GUIDANCE AND CONTROL 1991

CONFERENCE COMMITTEE

James McQuerry, Chairperson Ball Aerospace Systems Group	(303) 939-6102
Arlo Gravseth, Co-Chairperson Martin Marietta Astronautics Group	(303) 977-6147
Dick Zietz, Co-Chairperson Martin Marietta Astronautics Group	(303) 971-8539
Don Parsons, AAS National Martin Marietta Astronautics Group	(303) 971-7591
Pam Alstott, Registration and Finance Martin Marietta Astronautics Group	(303) 977-8322
Bud Gates, AAS National Steering Committee Martin Marietta Astronautics Group	(303) 977-1038

COMMITTEE MEMBERS

Pete Abbott	Martin Marietta Astronautics Group
Bob Culp	University of Colorado (Boulder)
John Durrett	Martin Marietta Astronautics Group
Larry Germann	Ball Aerospace Systems Group
Carl Henrikson	Ball Aerospace Systems Group
Terry Kelly	Ball Aerospace Systems Group
Bob Lewis	Ball Aerospace Systems Group
Lou Morine	W.J. Schafer Assoc., Inc
Paul Shattuck	Martin Marietta Astronautics Group

EDITORS
Dr. Robert D. Culp — University of Colorado
James P. McQuerry — Ball Aerospace Systems Group

Thanks are due Diane M. Massey, Univelt, Incorporated, for final preparation of the manuscript for publication.

Front Cover Illustration:

Ball Corporation has been awarded a contract by Canada's Spar Aerospace Limited to build a spacecraft bus and perform associated support services for the RADARSAT remote-sensing satellite program worth approximately $55M.

Being built for the Canadian Space Agency, RADARSAT will put a synthetic aperture radar in space to: (1) monitor and predict the location and type of ice in the Arctic Ocean in order to aid navigation; (2) assist in management of offshore oil, gas and mineral exploration and oil slick monitoring; (3) receive remote-sensing data for use in the management of agriculture, forest and hydrology resources; (4) acquire data to update the Canadian geological map; and, (5) create a three-dimensional map of the world for geological exploration (contact: Alexander Bracken 303-460-2259).

Frontispiece:

Technicians at Ball Corporation's aerospace systems group prepare the Oriented Scintillation Spectrometer Experiment (OSSE) for testing in the thermal vacuum chamber. The chamber simulates both temperature extremes and the airless environment of space. Composed of four gamma ray detectors, OSSE was built by Ball for the Naval Research Laboratory and is one of four major instruments on the Gamma Ray Observatory. A Shuttle launch is scheduled for Spring, 1991 (contact: Alexander E. Bracken 303-460-2259 or Bernadette C. Stechman 303-939-6433).

AN AMERICAN ASTRONAUTICAL SOCIETY PUBLICATION

GUIDANCE AND CONTROL 1991

Volume 74
ADVANCES IN THE ASTRONAUTICAL SCIENCES

Edited by
Robert D. Culp
James P. McQuerry

Proceedings of the Annual Rocky Mountain Guidance and Control Conference held February 2-6, 1991, Keystone, Colorado.

Published for the American Astronautical Society by
Univelt, Incorporated, P.O. Box 28130, San Diego, California 92198

AAS Publications Office
P.O. Box 28130
San Diego, California 92198

Affiliated with the American Association for the Advancement of Science
Member of the International Astronautical Federation

First Printing 1991

Library of Congress Card No. 57-43769

ISSN 0065-3438

ISBN 0-87703-334-X (Hard Cover)
ISBN 0-87703-335-8 (Soft Cover)

Published for the American Astronautical Society
by Univelt, Incorporated, P.O. Box 28130, San Diego, California 92198

Printed and Bound in the U.S.A.

FOREWORD

HISTORICAL SUMMARY

The Annual Rocky Mountain Guidance and Control Conference began as an informal exchange of ideas and reports of achievements among local guidance and control specialists. Since most area guidance and control experts participate in the American Astronautical Society, it was natural to gather under the auspices of the Rocky Mountain Section of the AAS.

In 1977, Bud Gates, Don Parsons, and Bob Culp organized the conference formally and began the annual series of meetings the following winter. In March, 1978, the Annual Rocky Mountain Guidance and Control Conference met at Keystone, Colorado. It has met there annually ever since. The fourteenth Conference was held this year.

The first Conference was chaired by Bud Gates, co-chaired by Bob Culp, with arrangements by Don Parsons. Bob Culp was Section Chairman. The local chairmen were Bob Barsocchi, Carl Henrikson, and Lou Morine; session chairmen were Sherm Seltzer, Pete Kurzhals, and Lou Herman. Other members of the original organizing committee were Ed Euler, Joe Spencer, and Tom Spencer.

A tradition from the beginning has been the opening banquet: this provides a chance to meet friends, renew acquaintances, and lay plans for the five-day meeting. It is an elegant feast, still the evening is marked by informality and good cheer. A general interest speaker has become a popular feature. These have been:

1978 Sherman Seltzer, NASA MSFC, told a joke

1979 Sherm Seltzer, Control Dynamics, told another joke

1980 Andrew J. Stofan, NASA Headquarters, "Recent Discoveries through Planetary Exploration"

1981 Jerry Waldvogel, Cornell University, "Mysteries of Animal Navigation"

1982 Robert Crippen, NASA astronaut, "Flying the Space Shuttle"

1983 James E. Oberg, author, "Sleuthing the Soviet Space Program"

1984 W. J. Boyne, Smithsonian Aerospace Museum, "Preservation of American Aerospace Heritage: A Status on the National Aerospace Museum"

1985 J. B. Irwin, NASA astronaut (retired), "In Search of Noah's Ark"

1986 Roy Garstang, University of Colorado, "Halley's Comet"

1987 Kathryn Sullivan, NASA astronaut, "Pioneering the Space Frontier"

1988 William E. Kelley and Dan Koblosh, Northrop Aircraft Division, "The Second Best Job in the World, The Filming of *Top Gun*

1989 Brig. Gen. Robert Stewart, U.S. Army Strategic Defense Command, "Exploration in Space: A Soldier-Astronaut's Perspective"

1990 Robert Truax, Truax Engineering, "The Good Old Days of Rocketry."

1991 Rear Admiral Thomas Betterton, Space and Naval Warfare Systems Command, "Space Technology: Respond to the Future Maritime Environment"

The Rocky Mountain Section of the Society established a broad-based Conference Committee, the Rocky Mountain Guidance and Control Committee, chaired ex-officio by the next Conference Chairman, to run the Annual Conference. The Conference has been a success from the start. Currently, the meeting attracts over 200 of the nation's top specialists in space guidance and control:

Year	Conference Chairman	Attendance
1978	Robert L. Gates	83
1979	Robert D. Culp	109
1980	Louis L. Morine	130
1981	Carl Henrikson	150
1982	W. Edwin Dorroh, Jr.	180
1983	Zubin Emsley	192
1984	Parker S. Stafford	203
1985	Charles A. Cullian	200
1986	John C. Durrett	186
1987	Terry Kelly	201
1988	Paul Shattuck	244
1989	Robert A. Lewis	201
1990	Arlo Gravseth	254
1991	James McQuerry	256

In addition to providing for an annual exchange of the most recent advances in research and technology of astronautical guidance and control, the Conference brings in a guest speaker of special prominence to present a lecture series in one specific area of current interest. These Speakers and their topics have been:

1978 Professor Daniel Debra, Stanford University, "Navigation"

1979 Professor William L. Brogan, University of Nebraska, "Kalman Filters Demystified"

1980 Professor J. David Powell, Stanford University, "Digital Control"

1981 Professor Richard H. Battin, Massachusetts Institute of Technology, "Astrodynamics - A New Look at Old Problems"

1982 Professor Robert E. Skelton, Purdue University, "Interactions of Dynamics and Control"

1983 Professor Arthur E. Bryson, Stanford University, "Attitude Stability and Control of Spacecraft"

1984 Dr. William B. Gevarter, NASA Ames, "Artificial Intelligence and Intelligent Robots"

1985 Dr. Nathaniel B. Nichols, The Aerospace Corporation, "Classical Control Theory"

1986 Dr. W. G. Stephenson, Science Applications International Corp., "Optics in Control Systems"

1987 Professor Daniel Debra, Stanford University, "Guidance and Control: Evolution of Spacecraft Hardware"

1988 Professor Arthur E. Bryson, Stanford University, "Software Application Tools for Modern Controller Development and Analysis"

1989 Professor John L. Junkins, Texas A&M University, "Practical Applications of Modern State Space Analysis in Spacecraft Dynamics, Estimation and Control"

1990 Professor Laurence Young, Massachusetts Institute of Technology, "Aerospace Human Factors"

1991 The Low-Earth Orbit Space Environment
Professor George W. Rosborough, University of Colorado, "Gravity Models"
Professor Raymond G. Roble, University of Colorado and NCAR, "Atmospheric Drag"
Professor Robert D. Culp, University of Colorado, "Orbital Debris"
Dr. James C. Ritter, Naval Research Laboratory, "Radiation"
Dr. Gary Heckman, National Oceanic and Atmospheric Administration, "Magnetics"
Dr. William H. Kinard, NASA Langley Research Center, "Atomic Oxygen"

In recent years, under the leadership of Don Parsons, the Rocky Mountain Section of the AAS has strongly supported students interested in the aerospace program. Every year approximately twenty college students are invited as guests of the Conference. They obtain a valuable, first-hand exposure to the state of the art in space guidance and control. The Conference has been a major sponsor of the University of Colorado's student-run Get-Away-Special Space Shuttle Experiment. Finally, the Section every year gives $1000 to two scholarships at the University of Colorado: one in Aerospace Engineering Sciences, and one in Electrical and Computer Engineering. These are recognized at the opening banquet the year following the award. Since they have been given, the winners have been:

Year	AERO	ECE
1981	Jim Chapel	
1982	Eric Seale	
1983	Doug Stoner	John Mallon
1984	Mike Baldwin	Paul Dassow
1985	Bruce Haines	Steve Piche
1986	Beth Swickard	Mike Clark
1987	Tony Cetuk	Fred Ziel
1988	Mike Mundt	Brian Olson
1989	Keith Wilkins	Jon Lutz
1990	Robert Taylor	Greg Reinacker

These scholarship winners are honored guests of the AAS at this Conference. The Rocky Mountain Section has made a commitment to continue this support through a $20,000 endowment, which will assure the existence of these scholarships indefinitely.

In 1983, Don Parsons, Joe Spencer, Ed Dorroh, and Bob Culp, at the request of the National Board of the American Astronautical Society, prepared a charter which established the Annual Conference as the AAS Rocky Mountain Guidance and Control Conference. The Conference will be sponsored by the National Organization for the AAS, will meet annually at Keystone, Colorado, and will be run by the Rocky Mountain Section.

A National AAS Guidance and Control Technical Committee, with national representation, was established to provide oversight to the local conference committee. W. Edwin Dorroh, Jr., was the first chairman of the AAS Guidance and Control Technical Committee. At the 1985 Conference, Robert L. Gates took over as the chairman of the National AAS Guidance and Control Committee. This committee meets every year at the Keystone Conference, and also at either the summer Guidance and Control Conference, or the fall AAS Annual Meeting.

Thus, the AAS Rocky Mountain Guidance and Control Conference continues as the premier conference of its type anywhere. As a National Conference sponsored by the AAS, it promises to be the preferred idea exchange for guidance and control experts for years to come.

The Conference Committee extends congratulations to the 1991 Conference Chairman, James McQuerry, for an excellent conference and a job well done. His report follows.

Robert D. Culp
University of Colorado
Boulder, Colorado

PREFACE

The fourteenth annual Rocky Mountain Guidance and Control Conference, sponsored by the AAS, began Saturday evening February 2, with the opening banquet. The guest speaker, Rear Admiral Thomas Betterton the Assistant Commander for Space Technology Space and Naval Warfare Systems Command, presented a timely speech on "Space Technology: Respond to the Future Maritime Environment". His presentation focused on the historical development of the Navy's Space Program. In light of the on-going Persian Gulf Crisis, the speech provided a significant timely kickoff to the conference.

The Sunday tutorial session was the culmination of a many year goal of the the conference committee. The focus of the tutorial was on the Low-Earth Orbit Space environment. Six separate presentations were given, divided between the morning and evening sessions. This session contained guidance from both a theoretical and a practical recent experience viewpoint. Many software programs were made available by the authors in addition to their fine papers.

The first technical session, on Monday morning, Advances in Guidance, Navigation and Control, was chaired by Ken Cox of NASA at the Johnson Space Center and co-chaired by Major Wiley Larson of the U.S. Air Force Academy. The seven papers presented in this session focused on the NASA and DOD space initiatives in applying technology advances to avionics for guidance, navigation and control. Strategic philosophy, architecture, development, intermediate bridging tasks, and technology and hardware developments were presented to provide insight for definition and development of future systems.

The second session on Monday was the annual storyboard displays. This session is one of the highlights of the conference since attendees have the opportunity to informally discuss the authors' hardware and other information in a relaxed social environment. The displays included examples of G&C hardware, prototyping advancements, and results from recent flight experiences. A buffet, sponsored by several corporate members and participants, was provided during this three-hour session.

Session three on Tuesday morning, Challenges and Approaches to Space-Borne Pointing Control, was chaired by Col. Roland Worrell III of the Space Defense Initiative Office and co-chaired by Robert Laskin of the Jet Propulsion Laboratory. This session covered the integrated approach to pointing of space systems including advanced analytical techniques, advanced test beds, and joint controls-structures-payload design, analysis and verification.

Session four on Tuesday evening, International space programs, was chaired by Jacques Busseuil of Aerospatial/Strategic and Space Systems Division and co-chaired by Lyle Jenkins of NASA at the Johnson Space Center. Considering that the 1990s will be a period of unprecedented international space activity, this session shared insight within the international guidance and control community of both current and planned programs. Activities in Germany, France, Canada, and the U.S.A. were presented in seven very well received papers.

The final session, Recent Experiences, was held on Wednesday morning. The session was chaired by Doug Broome of NASA Headquarters and co-chaired by Bernard Kaufman of the Naval Research Laboratory. This annual session covered successes and failures on recent programs including: CRESS, Hubble Space Telescope, Magellan, Pegasus, LACE and LDEF. Lessons learned topics included guidance and control results, sensor hardware performance and material effects in space.

The annual meeting of the AAS Guidance and Control Technical Committee was held concurrent with the storyboard session under the chairmanship of Bud Gates. The committee reviewed the format and content of the conference and offered many insightful recommendations for the 1992 conference, which will be held on February 1-5, 1992 at Keystone Colorado. Next year's chairperson will be Dick Zeitz, Martin Marietta Astronautics Group, P.O. Box 179, MS F4083, Denver, Colorado 80201, Tel: (303) 971-8539. Co-chairperson will be James McQuerry, Ball Aerospace Systems Group, P.O. Box 1062, MS CO-3, Boulder, Colorado 80306, Tel: (303) 939-6102.

The success of this year's conference was the result of the continued support of the guidance and control community as exemplified by the record registered attendance of 244 persons. As I stated at the conference opening remarks, this conference was the result of the dedicated work of 26 committee members, 8 chair and co-chair persons, 6 tutorial speakers, 1 dinner speaker and 127 authors and co-authors. The guidance and control community should be proud of its dedicated support in bringing this conference to its present stature.

If you have any suggestions for next years conference, please contact either Dick or myself as the 1992 planning cycle begins in April. I would like to thank all of you for making this conference a memorable success and hope to see each of you at Keystone next year.

James P. McQuerry
Ball Aerospace Systems Group
Electro-Optics and Cryogenics Division
Chairman, 1991
AAS Guidance & Control Conference

CONTENTS

Page

Section I
ADVANCES IN GUIDANCE, NAVIGATION AND CONTROL

SESSION I

Chairperson:	Ken Cox NASA Johnson Space Center
Co-Chairperson:	Major Wiley Larson U.S. Air Force Academy
Local Chairperson:	Kirk Stewart Ball Aerospace Systems Group

The following paper was not available for publication:

AAS 91-021 "Space Avionics Architecture", E. S. Chevers, NASA Ames Research Center

The following paper numbers were not assigned:

AAS 91-027 to -029

AAS 91-020

STRATEGIC AVIONICS TECHNOLOGY PLANNING

Kenneth J. Cox and Don C. Brown*

NASA experience in development and insertion of technology into programs has led to a recognition that a Strategic Plan for Avionics is needed for space. In the fall of 1989 an Avionics Technology Symposium was held in Williamsburg, Virginia. In early 1990, as a followon, a NASA wide Strategic Avionics Technology Working Group was chartered by NASA Headquarters. This paper will describe the objectives of this working group, technology bridging, and approaches to incentivize both the federal and commercial sectors to move toward rapidly developed, simple, and reliable systems with low life cycle cost.

INTRODUCTION

In the fall of 1989 a Strategic Transportation Avionics Technology Symposium was convened at Williamsburg, Virginia to promote a dialogue between avionics technology producers and users, including 192 NASA, government, industry, and academia participants. (Ref. 1)

NASA organized, in early 1990, a Strategic Avionics Technology Working Group (SATWG) to develop avionics technology planning in the civil space arena. Presently, all nine NASA field centers are participants in the SATWG.

Figure 1 shows the chief goal of the strategic avionics plan which is to develop effective technology requirements that will span multiple programs and be phased according to a master pre-planned product improvement strategy.

Customer driven technologies will be emphasized and a process developed to decide priority between the development of new technology, and the application of existing technology.

New ways of "doing business" must be developed to focus and harness the strengths of NASA institutional engineering and operations, and to emphasize team building, communication and a cooperative environment. "Gain sharing" or seeing what is best for NASA and industry must replace isolated attitudes.

ORGANIZATION

Strategic planning for avionics across multiple NASA programs is a recent need and has been brought about by missions that are characterized by factors such as; long duration, complex interfaces with multiple vehicles and surface systems, space based checkout, remote in-flight training, vehicle health management, and compatibility between manned and unmanned precursor designs. It is emphasized that both the flight segment and the ground infrastructure should be viewed as an integrated system. Another partitioning viewpoint, at a systems level, is to integrate the vehicle segment and the information systems segment.

* NASA Johnson Space Center, NASA Road 1, MS-EG, Houston, Texas 77058.

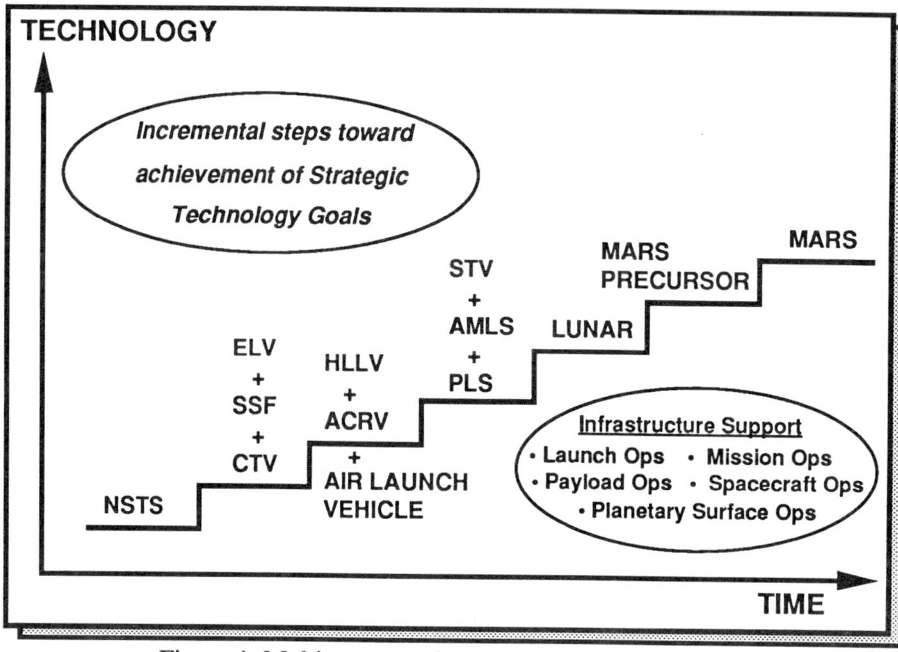

Figure 1 Multi-program Strategic Avionics Planning

The approach taken in this planning to establishing such a broadly based forum was like a good campus sidewalk designer who observes "where the paths in the grass are" before laying down any concrete. This is an essential principle of Total Quality Management, to try things on a small scale to validate its value added content before institutionalizing it while employing concurrent engineering, enhanced communication, teamwork and team building.

The present SATWG organization that has evolved is shown in Figure 2.

OBJECTIVES

Strategic Avionics Technology Working Group

The SATWG charter provides for a forum to develop a space strategic avionics technology plan that will establish: technology strategies and goals; long-range infrastructure elements to support future, existing and developing programs; and guidelines for functional commonality of avionics architectures. Figure 3 shows the scope of the categories of vehicles and systems that are being considered.

The primary objectives of the SATWG are to develop new processes that stress customer focus, foster technical interchange between NASA headquarters codes, technologists, advanced developers, and operators, employ concurrent engineering and common tool environment principles, and which produce comprehensive and complete forums to serve as advisors for NASA programs.

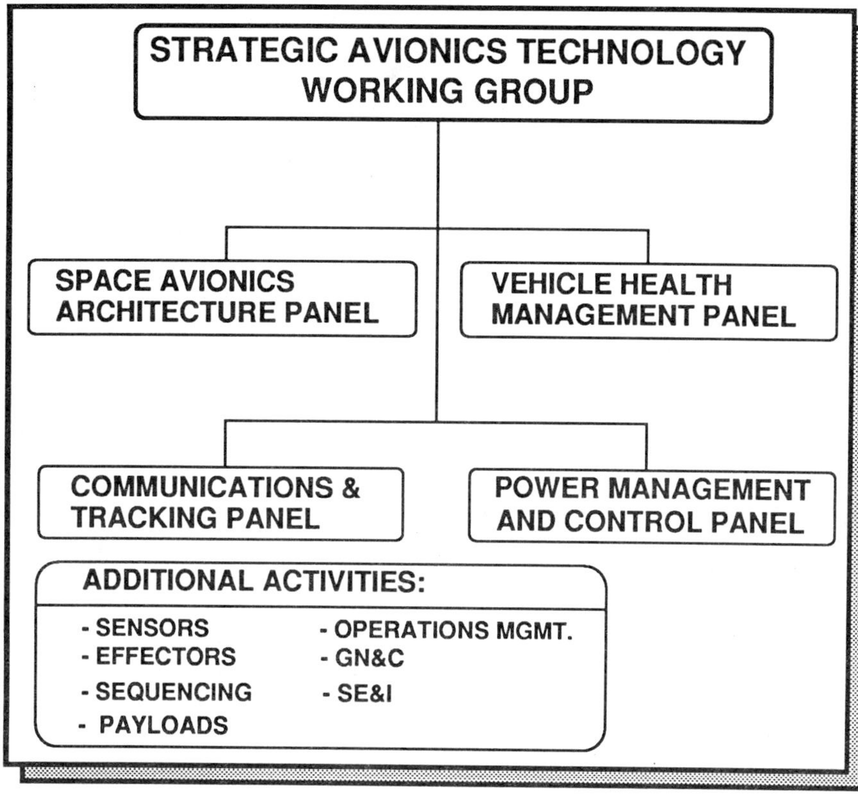

Figure 2 Strategic Avionics Technology Working Group Organization

Space Avionics Architecture Panel

The Space Avionics Architecture Panel will develop an advanced avionics architecture technology plan for space transportation and exploration programs which emphasizes lower life cycle cost and more efficient DDT&E cost. System architecture, software and hardware standards for open architecture will be defined. A forum for avionics architecture technology interchange between NASA, aerospace industry partners, DoD and the commercial sector will be established.

Vehicle Health Management Panel

The Vehicle Health Management Panel will provide a focus and technical interchange for automated vehicle health monitoring and checkout activities. System requirements and integration strategies will be developed for launch and space vehicles. Test and demonstrations of technology and systems concepts will be provided. The panel will provide a forum for technology interchange between NASA, aerospace partners, DoD and the commercial sector and will provide consulting support for new NASA programs.

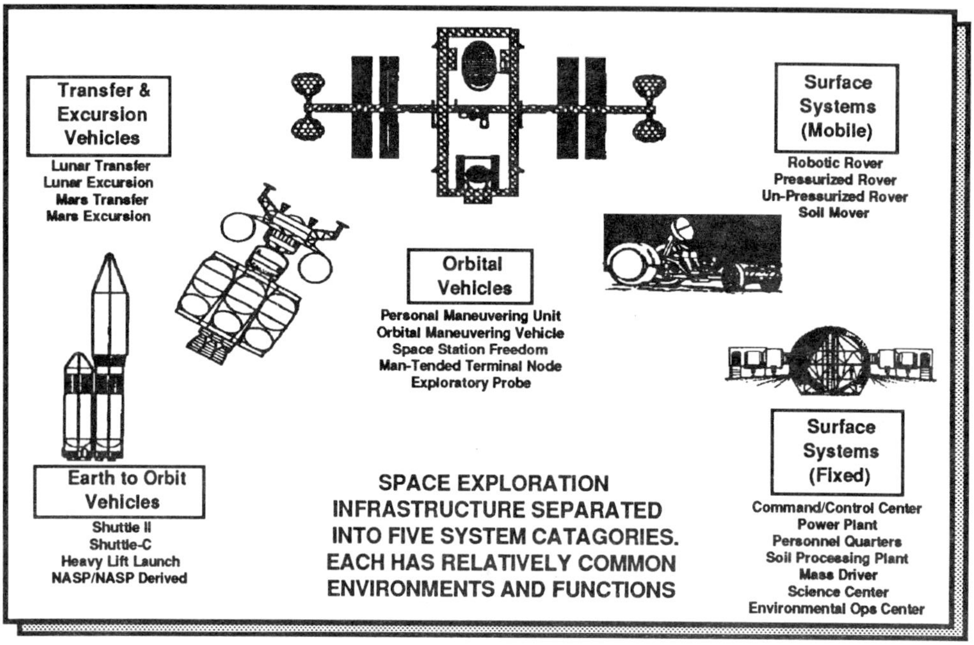

Figure 3 Scope of Avionics Systems Catagories

Communications and Tracking Panel

The Communications and Tracking Panel will develop top level communications and tracking systems requirements, systems architecture requirements and standards which employ open architecture approaches, modularity, and interface standardization. Strategies for utilization of common elements will be investigated. A forum for technology interchange between NASA, aerospace partners, DoD and the commercial sector and consulting support for new NASA programs will be provided.

Power Management and Control Panel

The Power Management and Control Panel will develop advanced integrated electrical power system technologies and strategies to ensure fault tolerant power for all mission phases. The panel will investigate technology for improved wiring and connectors. The panel will provide a forum for technology interchange between NASA, aerospace partners, DoD and the commercial sector and will provide consulting support for new NASA programs. An improved technology insertion process to reduce perceived risk, lower operational and life-cycle costs and maximize system operability and power availability will also be developed.

CUSTOMER DRIVEN TECHNOLOGY PROCESSES

"Technology Bridging" is a term developed to describe a cooperative technology development process between NASA technology centers, operations centers, program managers and the commercial sector.

Technology bridging projects are characterized by end-to-end concurrent engineering processes and environments, broadly based technologies that support multiple programs, NASA program management and institutional customer focus, and technology that is cost driven as well as performance driven.

Projects selected for "bridging" must provide significant evidence that the associated technology either enables mission objectives, reduces operational costs, mitigates obsolescence or enhances capability. Traditionally, performance enhancement has most often been the "driver" to incorporate new technology. The Space Exploration Initiative will require greater emphasis on additional objectives. For example, while current avionics technology can support a lunar or mars mission with few major advances, one of the primary motivators to invest in advanced avionics is to enhance the rewards that accrue from avionics support to fluids, propulsion and mechanical systems which will allow much more automated preflight checkout and health monitoring. Affordability is as valid an "enabler" as the ability to autonomously land on a distant planet.

In summary, "technology bridging" is a "process" that supports current NASA objectives to produce compatible, integrated avionics designs that will support the needs of multiple programs, lead to improved life cycle costs and take advantage of the many pockets of technical excellence that reside in NASA technology and operations field centers and associated industry advanced development centers.

Two technology bridging programs have been initiated in FY91: Electrical Actuation (EA) and Autonomous Guidance, Navigation and Control (AGN&C).

The Electrical Actuation program is jointly sponsored by JSC/LeRC/MSFC/KSC and will build upon the base for EA established by the Advanced Launch System technology program. The objective is to provide compelling technology demonstrations that will initiate the transition from fluid to electrical based high power actuation systems. Aerosurface control, thrust vector control (TVC), propulsion control valves (PCV) and many small actuators will benefit greatly by transition to electrical actuation. Automatic checkout, whether surface or space based will become routine. Generic electronic controllers capable of controlling a range of EA technologies and power ranges will be developed. Advanced scaleable motor designs for "soft-switched", field oriented pulse population control, and "hard-switched" pulse width modulated designs will be developed to allow application from primary flight control functions to surface based machines.

The Shuttle prime contractor, Rockwell, will serve as the architect for the advanced EA systems to ensure that Space Shuttle requirements are met. LeRC will manage EA technology from the ALS contractors, General Dynamics and Boeing, and will transfer ALS experience. An engineering model redundant design will be developed. KSC will develop new efficient checkout methods that take advantage of EA and will quantify benefits over existing Shuttle technology. MSFC will lead the application of EA to TVC and PCV's. JSC will provide customer interface with the benefiting programs, explore small horsepower EA applications (<10 HP), and demonstrate integration between EA with a power conditioning system driven by the energy system.

A significant government/industry coordination meeting on electrical actuation was held at General Dynamics Space Systems Division, San Diego, Ca. on December 6, 1990. The government was represented by JSC, LeRC, MSFC, KSC, SSC, NASA SRB project manager, and the ALS/JPO. Industry was represented by Boeing, Martin Marietta,

Lockheed, Rockwell, Sundstrand, Allied Signal, Eldec, General Electric, Parker-Hannifin, Honeywell, Moog, and HR Textron. The government presented overviews of EA initiatives in the arenas of NASA, civil and military transportation and ALS. Industry gave brief overviews of EA IR&D initiatives and test facilities.

There was a strong consensus to maintain the dialogue initiated between government and industry regarding electrical actuation barriers, technical challenges, test articles and demonstration facilities.

Figure 4 shows the elements of EA and the key engineering trades that will be done to select technologies for development and demonstration.

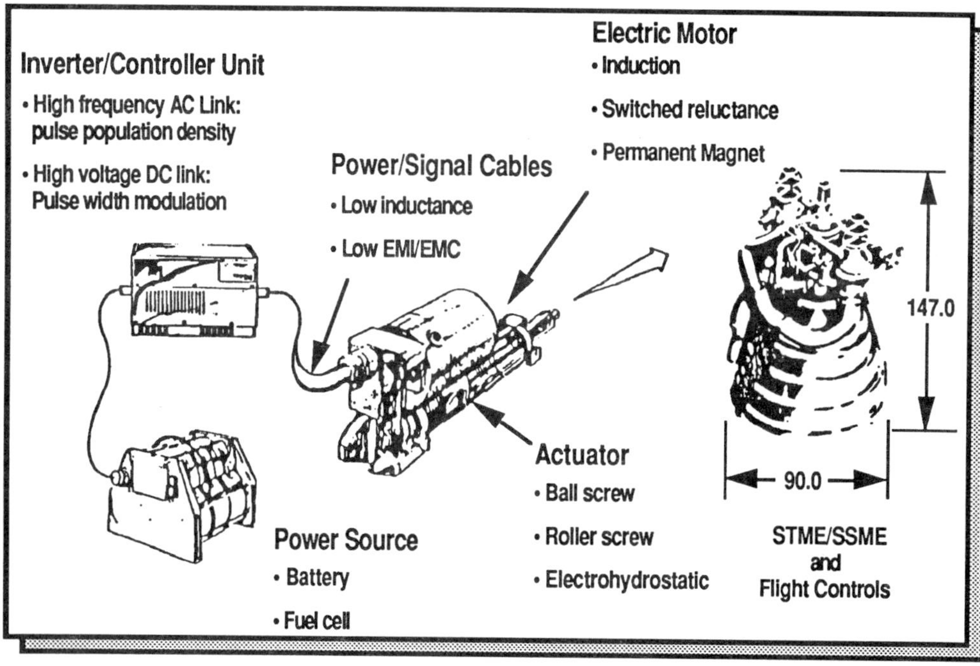

Figure 4 Elements of Electrical Actuation Systems

The AGN&C program is jointly sponsored by JSC/MSFC/LaRC/KSC. The objective is to develop and demonstrate lidar based wind sensing, more robust GN&C algorithms, and improved processes for incorporation of environmental data into Shuttle and future onboard systems. JSC will lead GN&C design changes, and application to operational processes. MSFC and LaRC will provide Lidar expertise and test articles and KSC will update operational techniques. Expected benefits from this technology program include more autonomy and robustness for ascent flight control, more efficient interface of environment constraint data and flight system preparations, and technical recommendations to new programs for advanced Autonomous GN&C systems. Figure 5 shows some of the elements of this technology initiative.

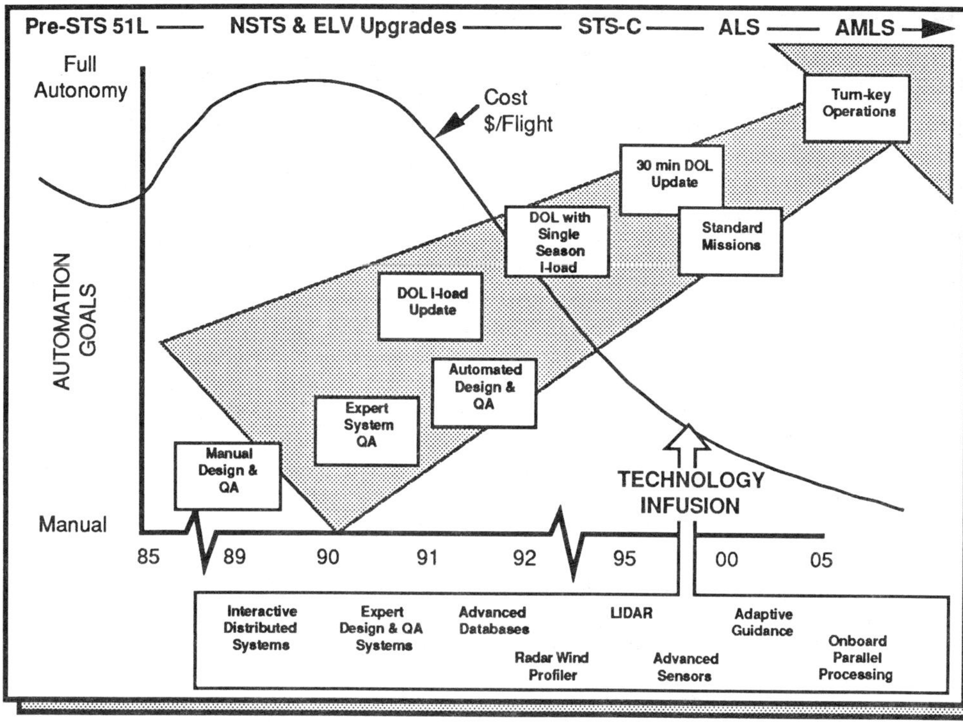

Figure 5 Evolution of Autonomous GN&C

In addition to "technology bridging", a concept called "technology transition" has also evolved from the SATWG. Technology transition is a process that emphasizes ongoing involvement of technology centers even into the operational deployment of the technology. Operations centers jointly sponsor technology bridging projects with technology centers until a program incorporates the technology into an advanced development phase. Program advocacy in the latter phases of technology bridging is needed to provide funding support for major demonstrations. Technology and operations centers continue to participate with programmatic advanced development projects, in a consulting fashion, until the technology is fully deployed.

EFFECTIVE AVIONICS SUPPORT TO SE&I

Shuttle and Space Station Experience

The practical experience of designing, building and verifying the Space Shuttle and planning for Space Station, has led to a strong consensus that mainline emphasis for advanced avionics should include open architectures that are scaleable, compliant with standards and utilize modular system concepts, trade studies of centralized versus distributed architectures, and development of flexible avionics facilities and test beds capable of multi-program support.

Application of Avionics Integration Methods to Non-avionics Systems

The principles of "horizontal integration" are well established to avionics practitioners who have had to integrate complex hardware and software and operational procedures into systems. Most of the non-avionics systems have emphasized "hardware" that performs an end function that may be developed and checked out somewhat independently of the remainder of the spacecraft. Avionics support to SE&I for fluids, propulsion and mechanical systems will facilitate automated checkout and health monitoring for systems that have traditionally been major contributors to intensive manual test methods requiring significant ground support equipment. Avionics horizontal integration support will permit these non-avionics systems to be sufficiently intelligent that the vehicle will literally contain the on-board checkout infrastructure to test itself. This is vital for future space based systems.

The application of avionics SE&I services to fluid, propulsion and mechanical systems has potentially high payoffs. Interfaces between vehicles will be made compatible by developing and utilizing functionally common tool sets that operate out of common data bases. Interfacing data will be defined and controlled by the developers of each system but will be made readily available to other systems developers.

As an aside, this experience was applied in the conceptual phase of the Space Station and has resulted in an approach to systems integration that provides functionally equivalent data management system test sets (DMS kits) to system developers so that their systems hardware and software may be developed and integrated to the basic spacecraft well in advance of the traditional end-to-end integration facility that is utilized for the Space Shuttle. DMS kits are configured with functionally equivalent spacecraft hardware, which will allow the same test procedures and software to follow the vehicle from DDT&E through on-orbit deployment and operations.

Need for Advanced Integration Tool Development

At a total system level, tools need to be developed that will balance the requirements of performance, economics and surface processing to yield "hard" data on the sensitivities of design features to operational features, such as launches per year, and costs of acceleration of certain paths in DDT&E. Several versions of this "Advanced Simulation Tool" (ASIM) have been developed by the ALS contractors and have produced excellent cost estimation data.

Avionics as an Enabling Technology

When avionics is viewed in a global sense, it is seen as an enabling technology for management of exploration initiatives operational costs and as a key element in containing and reducing the operational costs of existing vehicles. Space based systems will only be practical if all systems are capable of automated checkout in an operational context, not as standalone subsystems. Vehicle systems must be integrated into the overall information management system so that ongoing data reflecting trends and state of health are readily available.

INNOVATIVE INCENTIVIZING TO PROMOTE MORE COST EFFECTIVE AVIONICS

Cost effective avionics processes will lead to an approach that is characterized by "top down/bottom up/middle in". The "Middle In" function is synonymous with "horizontal integration", and emphasizes: (1) understanding "what's out there", (2) forecasting inevitable future available technology, (3) developing technology needs, and (4) deciding to what extent to invest in new technology as opposed to applying existing technology.

These new business approach processes include the three traditional phases of avionics advanced development; avionics design, development, test and evaluation (DDT&E), and avionics operations. What's unique about this new paradigm, is the integration and interdependencies of each phase with the others and the incorporation of industry as an equal partner in the planning process.

Avionics Advanced Development

The major trends for this phase include becoming customer focused, including both flight and ground infrastructure, planning smaller incremental projects, applying across multiple programs, and integrating advanced development with DDT&E, and operations.

DDT&E

The major trends for this phase include expansion from a one time build it, test it and fly it, role to an integral part of an ongoing continuous improvement process that will be repeated over the life of the program, which may span 30 years or more. New tools must be developed that are capable of providing closed loop metrics and assessment of performance/risk management, economics/validation of cost estimation and operations/process effectiveness. Strategic avionics plans that led to the original development of the system to which DDT&E is being applied must include this continuous improvement process. Such insight will allow timely and cost effective updates to be considered, given flight test data.

Operations

The major trends for this phase include migration from surface based control centers to more autonomous onboard control. This is essential for missions that exhibit long communications trip times. There will also be a move toward control centers that can concurrently service multiple missions. Long mission durations will require that traffic management, debris avoidance and onboard training be supported.

Incentives

Specific incentives include facilitation of contractor and NASA teaming. An example of effective contractor teaming is the NASP consortium, which fosters concepts of equality, equitability of work, cooperation within competitive boundaries and working toward a common goal.

Internally, NASA program managers could be incentivized to reduce program costs by allowing them to retain a part of realized cost savings to fund additional continuous improvement initiatives. Technologist's participation in "technology bridging" projects will allow them to see a practical application of their ideas. Institutional engineering and operations will benefit by providing applied technology, to solve real-world needs, that is developed in a timely manner.

An effective strategic avionics technology plan and process supports NASA's desire to move toward a work force that is more empowered to produce timely change to their work environment. Total Quality Management methodologies such as Quality Function Deployment, Design of Experiments, etc. may be employed to establish comprehensive, systematic ways and metrics to explore engineering and economic "trades" between possible technologies initiatives and between choices to implement in the ground or flight infrastructure.

CONCLUSIONS

The Exploration Initiative and technology upgrades to existing vehicles will become more feasible if new methods are developed to control and manage operational costs and to mitigate and reduce existing costs.

A significant study by General Dynamics Space Systems Division (GDSS), commissioned by the SATWG Space Avionics Architecture Panel, was discussed at a SATWG meeting at GDSS on November 14-16, 1990. This systematic review of common avionics requirements across the space avionics infrastructure, including surface systems, will be expanded to provide insight and consensus by the SATWG NASA and commercial sector community to provide a base for future technology definition studies and development of space avionics standards and guidelines.

Technology gaps identified by this study include: (1) Avionics Hardware - robust fault tolerant avionics components including cold sparing, non-intrusive built-in test, and improved test methods and design verification for complex systems, (2) Software technology - Standard, flexible operating systems, high coverage/low overhead redundancy management, minimization/automation of verification and validation, (3) Testability - Fault/Trend analysis for incipient failures, autonomous checkout. As an example, Figure 6 shows a projected trend for advancement of autonomous checkout.

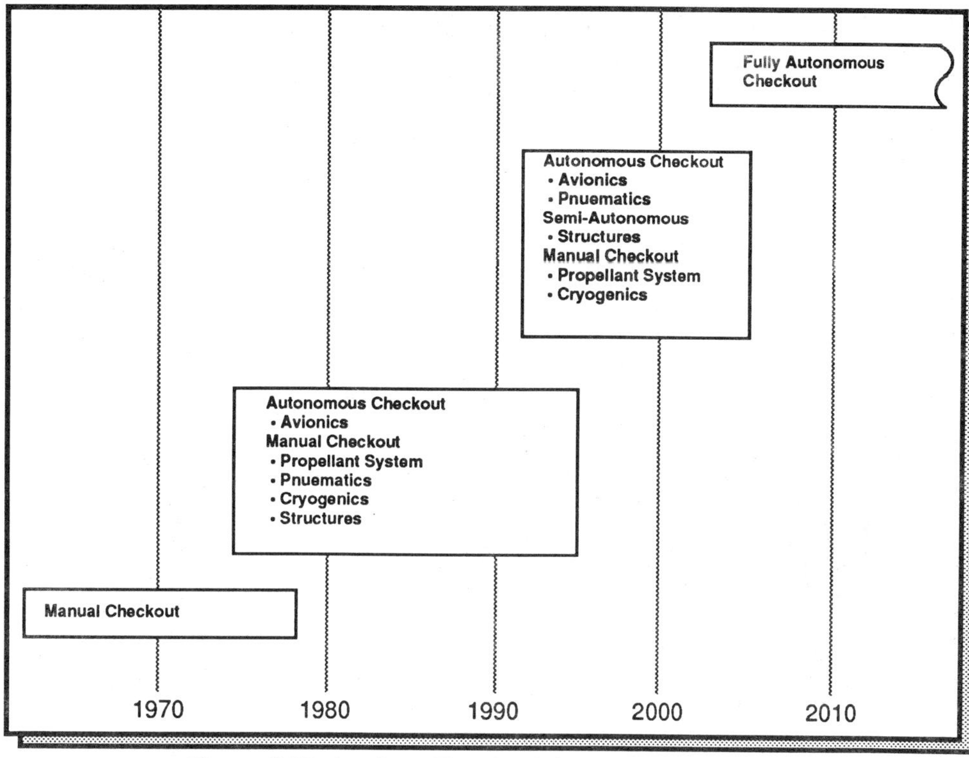

Figure 6 Technology Trends of Autonomous Checkout

A major challenge is to determine how to add fault tolerance without degrading system performance and increasing system complexity.

In conclusion, Figure 7 shows a vision for the day when rapidly developed, low cost, scaleable, simple, reliable systems will be commonplace throughout the space avionics infrastructure.

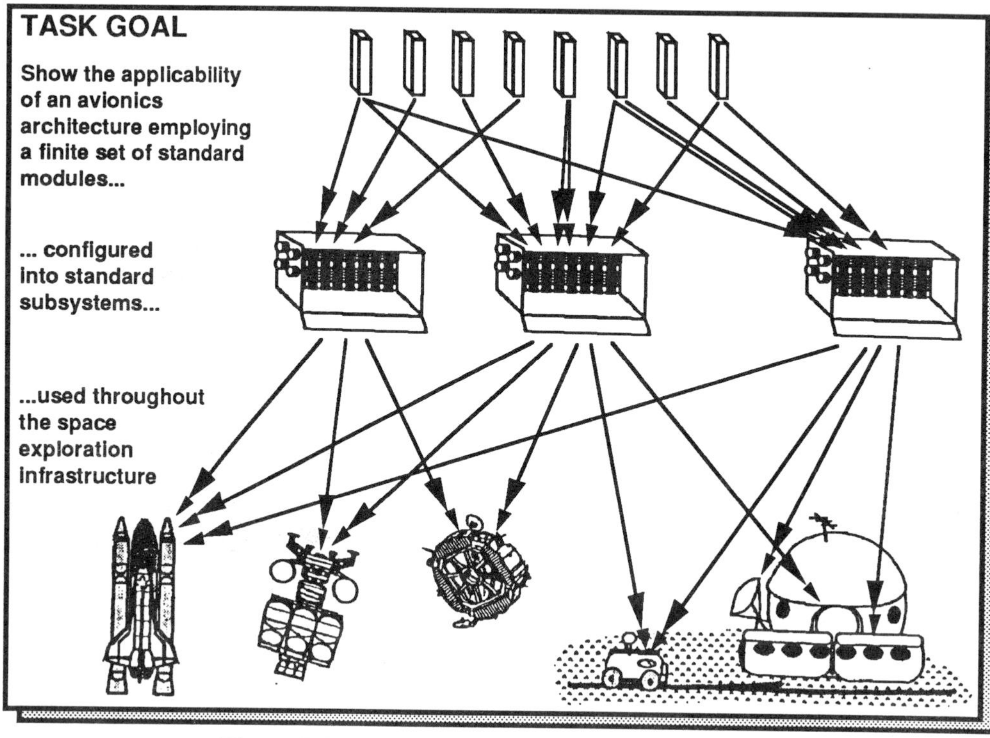

Figure 7 Multi-program Open Avionics Architecture

REFERENCES

1. Cox, Dr. Kenneth J., "Space Transportation Avionics Technology Symposium", NASA Conference Publication 3081, Vol. 1,2, November 7-9, 1990.

2. Nordstrom, John, "Space Avionics Requirements Study", NAS8-37588-TD006, November 1990, General Dynamics Space Systems Division.

ACKNOWLEDGMENT

John DiBattista/NASA-HQ/Office of Aeronautics and Exploration Technology, Paul Herr/NASA-HQ/Office of Space Flight, and a significant number of personnel from the NASA field centers together with the integrating space prime contractors have made substantial contributions to this effort.

AAS 91-022

AUTONOMOUS GUIDANCE, NAVIGATION AND CONTROL

A. J. Bordano*, G. G. McSwain* and S. T. Fernandes†

A review of the Autonomous Guidance, Navigation and Control (GN&C) Bridging Program sponsored by the NASA Office of Space Flight is presented. In this program, a four-center NASA team will develop and demonstrate mature technology applicable to autonomous GN&C systems. Application will be targeted primarily at the National Space Transportation System (NSTS) and other NASA programs, although application to future space vehicles will also be investigated. The program will be coordinated closely with the NSTS program office to ensure that the technology being developed will address real operations, cost, safety or performance issues.

Investigators will focus on four main areas of study: launch and landing day winds assessment, ascent guidance, ascent load relief and systems failures during ascent. The goal is to identify the features of a practical, cost-effective autonomous ascent GN&C system (which includes ground as well as onboard elements) which meets the requirements of the Shuttle as well as follow-on launch vehicles.

The overall program plan is presented, followed by a brief review of the space shuttle ascent GN&C system. Preliminary results from each of the main areas of study are then presented.

Significant early results from this study indicate that a properly designed GN&C system which relies on a ground-computed atmospheric steering profile can achieve near optimum performance and be very cost effective. Preliminary results also show that a steering profile based on a pre-launch wind sounding can eliminate the need for a complex load relief function.

Introduction

This paper summarizes the Autonomous GN&C Bridging Program. The goals of this program are to define ground and onboard changes that will reduce STS operations costs, make STS operations more reliable and more flexible, extend the list of credible failures that the shuttle can adapt to, improve abort capability, improve launch availability and improve ascent performance. The intent of the activity is to identify, develop and demonstrate technology that can have an immediate, quantifiable benefit to the STS program, and that may be adapted to other current and future programs. The activity is divided into four study areas, shown below with the hoped-for benefits:

1. Day of Launch and Landing Winds Assessment
   - Operations cost reduction
   - Increase operations flexibility and reliability

* NASA Johnson Space Center, NASA Road 1, Houston, Texas 77058.

† McDonnell Douglas Space Systems, Co., Houston, Texas.

2. Guidance Improvements
   - Operations cost reduction
   - Increase operations flexibility and reliability
   - Improve abort capability
   - Improve ascent performance

3. Improved Ascent Load Relief
   - Improve launch availability

4. Adaptive GN&C for Systems Failures
   - Improve failure adaptivity
   - Increase operations flexibility and reliability

Assembled in this program are four NASA centers - Johnson Space Center (JSC), Kennedy Spaceflight Center (KSC), Marshall Space Flight Center (MSFC) and Langley Research Center (LaRC) and their respective contractor support, each bringing specific skills and expertise to bear on the launch vehicle problem. KSC is providing operations and logistics support as well as providing a site for the wind measurement demonstrations. MSFC and LaRC are developing and demonstrating laser Doppler (lidar) wind measurement systems for use in launch and landing operations. JSC is providing operational requirements for the lidar and conducting the GN&C studies. Each center is making use of contractor support as well as in-house resources.

The planned time table includes performing analysis, assessments and failure scenario identification during 1991. Technology demonstrations will be performed in late 1991 and continuing through 1992. Technology recommendations will be made to the STS program office during 1992 and 1993.

Proper application of avionics technology can be an effective lever for reducing operations costs and increasing the operational flexibility of a launch vehicle program, particularly in the case of the space shuttle. The following two sections describe the space shuttle ascent GN&C system and summarize the issues associated with preflight analysis and mission planning as it pertains to GN&C. Some of the reasons for this leverage will be made clear in this discussion.

Space Shuttle Ascent GN&C

The elements of the Space Shuttle Avionics system used in ascent GN&C consist of the hardware and software needed to steer the vehicle safely and reliably to its orbital targets. The hardware and software are broken out by major function below, beginning with elements common to all functions.

Common Elements

The space shuttle ascent GN&C system consists of a number of elements whose assets are shared between all three major functions. First on the list, of course, are the five IBM-built AP-101 general purpose digital computers (GPC's) which run the GN&C software. Four of these run identical software and make up a "voting" set. These GPC's, their software and other associated hardware are called the primary avionics subsystem (PASS). The fifth GPC runs a different, but functionally equivalent set of software and communicates with the various sensors and effectors through a different set of channels. This system is called the backup flight system (BFS). It was originally designed to protect against generic coding errors in the PASS, but it has evolved into a specialized system that contains some unique functionality. For instance, it communicate with all 24 aft reaction control system (RCS) jets directly and contains special software to drive certain displays. Subsequent descriptions will address only the PASS.

The software running on the GPC's is event-driven, with events originating both externally (ie: an engine failure) or internally (ie: a guidance mode discrete). The various software modules are either scheduled to run in one of the three executive threads (high frequency exec - HFE, which cycles at 25 Hz, mid frequency exec - MFE, at .96 Hz and low frequency exec - LFE, at .25 Hz), or are invoked by the dispatcher (which, in turn, runs in the HFE) as needed.

Navigation

The navigation subsystem consists of various sensors and the software modules dedicated to processing the signals from those sensors, as well as software which calculates special parameters for use in the guidance and control modules and the redundancy management (RM) module.

The sensors used in the ascent phase of flight are the inertial measurement unit (IMU), rate gyro assembly (RGA) and accelerometer assembly (AA). The shuttle uses three model KT-70 all attitude, four-gimbal inertial platforms manufactured by the Kearfott division of Singer, which are located in the orbiter avionics bay. The Orbiter is equipped with four Northrop RGA's (each containing three single DOF rate gyros), located in the wing well at the aft bulkhead. Each SRB is equipped with three RGA's (containing two single DOF rate gyros). These are used exclusively during first stage flight. The orbiter is equipped with four AA's, each one capable of measuring both normal and lateral acceleration.

The software modules in the navigation subsystem perform the following tasks: vehicle state computation, the estimation of the gravity vector, and user parameter processing. There are additional nav software modules, but they support onorbit and entry flight phases and are not described here.

Guidance

The guidance subsystem is made up entirely of software. During atmospheric powered flight, the onboard guidance function consists of a table look-up of inertial pitch and yaw attitude commands as a function of relative velocity. This function is part of the guidance-control steering interface (G/C Steer) software. Second stage guidance consists of the various modules of the powered explicit guidance (PEG) software, which iteratively computes a thrust vector pointing and turning rate solution based on a linear tangent law.

Control

The flight control subsystem consists of the ascent thrust vector control digital autopilot (Ascent TVC DAP) software, software which reconfigures the DAP for various flight phases by looking up gains and toggling discretes as appropriate (Ascent Recon), software interfaces to the effectors(SOP's), and the effectors themselves, along with the necessary signal conversion hardware and circuitry. The shuttle uses hydraulic actuators to gimbal the space shuttle main engines (SSME's) and the SRB's for thrust vector control. The SSME's can also be throttled. The ascent DAP has the ability to command the orbiter elevons, but these are set only to follow a preflight computed schedule selected to minimize wing loads.

Mission Operations Cost Background

According to the Liquid Rocket Booster (LRB) study conducted in 1988, JSC mission operations comprise a significant part (29.8%) of the NSTS total launch cost per flight of approximately $258.4 M. Mission Operations Directorate (MOD) activities make up 72% of the total JSC operations cost resulting in a cost per flight of $32.3 M for MOD mission operations tasks.

The MOD mission operations cost is primarily driven by the staffing and associated facilities required to support the desired flight rate with the current mission design template. This template includes an Assessment Flight Profile (AFP), a Conceptual Flight Profile (CFP), and typically two Operational Flight Profile (OFP) cycles. The AFP and CPF, which are performed twelve months or more before launch, usually are not manpower-intensive. The OFP cycles, which start at L-9 months and L-3.5 months, are manpower intensive activities which include flight software initialization load (I-load) design and verification, product deliveries including Range Safety, MCC and SMS training, and Flight Readiness Review (FRR) support. Additional mission operations activities not included in the design template are the Day-of-Launch (DOL) support for the Launch Systems Evaluation Advisory Team (LSEAT) and post-flight analysis.

Only two OFP cycles are planned, but any significant change to the groundrules and constraints (i. e., manifest changes and launch slips) can cause additional cycles to be performed because mission design is seasonally-dependent and many of the abort designs are weight-dependent. Approximately 60% of the flights from STS-26 through STS-43 had significant changes to the groundrules and constraints between the OFP Cycle 1 and the final flight cycle. Additional design work is also caused by carrying multiple manifest options - which must be treated as different missions - through the design cycles, and by the current procedure of having 8 alternate first-stage I-load sets to accommodate launch day wind variations. In addition to the formal design cycles, mission changes are reflected in a steady stream of Shuttle Operational Data Supplements (SODS's), averaging twelve per flight, which must be evaluated for potential impacts.

The significant cost items within each OFP cycle are the I-load design, the detailed verification of the flight software load, and the systems analysis and verification (loads assessments, launch probability, thermal margins, control and stability assessments, etc.) performed to support the flight readiness review. Any problems found during the system verification process can precipitate a new OFP cycle or, if found soon enough, can cause the current cycle to halt and restart with new groundrules and constraints. The I-load design tends to use manual iteration on 3 degree-of-freedom (DOF) or 6 DOF simulations. Often "tweaking" is required to match 6 DOF simulation responses when 3 DOF design is performed. Since nominal design, engine-out and adaptive guidance/throttling (AGT) designs, and abort designs usually occur sequentially, any delay encountered early in the nominal design cycle ripples through the remaining areas. In this case, the abort designs usually get squeezed and overtime may be required to complete the designs in time for final verification and training. Another complication is that iteration may be required between the nominal/engine-out designs and the aborts, typically RTLS, to ensure acceptable abort trajectories.

Detailed verification of the flight software involves multiple facilities, such as the Software Production Facility (SPF) and the Shuttle Avionics Integration Facility (SAIL), and multiple participants (NASA and contractors). For a new software build, an extensive matrix of test cases must be run to verify the many logic paths. New software builds are not typically driven by specific mission requirements, but rather by generic capability improvements and the introduction of new flight modes. Mission requirements are met instead by selecting or designing an appropriate set of I-load values.

Any unique mission (new inclination, orbit altitude, etc.) increases the amount of design work considerably. MOD figures indicate that the flight-specific man-hours (and cost) required may double or triple for a unique mission such as 51-F. Additional abort , post-MECO, Range Safety, and dispersion analyses may be required for these types of mission. Usually, analysis for unique missions starts well ahead of the official start of OFP Cycle 1 so that problem areas can be identified.

FUNCTION	DESCRIPTION	# I.loads	Today	DOLILU	SII	COMMENTS
Nominal	Pitch	30	Unique	Upload	Upload	
Guidance	Yaw	30	Unique	Upload	Upload	Inclination Dependent
	Roll	15	Fixed	Fixed	Fixed	
	Velocity Tbl	30	Unique	Unique	Fixed	Changes to Mach
	Throttle Level	4	Unique	Unique	Upload	Probably Fixed
	Throttle Vel	4	Unique	Unique	Upload	Probably Fixed
	AGT	10	Unique	Unique	Fixed	
Nominal	Nz Table	20	Unique	Unique	Fixed	delta Nz enabled
Flight	SRB Trims	20	Unique	Unique	Fixed	delta trim enabled
Control	Velocity Table	19	Unique	Unique	Fixed	Changes to Mach
	Elevon Sched.	16	Fixed	Fixed	Fixed	
	Elevon Vel.	7	Unique	Unique	Fixed	Changes to Mach
Engine Out	Pitch	60	Unique	Upload	Fixed	Changes to deltas
Guidance &	Misc	8	Fixed	Fixed	Fixed	Constraint Dependent
Flight Control	Trins & Nz's	40	Fixed	Fixed	Fixed	Constraint Dependent
Contingency Guid	delta Pitch,Misc	26	Fixed	Fixed	Fixed	Constraint Dependent
Season Ind. Iload	Wind	10			Upload	
New	Mach Curve	2			Upload	Probably Fixed
Unique/Upload/Fixed		339	234/0/105	114/120/105	0/80/271	

Figure 1: Flight Design Simplification

Flight design and verification costs for mission reconfiguration work are clearly dependent on the number of I-loads which must be changed on a mission-by-mission basis. Presently, investigations are underway to reduce the number of these flight-specific or flight-unique I-loads. Investigations are centered around developing a ground support system which can shape the launch trajectory on the day-of-launch (DOL) and upload the appropriate I-loads to the vehicle. In addition, certain changes to the flight software can help make other I-loads less sensitive to the seasonal variations in the environment, and so these could become fixed, rather than unique. Figure 1 shows the path of progress from today's flight software), through the soon-to-be implemented day-of-launch system (DOLILU), and on to a new version called seasonal independent I-loads (SII). As can be seen, the number of mission unique I-loads needed to configure the ascent GN&C will drop to about 80, and these will be computed on the day of launch.

Day of Launch and Landing Winds Assessment

The wind profile that the launch vehicle will encounter during atmospheric flight is the most influential environmental variable affecting a given mission. Surface winds, winds aloft and weather at the landing site are also important for the Shuttle, as they would be for any aircraft. The space shuttle is guided by pre-selected pitch and yaw programs during atmospheric flight. These programs are selected based on simulations made with statistical mean wind profiles for the month in which the flight is planned. Launch day winds which are significantly different from the design profile may cause excessive structural loads on the vehicle. Thus, it is extremely important to make accurate wind soundings as close as possible to the launch time, so that an informed Go/No-go decision may be made.

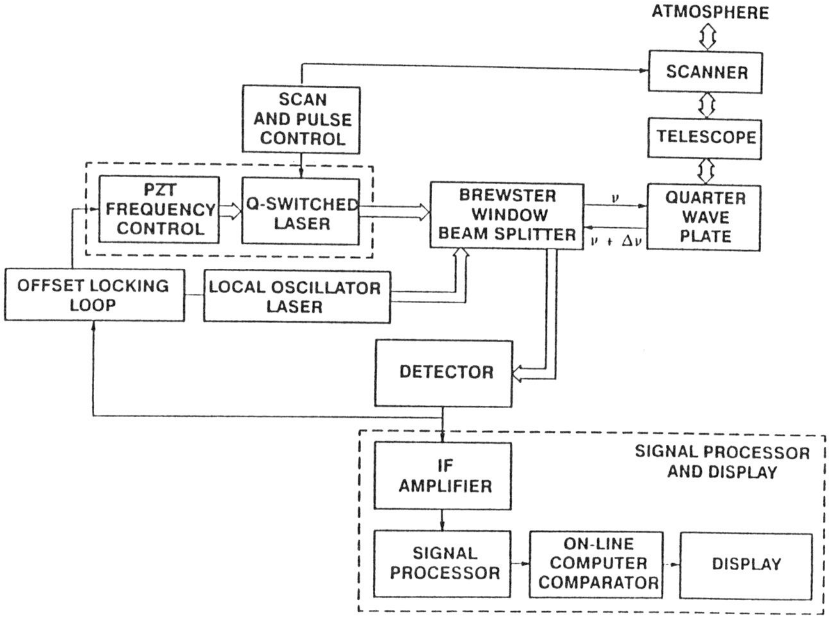

Figure 2: Lidar Wind Sounding System

Balloon-based wind soundings

Wind soundings for both ascent and entry are made with Jimsphere weather balloons. The balloons are released from a point in the vicinity of the launch pad or landing site and rise to maximum altitude of approximately 18 km. Radar installations at the range site track each balloon as it rises, obtaining wind speed and azimuth measurements at 25 m intervals. These measurements are sent to JSC for use in trajectory simulations to assess the effect on the planned launch. This system has two main disadvantages: First, the balloon takes approximately one hour to rise to the maximum altitude, which forces the balloon release to be made up to two hours before launch. Second, the balloon drifts with the wind, and therefore does not measure the wind profile along the launch or landing trajectory. These spatial and temporal differences can sometimes be significant. The predicted touchdown point for STS-31 was nearly 1000 feet in error because of inaccuracies in the measured wind.

Lidar-based wind soundings

Research into the use of laser radar (lidar) to measure winds has been ongoing for a number of years. Figure 2 shows a diagram of the signal processing technique employed by an optical heterodyne lidar wind sounding system[1]. Working lidar systems (based on CO_2 lasers) exist today that can measure winds at altitudes up to 3 km. Lidar systems fundamentally only make range-rate measurements, but researchers have developed techniques to enable a single unit to construct a three-dimensional wind field (figure 3). A lidar unit can make a complete wind sounding in about two to three minutes, and systems built around larger, higher powered CO_2 or solid state lasers appear to have the potential of reaching the 18 to 20 km altitudes required for launch assessment.

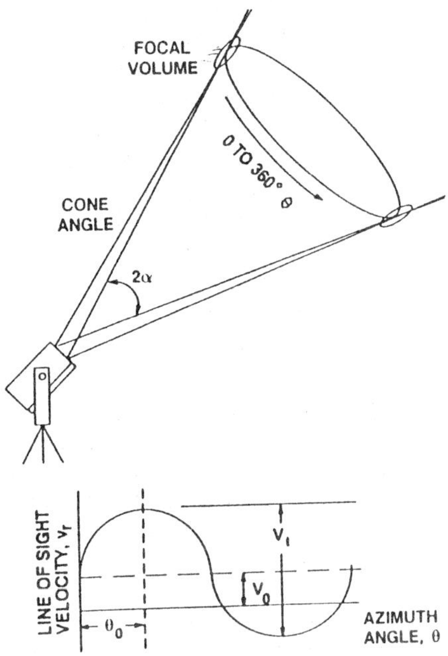

THE INSTANTANEOUS LINE-OF-SIGHT COMPONENT OF VELOCITY WITHIN THE SENSING VOLUME AS MEASURED BY THE LDV, v_r , IS GIVEN BY

$$v_r = v_h \sin\alpha \cos(q - q_0) + w \cos\alpha$$

WHERE:

w = WIND COMPONENT OUT OF CONE
= $V_0/\cos\alpha$

v_h = WIND COMPONENT ACROSS CONE
= $V_t/(2\sin\alpha)$

q_0 = DIRECTION OF ACROSS-CONE WIND COMPONENT

FOR SMALL CONE ANGLES, THE VAD IS MUCH MORE SENSITIVE TO THE DOWN-THE-CONE COMPONENT THAN THE ACROSS-THE-CONE COMPONENT OF THE WIND; i.e.:

α	1/cos α	1/2 sin α
30°	1.155	1.00
15°	1.036	1.93
6°	1.005	4.78
3°	1.001	9.61

Figure 3: Lidar Wind Measurement Technique

The use of a lidar system for launch and landing wind assessment can be beneficial to the STS program. Wind soundings could be made much closer to launch, because they could be completed in such a short time. The spatial differences could also be reduced by locating the lidar system close to the launch or landing site and by configuring a set of lidars to measure wind data along the actual shuttle trajectory. Current loads constraints incorporate allowances for these sources of uncertainty. The reduction of these uncertainties can improve both flight safety and launch availability.

The ability to make a sounding in a very short time may also be beneficial in that new soundings and simulations could be made after a short launch hold is declared, to maintain a more up-to-date assessment of the launch environment. Another benefit of using lidar is the ability to make a sounding right at launch, to be used for postflight analysis. Balloons cannot be used for this because the range radars are all trained on the vehicle itself at launch time.

Requirements definition for lidar wind sounding device

All four NASA centers are conducting analysis to define the operational requirements for a lidar wind sounding device. Table 1 lists some of the more important factors to consider and approach being taken to determine what these requirements are. The relationship between data frequency and sample interval deserves some clarification.

Requirement	Est./How determined
Schedule for Wind Soundings	? / Ops Procedures
Time to Complete One Sounding	10 min / Ops Proc.
Accuracy	0.5 m/s / analysis
Frequency (points per sounding)	Every 25m / analysis
Granularity (alt. range per point)	150m max. / analysis
Minimum and Maximum Altitude	30 m to 20 km
Two or Three Dimensional Wind Field	Currently 2D / anal.
Data Format and Transfer Method	Compatible w/current
Ability to Measure Along Trajectory	? / analysis
Ability to Measure in Launch Weather	Flight Rules / Demos
Hardware Reliability	Flight Critical
Operational Maturity	Operator skill req.

Table 1: Lidar Operational Requirements

Preliminary analysis on the sample interval requirement indicates that when wind measurements are averaged over sample intervals above about 200m, computed loads estimates begin to diverge from those estimated over smaller sample intervals. Figure 4 shows this effect for one of the load "indicators" calculated as part of the shuttle pre-launch analysis. This indicator represents External tank structural loads and is sensitive to dynamic pressure, angle of attack and angle of sideslip.

Lidar demonstrations are planned at KSC and Edwards during 1991. Comparisons of lidar measurements to balloon, Doppler radar, and possibly shuttle training aircraft (STA) based wind data will be made. The primary purpose is to demonstrate the operational maturity of the lidar as a wind sounding device and to identify critical issues so that the system, when developed, will meet STS program requirements.

Guidance Improvements

Space Shuttle guidance software can be grouped roughly into three categories: atmospheric powered flight, vacuum powered flight, and abort specific. We examined each of these categories, identifying possible improvement areas which would realize the greatest operational benefits, as outlined above.

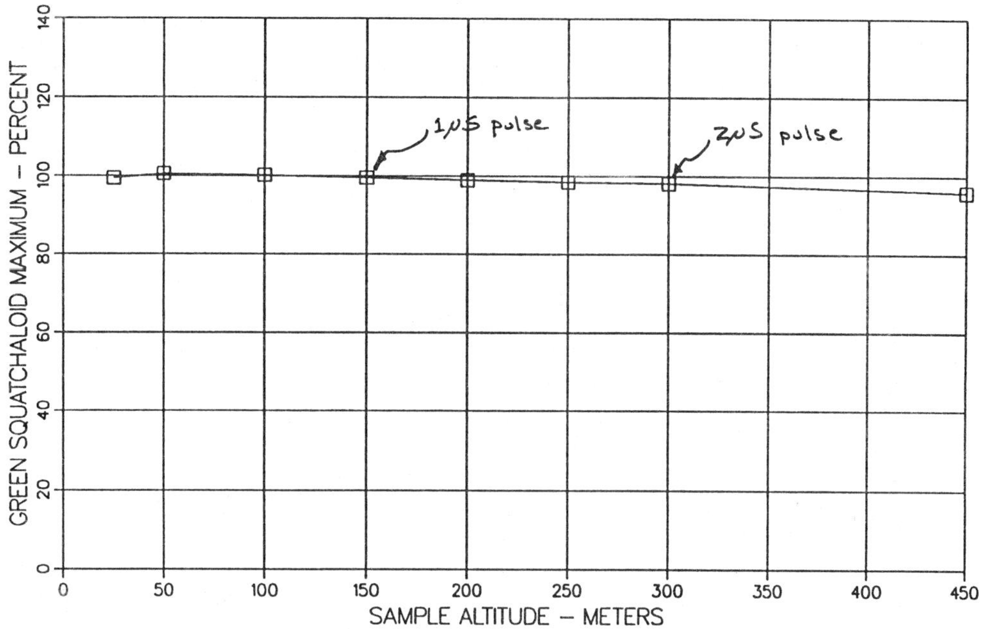

Figure 4: Effect of Sample Size on Load Indicator

Vacuum Powered Flight

The Space Shuttle Powered Explicit Guidance (PEG) algorithm has performed well for a wide variety of missions, requires virtually no pre-mission I-load changes except for new targets, and achieves near-optimum fuel consumption for nominal ascent missions. Although possible improvements have been identified, none of these had the potential to realize significant operational benefits to the STS program.

Atmospheric Powered Flight

The Space Shuttle, like all other current launch vehicles, uses pre-selected pitch and yaw programs during the atmospheric phase of flight. This essentially open loop technique is modified by a load relief feature which compares measured normal and lateral acceleration with a pre-selected profile and adds this error signal to the regular attitude and attitude rate error signals. On the basis of this signal, the space shuttle can weathercock into sufficiently large scale wind features and thus alleviate aerodynamic loads.

An improvement over this procedure, now in place in most launch vehicle programs and due to be incorporated in the Shuttle program, is the technique of modifying the pitch and yaw programs on the basis of wind soundings made close to launch time. These modified programs more closely match the wind to be encountered during the flight and enable performance targets to be met as well as matching trajectory constraints such as maximum dynamic pressure constraints. Figure 5 shows the process as it is being implemented in the Space Shuttle program.

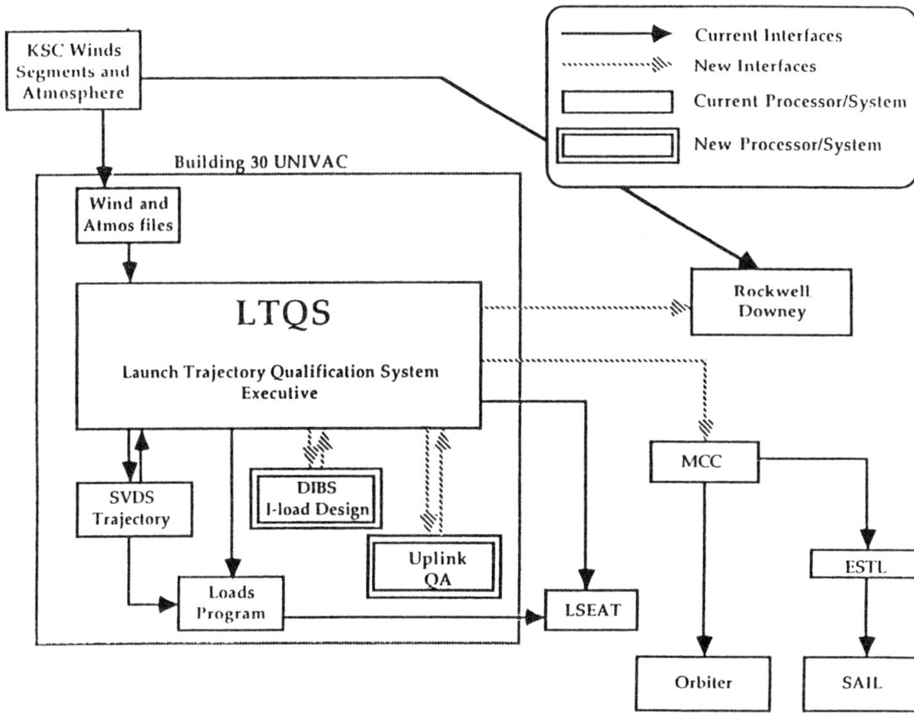

Figure 5: Space Shuttle Day of Launch I-load Update Process

Possible improvements to this method would include in-flight adaptive techniques which continuously compute and steer to an optimal flight path, taking into account dispersed environment and system performance. Wind measurements would still need to be made, possibly loaded onboard and ground-based analysis would still be required to support a Go/No-go decision, but such a method would potentially improve performance.

In order to determine the maximum possible improvement to be gained from such a technique, comparisons were made of optimized trajectories computed using various dynamic pressure constraints. The Space Shuttle trajectory is constrained by a maximum allowable dynamic pressure which is lower than what an unconstrained optimum would fly. Figure 6 shows the various dynamic pressure constraint lines used, and what they represent in terms of dispersion coverage. It also shows the increased performance resulting from these changes, measured in terms of additional propellant weight margin (which could be converted into additional payload). As shown, the most significant increase results from eliminating the reserve which accounts for the fact that the pitch and yaw programs are selected using a mean monthly wind. This reserve is effectively eliminated by using a wind sounding within two hours of actual launch to modify the pitch and yaw programs. The other reserves (two hour wind persistence and other systems dispersions) deliver much smaller performance increases, and probably would not be completely eliminated in a final, operational adaptive atmospheric guidance.

However, other steering techniques are being investigated (gamma guidance hybrids and modifications which would allow PEG to start earlier) which may reduce trajectory dispersions at staging, and thus improve overall system performance.

Abort Specific Powered Flight

The RTLS abort trajectory is shaped to meet a series of constraints. Figure 7 shows the main features of the RTLS trajectory and associated guidance modes. One area open for improvement is the fuel dissipation phase. Out of the design requirements of the RTLS, an unusual paradox arose; the RTLS consists of a fuel dissipation phase which steers the vehicle downrange in order to waste fuel, followed by a turnaround and flyback phase which make use of a fuel optimal steering law. This apparent contradiction was actually driven by two important operational considerations. First, prior to the second shuttle flight, in which the new Transoceanic Abort Landing (TAL) guidance mode was devised, it was very important to extend the availability of the RTLS to as late a time as possible. These late RTLS cases included no fuel dissipation phase, relying on the fuel optimal steering to make it back within gliding range of the KSC runway. Second, the limitations in onboard memory did not allow a dedicated, unique guidance algorithm to be used for the RTLS. Since the TAL abort mode is now preferred over the RTLS, and since GPC upgrades are planned, it is now feasible to revisit these decisions and possibly make some improvements in the RTLS.

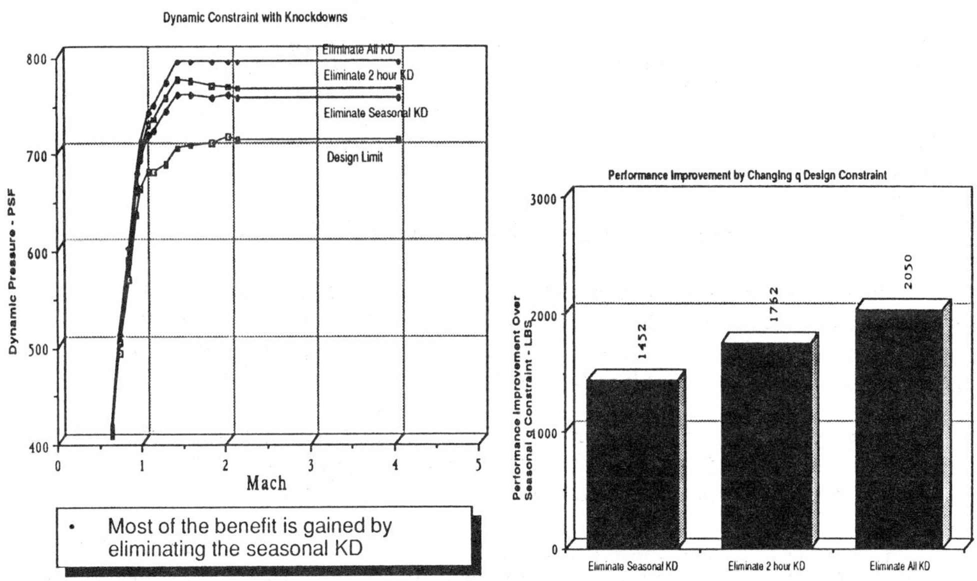

Figure 6: Potential Performance Gains from Relaxing Dynamic Pressure Constraint

Other possible areas of improvement are the Powered Pitchdown (PPD) maneuver and the turnaround phase. Typical fuel optimal trajectories for RTLS reach their range, velocity and flight path targets at a large angle of attack (alpha). Since separation from the ET must occur at alpha's around -4 degrees and since proper control of the orbiter-ET combination can be maintained only at small alpha's at the dynamic pressures typically encountered at separation, the PPD is used to correct the vehicle attitude once the target conditions are reached and before the main engines are shut down. This maneuver is a simple open-loop pitch program and can result in undesirable target dispersions.

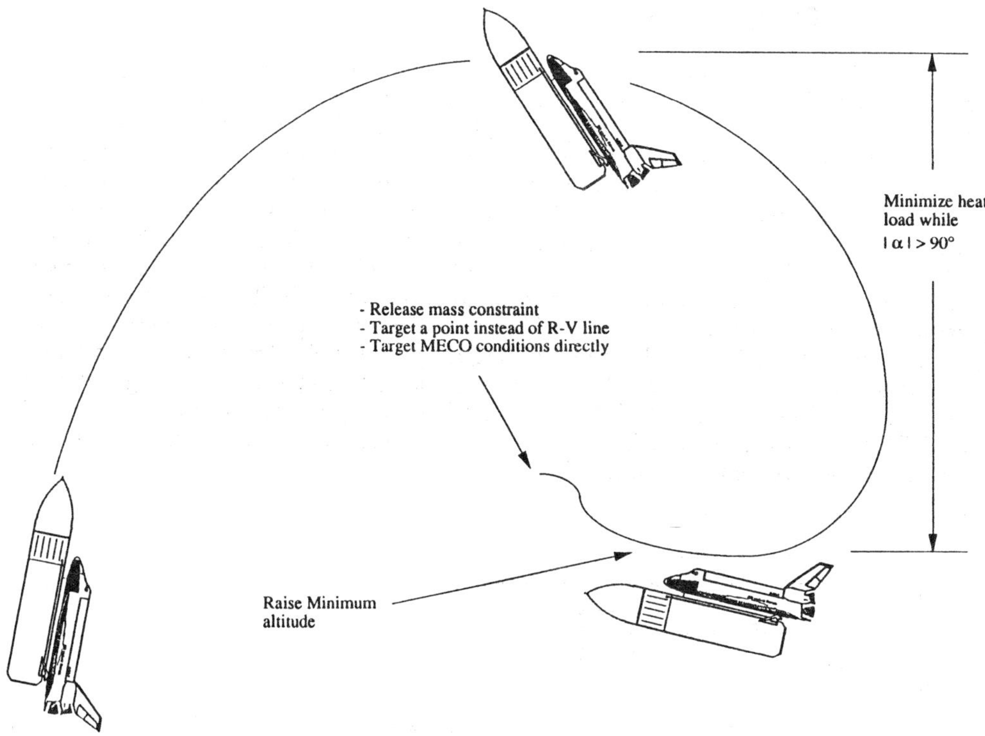

Figure 7: Space Shuttle RTLS Trajectory

The turnaround phase uses an in-plane maneuver to arrest the vehicle's downrange motion and begin flying back toward the KSC runway. This maneuver results in the orbiter-ET combination flying backwards at high speed through the SSME plumes and dropping very low into the atmosphere, which causes heating concerns.

In this study we are investigating a number of potential guidance improvements, including alternate fuel dissipation algorithms, active targeting schemes and out-of-plane maneuvers, to increase the adaptability and flight safety of this abort.

A variational calculus method, called the Sequential Gradient / Restoration Algorithm (SeGRAm), is being used to help determine the optimal control law for the RTLS abort. SeGRAm can compute optimal throttle and thrust pointing profiles for a set of in-flight constraints (minimum altitude, maximum heat rate and total heat load) and PPD conditions (altitude, flight path angle, mass, velocity) over a range of RTLS abort trajectories, providing a basis to measure the effectiveness of new algorithms.

In the current mechanization of the RTLS fuel dissipation phase, the shuttle flies a constant inertial pitch attitude after the engine failure to maintain altitude while continuing downrange until enough fuel has been burned that completing the flyback will leave a minimal amount at engine cutoff. This inertial attitude is computed prelaunch and is a function of velocity, altitude rate and engine failure time.

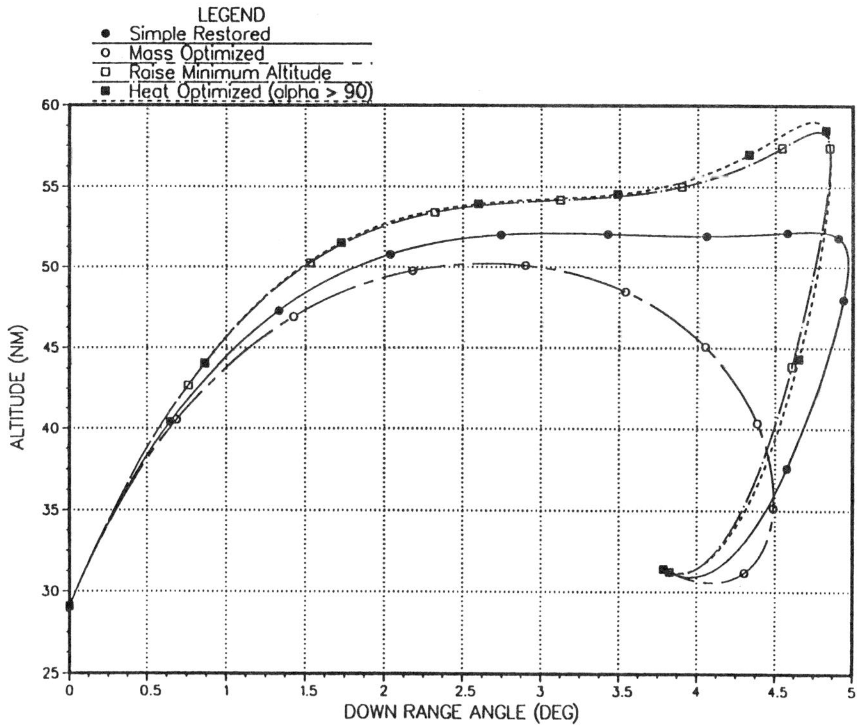

Figure 8: Various Optimizations of RTLS Trajectory

Preliminary in-plane optimizations have been performed using various penalty functions. The results are compared with the mass-optimized solution (for which PEG is an approximate mechanization) in figure 8. One obvious conclusion is that aerodynamic heating can be controlled by flying a higher trajectory, but that large negative altitude rates would be necessary to bring the vehicle to the desired engine cutoff conditions. In addition, these trajectories also require the vehicle to fly backwards for some period of time. Preliminary results with out-of-plane trajectories look promising. Small flight path angles can be maintained throughout the trajectory, altitude is kept above desired minimum values and vehicle attitude is more benign than in the in-plane maneuver. Comparisons are shown in figures 9 and 10.

A bi-linear tangent algorithm is being investigated as a possible mechanism to produce such a trajectory. It has the benefit of additional control variables (over the linear tangent approximated in PEG) which can be used to implement a non-linear turning rate, allow for additional terminal targets and eliminate the open loop phases of the trajectory.

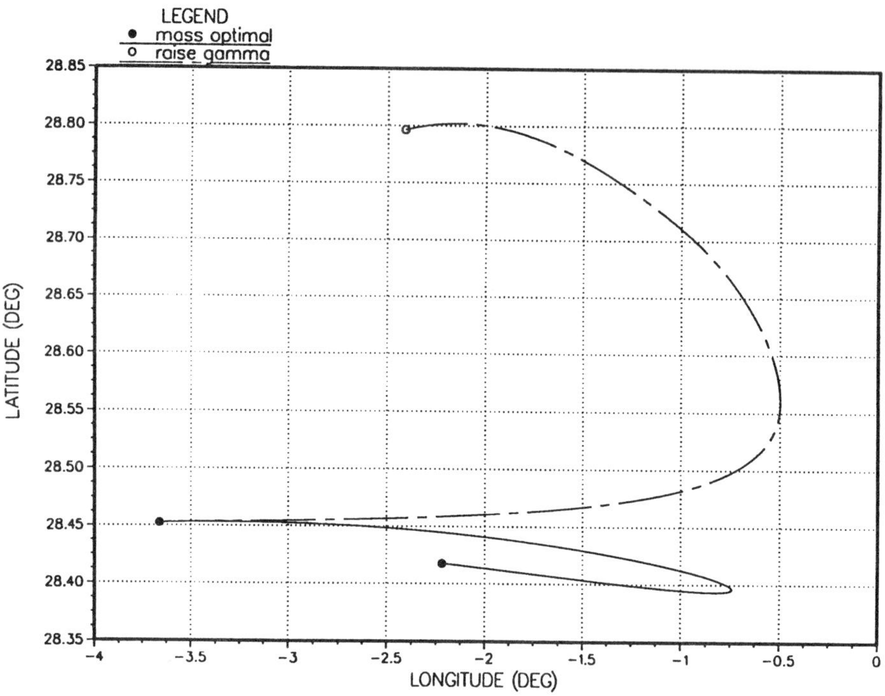

Figure 9: Out-of-plane RTLS Solution

Designing the fuel dissipation phase, as well as other open loop phases such as PPD, is an iterative procedure that is done months before launch. In addition to onboard changes, we are looking at ways to functionalize the parameters associated with these phases so that the design becomes mission independent. This involves for example, adding weight, altitude and downrange velocity to altitude rate and total velocity as independent variables in the evaluation of the fuel dissipation attitude. These modifications will reduce the cost and effort of the design as well as enhance the reliability.

Improved Ascent Load Relief

STS flight control incorporates a load relief algorithm that is designed to maintain a reference normal acceleration profile in the presence of environmental and vehicle systems dispersions. This feature is intended to keep structural loads within red-line limits so that a launch can proceed even if the winds on the day of launch are different than those for which the ascent profile was designed. Figure 11 shows a diagram of the pitch channel[2].

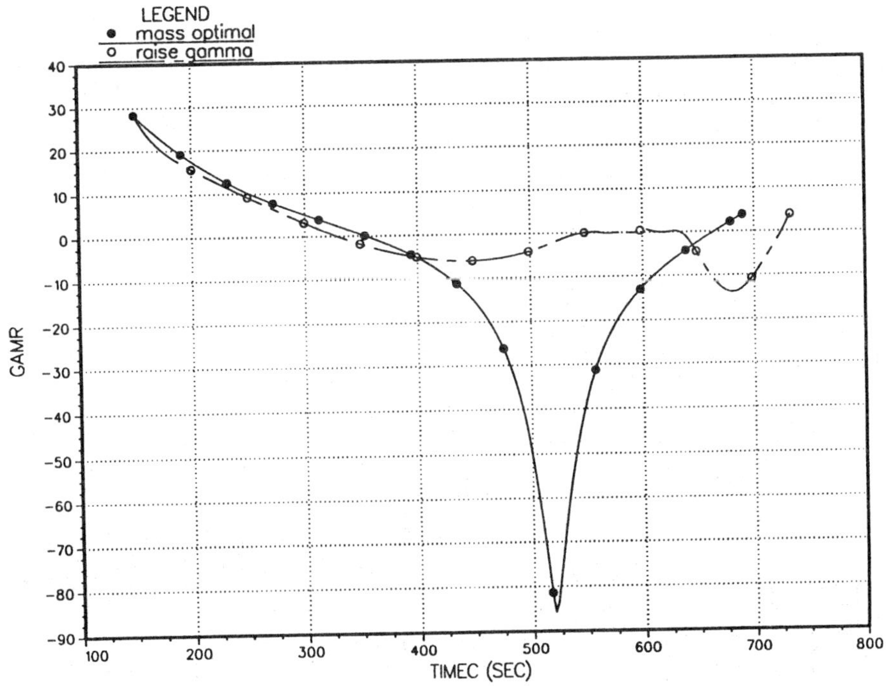

Figure 10: Out-of-plane RTLS Solution - Flight Path Angle History

The main problems associated with this load relief scheme center around the characteristics of the sensor itself and on the effect the load relief feedback loop has on the stability characteristics of the Shuttle. The accelerometer assembly unit tends to produce a noisy signal which contains a significant component due to the elastic bending modes of the shuttle. The signal must be heavily filtered to be useful for load alleviation. The additional loop must respond as quickly as practical to have the intended effect in the wind measured wind profiles, and this results in a reduction in the overall rigid body gain margin. Possible improvements include: modifying the filtering of the accelerometer signals, using angle of attack and sideslip sensors instead of the accelerometer measurements, and simply turning the load relief feature off. There are a number of operational issues associated with this question and these are outlined below.

Background: Launch Availability

Before the Shuttle program commits to a launch, detailed assessments of the predicted trajectory are made to ensure that the vehicle and its systems can operate successfully in the environment at the time of launch. Chief among the variables which influence the trajectory on launch day is the wind profile. As described earlier, balloon soundings are made at intervals prior to launch and these are fed into computer simulations of the trajectory, which are then used to estimate structural loads, performance and other parameters. These estimates form the basis for the decision to continue with the launch countdown, or to hold or scrub the launch. Shuttle launches have twice been scrubbed solely due to high predicted loads caused by adverse wind profiles. A number of other launches had high predicted loads, but were scrubbed for other reasons. Many launches have had predicted loads high enough to cause concern, prompting detailed reviews of the loads calculations before a decision was made to proceed with the launch. The ability of the Shuttle to proceed with a launch in the presents of environmental dispersions such as winds contributes to what is known as launch availability. Launch availability is characterized as the probability that a launch can proceed on a given date with a given payload and destination.

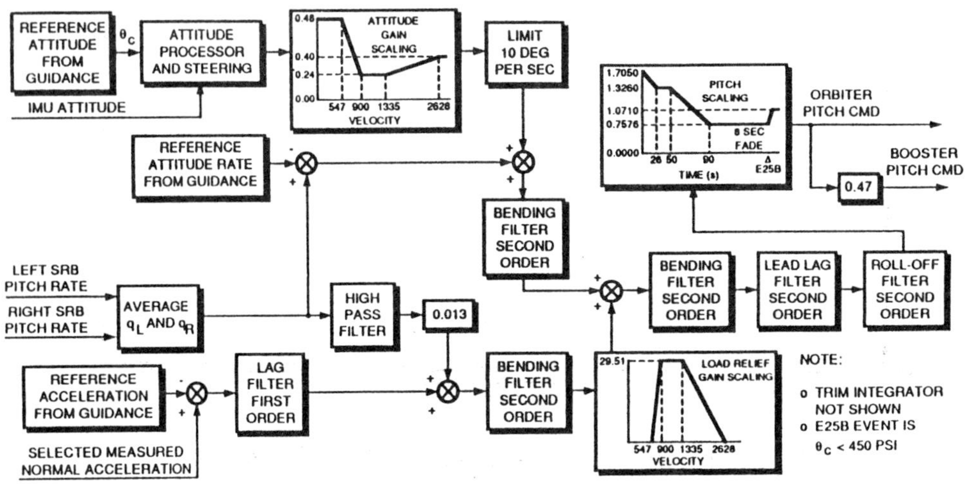

Figure 11: Space Shuttle Flight Control - Pitch Channel

Candidate improvements to the Shuttle GN&C are measured partly by their effect on launch availability. To estimate launch availability, simulations are made with a series of 150 winds measured in a particular month, chosen to represent the statistics of the actual population. Trajectory data from these winds are then used to drive a loads model, and the results of this model are time histories of approximately 200 load indicators, each representing a particular structural member of the launch vehicle. Critical values for each of these load indicators are defined by the shuttle program, and the percentage of trajectories in which all indicators remain below these critical values is an estimate of launch availability, at least as far as the influence of winds is concerned. Load relief algorithm improvements should reduce the peak values of these indicators and so improve the launch availability.

Load Relief On vs. Off

The simplest possible change to the current system is to disable the load relief feature. Since the primary purpose of load relief is to improve launch availability, estimates were made of the effect of load relief on launch availability. These estimates were made based on the current procedure of shaping pitch and yaw profiles using a mean monthly wind profile, and shaping to the measured wind profiles forming the 150 wind set for a particular month. Figure 12 shows the estimates for these two procedures and the effect of load relief (in both pitch and yaw and in pitch only) on each. Clearly, when trajectories are shaped to the mean monthly wind, the load relief feature improves launch availability substantially. However, when measured winds are used, load relief noticeably *reduces* launch availability. The reason for this is that the load relief feedback causes increased attitude oscillations in a frequency range of 0.1 to 0.2 Hz. Figures 13 through 16 show the vehicle response in attitude to synthetic, sinusoidal winds of various wavelengths, spanning the critical frequencies mentioned above. Response with load relief on reaches a peak gain of about 2. Also evident in these figures is the fact that load relief has no effect at all on small scale, high frequency wind features. Figures 17 and 18 compare the attitude response to a selected measured wind, and show that this behavior is evident in real winds as well.

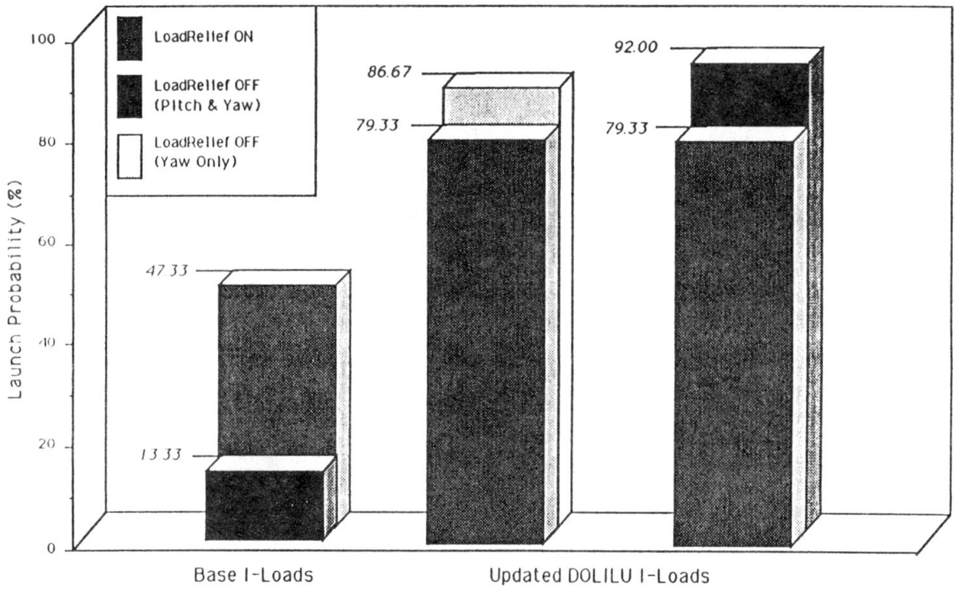

Figure 12: Launch Probability Comparisons

The increased amplitude of attitude excursions causes increased aerodynamic loading. A measure of aerodynamic loads is the product of dynamic pressure and either angle of attack or angle of sideslip, q-alpha and q-beta. Figures 19 and 20 compare envelopes of these parameters for all 150 winds with load relief on and off. In figure 19, which represents the trajectories shaped for the mean monthly wind, The envelopes for load relief on are shown to be larger, but are more nearly centered in the constraint corridor than the envelopes for load relief off. This is the desired effect of load relief: it weathercocks the vehicle into large scale wind features. However, the process of shaping the trajectory to the measured wind also accomplishes this same result, and we are left only with the adverse load relief effect of increasing the attitude excursions, as shown in figure 20.

Studies were made of the effect of load relief in the presence of an engine failure, SRB actuator failure, stuck throttle and aero model dispersion. In all these cases, turning load relief off revealed no degradation in overall system response.

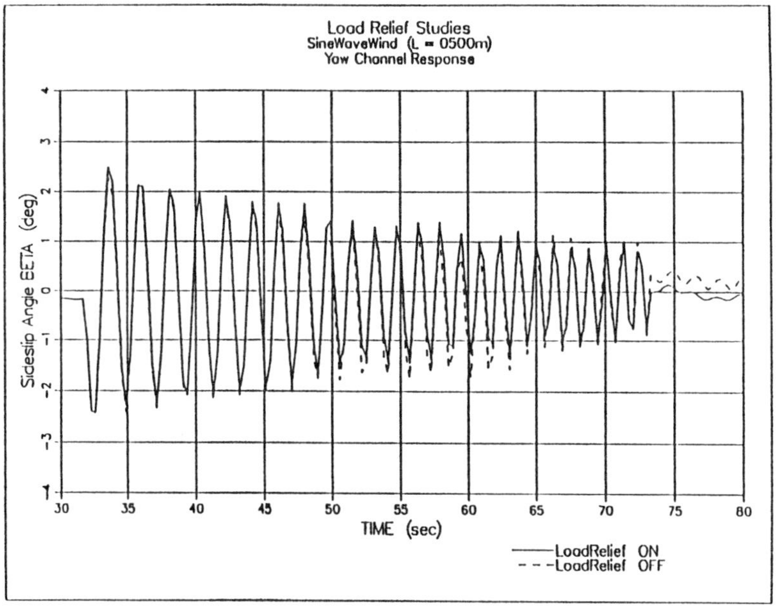

Figure 13: Flight Control Response - 500 m Wavelength

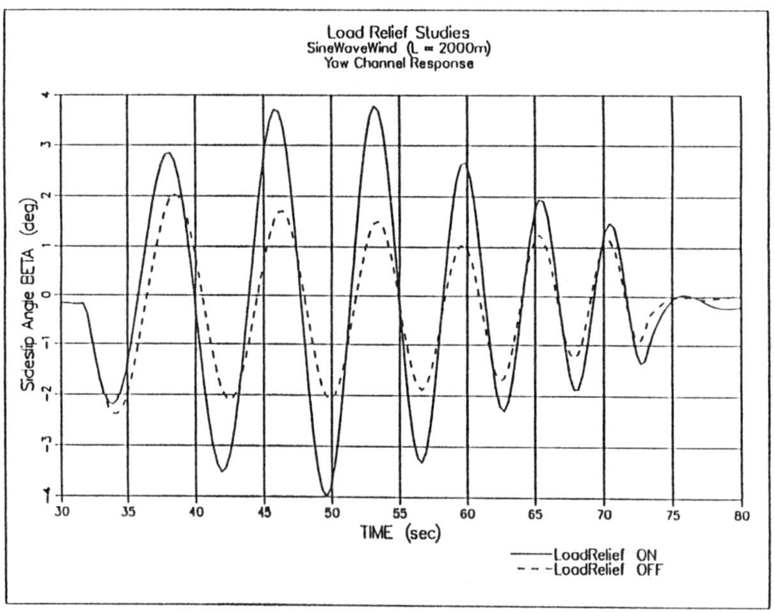

Figure 14: Flight Control Response - 2000 m Wavelength

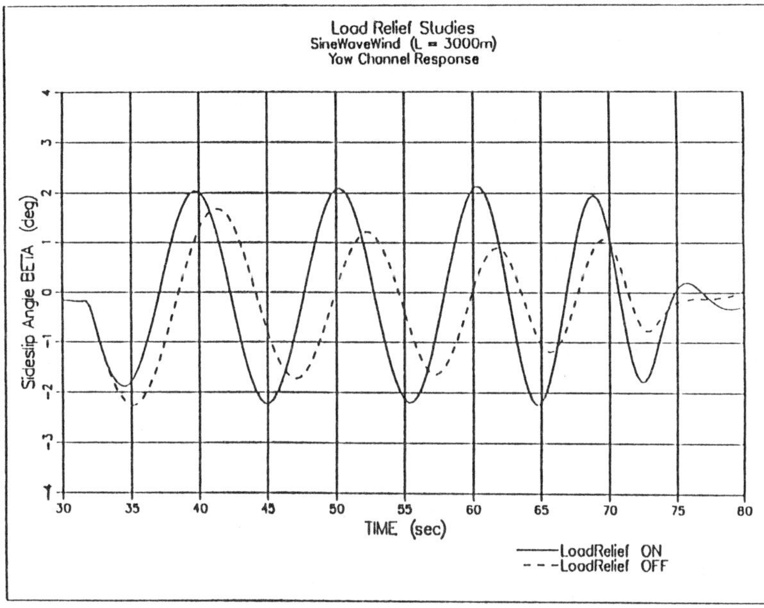

Figure 15: Flight Control Response - 3000 m Wavelength

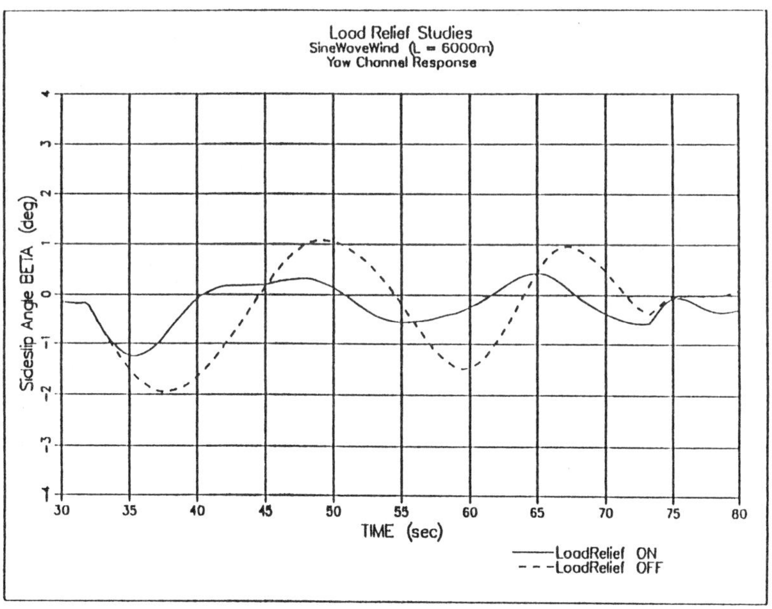

Figure 16: Flight Control Response - 6000 m Wavelength

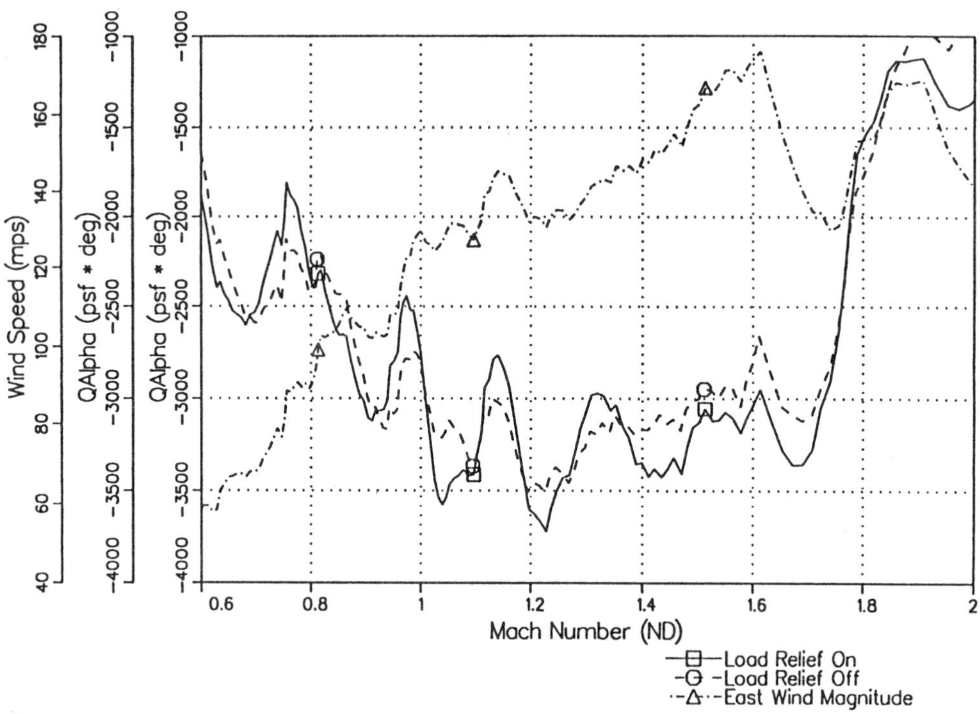

Figure 17: Pitch Channel Response - Measured Wind

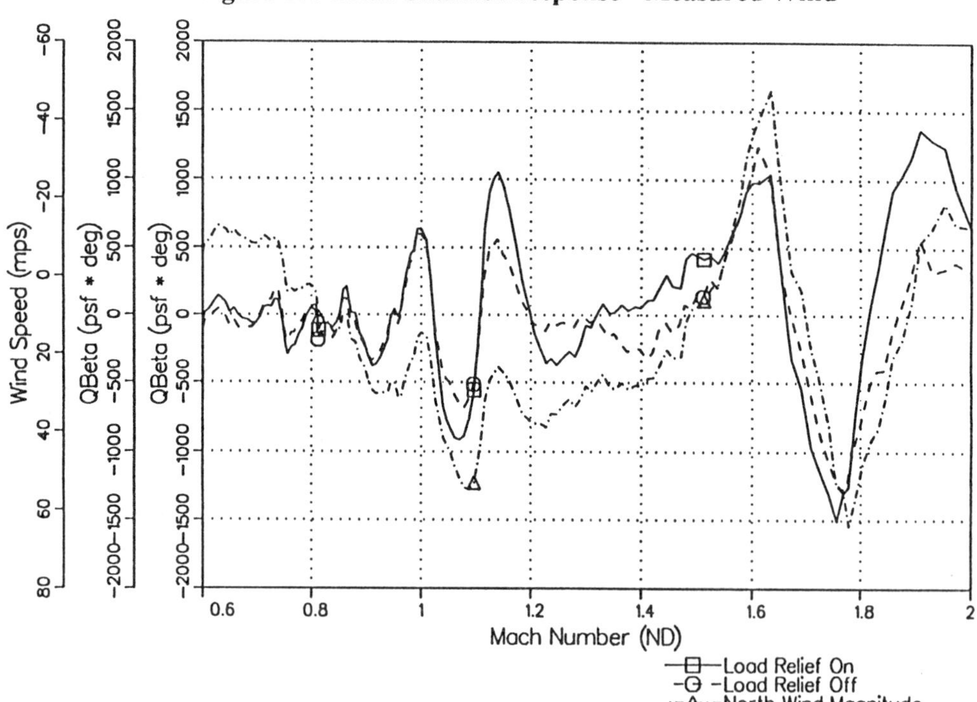

Figure 18: Yaw Channel Response - Measured Wind

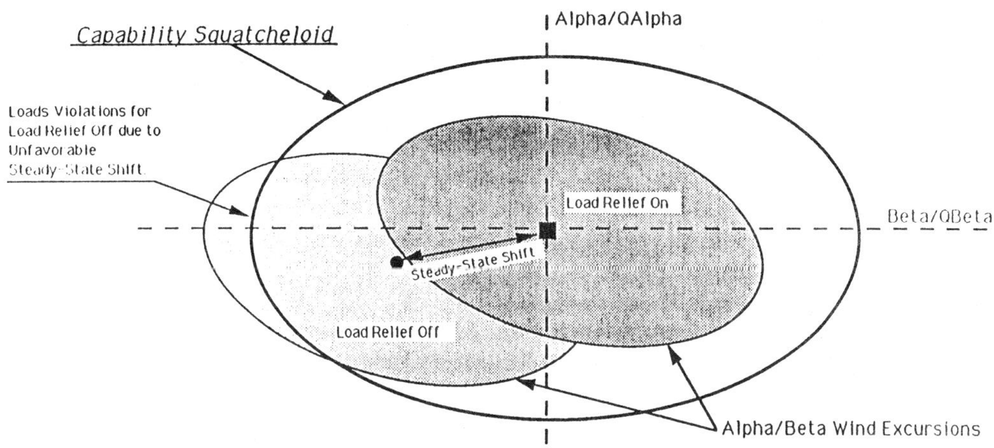

Figure 19: Effect of Load Relief without Day of Launch Update

Alternate Sensors for Load Relief

A possible alternative to turning load relief off would be to substitute sensors that directly or indirectly measure angle of attack and angle of sideslip instead of normal and lateral acceleration. This study will assess the feasibility of incorporating new sensors to measure the vehicle state and environment and incorporating those measurements into the flight control system to reduce loads and improve performance and launch availability.

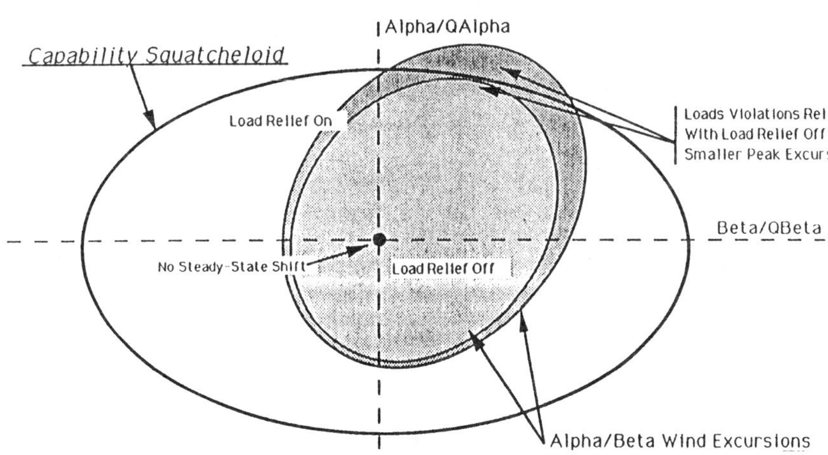

Figure 20: Effect of Load Relief with Day of Launch Update

New technology for directly measuring the angle of attack of the ascent vehicle, including such possibilities as using Global Positioning System (GPS) signals and using lidar data will be examined. STS requirements such as accuracy and update frequency will be determined, and the new technologies will be assessed on this basis.

This study will also quantify the improvements which could be realized by incorporating angle-of-attack feedback into the control algorithm in various ways. The goal will be to determine the characteristics and benefits of a system based on a realistic set of requirements for the sensor (as determined above).

Benefits of using additional accelerometers as feedback elements or different positioning of the existing ones will be examined. Current state-of-the-art in acceleration measurement will also be reviewed to see if there is any benefit to be gained by using updated hardware.

Adaptive GN&C for Systems Failures

A significant amount of STS mission reconfiguration and systems analysis work is devoted to making sure systems failures (such as an SSME failure) are handled properly. Improvements in the way the onboard GN&C reacts to such failures could reduce or eliminate portions of this work.

In addition, there is a list of systems for which safe recovery from a failure is unlikely with the current algorithms and implementations. GN&C improvements which would result in removing items from this list (the *Critical Items List*) would improve flight safety and simplify the Flight Readiness Review process.

Background: Failure Scenario Identification

This study will review and examine known and documented systems and scenarios in which failures would affect the operation of the flight control hardware and software. If possible, these failure will be classified according to the likelihood of occurrence, the specific effect on the flight control and whether it is possible to recover from the failure. Further analysis will be targeted to failures which have the highest likelihood of occurrence and where a solution is thought to be possible.

Sources for the failure scenarios will be the published STS Critical Items List, Volume X, and documentation on sources of systems dispersions.

Alternate Gain Synthesis Techniques

Study plans call for re-examining the current flight control gains for potential improvement, using modern techniques in gain synthesis (LQR and H-infinity). The goal will be to identify alternate gain sets which are more robust with respect to specific failures and which deliver better performance in conjunction with other proposed modifications to the GN&C (such as load relief modifications and guidance modifications currently under study to support day-of-launch trajectory shaping).

Special emphasis will be given to examining new algorithms which provide protection against the target failures automatically, which no special failure logic required. In addition, special purpose algorithms which are designed to respond to failures not currently handled will be investigated.

Summary

The Autonomous GN&C Bridging Program is providing an effective method for moving new and developing technology into an operational space launch program. By identifying and focusing at a detailed level on real operational problems, and by making the up front effort to quantify the expected benefits, a real payoff in space shuttle operations is expected. In addition, this work should provide insight into the application of similar technology to other space programs, both current and future.

REFERENCES

1. Targ, R., "Lidar Remote Atmospheric Sensing for Advanced Launch System Predictive Guidance and Structural Damping", NASA Contract NAS1-18444, January 1990.

2. Lyons, J.T. & Sorrells, J.E., "Current State of Practice in Guidance Navigation and Control for Space Launch Vehicles", WRDC-TR-89-3111, November 1989.

AAS 91-023

A SCALABLE ARCHITECTURE FOR AN OPERATIONAL SPACEBORNE AUTONAV SYSTEM

L. Jane Hansen*

To meet the expanding and diverse requirements of spaceborne processing applications, modular, reconfigurable, and scalable systems must be developed. A scalable system allows the user to vary capacity and complexity within cost constraints (size, weight, power, and price) and in terms of both software and processing hardware. The software system must accommodate varying types of hardware inputs, and will drive the processing requirements of the system in which it executes. Thus, the software design is critical to the success of the scalable system in meeting the top-level functional needs of future spacecraft. Additionally, the creation and acceptance of common processing modules and interfaces as a means of achieving spacecraft computer standardization is needed to further facilitate the development of scalable systems.

This paper discusses the concepts of *standardization, commonality,* and *reuse*, as they help to achieve the goals of low cost scalability. To illustrate these concepts, we will use the Microcosm Autonomous Navigation System (MANS) software development process as an example. MANS provides a scalable architecture for an operational spacecraft autonav system with varying levels of functionality based on the chosen configuration and the addition or deletion of software and hardware modules. The potential use of Ada library modules in the operational flight software, as well as the creation of new modules for these libraries, has been key to the development. A modular computer hardware architecture developed around the MIL-Std-1750A processor, and well-partitioned software modules, allow the user to scale the system to meet requirements at minimal costs.

We examine the key design decisions made during the MANS development, highlighting data critical to these decisions. We discuss the various steps of the development process as they relate to scalability. Finally, we present "lessons learned" from the development of one of the first scalable systems built for spaceflight.

INTRODUCTION AND BACKGROUND

This section introduces the concepts associated with scalable systems, the backbone of which is the standardization of hardware and software elements and interfaces.

* Director of Software Department, Microcosm, Inc., 2601 Airport Drive, Suite 230, Torrance, California 90505.

Additionally, it provides an overview of the Microcosm Autonomous Navigation System (MANS)† used as an example throughout the paper.

Standardization

The costs associated with embedded computer system development have been increasing since the 1960s. This increase in cost can be directly related to an increase in system and spacecraft capability. As computer systems become more powerful, they provide the basis for more powerful software, thereby increasing system functional capability. However, since spacecraft systems require very high reliability, new approaches are not quickly adopted. Thus, the potential increase in space system capability based on newly available processing systems and more complex software has not been realized. Because most space systems are built in one- and two-quantity lots, economies of scale are hard to achieve for costs of hardware production associated with computer processors and software development. An emphasis on system standardization in both processor hardware and software will allow the space community to overcome current cost penalties while providing the benefits of more complex systems scalable to need levels.

System and processing architectures have been developed which will achieve standardization for embedded spacecraft computer systems. Standardization implies the use of "common" elements by all spacecraft suppliers; however, this is not necessarily the precise implementation suggested. Standardization hopes to achieve reduced costs through high quantity production runs of hardware, the reuse of previously developed software modules, and greatly reduced development time. Standardization also leads to the implementation of scalable systems where hardware and software modules can be configured to meet the specific needs of any mission. Thus, standardization will allow for development of more capable systems in scalable packages.

Hardware Standardization

The concept of processor hardware standardization is based on several assumptions. We first confine standardization to functionally equivalent processing tasks. Thus, signal processing applications should not attempt to use standard elements designed to meet the needs of data processing applications. Standard elements should be identified in each of the functional processing categories required for spacecraft operations. An example of the partitioning of functional categories is as follows:

- Real-time Data Manipulation, Logic, and Mathematics
- Non-real-time Data Manipulation, Logic, and Mathematics
- Input, Output, and Communications links
- Payload Processing

From this partitioning, common hardware modules can be created to meet the varying needs of applications in each category. Any one application may only require a subset of the common hardware modules available to that type of software. Thus, the system becomes scalable in terms of capabilities required.

---

† The MANS effort depicted herein is monitored and sponsored in part by the United States Air Force, Air Force Systems Command, Phillips Laboratory, Kirtland AFB, NM 87117-6008. Microcosm, Inc. retains patent rights to the MANS software.

Most processors available for use in space today are built by individual suppliers to meet specific requirement specifications. Although the systems have similar elements, there has been little progress toward achieving standardization for interfaces or functionality. This is to be expected, since interface standardization is dependent upon and can only occur subsequent to the creation and acceptance of a common, standard backplane. Such a backplane, which can be used for all processors supporting multiple functional software groups, represents a major development activity. The implication here is that processor developers and suppliers must share their knowledge to create common technology specifications.

The commercial world has embraced the idea and has managed to achieve a high level of standardization. When IBM created the IBM Personal Computer from an Intel 8088 processor and their own backplane, they also made their standards known (by creating what is called an *open architecture*) so that others could develop *clone* systems and performance-enhancing peripherals. IBM chose an 8-bit processor so that existing 8-bit peripherals could be easily used. In some respects, IBM created its own competition. However, they also created a multitude of software application packages targeted for the major PC hardware market, consisting of their computers and clone systems. In so doing they may have increased their market share more than if they had kept their system design proprietary and created a smaller, captive-user audience.

The government is now trying to create similar standards. For example, the JIAWG (Joint Integrated Avionics Working Group—consisting of Air Force, Navy, and Army representatives) has selected the *PI bus* as their standard backplane. The specification for the PI bus is common across the industry, and the goal is to develop modules which everyone will use to work with that backplane. While there is potentially greater initial investment in designing all the modules which might be required by any system to meet the needs of various spacecraft functions, the payoff is in the long run reduction of maintenance and replacement costs. The Air Force PAVE PILLAR and PAVE PACE programs are working to further define common modules and to create large-scale architectures to accommodate them for aircraft avionics.

Software Standardization

Standardization for software is similar to hardware. Interfaces and functionality must be well defined so that standardized elements can be created. As with the hardware functional categories above, software functional categories can be created as follows:

- Control System Software (Orbit and Attitude)
- Navigation Software - Orbit and Attitude Determination
- Input, Output, and Communications links
- Payload Processing
- Computer System Control (Executive, Kernel, Utilities)

Within these categories, many software modules may exist; even multiple approaches to similar problems can be developed and maintained. These modules are stored in reuse libraries. The Government has been strongly supporting the development of reusable software libraries, particularly in Ada. The intent of these libraries of reusable modules is to allow software concepts, designs, and even code, to be reused by other developers, much as hardware chips are reused by computer manufacturers today. This idea is often referred to as *software I.C.* (integrated circuit). Software is stored in the libraries by type, based on requirements, performance attributes, documentation availability, and other criteria. Table 1 shows some of the reuse libraries specifically for Ada-based products available today.

Table 1
POTENTIAL ADA REUSE LIBRARIES

Library/Organization	Point of Contact	Attributes
Ada Software Repository (ASR)	Bob Kitchin P.O. Drawer 100 Building T148 White Sands Missile Range, NM 88002	• Government Sponsor • Oldest and largest Ada software library • Will take software from anyone who offers
Common Ada Missile Packages (CAMP)	CAMP Information Data and Analysis Center for Software (DACS) RADC/COED Griffiss AFB, NY 13441-5700	• Government Sponsor • Navigation and Guidance • Scientific • Includes test data
Booch Components	Wizard Software 835 S. Moore Street Lackwood, CO 80226	• Commercial - Costs $ • Educational Discount • 501 Components • Large number of smaller parts
Grace Components	EVB Software Engineering 5303 Spectrum Drive Frederick, MD 21701	• Commercial - Costs $ • 275 Components • Smaller number of larger parts

Data Source: Don Firesmith, Advanced Software Technology Specialists

The MANS System

The autonav system which Microcosm has developed is called MANS, the Microcosm Autonomous Navigation System. MANS provides the following data onboard, in real-time, at 250-millisecond intervals:

- Spacecraft position and velocity
- Spacecraft attitude and attitude rate
- Sun direction in spacecraft coordinates
- Subsatellite location
- Ground look-point for any axis defined in spacecraft coordinates

This is achieved by processing data from any combination of the following input sensors: one or two Barnes Engineering Dual Cone Scanners, gyroscopes, acclerometers, GPS receiver, and star sensor. Accuracy and coverage will depend on the sensor configuration and orbit conditions. With the baseline complement of two Dual Cone Scanners and no other sensors, MANS will provide position accuracy of ±100 m (3 sigma) and attitude accuracy of ±0.03 degrees (3 sigma) in low-Earth orbit. MANS will be flight tested in 1992 to validate its on-orbit performance.

The MANS software receives sensor inputs from the Dual Cone Scanner (DCS) sensor. The DCS has two components: the sensor head, shown in Fig. 1, and the associated electronics box. The sensor uses a pyroelectric detector operating in the infrared for Earth sensing, and a silicon detector operating in the visible for sensing the Sun and Moon. The DCS provides independent measurement data from which MANS derives the angles between the Earth, Sun, and Moon, as well as distance to the Earth. As illustrated in Fig. 2, the sensor receives a total of seven independent measurements: two components of the direction to the Sun, two components of the direction to the Moon, and three components of the direction and distance to the Earth. Using these seven independent measurements, MANS solves simultaneously for both the attitude and position of the spacecraft.

Photo courtesy of Barnes Engineering

Fig. 1. The Barnes Engineering Dual Cone Scanner (DCS).

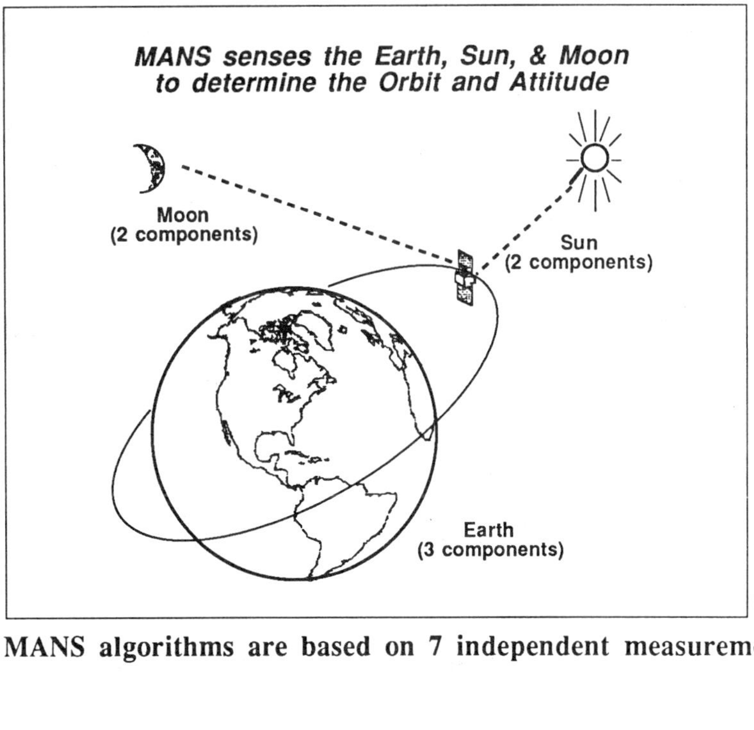

Fig. 2. MANS algorithms are based on 7 independent measurements.

The elements of the MANS Software, the Computer Software Components (CSCs), are shown in Fig. 3. Fig. 4 shows the general structure of the MANS software in terms of data flow. The processing begins by obtaining measurements from available sensors, validating the measurements through a variety of independent data checks—either internal to the sensor or between sensors—and then filtering the output to achieve the most accurate solution available at any time.

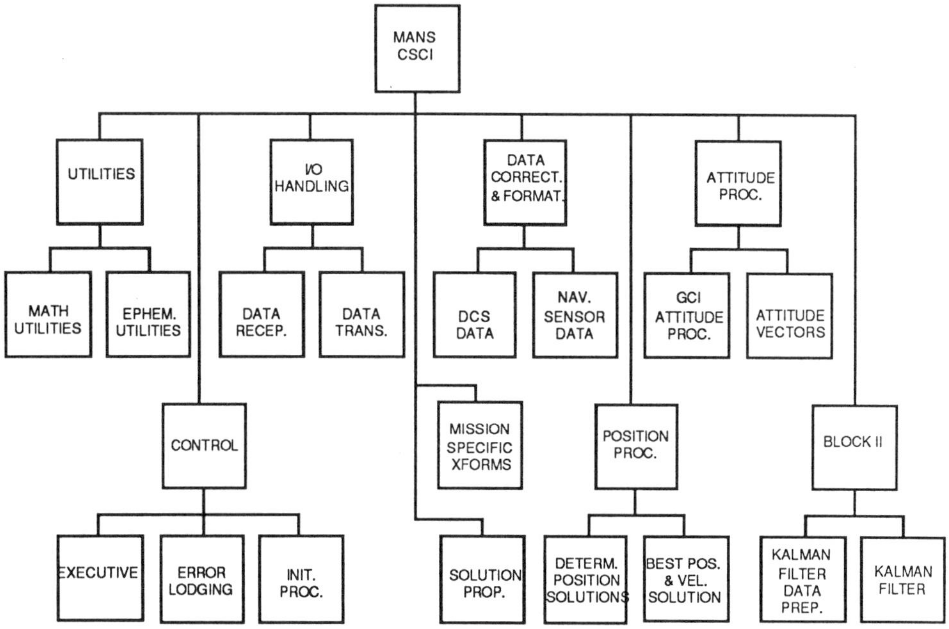

Fig. 3. MANS Computer Software Components (CSCs).

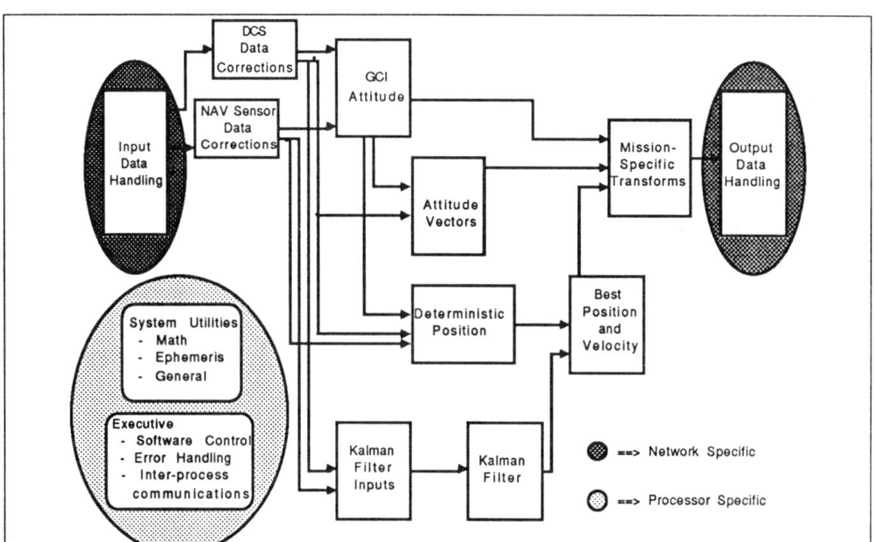

Fig. 4. MANS Software Data Flow.

MANS is fully autonomous. Failures in any of the hardware components are detected by the MANS software and can be accommodated in real time onboard by using whatever residual sensing is available. If no real data is available, MANS will propagate its best state solution as long as reasonable. The corollary to this is that both sensing and processing hardware can be eliminated from the system without necessarily changing the software structure or the mechanization of the system. This means that the selection of hardware for a particular system configuration can be based on the required accuracy, cost and weight constraints, and the level of functional redundancy necessary for each individual mission. Thus both external inputs and the associated software processing are scalable to meet specific mission needs.

MANS SOFTWARE DESIGN

During the initial development of MANS, we investigated approaches for data validation and measurement quality assessment. We evaluated processor configurations to determine which would provide the most long-term flexibility with the processing speed and numerical precision required to accommodate the mathematically intensive algorithms used in the MANS software. Ada was chosen for the MANS software development, for several reasons. Ada is broadly accepted within the Department of Defense, portable among varying systems, and very structured, which permits good large-scale development. It is also easy to maintain, and there are several reusable software libraries available.

Data Quality

As indicated, the DCS provides MANS with seven independent measurements for each of the DCS sensors flown. The gyroscope and accelerometer add additional information about spacecraft motion. GPS also provides position data. Finally, star trackers can provide high fidelity information about spacecraft attitude; however, they are typically expensive and often heavy. With all of this information available to MANS but not necessarily required, the navigation equations are overdetermined. Because of this, data checks can be made between measurements from a single sensor, measurements between two of the same type of sensor, and measurements from different sensor types. Data quality flags are attached to each measurement, first in its rawest form, and at higher levels as the data flows through the system.

The data quality flags allow the MANS software to operate at the highest fidelity available at any one time. In other words, graceful degradation has been achieved in the event of sensor failures, blockages or data drop out, control system failure or rapid re-orientation, filter divergence, and system restart. Because the data quality flags are passed throughout the various CSCs which use them, the data is also ultimately available to the user. In addition, the thresholds at which each of these flags is set can be modified within MANS to meet each user's needs and requirements.

MANS Processor Selection

The logic associated with creating and manipulating data quality flags created one type of requirement for the MANS processor. Additional requirements were found in MANS' manipulation of many angles, implying a need for an extensive use of transcendental functions. Most important for processing was the need to operate rapidly on large quantities of information to provide results in real time. Real time, in this case, means processing data quickly enough to be useful for a typical spacecraft control system. These requirements led us to a 16- or 32-bit processor with a throughput rating of over 1 MIPS (millions of instructions per second). Table 2 shows the processors we considered.

Table 2
PROCESSORS CONSIDERED IN THE MANS TRADE STUDIES

Processor	Positive Attributes	Negative Attributes
80C85	• Radiation hardened	• 8-bit internal and external data bus • Low throughput
80C86	• Radiation hardened • 16-bit internal and external data bus	• Low throughput
MIL-Std-1750A	• 16-bit internal and external data bus • Radiation hardened • Available with over 1 MIPS • Quick context switching	• Requires memory management unit for memory above 64K words
R3000	• 32-bit RISC architecture • High speed	• Not radiation hardened
Hybrid/custom	• Would meet all requirements exactly	• Non-recurring costs • Potential schedule impacts

Further considerations for the MANS processor included space-qualification, reliability and history of use, long-term support by suppliers, and ease of transition into "next generation processors." Because of the language choice, real-time kernel availability and supplier support were of great interest. Also of importance was the availability of cross compilers for VAX/VMS systems where high quality development environments exist. Table 3 shows survey information on cross-compilers and kernels.

Table 3
ADA CROSS COMPILERS AND KERNELS (VAX TO 1750A)

Supplier/Product	Attributes
Ready System / ARTX	• VRTX, original kernel—supports 64K word program • ARTX (Ada version of VRTX) is relatively immature • uses non-Ready Systems front-end
Tartan	• Supports single 64K word program • Compiler optimized for efficiency
TLD / SPK	• Mature compiler and kernal • Kernel optimized for speed • Supports single 64K word program
TLD / MPK	• Contains "classic" scheduler program underneath Ada function library • Supports up to 15 64K word programs • Immature kernel—not often used

A Mil-Std-1750A processor was selected for several reasons. First, the 1750A provides 16-bit processing at relatively high throughput. The 1750A is a standard used extensively in aircraft avionics and then transitioned into the spacecraft arena. Several companies produce 1750A processors which are radiation hardened for space use. The 1750A includes a definition of floating point instructions (providing mathematical functions), and has memory expansion capabilities in 64K word slices. Additionally, the new Mil-Std-1750B specification, soon to be released, is a natural extension of the 1750A system and will be easy to transition into. Finally, many Ada cross compilers and kernels are available for Mil-Std-1750A processors. The MANS 1992 flight test will use the Honeywell Generic VHSIC Spaceborne Computer (GVSC) chipset and their well-established SANDAC packaging.

Reuseable Software Modules

To start the process of looking for software modules in the established reuse libraries, Microcosm established a list of potential software elements, defined in the MANS concept, which might be available in each of the libraries. This list is shown in Table 4. We began to contact reuse libraries (listed in Table 1), to find out if the software modules were available. Generally, people were happy to help. Unfortunately, the data available in the libraries was not always very useful.

Table 4
MANS CANDIDATE ELEMENTS FOR REUSE SOFTWARE

MANS CSC	Attributes
Math Utilities	• Relatively generic functions • High fidelity required
Ephemeris	• Straightforward implementation of look-up tables
Kalman Filter	• Generic ones are available • Most Kalman filters must be tuned for application-specific use • Measurement and state models are application specific

A serious problem with much of the software available in reuse libraries is that it does not include either original requirements or test criteria/test drivers, or both. Without these elements, it is very difficult to ascertain the following critical information:

- If the module will provide the exact function you are seeking
- If the module can be modified to meet your needs
- The level of testing originally performed on the module
- If retest is required—how and what to test for to ensure errors were not introduced into the code

In some cases, software modules were available in Fortran or other languages to perform the functions we were searching for; however, because the original designs were not modular in nature, the particular software we wanted to extract was buried in extraneous code. The inputs and outputs were often ill-defined, and the "calling sequence" was frequently cryptic. Finally, the availability in terms of calender time for some of the libraries was lengthy. Most Government-based libraries are not well funded, so tapes for users are made once per month or once every six weeks. Also, many of the libraries either charged a fee of up to $200 for media, or required that the user provide their own. Providing media to the library supplier further reduces time available for development.

We ultimately decided that it would be equally or more time consuming to recreate the missing information as it would be to start from scratch. However, we did use the Charles Stark Draper Laboratory (CSDL) Ephemeris software (developed originally by JPL and modified by the U.S. Naval Observatory, then CSDL) as the design basis for our own. This CSDL software was originally developed in Fortran, but because of the tabular nature of the data, was relatively easy to port into Ada. CSDL had also developed a very comprehensive set of test data which we could use directly. This was the only instance of actual software reuse in the operational flight software development for the MANS project.

In creating our own libraries, we have attempted to make use of the lessons learned from this search. We include both requirements and test drivers for all baselined software. In essence we have created our own reuse library for autonomous navigation and attitude determination functions, so that in the future when requirements are modified we can reuse large portions of our software.

Criteria for Software Partitioning

Software partitioning has become increasingly important to the software design process as new architectures for both processors and systems are being evaluated. At the system level, the severing of traditional boundaries for subsystem elements is occurring more as common modules are being developed by the JIAWG (Joint Integrated Avionics Working Group). For example, signal processing for all types of sensor units is being combined through evaluation of the use of common signal processing modules. Likewise, with the emphasis on reuseable software modules, software partitioning internal to a functional element is being looked at strongly.

To increase the reuseability of each module for MANS, Microcosm searched for a software partitioning which would foster the most applicability in the future. Specifically, we chose to partition tasks into hardware- or processor-dependent and independent elements, network- or interface-dependent and independent elements, and mission-specific elements. Fig. 4 shows the partitioning of the MANS software for these criteria, with respect to the data flow.

Notice that we identified only the MANS software executive and system utilities as hardware- or processor-dependent. These are the functions which directly interact with the hardware or processor to provide higher-level functions to the application software. These functions were developed in Ada; however, they were allowed to make use of system built-in-functions and system-level utilities. Other application software was required to provide a generic and thus portable design for each design decision. Mission-specific data transformations and units conversions were grouped after the general or core processing but before the network-specific processing. In this way, the output data structure can be easily changed while the network or interface stays a separable issue.

The MANS I/O Handler was based on the MIL-STD-1553B interface selected for our flight test, scheduled in 1992. Inside this shell of interface software, MANS calculates full fidelity solutions and only packs and unpacks data when necessary, based on system modes. The I/O Handler is the only place where any 1553 attributes have been considered or accommodated. Thus, when another network is selected, the timing and data packing issues associated with that specific network can be accounted for in the I/O Handling software.

Finally, we evaluated the utility of partitioning application-specific units so that varying levels of accuracy could be achieved within the system architecture. The idea was not only feasible but also a very scalable approach to building the system. By partitioning software between attitude and position determination, between deterministic and Kalman Filtering solutions, as well as allowing varying data averaging times, the system will provide varying levels of fidelity and accuracy without modifying the software.

In summary, the MANS software has been partitioned in terms of processor dependencies, in terms of network and interface requirements, and to allow for varying levels of fidelity and accuracy. This has allowed us to develop a robust system scalable for any situation. The system level concept of MANS as a scalable architecture is illustrated in Fig. 5. Required elements are highlighted. These are the software pieces needed to operate every level of the scalable MANS. These might be considered "overhead" much like the power supply, backplane, and cooling plates are "overhead" for the processor.

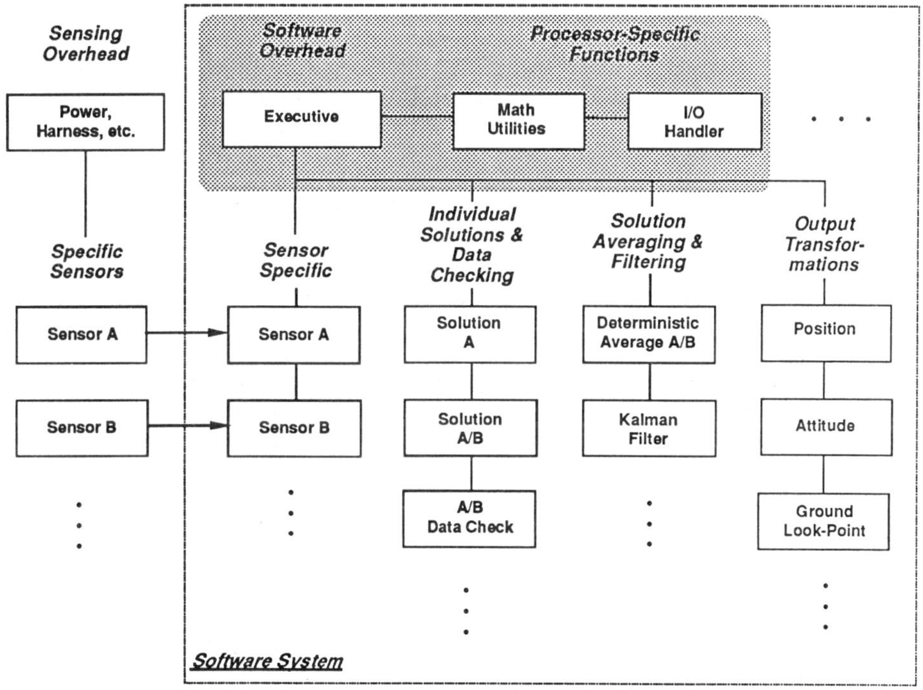

Fig. 5. MANS Software Partitioning.

From the data processing elements, you may select one or more from the data corrections, the deterministic solutions with one or more averaging filters, and choose whether or not to use the Kalman filters. In this way varying levels of fidelity, and ultimately of accuracy, can be achieved without changing the software structure. The concept of standardization centers around well-defined interfaces and functions which enable the MANS software to provide this scalable operational capability.

LESSONS LEARNED

Work on the MANS software development activities has provided several valuable lessons for future development. Many of these, such as reuse libraries and software partitioning, were discussed in the previous sections. With the MANS software, Microcosm attempted to reduce system errors by involving all disciplines in all of the general discussions. Thus, software engineers were included in the algorithm design discussions and analysts attended software design walk-throughs. Testability and maintainability were accepted goals addressed at all levels of design. Finally—though not a universally transferable characteristic—the informality of a relatively small company such as Microcosm stimulated exchange of information, as long as results or significant outputs were well documented.

Mil-Std-1750A Processor

The Mil-Std-1750A processor, for which Microcosm has targeted the first MANS software development, is from an early production run of a newly developed VHSIC 1750A processor line. This brought about several unexpected problems. First, the kernel, based on the TLD MPK, was customized for our program so that device and network drivers, as well as other program-specific needs, could be provided. Additionally, the cross-compiler was not modified to include the processor's built-in-functions (BIFs) nor did it accommodate supplier-specific 1750 extensions. Thus, the kernel was not "off-the-shelf' as advertised and the development environment was not fully matured. Also, as expected, several "features" of the processor were discovered first by the developers, when advertised functions worked differently than expected. With time, all of the companies involved worked together to resolve these issues.

In addition to the problems associated with development on a low production number processor, Microcosm found several general issues associated with Mil-Std-1750A processors. The Mil-Std-1750A specification is nearly 15 years old. The system was specified in detail in some places, and not at all in others. Data word size and addressable memory space is specified in detail, based on the general capabilities of the late 1970s. For example, the longest native type length available is 48 bits. This represents extended precision for floating point values. With personal computers today providing 64-bit and higher accuracy, this limitation, though specified for the 1750A, may not be expected. Lack of understanding for each aspect of the 1750A specification may affect the accuracy of the software systems which can be achieved using the product.

Also, the original addressable memory space for a Mil-Std-1750 processor was 64 K bytes for executable code, and a comparable amount for data storage. Many suppliers have augmented their systems per the 1750A specification with memory management units which allow up to sixteen of these 64 KB memory spaces to be addressable. While most software solutions can be engineered to execute within 64 KB or one address state, this limitation is just one more constraint placed on the software developer. On the other hand, this constraint also aids in "forcing" software partitioning.

Finally, mathematical functions are not entirely specified by the Mil-Std-1750A specification. Math functions are included in the description, yet specific implementations and their associated accuracies are not specified. Thus, each supplier is at liberty to provide their own math capabilities either through attached co-processor chipsets or with BIFs. The VHSIC 1750A used for the MANS development provided BIFs for the mathematical functions. Unfortunately, many of the transcendental functions did not achieve the accuracy afforded by the already limited 48-bit word length.

Software Sizing Accuracy

Microcosm originally estimated the MANS software to be about 50,000 lines of code (using semi-colons as a basis for counting), requiring two address states or 128 KB of memory for execution. We calculated these numbers by breaking the predicted MANS functional modules into the smallest pieces identified, and estimating sizes based on comparison with similar, already developed, modules. As many experts recommend, in most cases the numbers were doubled, tripled and even quadrupled, so that processor requirements would not be a limiting factor in the software development. Data words are counted at about one-quarter of an executable statement since time and energy are expended in defining and developing these elements. We modified the MANS estimates as the development progressed, and have yet to collect the final numbers. Currently, the predictions are 20,000 lines of code executing in three address states.

Several interesting things become apparent about the sizing process throughout the development. First, sizing estimates associated with well-defined functional elements and for which any algorithm information was available should not have been made artificially larger. These elements were often over-sized. Also, for elements where the algorithms are unknown and the definition is not clear, estimates are not very reliable. In two cases—the ephemeris generation and the Kalman filter design—Microcosm substantially underestimated the size requirements. Both of these functions were ultimately implemented using very different approaches than those identified during the initial sizing effort. The new approaches were accepted because of data storage optimization for the ephemeris and to increase accuracy and fidelity in the Kalman filter.

The sizing estimates for computer processor control were estimated very closely; however, identifying an approach to control is required to make these estimates. This locks the developers into a pre-defined scheme during implementation. One approach is often as good as another, but being locked into an approach is rather restricting. General utility function estimates were relatively low. This is primarily because the number of utility functions increased dramatically from estimation to implementation. In the case of the mathematical utilities, the number and complexity of the functions increased as algorithms became more complex. Additionally, applications programming engineers (who were not as familiar with the mathematical functions as the computer system programming engineers) found that it was less time consuming and more reliable to have all mathematical processing provided through the math utilities. While this is ultimately more reliable, it puts a heavy burden on the utilities and their developers to provide large, sometimes suboptimal, software packages.

SUMMARY AND CONCLUSIONS

MANS, the autonav system which Microcosm has developed, provides an example of various elements of the software development process which can be tailored to produce a scalable architecture. By focusing on scalability, Microcosm developed MANS, which provides a scalable architecture for spacecraft autonomous navigation and control. MANS produces fully autonomous, real-time position and attitude fixes at 250-millisecond intervals, in varying levels of fidelity, on-board the spacecraft. Ada modules in the operational flight software have been used to create a library of functions which can be combined in different configurations to create the desired system. A modular computing architecture, developed around the Mil-Std-1750A processor and well-partitioned software modules, has allowed MANS to perform as a truly scalable system.

BIBLIOGRAPHY

B. W. Boehm, *Software Engineering Economics*, New Jersey: Prentice-Hall, Inc., 1981.

Donald Firesmith, Seminar notes: "Ada for Program Managers," Advanced Software Technology Specialists, January, 1989.

Alan S. Fisher, *CASE Using Software Development Tools*, New York: John Wiley and Sons, 1988.

John A. Stankovic and Krithi Ramamritham, "Tutorial: Hard Real-Time Systems," IEEE Catalog Number EH0276-6, 1988.

C. R. Vick and C. V. Ramamoorthy, eds., *Handbook of Software Engineering*, New York: Van Nostrand Reinhold Co., 1984.

James R. Wertz and Wiley J. Larson, *Space Mission Analysis and Design*, Dordrecht, The Netherlands: Kluwer Academic Publishers, 1991.

AAS 91-024

MITIGATING MULTIPATH ERROR IN GPS BASED ATTITUDE DETERMINATION

Clark E. Cohen* and Bradford W. Parkinson†

The Global Positioning System (GPS) offers the potential for significant cost savings in applications where inertial guidance has traditionally been the standard approach. Usually the velocity and position accuracies are the key selling points for GPS. Often overlooked is the capability of also using GPS to determine vehicle attitude.

Attitude is determined by differential measurements of GPS carrier phase between pairs of antennae. A major disturbance of phase measurements comes from unwanted reflections of the carrier (multipath). Reducing the causes or symptoms of multipath will enable differential phase receivers to achieve significantly higher accuracy. For example, with a 2 m antenna separation and a 2 Hz output rate, the method presented here has the capability to reduce the rms orientation error to less than one minute of arc with existing receivers.

A new calibration technique is described which models out multipath error emanating from the vehicle itself. The technique is readily applicable to flight vehicles (and perhaps certain classes of ground vehicles), where phase measurement errors due to multipath should be repeatable as a function of satellite direction relative to the vehicle body frame. Experimental results are presented to substantiate the accuracy improvements made possible by the technique.

INTRODUCTION

The Global Positioning System (GPS) is revolutionizing navigation. As the GPS satellite constellation nears completion, coverage has become continuous, extending globally from the surface of the Earth to well above low Earth orbit. To date there are 16 satellites in orbit and operational. *As such, GPS offers enormous potential for flight vehicles.*

Although GPS has traditionally been thought of primarily as a sensor for navigation, it offers the additional capability of providing an *attitude reference*. Such a reference is based on measuring the *differential carrier phase* of the GPS signal between separate antennae on a vehicle.

* Ph.D. Student, Department of Aeronautics and Astronautics, Stanford University, Stanford, California 94305.

† Professor, Department of Aeronautics and Astronautics, Stanford University, Stanford, California 94305.

Since GPS receivers are comprised purely of electronic components, they have the potential to provide considerable cost, weight, size, and reliability advantages. Furthermore, the appealing commonality between translational and rotational sensing afforded by GPS presents additional opportunities for cost savings.

In many applications, combined attitude determination and navigation through GPS can augment or replace inertial guidance systems. A key advantage is that the GPS measurements of position and attitude are not subject to the long term drift associated with inertial devices.

In space applications, GPS has brought about new capability in GN&C, including the arena of autonomous operations. The Gravity Probe B spacecraft being developed by Stanford University and NASA to measure effects of General Relativity on orbiting gyroscopes will use GPS in a closed-loop translation control system for precise initial orbit insertion.[1] The spacecraft, equipped with several GPS antennae, also presents an opportunity for providing attitude determination test data in orbital applications. The Gravity Probe B is just one example of how GPS can potentially reshape spacecraft GN&C.

The state of the art in GPS based attitude determination has been advancing.[2-5] In many cases, multipath is a limiting factor for accuracy. The technique described in this paper is a way by which the accuracy of GPS based attitude determination may be improved by reducing the disturbances from multipath. In applications where the system bandwidth is not driving the error budget, multipath becomes the largest source of error. This paper reports a method by which this error source may be significantly reduced.

THEORY OF THE METHOD

Foundations of Attitude Determination

GPS based attitude determination is founded on measurements of differential carrier phase between two or more GPS antennae. As shown in Fig. 1, a planar wave front approaching from a GPS satellite reaches one antenna on a vehicle-mounted baseline before the other. Measurement of the *difference* in carrier phase, $\Delta\varphi_{ideal}$, between the two antennae (modeled as points) yields the projection of the antenna baseline vector, **a**, along the satellite line of sight, $\hat{\mathbf{s}}$ as follows:

$$\Delta\varphi_{ideal} = 2\pi \frac{\mathbf{a}\cdot\hat{\mathbf{s}}}{\lambda} + b - 2\pi k \qquad (1)$$

The measured phase angle is also subject to an additive line bias, b, and an integer wavelength ambiguity, k. The line bias arises from delays in the receiver hardware, such as differences in cable lengths between the two antennae. The L_1 GPS frequency at L-band corresponds to a wavelength, λ, of 19 cm. With multiple baselines and GPS satellites, the 3-axis vehicle attitude, line biases, and integer ambiguities are all observable through the phase measurements.

Achieving High Accuracy with Carrier Phase

Of the GPS signal characteristics, use of the relative *carrier phase* for attitude determination provides the best accuracy, especially since the signals from each antenna are referenced to the *same oscillator* in the receiver. Operating differentially, such receivers offer unparalleled common-mode noise rejection at low cost. Furthermore, many of the error sources present in GPS based navigation (such as ionospheric effects and Selective Availability -- intentional degradation of the satellite signal for non-military users) simply cancel out.

Using GPS, ground surveying (using two separate receivers) is now performed routinely to an accuracy better than one centimeter (or a few parts in 10^6). GPS based attitude determination is remarkably similar to surveying and also has the potential to improve this surveying accuracy to *sub-millimeters*.

Relative position measurement error may be expressed in terms of phase angle or distance. In attitude determination, the angular accuracy is approximately determined by the relative antenna position error divided by the baseline length. Hence, longer baselines yield proportionately higher accuracies.

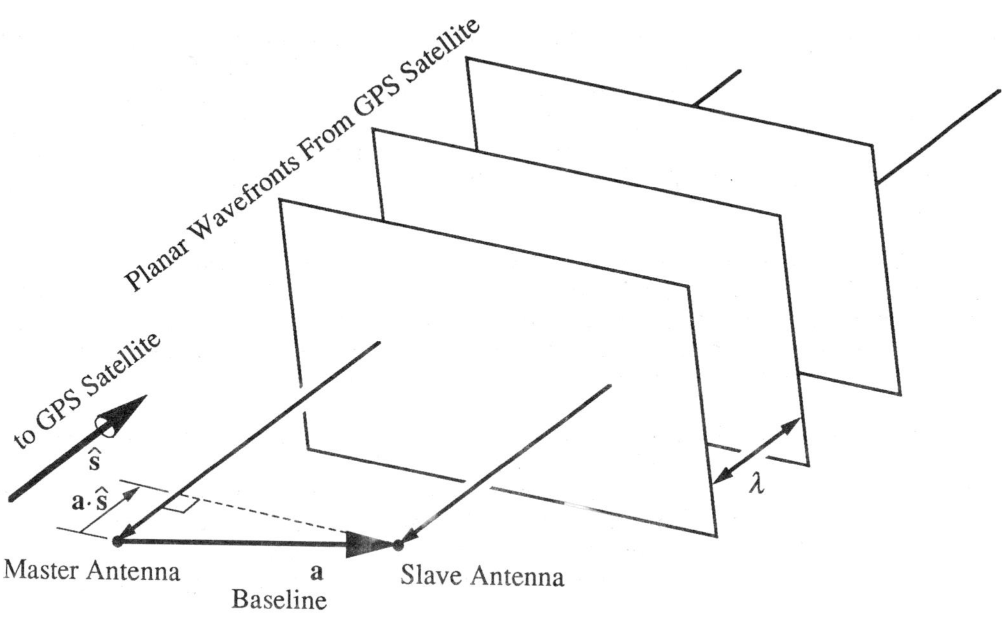

Fig. 1 Antenna Geometry for Attitude Determination

Wave fronts from the GPS satellites appear as plane waves, reaching one antenna before another. The carrier phase between the two antennae is proportional to the projection of the antenna baseline onto the line of sight to the satellite. With multiple baselines and satellites, the platform orientation may be determined.

A Model of Multipath

In most cases multipath is the dominant error source in relative carrier phase measurements. Perturbations to the measured carrier phase in typical multipath environments are observed to peak around 40-50 deg. In all but the most pathological cases, larger deviation (in conjunction with deep fading) is precluded for higher signal levels, such as when there is a direct line of sight between the antenna and the GPS satellite.

When an antenna is mounted on a vehicle, one may consider the vehicle as part of the antenna. The key idea behind the work presented here is as follows: the multipath arising from a "far field" plane wave emanating from any given direction with respect to the vehicle body frame should be *repeatable*. In reflections the principle of superposition applies. Furthermore, receivers should have a repeatable response independent of signal modulation, since the carrier may be reconstructed by demodulating the signal with its original pseudo-random sequence. Therefore, the correction to the repeatable component of multipath should be independent of the signal strength and the specific source satellite. Doppler shift is expected to be a negligible effect even at its nominal maximum magnitude of 10^{-5} (assuming that the vehicle flies at orbital velocities).

Note that the phase perturbations originating from a variety of sources may be conveniently lumped into the same repeatable category. Specifically, antenna phase center deviation (individual antenna response), inter-antenna interaction, and conventional multipath reflection all combine together to yield a repeatable phase response. In fact it is arguable that these separate phenomena are indeed *indistinguishable* if one simply expands the definition of the antenna to include the vehicle as well.

It is postulated, therefore, that it should be possible to *calibrate* out the multipath error for a flight vehicle provided that the only sources of reflections in the immediate vicinity are due to the vehicle itself. Note that reflections arriving at the receiver by a difference in path length exceeding one pseudo-random noise code chip (300 m) will be suppressed by the autocorrelation characteristics of the code sequence.

In addition to the ideal point model (1) of the antennae, the measured differential phase, $\Delta\varphi_{measured}$, will be perturbed by the repeatable multipath component of phase error, $\delta\varphi(\theta,\phi)$. The multipath perturbation is assumed to be a single valued function of the angular position of the satellite (θ,ϕ) in spherical coordinates with respect to the local body frame of the vehicle. Multipath error not corrected in the model and receiver random noise are lumped into the remaining error, $\Delta\varphi_n$. The measured differential phase is then

$$\Delta\varphi_{measured} = \Delta\varphi_{ideal} + \delta\varphi(\theta,\phi) + \Delta\varphi_n \qquad (2)$$

This formulation departs from the traditional treatment of multipath error as a random process. Although its signature is typically zero-mean and band limited,[6] in fact, multipath from the vehicle itself is *not* a random process. Much of the desired information in the signal is *still contained* in the measurements disturbed by multipath.

Spherical Harmonic Model

Although there are a number of ways to model the repeatable characteristic, spherical harmonics were chosen as the calibration function. This form of representation is desirable because

- few coefficients are required to represent the smoothly varying function
- the set of body frame direction vectors covers the spherical domain
- for calibration, the representation must "interpolate" between satellite data arcs
- there is no projection distortion

The functional form of the spherical harmonic representation is

$$\delta\varphi(\theta,\phi) = \sum_{l=1}^{n}\left(J_{l0}P_{l0}(\cos\theta) + \sum_{m=1}^{l} P_{lm}(\cos\theta)\left(C_{lm}\cos m\phi + S_{lm}\sin m\phi\right)\right). \tag{3}$$

For an nth order model of the multipath error, the coefficients, J_{l0}, C_{lm}, and S_{lm}, describe an approximation to $\delta\varphi(\theta,\phi)$. In the expression, $P_{lm}(x)$ functions are Legendre polynomials. Thus each baseline pair of antennas has a set of coefficients to describe the vehicle multipath specific to that pair. The capacity of this functional form to follow variations of increasing complexity is governed by the model order, n.

EXPERIMENTAL RESULTS

To demonstrate the feasibility of the multipath calibration technique, a GPS receiver was statically employed to collect data over a large range of satellite directions. Ideally, to build up a multipath model, one would accumulate measurements from plane wave signals originating from every possible direction in the vehicle body frame. In practice, measurements may consist of "slices" of the celestial sphere to create a mesh of sufficient density to solve for the coefficients of the spherical harmonic model.

GPS Receiver Hardware

A Trimble Navigation TANS-based GPS receiver, configured to provide differential phase measurements, was employed to collect the experimental data. The architecture of the receiver is shown in Fig. 2. Signals from each antenna are amplified and down-converted in the RF front-ends before the GPS carrier phase is differenced by any one of six receiver channels. Each channel is capable of tracking a separate satellite to yield continuous and simultaneous differential phase measurements from each satellite over the same baseline. Although only a single baseline was used for the experiments reported here, other baselines can be added by replicating the slave antenna, RF front-end, and receiver channel set.

Of crucial importance is that a single oscillator be used to generate all of the timing signals in the entire receiver. Phase noise emanating from the oscillator or synthesizer affects *both* signal lines equally and thus cancels out when the differential phase is determined. The common clock enables this receiver to achieve raw phase measurement accuracies below 1 deg rms (500 μm) with a 500 ms averaging time.

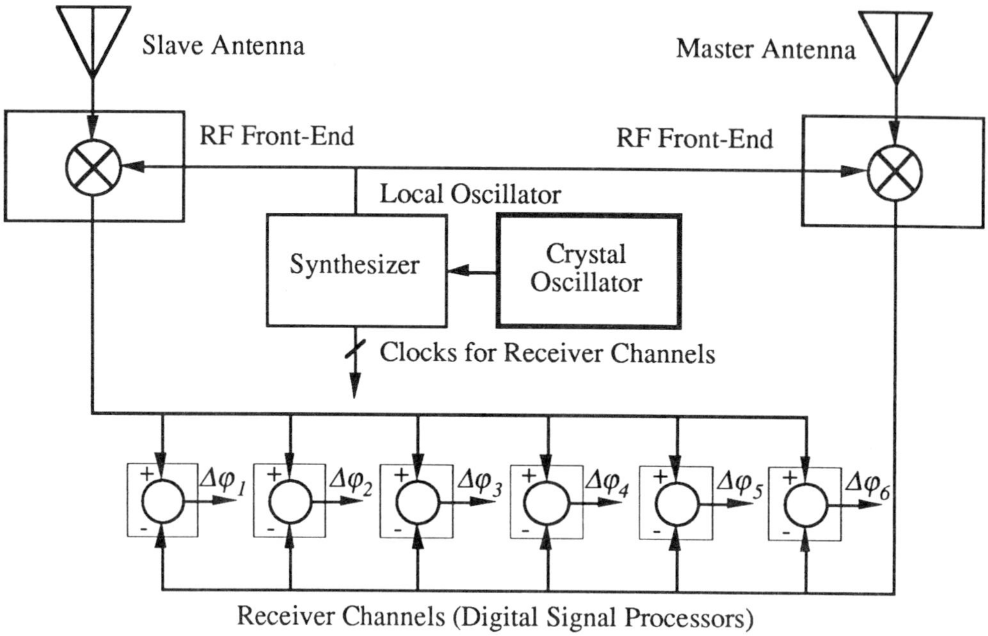

Fig. 2 Architecture of the Trimble Differential Receiver

Down-conversion and relative phase measurements are based on the same oscillator. Any phase noise in the local oscillator or timing signals cancels out in the differential operation of the receiver, allowing the design to achieve superior measurement accuracy. Six digital signal processors lock on to GPS signals, providing continuous differential carrier phase measurements on up to six separate satellites simultaneously. Additional baselines may be added by replicating the slave antenna, RF front-end, and receiver channel set.

Antenna Configuration: Analogy to a Vehicle in Flight

In lieu of actual calibrations of vehicles in flight, a ground-based analog was developed. By using a large (3λ radius) horizontal ground plane, shown in Fig. 3, it was attempted to attenuate radiation coming up from below the antenna. To the extent that this was successful, the antenna was rendered immune to its multipath environment for azimuthal rotations when the horizontal ground plane was the highest object around. For the experiments performed here, the antenna was placed on the roof of a building such that it was the highest object for at least 500 m in every direction. In this respect, the analogy corresponds to a flight vehicle operating well away from other reflective objects. With the ground-plane and antenna pair alone, the residual "noise floor" due to multipath was 4.9 deg rms of carrier phase (2.6 mm relative position error).

To intentionally create multipath errors, a rectangular sheet metal reflector was mounted on the ground plane as shown in the figure. The analog to attitude motion of a flight vehicle is *rotation of the entire ensemble* (consisting of the antennas, ground plane, and reflector) about the zenith line. The observed result is that the perturbations in carrier phase relative to the ensemble body frame *moved with the body* about its azimuthal rotations.

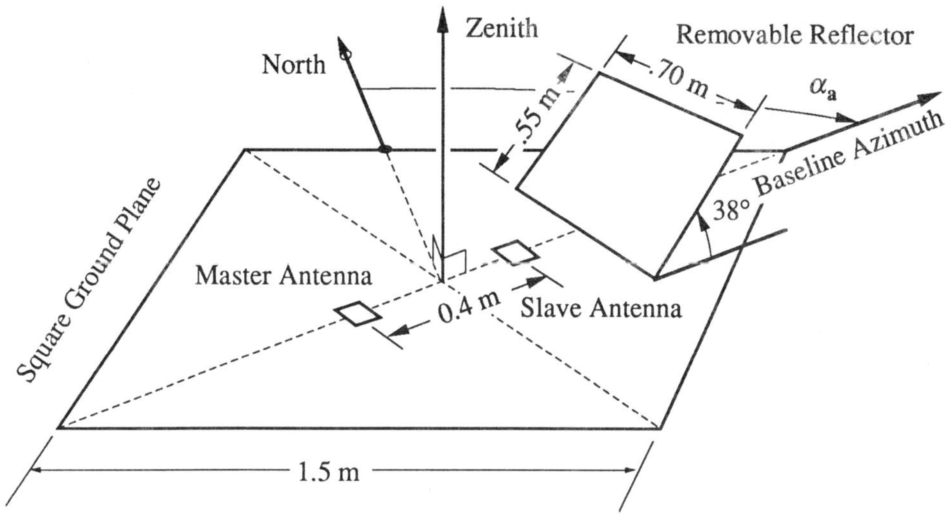

Fig. 3 Test Antenna Configuration and Perturbing Reflector

The horizontal ground plane was used to isolate the antennae from the multipath environment below the setup. A detachable sheet metal reflector was positioned to intentionally induce multipath disturbances. The entire ensemble was rotated about the zenith to various azimuth angles, α_a, with respect to true north.

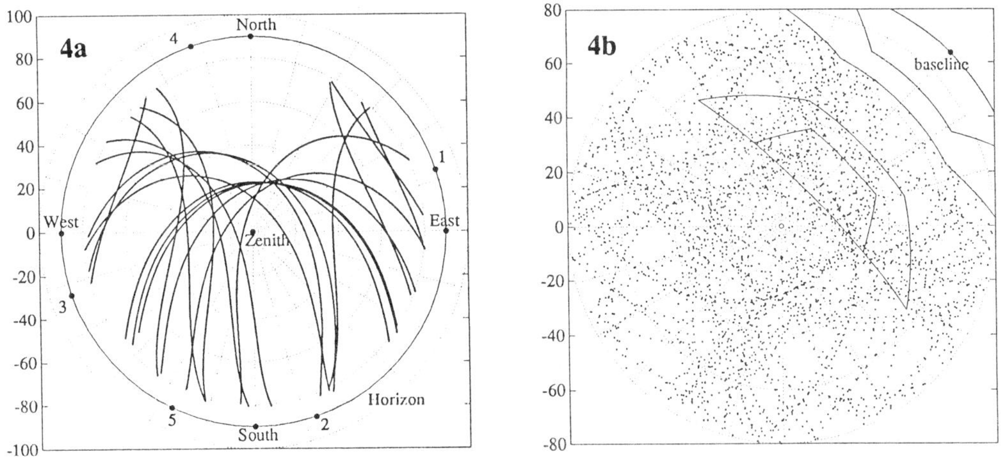

Fig. 4 Tracks of GPS Satellites across the Celestial Hemisphere

Over a 24 hour period, the repeating ground track of the GPS constellation traces out arcs (4a) in the sky. The antenna baseline was oriented along any of the 5 indicated azimuths. In the antenna body frame (4b), data from four 24 hour runs separated by 90 deg azimuthal rotations of the antenna ensemble were combined together for a single calibration fit. To reduce the size of the data set, phase measurement residuals along each arc were spatially prefiltered and thinned out to the density shown. Based on a simple ray-tracing model, the regions of the sky blocked from each antenna by the reflector are shown along with the regions of the sky each antenna sees reflected through the sheet metal plate.

GPS Satellite Sky Coverage

To build up a model of the phase perturbations, the antenna ensemble was left motionless for 24 hours in various azimuthal positions. The orbital motion of the GPS satellites in the sky naturally varied the directions of incidence of the GPS signals. Fig. 4a shows the cumulative paths of the satellites which repeat every 24 hours. Thus over a 24 hour period, a measure of the multipath phase perturbation is available over the entire mesh depicted in the figure.

The spherical harmonic model also acts as an angular (spatial) filter. The higher the model order, n, the greater the angular detail (angular or spatial frequency) the model can approximate. A rough rule of thumb is that the smallest variation with angle is roughly 180 deg/n. Therefore, gaps in the regions of the sky shown in Fig. 4a (excluding the region in the north where no satellites fly) preclude a higher order model than about 5. (For higher order, n, the best fit would not be accurate in the uncovered regions since there would be no data to shape the estimate).

To circumvent this limitation, measurements were collected with the antenna rotated among a total of five azimuthal positions (shown in Fig. 4a). Fig. 4b shows the resulting collection of data points consisting of positions 1-4. Note that the areas which remain uncovered are much smaller, suitable to increasing the model order to around n=18. To reduce the size of the data set, the measurements were spatially prefiltered before the spherical harmonic fit.

Also shown for reference is a simple ray-tracing model of the regions in the sky over which the reflector would influence the phase measurements. However, since the reflector is less than three wavelengths in dimension, diffraction effects severely limit the validity of the ray-tracing model.

Experimental Data

For a given 24 hour data run, the phase residuals were obtained by solving (1) with a least-squares fit to all the measurement data. We define the solution, **a**, to this equation as the *differential phase center* of the antenna pair. Phase residuals (constructed by subtracting the resulting $\Delta\varphi_{ideal}$ from the actual measurement) are exceptionally repeatable, as shown in Fig. 5 and as documented previously.[6,7] As each satellite retraces its sky track each day, the signature in the phase residuals is almost identical. Furthermore, the signatures of *different satellites* following closely spaced arcs at different times of day are also exceptionally repeatable as shown in Fig. 6. Implicit in this observation is verification that the multipath perturbations are independent of carrier modulation code and doppler shift.

For comparison, 8th order fits of (3) were performed on the phase measurement residuals which were produced both with and without the rectangular reflector affixed to the antenna (Fig. 7). To illustrate the effectiveness of the calibration, Fig. 8 shows the measured phase residuals from a completely *independent* set of data taken at position 5 against the calibration accumulated over positions 1-4. For reference, the phase residuals in Figs. 5, 6, and 8 came from the regions in the antenna body frame celestial sphere shown in Fig. 9.

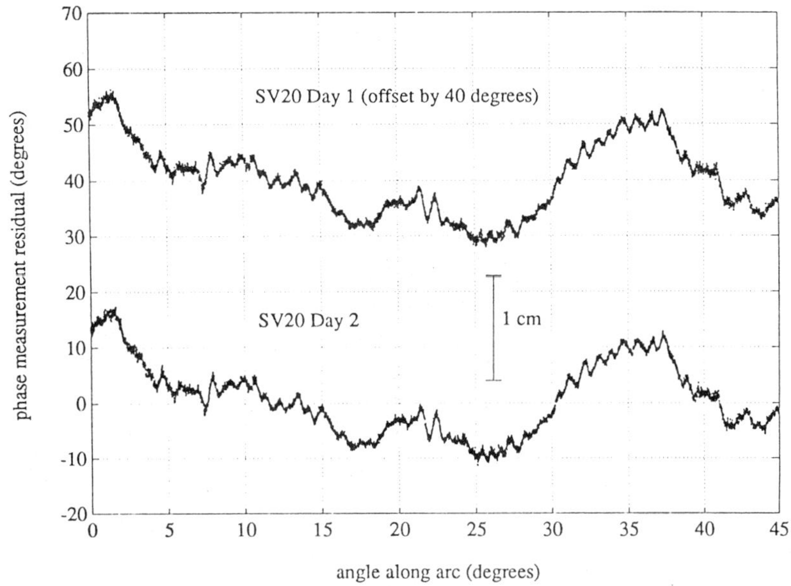

Fig. 5 Repeatability of Phase Residuals Perturbed by Multipath

From one day to the next, the multipath signature from a given satellite is exceptionally repeatable. Since the sky track of the satellite repeats every 24 hours, the Space Vehicle (SV) transmission arrives at the antenna from the same direction the following day. An averaging time of 500 ms was used for all the data presented.

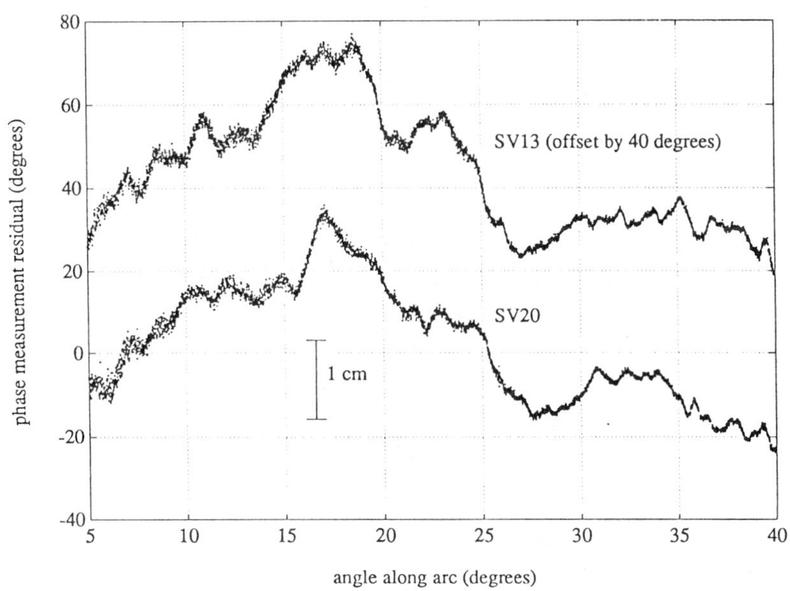

Fig. 6 Repeatability of Residuals among Different Space Vehicles

The exceptional repeatability of the phase measurement residuals is largely independent of satellite-specific modulation code and doppler shift as indicated by these adjacent sky tracks of different satellites. Note that SV20 is setting over the interval while SV13 is rising. Data points on the left are closer to the horizon.

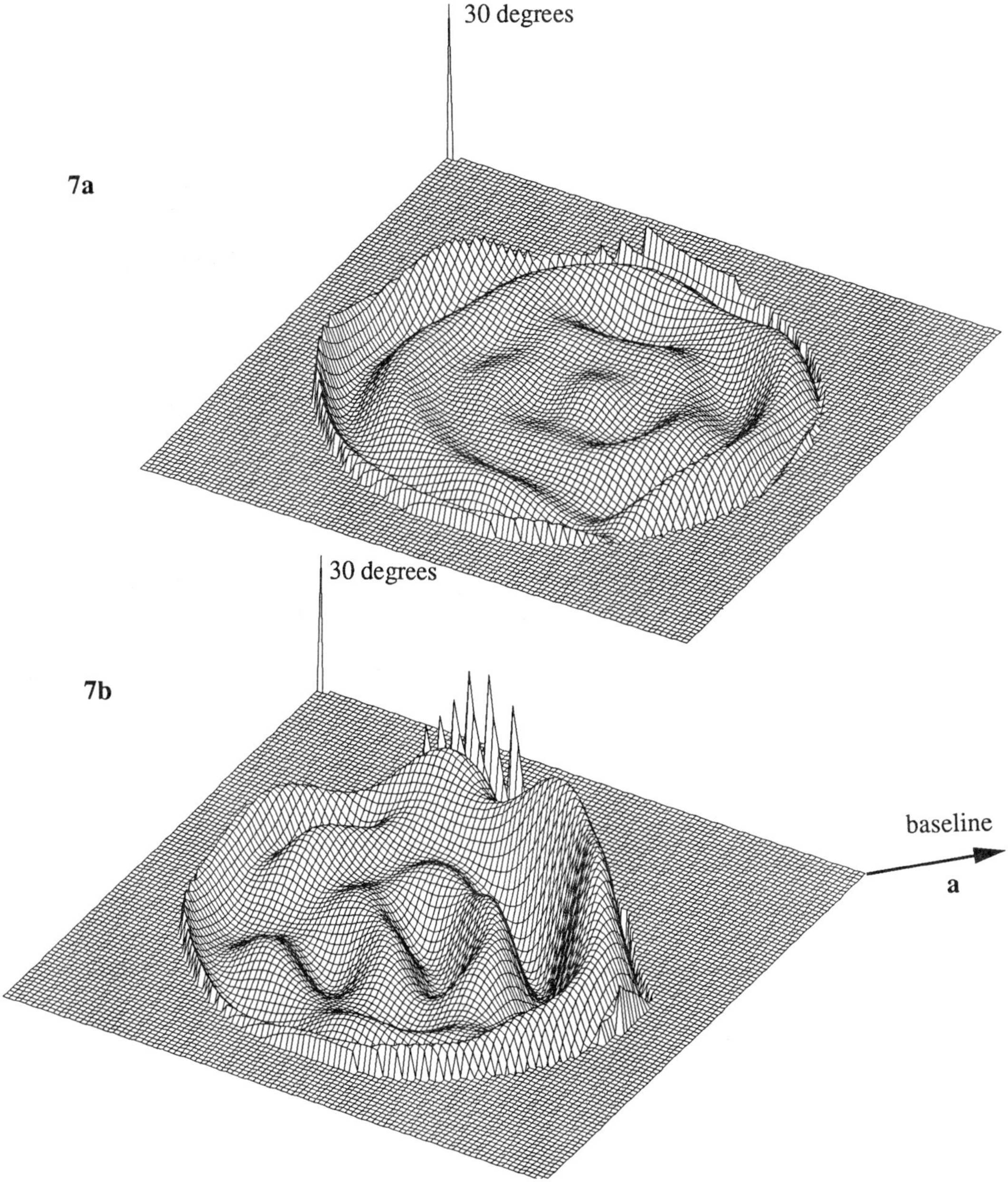

Fig. 7 Best Fit Multipath Calibration (8th Order Model)

This mesh plot depicts the calibration model error (represented as height) for the antenna pair both *with* (7b) and *without* (7a) the reflector disturbance. This representation could be overlaid on Fig. 5, since the horizontal dimensions correspond to direction in the antenna body frame. The phase correction ranges between ±9 deg without the reflector and from -38 to +27 deg with the reflector. The portion of the sky blocked by the reflector is masked off. Two 24 hour runs at azimuthal positions 1 and 2 were used for the fit without the reflector. Four 24 hour runs at azimuthal positions 1-4 were used for the fit with the reflector.

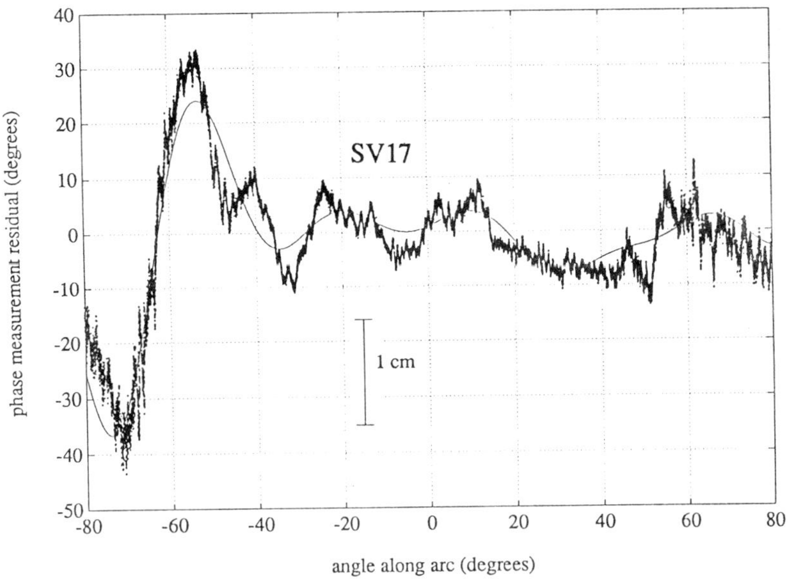

Fig. 8 Calibration Results with Reflector

The continuous 8th order calibration curve from data taken at azimuthal positions 1-4 is shown against a completely *independent* set of phase measurement residuals taken at position 5.

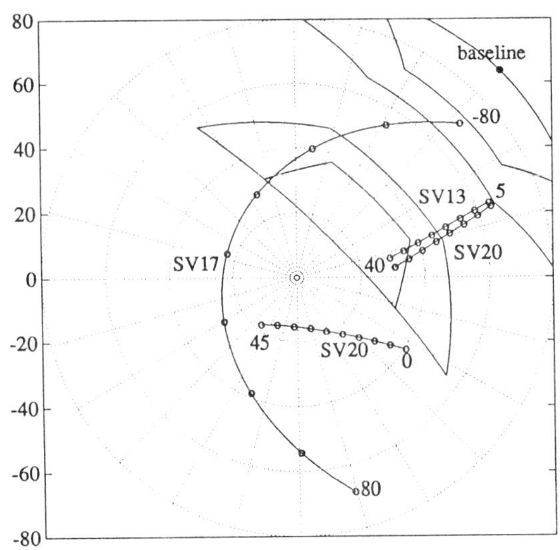

Fig. 9 Sky Map of Phase Measurement Residuals

The satellite positions for the phase measurement residuals of Figs. 5, 6, and 8 are shown with respect to the antenna body. The circles correspond to the abscissa grid lines in the referenced figures.

Intentionally introducing multipath with the reflector doubled the uncorrected error residuals. This disturbance raised the residuals above the 4.9 deg rms minimal multipath "noise floor" to more than 10 deg rms (5.2 mm). Applying the calibration reduced the error to below 6 deg rms (3.2 mm), almost fully recovering the original accuracy achieved without the disturbance.

The residuals resulting from the independent data set are plotted against model order in Fig. 10. In general, the model order n required to handle a vehicle dimension, L, can be loosely approximated as $2L/\lambda$ (around n=12 for the reflector dimension used here). Reflections from objects farther away than the antenna contribute to the higher frequency components of the multipath signature. As can be seen from the plot, the calibration model offers diminishing returns above 8th order. The "noise floor" conceals any additional improvement.

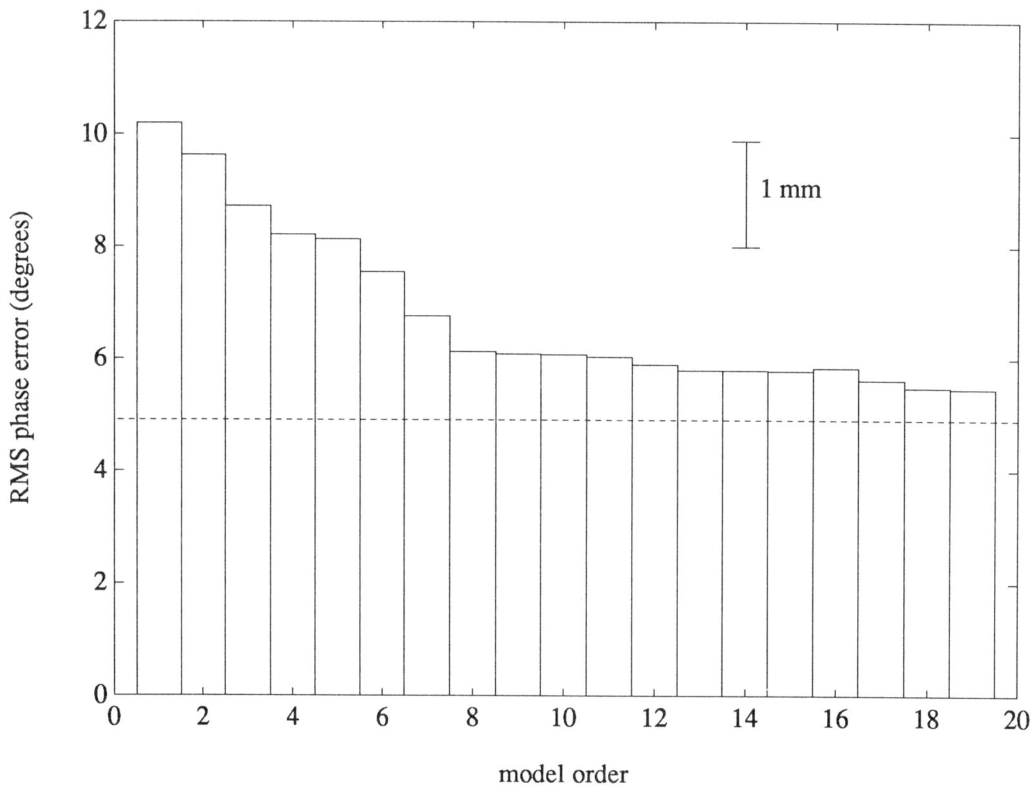

Fig. 10 RMS Error Remaining after Calibration

This plot shows the rms error remaining after the subtraction of the calibration data taken at positions 1-4 from the *independent* phase measurement residuals taken at position 5. Up through model order 8, accuracy improves steadily. The dashed line indicates the 4.9 deg rms residual error for measurements taken without the reflector disturbance. Almost all of the accuracy lost as a result of the reflector disturbance is recovered. In the air or space environment, well away from other reflective objects, it is expected that the "noise floor" seen here at higher orders will drop considerably.

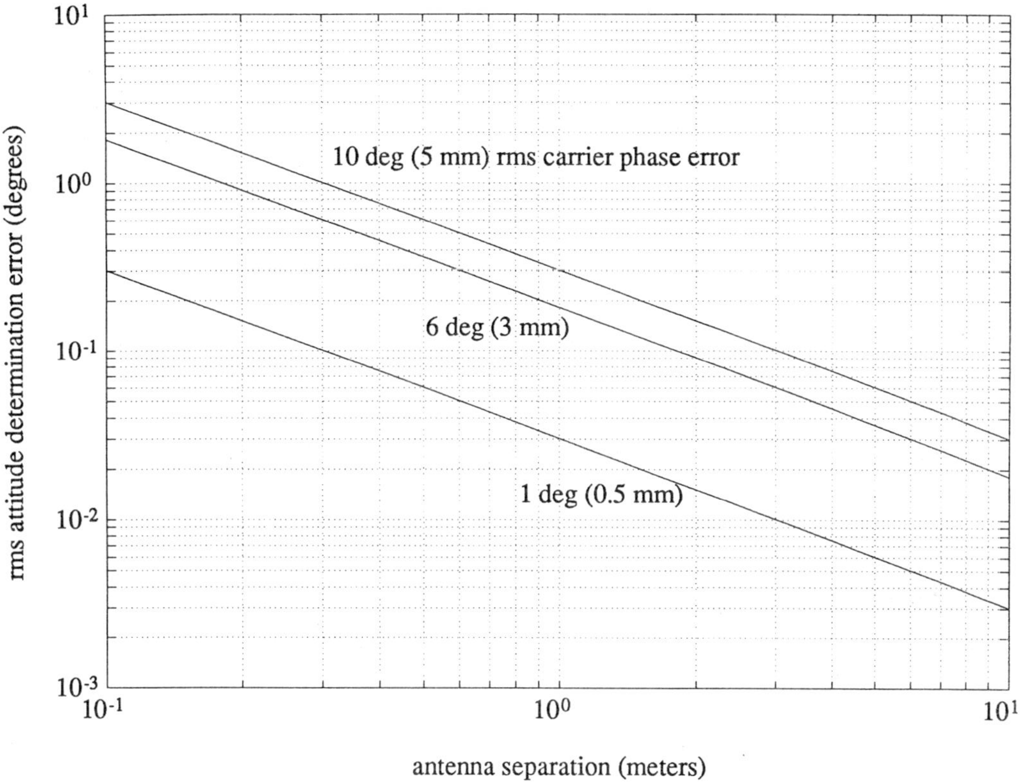

Fig. 11 Expected Flight Vehicle Performance

The attitude determination error is given by the differential phase measurement error divided by the baseline length. Based on experimental data, the three curves shown here (in improving order) correspond to uncalibrated error, calibrated error, and expected calibrated error in flight. The lowest curve is based on the demonstrated experimental receiver accuracy when set at an averaging time of 500 ms.

Extrapolation of Results to Flight Vehicle Performance

For experiments in flight vehicles operating well away from other reflective objects, one would expect the "noise floor" of Fig. 10 to disappear. In principle, multipath need not limit the angular accuracy that can be achieved. Instead, the GPS carrier-to-noise ratio must ultimately drive the accuracy and bandwidth performance.[4,5] The inherent measurement accuracy of the receiver used for these experiments is at least an order of magnitude better than the disturbance level due to multipath. Fig. 11 shows how this demonstrated receiver performance may be extrapolated to the air and space environment. For an averaging time of 500 ms, the 1 deg rms (0.5 mm) phase measurement accuracy translates into attractive accuracies with reasonable baselines for many applications.

The impact of body angular rates on attitude determination accuracy depends on the angular acceleration disturbances to the vehicle. With suitable carrier tracking, constant angular rates should not degrade accuracy. For a given disturbance environment, the trade-off between accuracy and bandwidth will be governed by the carrier-to-noise ratio.

PRACTICAL IMPLICATIONS

The calibration technique outlined here is essentially a correction to the ideal phase based on the antenna point model (1). One should note that the correction applied to each signal is a function of satellite direction in the body frame, which is actually a quantity sought in the attitude determination process. Fortunately, since the coarse attitude without the correction is already known, only one iteration is generally required. Furthermore, if the vehicle attitude is changing slowly, the corrections need not be updated frequently.

Initial calibration to determine the coefficients before flight could be performed in a number of ways, one of which is with an anechoic chamber. Where practical, the calibration could also be performed through simulation. In some circumstances, in-flight calibration is possible if some external truth reference, such as an IMU, is used for initial testing. Vehicle articulation could constitute an error source with this technique, although it could be accounted for by adding additional dimensions to the calibration function.

An important practical consideration associated with the technique is the model order, n. The number of coefficients for an nth order model is $(n+1)^2$. Storage and computation requirements mount quickly as model complexity is increased. Recall, however, that the model order is roughly proportional to the vehicle dimension. For small vehicle dimensions (on the order of a few wavelengths of the GPS signal) the calibration model should be extremely effective and efficient in improving accuracy. For larger vehicles (above ~2 meters), one may choose to exploit the larger vehicle dimension and employ a longer baseline length. In either case, the disturbance of multipath is attenuated with a resulting improvement in accuracy.

APPLICATIONS

Differential phase measurement capability opens up a bountiful array of new applications in aerospace. While attitude determination of flight vehicles using GPS is advancing, other potential uses of differential carrier phase lie ahead as well.

GPS based attitude determination in space may emerge in a number of roles, ranging from primary to backup operation. Since it provides unrestricted 3-axis attitude and position, the method is especially appealing for maneuverable vehicles, including the Gravity Probe B, Orbital Maneuvering Vehicle, and the Space Shuttle. GPS also offers new capability in booster guidance. Similar potential exists for Smallsats. GPS can play a role in initial attitude acquisition such as bringing a guide star within the field of view of a star tracker

In aircraft, GPS can be used for in-flight inertial alignment, cross-check, flight control, or redundant attitude determination. It also offers the capability to provide instantaneous local wind velocity measurements. Subtracting the local airspeed vector (determined by vehicle attitude and pitot tube measurements) from the inertial velocity (determined by GPS) yields the local wind velocity *vector*. This capability may aid in the detection of wind shear.

Vehicle flexibility may yield an important new class of applications for GPS differential phase measurements. Stringing GPS antennae along the length of a flexible space structure (such as the Space Station) could provide an excellent means of estimating the vibrational mode shapes and excitation. In the field of aeroelasticity, a similar arrangement could allow in-flight dynamic model verification. Space tethers are another area of application.

SUMMARY

A new calibration technique based on the principle of superposition was applied to correct for the disturbance of multipath on GPS based attitude determination. A pair of antennae mounted on a 3λ radius horizontal ground plane placed higher than any other object within 500 m was employed as an analog to a vehicle in flight. To create a real world situation, a reflector was placed near the antennae which effectively doubled the rms multipath error from 5 deg (2.6 mm) to more than 10 deg (5.2 mm). Applying the calibration model to the measurements brought the error back down to 6 deg (3.2 mm) -- almost the entire accuracy of the measurements without the reflector disturbance.

CONCLUSION

Extrapolating the experimentally demonstrated performance improvement to the air or space environment, one can reasonably expect to improve the accuracy of GPS based attitude determination by more than an order of magnitude in many applications. Multipath should give way to GPS carrier-to-noise ratio as the driving error source. Assuming a baseline length of 2 m and a 500 ms averaging time, the receiver used in these experiments can achieve an accuracy better than *one minute of arc.*

ACKNOWLEDGEMENT

The authors gratefully acknowledge Trimble Navigation, Ltd. for providing the GPS receiver used in the experiments and for assistance in performing this research.

REFERENCES

1. P. Axelrad and B.W. Parkinson, "Closed Loop Navigation and Guidance for Gravity Probe B Orbit Insertion", *Journal of the Institute of Navigation*, Vol. 36, No. 1, Spring, 1989, pp. 45-61.

2. L.R. Kruczynski, P.C. Li, A.G. Evans, B.R. Hermann, "Using GPS to Determine Vehicle Attitude", *Proceedings of the International Technical Meeting*, Institute of Navigation, Colorado Springs, Colorado, September, 1988.

3. L.R. Kruczynski, P.C. Li, A.G. Evans, B.R. Hermann, "Using GPS to Determine Vehicle Attitude: USS Yorktown Test Results", *Proceedings of the International Technical Meeting,* Colorado Springs, Colorado, September, 1989.

4. I.M. Weiss, B.E. Brumbaugh, and J.S. Avrin, "Attitude Determination using GPS Measurement Techniques", *Proceedings of the 11th annual AAS Guidance and Control Conference*, Keystone, Colorado, 1988.

5. J. Rath and P. Ward, "Attitude Estimation using GPS", *Proceedings of the National Technical Meeting*, Institute of Navigation, San Mateo, California, January 1989.

6. Y. Georgiadou and A. Kleusberg, "On Carrier Signal Multipath Effects in Relative GPS Positioning", *Manuscripta Geodaedica*, Vol. 13, No. 1, 1988, pp.1-8.

7. I. Nesbø, "Applications for GPS Determined Attitude for Navigation", Institute of Navigation, Colorado Springs, Colorado, September, 1988.

AAS 91-025

AUTONOMOUS, ALL-STELLAR ATTITUDE DETERMINATION EXPERIMENT: GROUND TEST RESULTS

H. L. Fisher, T. E. Strikwerda, C. C. Kilgus, L. J. Frank and M. D. Shuster*

An all-stellar instrument has been designed to demonstrate five-arc-second autonomous attitude determination on a NASA Spartan spacecraft. The instrument includes a Charge-Coupled Device (CCD) star camera that provides centroid measurements of up to five simultaneous star images with an update rate of 10 per second. At slue rates less than 0.1 degrees per second the camera accuracy varies from 5 to 15 arc seconds per image, depending on star magnitude. The instrument processor matches measured stars with a 9000 entry catalog and then computes a single-frame and Kalman-filter estimate of Spartan attitude and attitude rate each second. By using multiple star images for attitude determination, an accuracy of a few arc seconds can be obtained. The performance of the camera and processing techniques have been studied in an extensive modeling and test program including star measurements taken under realistic conditions at the Jet Propulsion Laboratory Table Mountain Observatory. Detailed results of these tests are presented.

INTRODUCTION

The star-tracker instrument (START) described in this paper was developed to test a new CCD-based star camera and a real-time technique for star identification and attitude determination. Both the instrument and star identification methods have been described in previous papers;[1, 2] these will be reviewed briefly here. In this paper we concentrate on several methods which have been developed for evaluating the accuracy of the camera's star measurements and the performance of the attitude determination techniques when observing real stars. Algorithms for evaluating the data and results from several test cases are presented and discussed.

Tests under realistic conditions are useful, because camera performance tests in the laboratory are limited typically to one test source or a grid of sources and precise motion of the test sources or camera is difficult. Test sources may not accurately simulate the true image profile, size or spectral characteristics. In addition, the software for attitude determination cannot be adequately tested with these sources; simulated data or actual star observations are required.

The flight test of the instrument is expected in 1993. The Spartan spacecraft, built by NASA Goddard Space Flight Center and funded by the U. S. Air Force, will be deployed from the shuttle for a 40-hour preprogrammed mission that will include sluing and staring observations.[1] The instrument will then be retrieved for possible upgrades leading to sub-arc-second performance.

* The Johns Hopkins University Applied Physics Laboratory, Johns Hopkins Road, Laurel, Maryland 20723-6099.

STAR CAMERA INSTRUMENT

The START instrument consists of a CCD star camera and its electronics and processor package, the instrument flight processor for data collection, attitude determination and instrument control, the power distribution system, and support structure.

CCD Star Camera

The CCD camera, built by Hughes Danbury Optical Systems, has a 7.2 deg × 9 deg field of view (FOV) and a 256 pixel × 403 pixel CCD array. Ten frames per second of digitized CCD data are sent to the camera processor, which computes image plane coordinates for up to five stars in each frame, corrects for various effects such as radial distortion, and sends the data to the flight processor. In the event there are more than five stars in one frame, an algorithm attempts to select five stars that are somewhat distributed throughout the FOV.

Instrument Flight Processor

The instrument processor has three main functions. Its highest priority is to collect all camera data, format it, and send it to the Spartan. The second function is to receive and act on preprogrammed commands from Spartan for instrument control. When there is time remaining after these tasks finish, the processor is used by attitude-determination software to process as many camera frames as possible. This task determines attitude by identifying stars in the FOV and then computing the optimal attitude estimate from the inertial frame to camera frame. The attitude results are stored on the tape recorder along with the camera data for later ground analysis; they are not used in the Spartan control system.

Star Identification

The technique for star identification follows closely that of Junkins and Strikwerda.[3] First, the angle between a pair of measured stars is compared with the angle between candidate pairs of catalog stars. This comparison is repeated using different pairs of measured stars until a possible match is found. The rotation matrix from the catalog frame to the camera frame is then computed (see below) and image-plane coordinates of other catalog stars are computed and checked to determine if they match other measured stars. Matches of at least three stars are required to confirm the identification. The attitude estimate can then be refined using all the identified stars.

The on-board catalog for this mission contains about 9000 stars. In sparse regions of the sky it is complete to approximately 6.0 visual magnitude. In dense regions some faint stars in this range have been eliminated. Although 9000 stars appear to be more than enough, it must be kept in mind that in dense regions the camera's star selection algorithm may not always select the brightest stars, but rather those with a good spatial distribution. Thus, to better "guarantee" a match for most camera orientations, we need a large catalog. In addition, we plan to test the camera and software performance when the CCD upper threshold is set to a lower value (fainter magnitude). This will require fainter stars in the catalog.

Our attitude determination software requires an *a priori* attitude estimate to avoid having to search the entire catalog. This estimate is provided by preprogrammed commands from Spartan's maneuver timeline.

Attitude Determination

If measured stars are matched with catalog stars, several methods exist to calculate the transformation, or camera attitude, between the catalog frame and the camera frame. The most intuitive is the Triad method,[4,5] a simple algorithm which estimates attitude from only two stars. A more general algorithm is QUEST,[5] which determines the attitude using any quantity of star-direction data.

Successive attitude estimates can be used to estimate the camera angular rate. Again, two methods exist for this calculation. The straightforward technique is to compute finite differences of the attitude and divide by the time. However, this method can be very inaccurate if the time interval is short. A Kalman filter was developed to produce a more accurate estimate of attitude and attitude rate.[6]

GROUND TEST METHODS

Two methods were developed for evaluating the camera and attitude determination performance with real star data: Earth-fixed tests and sluing tests. Our first Earth-fixed tests were conducted at the Applied Physics Laboratory (APL) and were valuable for end-to-end functional testing of the overall instrument and for performance testing of the camera and attitude software. We later tested the camera at Jet Propulsion Laboratory's Table Mountain Observatory (TMO) using both Earth-fixed and sluing tests.

Earth-Fixed Tests

Earth-fixed tests consist of holding the camera in a fixed, vertical orientation and letting the stars drift through the FOV. The constant rotation of the Earth causes the stars to pass through the FOV slowly and smoothly. Using the analysis discussed in a later section, it is possible to examine errors on both a pixel and FOV scale as well as frame-to-frame accuracy of the camera and attitude determination. Earth-fixed tests are relatively easy to conduct and quite useful for detecting high-spatial-frequency errors.

Our original test configuration used for testing the entire instrument consisted of the camera in a special enclosure with an optical window and a nitrogen purge port, instrument flight processor, and the instrument ground support equipment (GSE). The GSE serves primarily to emulate the Spartan for both instrument commanding and data logging. The camera, in its enclosure, was placed outside and pointed vertically to minimize errors due to atmospheric refraction. Meanwhile, the instrument processor and GSE collected and processed data. This test verified overall function, control, and measured throughput of camera, flight processor, and GSE. With this arrangement we were able to operate the instrument in a realistic way and verify that the attitude determination software was properly identifying stars and computing attitude.

Our camera performance evaluation was hampered, however, by poor atmospheric transparency and by being limited to only one slue rate (Earth rate). The rate of star motions is far too slow to be representative of most space applications of this technology. These limitations led us to develop more comprehensive tests and analysis techniques for the second method of testing at TMO.

Sluing Tests

The second phase of testing was carried out at TMO. The camera and enclosure were mounted to the side of a 24-inch telescope which provided a stable mounting, when the drive was off, for the Earth-fixed observations, and, with its sidereal and sluing capability, provided various slue rates about one or both axes to simulate motion of a spacecraft. Data were collected with a portable computer (eliminating the need to transport the entire instrument and GSE) and all data processing and analysis were performed off-line.

The TMO tests were conducted primarily for camera performance evaluation but also provided additional data for analyzing the attitude determination methods. Attitude performance will change as camera performance degrades with faster slue rate. However, testing the camera on a moving platform presents special problems not encountered with Earth-fixed tests. One of the primary concerns was how steady the telescope would be as a platform. Although direct measurements were not made, visual checks through the telescope indicated that the telescope drive motion was quite smooth. Also, from the data that has been processed, position errors between stars do not appear to be correlated.

Another problem is atmospheric refraction, both total and differential. While most of our data segments were short and taken with the telescope nearly vertical to minimize refraction effects, the higher rate slues subtended a sufficiently large arc that refraction could be a problem. No corrections were made for refraction although it could be handled without too much additional processing.

The Table Mountain experiments also presented the opportunity to test the camera sensitivity at a site with very good atmospheric transparency, or low extinction. Collecting data at a low extinction site gives a better indication of how the camera will perform in space. The extinction at TMO is typically 0.1 magnitude, which is very good. We believe extinction was at least that low during our tests, based on our extinction data. To further characterize the camera sensitivity several minutes of data were collected with and without the protective glass window on the enclosure. Without the window the signal increased by approximately 0.20–0.25 magnitude.

TEST-DATA PROCESSING

Data processing for these ground tests requires several steps: 1) identification of the measured stars, 2) computation of the single-frame and Kalman-filter estimates of attitude, 3) estimation of the camera slue axes and rates to obtain a reference attitude history, 4) computation of ideal star positions or paths on the CCD based on the calculated reference attitude, and 5) computation of residuals between measured and computed positions. The star position residuals can reveal interpolation errors, quantization errors and large scale image plane errors. Comparisons between the reference attitude and single-frame or Kalman-filter attitude estimates shows attitude determination performance.

Star Identification

Except for the end-to-end functional tests of the entire instrument, data processing is performed offline. The first step is to identify stars as mentioned above.[3] The output of this process is a table consisting of six columns and n rows, one for each frame. Each row of the table consists of the frame time reported by the camera, followed by as many as five catalog star identification numbers, one for each measured star, if it was identified in that frame. This table was the point of reference for the following processing steps and avoided the problem of having to reexecute the star-identification program when different reduction techniques or parameters are tested.

Data-Processing Method

A method to estimate the errors in the star measurements and in the attitude estimates from the on-board software is highly desired. Ideally, this method would be able to independently measure the attitude of the camera in order to obtain a true reference from which to measure the errors. Unfortunately, there was no method available to measure the attitude of the camera to an accuracy greater than that provided by the measurements themselves. Therefore, the data itself must be used somehow to estimate a reference attitude that is, hopefully, more accurate than the estimate from the on-board attitude-determination software. To do this, models of the camera motion were constructed that exploit the characteristics of the tests. For the Earth-fixed case, it was assumed that the camera is rigidly attached to the Earth, and therefore the motion of the camera is due solely to the rotation of the Earth, which is well known. For the telescope slue case, it was assumed that the telescope drive rate is constant, and that the motion of the camera is due to the combined rotation of the Earth and of the telescope drive. The tests were carefully set up and performed to eliminate as much as possible any vibrations or other disturbances that would violate these assumptions.

A maximum-likelihood estimator has been developed for the camera motion. Given a set of n star measurements, the kth star measurement obeys the following relation

$$\hat{\mathbf{W}}_k = R_{\text{Cam}\leftarrow I}(t_k)\,\hat{\mathbf{V}}_k + \Delta\hat{\mathbf{W}}_k \quad , \tag{1}$$

where $\hat{\mathbf{V}}_k$ is a unit vector representing the star direction in the inertial coordinate system (obtained from the star catalog), $\hat{\mathbf{W}}_k$ is the corresponding unit vector representing the measured direction of the star in the camera coordinate system, t_k is the time of the measurement, $R_{\text{Cam}\leftarrow I}(t_k)$ is the attitude matrix at that time, and $\Delta\hat{\mathbf{W}}_k$ is the measurement error, which is assumed to be Gaussian with

$$E\{\Delta\hat{\mathbf{W}}_k\} = 0 \quad , \quad E\{\Delta\hat{\mathbf{W}}_k \Delta\hat{\mathbf{W}}_k^T\} = \sigma_k^2\left(I - \hat{\mathbf{W}}_k\hat{\mathbf{W}}_k^T\right) \equiv R_k \quad . \tag{2}$$

Although some of the error components, such as quantization error, are not Gaussian, the total error is reasonably close to Gaussian.

For the Earth-fixed case, the attitude can be written as a sequence of two rotations: a rotation from the inertial system to an Earth-fixed system, and a rotation from the Earth-fixed system to the camera coordinate system, as follows:

$$R_{\text{Cam}\leftarrow I}(t) = R_{\text{Cam}\leftarrow E}\, R_{E\leftarrow I}(t) \quad . \tag{3}$$

The first rotation is time-varying due to the Earth's rotation. The second is assumed to be constant, i.e., the camera is assumed to be rigidly fixed to the Earth. Any Earth-fixed coordinate system may be chosen, which need not be aligned with any "natural" axes of the Earth. In fact, a somewhat unnatural choice will be made in order to simplify the equation. With this in mind, Eq. (3) is rewritten with the rotation matrices expressed as functions of rotation vectors, as follows:

$$R_{\text{Cam}\leftarrow I}(t) = R^T(\boldsymbol{\theta})\, R(\boldsymbol{\omega}_E(t - t_o)) \quad , \tag{4}$$

where $\boldsymbol{\omega}_E$ is the angular velocity vector of the Earth's rotation, t_o is the time of the first measurement, and $\boldsymbol{\theta}$ is a constant rotation vector representing the rotation from the camera coordinate system to the Earth-fixed system. In writing this equation the Earth-fixed coordinate system has been implicitly chosen to be the one which is aligned with the inertial coordinate system at time t_o. Thus, the z-axis of our Earth-fixed system is aligned with the polar axis, as usual, but the x-axis is not necessarily aligned with Greenwich. Also, $R^T(\boldsymbol{\theta})$ is necessarily the star camera attitude at time t_o.

Substituting the kinematic model of Eq. (4) into the measurement model of Eq. (1) results in

$$\hat{\mathbf{W}}_k = R^T(\boldsymbol{\theta})\, R(\boldsymbol{\omega}_E(t_k - t_o))\,\hat{\mathbf{V}}_k + \Delta\hat{\mathbf{W}}_k \quad , \tag{5}$$

which can be written as

$$\hat{\mathbf{W}}_k = R^T(\boldsymbol{\theta})\,\hat{\mathbf{V}}_k^E + \Delta\hat{\mathbf{W}}_k \quad , \tag{6}$$

where

$$\hat{\mathbf{V}}_k^E = R(\boldsymbol{\omega}_E(t_k - t_o))\,\hat{\mathbf{V}}_k \tag{7}$$

is the direction of the star in the Earth-fixed coordinate system. From Eq. (6) then, we see that the attitude estimation problem consists of finding the rotation matrix $R^T(\boldsymbol{\theta})$ which best transforms the set of vectors $\hat{\mathbf{V}}_k^E$ into the set $\hat{\mathbf{W}}_k$. This can be accomplished with the QUEST algorithm.[5]

Once $R(\boldsymbol{\theta})$ has been estimated, the attitude at any time can be computed using Eq. (4). This attitude can be used as a reference for computing errors in the attitude estimates provided by the on-board software.

The reference attitude can be used also to obtain estimates of individual star measurement errors. Note that $R^T(\boldsymbol{\theta})\hat{\mathbf{V}}_k^E$ would be the position in the camera system of the kth star, if our attitude estimate was error free. Thus $\hat{\mathbf{W}}_k - R^T(\boldsymbol{\theta})\hat{\mathbf{V}}_k^E$ is an estimate of $\Delta\hat{\mathbf{W}}_k$, the error in the kth measurement. Because the camera is likely to measure a given star multiple times as it moves across the field of view, a time history of the measurement errors for that star can be constructed.

Camera Slue Analysis

The Earth-fixed case is extended to the single-axis-telescope-slue case by adding another intermediate reference frame, one that is fixed in the telescope. The telescope is assumed to cause the telescope to rotate with a constant angular velocity with respect to the Earth. Thus, another time-varying rotation is added to the sequence of rotations as follows,

$$R_{\text{Cam}\leftarrow I}(t) = R_{\text{Cam}\leftarrow\text{Tel}}\, R_{\text{Tel}\leftarrow E}(t)\, R_{E\leftarrow I}(t) \tag{8}$$

$$= R^T(\boldsymbol{\theta})\, R(\boldsymbol{\omega}_{\text{Tel}}(t - t_o))\, R(\boldsymbol{\omega}_E(t - t_o)) \quad , \tag{9}$$

where $\boldsymbol{\omega}_{\text{Tel}}$ is the angular velocity of the telescope drive with respect to the Earth. Again, this equation is substituted into Eq. (1) and the catalog star positions transformed from the inertial frame to the Earth-fixed frame, resulting in

$$\hat{\mathbf{W}}_k = R^T(\boldsymbol{\theta})\, R(\boldsymbol{\omega}_{\text{Tel}}(t_k - t_o))\, \hat{\mathbf{V}}_k^E + \Delta\hat{\mathbf{W}}_k \quad . \tag{10}$$

We now have a set of nonlinear equations in both $\boldsymbol{\theta}$ and $\boldsymbol{\omega}_{\text{Tel}}$, and we cannot use the QUEST algorithm as in the Earth-fixed case. Instead, we will use a perturbation method to arrive at a solution. First, let us rewrite the remaining rotation matrices as matrix exponentials,

$$\hat{\mathbf{W}}_k = e^{-[[\boldsymbol{\theta}]]}\, e^{[[\boldsymbol{\omega}_{\text{Tel}}]](t_k - t_o)}\, \hat{\mathbf{V}}_k^E + \Delta\hat{\mathbf{W}}_k \quad , \tag{11}$$

where matrices of the form $[[\,\mathbf{U}\,]]$ are anti-symmetric matrices such that $[[\,\mathbf{U}\,]]\,\mathbf{V}$ is equal to the vector cross product $\mathbf{V} \times \mathbf{U}$ (or $-\mathbf{U} \times \mathbf{V}$). Given initial estimates of $\boldsymbol{\theta}$ and $\boldsymbol{\omega}_{\text{Tel}}$, denoted by $\boldsymbol{\theta}^*$ and $\boldsymbol{\omega}^*_{\text{Tel}}$, we can define perturbations $\delta\boldsymbol{\theta}$ and $\Delta\boldsymbol{\omega}_{\text{Tel}}$ such that

$$\boldsymbol{\theta} = \delta\boldsymbol{\theta} \otimes \boldsymbol{\theta}^* \tag{12}$$

$$\boldsymbol{\omega}_{\text{Tel}} = \boldsymbol{\omega}^*_{\text{Tel}} + \Delta\boldsymbol{\omega}_{\text{Tel}} \quad , \tag{13}$$

where $\otimes$ denotes rotation-vector composition, so that rotation $\boldsymbol{\theta}$ corresponds to the rotation represented by $\boldsymbol{\theta}^*$ followed by the rotation represented by $\delta\boldsymbol{\theta}$. Equivalently, in terms of rotation matrices,

$$e^{[[\boldsymbol{\theta}]]} = e^{[[\delta\boldsymbol{\theta}]]}\, e^{[[\boldsymbol{\theta}^*]]} \quad . \tag{14}$$

Equation (11) can now be expanded in a first-order Taylor series

$$\hat{\mathbf{W}}_k = e^{-[[\boldsymbol{\theta}^*]]}\,(I - [[\delta\boldsymbol{\theta}]])\left(e^{[[\boldsymbol{\omega}^*_{\text{Tel}}]](t_k - t_o)} + \sum_{i=1}^{3} \left.\frac{\partial e^{[[\boldsymbol{\omega}]](t_k - t_o)}}{\partial \omega_i}\right|_{\boldsymbol{\omega}^*_{\text{Tel}}} \Delta\omega_{\text{Tel}\,i}\right) \hat{\mathbf{V}}_k^E$$
$$+ \Delta\hat{\mathbf{W}}_k \quad . \tag{15}$$

Expanding, and keeping only first-order terms,

$$\hat{\mathbf{W}}_k = \hat{\mathbf{W}}_k^* + e^{-[[\boldsymbol{\theta}^*]]}\left[\left[e^{[[\boldsymbol{\omega}^*_{\text{Tel}}]](t_k - t_o)}\hat{\mathbf{V}}_k^E\right]\right] \delta\boldsymbol{\theta}$$
$$+ \sum_{i=1}^{3}\left(e^{-[[\boldsymbol{\theta}^*]]} \left.\frac{\partial e^{[[\boldsymbol{\omega}]](t_k - t_o)}}{\partial \omega_i}\right|_{\boldsymbol{\omega}^*_{\text{Tel}}} \hat{\mathbf{V}}_k^E\right) \Delta\omega_{\text{Tel}\,i} + \Delta\hat{\mathbf{W}}_k \quad , \tag{16}$$

where

$$\hat{\mathbf{W}}_k^* \equiv e^{-[[\boldsymbol{\theta}^*]]} e^{[[\boldsymbol{\omega}_{\text{Tel}}^*]](t_k - t_o)} \hat{\mathbf{V}}_k^E \quad , \tag{17}$$

and we have used the fact that $[[\mathbf{U}]]\,\mathbf{V} = -[[\mathbf{V}]]\,\mathbf{U}$. Thus, an effective measurement can be defined as

$$\mathbf{Z}_k = H_k\, \delta\mathbf{X} + \Delta\hat{\mathbf{W}}_k \quad , \tag{18}$$

where

$$\mathbf{Z} = \hat{\mathbf{W}}_k - \hat{\mathbf{W}}_k^* \quad , \tag{19}$$

$$\delta\mathbf{X} = \begin{bmatrix} \delta\boldsymbol{\theta} \\ \Delta\boldsymbol{\omega}_{\text{Tel}} \end{bmatrix}_{6\times 1} \quad , \tag{20}$$

$$H_k = [\, H_{\theta,k} \quad | \; H_{\omega_{\text{Tel}},k} \,]_{3\times 6} \quad , \tag{21}$$

$$H_{\theta,k} = [[\,\hat{\mathbf{W}}_k^*\,]]\, e^{-[[\boldsymbol{\theta}^*]]} \quad , \tag{22}$$

$$H_{\omega_{\text{Tel}},k} = e^{-[[\boldsymbol{\theta}^*]]}\, [\, \mathbf{U}_{1,k},\, \mathbf{U}_{2,k},\, \mathbf{U}_{3,k} \,] \quad , \tag{23}$$

and

$$\mathbf{U}_{i,k} \equiv \left. \frac{\partial e^{[[\boldsymbol{\omega}]](t_k - t_o)}}{\partial \omega_i} \right|_{\boldsymbol{\omega}_{\text{Tel}}^*} \hat{\mathbf{V}}_k^E \quad . \tag{24}$$

Since the unit vector measurements have only two degrees of freedom, the measurements must be reduced to minimum dimension by using only the x and y components. The reduced measurement is given by

$$\mathbf{Z}_k' = H_k'\, \delta\mathbf{X} + \Delta\hat{\mathbf{W}}_k' \quad , \tag{25}$$

where

$$\mathbf{Z}_k' = M\, \mathbf{Z}_k \quad , \quad H_k' = M\, H_k \quad , \quad \Delta\hat{\mathbf{W}}_k' = M\, \Delta\hat{\mathbf{W}}_k \quad . \tag{26}$$

and

$$M \equiv \begin{bmatrix} 1 & 0 & 0 \\ 0 & 1 & 0 \end{bmatrix} \quad . \tag{27}$$

All of the measurements, $\mathbf{Z}_k$, $k = 1, \ldots, n$, can be stacked to form a single measurement

$$\mathbf{Z} = H\, \delta\mathbf{X} + \boldsymbol{\varepsilon} \quad , \tag{28}$$

where

$$\mathbf{Z} = [\, \mathbf{Z}_1'^T, \ldots, \mathbf{Z}_n'^T \,]^T \quad , \tag{29}$$

$$H = [\, H_1'^T, \ldots, H_n'^T \,]^T \quad , \tag{30}$$

$$\boldsymbol{\varepsilon} = [\, \Delta\hat{\mathbf{W}}_1', \ldots, \Delta\hat{\mathbf{W}}_n' \,]^T \quad . \tag{31}$$

The minimum variance estimate of $\delta\mathbf{X}$ is then

$$\delta\mathbf{X} = \left(H^T R^{-1} H\right)^{-1} H^T R^{-1} \mathbf{Z} \quad , \tag{32}$$

where

$$R = \begin{bmatrix} R'_1 & 0 & \dots & 0 \\ 0 & R'_2 & \dots & 0 \\ \vdots & \vdots & \ddots & \vdots \\ 0 & 0 & \dots & R'_n \end{bmatrix}_{2n \times 2n} , \tag{33}$$

$$R'_k = M\, R_k\, M^T \quad . \tag{34}$$

This perturbation can be applied to the prior estimate to obtain a new estimate of θ and ω_{Tel}. The entire process can be iterated until $\delta \mathbf{X}$ becomes sufficiently small. Usually only two or three iterations were required for the cases that we have examined. Once the final state estimate is computed, the attitude at any time is obtained from Eq. (9). An estimate of the error in the kth measurement is obtained by using the final state estimate to compute $\mathbf{Z}_k$.

RESULTS

We discuss results of three types of observations: a slue slightly faster than sidereal rate, an Earth-fixed case, and a 0.1 deg/second slue.

Delta Sidereal Slue Results

The first results that we will examine are for the star camera mounted on the telescope and slued in right ascension at a rate slightly different than sidereal rate. The rate was such that the stars moved the distance of two pixels in the x-direction during the data segment. Every sixth data frame (i.e., one frame every 0.6 sec.) was processed by the (ground-based) on-board attitude determination software to identify the stars and estimate attitude. A reference attitude history was computed and individual star residuals determined using the technique presented earlier. Figure 1 shows the residuals for star 5303 (5303 is the catalog identification number for the star). There are clearly two cycles of a sinusoidal trend in the x-residuals, with an amplitude of 5 arc seconds. This is almost certainly due to centroid interpolation error. Telescope drive errors have been ruled out as a possible cause, because residuals for another star in the same experiment have similar sinusoidal behavior but different phase. Also, residuals for a dimmer star have non-sinusoidal behavior.

Earth-Fixed Case

The next results are for an Earth-fixed case in which the camera was oriented to point toward the zenith. Data were collected for approximately 18 minutes. Every tenth data frame (i.e., one frame per second) was processed by the on-board attitude determination software to identify the stars and estimate attitude.

Figure 2 shows the tracks of the stars that were identified. A reference attitude history was computed and individual star residuals and attitude errors determined using the technique presented earlier. Figure 3 shows the residuals for star 3462. This star left the FOV at about $t = 10000$ sec, which is why there is no data after that time. The last several samples have a large x-error, probably because a portion of the star image was lying off of the usable area of the CCD. This "edge effect" is apparent in most of the cases in which a star leaves or enters the FOV. In addition to the edge effect, there may be a bias in both the x and y residuals that could be due to catalog error. Residuals for another star (5170) for this case are shown in Fig. 4. This star entered the FOV at about $t = 9650$ sec, at which point there appeared an edge effect error in x. There is also a slight downward trend in the x-residuals which may be the result of uncompensated radial distortion due to star color. The y- residuals have a somewhat sinusoidal behavior which is probably due to centroid interpolation error, since the star motion had a slight component in the y-direction which caused the star to move about 1 pixel along the y-axis.

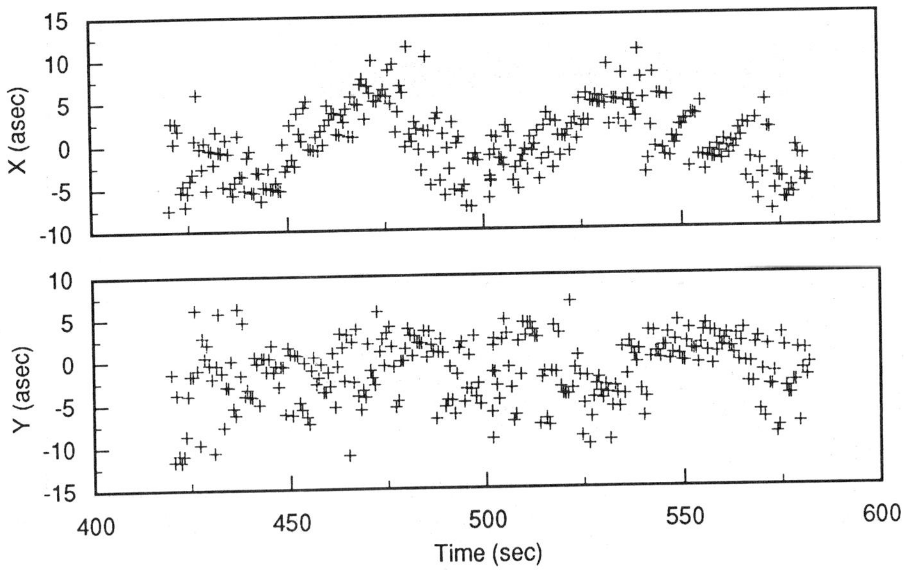

Fig. 1 Residuals for star number 5303 which moved the distance of two pixels in the x-direction. The cyclic variation in the x-residuals is probably due to interpolation error.

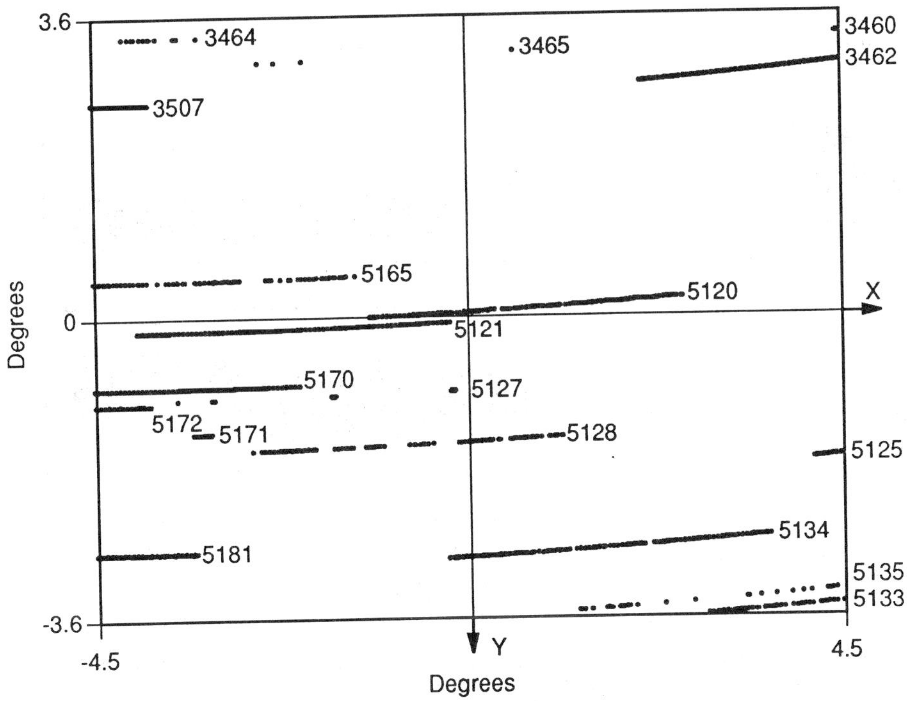

Fig. 2 Paths of stars through the FOV for an Earth-fixed case.

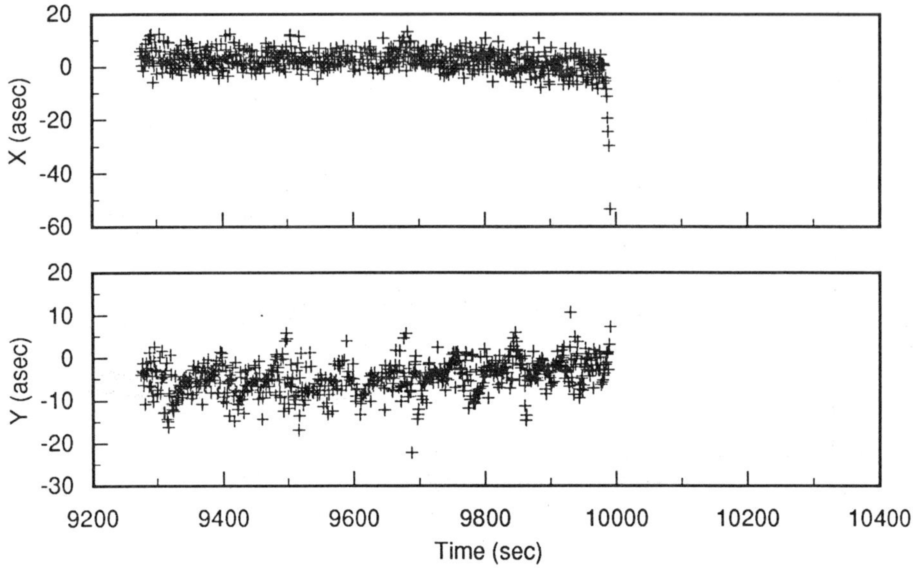

Fig. 3 Residuals for star number 3462. Note the large x-error when the star left the FOV.

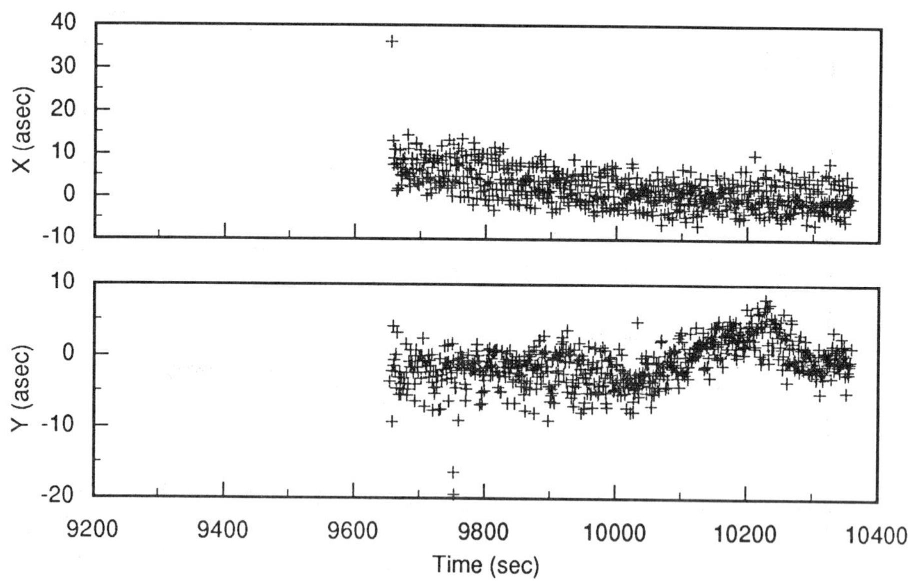

Fig. 4 Residuals for star number 5170. The slow variation in the y-residuals occurs as the star transits one row of the CCD.

In addition to the individual star residuals, we have also examined the attitude estimates from the on-board software for the Earth-fixed case. Both the single-frame attitude and the Kalman-filter attitude were analyzed by computing the small angles characterizing those attitudes as measured from the reference attitude. Results for the single-frame attitude are shown in Fig. 5. The middle portion of the data segment appears to be quieter than the beginning and end. During the middle portion there were three stars (5170, 3462, 5121) in the third to fourth magnitude range that were measured consistently. There were fewer good stars being measured at the beginning (before 5170 entered the FOV) and at the end (after 3462 left the FOV). Also, there is a noticeable spike in the attitude error at about $t = 10000$ sec, which is probably due to the edge effect error of star 3462 as it left the FOV. Not visible from this figure is the fact that between $t = 9500$ sec and $t = 9650$ sec there were many frames for which there were not enough stars identified for a confirmed match (i.e., less than three were identified). This may be because one of the stars being measured during that time may have been absent from the catalog, and the errors for others may have been too large. Except for this problem, though, the single-frame attitude estimates seem to be fairly good. Sample standard deviations of the errors were computed to be 4.5, 3.7, and 83. arc seconds, for the x, y, and z axes of the camera, respectively. These numbers are about what was expected given the errors in the individual star measurements. The z-axis (roll) is along the camera boresight. The large value for the z-axis attitude error is consistent with position errors of 5–10 arc seconds for stars several degrees from the camera boresight. Figure 6 shows the errors in the Kalman-filter attitude estimate for the same case. There is some improvement over the single-frame errors in all three axes, but the greatest improvement is in roll (z-axis).

The standard deviations for the filter errors were 3.9, 3.3, and 49. arc seconds. There is some structure visible in the filter results, and to a lesser extent in the single-frame results. Although this could be due to uncompensated motion of the camera, it is more likely the result of the systematic errors in the star measurements, such as interpolation error, which change as the stars move across the FOV and also as different stars are selected.

Slue (0.1 deg/sec) *Case*

Figure 7 shows the single-frame attitude errors for a slue case. The slue was at 0.1 deg/sec in right ascension and was about three minutes long. One frame of data per second was processed. Three periods are labeled along the time axes of the plots. During the first period there were always three or four bright stars (magnitude range 3.0–4.0) in each frame. The attitude results during this period were, accordingly, very good. Then, during the second period, the brighter stars began to drop out and were replaced by dimmer stars, with more toggling (the camera selected different stars in each frame). This caused a noticeable degradation in the attitude accuracy. At the transition from the second to the third period, the number of bright stars was reduced from two to one. The single remaining bright star was measured for the remainder of the third period. However, it was moving parallel to and slowly closer to the edge of the FOV, so that it had a large edge effect error during the last several seconds of the data segment. The other stars measured during the third period were fainter. In addition, all of the stars (including the bright one) measured during that period were close together, with the largest separation of any two being about 3 degrees. All of these problems caused the attitude errors to grow even worse during the third period, especially the roll errors. Averaged over the entire data segment, however, the attitude performance was quite good. The single-frame sample standard deviations were 4.0, 4.8, and 79. arc seconds for the x, y, and z errors, respectively. Standard deviations for the Kalman-filter errors were 3.5, 3.5, and 52. arc seconds. As in the Earth-fixed case, the filter roll error shows the most improvement over the single-frame results.

The attitude errors for the 0.1 deg/sec slue case did not seem to be any larger than the Earth-fixed case. However, for a 1.0 deg/sec slue case that we processed, the errors were about twice as large.

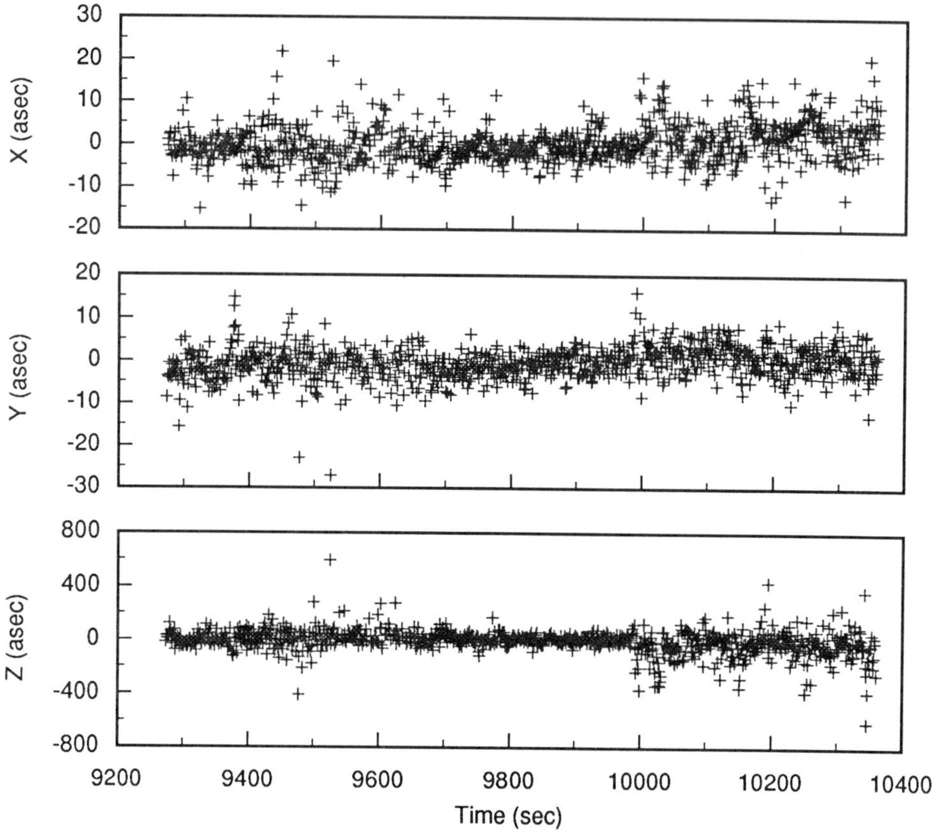

Fig. 5 Single-frame attitude errors for an Earth-fixed case. The errors are smaller in the middle, when a consistent star pattern was observed.

SUMMARY OF TEST RESULTS

Star Position Accuracy

The test data reveal much of the fine structure of star position errors. There are several sources of error which likely dominate. One is the quantization error (eight bits) of the camera's analog-to-digital converter for the CCD readout. This is the dominant error for faint stars. Another source is the interpolation error, which is a function of the image location with respect to the pixel and of star image size and shape. There may also be large spatial-frequency errors due to the limited precision in the lenticular radial-distortion correction. In addition, the calibration is applicable to only one particular spectral distribution, stars of different spectral type requiring different corrections. There is also a temporal error caused by photon and electronic noise in the system. Some of these errors are sensitive to the specific camera design and will be improved as the design matures.

There are other contributions to the residuals besides camera measurement errors. Biases in the residuals may be due to catalog errors. Uncompensated motion of the camera, caused by telescope drive errors for example, also add to the residuals. However, this can be detected by looking for correlation between the residuals of all the stars in the FOV. Atmospheric refraction also contributes to the star

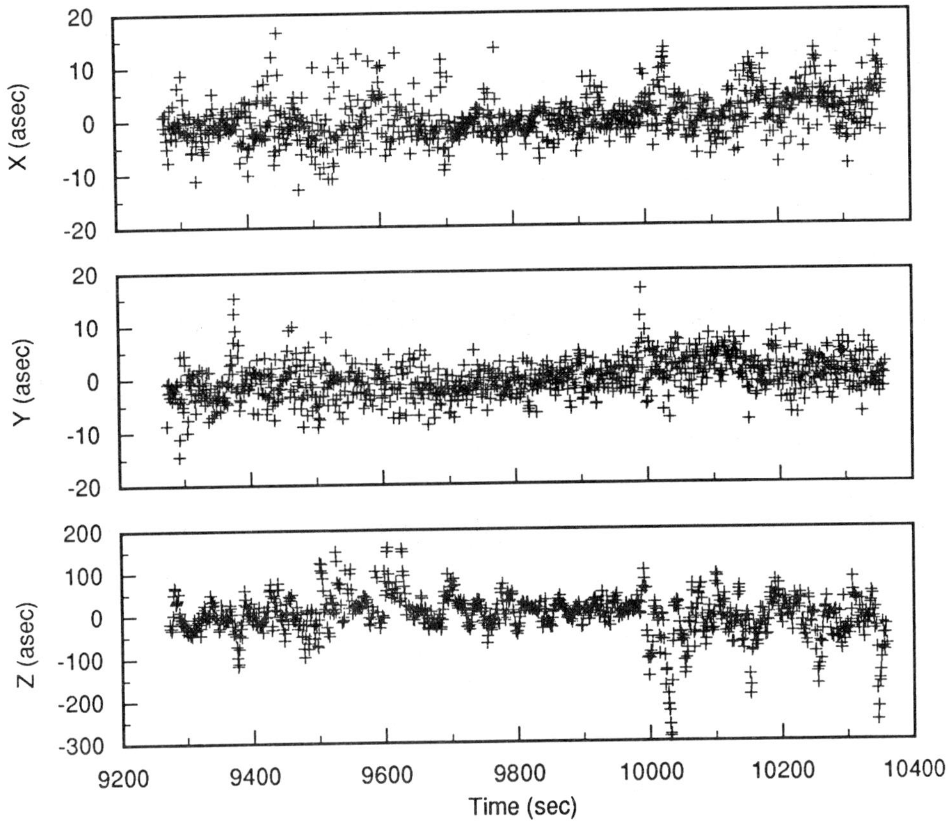

Fig. 6 Kalman-filter attitude errors for an Earth-fixed case. The roll (z) errors show the greatest improvement over the single-frame results shown in Fig. 5.

position error and is as yet uncompensated. Total refraction varies from zero at the zenith to about 35 arc seconds at a zenith angle of 30 degrees. Differential refraction within the FOV could cause biases in the residuals. Changes in total refraction due to changing optical path length as a function of star camera elevation may have more complicated effects.

From a large number of observations and conditions, we have developed an accuracy versus magnitude relationship, shown in Fig. 8. Each point represents the standard deviation of the residuals for a single star from a single data segment (since the stars were not all measured an equal number of times, each point represents different amounts of data). The fainter stars have larger errors, as expected. The ordinate of Fig. 8 is instrument magnitude as determined from CCD spectral response. An approximate visual-magnitude scale can be made by adding 0.5 to the instrument magnitude. The stars were attenuated by atmospheric extinction and absorption by the enclosure window and would appear brighter in space by about 0.30–0.35 magnitudes. Thus, the position errors associated with observations *in space* in stars of fifth to sixth magnitude would be reduced by 3 to 4 arc seconds.

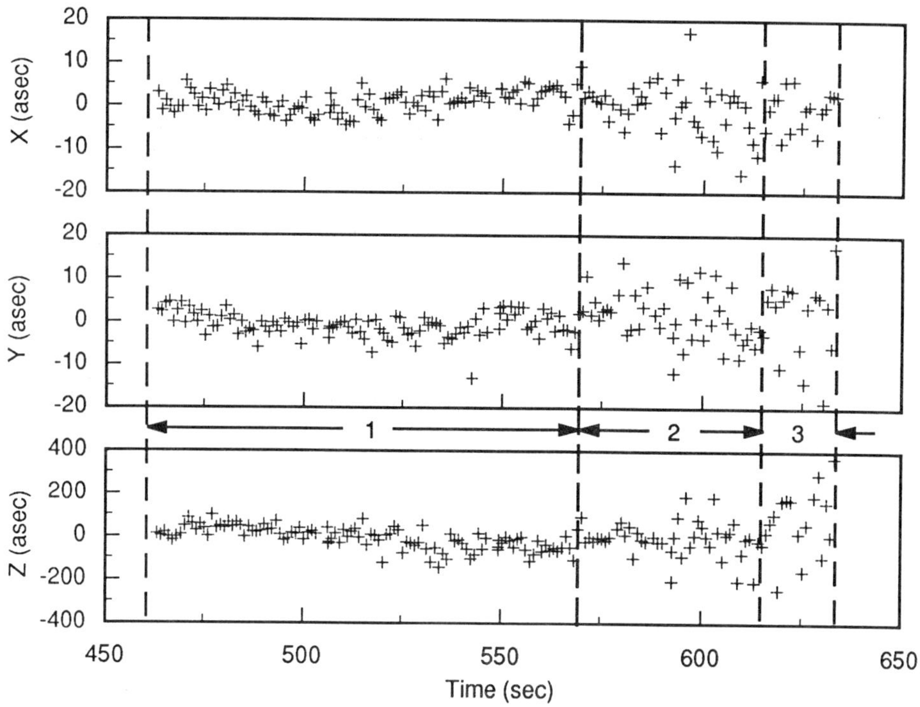

Fig. 7 Single-frame attitude errors for a 0.1 deg/sec slue. The larger errors at the end are caused by a rapidly changing star pattern.

Attitude Accuracy

The single-frame attitude accuracy is determined by the size of the errors of the individual stars in the frame. Ideally, the attitude accuracy about the two axes orthogonal to the boresight should be greater than the individual star accuracy by a factor equal to the square-root of the number of stars used to compute the attitude. As can be seen from the cases discussed here, individual position errors were about 5 to 15 arc seconds, depending on the star brightness, while the single-frame attitude accuracy, when using 3–5 stars, was about 4–5 arc seconds about the x- and y-axes. Obviously, there are many systematic errors which may affect this rule but these tests verify the power of the overall technique of using multiple stars for attitude determination.

Our tests also indicate additional improvement is possible by filtering the single-frame attitude estimates, particularly in roll. However, it is more difficult to quantify this and we have not processed a sufficient quantity of data from which to draw meaningful conclusions.

Camera Improvements

The frame-to-frame attitude "jitter" evident in several of the cases discussed above could be reduced in some cases if the camera tracked the same stars across the FOV rather than selecting five stars on its own. The present algorithm sometimes produced a frequently changing star pattern; consequently, different systematic position errors for each star will produce jitter in the attitude solution. Ensuring that "good" stars are tracked, however, would probably require that the instrument processor, with

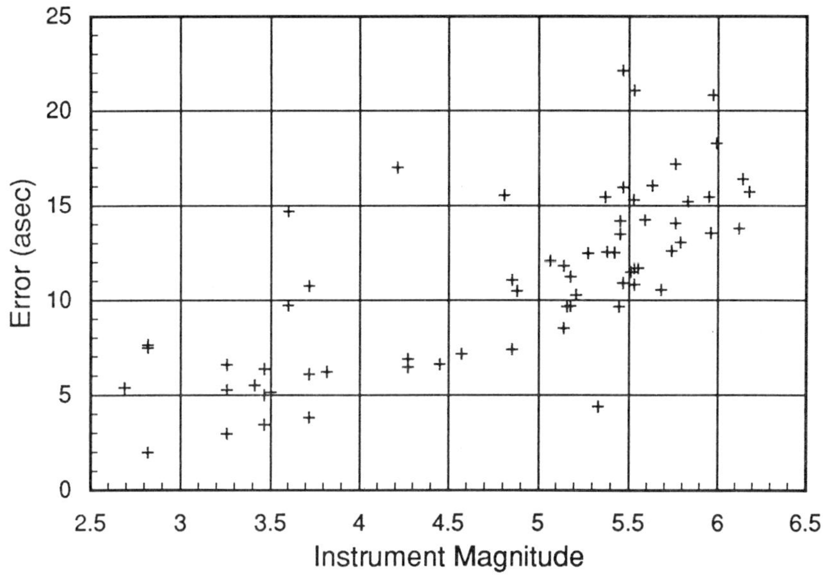

Fig. 8 Distribution of total position error for stars measured in these tests.

access to the star catalog and knowledge of the spacecraft rates, determine which stars to track; this obviously requires additional processing and real-time control of the camera.

We have detected larger position errors as stars near the edge of the FOV. We have not yet determined the cause, but in any case, one simple solution would be for the camera processor or our data-processing software to delete stars near the edge from consideration.

A number of improvements in camera design are possible. Cameras will use CCD arrays of larger format, higher resolution analog-to-digital converters, and more powerful processors.

Future Tests and Analysis

Only a small amount of data from our tests has been processed to the level discussed in this paper. Although we have learned a great deal, there are other cases, such as other slue rates and extinction data, which may provide insight for performance we might expect in a space experiment.

One refinement we will incorporate is correction for atmospheric refraction in our data-reduction software. We have the information needed (time and angles) to compute the mean refraction correction, which should be sufficient for our needs, but not the temperature and pressure for higher-accuracy correction.

The instrument star catalog can be validated using the test data. There were several instances of detected stars that apparently were not in our catalog or catalog stars that were not detected, even in sparse regions. The catalog was edited to remove faint stars in dense regions and to remove closely spaced doubles that would cause confusion in the star identification process; this editing should be verified.

As star cameras improve in accuracy, more accurate testing facilities must also be provided. The test methods presented here may also have to be improved and conditions monitored more closely.

Tests with real stars may continue to be a valuable part of testing both the cameras and attitude-determination software.

CONCLUSIONS

The test methods and results presented in this paper demonstrate the value of testing CCD star cameras and attitude determination software with real stars and precision sluing. Earth-fixed and sluing tests provide a measure of camera performance that is difficult or impossible to produce in the laboratory. Frame-to-frame position errors can be detected very easily as stars move slowly through the field of view. Bias and larger spatial errors can also be detected. We have shown that it is possible to identify stars and determine attitude accurately in a variety of conditions. Attitude accuracy is improved by using multiple stars in each frame, increasing the accuracy, in our case, from 5–15 arc seconds, for a single star, to 4–5 arc seconds. for a stationary or slowly moving camera.

Our methods for testing worked very well. The Earth-fixed tests require limited test equipment: the camera, an enclosure and stable mount, data-collection computer, and a dark, clear site. Sluing tests require, in addition, a precision platform such as an observatory telescope or rotary table.

ACKNOWLEDGEMENTS

The research and development reported here were carried out under contract to the US Army Engineer Topographic Laboratories, Fort Belvoir, Virginia. The support and encouragement of Connie Gray is gratefully acknowledged. We are grateful to Allan Eisenman of JPL and the staff of JPL's Table Mountain Facility for their assistance during our tests.

REFERENCES

1. GRAY, C., and STRIKWERDA, T. E. "Satellite Attitude Determination Using a CCD Star Camera," *Annual Convention of the Am. Congress on Surveying and Mapping and Am. Soc. of Photogrammetry and Remote Sensing*, March 1988.

2. STRIKWERDA, T. E., and FISHER, H. L., "A CCD Star Camera Used for Satellite Attitude Determination," Proceedings, Summer Computer Simulation Conference, Seattle, Washington, July 1988.

3. JUNKINS, J. L., and STRIKWERDA, T. E., "Autonomous Star Sensing and Attitude Estimation," *Rocky Mountain Guidance and Control Conference*, Paper AAS 79-013, February 1979.

4. BLACK, H. D., "A Passive System for Determining the Attitude of a Satellite," *AIAA Journal*, Vol. 2, No. 7, pp. 1350–1351, July 1964.

5. SHUSTER, M. D., and OH, S. D., "Three-Axis Attitude Determination from Vector Observations," *Journal of Guidance, Control and Dynamics*, Vol. 4, No. 1, pp. 70–77, Jan- Feb, 1981.

6. FISHER, H. L., SHUSTER, M. D., and STRIKWERDA, T. E., "Attitude Determination for the Star Tracker Mission," *AAS/AIAA Astrodynamics Specialist Conference*, Paper AAS 89–365, August 1989.

AAS 91-026

THE MINIMUM CONTROL AUTHORITY OF A SYSTEM OF ACTUATORS WITH APPLICATIONS TO GRAVITY PROBE-B

Peter Wiktor and Dan DeBra*

The forcing capabilities of systems composed of many actuators are analyzed in this paper. Multi-actuator systems can generate higher forces in some directions than in others. Techniques are developed to find the force in the weakest direction. This corresponds to the worst case output and is defined as the "minimum control authority." The minimum control authority is a function of three things: the actuator configuration, the actuator controller and the way in which the output of the system is limited. The following three output limits are studied: 1) fuel flow rate, 2) power and 3) actuator output. The three corresponding actuator controllers are derived. These controllers generate the desired force while minimizing either fuel flow rate, power or actuator output. It is shown that using the optimal controller can substantially increase the minimum control authority. For example the minimum control authority of a system limited by actuator output can be increased by 40 % if the controller minimizes actuator output instead of power. The techniques for calculating the minimum control authority are applied to the Gravity Probe-B spacecraft thruster system. This example shows that the minimum control authority is a useful design tool. It can be used to design the individual actuators, choose an actuator configuration, choose an actuator controller and study redundancy.

1 Introduction

When choosing a motor or thruster or any other type of actuator, it is important to make sure that it generates enough force to overcome the applied loads. A conservative approach is to calculate the biggest foreseeable load and then choose an actuator which

* Stanford University, Stanford, California 94305.

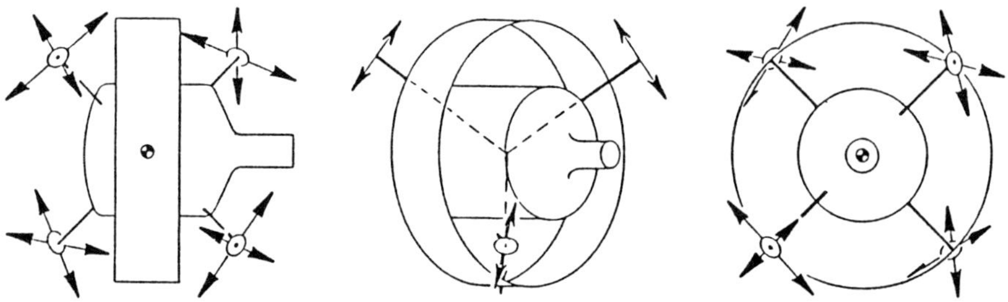

Figure 1: Three views of the Gravity Probe-B spacecraft thruster system.

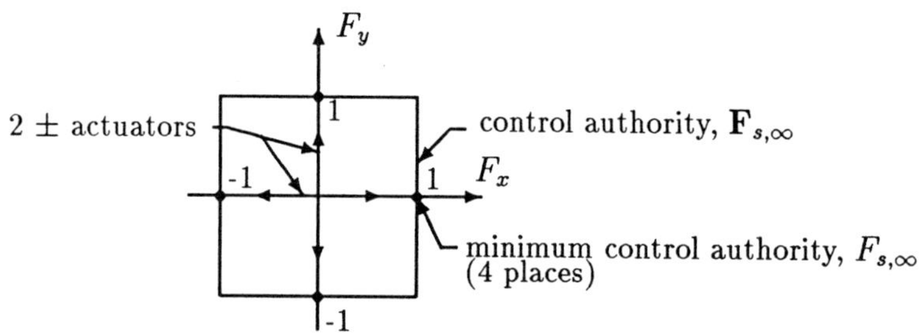

Figure 2: Control authority of a system of two mutually perpendicular actuators. Each actuator has a peak output of 1.

can overcome this load under worst case conditions. We call the worst case output of an actuator the minimum control authority.

In this paper we are concerned with systems made up of more than one actuator. The Gravity Probe-B (GP-B) spacecraft [10, 8], for example, will have a system of thrusters providing forces for attitude and translation control [Fig. 1]. For this multi-actuator system we want to make sure that the actuators can generate enough force to overcome the applied loads; therefore, we need to know the minimum control authority.

A multi-actuator system can generate higher forces in some directions than in others. In other words the system is stronger in some directions and weaker in others. The minimum control authority is the magnitude of the output force in the weakest direction. For example, consider a system made up of two mutually perpendicular actuators as depicted in Fig. 2. If each actuator can generate a force of at most one, then the square in Fig. 2 depicts the maximum force which can be generated by the actuator system in any direction. We call this square the control authority. This system can generate higher forces at ±45° than straight up and down or side to side. The worst case output, or minimum control authority, corresponds to the points on the square which are closest to the origin. In this simple example we are able to determine the minimum control authority by looking at a diagram. In general an actuator system is capable of generating three components of force and three components of moment.

The control authority is a surface in six dimensional space and therefore it is impossible to determine the minimum control authority graphically. A mathematical solution is required. General techniques to calculate the minimum control authority of a system of actuators are derived in this paper. Techniques to calculate the direction of the minimum control authority (the weakest direction) are also derived.

The minimum control authority is a single number which quantifies the performance of an actuator system. Since the minimum control authority corresponds to the worst case output, it is arguably the most important performance measure. If the minimum control authority is less than the applied load then there is a good possibility that the actuator system won't work. An actuator system with a higher minimum control authority is typically better than a system with a lower one. The minimum control authority therefore is a quantifiable way of judging different candidate actuator system designs. The minimum control authority also lends itself to a natural definition of redundancy. If the minimum control authority of a system of actuators under a worst case actuator failure is higher than the applied loads then the system is redundant.

The minimum control authority of a system of actuators depends on three factors: 1) the actuator configuration, 2) the way in which the actuators are controlled and 3) the way in which the actuator system's output is limited. These factors are defined as follows:

Configuration The number, location and orientation of the individual actuators define the actuator configuration.

Control The outputs from the individual actuators are force vectors. The sum of these vectors results in the output force of the system. The actuator controller chooses the appropriate actuator outputs which result in a desired output force. For redundancy or for some other reasons, a system may have more actuators than the minimum required to generate the desired force. In this case many different actuator controllers can be used. We define three specific actuator controllers in this paper; each controller generates the desired force and additionally minimizes one of the following three quantities: 1) actuator output, 2) power or 3) flow rate. The minimum power controller is the easiest to implement. Under certain conditions however, it makes sense to use one of the other two controllers. For example, if there is a limit on available flow rate then the minimum flow rate controller yields the highest output force and therefore the highest minimum control authority. As we show in Sect. 5 the increase in the minimum control authority can be quite significant under certain circumstances.

Output limit Any actuator system has a limit on the amount of force which it can generate. There are three important ways in which this force may be limited. There may be a limit on: 1) the peak force output from the individual actuators, 2) the power available to all of the actuators or 3) the total flow rate of propellant available to the actuators. In Sect. 2.3 we give a concise mathematical definition and examples for each one of these limits.

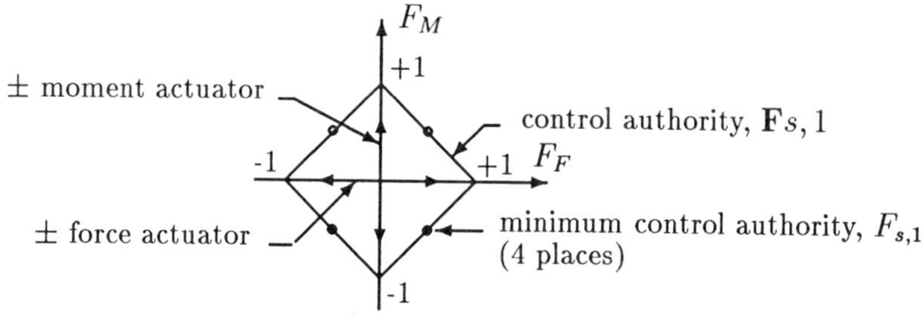

Figure 3: The control authority of a system consisting of one force actuator and one moment actuator. The flow rate is limited to a peak value of 1.

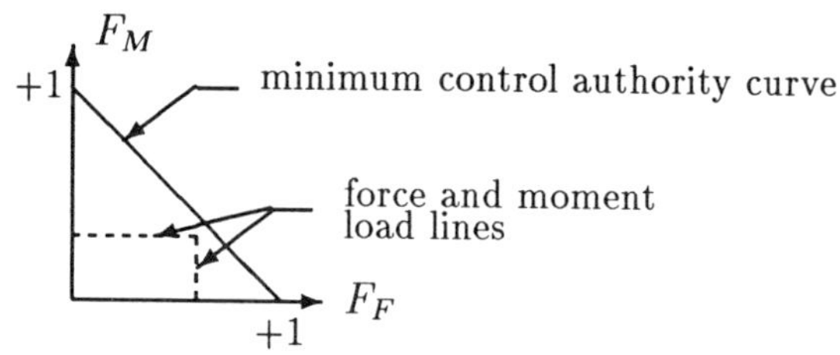

Figure 4: The minimum control authority curve of a system consisting of one force actuator and one moment actuator. The flow rate is limited to a peak value of 1. The dotted lines correspond to the applied force and moment loads on the system.

Some actuator systems, like the GP-B thruster system for example, can generate both forces and moments simultaneously. There is a trade off between the two. As the amount of force generated by the system increases then the ability to generate moments decreases. To analyze this situation we have developed the minimum control authority curve. As a simple example consider a system of two actuators again, one generating a force and the other a moment. If the total flow rate available to these actuators is limited to a peak value of 1 then the diamond in Fig. 3 illustrates the control authority for this case. The minimum control authority curve is depicted in Fig. 4. As force increases then the ability to generate a moment decreases along the diagonal line. By plotting the worst case force and moment load lines on the same plot (dotted lines) we can tell if the actuator system will work. If the worst case load lines lie below the minimum control authority curve then we know that the actuator system can overcome these loads even in its weakest direction.

As a more practical example consider the GP-B thruster system again. The solid lines in Fig. 12 depict the minimum control authority curves for different output limits and actuator controllers. The dotted lines depict the worst case force and moment loads due to atmospheric drag and Earth's magnetic field. Those curves which lie above the load lines correspond to actuator systems which can handle the loads. The distance from the load lines to the minimum control authority curves is the margin of safety.

A rigorous mathematical definition of the minimum control authority is presented in the next section. Following this, various techniques to calculate the minimum control authority for the three basic types of controllers are derived. A general technique for calculating the minimum control authority curve is presented next. These techniques are then applied to the GP-B thruster system to illustrate how useful the minimum control authority curve is in analyzing real world actuator systems. The definitions of some vector norms are presented in the Appendix.

2 Problem Definition

A rigorous mathematical definition of the minimum control authority is developed in this section, which is organized under the following headings: 1) Actuator Configuration, 2) Control, 3) Control Authority and 4) Minimum Control Authority.

2.1 Actuator Configuration

The force generated by the combined outputs of the individual actuators is,

$$\mathbf{F} = A\mathbf{T}. \tag{1}$$

In general, an actuator system can generate both forces and moments. The 6×1 generalized force vector, $\mathbf{F}$, is composed of 3 components of force, $\mathbf{F}_F$, and 3 components of moment, $\mathbf{F}_M$. If n is the number of actuators then $\mathbf{T}$ is an $n \times 1$ vector of the magnitudes of the force exerted by the individual actuators and A is the $6 \times n$ actuator configuration matrix which defines the position, orientation and scale factors of the actuators. For example consider the thruster configuration depicted in Fig. 5 consisting of three actuators radiating from a single point. Equation (1) for this case is,

$$\mathbf{F} = \begin{bmatrix} \mathbf{F}_F \\ \mathbf{F}_M \end{bmatrix} = \begin{bmatrix} 1 & -.5 & -.5 \\ 0 & .87 & -.87 \\ 0 & 0 & 0 \\ 0 & 0 & 0 \\ 0 & 0 & 0 \\ 0 & 0 & 0 \end{bmatrix} \begin{bmatrix} T_1 \\ T_2 \\ T_3 \end{bmatrix}. \tag{2}$$

2.2 Control

Given a desired generalized force, $\mathbf{F}$, and a configuration matrix, A, the actuator control problem consists of solving the set of simultaneous linear equations, $\mathbf{F} = A\mathbf{T}$, for the actuator command vector, $\mathbf{T}$. In general if there are more than six actuators then there are an infinite number of solutions to the under determined set of equations. In this case a specific solution which also satisfies some other condition can be found. In this paper we find the three solutions, $\mathbf{T}_s$, which satisfy $\mathbf{F} = A\mathbf{T}$ and in addition minimize the $s = 1, 2, \infty$ norms of $\mathbf{T}$. See the Appendix for definitions of these three vector norms. In mathematical terms we find the following three solutions, $\mathbf{T}_s$:

given $\mathbf{F}$ and A,

$$\mathbf{T}_s \stackrel{\text{def}}{=} \arg\min_{\mathbf{T}} \|\mathbf{T}\|_s \tag{3}$$
$$\text{such that } \mathbf{F} = A\mathbf{T}, \text{ where } s = 1, 2, \infty.$$

The techniques used to calculate these three solutions are defined in Sect. 2.4. The notation, arg min, in (3) means that $\mathbf{T}_s$ is that specific $\mathbf{T}$ which minimizes the s-norm of $\mathbf{T}$, subject to $\mathbf{F} = A\mathbf{T}$.

2.3 Control Authority

By control authority we mean the maximum force which can be generated by an actuator system in any direction. The control authority depends on the actuator configuration, A, the type of controller, $\mathbf{T}_s$, and on the way in which the output of the system is limited, $\|\mathbf{T}_s\|_p \leq m_p$, $p = 1, 2, \infty$. If $\mathbf{F}$ is 6×1 then the control authority is a surface in 6 dimensional space. The control authority, $\mathbf{F}_{s,p}$, is defined as the set of forces,

$$\begin{aligned} \mathbf{F}_{s,p} &\stackrel{\text{def}}{=} A\mathbf{T}_s && (4) \\ \text{such that } \|\mathbf{T}_s\|_p &= m_p && (5) \\ \text{where } \mathbf{T}_s &\stackrel{\text{def}}{=} \arg\min_{\mathbf{T}} \|\mathbf{T}\|_s \\ \text{and } s, p &= 1, 2, \infty. \end{aligned}$$

The three p-norms in (5) correspond to the three most common ways in which the output of an actuator system may be limited: 1) Actuator Force Limited ($p = \infty$), 2) Flow Rate Limited ($p = 1$) and 3) Power Limited ($p = 2$). These three limits are discussed individually in the following subsections.

2.3.1 Actuator Force Limited

There may be a limit on the amount of force which the individual actuators can generate. This is characterized as a limit on the ∞-norm of $\mathbf{T}_s$. The square in Fig. 2 for example is the control authority of a system of two actuators, each of which has a peak output of 1, $||\mathbf{T}_s||_\infty = 1$.

2.3.2 Flow Rate Limited

A system with a finite flow rate is characterized by a limit on the 1-norm of the actuator output vector, $||\mathbf{T}_s||_1 = m_1$. The diamond in Fig. 3, for example, is the control authority of a system of two actuators limited by a total flow rate of 1, $||\mathbf{T}_s||_1 = 1$.

2.3.3 Power Limited

A limit on the net power available to a system of actuators is characterized by an upper bound on the 2-norm of $\mathbf{T}_s$. For example, the control authority of a system of two actuators limited by a peak power of 1, $||\mathbf{T}_s||_2 = 1$, is a circle with radius 1.

controller type	*output limit*		
	flow rate	power	actuator output
minimum flow rate, $\mathbf{T}_1$	$F_{1,1}$	—	—
minimum power, $\mathbf{T}_2$	$F_{2,1}$	$F_{2,2}$	$F_{2,\infty}$
minimum actuator output, $\mathbf{T}_\infty$	—	—	$F_{\infty,\infty}$

Table 1: Combinations of output limits and controllers.

2.4 Minimum Control Authority

The minimum control authority, $F_{s,p}$, is the shortest distance to the control authority surface,

$$F_{s,p} \stackrel{\text{def}}{=} \min \|\mathbf{F}_{s,p}\|_2, \quad s,p = 1,2,\infty. \tag{6}$$

The actual vector to the closest point on the control authority surface is,

$$(\mathbf{F}_{s,p})_{\min} \stackrel{\text{def}}{=} \arg\min \|\mathbf{F}_{s,p}\|_2, \quad s,p = 1,2,\infty. \tag{7}$$

We call $(\mathbf{F}_{s,p})_{\min}$ the minimum control authority vector.

For the two actuators in a plane [Fig. 2] for example, the places where the square intersects with the horizontal and vertical axes are closest to the origin. The minimum control authority, $F_{s,\infty}$, is the distance to these points and $(\mathbf{F}_{s,\infty})_{\min}$ are the vectors from the origin to these points.

3 Calculating the Minimum Control Authority

In this section we present the techniques used to calculate the minimum control authority of a system of actuators. Since the minimum power solution is the easiest to compute, it is commonly used even if the actuator system is not limited by available power. Systems which are limited by actuator output or flow rate for example often use the minimum power controller. For the minimum power controller, therefore, we show how to compute the minimum control authority for all three cases: 1) flow rate limited, 2) power limited and 3) actuator output limited. For the minimum flow rate controller we show how to calculate the minimum control authority only for the case when the flow rate is limited. Similarly only the actuator output limited case is discussed for the minimum actuator output controller. The specific combinations of controllers and output limits which we analyze in this section are summarized in Table 1.

3.1 Minimum Power Controller

3.1.1 Control

The solution to $\mathbf{F} = A\mathbf{T}$ which minimizes power, $\mathbf{T}_2 = \min \|\mathbf{T}\|_2$, is the simplest to implement since the actuator command vector, $\mathbf{T}_2$, is simply a linear function of the desired force, $\mathbf{F}$:

$$\mathbf{T}_2 = A^{+}\mathbf{F}. \tag{8}$$

The matrix A^+ is the 'pseudo-inverse' of A and is equal to, $A^+ = A'(AA')^{-1}$. The pseudo-inverse can also be conveniently found using the function 'pinv' in Matlab [6] which uses a numerically stable algorithm based on the singular value decomposition. For a given actuator configuration, A, the pseudo-inverse, A^+. has to be computed only once. Given A^+, computing the actuator command vector, $\mathbf{T}_2$, involves only the simple matrix multiplication in (8).

3.1.2 Minimum Control Authority

Not only is the minimum power controller the easiest to compute, it also makes it easy to compute the minimum control authority. Recall from (4) and (5) that the control authority is the set of forces, $\mathbf{F}_{s,p} = A\mathbf{T}_s$ satisfying some p-norm on $\mathbf{T}_s$, $\|\mathbf{T}_s\|_p = m_p$. From (8) we know that the minimum power solution, $\mathbf{T}_2$, can be written as a function of $\mathbf{F}$, $\mathbf{T}_2 = A^+\mathbf{F}$. Therefore the minimum control authority is,

$$\begin{aligned} \mathbf{F}_{2,p} &\stackrel{\text{def}}{=} \min_{\mathbf{F}} \|\mathbf{F}\|_2 & (9) \\ \text{such that } \|A^+\mathbf{F}\|_p &= m_p, \quad p = 1, 2, \infty. & (10) \end{aligned}$$

Equations (9) and (10) indicate that if the minimum power controller is used then the minimum control authority depends only on the input output properties of the pseudo-inverse matrix, A^+. The input to A^+ is $\mathbf{F}$ and the output is $A^+\mathbf{F}$. Equations (9) and (10) state that the minimum control authority corresponds to the force with the smallest 2-norm which results in an output with a p-norm of $\|A^+\mathbf{F}\|_p = m_p$. In other words, if the size of the input is measured by the 2-norm, $\|\mathbf{F}\|_2$, and the size of the output is measured by the p-norm, $\|A^+\mathbf{F}\|_p$, then the minimum control authority corresponds to the force which makes the ratio of $\|A^+\mathbf{F}\|_p$ to $\|\mathbf{F}\|_2$ as big as possible. We call this ratio, $A^+_{2,p}$:

$$A^+_{2,p} \stackrel{\text{def}}{=} \max_{\mathbf{F}\neq\mathbf{0}} \frac{\|A^+\mathbf{F}\|_p}{\|\mathbf{F}\|_2}. \tag{11}$$

The scalar, $A^+_{2,p}$ is a matrix norm [5] and corresponds to the peak gain of the matrix, A^+ when the size of the input is measured by the 2-norm and the size of the output is measured by the p-norm. In our case the size of the input is the minimum control authority, $\|\mathbf{F}\|_2 = F_{2,p}$, and the size of the output is $\|A^+\mathbf{F}\|_p = m_p$. From (11) therefore, $A^+_{2,p} = m_p/F_{2,p}$, so the minimum control authority is,

$$F_{2,p} = \frac{m_p}{A^+_{2,p}}, \quad p = 1, 2, \infty. \tag{12}$$

Therefore, in order to calculate the minimum control authority when the minimum power controller is used, we must first find the matrix norm, $A^+_{2,p}$. As we show in the following three subsections, finding $A^+_{2,p}$ for the three cases when $p = 1, 2, \infty$ is relatively easy.

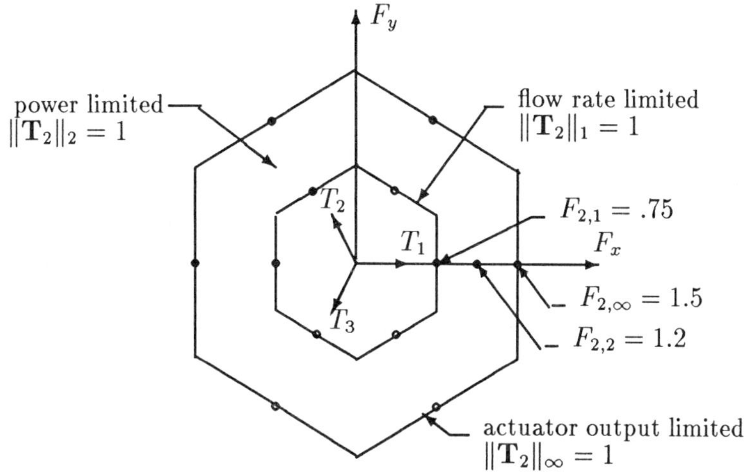

Figure 5: The control authorities of a system of three actuators subject to a limitation on flow rate, power or actuator output. The minimum power controller is used in all 3 cases.

Power Limited, p = 2

In this subsection we show how to calculate the minimum control authority of a system of actuators if the minimum power controller, $\mathbf{T}_2$, is used and if the actuator system has a limit on the total power available, $\|\mathbf{T}_2\|_2 = m_2$. From (12) we know that we must first find $A^+_{2,2}$. It is a well known fact that the peak gain of the matrix, A^+, when the sizes of both the input and output are measured by the 2-norm is the maximum singular value, $\sigma_1(A^+)$, of A^+ [5]. Therefore, the matrix norm is $A^+_{2,p} = \sigma_1(A^+)$. From (12) the minimum control authority is,

$$F_{2,2} = \frac{m_2}{\sigma_1(A^+)}. \tag{13}$$

The maximum singular value, $\sigma_1(A^+)$ can easily be found by typing $[U, S, V] = \text{svd}(A^+)$ in Matlab [6]. The direction of the minimum control authority corresponds to the first column in V, $\mathbf{V}_1$; therefore, $(\mathbf{F}_{2,2})_{\min} = \pm F_{2,2}\mathbf{V}_1$.

Example As a simple example consider three actuators 120° apart in a plane as depicted in Fig. 5. The actuator configuration matrix, A, and the pseudo-inverse matrix, A^+, are respectively,

$$A = \begin{bmatrix} 1 & -.5 & -.5 \\ 0 & .87 & -.87 \end{bmatrix}, \quad A^+ = \begin{bmatrix} .67 & 0 \\ -.33 & .58 \\ -.33 & -.58 \end{bmatrix}. \tag{14}$$

If the power available to the actuator system is limited to $\|\mathbf{T}_2\|_2 \leq 1$, then the control authority is depicted by the circle in Fig. 5. The maximum singular value of A^+ is,

$\sigma_1(A^+) = \sqrt{2/3}$, the minimum control authority from (13) is $F_{2,2} = \sqrt{3/2}$. Since all of the singular values of A^+ are equal to $\sqrt{2/3}$, the minimum control authority is the same in all directions as illustrated by the circle in Fig. 5.

Actuator Output Limited, $p = \infty$

Here we show how to compute the minimum control authority of an actuator system whose output force is limited by the force which can be generated by the individual actuators. Even though the minimum power controller does not result in the greatest possible minimum control authority for this type of actuator system (see Fig. 7) it is still used very frequently since it is by far the easiest to compute. From (12), the minimum control authority for this case is

$$F_{2,\infty} = \frac{m_\infty}{A^+_{2,\infty}}. \tag{15}$$

The matrix norm, $A^+_{2,\infty}$, is one of the easiest to compute. It is equal to the maximum 2-norm of the rows of A^+. If $\mathbf{a}'_i$ are the rows of A^+,

$$A^+ = \begin{bmatrix} \mathbf{a}'_1 \\ \vdots \\ \mathbf{a}'_n \end{bmatrix}, \tag{16}$$

and if $\mathbf{a}'_m$ is the row with the largest 2-norm, $\mathbf{a}'_m \stackrel{\text{def}}{=} \arg\max_i \|\mathbf{a}'_i\|_2$ then $A^+_{2,\infty} = \|\mathbf{a}_m\|_2$. From (15) therefore, the minimum control authority is,

$$F_{2,\infty} = \frac{m_\infty}{\|\mathbf{a}_m\|_2}. \tag{17}$$

The minimum control authority is in the same direction as the row with the largest 2-norm:

$$(\mathbf{F}_{2,\infty})_{\min} = \pm F_{2,\infty} \frac{\mathbf{a}_m}{\|\mathbf{a}_m\|_2}. \tag{18}$$

Example Consider three thrusters in a plane again [Fig. 5]. If the output of the individual thrusters is limited to 1, $\|\mathbf{T}_2\|_\infty \leq 1$, and if the minimum power controller is used, then the control authority is the large hexagon in Fig. 5. The pseudo-inverse, A^+, is again given by (14). The minimum control authority depends on the row with the largest 2-norm in A^+ which in this case is $\|\mathbf{a}_m\|_2 = 2/3$. From (17) the minimum control authority is $F_{2,\infty} = 3/2$. From (18) the directions corresponding to the minimum control authority are,

$$(\mathbf{F}_{2,\infty})_{\min} = \begin{cases} \pm[\ 1.5\ \ 0.0]' \\ \pm[-.74\ \ 1.3]' \\ \pm[-.74\ \ -1.3]'. \end{cases}$$

These are the points on the larger hexagon in Fig. 5 which are closest to the origin.

Flow Rate Limited, $\mathbf{p = 1}$

If the flow rate to the actuator system is limited but the minimum power controller is used then calculating the minimum control authority involves finding $A^+_{2,1}$ from (12). The technique for finding $A^+_{2,1}$ is similar to the technique described above for finding $A^+_{2,\infty}$ with the additional complication that we must first define the following matrix, Q:

Definition 3.1 *If n is the number of actuators, let Q be the $2^{n-1} \times n$ matrix formed by counting in binary to 2^{n-1} with the 0's replaced by $+1$ and the 1's replaced by -1. For example, if $n = 3$ then,*

$$\begin{bmatrix} 0 \\ 1 \\ 2 \\ 3 \end{bmatrix} \underset{binary}{\longrightarrow} \begin{bmatrix} 0 & 0 & 0 \\ 0 & 0 & 1 \\ 0 & 1 & 0 \\ 0 & 1 & 1 \end{bmatrix} \underset{\substack{0\to 1 \\ 1\to -1}}{\longrightarrow} \begin{bmatrix} 1 & 1 & 1 \\ 1 & 1 & -1 \\ 1 & -1 & 1 \\ 1 & -1 & -1 \end{bmatrix} \stackrel{\text{def}}{=} Q \tag{19}$$

The matrix norm, $A^+_{2,1}$, is equal to the maximum 2-norm of the rows of QA^+. More rigorously, if $nr = 2^{n-1}$ and $\mathbf{q}'_i A^+$ are the rows of QA^+:

$$QA^+ = \begin{bmatrix} \mathbf{q}'_1 A^+ \\ \vdots \\ \mathbf{q}'_{nr} A^+ \end{bmatrix}, \tag{20}$$

and if $\mathbf{q}'_m A^+$ is the row with the largest 2-norm, $\mathbf{q}'_m A^+ = \arg\max_i \|\mathbf{q}'_i A^+\|_2$ then $A^+_{2,1} = \|\mathbf{q}'_m A^+\|_2$. From (12) the minimum control authority for this case is,

$$F_{2,1} = \frac{m_1}{\|\mathbf{q}'_m A^+\|_2}, \tag{21}$$

and the direction of the minimum control authority is the same as the vector, $\mathbf{q}'_m A^+$,

$$(\mathbf{F}_{2,1})_{\min} = \pm F_{2,1} \frac{\mathbf{q}'_m A^+}{\|\mathbf{q}'_m A^+\|_2}. \tag{22}$$

Example Consider the same configuration of three actuators as in the last two examples [Fig 5]. This time the system output is limited by a maximum flow rate of 1 being available to the actuators, $\|\mathbf{T}_2\|_1 = 1$. To calculate the minimum control authority we must first define the matrix, Q,

$$Q = \begin{bmatrix} 1 & 1 & 1 \\ 1 & 1 & -1 \\ 1 & -1 & -1 \end{bmatrix}.$$

The matrix, QA^+, is therefore,

$$QA^+ = \begin{bmatrix} 0 & 0 \\ .67 & 1.15 \\ .67 & -1.15 \\ 1.33 & 0 \end{bmatrix}.$$

The last 3 rows of QA^+ have a 2-norm of, $\|\mathbf{q}'_m A^+\|_2 = 4/3$. Therefore, from (21) the minimum control authority is $F_{2,1} = 3/4$ and the directions corresponding to the minimum control authority from (22) are:

$$(\mathbf{F}_{2,1})_{\text{min}} = \begin{cases} \pm[.38 \;\; .65]' \\ \pm[.38 \;\; -.65]' \\ \pm[.75 \;\; 0]'. \end{cases}$$

The control authority for this case is the smaller hexagon in Fig. 5. The minimum control authority corresponds to the points on the hexagon which are closest to the origin.

3.2 Minimum Flow Rate Controller

3.2.1 Control

The minimum flow rate controller yields the greatest possible output force for a system which has a limit on the total flow rate supplied to the actuators. The solution, $\mathbf{T}_1$, which minimizes the flow rate can be found by solving the following linear program [9]:

$$\mathbf{T}_1 = \arg\min_{\mathbf{T}} \mathbf{I}'\mathbf{T} \tag{23}$$

$$\text{subject to, } A\mathbf{T} \geq \mathbf{F}, \;\; \mathbf{T} \geq \mathbf{0}, \tag{24}$$

where $\mathbf{I}$ is a vector of ones and $\mathbf{0}$ is a vector of zeros of the same length as $\mathbf{T}$. In words, equation (23) says "minimize the 1-norm of $\mathbf{T}$" and (24) says "subject to the constraints that $\mathbf{T}$ generates at least the desired force, $A\mathbf{T} \geq \mathbf{F}$, using only one-sided actuators, $\mathbf{T} \geq \mathbf{0}$."

Linear programs like (23) and (24) can easily be solved using the simplex method [9]. Many computer programs which implement the simplex method are available. In general, the simplex method converges very quickly and therefore can be used for real time control for many applications. For example, we used the simplex method to solve for $\mathbf{T}_1$ in the GP-B thruster system. For a thruster system consisting of 18 individual thrusters we were able to solve for 1000 completely random generalized force commands in 80 seconds. This corresponds to an update rate of 12.5 Hz. The calculations were done on a Sun-3 computer which is rated at 12 million instructions per second (mips). The program, LSSOL [4], was used which implements the Simplex method in Fortran. Although 12.5 Hz is fast enough for the GP-B translation and attitude control systems it is not nearly as fast as the simple matrix multiplication of the minimum power controller (8). This controller runs about 100 times faster on the same computer.

3.2.2 Minimum Control Authority, Flow Rate Limited

Calculating the minimum flow rate controller requires solving a linear program. To find the minimum control authority of a system using this type of controller, a nonlinear program must be solved. In this section we show how to calculate the minimum control authority of a system which has a limit on flow rate and which uses the minimum flow rate controller. In other words we wish to find $F_{1,1}$ which is the shortest distance to the control authority surface,

$$F_{1,1} = \min \|\mathbf{F}_{1,1}\|_2. \tag{25}$$

The control authority, $\mathbf{F}_{1,1}$, is the set of output forces which are attainable by the system if the minimum flow rate controller, $\mathbf{T}_1$, is used and if the system is operating at its maximum flow rate limit,

$$\|\mathbf{T}_1\|_1 = \mathbf{I}'\mathbf{T}_1 = m_1. \tag{26}$$

From the previous subsection we know that $\mathbf{T}_1$ satisfies the linear program defined in (23) and (24). The dual [9] to this linear program is

$$\max \quad \mathbf{y}'\mathbf{F} \tag{27}$$

$$\text{such that } A'\mathbf{y} \leq \mathbf{I} \text{ and } \mathbf{y} \geq \mathbf{0}. \tag{28}$$

The duality theorem of linear programming states that for an optimal solution the minimum of the primal problem (23) and the maximum of the dual (27) have the same value [9], $\mathbf{y}'\mathbf{F} = \mathbf{I}'\mathbf{T}_1$. From (26) this becomes

$$\mathbf{y}'\mathbf{F} = m_1. \tag{29}$$

The beauty of the duality theorem is that it allows us to replace the requirement that $\mathbf{T}_1$ solves a linear program with the simple scalar constraint of (29). Combining (25), (29), (24), (28) and (26) results in the following formulation for the minimum control authority:

$$\begin{aligned} F_{1,1} &= \min_{\mathbf{F}} \|\mathbf{F}\|_2 \qquad (30) \\ \text{such that } \mathbf{y}'\mathbf{F} &= m_1 \\ A\mathbf{T} &\geq \mathbf{F} \\ A'\mathbf{y} &\leq \mathbf{I} \\ \mathbf{I}'\mathbf{T} &= m_1 \\ \mathbf{T}, \mathbf{y} &\geq \mathbf{0}. \end{aligned}$$

The unfortunate aspect of this minimization problem is that the constraint, $\mathbf{y}'\mathbf{F} = m_1$ involves the product of two variables which is nonlinear. Finding the minimum control authority therefore requires minimizing a quadratic objective function (30) subject to a nonlinear constraint. Nonlinear programs of this type are readily solved using the program MINOS [7].

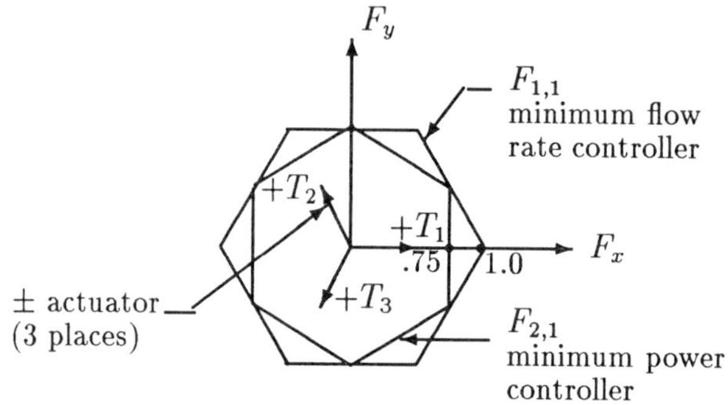

Figure 6: The control authorities of a system of three actuators using either the minimum power or minimum actuator output controller. The flow rate is limited to 1 in both cases.

3.2.3 Example

Consider three actuators, 120° apart in a single plane as in Fig. 6. The two hexagons in the figure correspond to the control authority of the system if the total flow rate is limited to 1, $\|\mathbf{T}_s\|_1 \leq 1$. The larger hexagon is the control authority if the minimum flow rate controller is used, $s = 1$, and the smaller hexagon corresponds to the minimum power controller, $s = 2$. The figure graphically illustrates the increased output which can be achieved by optimizing the controller. The minimum control authority for the minimum flow rate controller is $F_{1,1} = 0.87$ compared to $F_{2,1} = 0.75$ for the minimum power controller. The minimum control authority can be found by solving the nonlinear program (30) using MINOS.

3.3 Minimum Actuator Force Controller

3.3.1 Control

A system which is limited by the outputs of its individual actuators produces the largest possible force if the minimum actuator force controller is used. This controller is the solution to the following linear program:

$$\mathbf{T}_\infty = \arg\min \; [\mathbf{0}' \; 1] \begin{bmatrix} \mathbf{T} \\ z \end{bmatrix} \tag{31}$$

$$\text{subject to} \quad \begin{bmatrix} -I & \mathbf{I} \\ A & \mathbf{0} \end{bmatrix} \begin{bmatrix} \mathbf{T} \\ z \end{bmatrix} \geq \begin{bmatrix} \mathbf{0} \\ \mathbf{F} \end{bmatrix} \tag{32}$$

$$\begin{bmatrix} \mathbf{T} \\ z \end{bmatrix} \geq \mathbf{0}, \tag{33}$$

where I is the identity matrix, $\mathbf{I}$ is a vector of ones, $\mathbf{0}$ is a vector of zeros and z is a scalar. Another way of stating (31), (32) and (33) is

$$\begin{aligned} \min \quad & z & (34)\\ \text{subject to } \mathbf{T} & \leq \mathbf{z} & (35)\\ A\mathbf{T} & \geq \mathbf{F} & (36)\\ \mathbf{z}, \mathbf{T} & \geq \mathbf{0}, & (37) \end{aligned}$$

where $\mathbf{z}$ is a vector of z's the same length as $\mathbf{T}$. This linear program finds the solution to $\mathbf{F} = A\mathbf{T}$ with the smallest ∞-norm.

3.3.2 Minimum Control Authority, Actuator Output Limited

This case is analogous to the minimum flow rate controller where again the minimum control authority is the solution to a nonlinear program. The minimum control authority of an actuator system whose output force is limited by the outputs of the individual actuators and which uses the minimum actuator output controller is,

$$\begin{aligned} F_{\infty,\infty} &= \min_{\mathbf{F}} \|\mathbf{F}\|_2 & (38)\\ \text{such that } \|\mathbf{T}_\infty\|_\infty &= m_\infty, & (39) \end{aligned}$$

where $\mathbf{T}_\infty$ is the minimum actuator output controller and therefore satisfies the linear program in (31), (32) and (33). The dual to this linear program is:

$$\begin{aligned} \max \quad & [\mathbf{y}]' \begin{bmatrix} 0 \\ \mathbf{F} \end{bmatrix} & \\ \text{subject to } \begin{bmatrix} -I & A' \\ \mathbf{I}' & \mathbf{0}' \end{bmatrix} \mathbf{y} & \leq \begin{bmatrix} 0 \\ 1 \end{bmatrix} & (40)\\ \mathbf{y} & \geq \mathbf{0}. & (41) \end{aligned}$$

From the duality theorem of linear programming [Sect. 3.2.2],

$$[\mathbf{0}'\ 1] \begin{bmatrix} \mathbf{T} \\ z \end{bmatrix} = \mathbf{y}' \begin{bmatrix} 0 \\ \mathbf{F} \end{bmatrix}. \qquad (42)$$

Combining (38), (42), (32), (40), (39), (33) and (41) the minimum control authority is:

$$\begin{aligned} F_{\infty,\infty} &= \min \|\mathbf{F}\|_2 & (43)\\ \text{subject to } [\mathbf{0}'\ 1] \begin{bmatrix} \mathbf{T} \\ z \end{bmatrix} &= [\mathbf{y}]' \begin{bmatrix} 0 \\ \mathbf{F} \end{bmatrix} & \\ \begin{bmatrix} -I & \mathbf{I} \\ A & \mathbf{0} \end{bmatrix} \begin{bmatrix} \mathbf{T} \\ z \end{bmatrix} &\geq \begin{bmatrix} \mathbf{0} \\ \mathbf{F} \end{bmatrix} & \\ \begin{bmatrix} -I & A' \\ \mathbf{I}' & \mathbf{0}' \end{bmatrix} [\mathbf{y}] &\leq \begin{bmatrix} \mathbf{0} \\ \mathbf{1} \end{bmatrix} & \\ z &= m_\infty & \\ \mathbf{T}, \mathbf{y}, z &\geq \mathbf{0}. & \end{aligned}$$

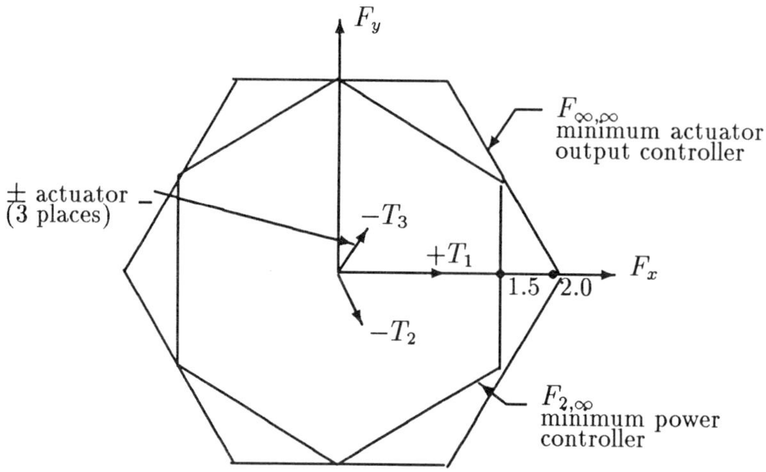

Figure 7: The control authorities of a system of three actuators using either the minimum power or minimum actuator output controller. The peak output of the individual actuators is limited to 1 in both cases.

The constraint in (42) is nonlinear which means that finding the minimum control authority, $F_{\infty,\infty}$, requires minimizing a quadratic objective function, (43), with a nonlinear constraint. Again this type of problem is readily solved using the program MINOS [7].

3.3.3 Example

Once again consider three actuators in a plane, Fig. 7. This time the two hexagons correspond to the control authority of the actuator system if the individual actuators are limited by their output force, $||\mathbf{T}||_\infty \leq 1$. The larger hexagon shows the increased control authority obtained by minimizing actuator output (min $||\mathbf{T}||_\infty$ controller). The smaller hexagon corresponds to the minimum power controller. The minimum control authority vector for the minimum power controller is $(\mathbf{F}_{2,\infty})_{min} = [1.5 \quad 0]'$. The actuator outputs which yield $(\mathbf{F}_{2,\infty})_{min}$ and which minimize power are $\mathbf{T}_2 = A^+(\mathbf{F}_{2,\infty})_{min}$. Therefore, from (14) $[T_1 \quad T_2 \quad T_3] = [1 \quad -.5 \quad -.5]'$. The three actuator outputs, $\mathbf{T}_2$, are shown in Fig 7. Notice that $\mathbf{T}_2$ satisfies the constraint, $||\mathbf{T}_2||_\infty \leq 1$. The vector, $\mathbf{T}_\infty = [1 \quad -1 \quad -1]'$ also satisfies $||\mathbf{T}_\infty||_\infty \leq 1$ but yields the higher output, $\mathbf{F}_{\infty,\infty} = A\mathbf{T}_\infty = [2 \quad 0]'$. The minimum actuator output controller, $\mathbf{T}_\infty$, does the smarter thing in this case: it turns each actuator on full blast. It uses more power than $\mathbf{T}_2$ but, yields a higher output. The minimum control authority for the minimum actuator output controller is $F_{\infty,\infty} = 1.73$ and can be found by minimizing (43) in MINOS.

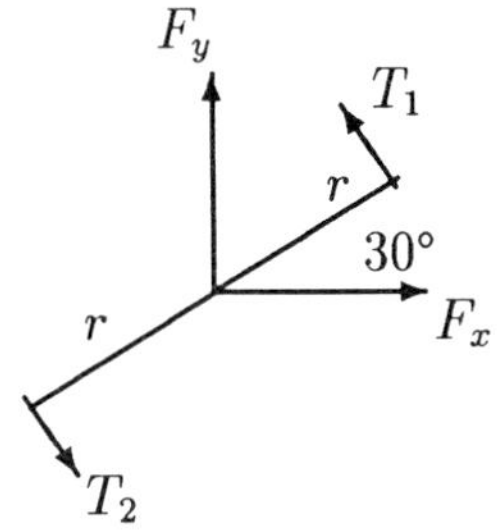

Figure 8: Planar configuration of two actuators

4 Minimum Control Authority Curve

As the force generated by an actuator system increases, its ability to generate moments decreases. The minimum control authority curve is a graphical representation of this fact. For any *force*, the minimum control authority curve gives the worst case or minimum control authority *moment* which can be generated.

The loads which an actuator system is subjected to are often conveniently expressed in terms of forces and moments. By plotting the force and moment load lines on the same plot as the minimum control authority curve we can tell whether the system is strong enough to overcome the load. If the minimum control authority curve is above the load lines then we know that the actuator system can handle the loads under worst case conditions [Fig. 4].

Finding the minimum control authority involves normalizing the actuator configuration matrix, A, before applying the techniques described in Sect. 2.4. For example, consider the actuator configuration depicted in Fig. 8 consisting of two actuators each located a distance r from the center of mass. The actuator configuration matrix for this case is

$$\mathbf{F} \stackrel{\text{def}}{=} \begin{bmatrix} \mathbf{F}_F \\ \mathbf{F}_M \end{bmatrix} \stackrel{\text{def}}{=} \begin{bmatrix} -.5 & .5 \\ .87 & -.87 \\ 0 & 0 \\ 0 & 0 \\ 0 & 0 \\ r & r \end{bmatrix} \begin{bmatrix} T_1 \\ T_2 \end{bmatrix} \stackrel{\text{def}}{=} A\mathbf{T}. \tag{44}$$

where $\mathbf{F}_F$ and $\mathbf{F}_M$ are 3×1 vectors corresponding to force and moment outputs respectively. Equation (44) can be normalized by dividing the last 3 rows of both sides by $\hat{r}$,

$$\hat{\mathbf{F}} \stackrel{\text{def}}{=} \begin{bmatrix} \mathbf{F}_F \\ \hat{\mathbf{F}}_M \end{bmatrix} \stackrel{\text{def}}{=} \begin{bmatrix} -.5 & .5 \\ .87 & -.87 \\ 0 & 0 \\ 0 & 0 \\ 0 & 0 \\ r/\hat{r} & r/\hat{r} \end{bmatrix} \begin{bmatrix} T_1 \\ T_2 \end{bmatrix} \stackrel{\text{def}}{=} \hat{A}\mathbf{T}. \tag{45}$$

The scalar $\hat{r}$ can be any number. The vector, $\hat{\mathbf{F}}$, is now a normalized generalized force vector. All of the components of $\hat{\mathbf{F}}$ have units of force. The matrix, $\hat{A}$, is a normalized version of A.

All of the techniques derived in Sect. 2.4 for finding the minimum control authority still apply to the normalized actuator configuration matrix $\hat{A}$ but now we can weight the relative importance of force and moment by adjusting the parameter $r/\hat{r}$. If we make $\hat{r}$ small for example, then $r/\hat{r}$ is big. This is equivalent to a thruster system with actuators mounted on long moment arms. Such a system can generate larger moments than forces. In other words moment outputs are weighted heavier and force outputs are smaller by comparison. The minimum control authority therefore favors force outputs. In the limit as $r/\hat{r}$ get really big, $r/\hat{r} \rightarrow \infty$, the minimum control authority has no moment component, $\|\mathbf{F}_m\|_2 \rightarrow 0$.

The technique for computing the minimum control authority curve involves the following steps:

1. Normalize the configuration matrix, A, by $\hat{r}$.

2. Compute the minimum control authority vector, $(\mathbf{F}_{s,p})_{\min}$, using any of the techniques in Sect. 2.4 depending on the type of output limit and method of control.

3. Compute the force magnitude, $\|\mathbf{F}_F\|_2$, and the normalized moment magnitude, $\|\hat{\mathbf{F}}_M\|_2$ from $(\mathbf{F}_{s,p})_{\min}$. Note, $(\mathbf{F}_{s,p})_{\min} = [\mathbf{F}'_F \ \hat{\mathbf{F}}'_M]'$.

4. Find the true moment magnitude, $\|\mathbf{F}_m\|_2 = \hat{r}\|\hat{\mathbf{F}}_M\|_2$.

5. Plot the point, $\|\mathbf{F}_M\|_2$ vs $\|\mathbf{F}_F\|_2$.

6. Repeat the procedure for various $\hat{r}$ between 0 and ∞ to fill in the minimum control authority curve.

If this procedure is followed the minimum control authority curve will generally have a break in the middle as depicted by the dotted line in Fig. 9. The discontinuity occurs at $r/\hat{r} = 1$. It is due to the fact that if $r/\hat{r} < 1$ then the minimum control authority is mostly in the moment direction but for $r/\hat{r} > 1$ it favors the force direction. Since the two points at the discontinuity, points a and b in in Figure 9, both correspond to $r/\hat{r} = 1$ they consequently have the same minimum control authority, $F_{s,p}$. The discontinuity can therefore be completed or 'filled in' with the line of constant minimum control authority,

$$\|\mathbf{F}_M\|_2 = \sqrt{F_{s,p}^2 - \|\mathbf{F}_F\|_2^2}. \tag{46}$$

Equation (46) is the dotted line in Figure 9. It smoothly connects the discontinuity between the points, a and b. Minimum control authority curves for the GP-B thruster system are shown in Fig. 12.

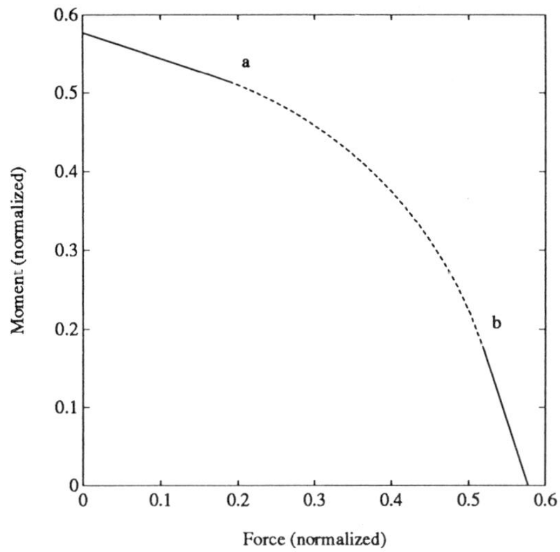

Figure 9: The discontinuity in the minimum control authority curve.

5 Applications to Gravity Probe-B

We now analyze the minimum control authority of the Gravity Probe-B thruster system as a means of illustrating how the techniques developed in the previous sections can be applied to a real world actuator system. The GP-B thruster system has to overcome a translational force predominantly due to atmospheric drag and a magnetic moment induced by Earth's magnetic field. The force and moment loads can come from any direction relative to the spacecraft, therefore the thruster system must provide 6 degree of freedom control.

Recall from Section 1 that there are three things which affect the minimum control authority of a system of actuators: 1) the actuator configuration, 2) the way in which the output is limited and 3) the actuator controller. We now describe how each one of these factors influences the minimum control authority of the GP-B thruster system.

Actuator Configuration To keep things simple we consider only one configuration of thrusters: 16 thrusters clustered in groups of four at the four vertices of a regular tetrahedron [Fig. 1]. The thrusters in each cluster lie in a single plane and are 90° apart [Fig 2]. Although the actuator configuration is fixed in this example we still analyze the minimum control authority under a worst case thruster failure. In this sense we analyze two different actuator configurations.

The actuator configuration which we analyze in this example is a bit academic in the sense that it doesn't take into account the location of solar panels and other equipment on the spacecraft. As the design of the GP-B spacecraft becomes more mature the position and orientation of the thrusters can be picked to maximize the minimum control authority while avoiding interference with the spacecraft and its equipment.

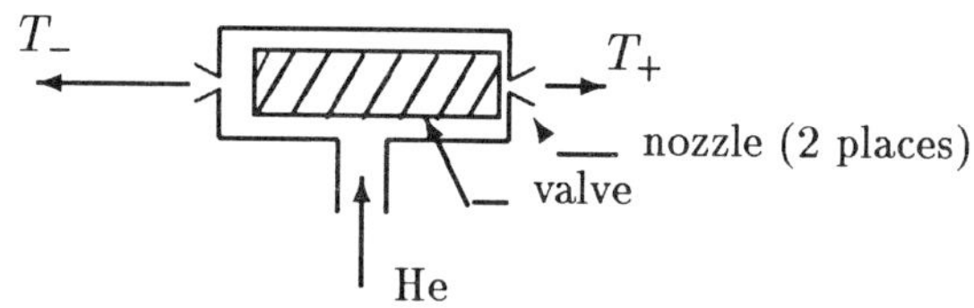

Figure 10: Differential thruster.

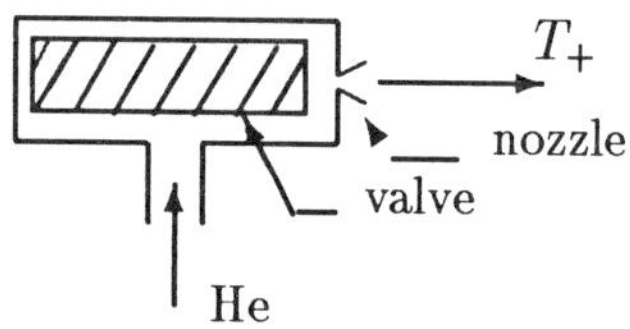

Figure 11: One-sided thruster.

Output Limit There are three ways in which the output of the GP-B thruster system may be limited. There may be a limit on 1) actuator force output based on differential thrusters, 2) flow rate based on heat leak and 3) actuator force output based on throat area and supply pressure. We analyze the minimum control authority for all three cases.

Actuator output limit, differential thrusters The GP-B thrusters use the continuously boiling off helium gas from a cryogenic cooling system as their propellant. A 'differential' thruster which vents this gas continuously is shown schematically in Fig. 10 [2]. The thruster has two opposing nozzles. As the valve closes off one nozzle the other side opens up more; a net force is generated but the total mass flow rate remains constant. If the total mass flow rate is w_n kg/sec and if there are n differential thrusters then the mass flow rate through each thruster is w_n/n kg/sec. The peak force per thruster is, therefore, $v_e w_n/n$ N where v_e is the exhaust gas velocity and has units of force per mass flow rate, N/(kg/sec)). The control authority of a system of differential thrusters is the set of forces,

$$\mathbf{F}_{s,\infty} = A\mathbf{T}_s \text{ such that } \|\mathbf{T}_s\|_\infty = v_e w_n/n. \tag{47}$$

Flow rate limit, heat leak A standard 'one-sided' thruster [Fig. 11] can not generate both positive and negative forces and therefore satisfies the constraint, $\mathbf{T} > \mathbf{0}$. Using one-sided thrusters it is possible to design a thruster system for the GP-B spacecraft which not only generates the desired forces but which also keeps the total flow rate constant [11]. If the total flow rate is not allowed to exceed w_n kg/sec then the control authority is the set of forces,

$$\mathbf{F}_{s,1} = A\mathbf{T}_s \text{ such that } \|\mathbf{T}_s\|_1 = v_e w_n. \tag{48}$$

As before, v_e is the exhaust gas velocity.

controller type	*output limit*		
	flow rate	actuator output, differential thrusters	actuator output, one-sided thrusters
	$\|\|\mathbf{T}_s\|\|_1 = v_e w_n$	$\|\|\mathbf{T}_s\|\|_\infty = v_e w_n/n$	$\|\|\mathbf{T}_s\|\|_\infty = v_e k w_n$
$\mathbf{T}_2 = \min \|\|\mathbf{T}\|\|_2$	Line 2	Line 1	Line 3
$\mathbf{T}_\infty = \min \|\|\mathbf{T}\|\|_\infty$	—	—	Line 4

Table 2: The combinations of output limits and controllers which are analyzed for the GP-B thruster system.

parameter	*value*
exhaust gas velocity	$v_e = 1274$ m/s
total flow rate	$w_n = 5 \times 10^{-6}$ kg/s
fraction of total flow rate through one thruster	$k = 0.33$
distance from spacecraft center of mass to thruster	$r = 1$ m

Table 3: Parameter values of the GP-B thruster system.

Actuator output limit, supply pressure and throat area If one-sided thrusters are used [Fig. 11] then the peak output force of each thruster depends on the supply pressure and the nozzle throat area. For GP-B the nozzle throat area can be chosen so that some fraction, k, of the total flow rate, w_n, can go through one nozzle [11]. In this case the control authority is the set of forces,

$$\mathbf{F}_{s,\infty} = A\mathbf{T}_s \text{ such that } ||\mathbf{T}_s||_\infty = v_e k w_n. \tag{49}$$

Actuator Controller Due to its simplicity we analyze the minimum control authority corresponding to the minimum power controller for all three of the output limits listed above. We also compute the minimum control authority for the minimum actuator output controller for the two cases when actuator output is limited, $||\mathbf{T}_\infty||_\infty \leq v_e w_n/n$ and $||\mathbf{T}_\infty||_\infty \leq v_e k w_n$.

Table 2 summarizes the combinations of output limits and controllers which we analyze for the GP-B thruster system. The minimum control authority curves for the four cases in Table 2 are displayed in Fig. 12. These curves are based on the parameter values listed in Table 3. The worst case force and moment loads are also displayed in Fig. 12. Here is a description of each line in the figure:

Line 1 The minimum control authority curve of a system of differential thrusters using the minimum power controller,$||\mathbf{T}_2||_\infty = v_e w_n/n$. The configuration of 8 differential thrusters which we are analyzing [Fig 1] happens to be an optimal configuration [12]. This means that if the minimum power controller is used then there are no configurations of differential thrusters which yield a higher minimum control authority. One way of increasing the minimum control authority of a system of differential thrusters is by using the minimum actuator output controller.

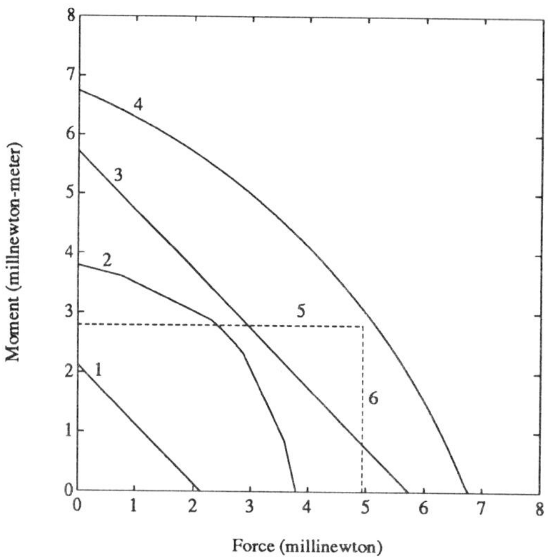

Figure 12: Minimum control authority curves for the Gravity Probe-B thruster system with no thruster failures.

Line 2 The minimum control authority curve of a system of one-sided thrusters using the minimum power controller and with the output limited by a total flow rate of $||\mathbf{T}_2||_1 = v_e w_n$. The minimum control authority can be increased by using the minimum flow rate controller but only by 6%. The minimum control authority of the GP-B thruster system can also be increased by increasing the overall flow rate above w_n kg/sec [Line 3]. The GP-B thruster system can handle increased flow rates for short periods of time or for extended periods of time with the use of a heater [11].

Line 3 The minimum control authority of a system of one-sided actuators using the minimum power controller and limited by pressure, $||\mathbf{T}_2||_\infty = v_e k w_n$. The value of k was chosen so that Line 3 is greater than Line 2. The value k can be made bigger to give a wider dynamic range but this is at the expense of higher thruster errors. Line 3 illustrates two important points about the minimum control authority curve:

1. The minimum control authority curve is a good design tool. In this case, for example, the value k is a function of the throat areas of the thrusters. The minimum control authority therefore gives us a rational basis for choosing the GP-B thruster throat areas. In general, the actuators of any system can be designed to yield a desired minimum control authority.

2. The overall minimum control authority curve of a system which has more than one output limit is the *intersection* of the minimum control authority curves corresponding to the individual output limits. For example if k is

smaller so that Line 3 falls below Line 2 in some places, then the overall minimum control authority curve is the boundary of the intersection of the areas under these two lines. This means that for some directions the overall flow rate drops below w_n kg/sec.

Line 4 The same thrusters and output limit as Line 3 but here the minimum actuator output controller is used. Line 4 illustrates the significant increase in minimum control authority which can be achieved by optimizing the controller (approximately 40% increase). Line 4 is above the worst case force and moment loads, Lines 5 and 6, indicating that the GP-B thruster system can overcome the worst case loads if the optimal controller is used.

Line 5 The worst case moment load on the GP-B spacecraft based on the moment induced by Earth's magnetic field. This load line also includes moments produced by aerodynamic drag, solar radiation pressure, gravity gradient and principal axis misalignment. The aerodynamic drag is based on the density at a 650 km orbit on the worst day in the year 2001 (peak of 11 year solar cycle).

Line 6 The worst case force load on the GP-B spacecraft based on the aerodynamic drag at a 650 km orbit on the worst day in the year 2001 (peak of 11 year solar cycle). The load line also includes forces due to solar radiation pressure, gravity gradient and center of mass offset.

The GP-B thruster system illustrates how useful the minimum control authority curve is as a design tool for actuator systems. For example:

- It can be used to study the redundancy of an actuator system.
- It is a measure of the performance of an actuator system and as such can be used to trade-off one actuator system design against another.
- It can be used to design the individual actuators. The output of the individual actuators can be adjusted to yield a desired minimum control authority.
- It can be used to compare different actuator controllers. A decision can then be made whether the additional complexity of the minimum flow rate or minimum actuator output controllers is justified by their increased output. Even if these more complicated controllers are not used it is still useful to know how close the output of the minimum power controller is to the optimum.

6 Acknowledgments

The authors wish to acknowledge J-H. Chen of CSIST in the Republic of China. In his doctoral dissertation at Stanford University [2], J-H. Chen investigated various thruster configurations for the Gravity Probe-B spacecraft. He identified the minimum control authority as an important performance measure of a thruster system and wrote a computer program which automatically converges on a thruster configuration which maximizes the minimum control authority.

The authors also wish to acknowledge Michael Saunders of the Department of Operations Research at Stanford University. He generously provided the optimization program, MINOS [7], which we used to find the minimum control authority for the minimum flow rate and minimum actuator output controllers. He also provided invaluable advise on formulating the nonlinear programs for these problems.

The work performed by Peter Wiktor is in partial fulfillment of the degree of Doctor of Philosophy at Stanford University. This work is supported by NASA contract NAS8-36125.

7 Appendix: Vector Norms

Norms are a measure of the size of a vector. A useful class of vector norms are the p-norms defined by

$$||\mathbf{T}||_p \stackrel{\text{def}}{=} \left(\sum_{i=1}^{n} |T_i|^p\right)^{1/p} \qquad p \geq 1 \tag{50}$$

The following three p-norms are used in this paper:

1-norm The sum of the absolute values of the elements of $\mathbf{T}$,

$$||\mathbf{T}||_1 \stackrel{\text{def}}{=} \sum_{i=1}^{n} |T_i| \tag{51}$$

For example, if $\mathbf{T} = [1 \quad -2]'$ then $||\mathbf{T}||_1 = 3$.

2-norm The square root of the sum of the squares of the elements of $\mathbf{T}$,

$$||\mathbf{T}||_2 \stackrel{\text{def}}{=} \left(\sum_{i=1}^{n} T_i^2\right)^{1/2} = \left(\mathbf{T}^T\mathbf{T}\right)^{1/2}. \tag{52}$$

This is also known as the Euclidean norm and is what is commonly thought of as the length or magnitude of a vector. For example, if $\mathbf{T} = [1 \quad -2]'$ then $||\mathbf{T}||_2 = \sqrt{5}$.

∞-norm The maximum absolute value of $\mathbf{T}$,

$$||\mathbf{T}||_\infty \stackrel{\text{def}}{=} \max_i |T_i| \tag{53}$$

For example, if $\mathbf{T} = [1 \quad -2]'$ then $||\mathbf{T}||_\infty = 2$.

References

[1] J.S. Bull. Precise attitude control of the Stanford relativity satellite. *Stanford University Ph.D. Thesis*, SUDDAR 452, 1973.

[2] J-H Chen. Helium thrustor propulsion system for precise attitude control and drag compensation of the Gravity Probe-B satellite. *Stanford University Ph.D. Thesis*, SUDDAR 538, 1983.

[3] J-H Chen and D.B. DeBra. A continuous proportional low-thrust propulsion system. *XIth IFAC Symposium on Automatic Control in Aerospace*, July 1989.

[4] P.E. Gill, S.J. Hammarling, W. Murray, M.A. Saunders, and M.H. Wright. User's guide for LSSOL: a Fortran package for constrained least-squares and convex quadratic programming. 1986.

[5] G. H. Golub and C. F. Van Loan. *Matrix Computations.* The Johns Hopkins University Press, Baltimore, Maryland, 1983.

[6] C. Moler, J. Little, and S. Bangert. *Matlab User's Guide.* The MathWorks, Inc., Sherborn, MA, 1990.

[7] B.A Murtagh and M.A Saunders. MINOS 5.1 users's guide. 1987.

[8] L.I. Schiff. Motion of a gyroscope according to Einstein's theory of gravitation. *Proceedings of the National Academy of Sciences*, 46(871), 1960.

[9] G. Strang. *Linear Algebra and its Applications.* Harcourt Brace Jovanovich, Inc., 1989.

[10] J.P. Turneaure, C.W.F. Everitt, and B.W. Parkinson. The Gravity Probe B Relativity Gyroscope Experiment: Approach to a flight mission. *Proceedings of the Fourth Marcel Grossman Meeting on General Relativity*, 1986.

[11] P.J. Wiktor. The control of a liquid helium propulsion system with applications to Gravity Probe B. *AIAA and ASME 27th Joint Propulsion Conference*, June 1991.

[12] P.J Wiktor, J-H Chen, and D.B. DeBra. Optimal thruster configurations for the GP-B spacecraft. *XIth IFAC Symposium on Automatic Control in Aerospace*, July 1989.

Section II
GUIDANCE AND CONTROL STORYBOARD DISPLAYS

SESSION II

Chairperson: Ron Rausch
Martin Marietta Astronautics

Co-Chairperson: Mary Sullivan
Ball Aerospace Electro-Optics/
Cryogenics Division

The following papers were not available for publication:

AAS 91-030	"Preliminary Design and Analysis of a Counter-Balanced Lunar Base Construction Crane", Martin Mikulas, Jr., Li Farn Yang, and Peter Withnell, University of Colorado
AAS 91-031	"A Shape Memory Material Actuator for Adaptive Truss Applications", B. J. Maclean, J. L. Draper, and M. S. Misra, Martin Marietta Astronautics
AAS 91-032	"Simple Mars Walker: Walking Beam", Steve Price, Martin Marietta Astronautics.
AAS 91-033	"Performance and Guidance Improvements for Advanced Kinetic Energy Weapons", Thomas Webber, Lt. Michael Dickey, Astronautics Laboratory, Edwards AFB
AAS 91-034	"Laboratory Demonstration of SAVI Magnetic Suspension Technology", Brian J. Hamilton, Delano R. Carter, Honeywell, Inc.
AAS 91-036	"Figure Maintenance Sensor, Precision Segmented Reflectors", Noble M. Nerheim, Jet Propulsion Laboratory
AAS 91-037	"Testing the New Generation Stellar Attitude Sensors", James P. McQuerry, David D. Wagner, Ball Aerospace Electro-Optics/Cryogenics Division
AAS 91-039	"Six-Degree of Freedom Magnetically Suspended Fast-Steering-Mirror (MSFSM) Tracks Atmospheric Turbulence While Providing Focusing and Collimation Travel", Avanindra A. Gupta, Lawrence M. Germann, Jim Medbery, Ball Aerospace Electro-Optics/Cryogenics Division
AAS 91-040	"Relay Mirror Experiment, A Ground Based Laser Pointing and Tracking Demonstration", Jeff Dierks, Ball Aerospace Space Systems Division, Joel Anspach, Applied Technology Association, Steve Katapski, USAF, Weapons Laboratory
AAS 91-044	"Two-Axis RAMS Prototype", Curtis L. Schroeder, Ball Aerospace Electro-Optics/Cryogenics Division
AAS 91-045	"Applicability of Real-Time Controllers for Satellite Simulations", Doug Case, Fred Tubb, Integrated Systems, Inc.
AAS 91-049	"A Magnetically Suspended Internally Redundant Earth Sensor Assembly", Richard A. Swendsen, Kevin J. Taylor, TRW Space & Technology Group
AAS 91-080	"Real-Time GN&C Rapid Prototyping", Frank Barone, Ron Eicher, Bob Zermuehlen, Satish Anand, Martin Marietta Astronautics
AAS 91-081	"Flight Results of Attitude Matching Between Space Shuttle and Inertial Upper Stage (IUS) Navigation System", Alfred J. Treder, Keith L. Meldahl, Boeing Aerospace & Electronics
AAS 91-082	"Line-of-Sight Stabilization of a Flexible Planar Trust Using Mechanical and Optical Feedback", Capt. M. Scott Trimbolie, Lt. Rick Adams, USAF Academy

AAS 91-035

MICROGRAVITY FLIGHT TESTING OF A LABORATORY ROBOT

Elaine M. Hinman*

Robotics will be necessary for productivity in space-based laboratories and processing facilities. Robots used for these functions will require a higher degree of precision and dexterity than current space positioning arms.

Although computer simulations are plentiful, very little actual data on manipulator performance characteristics in low gravity exists. This project attempts to combine microgravity data with dynamic computer simulation to drive out performance characteristics and beneficial control strategies for dexterous and high-precision small robotic systems.

A manipulator designed for laboratory use has been modified and incorporated into a materials transfer workcell. Accelerometers have been mounted at the base and end effector of the arm to take measurements during robotic motions. This workcell has been flown on the KC-135 Microgravity Simulator. Concurrently with these tests, a computer dynamic simulation using the TREETOPS package has been developed. Results from each are compared, and as new data from flights is taken and analyzed, the fidelity of the simulation is being improved. Control strategies to minimize the effects of the robot's motion on the item being manipulated may be developed in software and compared with flight results. Preliminary statements regarding robot and workcell configuration for space processing applications may be made based on the information gathered from the flights and dynamic simulation.

While the information gained from this project will be useful to future space automation development in general, specific robotic design criteria and workcell processes may also be defined for individual projects as their needs are specified.

HARDWARE SYSTEM

The flight system hardware includes a modified Zymark Zymate II with the Zymate System V controller. A Commodore Amiga 2500 provides supervisory control of the robot through the PC/AT Bridgeboard, which is connected to the Zymate controller. The Amiga's multi-tasking capability allows accelerometer data to be taken as the robot performs its trajectory. A Sunstrand accelerometer is

* NASA/Marshall Space Flight Center, EB24, Huntsville, Alabama 35812.

mounted on the end effector of the robot in one of three orientations for measuring along the X, Y, and Z axes. This system is shown in the attached diagrams.

The initial flight system configuration incorporated a UMI-RTX laboratory robot with a PC compatible controller. This system was upgraded for the higher-precision robot and the multi-tasking computer of the current system.

FLIGHT ENVIRONMENT

The KC-135 Microgravity Simulator provides a means of low-cost access to the microgravity environment. This is useful for researchers involved in proof-of-concept, initial testing, and system verification for spaceflight.

The KC-135 aircraft is similar to a commercial Boeing 707. It is a four engine turbojet, with the cargo area made available for researchers and their experiments. Several experiments will share a flight.

The aircraft flies a parabolic flight path, which provides approximately 23 seconds of microgravity followed by a 1.8g pullout. In a day's flight, 25 to 40 parabolas are flown. Typically, an experiment flies several days in a week to maximize access to the plane. This approach allows modifications to be made to the experiment -- either small changes during flight, or larger changes before the next day's flight.

SOFTWARE SIMULATION

The TREETOPS software package was used in developing the dynamic model of the Zymate flight unit. This program was designed for use in dynamic analysis and control system development.

TREETOPS is designed to model and simulate multi-body structures which have a tree topology. A Lagrangian formulation of the equations of motion is used. TREETOPS may be used to produce a time history of interactions which take place during the motion of the system modeled.

A system to be modeled is broken into separate 'bodies' connected by 'hinges' of defined degrees of freedom. Hinges may be of several types, with varying spring and damping values assigned to them. 'Sensors' and 'actuators' may be placed on the bodies to initiate motion and measure (calculate in software) motions and accelerations.

Each body is defined independently of the others in the tree. This gives TREETOPS greater flexibility in the types of systems it may model. For example, TREETOPS can incorporate NASTRAN information about a body into the overall system simulation.

Plots of the simulated end effector's motion are shown. A diagram of defined bodies and joints for the system is also shown.

ACKNOWLEDGEMENTS

I would like to thank Dr. Gary Workman, Guy Smith, Houcheng Li, and David Gilliam of the University of Alabama in Huntsville for their work in support of this experiment.

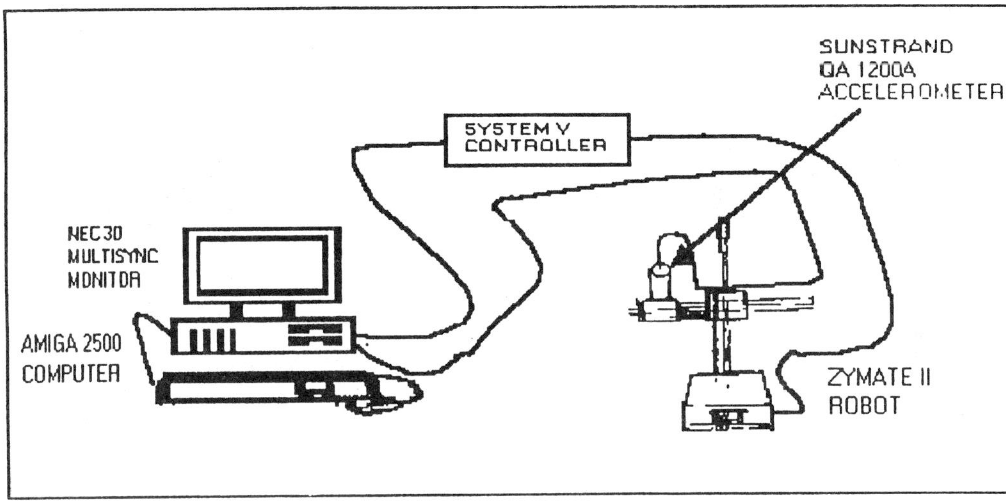

Figure 1 Zymate Robot System Components for Study of Robot Dynamics in Reduced Gravity

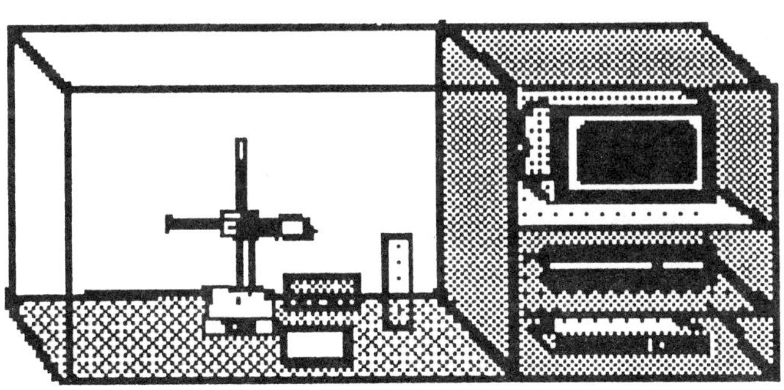

The zymate robot is programmed to perform materials transfer processes during parabolic flight on NASA's KC-135. Accelerometer measurements allow researchers to obtain data on the actual g levels occurring at the end-effector. This data will enable design and development of robotic devices with minimum g-levels affecting materials processing crystal growth experiments.

Figure 2 KC-135 Experiment for Robotic Materials Transfer in Low Gravity Using Zymate Robot

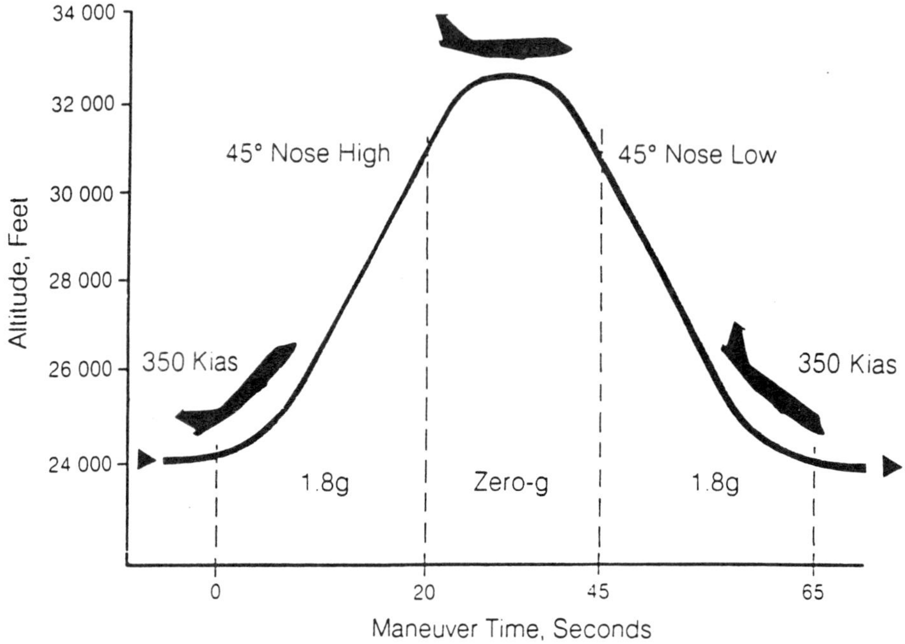

The above diagram shows a typical zero-g maneuver. However, the maneuver can be modified to provide any level of g-force less than one g. Some typical g-levels used on different tests and the corresponding time for each maneuver are as follows:

Negative-g (-0.1g)	Approximately 15 seconds
Zero-g	Approximately 25 seconds
Lunar-g (one-sixth g)	Approximately 30 seconds
Martian-g (one-third g)	Approximately 40 seconds

Figure 3

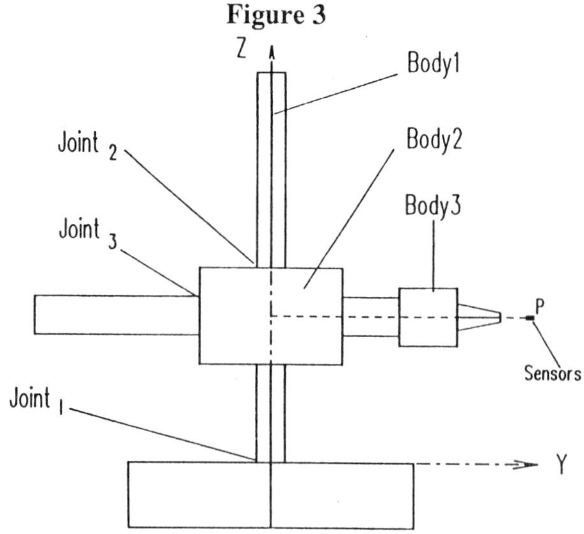

Figure 4 The Simulation Model of the Zymate Robot as Defined for Use 'TREETOPS'

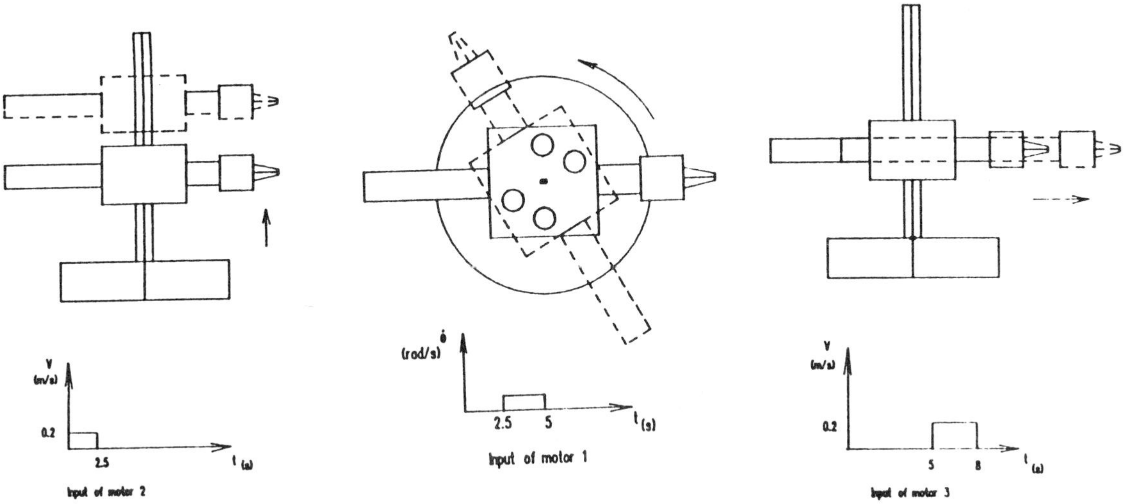

Figure 5 Illustration Showing Sequences of Motion for Zymate Robot

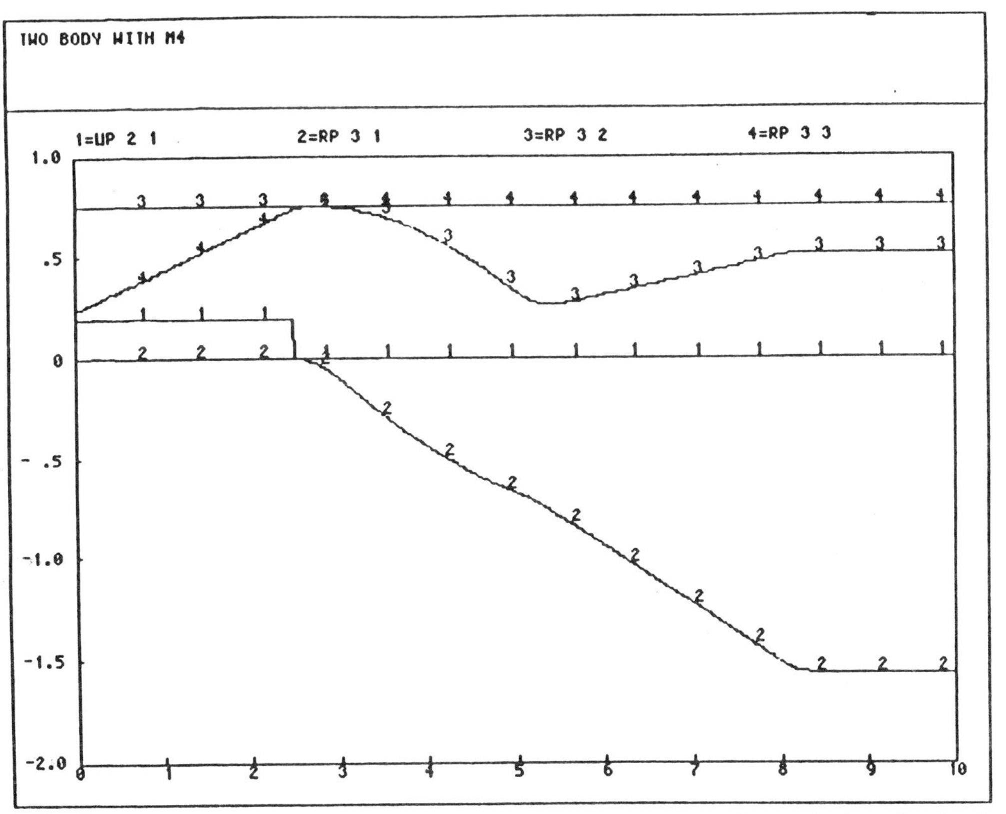

Figure 6 TREETOPS plot Showing Motion of End-Effector for a Typical Trajectory: $P_X = 2, P_Y = 3, P_Z = 4$

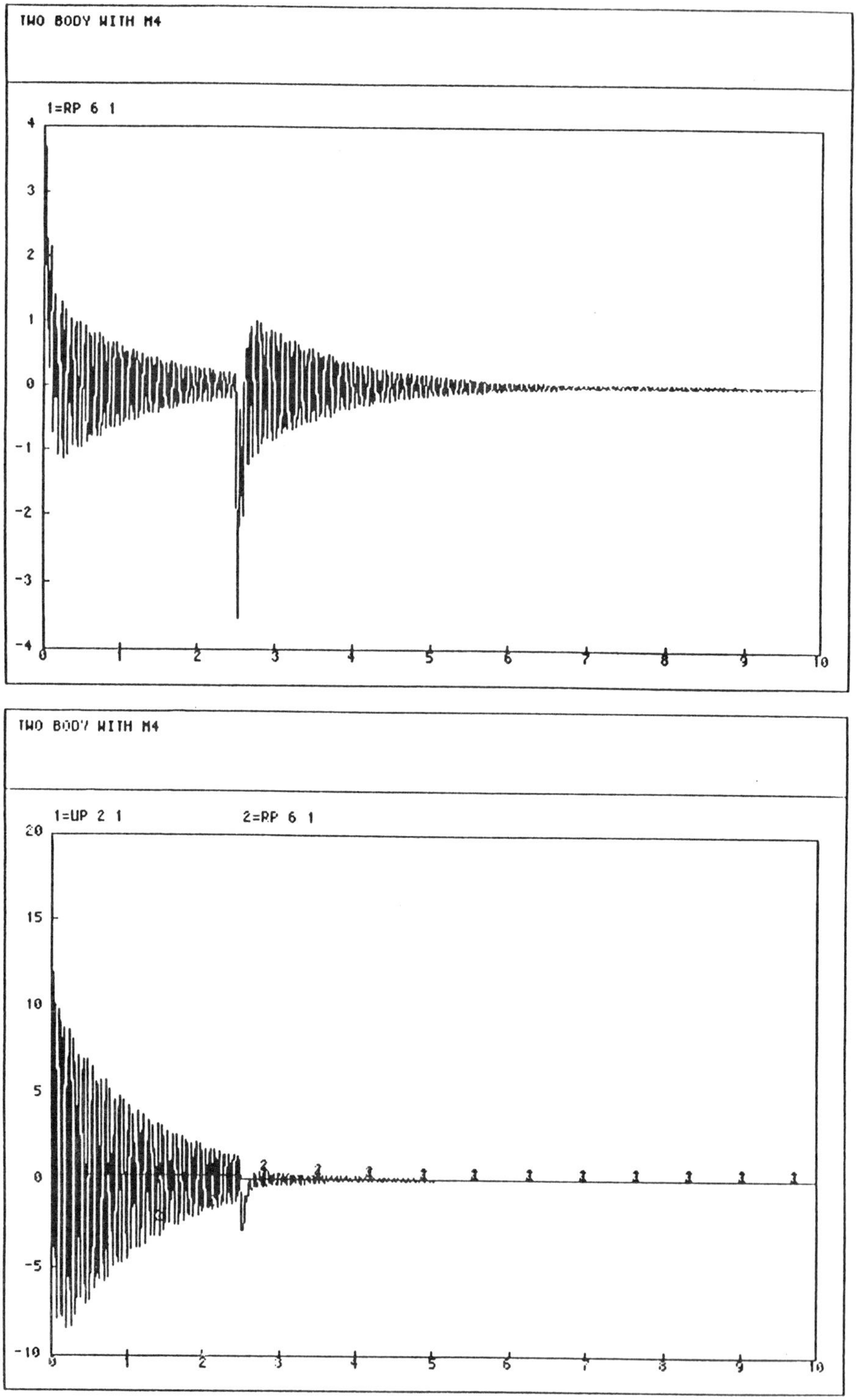

Figure 7 The Output of Z Direction Acceleration Sensor at End-Effector for a Typical Trajectory Without Gravity

AAS 91-038

REACTION WHEEL CONTROL OF THE MSX SATELLITE

Terry Marshall*, Tracy Gunderman* and Fred Mobley†

The satellite for the Midcourse Space Experiment (MSX) will carry an array of photon sensors for measurement of radiation from cooperative missile launches into space. The MSX satellite is required to point these sensors at targets as they move from horizon to horizon in 10 to 20 minutes. The orientation of the entire satellite is controlled by four powerful reaction wheels to achieve rates of 1.6 deg/s (and more) and acceleration of 0.03 deg/s^2. A new and more powerful reaction wheel, developed to meet the torque and momentum storage requirements of the MSX, has a torque capability of 1.2 ft-lb and momentum storage of 150 ft-lb-s (at 4500 rpm) in a wheel assembly weighing 60 lb. Overall pointing accuracy of 0.1 deg is required, and pointing jitter must not exceed 1 arc s. The jitter requirement impacts the wheel bearing characteristics, but is not nearly as severe as that for Hubble Space Telescope (~0.003 arc s). Nevertheless, analysis of bearing-noise excitation of structural modes is necessary to confirm that jitter requirements are met. Four reaction wheels are arranged so that loss of any one wheel will not prevent control of the satellite.

INTRODUCTION

The primary purpose of the Midcourse Space Experiment (MSX) is to measure photon radiation from intercontinental-warhead-like devices during coasting flight well outside of the Earth's atmosphere. Rocket launch of warhead-like objects will be tracked by the MSX satellite; radiation from these devices will be measured by ultraviolet, visible, and infrared light detectors. The detector systems use fairly small field-of-view telescopes, so accurate tracking of the targets to ~0.1 deg is necessary. Furthermore, the targets are moving rapidly, sometimes with additional thrusting, and making large apparent angular traverse, perhaps as much as 180 deg. This presents a challenge to the designers of the attitude control system.

Fig. 1 shows the MSX satellite as presently conceived. It weighs 5200 lb, is 17 ft tall, and has two large solar panels with one degree-of-freedom of rotation. The array of four S-band receiving antennas is used for target pointing information with targets that have S-band transmitters, using phase information to determine angular pointing direction. Four Reaction Wheel Assemblies (RWA), mounted in the lower four corners at the base of the satellite, provide the torque for maneuvering satellite attitude. The large dewar in the center contains solid H_2 used for cooling the infrared detectors of the SPIRIT III telescope.

Fig. 2 shows a typical tracking maneuver. The target is first detected near the horizon, and is tracked throughout its trajectory. At the point of closest approach to the MSX satellite, we expect to see maximum RWA rotation rates. After completion of a tracking maneuver, the satellite returns to the parked mode orientation with X-axis vertical and solar panels oriented to the sun for recharging the batteries.

* Honeywell Inc., Satellite Systems Operations, 19019 N. 59th Avenue, Glendale, Arizona 85308-9650.

† Johns Hopkins University, Applied Physics Laboratory, Johns Hopkins Road, Laurel, Maryland 20723-6099.

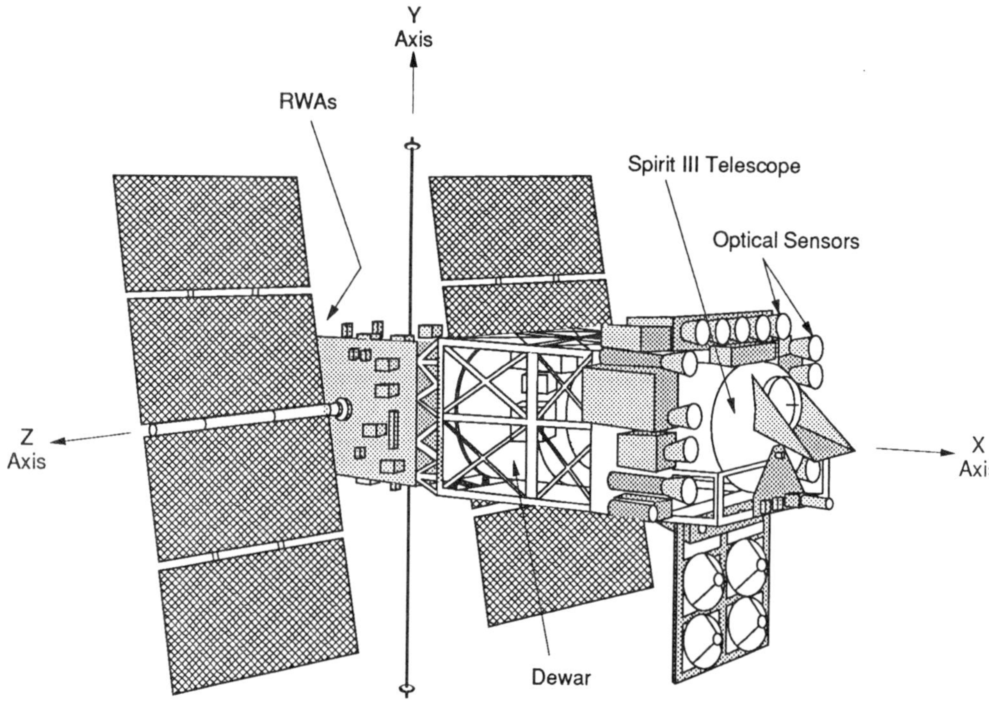

Fig. 1 MSX Satellite

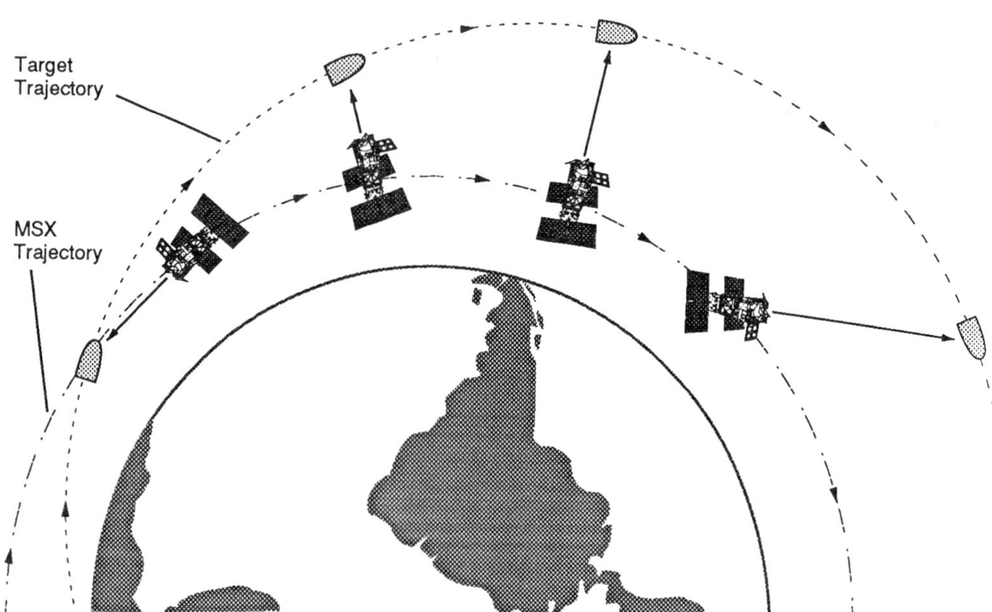

Fig. 2 Typical Tracking Scenario

A system of ring-laser gyros is used for continuous attitude determination. The initial conditions for the gyros are set by a single star tracker using readings on five stars simultaneously. Two three-axis gyro systems are used (for redundancy); these are mounted on a graphite fiber optical bench along with the star camera and a pair of auto-collimators, which detect the orientation of the SPIRIT III telescope. By combining simultaneous SPIRIT III and star-tracker detection of the same stars, we can make an in-orbit calibration of the SPIRIT III axis relative to the star camera's field-of-view. This calibration may be repeated as necessary.

Preliminary analysis of typical target-tracking scenarios showed that RWAs with angular momentum of 100 ft-lb-s and torque of 0.6 ft-lb can meet our requirements. We also have a requirement that the telescope pointing axis not jitter more than ±1 arc s in any 0.7-s time period. This requirement brings up the issue of spacecraft mechanical vibration from all possible sources, especially from the four RWAs. Although our requirement is not nearly as stringent as that of the Hubble Space Telescope (HST) (0.003 arc s), we are concerned about this problem.

A block diagram of the attitude control system for the MSX satellite is shown in Fig. 3. The attitude computer produces command voltages for each of the four RWAs. A command voltage nominally produces a proportional reaction torque along the wheel's rotational axis. Wheel command voltages will be adjusted at 0.5-s intervals. The four wheel speeds are made known to the attitude computer by the wheel tachometer signals. With these signals, the control equations can be corrected for the internal angular momentum of the wheels. The attitude computer also receives target coordinates and angular rates from another computer (the tracking processor).

The wheel speeds will be maintained at approximately 100 rpm while in the parked mode. Three magnetic torque rods are used to unload excess wheel momentum during rest periods in the parked mode.

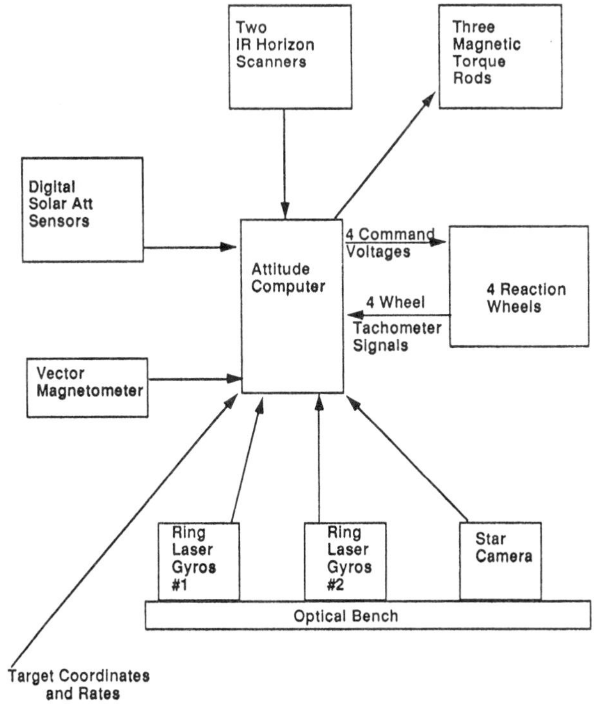

Fig. 3 Block Diagram - MSX Attitude Control System

Reaction Wheel Activity During a Typical Maneuver

Fig. 4(a) shows the body angular rates required to follow a typical target in a tracking maneuver[1]. Using our control laws, the reaction torque, from each of the four wheels shown in 4(b), will produce the required angular velocities. The resultant variation of wheel speed with time is shown in 4(c). The jitter in the reaction torques is due to inclusion of realistic gyro and star-camera noise.

Since only three wheels are required to accomplish the maneuver, the fourth wheel provides redundancy. It also produces non-uniqueness in the wheel torque and wheel speed versus time. We take advantage of this fact at the start of the tracking maneuver to keep all wheel speeds away from zero speed – a speed below the bearing's Elastohydrodynamic (EHD) regime, a point where lubricant film thickness is not fully developed. However, we can't prevent some speeds from going through zero during the actual maneuver.

REACTION WHEEL DESIGN

Satellite Performance and the MSX RWA

With the MSX satellite's maneuvering and target tracking scenarios defined, the role of RWA performance requirements has been established, with momentum, torque, and minimum rotating induced vibrations as key design drivers. The reaction torque requirement of 0.6 ft-lb and angular momentum capacity of 100 ft-lb-s, plus enhanced operation capability of both parameters, introduced a new midsize RWA previously unavailable. The new MSX RWA offers proven, high output-torque capability up to 1.2 ft-lb, while containing momentum-storage potential up to 150 ft-lb-s at 4500 rpm. The motor-drive design, based on the HST RWA, uses a DC motor and hybrid drive electronics uniquely packaged to improve thermal power dissipation paths. The inertial rotor design incorporates the rotor structure and bearing suspension system baseline design found in all Honeywell momentum devices, which deliver in-orbit, space-proven reliability and performance. Both the electronics and the inertial rotor assembly are integrated in a single, evacuated housing package. By offering large-wheel performance in a small package, a new level of RWA performance-to-real-estate density is achieved. See Fig. 5.

RWA Sizing Techniques

The MSX RWA design was created through a Honeywell-developed wheel optimization Computer Aided Design (CAD) program. The program's data base, consisting of many existing RWA designs, allows an accurate target configuration to be developed through iterative exercise within the CAD program directed through engineering judgment. The MSX RWA performance spreadsheet, outlined in Fig. 6, demonstrates an RWA configuration of specific power and weight projection, which was within a small percentage of actual measured parameters of the fabricated MSX RWA prototype.

RWA Design - Electronics

The block diagram, Fig. 7, shows the interface commands and telemetry signals, as well as the method of signal isolation in the RWA electronics. The torque command, after isolation, drives each of three motor-current control loops. The commutation of each loop is controlled by Hall-sensing devices that sense shaft position and also provide direction information, a tachometer pulse train, and the input to generate an analog wheel-speed signal.

RWA Design - Motor Drive

Normally the interaction of wheel diameter with maximum rpm is the tradeoff of major interest. In the MSX, however, the actual method of motor drive became a dominant factor.

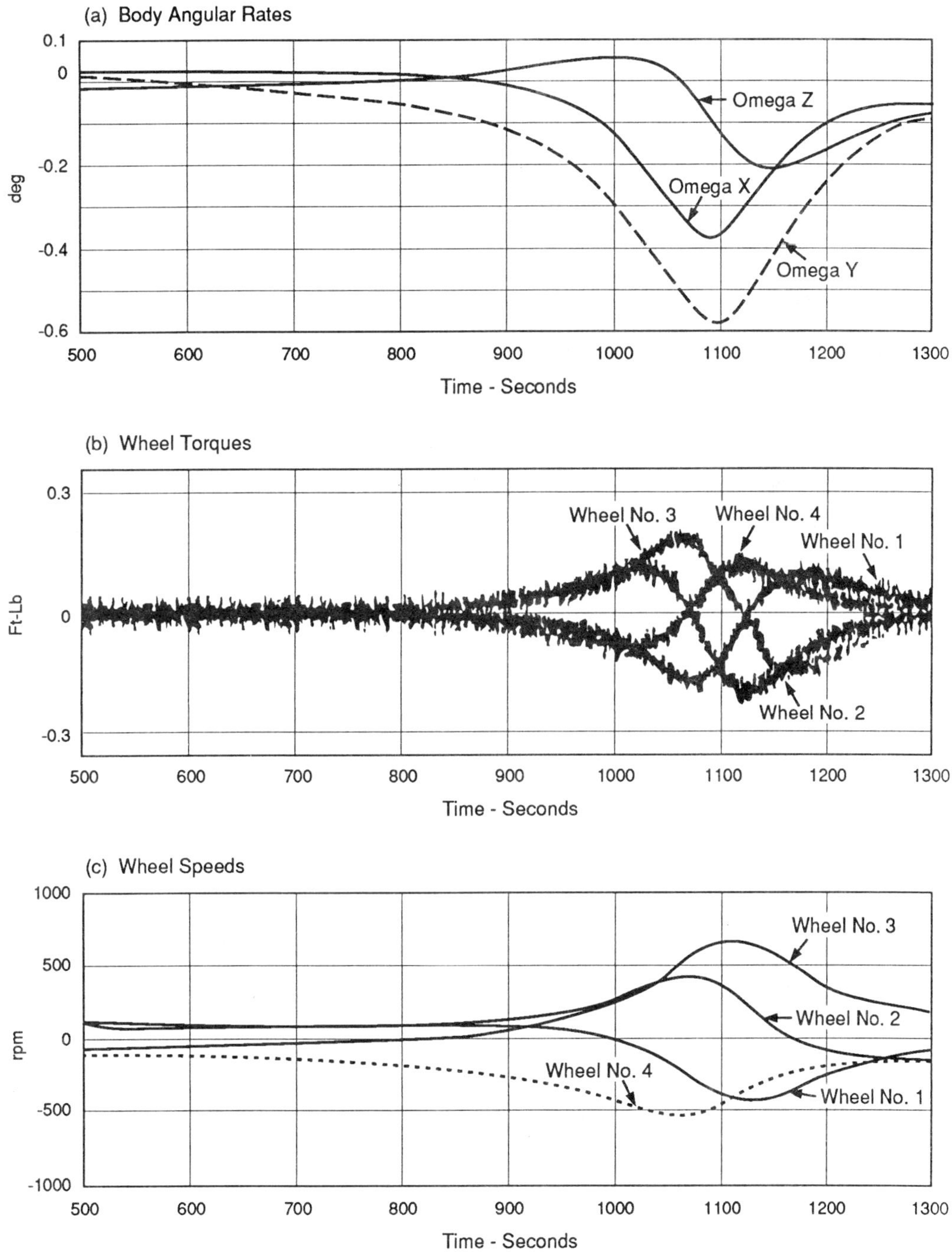

Fig. 4 Typical Maneuver Dynamics

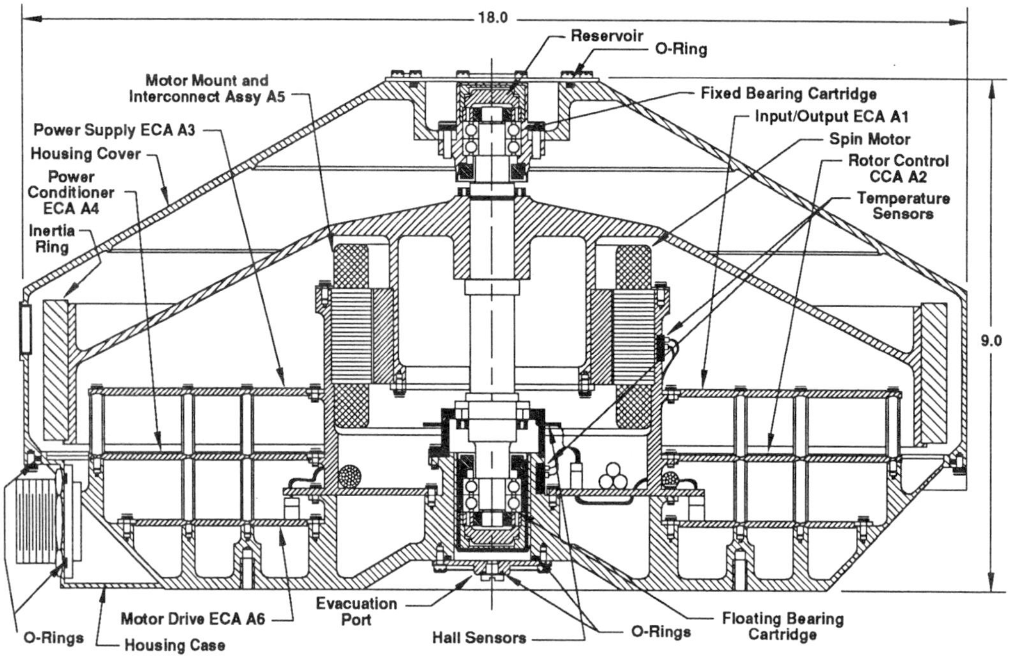

Fig. 5 MSX RWA Cross Section

RWA Weight and Power Estimate Rev 1.3 File: MSX 11/28/1990								
			Weight, lb	Power, W				
				Max	St. St.			
Torque Out (O-I)	230.0							
Momentum (f-lb-s)	100.0	Rotor	25.1			Brg Drag, oz-in.	2.4
PSD	0.125	Housing	13.4			Mtr Drag, oz-in.	0.6
Wheel Speed (rpm)	3000.0	Bearing	1.0			Brg Load, lb	1476.0
Rotor Dia (in.)	18.0	Motor	9.0	26.8	0.0	Mtr I, A	18.9
Mtr Design (km)	3.0	Resolver	0.0				
Min Bus Voltage	22.0	Electronics	12.3	247.8	8.2		
Digital I/O (1/0)	0.0	Total Wheel	60.8	274.7	8.2		
Rad Hard (1/0)	0.0	Shaft Pwr		517.0	6.6		
Mtr (2, 3Y, 3I, 2-3Y)	3.0	Total System	60.8	791.7	14.9		
Resolver ID (in.)	0.0	Pwr Cond	3.5	72.0	0.0	Incremental	
# Motor	1.0	LVPS	1.3	6.0	6.0	Pwr/Wt	17.0
# Resolver	0.0	I/O Card	0.8				
# Pwr Cond	1.0	Rotor Cont	1.6				
# Sec Supply	1.0	Drivers	5.0	169.9	2.2		
# I/O card	1.0						
# Rotor Control	1.0						
# Motor Driver	1.0						
# Hall Assy	1.0						

Fig. 6 Typical RWA Optimization

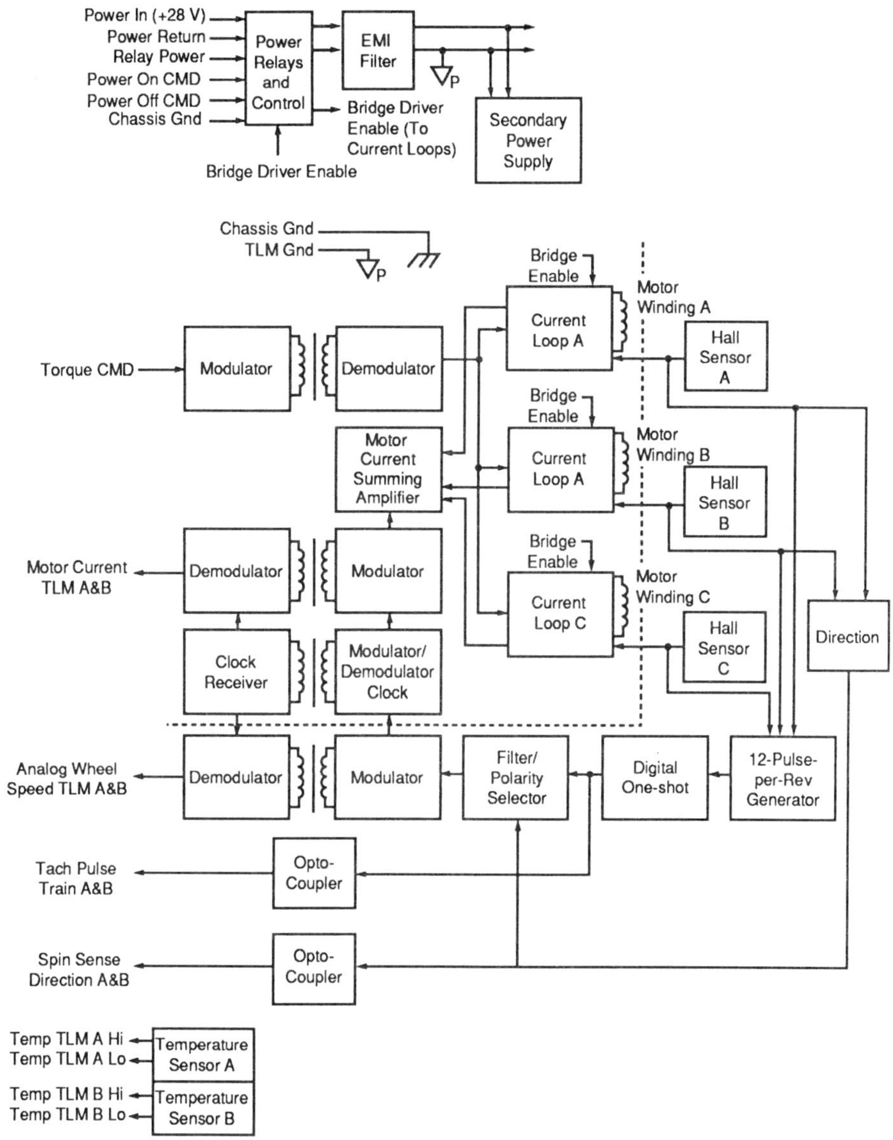

Fig. 7 RWA Functional Block Diagram

A motor's torque constant is directly related to its self-generated back-EMF constant. Therefore, by knowing the minimum bus voltage, the necessary motor coil current to produce the required torque can be found. In the case of MSX, to achieve the goal of 1.2 ft-lb, the coil current is 30 A when using the normal three-phase drive method shown in Fig. 8.

Reduced coil current, and therefore motor driver current, is possible by driving the three motor coils simultaneously (with 120-deg phase shifts), also shown in Fig. 8. By using this method, a current of 16 A is required to power individual coil and driver currents. This represents a considerable reduction in driver-transistor stress levels, as well as I^2R losses over conventional drive schemes, and was the reason for selecting this drive method.

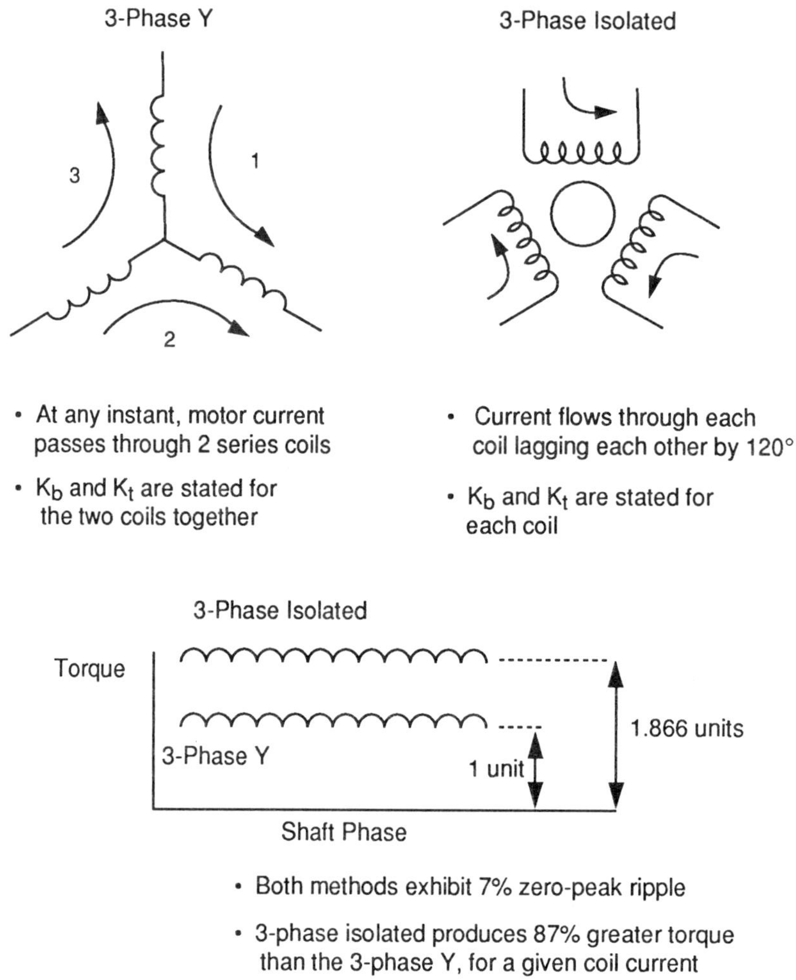

Fig. 8 Three Motor-drive Arrangements Compared

RWA Design - Motor Current Loop

The MSX RWA uses motor current control loops, which are enhanced to minimize the effects of self-generated back EMF. The back EMF can cause a gradually decreasing torque as the wheel speed increases.

To prevent this torque falloff, the MSX RWA design encloses the conventional control loop within an integrator, which senses feedback motor current and maintains a correct average motor current relative to the input torque (current) command.

RWA Structural and Rotor Bearing Suspension

The MSX RWA structural design uses a high-strength aluminum alloy in a conically shaped housing design for a high radial and axial stiffness-to-weight ratio. Similar strength-to-weight ratio techniques are employed in the bimetallic inertial rotor structure. A steel alloy rim, thermally fitted to an aluminum alloy web (in turn thermally fitted to a hollow stainless steel shaft), is used.

The housing and rotor structure designs utilize asymmetric finite-element modeling techniques to predict natural frequencies, modal shapes, critical stresses, and initial bearing loads.

The MSX RWA employs a space-proven, duplex 101 bearing system in a fixed- and floating-cartridge assembly suspension. With Honeywell-developed geometric enhancements and an inertially biased bearing cage design, extended long life is realized and proven with over 5.6 millions of hours of accumulated space-environment performance. The suspension system incorporates load-optimized capability through special bearing raceway refinements, which include precision bearing race curvature ratios for improved axial runout control essential in reducing satellite disturbances.

The lubrication transport mechanism employs a passive oil/grease labyrinth design with sacrificial oil reservoir assemblies for extended long-life operation. The bearing oil/grease chemical composition includes Tricresyl Phosphate (TCP) enhancement for metallic treatment to prevent premature wear of bearing components when operating at low speeds in the sub-EHD regime.

Electronic Packaging and Thermal Dissipation

Proper thermal management is needed for the significant RWA power consumption possibly required to support the vehicle's more difficult tracking scenario. The RWA's power distribution, shown in Fig. 9, presents the hybrid drive electronics as the main power consumer of the electronic package. Therefore, the electronics-package design focused on improving heat dissipation of this assembly by taking advantage of the RWA's enlarged mounting platform and locating the driver's metal board assemblies in the base of the RWA. This provided substantial primary dissipation through conduction and secondary dissipation by surface radiation to the inside of a black-anodized housing structure. Other important design approaches include thermal conductive mounting of all significant power-generating discrete devices, including the DC motor, while also using secondary heat dissipation through direct radiation to an emissivity-treated inertial rotor web. Through asymmetric and solid finite element modeling techniques,

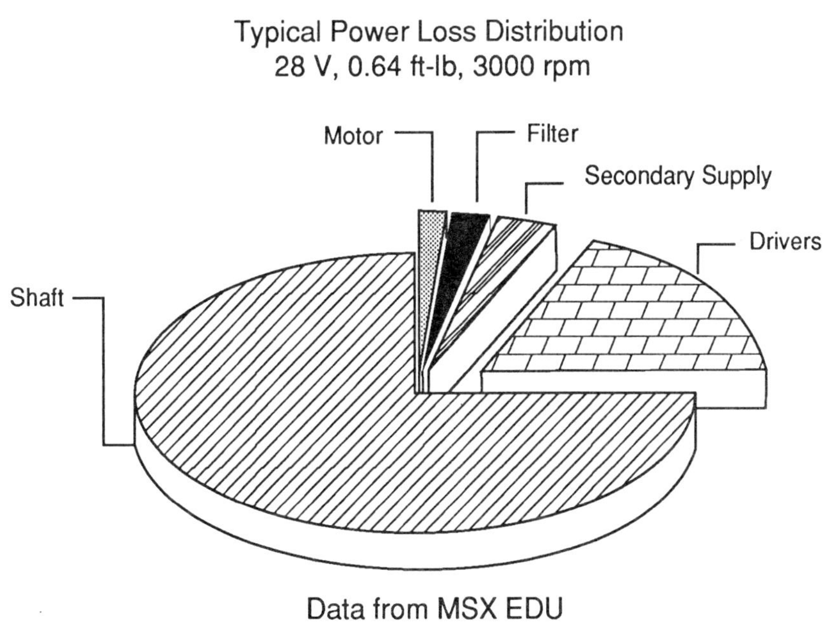

Fig. 9 MSX RWA Power Distribution

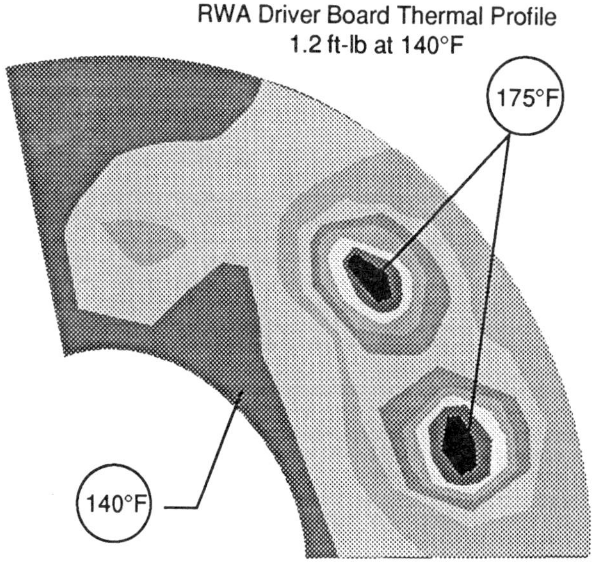

Fig. 10 MSX RWA Drive Electronics Thermal Gradient Under Worst-case Operating Conditions

the MSX RWA was designed for thermal dissipation under worst-case, enhanced-performance, power-generating operations (1.2 ft-lb and 140°F). An expected thermal gradient of the motor-drive assembly, based on a controlled RWA base temperature, is shown in Fig. 10. Note the sharp temperature gradient at the two hybrid-device locations.

THE MSX RWA PROTOTYPE

Operating Performance

After fabrication of the MSX RWA prototype, performance checkout operations were completed, with the electronic assemblies located external to the housing, except for the motor and Hall-sensing commutator assembly. This configuration allowed wheel operation and efficient troubleshooting with direct access to the electronic assemblies. Following checkout, the RWA was fully assembled with electronics integrated and successfully subjected to ambient, thermal-vacuum, and vibrational environmental testing conditions.

Output Torque

Fig. 11(a) shows the relative independence of output torque over the speed range of 100 to 3000 rpm, temperature range of -20 to +60°C, and bus voltage range of +22 to +34 VDC. This data was for a fixed torque command of 5 V (0.64 ft-lb).

Torque Linearity

Fig. 11(b) demonstrates linearity over a torque command range of -5 to +5 V, again over the operating ranges of bus voltage and temperature, while holding a fixed wheel speed of 1500 rpm. Torque command was adjusted in 1-V increments. Torque was found by measuring the time taken to change speed from 1450 to 1550 rpm, or 1550 to 1450 rpm.

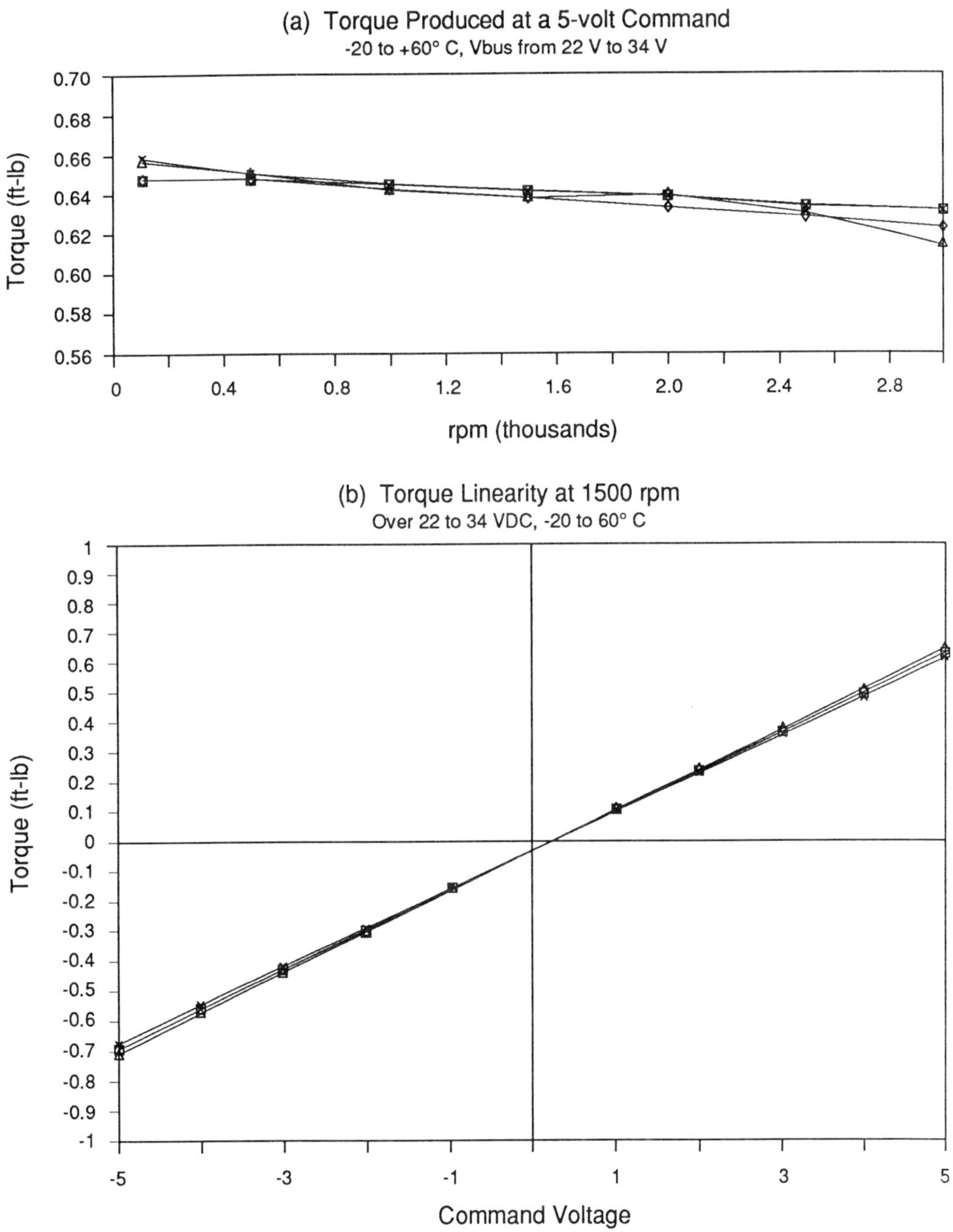

Fig. 11 MSX RWA Prototype - Normal Performance Range

Enhanced Performance Operation

As mentioned previously, a design goal to enhance operating capability beyond normal vehicle requirements was incorporated and verified. Fig. 12(a) demonstrates enhanced torque operation at 1.28 ft-lb, again showing variations caused by bus voltage and temperature variations. Note that the data for the 22-V bus condition is terminated at 2000 rpm, necessary because the larger load power (800 W) and voltage drops of the motor and drive circuits reduce normal "headroom" at low bus voltage. This maximum speed increases as the bus rises from 22 to 28 V.

Torque command linearity over bus voltage and temperature, shown in Fig. 12(b), shows consistent performance over command voltages of ±10 V. The 22-V bus case was not performed.

In addition to enhanced torque capability, higher wheel speeds are also possible. Torque characteristics are shown in Fig. 13 for speeds to 4500 rpm. As shown, the higher back EMF generated at 4500 rpm causes falloff at the 28-V bus at a speed of 4000 rpm.

The limitation relative to wheel speed and bus voltage has been noted. Another limitation to extended operation is that the input current should not exceed the derated design limit of 33 A.

MSX RWA Power Consumption

Power consumption was monitored closely throughout prototype testing to fully characterize the MSX RWA power envelope. A power equation was developed to compile the various RWA component contributions: shaft power delivered, EMI input filter, driver loss consisting of both switching and semiconductor voltage drop losses, motor winding resistance, and secondary power supply input power. The power equation required empirical influence to compensate for nonlinearities and secondary effects. These constants were developed by enclosing the power equation within a "hill-climbing" optimization strategy to obtain a best-fit curve to actual measured data. Some results, shown in Fig. 14, compare measured and calculated input powers for torque and detorque operating conditions over 28-V bus, 4-V command, and wheel speed.

The power equation, developed primarily for satellite vehicle simulation, allowed generation of an input-power locus over many operating conditions. Fig. 15 was created from the power equation to aid in a quick power-consumption guide as applicable to various vehicle tracking scenarios, both planned and unplanned.

Structural Performance Results

The MSX RWA prototype was structurally analyzed under flight vibration levels through the use of strategically located internal accelerometers. The transducer responses aided in assessing structural modal responses regarding natural frequencies and transmissibilities. The modal response results allowed calculation of the bearing loads expected during launch. Fig. 16 shows the RWA's accelerometer locations. The resulting bearing loads seen on the prototype indicate good response agreement versus modal and natural frequency predictions, and overall reduced transmitted bearing loads. In support of the calculated loads, bearing-component inspection following vibration exposure indicated no signs of bearing brinelling.

Thermal Performance Results

The MSX RWA prototype was subjected to acceptance-level thermal-vacuum environments to assess not only functional operation but also to determine its ability to adequately transfer heat from the high power-generating devices and ensure good thermal management under extreme operating conditions. Temperature sensors, located in a special configuration, allowed thermal measurement of the heat-generating devices with respect to the thermally controlled mounting base of the RWA. Thermal model analysis of the driver assembly, a key contributor, predicted a temperature differential from the RWA base of 35°F, under worst-case operating conditions (1.2 ft-lb and 140°F). Actual thermal response indicated a slightly increased temperature delta of 41.5°F. Overall RWA results indicate good thermal-management capability and healthy thermal margins for RWA/vehicle tracking operations.

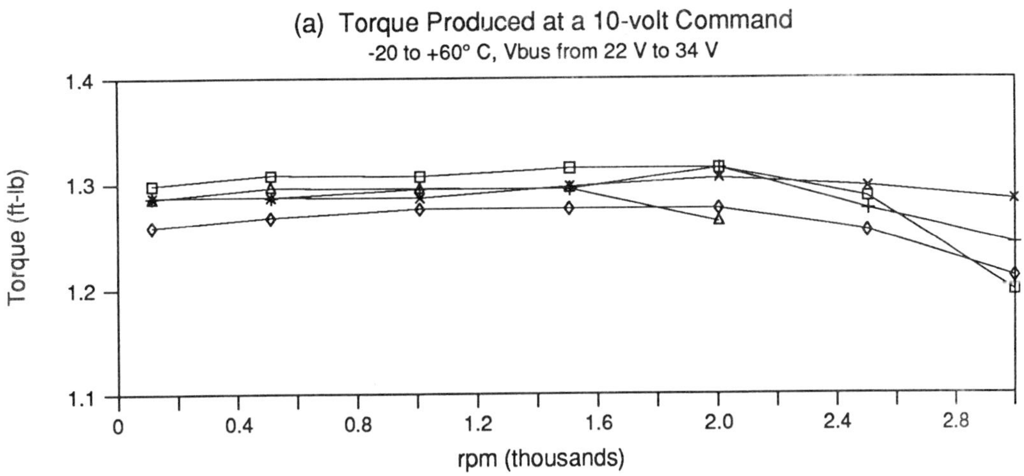

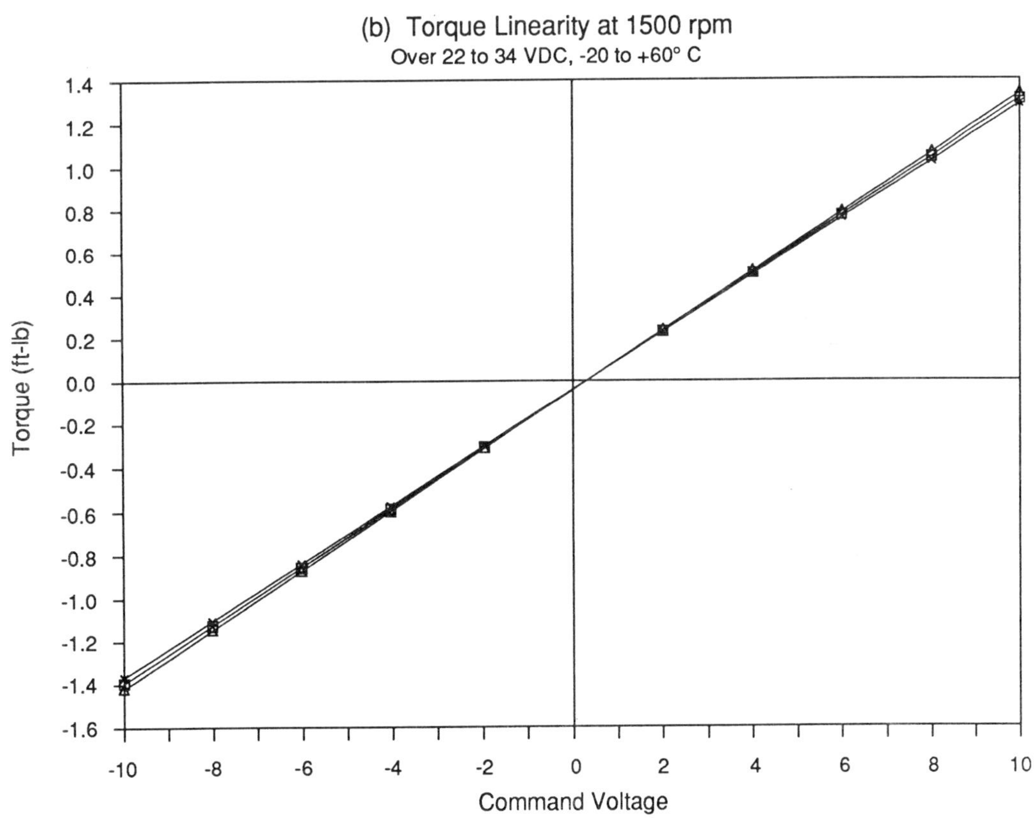

Fig. 12 MSX RWA Prototype - Enhanced Performance Range

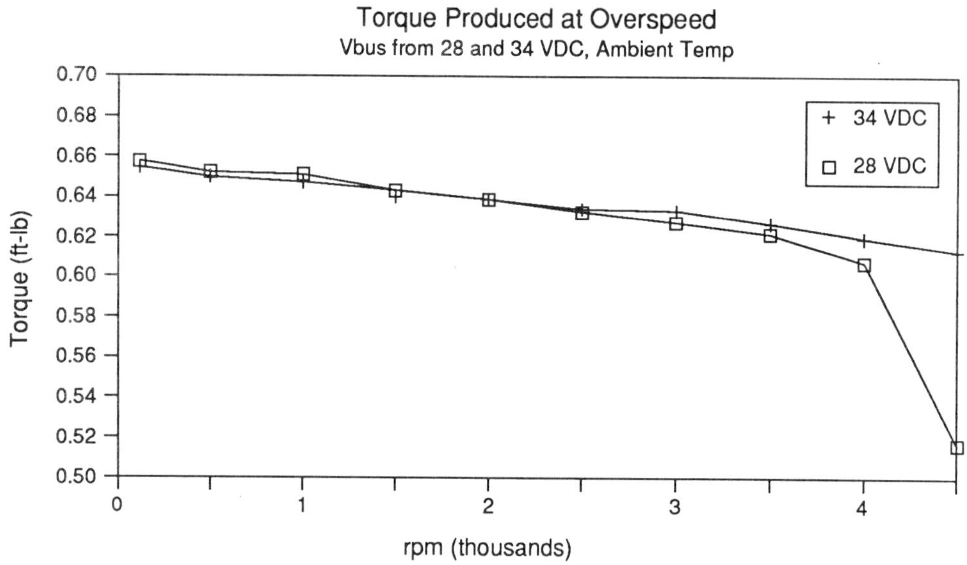

Fig. 13 MSX RWA Prototype - Enhanced Operation

BEARING-INDUCED VIBRATION AND EFFECT ON JITTER

Analysis of RWA-induced Vibration

Ball bearings produce mechanical vibration at a number of frequencies. Extensive testing of the reaction wheels for the HST determined that the *axial* excitation of the bearings was worse than "lateral", and that frequencies at the rotor speed and multiples of 2, 2.8, and 5.2 were the primary offenders.[2,3] Lower levels of random vibration at all frequencies were also present. In order to limit wheel-induced structural vibrations, a specification was placed on the maximum allowable wheel axial force from any single frequency or combination of frequencies to the curve shown in Fig. 17. This curve satisfies the formula $F = 0.001\ f^2$, where F is the peak axial force in lb, and f is the excitation frequency in Hz. The rationale for this formula is that, if the bearings produce ±45 μin. of oscillation (allowed by the bearing specification) and the rotor weighs 25 lb, then the force to produce a sinusoidal oscillation of 45 μin. is:

$$F = (\frac{45 \times 10^{-6}}{12}) \times \frac{25}{32.2} \times (2\pi f)^2 = 115 \times 10^{-6} f^2$$

If a magnification factor of 8 is applied at an assumed mechanical resonance of 25 Hz (in the rotor or mounting structure), then $F = 0.001 \times 25^2 = 0.625$ lb (peak). This is a substantial force and will be greater at higher frequencies. We believe this to be a conservative number: more severe than reality.

How would the satellite structure respond to this level of mechanical excitation? A NASTRAN model of the structure evaluated the angular pointing oscillation of the SPIRIT III telescope as excited by this level of wheel axial forcing[4]. Fig. 18 shows the Y- and Z-axis angular oscillations produced by *one* frequency of excitation at the level specified by Fig. 17. If 4 frequencies (at 1.0, 2.0, 2.8, and 5.2 times rotor speed) are imposed simultaneously, we get the response shown in Fig. 19. Wheel No. 2 produced the largest response of the 4 wheels, each analyzed individually, while Fig. 20 shows the excitation due to the static unbalance of one wheel.

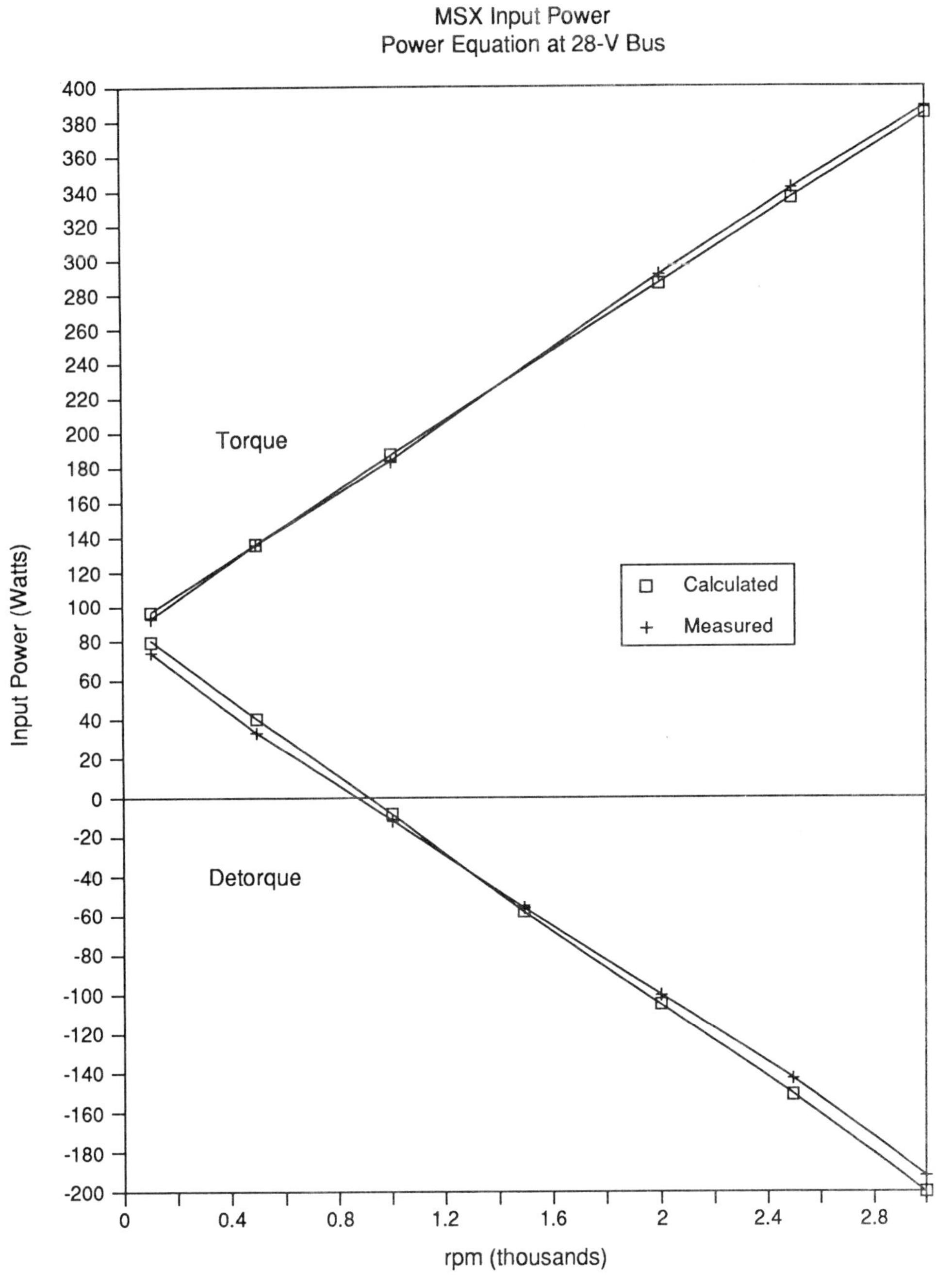

Fig. 14 MSX RWA Power Calculation Versus Actual Prototype Measurement

Power Equation:

$$P\_Total = P_{sec\_sup} + P_{mtr} + P_{switch} + P_{dvr+drop} + P_{emi} + P_{shaft}$$

Individual Terms:

$$P_{sec\_sup} = 7.3 \text{ watts (measured)}$$

$$P_{mtr} = (Imtr)^2 \, Rmtr$$

$$P_{switch} = Vbus \; |Imtr| \; 0.1 \, K_{switch}$$

$$P_{dvr\_drop} = 3 \, Vdrop \; \frac{|Imtr + 0.5 \, Delta\_I| + |Imtr - 0.5 \, Delta\_I|}{2}$$

$$P_{emi} = \left[\frac{|P_{sec\_sup} + P_{shaft} + P_{mtr} + P_{switch} + P_{dvr\_drop}|}{Vbus} \right]^2 Remi$$

$$P_{shaft} = 1.356 \, (-0.12) \, Vcmd \left[\frac{rpm}{60} \right] 2 \, pie$$

Where:

$$Vbak = \frac{|rpm|}{3000} \, 16.7 \qquad Duty\ Cycle = \frac{Vbak}{Vbus} \qquad Imtr = 1.64 \, Vcmd$$

$$Delta\_I = 0.29 \, Kcomm \, (Vbus - Vbak) \, \frac{100}{Lmtr} \, Duty\ Cycle$$

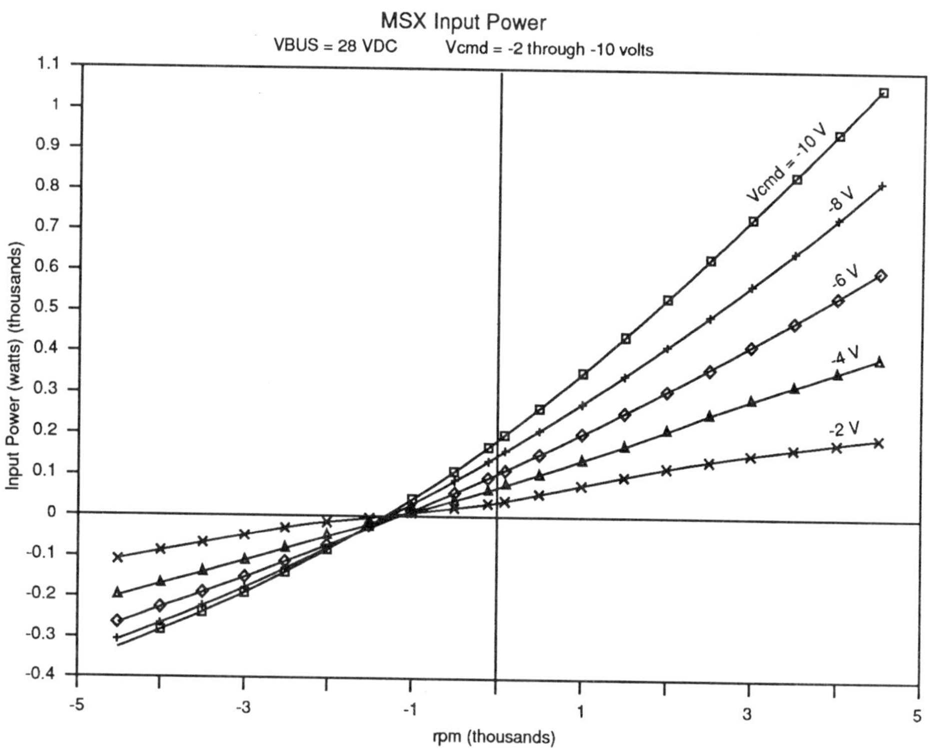

Fig. 15 MSX RWA Equation for Estimating Various Power Consumption Scenarios
(Data plotted for torque constant of 0.135 ft-lb/V)

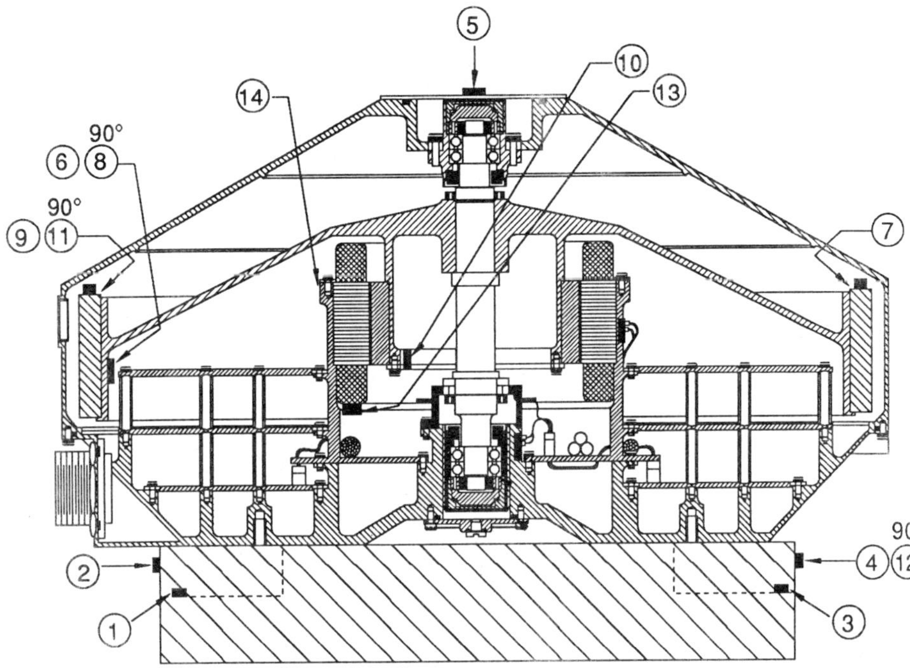

Fig. 16 MSX RWA Prototype Modal Response Accelerometer Layout

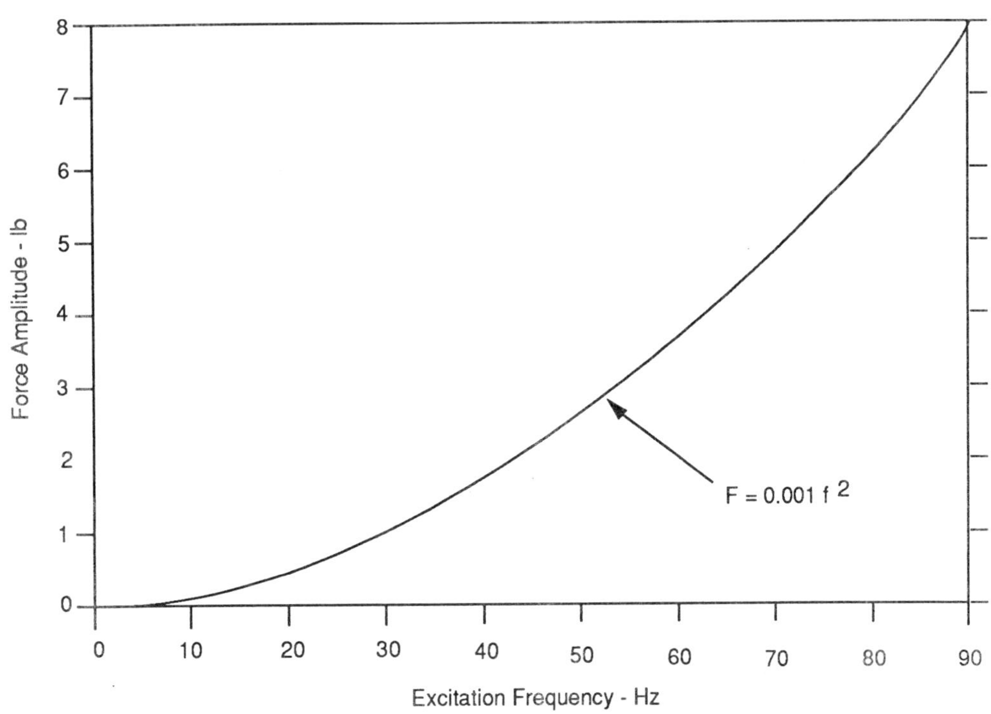

Fig. 17 Maximum Permitted Axial-force Excitation (by Specification) as a Function of Excitation Frequency

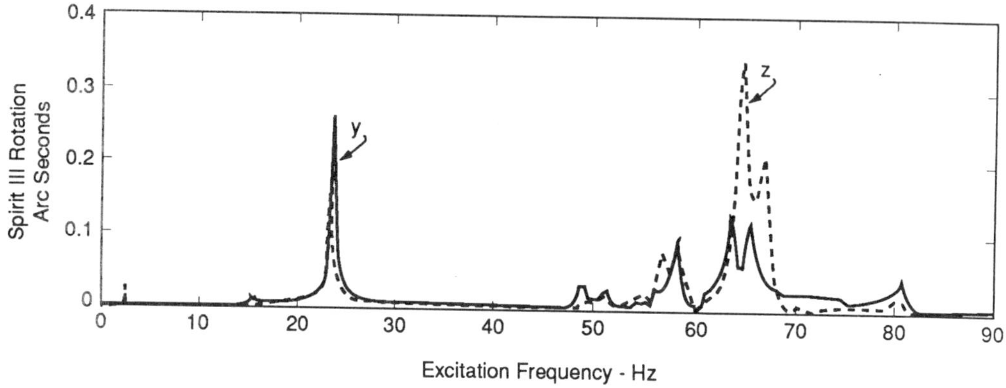

Fig. 18 SPIRIT III Response to One Wheel's Axial Sinusoidal Forcing Frequency

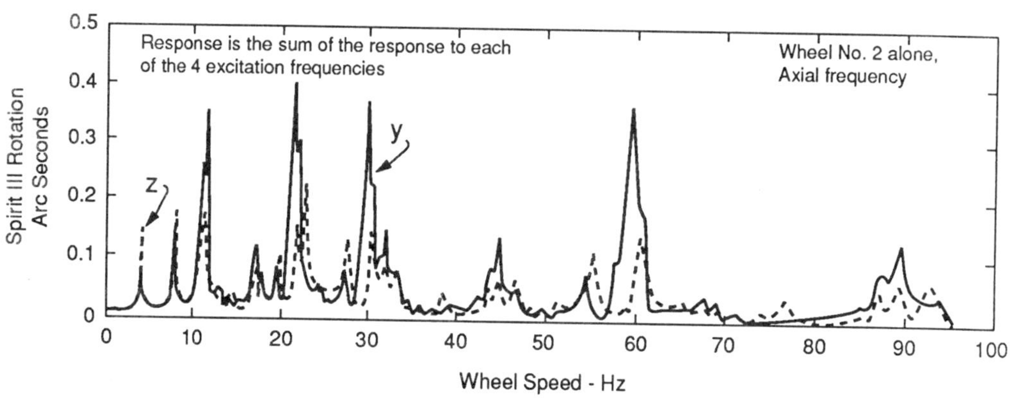

Fig. 19 SPIRIT III Response to 4-frequency Wheel Excitation

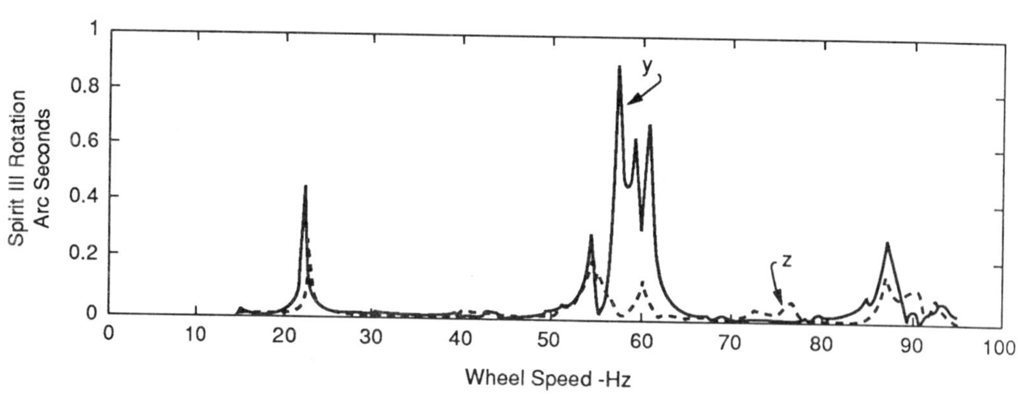

Fig. 20 SPIRIT III Response to Static Unbalance in Wheel No. 2

Since all 4 wheels will be running simultaneously at different and changing speeds, it is difficult to predict the combined result of these inputs. Fig. 21 shows an attempt to do this. In this case we let the wheel speeds vary during a tracking maneuver and plot the SPIRIT III angular excitation as a function of time. Included in this plot is excitation due to wheel static-unbalance force. Note that the peak angular rotation of SPIRIT III is only 0.015 arc s, much less than our allowed limit of 1 arc s. It appears that the 4 reaction wheels will not cause excessive jitter in the SPIRIT III pointing. Future analysis may include the response of *internal* elements of the telescope to the vibration environment, as well as the satellite structure considered here.

Axial Movement of the Flight Bearings

As discussed above, one of the major sources of wheel-generated vibration is axial movement of the bearing assemblies during wheel rotation. The bearing specification allows ±45 μin. of axial motion in the bearings. This might be caused by geometrical limitations in machining of the inner and outer bearing races. The bearing manufacturer, Miniature Precision Ballbearings (MPB), indicates that the actual flight bearings should perform much better than the specified limit. Special tests were run on our flight bearings by MPB to prove this point. Twelve bearing pairs were tested individually, with the results of the best and worst bearing pairs shown in Fig. 22. Bearing pair No. 316 shows a 35-μin. "bump" from the best-fit circle. Bearing No. 317 shows only a 5-μin. deviation from the best-fit circle. All bearing pairs are well within the specification limit of ±45 μin. Therefore, the wheel-induced vibration should be much less than predicted by the conservative analysis based on ±45 μin. axial excitation.

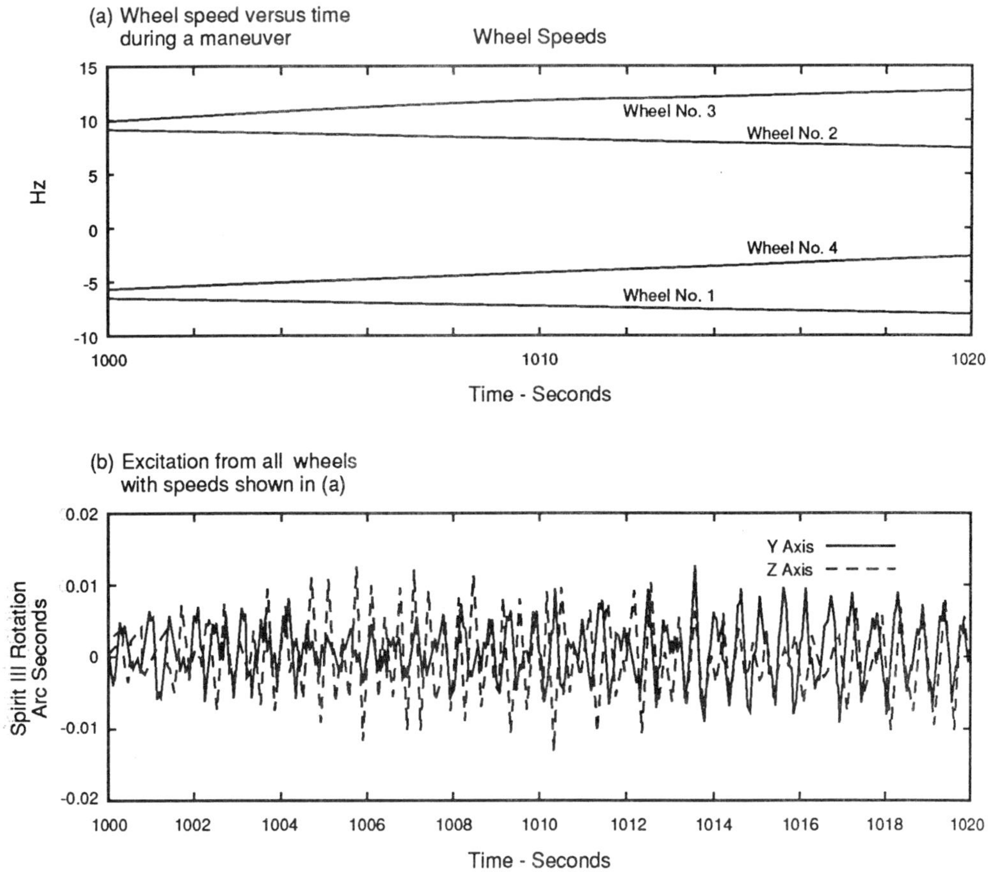

Fig. 21 SPIRIT III Combined Wheel Response to 4-frequency Excitation

Each radial division = 10 microinches of axial movement

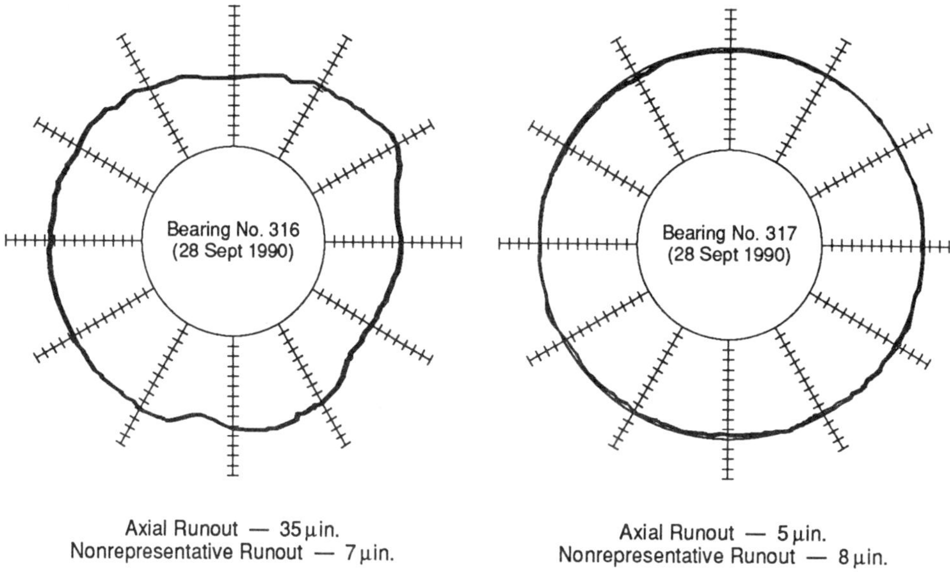

Fig. 22 Axial Movement Versus Rotation Angle for Bearing Pairs

CONCLUSION

The MSX presented an important challenge to the designers of the attitude control system and one of its key components – the RWA. Through proper technical attention given to key design parameters, a new and more powerful reaction wheel has been developed. The RWA promises to not only satisfy the MSX vehicle attitude control needs but also to provide enhanced performance capability for true mission versatility.

REFERENCES

1. Ray, J.C. "Typical MSX Maneuver Results," Applied Physics Laboratory Memo No. S3G-91-003, dated January 5, 1991.

2. Hasha, M.D., "Reaction Wheel Mechanical Noise Variation," Lockheed EM No. SSS218, dated 30 June 1986.

3. Blair, Mark A., and Vadlamudi, N., "Jitter Performance Test for the Hubble Space Telescope: Method and Results," International Modal Analysis Conference, Orlando, FL, February 1988.

4. Garner, J.P., "SPIRIT III Structural Response to Disturbances," Garner Associates Technical Report MSX11, dated August 29, 1990.

AAS 91-041

NON-COLLOCATED CONTROL OF AN INVERTED FLEXIBLE STRUCTURE

Steve Clark and Charles Fosha*

This paper summarizes an undergraduate project in the controls lab in the Electrical and Computer Engineering Department at the University of Colorado at Colorado Springs. The work was undertaken to produce a very inexpensive experiment in the control of flexible structures area and to provide the ground work for more advanced and intensive study in the field. The work involves investigation into the dynamics and active vibration control of a lightly damped beam. Both the force actuator and the non-collocated feedback sensor are very non-linear. The sensor output versus displacement is modelled by an odd ordered polynomial whose parameters are found using a least squares algorithm. The effect of the sensor output on system stability is investigated and limits on stability established. The sensor is then compensated in a piece-wise linear fashion using a non-linear circuit. The first three flexible modes were then modelled and the control law investigated. Since this is an undergraduate experiment, control of the first mode was attempted using a classical lead-lag filter. The intermediate findings and the final results are explained and evaluated.

EXPERIMENTAL HARDWARE

Beam

The focus of the experiment is an extruded aluminum beam. See Figure 1. The beam is pinned at one end and free at the other. Force actuation is accomplished with two hobby magnets attached approximately 1" from the pinned end and acted upon by two electromagnetic coils in a push pull fashion. The coils are activated by a transconductance amplifier built specifically for this experiment.

Sensors

End effector sensing is accomplished using a unique configuration of Hall effect sensors that detect the position of two additional magnets placed on the end of the beam. Extruded aluminum was chosen because it is stiff enough to position the beam vertically with the free end up but still exhibit flexible modes.

* University of Colorado, Department of Electrical Engineering, Colorado Springs, Colorado 80933.

Figure 1. Flexible Beam Control System

End point sensing had to be very inexpensive and suitably sensitive. Commercially available Hall effect sensors fit these criterion. They cost about $3.00 a pair and have proven to have more than adequate sensitivity. They are very linear with respect to the magnetic (B) field but the B field is very non-linear with respect to distance. See Figure 2. This proves to have a significant effect on performance and stability.

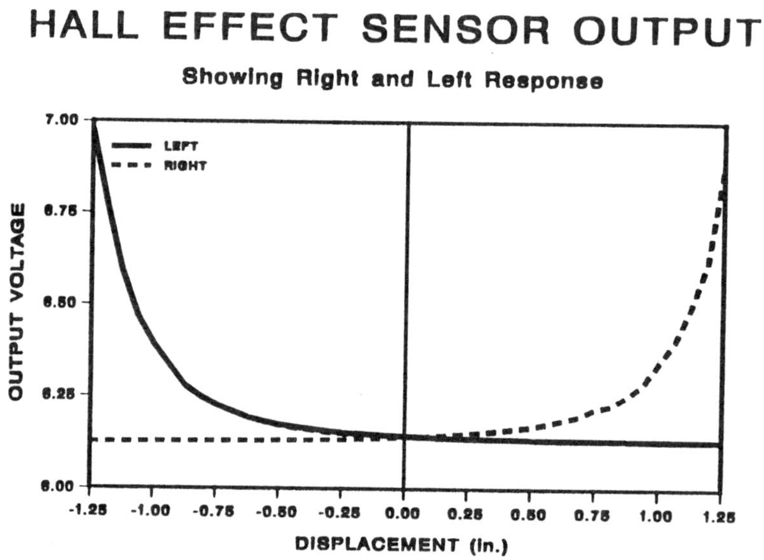

Figure 2.

The power actuator provides 1 amp/volt, but there is a very nonlinear force saturation in the actuator. This is investigated and the limits on linearity are established.

MATHEMATICAL MODEL

There are theoretically an infinite number of flexible modes for a continuous beam. Concentration on the directly coupled and the first three flexible modes will yield an adequate model for the purposes of this investigation. Excitation of the higher modes decay quickly and become insignificant compared with the lower modes, especially the first.

The system model was found using a gain-phase meter. System poles are located at the local maximums and the system zeros are located at the minimums. Modal damping is determined using the following derivation.

$$|x| = Kx_{res}$$

K can be any fraction of the output peak magnitude.

$$\left(\frac{K}{2\zeta}\right)^2 = \frac{1}{\left(1-\left(\frac{\omega}{\omega_n}\right)^2\right)^2 + \left(2\zeta\left(\frac{\omega}{\omega_n}\right)\right)^2}$$

$$\left(1-\left(\frac{\omega}{\omega_n}\right)^2\right)^2 + \left(2\zeta\left(\frac{\omega}{\omega_n}\right)\right)^2 - \left(\frac{2\zeta}{K}\right)^2 = 0$$

$$\left(\frac{\omega}{\omega_n}\right)^4 - 2\left(1-2\zeta^2\right)\left(\frac{\omega}{\omega_n}\right)^2 + 1 - \frac{2\zeta}{K} = 0$$

Solve for $\left(\frac{\omega}{\omega_n}\right)^2$.

And since $\zeta \ll 1$ we can ignore ζ^2 terms. This gives

$$\left(\frac{\omega}{\omega_n}\right)^2 = 1 \pm 2\zeta\left(\frac{1}{K^2}-1\right)^{\frac{1}{2}}$$

These two roots are sidebands of the resonant maximum or minimum:

$$\left(\frac{\omega_1}{\omega_n}\right)^2, \left(\frac{\omega_2}{\omega_n}\right)^2$$

Now finding the difference of the two sidebands:

$$\frac{\omega_2^2 - \omega_1^2}{\omega_n^2} = 4\zeta\left(\frac{1}{K^2} - 1\right)^{\frac{1}{2}} \approx 2\left(\frac{\omega_2 - \omega_1}{\omega_n}\right)$$

Solving for ζ:

$$\zeta = K_g\left(\frac{\omega_2 - \omega_1}{\omega_n}\right)$$

Where:

$$K_g = \frac{1}{2}\left(\frac{1}{K^2} - 1\right)^{-\frac{1}{2}}$$

The system is summarized in Table 1.

TABLE 1.

Mode i	Zero Frequency (Hz)	Zero Damping	Pole Frequency (Hz)	Pole Damping
0			16.5	real
1	4.7	.03	5.25	.01
2	38.5	.1	37.5	.2
3	100	.2	102	.2

The frequency and phase response are shown in Figures 3 through 6.

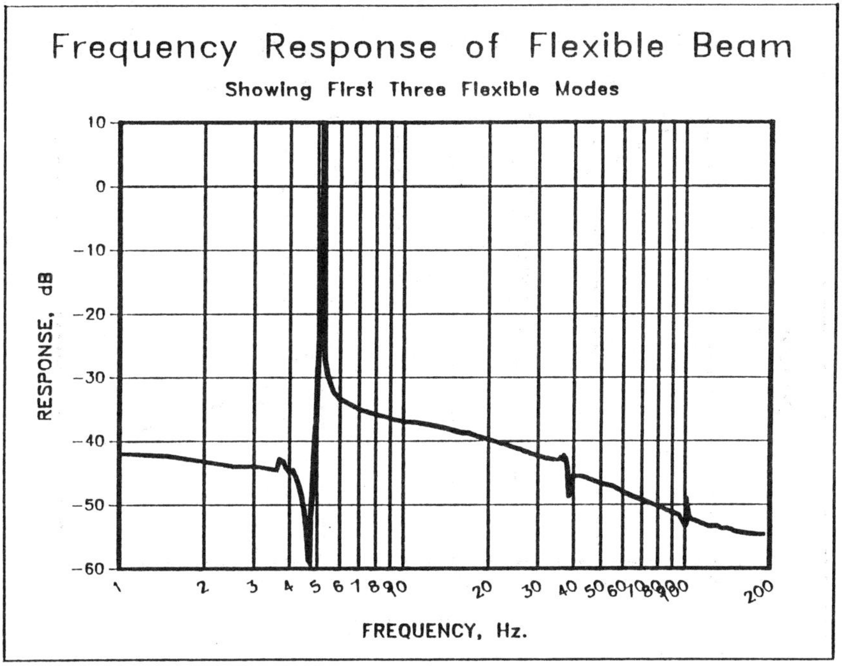

Figure 3.

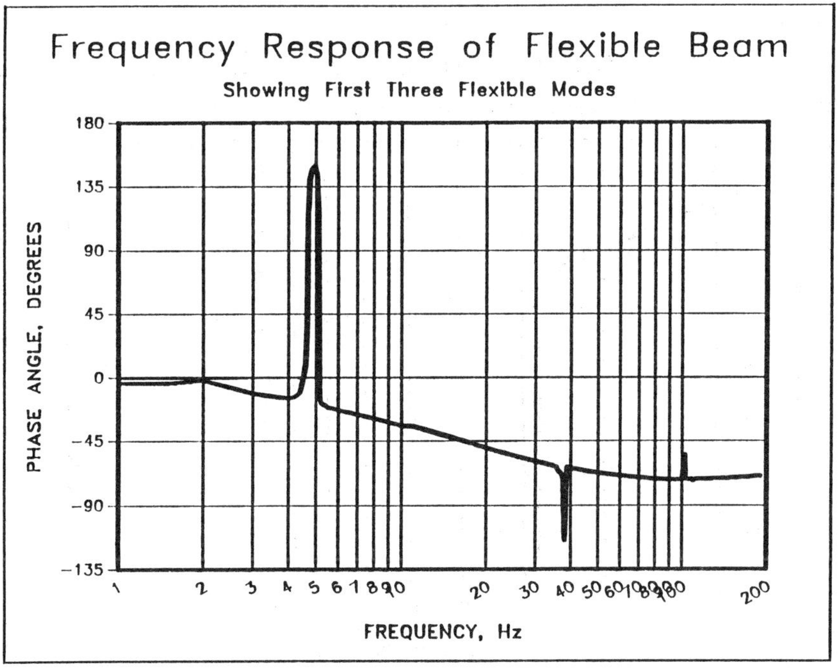

Figure 4.

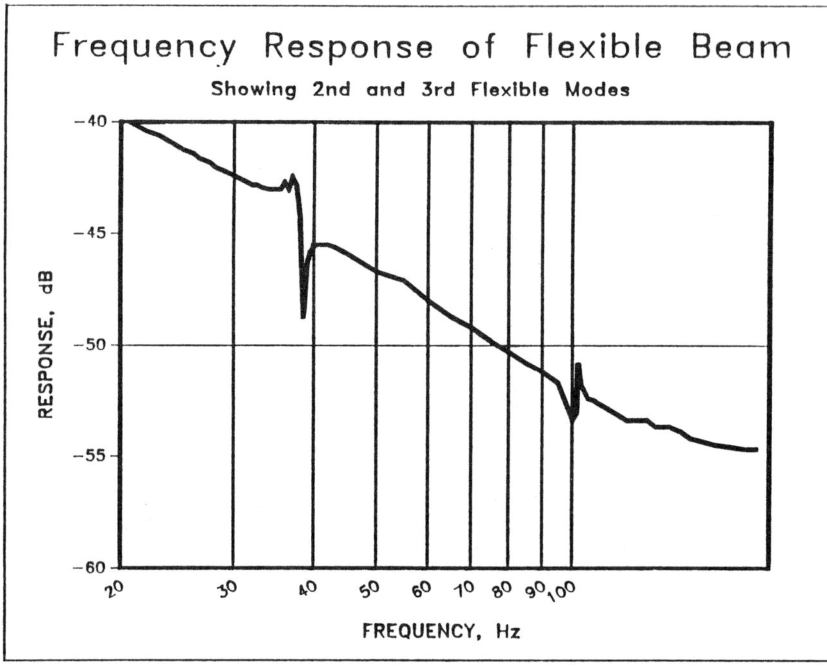

Figure 5.

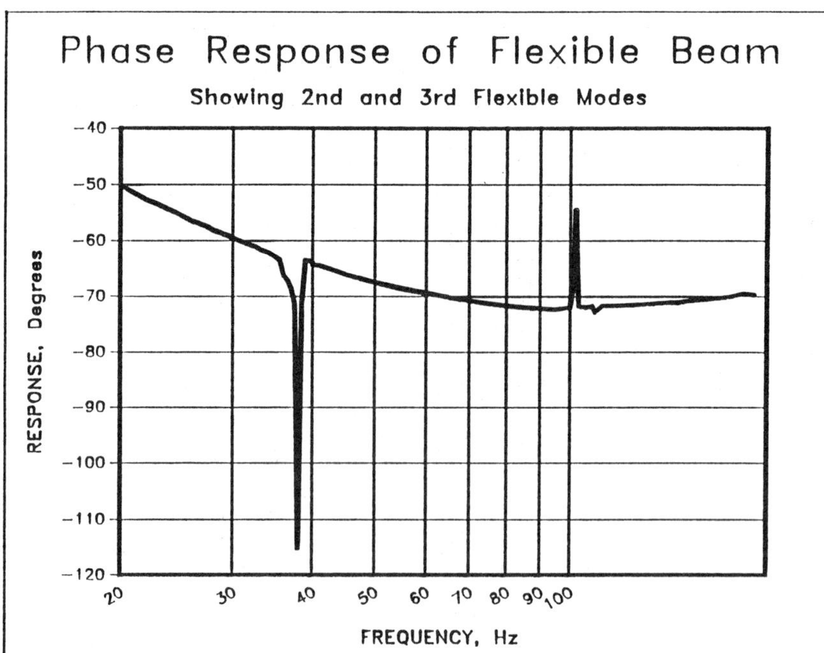

Figure 6.

SENSOR MEASUREMENTS

The output of the non-collocated sensors is very non-linear. Figure 7 shows the sensor circuit. Figure 8 shows the output of the sensor conditioning circuit. The circuit simply subtracts one output from the other along with gain and zero adjustments. The actual equation describing the input-output of the sensor circuit is difficult to accurately model. However, the output of the sensor circuit is similar to an odd-ordered polynomial . A fifth order polynomial fit to the sensor output is also shown on Figure 8.

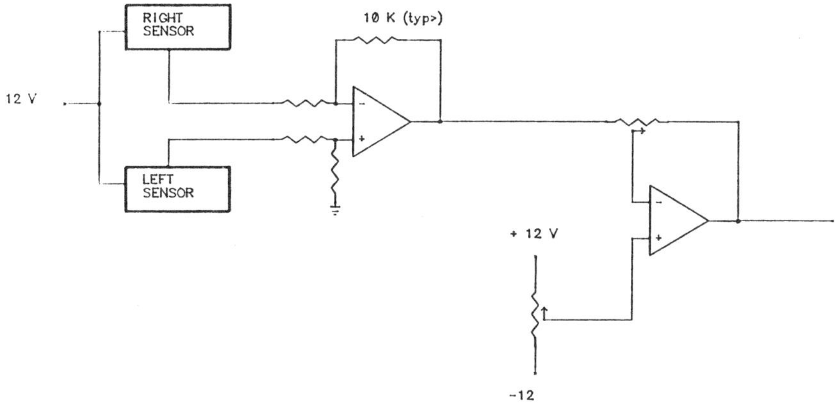

Figure 7

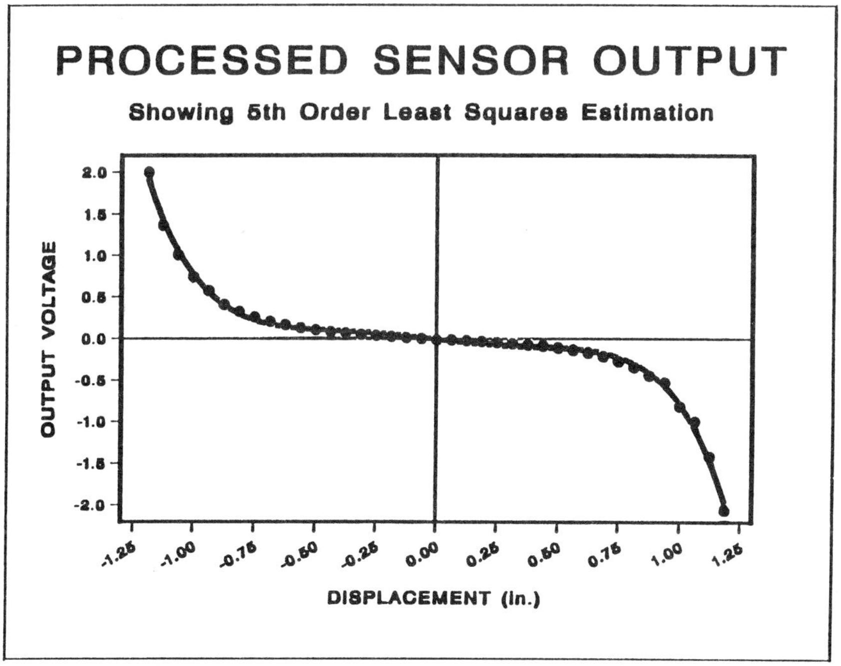

Figure 8.

COMPENSATOR DESIGN

The compensator is a simple lead-lag type with a zero at -10 and a double pole at -100. The overall gain of the compensator can be varied . The compensator schematic is shown on Figure 9.

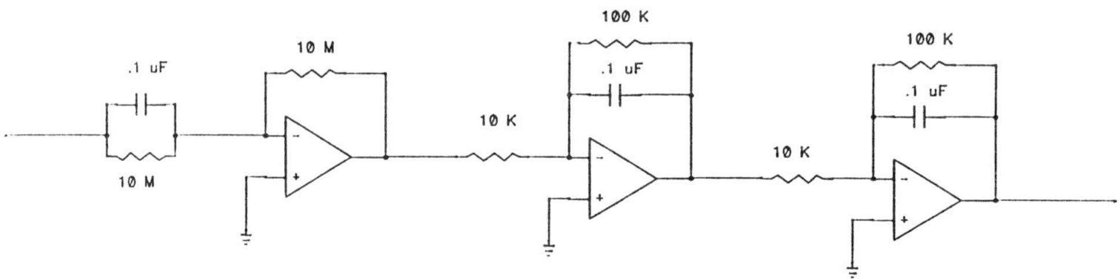

Figure 9.

EXPERIMENTAL RESULTS

Figure 10 shows the system and compensator poles and zeros. The upper half of the Figure 11 shows the impulse response of the uncompensated system. The lightly damped oscillatory response is clearly evident.

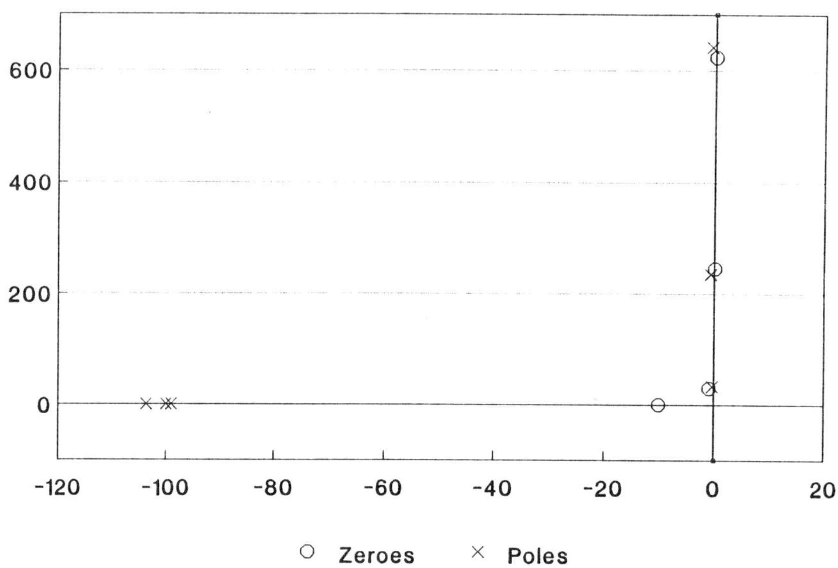

Figure 10.

The lower half of Figure 11 shows the impulse response with the compensator in the loop. The system is clearly more heavily damped. However, with too large an input, the system goes unstable showing its highly non-linear aspect.

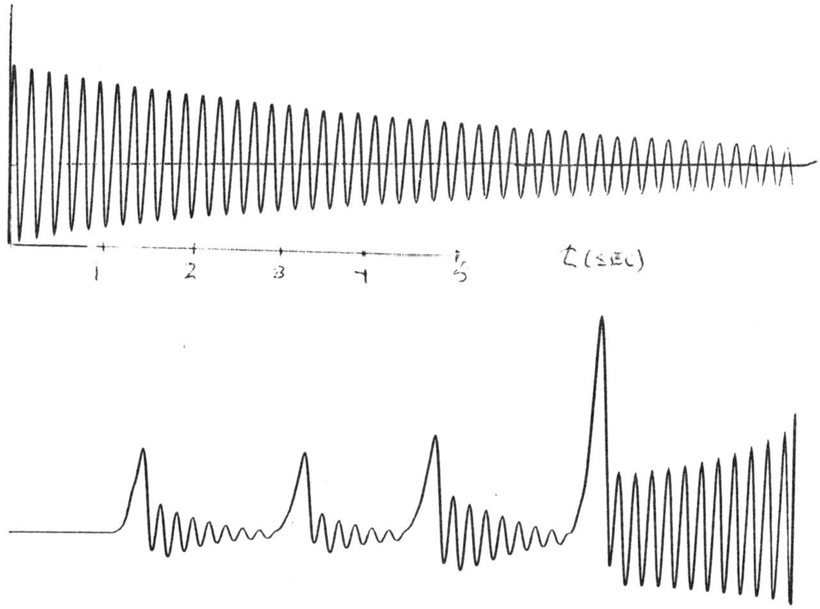

Figure 11.

CONCLUSION

A very inexpensive inverted flexible structure has been demonstrated which clearly shows the problems associated with the control of lightly damped systems. While the compensator used in the experiment is simple, the experiment can be used for identification and implementation of more sophisticated modern design techniques such as observors, linear state feedback and pole placements. These results will be presented next year.

AAS 91-042

A MAGNETIC BEARING MOMENTUM WHEEL FOR HIGH POINTING ACCURACY AND VIBRATION SENSITIVE SPACE APPLICATIONS

U. J. Bichler*

Magnetic Bearing Momentum Wheels are well suited as gyroscopic actuators in the attitude control system of spacecraft.
Their advantageous features are no stiction and low friction torque, a higher speed capability, no need for lubrication and the absence of a mechanical life limit.
Furthermore, a full actively controlled magnetic bearing wheel offers the following special features compared with ball bearing wheels or with mainly passive magnetic bearing wheels:

> An active tilting of the momentum vector with respect to the spacecraft ('vernier gimballing') enables to use the wheel as a control moment gyro and a cross momentum storage device. This enables a three axis attitude control of the spacecraft in a fine pointing range with only one wheel and revokes the need for a separate nutation damping device.

> Active Vibration Suppression, enabling isolation of unbalance vibrations of the wheel from the spacecraft. This meets the increasing demand for quiet wheels which can be be used in certain missions.

> Active damping of flexible structures through fully controllable translational bearing forces.

Possible applications are satellites with optical communication links, micro–gravity missions, space telescopes and high resolution earth observation satellites. In long term manned space missions low noise wheel's can reduce the effect of vibrations and the associated noise on the spacecraft environment and crew.

INTRODUCTION

Almost all spacecraft in orbit are equipped with ball bearing momentum or reaction wheels serving as actuators in the attitude control system. Amongst the various on board system components the wheels have been identified as one of the main sources of vibration noise due to residual unbalances, bearing imperfections, etc.

* Teldix GmbH, Bosch Telecom, Grenzhöfer Weg 36, W-D-6900 Heidelberg, Germany.

This disadvantage can be avoided by the use of magnetic bearing wheels having all the well known special advantages as higher speed, no contact and lubrication, no stiction etc., but permitting the acceptance of the increased electronics complexity with the inevitable impacts on mass, volume and reliability[1]. However, these impacts appear in a more favorable light, if the comparison is performed on a whole AOCS level.

In magnetic bearings – as far as no permanent magnets are involved for direct attraction/repulsion – the bearing forces between the statoric and the rotoric part are fully controllable. They can be designed frequency dependant and with a time varying characteristic for example. Unbalance forces are transmitted to the stator only at an amount proportional to the suspension current. Applying a special control law, the disturbance component in the suspension current can be attenuated while keeping an overall stable levitation of the rotor. This in fact isolates the unbalance forces from the stator and the satellite structure.

A second interesting feature is, that if a vernier gimballing capability is implemented, the use of such a magnetic bearing momentum wheel enables an active satellite nutation damping and a three axis attitude control for small roll and yaw angles with one rotating mass only, that is, a fine pointing system with an accuracy depending only on the performance of the attitude sensors. This may result in overall mass savings and performance enhancements.

Furthermore, it is possible to use the magnetic bearing wheel as actuator for the active damping of structure oscillations since active damping forces can be generated by moving the rotor within the mechanical gap.

In the following, this new magnetic bearing momentum wheel is described, and especially the low noise feature is discussed theoretically and proven by practical results.

THE TELDIX ELECTRODYNAMIC BEARING WHEEL MW-X

In the following, the features of the wheel this paper deals with are described.

Design

A cross section of the wheel is shown in Fig. 1. It has been designed to feature the following advantages:

> pancake profile
> low weight
> low volume
> high reliability
> vernier gimballing
> low noise

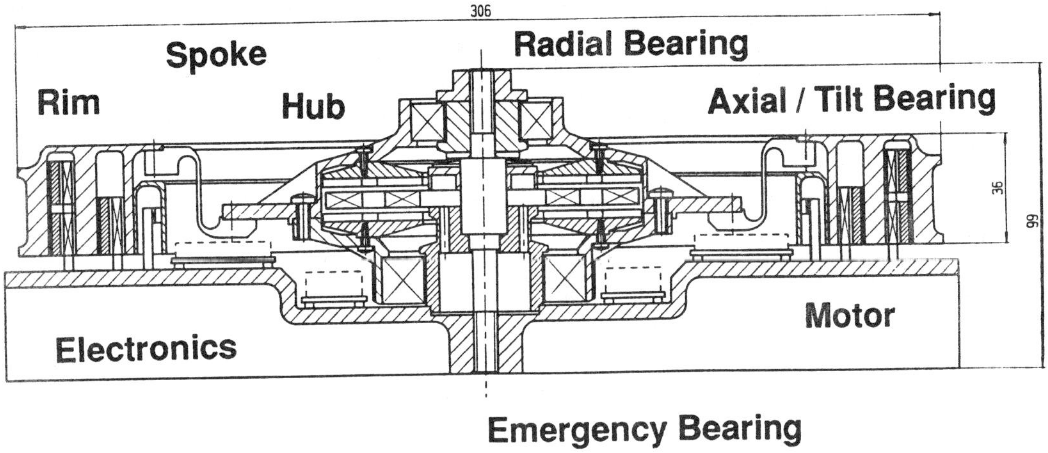

Fig. 1: Teldix 5–d.o.f. electrodynamic bearing wheel MW–X

Rotor

The rotor is basically composed of a rim which is connected to the hub by five titanium springs. These are designed to accommodate the different strains of hub and rim due to temperature and centrifugal stresses.

Most of the mass is concentrated at the outer rim of the rotor contributing to the required moment of inertia.

All five actively controlled degrees of freedom are of the electrodynamic (Lorentz) force generation type. This principle has many advantages with respect to controllability, low noise and the mass distribution between rotor and stator.

The axial and tilt loop control elements are located in the outer slot of the rotor rim. The slot is equipped with permanent magnets which produce a circumferential uniform magnetic field. Four epoxy embedded stator coils, each of them covering 90^0 of the circumference, are fitted in the slot. The current in each of these coils produces a force in axial direction. Applying equal currents to all coils produces axial forces. Tilt torques are generated by exciting two opposite coils with opposite currents. Since it is possible to generate the required forces and torques for the suspension by use of only three coils, sufficient redundancy is attained.

The radial bearings are located in the hub. A disk shaped coil system on the stator fits between two pole plates equipped with permanent magnets. Redundancy is assured by the use of two coils per radial axis.

Any deviation from the nominal rotor position is detected by a set of axial and radial position sensors, and – using five control loops – balanced by forces generated in the coils.

Vernier Gimballing

The rotor and thus the momentum vector can be tilted actively within a range of about 1^0. A special tilt control law assures, that the tilt responses are not deteriorated by gyroscopic effects, i.e. a tilting in one direction is decoupled from the orthogonal direction. This is demonstrated by the step responses shown in Fig. 2. Steps have been applied to one of the tilt axes and the responses of both have been measured at different speeds. In the above row it can be seen, that with a simple uncoupled tilt control law the quality of the desired response deteriorates with speed and even the uninvolved axis becomes disturbed. In the below row, the more sophisticated control law of the MW–X has been applied which removes all the disturbing effects. It is easy imaginable, that such a behaviour is much better suited for spacecraft attitude control purposes.

It is a special feature of the MW–X, that it is possible to exert gyro control torques for the attitude control of the S/C while the produced noise of the wheel is still low.

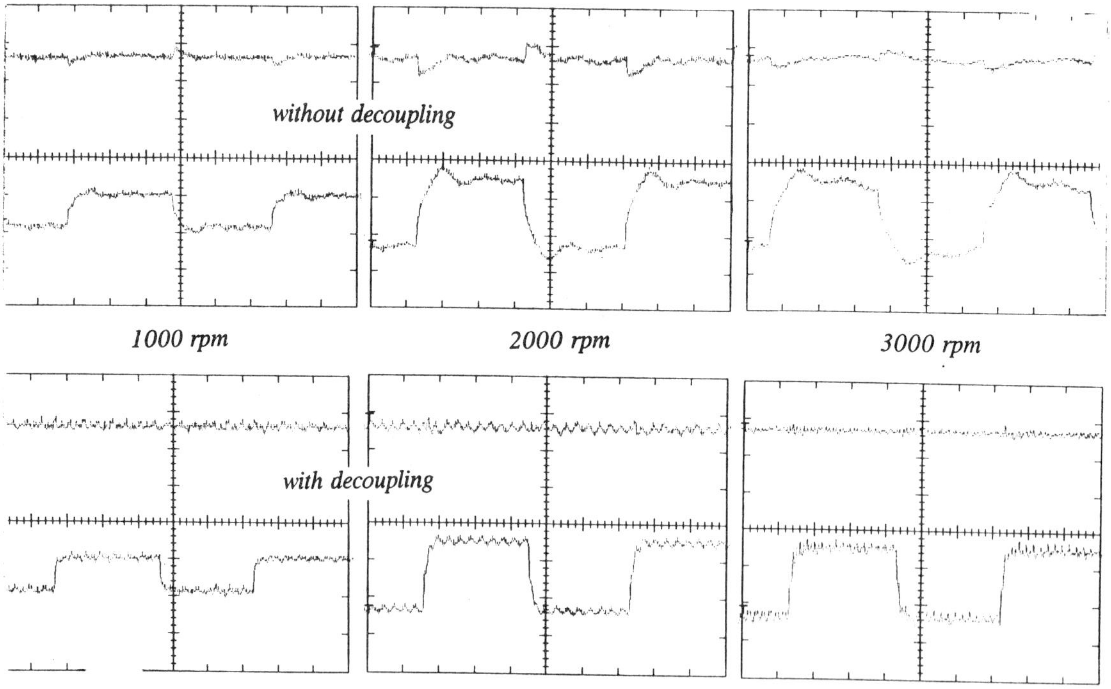

Fig. 2 Vernier Gimballing Step Responses

Low Noise

The MW–X is equipped with a special suspension control system, which keep the overall stability while attenuating the noise, even during a vernier gimballing activity.

The low noise behaviour is further improved by the electrodynamic force generation principle, which enables the rotor to rotate about its natural axis of inertia without transforming the resulting whirling motion into force and torque.

This results in an isolation of the disturbing forces and torques from the stator and the satellite structure.

This active vibration suppression system is described in detail below.

Bearing Stiffness

Since the MW–X is provided with an effective noise suppression control system, it has a very low bandwidth passive stiffness against externally impressed movements of the S/C. Under normal operating conditions, the wheel will keep its long term mid gap position. If, for instance, a thruster is fired, the MW–X switches automatically in a stiff mode, where it is securely kept within the gap. When the external disturbance is over, it returns automatically to the low noise condition.

Emergency Bearings

The task of the emergency bearings is to provide a non–disturbing landing and a smooth run–down exhibiting minimum noise figures in case of a failure in the suspension electronics.

The emergency bearing system consists of two ball bearings, which are located inside the hub of the rotor.

Tests performed with gravity compensation have shown, that the emergency bearing operation is safe and uncritical. The rotor runs down without any whirling or wobbling.

Locking Mechanism

A locking mechanism with re–lock capability keeps the rotor in a safe mid–gap position during launch. It is closed by applying compressed air and released by a redundant set of magnet valves.

Drive Motor

The drive motor is of the brushless d.c. type featuring high efficiency and excellent torque controllability. It is an ironless three–phase, six–pulse design which assures minimum torque ripple and avoids magnetic cogging torque.

The associated wheel drive electronics receives the rotor position from an eddy current sensor system and performs commutation switching. It incorporates a four quadrant converter and can operate either in torque or speed control mode.

In case of a power–supply break down, the motor is switched into an auto–generator mode to maintain the stable suspension during the break down period.

Technical Data

Diameter		340	mm
Height		120	mm
Rotor Mass		6.65	kg
Total Mass	(incl. electronics)	13	kg
Nominal Speed		6000 ... 10000	rpm
Angular Momentum		60 ...100	Nms
Gimballing Torque		3	Nm
Gimballing Angle		1.0	°
Slew Rate Capability		10	°/s
Angular Momentum Cross Component		1	Nms
Motor Torque		0.1	Nm
Drag Torque		< 0.01	Nm
Stiction Torque		–	Nm
Power Consumption:			
at 0 rpm		19	W
at 6000 rpm (Motor off)		23	W

DISTURBANCES IN MAGNETIC SUSPENSION SYSTEMS

Magnetic bearing wheels should be on the first glance ideally suited to exhibit very low noise figures, because they have no mechanical contact between statoric and rotoric parts. But unfortunately there are several possible sources of disturbing forces and torques in magnetic bearing wheels as well. Some of these sources are due to unfavorable design details and should be foreseen and avoided in time. Some others are of principle nature but can be overcome by control means. These sources will be introduced first and then possibilities on how to suppress the disturbances are given.

Unwanted disturbing forces and torques in magnetic bearings can have the following sources:

1. The main source is imbalance, which means that the rotational axis is shifted (resulting in static imbalance) or/and tilted (resulting in dynamic imbalance) against the natural axis of inertia. In a magnetic bearing this rotational axis is the axis of the position sensors measurement surface. Thus if the rotor spins about its natural axis, the measurement surface wobbles. In the subcritical speed range this leads to forces and torques which (try to) keep the rotor rotating about this axis. In the

supercritical speed range, the rotor rotates about its natural axis, but the wobbling sensor signal component also causes forces and torques.

2. Mechanical imperfections of the rotating parts as irregularities in a sensor surface, a pole piece or a permanent magnetic field.
3. Poorly damped bearing control loops, for example due to a passive nature, non-linearities, structural modes or coupling effects between the different loops.
4. Poorly damped gyroscopic oscillations as nutational and precessional motions, which means a whirling motion of the spin axis around the angular momentum vector, in connection with a type of bearing, that transforms motion into force.
5. Sensor types and locations, which do not reject harmonics produced by mechanical imperfections of the rotating measurement surface.

PRINCIPLE OF ACTIVE VIBRATION SUPPRESSION

In this paper, Active Vibration Suppression (AVS) means to enable the rotor rotating about its natural axis of inertia and to suppress all the resulting noise components from the whirling and wobbling sensor surface without affecting the suspension stability. It does not mean a suppression of the vibrational movement of the rotor surface to achieve a rotation about the geometrical axis[3,5] because then the full imbalance forces and torques would be present at the stator, just like in a ball bearing.

A simple suspension loop model is used to introduce and to discuss qualitatively the different proposed kinds of AVS. It is shown in Fig. 3. It consists of a controller, a power amplifier including the force actuator and the rotor as the plant.

The loop is depicted in a possibly unusual way with the disturbances due to unbalance movement of the measurement surface and other noise as input and the force as output.

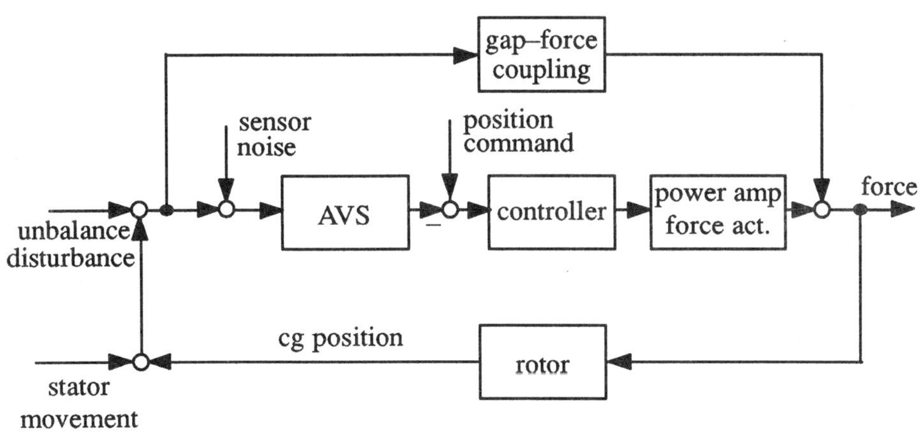

Fig. 3 Simple suspension loop model

AVS means, that the sensor disturbances must be filtered out before reaching the power amplifier and becoming force or torque. Thus a block 'AVS' is introduced in the loop between sensor and controller which must be designed to filter the disturbances out of the measurement signal while keeping the useful information for the suspension loop.

If there is a gap–force coupling in the bearing, the synchronous forces are produced in the bearing before it is possible to filter them out. This is true for all electromagnetic force generation principles, where a change of the gap causes a change of the (bias or permanent) flux and thus of the force. Then these forces must be identified and cancelled by inverse AVS force components. This seems to be more complicated as just to filter, so it is better to have no gap–force coupling. Thus the better choice for a low noise wheel is the electrodynamic principle, where the generated force is independent of the relative position or the relative motion of rotor and stator.

The sensor noise in a magnetic bearing consists of a synchronous component, i.e. of a frequency equal to the rotational speed, and harmonics thereof. Thus an AVS must be either selective and speed adaptive or it must cover the whole frequency range of the disturbances.

Several methods of active vibration suppression have been investigated, some of them speed tracking, others covering a broad frequency range, the following has been found the best:

Model Following Control (MFC)

The principle idea of the model following control is to use the output of an observer, simply a double integrator, for the control feedback. The observer output is well suited for low noise purposes, because it does not contain any disturbances. Fig. 4 shows the block diagram of the control loop with observer, which is fed with the current command signal. Since there is a linear relationship between current and force in an electrodynamic system, the observers output will be very close to the real position of the rotor.

To prevent the observer from long term drift effects, the low pass filtered difference between the real position signal and the model is used to correct the observers state variables. Thus from the real sensor signal only the lowest (near dc) frequency components are used and the remaining noise in the suspension loop is very small.

The AVS method has been investigated by linear analysis and time domain simulations of a single suspension loop as representative of the three translational d.o.f.'s and also of the coupled tilt loops.The result in terms of noise suppression gain is shown in Fig. 5. The baseline for the comparison was a suspension loop with a simple PD–controller of the same parameter set.

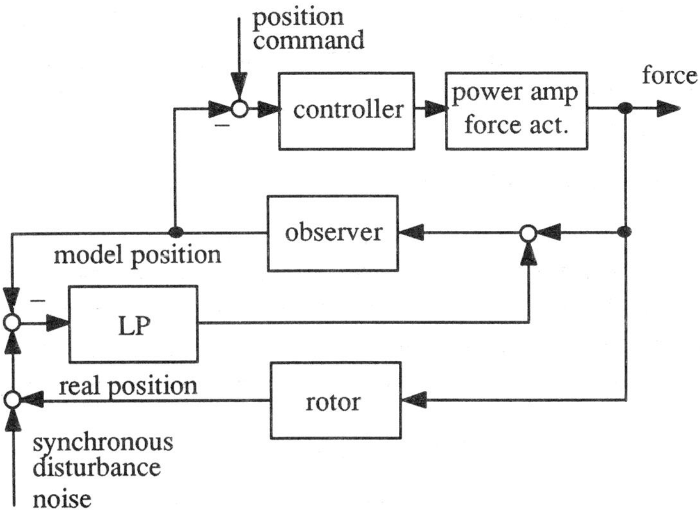

Fig. 4 Suspension loop with observer

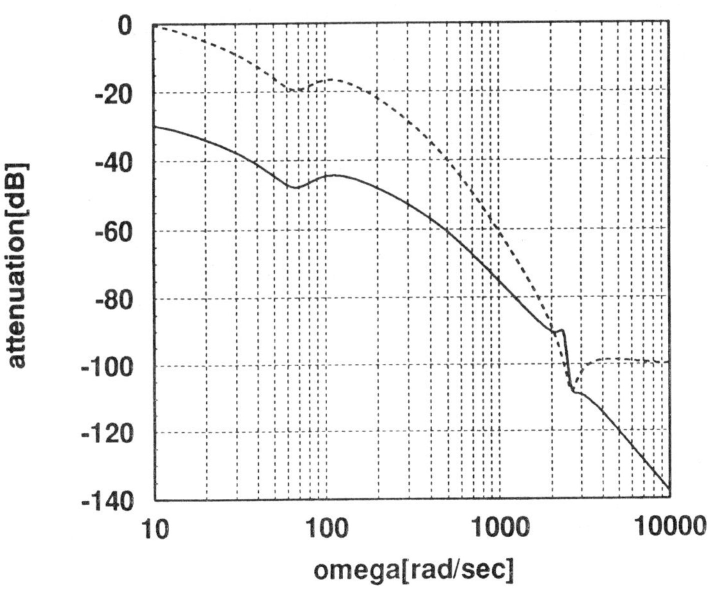

Fig. 5 Noise attenuation of the model following control

At a speed of 6000 rpm for instance the synchronous noise is predicted to be reduced by 65 dB and the higher harmonics up to 108 dB (5.').

The model following control has been chosen as the superior AVS principle, because it offers very good suppression results and it keeps the active stiffness, i.e. the wheel is usable as tilt actuator in the attitude control loop of spacecraft.

The obtained suppression values are considered as theoretical values, which will presumably not be reached in the real implementation, but can be taken as indicator, that an improvement of about 40 dB at least should be achievable practically at the nominal speed.

IMPLEMENTATION OF THE MFC

Two problems remain with this principle. Firstly there is the poor passive stiffness, i.e. the stiffness against externally impressed stator movements. If the satellite fires a thruster for example, the stator moves towards the rotor within fractions of a second and the suspension control cannot react fast enough. This problem can be solved with a second, faster feedback for the observer. This is depicted in Fig. 6.

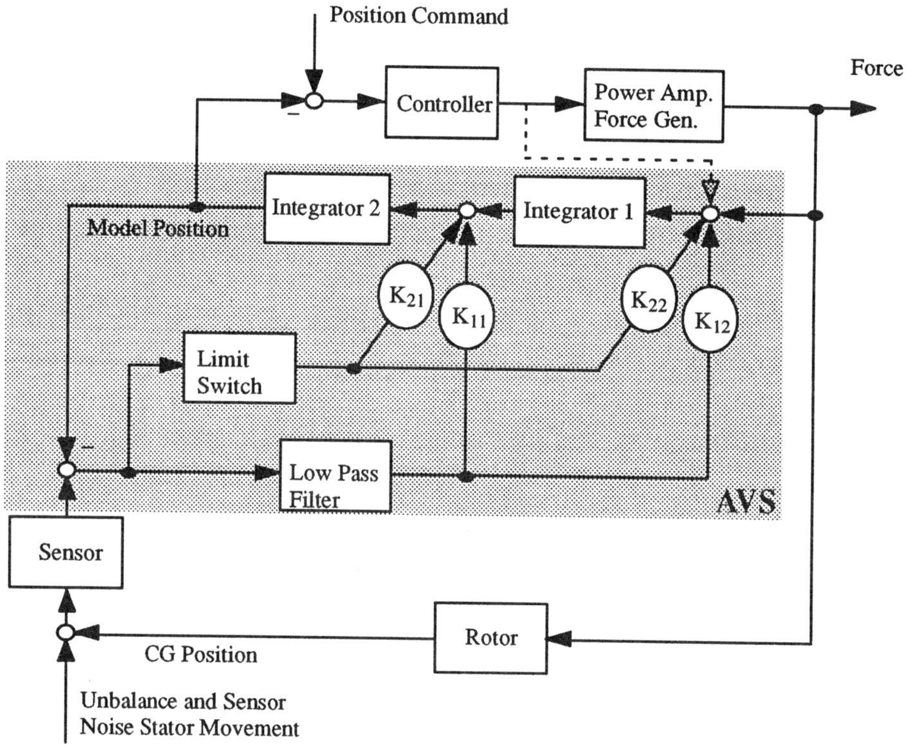

Fig. 6 MFC with touch down protection

The difference of the real position and the observer position is fed through a low pass filter to achieve the long term compensation of drift effects. To control the faster externally impressed stator movements, this difference is additionally fed to a limit switch, which for example detects if the difference exceeds half of the gap. If so, it uses a second, stronger feedback to force the observer position immediately to the real position and, with this, the suspension loop can react and prevent a touch down of the rotor. During the activity of the limit switch, the low noise quality is of course reduced, but there is no low noise condition during an external disturbance anyway.

The second problem is, that this method is not directly applicable in the gyro-coupled tilt loops, because then the model would not represent the real plant. The gyro-couplings must be either introduced into the observer too, or a decoupling network[2] must be used to distribute the controller outputs over both torque inputs. The second possibility is preferred, because it additionally improves the dynamic behaviour of the tilt loops. Then each controller sees the same plant as if the wheel would stand still and the observer remains the same as before. The overall tilt control system with MFC is shown in the block diagram Fig. 7. Additionally a nutational damping cross coupling is foreseen, because the damping of the nutational mode can become poor due to the vibration suppression.

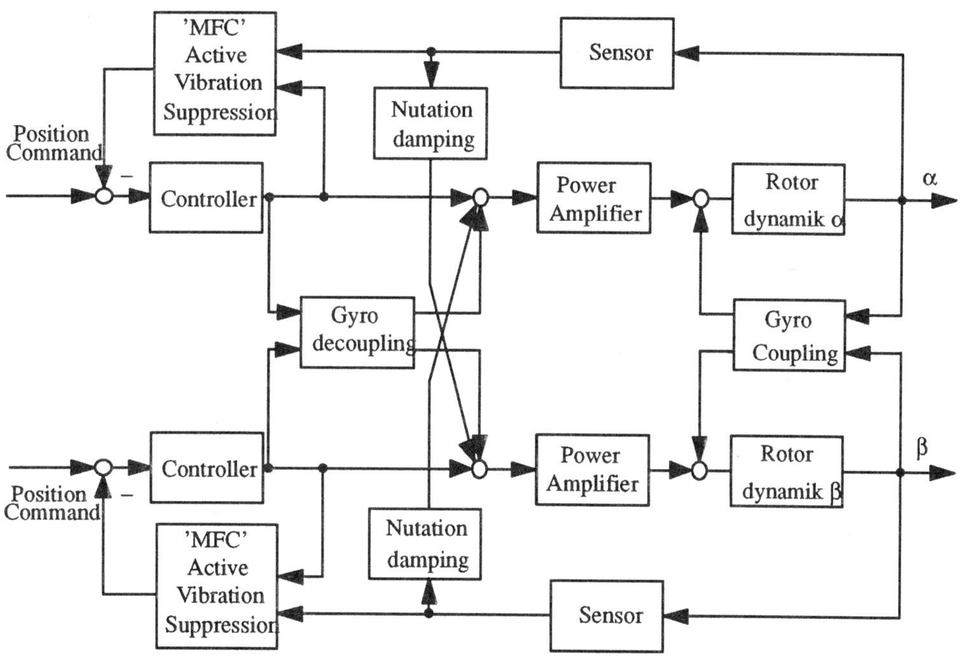

Fig. 7 MFC in the tilt loops

NOISE SUPPRESSION RESULTS

Some results achieved with the MFC in the tilt axes of the real wheel are shown in the next figures. Fig. 8 shows the force Fx in radial direction and the torque Tx around the x-axis that are induced into the wheels mounting base, for example the satellite, at a speed of 5400 rpm.

The forces and torques were measured with a six d.o.f. dynamometer. First, on the left hand side of the figure, the wheel is controlled by the conventional PD law of magnetic bearings. On the right hand side, the wheel is controlled by the MFC low noise control system. It can be seen, that the suppression works fairly well.

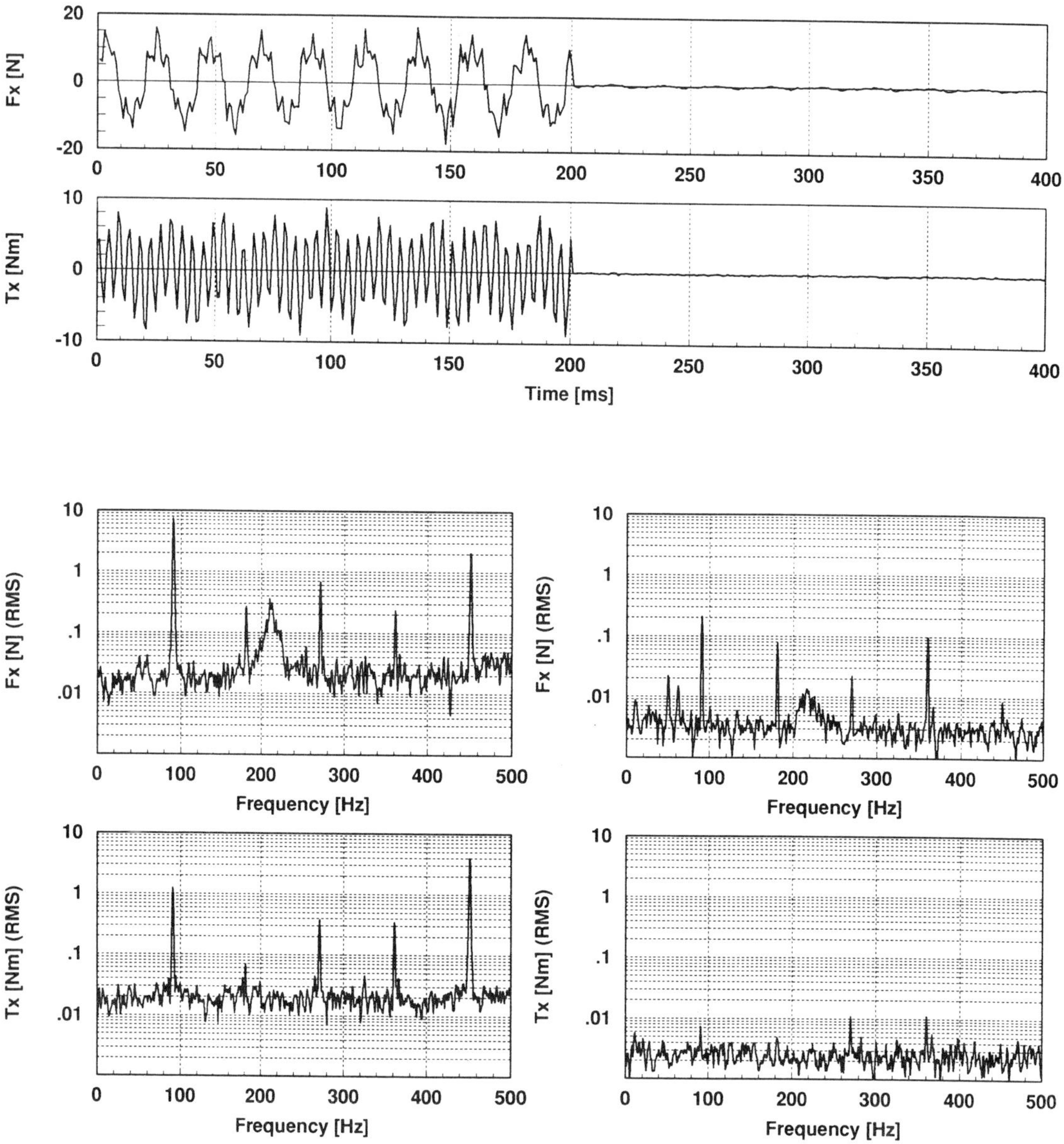

Fig. 8 Wheel Force and Torque at 5400 rpm with conventional (left) and MFC (right) controllers

Fig. 9 gives some insight in the dynamic behaviour of the system by depicting the responses of the wheel and the observer on a small step applied at the tilt angle reference input. The difference in the superimposed noise is evident and it is obvious that the observer output causes much less disturbance torque in the loop than the real position signal. Furthermore it is demonstrated that the responses are globally of the same time behaviour and thus the vibration suppression does not affect the active stiffness.

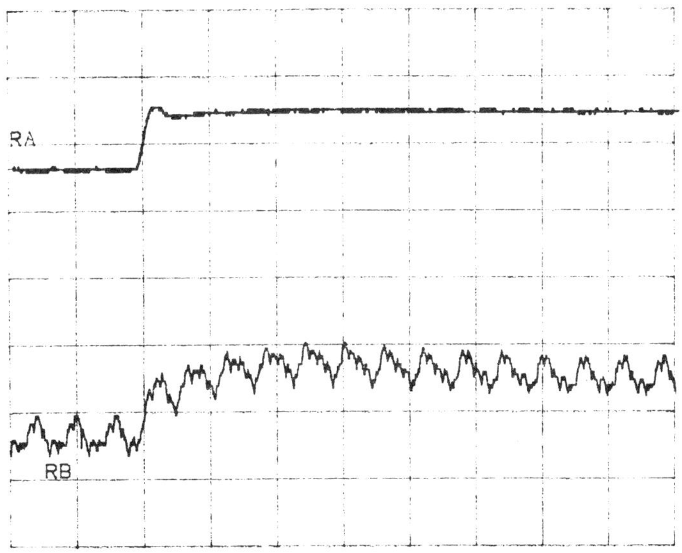

Fig. 9 Small signal step response of the tilt loop with MFC at 2000 rpm RA: observer , RB: sensor.

ACTIVE DAMPING FORCES

The wheel can be moved freely within the mechanical gap of the emergency bearings, because all translational axes are actively controlled. Thus forces can be generated by accelerating and decelerating the rotor. These forces can be used to damp oscillations of antennas, solar arrays or large structures, i.e the wheel with it's translational axes can perform on the side the task of an actuator within an active damping loop.

Fig. 10 shows the generation of sinusoidal forces in the axial and radial direction of the wheel.

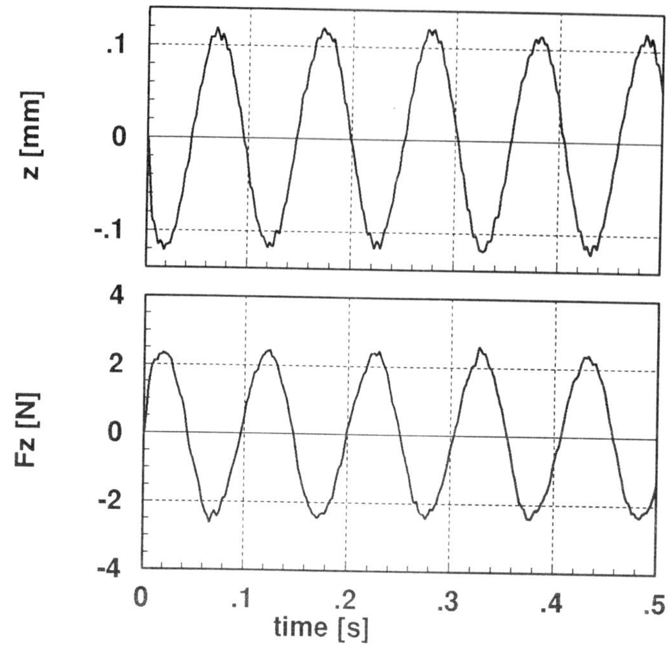

Fig. 10 Active Axial Force Generation

The maximum forces that can be generated are functions of the frequency and the gap. Fig. 11 shows the possible forces obtainable with the MW–X:

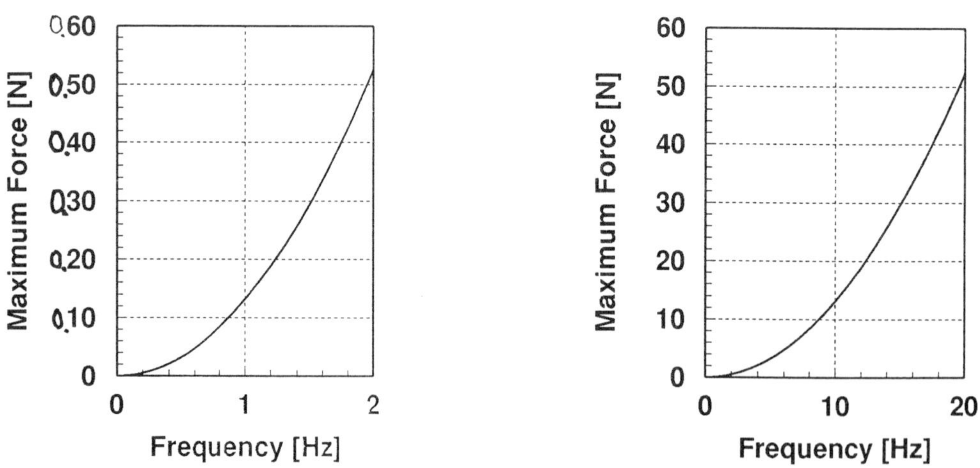

Fig. 11 Maximum Possible Active Translational Force

SUMMARY

The Teldix MW–X, a five d.o.f. magnetic bearing momentum wheel, has been described. The special features of the wheel, the vernier gimballing and the low noise control, have been explained and demonstrated by practical results.

It has been shown that the wheel is very well suited to be used as actuator in special space missions where low vibrational noise, a cross momentum storage capability or a fine pointing three–axis attitude–control with only one wheel would be advantageous. Even the active damping of structural oscillation are imaginable through the utilization of the active translational bearing forces.

An AOCS, build around this wheel, can be considered a real and interesting alternative since it offers and combines different features not obtainable with ball bearing wheels. Thus, for certain missions, a trade off analysis on system level may result in an overall advantage for the slightly bigger and heavier magnetic bearing wheel.

The absolute noise values shown in some figures are not meaningful, since the wheel is an laboratory bread board model. It was not balanced and no steps have been taken to reduce the hardware cause noise in the 1–g field. The only goal was to demonstrate the relative attenuation of the produced noise.

Later, in space, it will be a high quality wheel, well balanced, with high circular uniformity of the magnetic field and no steady state current to bear the 1–g load will flow. Then, the demonstrated noise suppression will further attenuate the residual noise of this wheel resulting in real low noise behavior.

ACKNOWLEDGEMENTS

The development of the magnetic bearing momentum wheel MW–X is jointly funded by Teldix and the European Space Agency (ESA).

ABBREVIATIONS

AOCS	Attitude and Orbit Control System
AVS	Active Vibration Suppression
cg	Center of Gravity
d.o.f.	Degree of Freedom
MFC	Model Following Control
MW–X	Magnetic (Bearing Momentum) Wheel – X Nms
PD	Proportional–Differential (Control Law)
S/C	Spacecraft

REFERENCES

1. Ball Bearing Versus Magnetic Bearing Reaction and Momentum Wheels as Momentum Actuators
   W. Auer
   AIAA 'Global Technology 2000', Baltimore 1980.

2. A 5 Degree of Freedom Electrodynamic–Bearing Wheel for 3–Axis Spacecraft Attitude Control Applications
   U. Bichler, T.Eckardt
   Proceedings of the First International Symposium on Magnetic Bearings, ETH Zurich, June 6–8, 1988.

3. Magnetic Bearing With Rotating Force Control
   H.M.Chen, M.S.Darlow
   Transactions of the ASME, VOL. 110, Jan. 1988

4. Synchronous Response Modelling and Control of an Annular Momentum Control Device
   R.Hockney, B.G.Johnson, K.Misovec
   NASA Contractor Report 4166

5. A Design of Robust Servo Controllers for an Unbalance Vibration in Magnetic Bearing Systems
   F. Matsumara, M. Fujita, C. Oida
   Proceedings of the First International Symposium on Magnetic Bearings, ETH Zurich, June 6–8, 1988.

6. Vorrichtung zur Kompensation synchroner Störeinflüsse bei einem magnetisch gelagerten Läufer.
   H.Habermann, M.Brunet, P.Joly
   Patent 2658668

AAS 91-043

FLIGHT TELEROBOTIC SERVICER PROTOTYPE SIMULATOR

Rob Schein, Linda Krauze, Craig Hartley, Alan Dickenson, Tom LaVecchia and Bob Working*

The Flight Telerobotic Servicer (FTS) will be a space-based, teleoperated robot designed to perform on-orbit assembly and servicing tasks. Its purpose is to act as a surrogate for a space-suited astronaut in order to reduce the amount of astronaut extra-vehicular activity required to build and maintain Space Station Freedom. It is scheduled to fly its first demonstration test flight (DTF-1) on the STS Orbiter by early 1993. In order to obtain timely astronaut and designer inputs into the system design, the project plans to employ a series of simulation tools providing successively greater fidelity. The first of the simulators, the FTS Prototype Simulator, has already been delivered to the NASA's Johnson Spaceflight Center. This simulator uniquely combines a real hand controller, a rapid prototype of the user interface, and a graphics workstation to produce what we call a "rapid simulator", a tabletop simulator based on rapid prototyping tools but having additional powerful and useful capabilities that allow users to conduct meaningful system analyses. The simulator software consists of kinematic modeling, manipulator control algorithms, graphics, and communications and executive software interacting with a highly realistic graphical representation of the STS Orbiter hardwired FTS panels and computer screens (operator interface). All the controls and displays on the control panel and simulated computer screen interact with the simulator operation just as real hardware would. For example, when the user actuates the power toggle switch, the computer screen comes on and appropriate status lights are illuminated. At the same time, on the graphics workstation, the operator sees graphical representations of the FTS system through various camera views available to the operator in the real system. When the operator moves the hand controller, the robot's arm moves and joint angles are displayed on the user interface prototype. This paper describes the system software and hardware and lessons learned in detail. The system will be demonstrated at the conference.

INTRODUCTION

The Flight Telerobotic Servicer (FTS) system is comprised of a series of three development and operational phases.The first phase will result in a single manipulator

* Martin Marietta Civil Space Systems Company, P.O. Box 179, Denver, Colorado 80201.

telerobot launched in the NSTS space shuttle cargo bay for a development test flight (DTF-1) designed to exercise and validate the telerobot functions, operations, and control systems. The FTS will be rigidly mounted to a support pallet along with a task panel which contains a variety of task elements. Fig. 1 depicts the configuration. The second phase will include a dual arm telerobot with a positioning and stabilization manipulator which will act as a "leg". This phase will culminate with a second space shuttle flight demonstrating proof of concept for Space Station tasks and operations. The third phase will deliver the fully operational FTS system for launch and subsequent residence on the Space Station. This paper will be geared toward the DTF-1 system configuration.

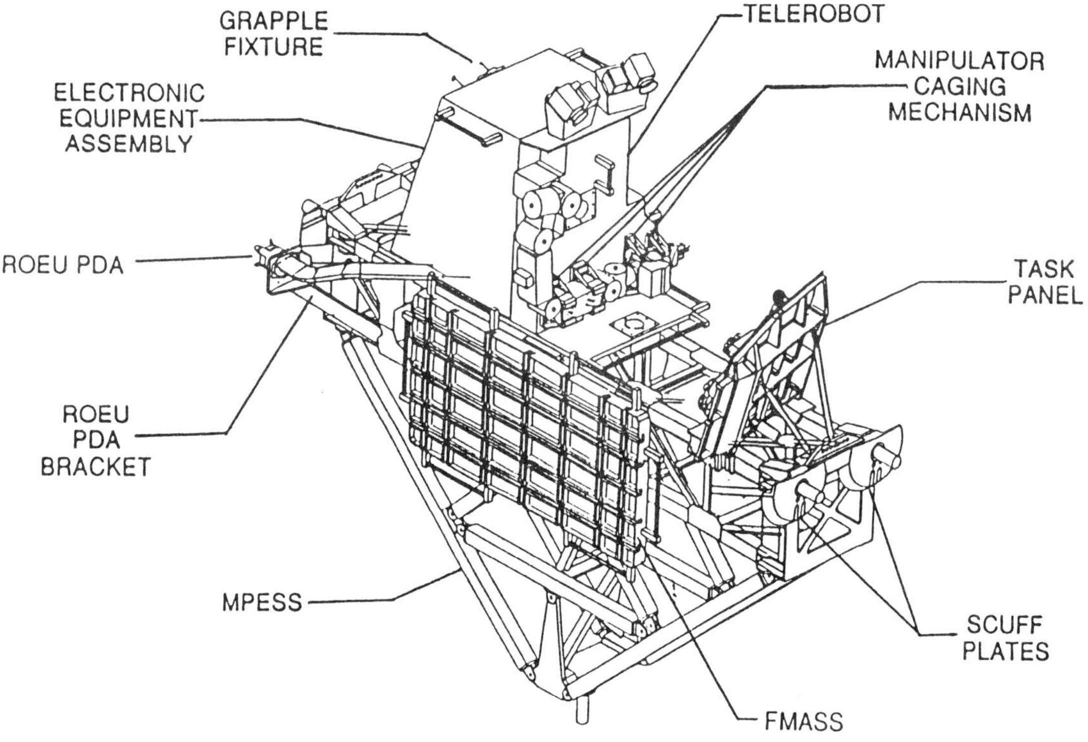

Fig. 1 - FTS Payload Bay Element

The manipulator is configured with seven degrees-of-freedom (DOFs) which include in the following order from the manipulator base to the end effector: shoulder roll, shoulder yaw, shoulder pitch, elbow pitch, wrist pitch, wrist yaw, wrist roll, as shown in Fig. 2. Six of the seven DOFs will be actively controlled to provide positioning of the end effector in each of six Cartesian positions and orientations. The seventh DOF, shoulder roll, will be used to provide the optimum kinematic solution for each of the tasks.

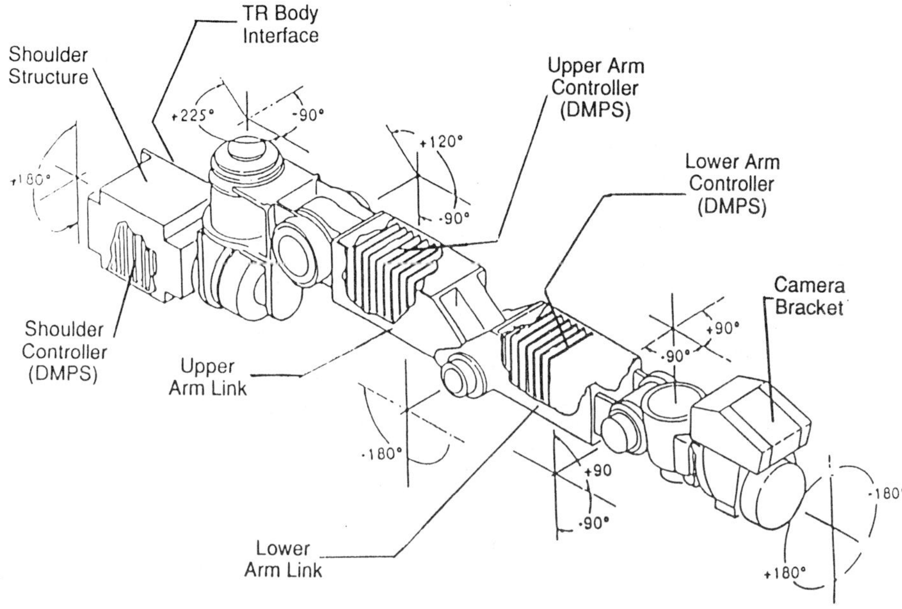

Fig. 2 - FTS Manipulator

FTS can be controlled by a variety of means. The obvious method of control is the teleoperation mode whereby an operator controls the movement of the manipulator via hand controller commands. The robot will also have autonomous control capabilities, however for DTF-1 these will be limited to tasks such as destow and stow, and some autonomous moves required for performance verification. A joint-by-joint backup control mode will be implemented for contingency operations.

The DTF-1 task elements provide for the capabilities demonstration of Space Station assembly tasks using simulated hardware which in some cases is scaled down for demonstration purposes. There are six task elements included, 5 of which are on the task panel, while one task requires two similar elements, one located on the telerobot body and the other is on the pallet in front of the body. The elements include a Space Station truss element section, a fluid connector section, a peg-in-the-hole task element, a contour board element, a removable mass, and two video positioning target assemblies. Each task will provide verification of performance requirements. These tasks can be seen on the task panel in Fig. 3.

The FTS prototype simulator was developed as the first simulation tool which combined a flight like hand controller with a graphics representation and controls and displays of the FTS system configuration. It provided the first look at the integrated kinematic operations of FTS. From the first day that the simulator was operational, valuable information pertaining to the design and operation of the entire system was produced. This paper describes the simulation from the hardware and software perspectives as well as describe the operational capabilities, and concludes with some results and lessons that have been learned about the FTS system.

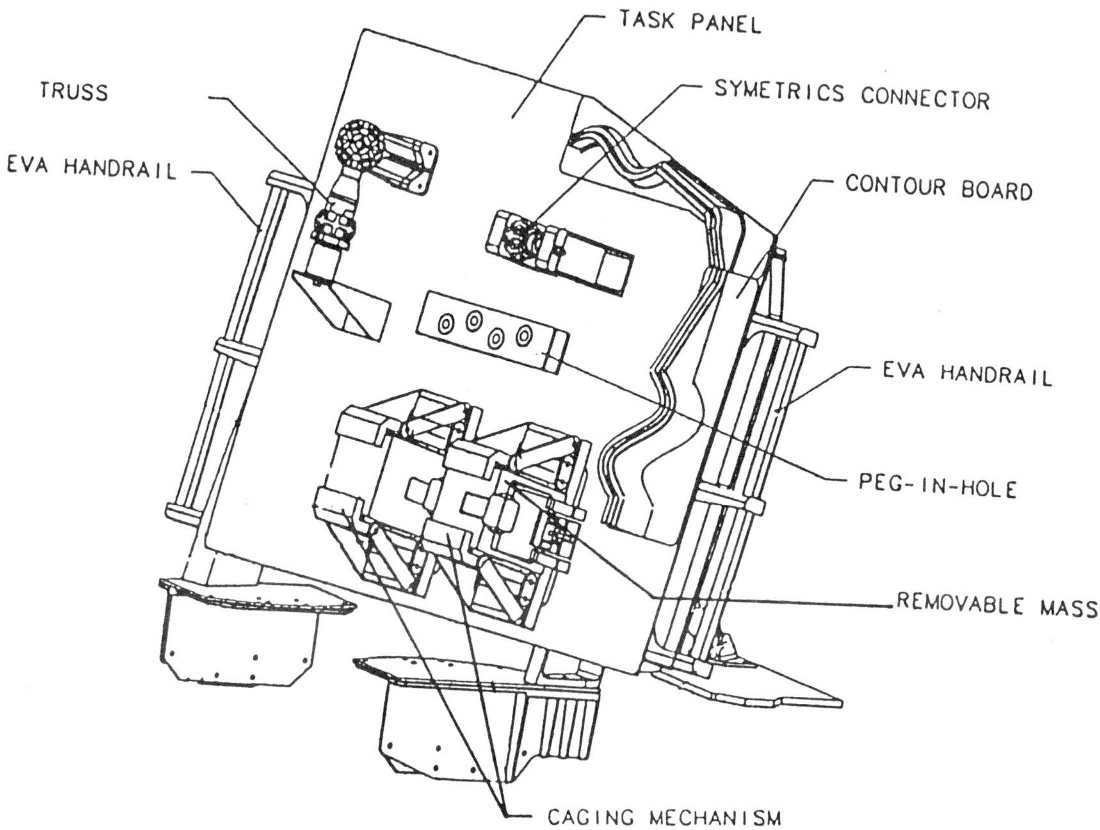

Fig. 3 - Task Panel Assembly

SIMULATION DESCRIPTION

Hardware Description

The prototype simulator hardware was configured to emulate as closely as possible the workstation to be used in the first FTS demonstration test flight, DTF-1, to be flown on the Space Shuttle Orbiter by early 1993. It included the following:

a) Kraft six DOF hand controller and associated electronics box,
b) Silicon Graphics Iris 4D/220 Unix workstation
c) Omega data communications conversion box
d) Apple Macintosh II ci 8/80 with a 21 in. Viking Moniterm color monitor.

The Kraft hand controller (Fig. 4) is the primary input for the user to drive the simulated FTS manipulator. Motion of the robot is a function of the position of the grip of the hand controller. Motion of individual joints of the hand controller may also be used to drive the joints of the FTS manipulator separately. Actuators at each of the hand controller joints allow forces to be fed back to the operator. This could be used to simulate contact forces or as a means of centering the hand controller in rate mode. Various control modes available in the prototype system are described in the section entitled Flight System Software Models. The grip of the hand controller also has a rocker switch and two pushbutton switches. The rocker is used to send signals to open and close the gripper of the simulated manipulator. One pushbutton switch is used to enable the functionality of the rocker and the second, momentary pushbutton switch is used as a "deadman" switch and enables the use of the hand controller. Motion of the hand controller is not translated into robot motion unless this switch is depressed.

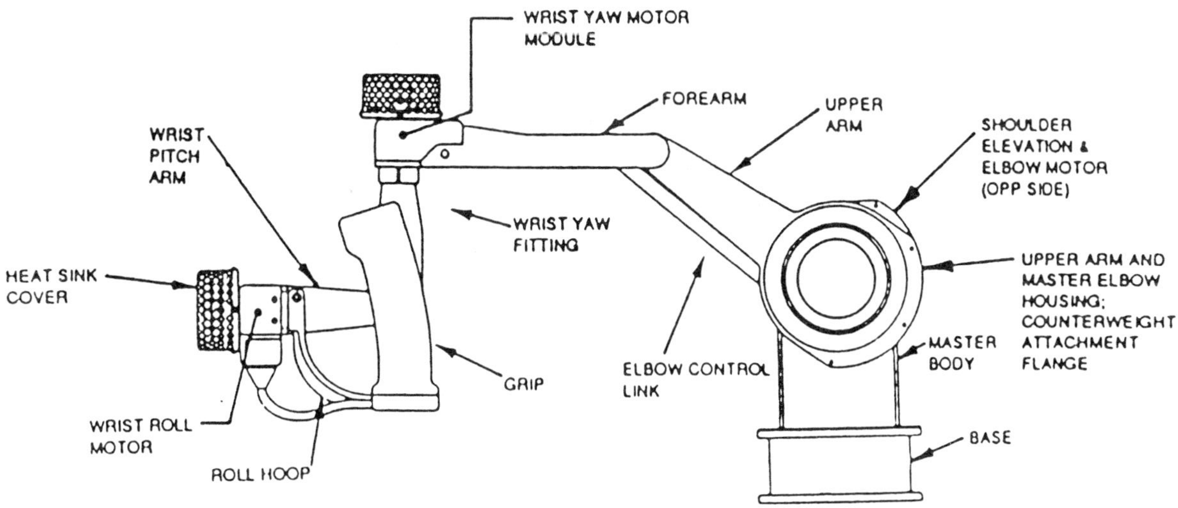

Fig. 4 - Hand Controller Configuration

Data read and written by the Kraft hand controller is formatted to conform to RS-485 standards. To communicate to a serial port on a computer the information must be translated to RS-232 format. This is the function of the Omega converter box.

The Iris 4D/220 is the simulator control computer. The code which communicates with the hand controller and the Macintosh resides here. This machine performs all the analytic computations and hosts the graphics display code and DTF-1 workcell models.

A 21 in. color CRT Macintosh display simulates the hardware which would be found on the space station workstation control and display panel. This display is fully functional, and is another path available to the operator to control the simulated robot.

The prototype simulator configuration is shown in Fig. 5, and a functional block diagram is shown in Fig. 6.

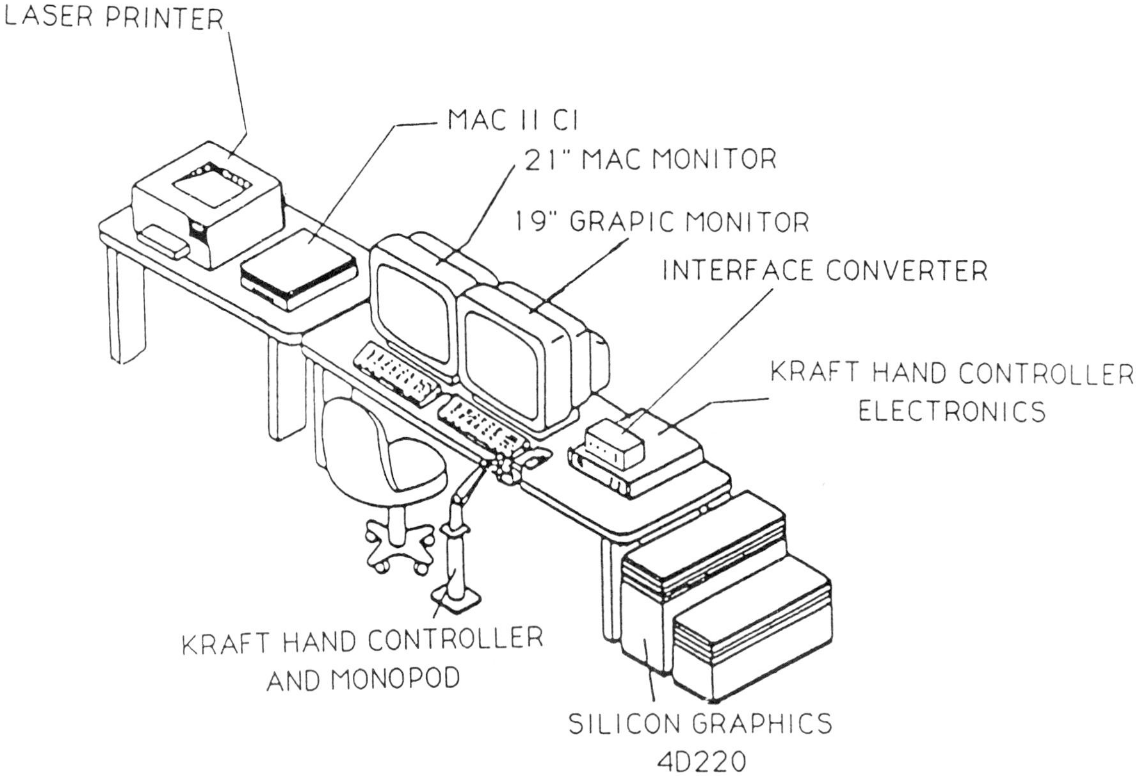

Fig. 5 - DTF-1 Prototype Simulator Configuration

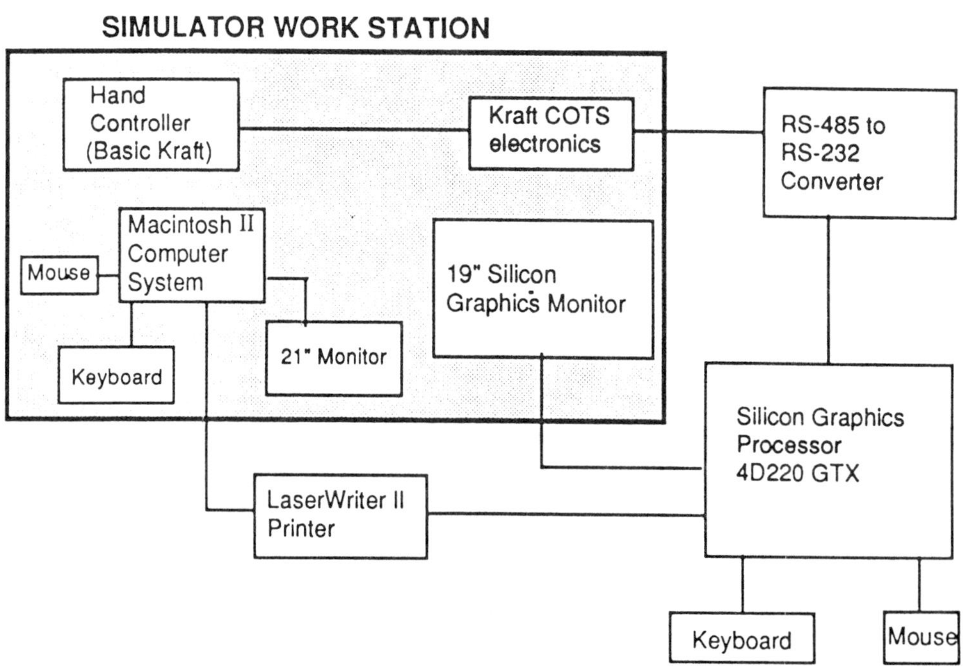

Fig. 6 - DTF-1 Prototype Simulator Functional Block Diagram

Flight System Software Models

The FTS flight system models are run by a pseudo-realtime executive resident on the Iris that was originally developed to run high fidelity system simulations in a non-realtime environment. It offers a high degree of flexibility to taylor fit the desired simulation to assess particular simulation requirements. Input and output can be controlled strictly via database management, providing a user friendly simulation environment. Simulation models may also be database selectable, offering the user the opportunity to select models of varying fidelity, and running them at user defined rates (i.e. analog models or dynamics can be run at fast frame rates, while digital control algorithms will run at their associated rates). As a new model, or a model of increased fidelity is developed, a simple database entry will notify the executive of its presence. In adapting the executive to run realtime, these features were maintained. A non-realtime high fidelity nonlinear simulation (System Performance Analysis Simulation, or SPAS) of the FTS system which includes dynamics, actuator and control loop models, has also been developed and is used in conjunction with the prototype simulator. This feature will be discussed later.

The FTS control system algorithms were developed prior to the flight algorithms and provide the functionality that the flight algorithms will have. These algorithms include a variety of manipulator teleoperated and autonomous control modes described below.

Position/Position Control

This control mode generates a manipulator Cartesian position command based directly on the position of the hand controller. The hand controller acts as a master while the manipulator follows the positional motion of the hand controller as the slave. The motion may be scaled so that large motions of the hand controller may result in small manipulator motions, as would be the case while performing dexterous tasks, or visa-versa, as would be when traversing large distances. This scaling may be performed on a per axis basis.

Hand Controller Indexing

Based upon the scale factor selected, the hand controller workspace may be far lower than that of the manipulator. Indexing is the method by which the hand controller workspace can be mapped into different areas in the manipulator workspace. The control algorithms will not revise position commands to the manipulator until the hand controller enable switch, located on the front of the hand grip under the index and middle fingers, has been activated. When this occurs, the present position of the manipulator is set equal to the current hand controller position. This allows the operator to extend the hand controller to the proximity of its limit, release the hand controller enable switch, reposition the hand controller to a suitable workspace location without commanding manipulator motion, and reactivate the enable switch, providing a more usable envelope. This function is required when moving the manipulator through motions greater than the scaled hand controller commands can generate, or simply to vary the operator's physical stance while performing a task.

Position/Derived Rate Control

This control mode generates a Cartesian rate command which is then integrated to form a Cartesian position command for the manipulator and it is proportional to the displacement of the hand controller. Upon activation of the hand controller enable switch, a software detente and deadband are formed at the current position of the hand controller. As the hand controller is deflected, a Cartesian rate command is generated in the same direction. The rate command can be adjusted simply by varying the deflection in any given direction. When the operator returns the hand controller to the deadband, or when the hand controller enable switch is released, the rate commands are zeroed and the manipulator stops. Like position/position control, the manipulator rate verses hand controller deflection curve can be scaled as selected by the operator.

Hand Controller Force Centering

This feature is used in the position/derived rate mode only. It is used to provide positive feedback of the relative rate requested. Forces are fed back through the hand controller joint actuators in the direction opposite to the requested rate (in the opposite direction of hand controller displacement). This force is also proportional to the deflection of the hand controller and may be implemented as a linear function, or a "dog leg" function where the forces will provide a different stiffness depending on displacement. This could provide a soft spring force for small deflections, and a stiff spring force when reaching the selected travel limits. This option as well as the associated force gains is operator selectable via the operator interface.

Gripper Operations

The grippers may be opened and closed by first activating the gripper enable switch located on the front of the hand controller hand grip, then by deflecting the rocker switch located on the back of the hand grip near the top. Deflecting the switch to the right will cause the gripper to close, while deflection to the left will result in opening the gripper.

Hand Controller Reference Frames

The coordinate frames which the hand controller can represent can be varied by the operator. Ideally, the reference frame selected should be one affixed to the view, either camera, or out the window, that the operator is using. This assures that any commands input by the operator produce motions which are obvious. While the operator is using the wrist camera, any motion input to the hand controller should create the same viewed motion. This is true for a head camera as well. The prototype simulator has two coordinate reference frames modelled: 1) End effector reference frame which is located at the center of the gripper fingers, and 2) Base reference frame which is located at the intersection of the shoulder roll axis and the shoulder yaw axis.

Inverse Kinematics

The inverse kinematics algorithm is used when the manipulator is under Cartesian control to compute a set of joint angles required to produce the commanded manipulator Cartesian position and orientation determined by one of the previously described Cartesian control modes. The inverse kinematics for a general 6-DOF manipulator can be represented by six nonlinear, transcendental equations with six unknowns. Unlike linear equations, there are no general algorithms which can be used to solve a set of nonlinear equations. In general, numerical methods are required to solve these equations. However, manipulator configurations exist for which analytic closed form solutions are possible. A sufficient (although not necessary) condition for the existence of a closed form inverse kinematic solution is for three consecutive joints axes to intersect at a point for all arm configurations. This condition exists for the FTS manipulator design (shoulder, elbow, and wrist pitch joints axes intersect at infinity).

Use of a closed form solution for the six actively controlled joints provides the most time efficient computation of the inverse kinematics within the control loop. The order in which the solution proceeds is: shoulder yaw, the sum of the three pitch angles, wrist yaw, elbow pitch, shoulder pitch, wrist pitch, and wrist roll. Imbedded in this module are a variety of system safety limits and checks. The first check determines whether the manipulator solution is outside of the workspace. This is called an envelope boundary check. Next, the joint rates are checked and limited if necessary. If any joint or joints exceed their associated joint rate limits, the joint with the maximum ratio of desired rate-to-rate limit will be used to scale the other joint rates thus maintaining coordinated Cartesian motion at the end effector. Finally, the joint desired joint angles are compared to their joint angle limits and if any joint exceeds its limit, the commanded joint angles are set to the last acceptable values calculated. This in effect stops the manipulator until the commanded motion drives the affected joint away from its limit.

The prototype simulator does use the flight algorithms for this module.

Autonomous Movements

The prototype simulator contains two types of autonomous movement or trajectory planning techniques. The first technique involves a sophisticated Cartesian trajectory path planner whose inputs are the initial (current) manipulator position, and the desired position. The planner then calculates maneuver times based on available torques and system mass properties, then generates the Cartesian commands required to perform the motion. These commands are then fed into the inverse kinematics routine which calculates the appropriate joint angle commands.

The second autonomous trajectory planner is much simpler than the first. This planner is user for destowing and stowing the manipulator. This movement must be performed in joint control space. Each joint is commanded separately in this mode. This is required because in order to enter and exit the caging mechanisms, the manipulator must be set to a specific pose, then maneuvered using the shoulder yaw joint only.

Hardwire Control

This function provides a backup method of controlling the manipulator given a control computer or data bus failure which prohibits nominal operations. The FTS system has a second set of motor windings which can be commanded via a control panel in the aft flight deck which is wired directly to these windings. The simulator provides this control mode via the Macintosh graphics interface. The operator can select each axis to be commanded individually and drive the axis in either the positive or negative directions.

Rapid Prototype of the User Interface

The rapid prototype of the DTF-1 hardwired control panel and computer screen consists of a single 1.1 megabyte Macintosh application program written using Silicon Beach Supercard 1.0 software. This program creates two primary windows on the large color monitor. Fig. 7 depicts the Macintosh display screen. One window consists of a 2/3 scale graphical representation of the DTF-1 hardwired control panel. This panel contains a number of 3-D graphical representations of the toggle switches, lighted pushbuttons, indicator lights, rotary selector switches and emergency stop pushbuttons. All switches are fully functional through Macintosh mouse interaction. For example, a click on the power toggle switch flicks the switch to "on", makes a switch actuation sound (through the Macintosh/ Supercard sound capability), and lights all appropriate indicator lights and computer screens. At the same time, the software sends notification of a change in the power switch toggle state to the Iris via a 38.4 kilobaud RS-232 link. A process running on the Iris then writes the switch change to an area of shared memory where it is acted upon, as appropriate, by the other simulation software processes on the Iris. The momentary toggle switches on the rapid prototype of the control panel work by switching an "on" command to the Iris while the mouse button is held down over the particular switch, and an "off" command on release of the mouse button. The graphical representations of the switches and their shadows change when they are activated to make them appear to be depressed or moved.

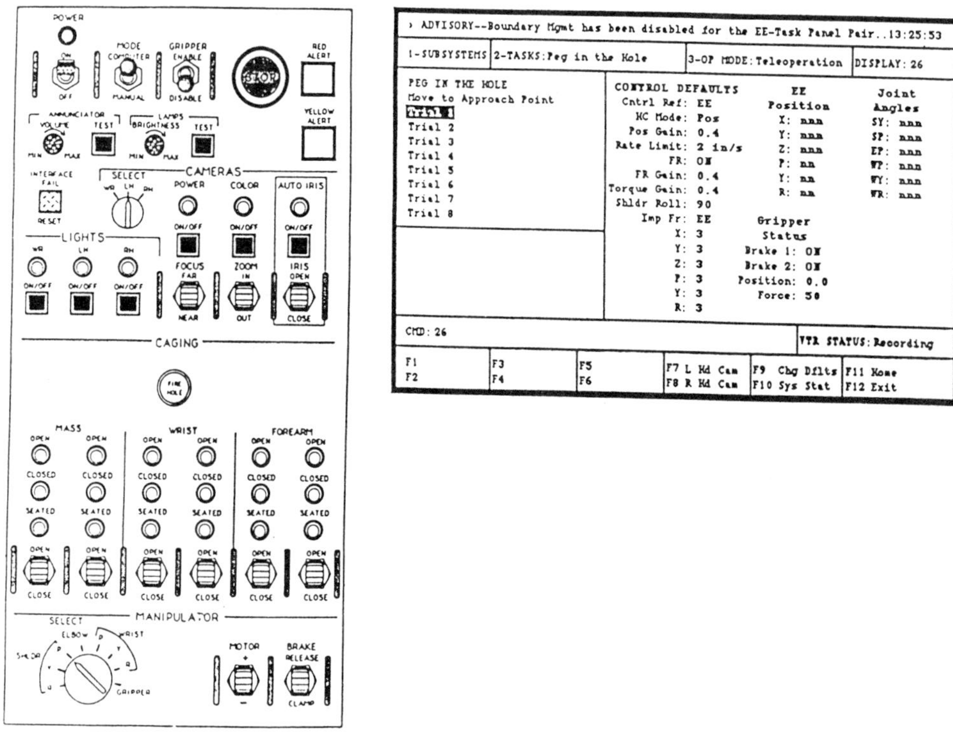

Fig. 7 - Prototype Simulator Controls and Display Panel

The second primary window on the Macintosh rapid prototype is a full-scale representation of a Grid computer screen that will be used to operate the FTS when it is in its (normal) computer control mode. This screen uses the same fonts, layout, and function keys (on the Macintosh keyboard) planned for the actual flight hardware and software. Among the most important displays on this screen are the realtime representations of FTS joint angles as the hand controller is actuated. These displays are sent to the Macintosh by the Iris communications process via the same RS-232 link described above. The Macintosh and Iris "handshake" so that the Macintosh asks for a second data set as soon as it finishes passing and displaying the first data set. The Macintosh effectively displays a complete set of joint angles for the FTS robot at about 5 Hertz. The operator can also use the simulated computer display to change control modes, set shoulder roll angles, and other typical functions.

In addition to the two windows containing the user interface rapid prototype, the Macintosh software also provides a variety of other functions that enhance the value of the prototype simulator. These include a comprehensive "help" capability consisting of a text window that pops up with appropriate information each time a user clicks on any screen object while holding down the "option" key. Another valuable feature is a detailed activity log that records the activity and time of each user interaction with the rapid prototype. The Macintosh software also provides a guided tour of all of the features of the prototype simulator, including practice using the hand controller and the Iris simulated camera views.

The combination of the Macintosh II ci (without cache), large color monitor, and Supercard software has proved to be an extremely powerful, flexible, and low-cost rapid prototyping environment. We were able to easily import and give function to existing Apple HyperCard and Claris MacDraw files. We were even able to completely revise the control panel layout in only a few hours after a major change in the project design.

Workcell (Graphics) Models

The graphical models displayed on the screen of the Iris computer accurately depict the DTF-1 workcell. This includes representations of the telerobot body, head and wrist cameras, manipulator joints and links, and end effector and gripper as well as the structure necessary for support in the shuttle payload bay. The DTF-1 task panel and elements are also modeled. Fig. 8 shows these workcell elements. The software to generate the graphical images was developed in-house. The manipulator forward kinematics are a part of this code as are transformations to various workcell camera locations. This allows the operator to simulate the camera views that will actually be available in the real system.

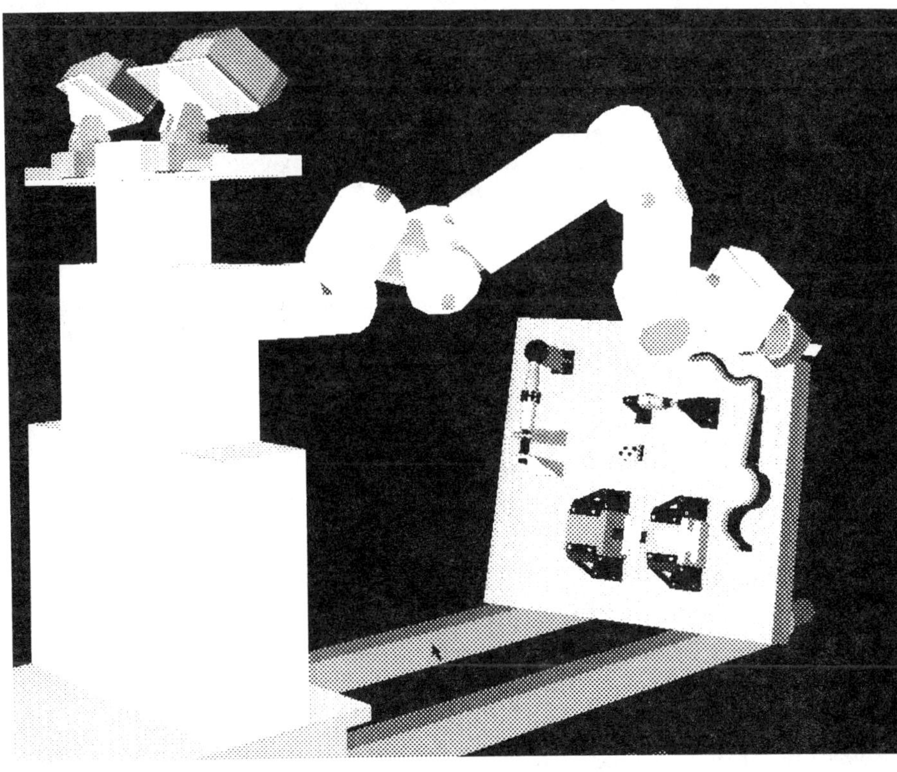

Fig. 8 - Graphics Workcell Elements

Data Flow and Communications

There are five software processes running simultaneously to provide the full system simulation. The Macintosh display driver resides on that system while the other four processes reside on the Iris computer. These are the graphics modeling and display process, called Grasp, the analysis and control software, called Satsim, the kraft hand controller communications software kraft_comm, and the Macintosh communications software mac_comm. Communication and coordination of these processes is accomplished via a block of shared memory containing system data which each process residing on the Iris has access to.

The process kraft_comm is continually sending requests for data to the hand controller and reading the data packets it sends back. Data sent to the Kraft includes torques to be applied to the hand controller joint actuators. Data from the Kraft contains hand controller joint positions and rocker and switch status. Information going to the Kraft is obtained from the shared memory block and data read from it is placed in this same block of shared memory.

The control software Satsim contains an executive to determine the around the loop timing and data update frequencies. It reads the hand controller data and performs the kinematic and control calculations needed for the mode of operation in use. It then places manipulator joint angles and gripper data into shared memory.

The graphics display process, Grasp reads shared memory and uses the information to place the manipulator, gripper and other workcell components in their correct locations at the next frame update.

Mac_comm is the pipeline through which the control and display panel emulated on the Macintosh is updated to reflect current system status. It also takes operator inputs to the display panel and places them in shared memory so subsequent control modes or parameters reflect these inputs. A schematic of this data flow is shown in Fig. 9.

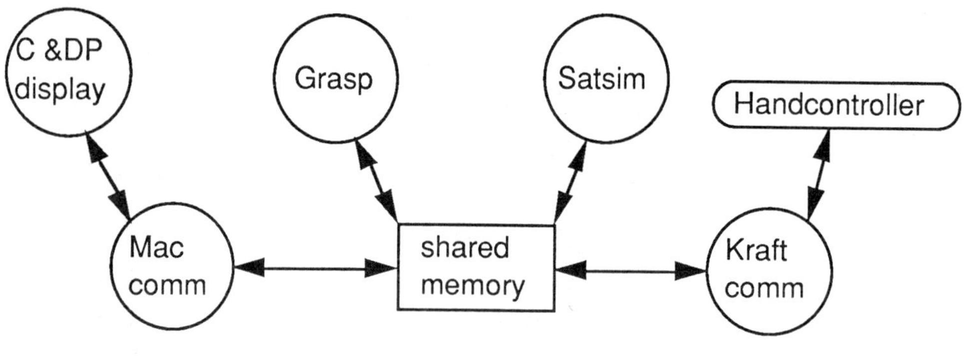

Fig. 9 - Prototype Simulator Data Flow

This shared memory method of communication allows processes running at different update frequencies to function properly and not be tied to the update rate of the slowest process. This would assure accurate system performance by letting the the control software Satsim run at the correct frequency on a machine with a marginal graphics display capability and slow refresh rates.

APPLICATIONS

There is much discussion as to the application of a graphics simulation tool such as the prototype simulator. The tool was originally conceived to provide crew familiarization with the FTS system in general, as well as to obtain crew feedback on operations, screen messages, menus, and controls and displays layout. Prior to development, it was realized that many system performance issues could be addressed using the simulator. This led to the development of a simulator which contained a much higher level of fidelity than had been originally envisioned. One must take care in determining which issues can be realistically addressed and apply the results accordingly. Certainly it is obvious that issues which require things like contact dynamics, or high fidelity lighting cannot be assessed with this type of a system. There are a number of items, however, that either have been addressed, or will soon be.

The prototype simulator was delivered in June, 1990 to the Johnson Space Center for crew evaluations. A copy of the simulator was maintained in Denver at the Martin Marietta Space Operations Simulation facility and has been upgraded as required to address performance simulations requirements.

Because the prototype simulator was the very first time anyone had a realtime, human-in-the-loop kinematic simulation of the FTS system including actual geometries, flight-like hand controller, and controls and displays, its uses were apparent immediately upon becoming operational. Prior to this time, only static kinematic analysis had been performed. The first issue that surfaced was the fact that joint limits were being reached quite often. An analysis was performed to determine whether or not this was a real concern. All of the tasks were performed and it was concluded that while joint limits obviously could be reached, they could be minimized or eliminated operationally.

More recently, a structural analysis was performed to determine a velocity limiting technique to assure that each actuator would be able to stop as it approached its limit. This resulted in a trapezoidal rate verses angle velocity limit scheme. This was quickly implemented in the simulator and assessed. It was determined that while the limiting scheme was annoying to the operator when joint angular limits were approached, it had no effect on the success in performing the task.

Vision studies were also performed using the prototype simulator. Items which were addressed included viewablity constraints, location of head cameras, camera field-of-view, boresighted vs. a tilted camera looking at the grippers, and use of payload bay and RMS cameras and out the window views to manually stow the manipulator and the removable mass. These issues were all addressed initially using this simulator and then verified using a variety of mockups and actual cameras.

One of the truly nice capabilities of the simulator is its use in conjunction with the high fidelity offline FTS simulation (System Performance Analysis Simulation, or SPAS). Scenarios can be simulated in realtime and commanded joint angles can be stored in a database file. This file can then be read in to SPAS for more detailed performance analysis. This method is used to provide system level rough estimates of power usage, thermal duty cycles, etc.

The simulator will be used from a mission operations standpoint to determine a "first cut" at timeline and procedures development. By including all of the functionality of the operator interfaces, the operator has to perform all of the required inputs to conduct a given operation.

In its original completed state, the operator interface was extensively used to develop control panel layouts and the computer screen contents.

ACKNOWLEDGEMENT

The authors wish to acknowledge the NASA Goddard Space Flight Center, responsible for the overall development of the Flight Telerobotic Servicer project. We would also like to thank the support crew at Martin Marietta's Space Operations Simulation Laboratory for their logistics, equipment, and facilities support.

AAS 91-046

SPACE TELEROBOTIC RESEARCH AND APPLICATIONS AT SPACE SYSTEMS/LORAL

Scott W. Tilley, Michael G. Hollars,
Colin M. Francis and Alfred H. Tadros*

Highly-valued assets in space should be maintained through servicing rather than expended after one use or failure. The benefits of using teleoperated or autonomous robotics for assembly and servicing have been amply demonstrated in nuclear material handling, underwater operations, and some space operations. Trades must be made over autonomy, operational flexibility, and required human interaction to design an effective robotic system. Space Systems/Loral studied spacecraft servicing for NASA and began a research and development (R&D) to analyze space manipulators and their operational control in a laboratory environment. Current research on teleoperator control focuses on utilizing impedance control on a kinematically redundant manipulator. Impedance control provides a unified approach for simultaneously controlling manipulator position and force interaction with the environment. A simple peg-in-hole insertion under teleoperator control was successfully demonstrated without bounce or undesirable force overshoots. Further features are being added to demonstrate force guided assembly tasks under supervisory control in the presence of large time delay.

INTRODUCTION

In space, economic or national security needs dictate that highly-valued assets should be maintained through servicing rather than expended after one use or failure. Grave concerns have been recently expressed over the gross disparity between required manned extra-vehicular (EVA) hours required for Space Station Freedom (SSF) maintenance versus the hours actually available. Robotic EVA, as opposed to or in conjunction with manned EVA, is recognized as a solution. A further issue is whether these robots should be supervised from the SSF, using precious intra-vehicular (IVA) time, or from the ground. The benefits of using teleoperated or autonomous robotics for assembly and servicing have been amply

* Space Systems/Loral, 3825 Fabian Way, M/S G-75, Palo Alto, California 94303. **Space Systems/Loral** was formerly **Ford Aerospace, Space Systems Division**.

demonstrated in nuclear material handling, under-water operations, and some space operations.

NASA is looking beyond the Shuttle Remote Manipulator System (RMS) to the Flight Telerobotic Servicer (FTS) to assist assembly, maintenance, and inspection tasks at the SSF. The NASA Satellite Servicer System (SSS) is also being studied for remote unmanned servicing missions at orbits accessible by an unmanned transport. Space Systems/Loral (SS/L) is conducting research into the systems-level integration of reliable and cost-effective technologies required to address the harsh realities of space operations:

- the need for safe, dexterous manipulation in unstructured environments (possibly along side of humans)
- provisions for teleoperator control with time-delay and very limited communication bandwidth
- efforts to minimize teleoperator workloads

Fig. 1 illustrates the trades that must be made when considering robotic system requirements such as autonomy, operational flexibility, and required human interaction. Each of these has associated benefits and costs. The critical role of systems engineering is to maximize this ratio over a range of tasks.

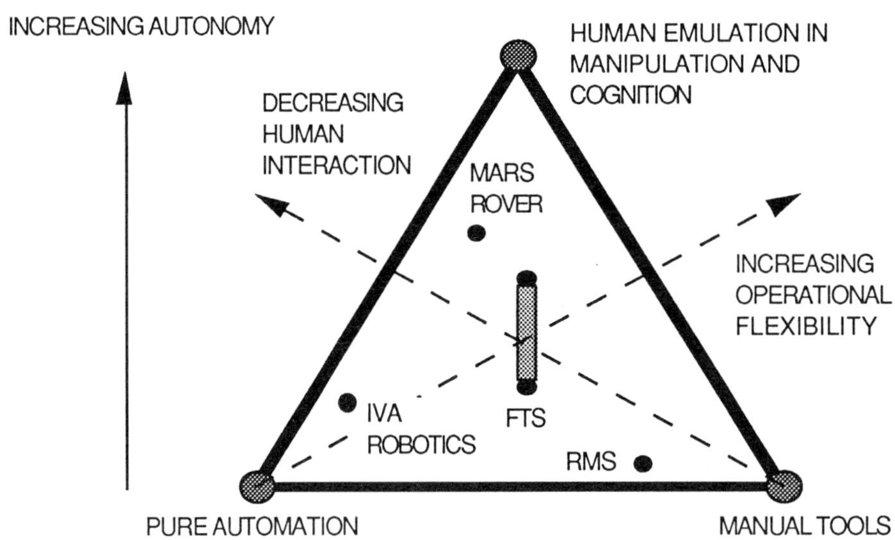

Figure 1: Trade-offs are required for effective robot systems designs.

ROBOTICS R&D HISTORY AT SPACE SYSTEMS/LORAL

Space System /Loral studies in spacecraft servicing date back to the mid-1980's with a series of NASA contracts investigating economics and technical aspects of LEO and GEO servicing. As the FTS program evolved from studies to hardware development, Space Systems/Loral began an internally funded research and development (R&D) program in space robotics. Research in control issues was built upon a close affiliation with Stanford University's Aerospace Robotics Laboratory [1] (due to proximity and hiring of graduates) headed by Prof. Robert H. Cannon, Jr. In 1988, Space Systems/Loral was awarded the contract for FTS end-effectors and associated hardware. The R&D continued in parallel and the analytical research and simulation in space manipulator design and operational control progressed into hardware experimentation in a laboratory environment, established in 1988.

First, an elemental building block of a manipulator, the rotary joint, was studied. Based on a preliminary design of a 2 meter, 7 degree-of-freedom (7 DOF) space manipulator, a prototype space qualifiable joint was designed, built, and tested. The compact joint, shown in Fig. 2, is capable of torques up to 3000 in-lbs and speeds of 45°/sec by utilizing DC motor and harmonic drive technologies. In order to achieve good output position and torque characteristics, these quantities were sensed and fedback for control. With proper logic to suppress vibration and Coulomb friction, joint control was achieved yielding an order of magnitude "softer" response than possible using traditional position control gains [2]. Desired position control bandwidth and accuracy requirements were simultaneously maintained. These basic 1 DOF tests laid the groundwork for impedance control in a 7 DOF manipulator.

Figure 2: Prototype joint tests are a critical element in manipulator impedance control.

CURRENT R&D EFFORT

Our current research on robot control focuses on implementing impedance control on a 7 DOF manipulator. Fig. 3 is a photograph of the key lab components. In the foreground is a Robotics Research K-1607HP redundant manipulator with a force/torque sensor mounted at the wrist. The graphical display of the manipulator at the operator workstation along with a 6 DOF hand controller is in the background. The equipment rack to the left contains a Sun UNIX-based development workstation along with the real-time control microprocessors. A variety of software modules reside on these processors to provide real-time data display and archiving, network communications, task macro execution and coordination, and other features. Fig. 4 is a graphical representation of the components and communication paths in the telerobotics lab.

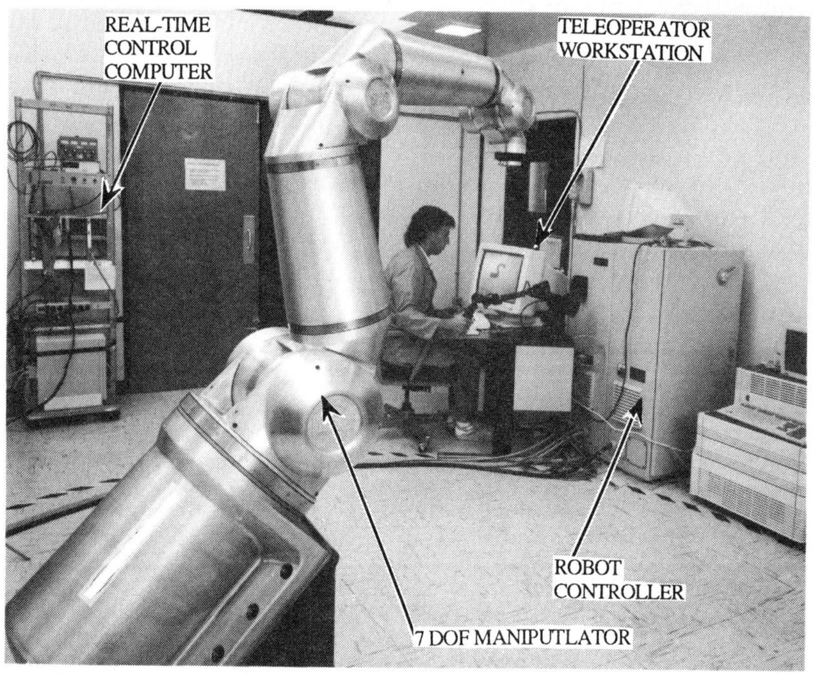

Figure 3: SS/L space robotics laboratory contains critical features of a telerobotic system.

Impedance control [3, 4, 5] is implemented to achieve dexterous manipulation because it provides a unified approach for controlling manipulator position and force interaction with the environment. The key idea is the matching of the mechanical impedance of the manipulator to the admittance of its payload or environment. This creates balanced position/force control without mode switching to obtain graceful workpiece contact and smooth force guided assembly.

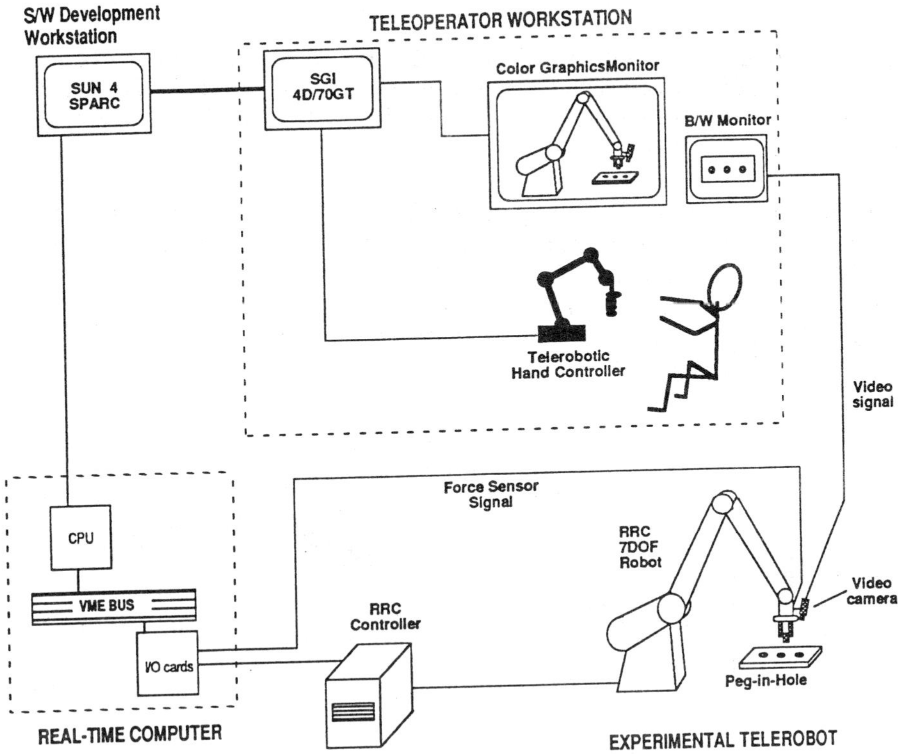

Figure 4: Key components and communication paths in the space robotics laboratory.

This concept was successfully developed in our lab and demonstrated as part of a teleoperated system. We initially used the existing, proven position control loop supplied with the manipulator and overlaid an impedance control loop. The force/torque sensor mounted at the tip provided input to the impedance control logic. Fig. 5 illustrates the key components in this experiment. We have also developed a torque based impedance control loop which by-passes the commercial robot control box and commands the individual joint torques directly. Utilizing a gravity compensation model, this controller reacts to forces applied to all parts of the manipulator, not just the end-effector.

The teleoperator workstation commands the manipulator in two ways. One method directly commands the Cartesian position of the end-effector, in local tool coordinates, while simultaneously allowing the impedance control to handle force interaction with the environment. Scalable gains allow the hand controller commands to be implemented as either gross or fine motions at the end-effector. A second command method drives the manipulator in a global Cartesian rate mode (yet still under impedance control) to cause large slews in world coordinates.

This integrated teleoperator system was used to demonstrate a simple peg-in-hole insertion under teleoperator control. The hole location is taught by the operator and registered in a database. The task was successfully achieved with an operator command sample rate of 4 Hz without contact bounce instability or excessive forces. End-effector forces were controlled at a 40 Hz sample rate.

Features are currently being added to command stored tasks, macros such as "insert peg", under supervisory control. Also, locally stored impedance field time histories can be used to allow tasks to be performed with minimal requirements on the communication bandwidth and no problems from time delay. With this approach, the force or position control stability problems from time delay are eliminated.

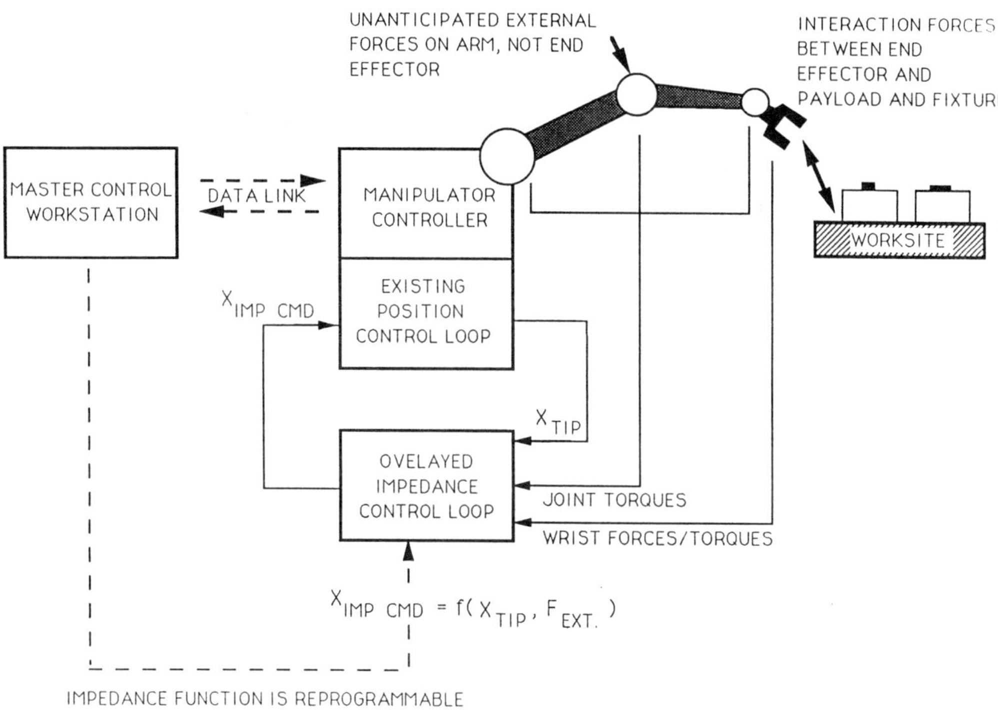

Figure 5: Impedance control block diagram for teleoperation demonstration.

FUTURE RESEARCH PLANS

Currently, the lab manipulator is being fitted with an end-effector similar to that on FTS Demonstration Test Flight 1 (DTF-1) for a series of tests on insertion forces, ORU compatibility, finger and tool usage, and operator task macros. A single black and white video camera will be mounted on the wrist to allow visual feedback to the operator (presently the operator is in direct line of sight of the robot) and yield important information on variables such as view angles and lighting. The operator will continue to issue move or macro commands in position only while the selected impedance field controls any force interactions to within safe levels.

In its new, larger facility the telerobotics lab is able to accommodate another robot for cooperative manipulation activities with larger scale spacecraft mockups. Machine vision will be implemented in a logical, progressive manner to augment the operator's own observations of video and computer generated images. The role of limited robot autonomy will slowly be expanded. The future development of the space robotics testbed will address issues such as:

- demonstrations with variable time-delay, automated task primitives, dynamic workpieces
- integration of proximity sensor and active collision avoidance through use of kinematic redundancy
- design of spacecraft subsystems for robot servicing

The last item is particularly important. As these space robotic systems mature, there must be a corresponding evolution in the spacecraft design process to allow for increased modularity and "robot friendly" interfaces. These interfaces will facilitate on-orbit refueling, subsystem changeouts, and payload upgrades - all items which increase the life or value of the spacecraft in orbit. Since Space Systems/Loral is a spacecraft manufacturer, these features are crucial for delivering cost-effective spacecraft systems to our customers in the future.

REFERENCES

[1] Hollars, Morse, Alexander, and Cannon. "Experiments in Advanced Control Concepts for Space Robotics: An Overview of the Stanford Aerospace Robotics Laboratory", presented at the AAS Dynamics and Control Conference in Keystone, Colorado, February, 1987.

[2] Tilley, Hollars, and Emerick. "Experimental Control Results in a Compact Space Robot Actuator", presented at the ASME Winter Annual Meeting in San Francisco, California, December 1989.

[3] Hogan. "Impedance Control: An Approach to Manipulation", ASME Journal of Dynamic Systems, Measurement, and Control, Vol. 107, March, 1985.

[4] Newman, Dohring, Farrell, Eismann, and Vold. "Preliminary Work in Impedance Control on a Kinematically Redundant Manipulator", presented at the ASME Winter Annual Meeting in San Francisco, California, December 1989.

[5] Chapel and Lawrence. "Stability Analysis for Alternative Force Control Schemes as Applied to Remote Space Teleoperation", presented at the AAS Dynamics and Control Conference in Keystone, Colorado, February, 1987.

AAS 91-048

IntelliSTAR^{TM} - AN EMERGING SPACECRAFT CONTROL ARCHITECTURE TO IMPROVE AUTONOMOUS CONTROL AND OPERATION

John Barry, Linas Raslavicius and Thomas P. Gathmann*

At last year's Guidance and Control Conference, Rockwell reported on IntelliSTAR^{TM}, an evolving architecture which improves autonomous control and operations for manned and unmanned space systems. This architecture, and the software being developed, will reduce dependence on ground systems, increase operator and astronaut productivity, and reduce life cycle costs. Just as important, this approach will allow scientists, ground operators, and astronauts to concentrate more on the space science mission than health and status of the equipment. Rockwell's IntelliSTAR^{TM}, the Intelligent Satellite Technology Automated Resource, is comprised of processes that address planning, scheduling, coordination, task execution and monitoring and an analysis function. These capabilities, along with existing resources of the current subsystems, can define the processing and decision making abilities for both manned and unmanned space systems. This paper describes current Rockwell space systems autonomy research and recent IntelliSTAR^{TM} advancements. This year, the paper focuses on the evolution, current research activities, and the newly developed spacecraft simulator that is being used to validate the architecture.

BACKGROUND

Satellite autonomy is a continuing goal for space systems, and has received increased emphasis in recent years in such documents as the Air Forces Satellite Architecture Study, Space and Missile Investment Strategy Study, the draft Autonomy Statement of Operational Need, the AIAA Autonomy Standards Initiative, and specific requirements in satellite requests for proposals. Similarly, NASA and DARPA are sponsoring programs to reduce space operations costs through increased autonomy and automation. These requirements are being driven by the need to reduce space systems life cycle costs.

PROBLEM

Various efforts are being promoted in the aerospace industry to attack the labor intensive and costly problems associated with command and control using artificial intelligence. These efforts usually involve developing solutions to specific portions of the space control problem involving scheduling, monitoring, interpreting, data for either the ground or planetary system. These approaches, while valid,

* Space Systems Division, Rockwell International, M/S AE09, P.O. Box 7009, Downey, California 90241-7009.

do not address the integration and framework necessary to efficiently command, control and operations of space systems. This lack of integration could result in fragmented and expensive implementations which are counterproductive. In contrast, the Rockwell approach develops an architecture which integrates both ground and space systems operations for manned and unmanned activities. This architecture is being designed to overlay existing systems and tools for integrations into a variety of applications. These applications can range from launch systems, manned and unmanned space platforms, planetary probes, and extraterrestrial outposts.

INCREASED EMPHASIS

Evidence of the increased emphasis in space systems autonomy is the focus of the American Institute of Aeronautics and Astronautics (AIAA) Autonomy Sub-Committee of Space Based Observations Committee on Standards (SBOS/COS). Rockwell shares the chair of the committee with Martin Marietta. This committee progress has been assisted by extensive government and industry participation. The AIAA initiated an effort to create new satellite autonomy and automation guidelines based on the 1981 autonomy guidelines developed by the Jet Propulsion Laboratory (JPL) for the Air Force Space Systems Division. Significant technology changes have occurred since 1981 which suggest a reexamination of the JPL guidelines.

The AIAA noted that there are many implementations of specific features of autonomy, but there exists very few integrated approaches. Additionally, the present approaches concentrate on identification and detection of known faults to specific, known problems. These faults are pre-determined through failure mode and effects analysis. Traditional satellite control theory concentrates on designing responses to these known and anticipated space system faults and anomalies. At a minimum, safing conditions are embedded within system and subsystem control logic to avoid catastrophic events resulting from anomalies. This embedded control can accommodate simple failures such as an solar orientation lock by designing in logic and hardware to sense and safe the problem. The AIAA Autonomy committee discovered that incipient failures that gradually build up as the result of multiple activities across subsystems presently require extensive ground monitoring, commanding and control. No systems exist that can anticipate, sense and either correct or provide recommendations to preclude the problem. Nor does a system exist that allows designers to anticipate the infinite anomalous conditions that could arise from such subsystem interaction. However, the IntelliSTAR ™ architecture and its supporting software system is being developed to monitor, anticipate and control complex subsystem and system problems.

IntelliSTAR ™ APPROACH

As published in the 1990 Guidance & Control Conference, IntelliSTAR ™ addresses the issues of both ground and orbital operations support associated with the processes of planning, scheduling, and coordination, task execution and monitoring and an analysis function. The Rockwell approach has the following two thrusts: systems engineering analysis and prototyping. The systems engineering part is concerned with the definition and refinement of requirements and top level considerations for the Rockwell spacecraft autonomy architecture. The prototyping research concentrates on features of the architecture that have been sufficiently defined. This prototype uses the requirements definition products to construct working portions of the architecture as well as a simulation testbed and user interface. This parallel approach effort allows a closed loop development cycle for rapid concept verification, reduced risk and cost, and increased product assurance. The IntelliSTAR ™ capabilities, along with the existing capabilities of existing subsystems, define the processing and decision making abilities of the spacecraft. The architecture can therefore be independent of specific Artificial Intelligence (AI) techniques and satellite design implementations.

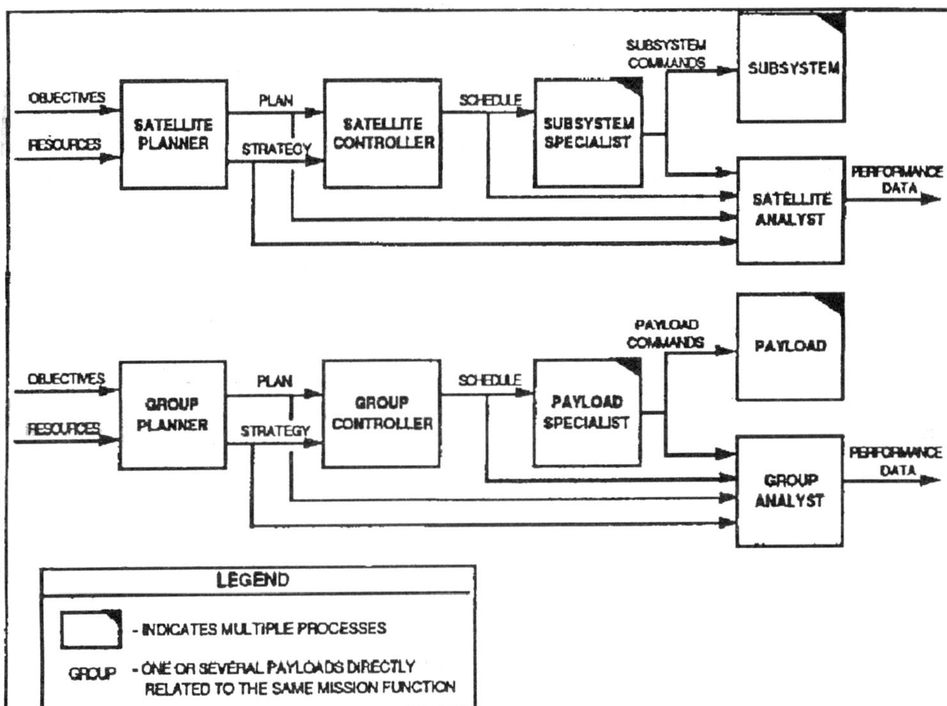

Figure 1 Rockwell IntelliSTAR Architecture

EVOLUTION OF INTELLISTAR ™ ARCHITECTURE

Since 1985, Rockwell has been building and refining its intelligent processing architecture. The overall architecture has resulted in an evolutionary refinement and validation of increasingly more complex prototypes. This architecture is based on a space systems mission operations cycle. This cycle is comprised of the following four supporting cycles: 1) planning; 2) scheduling; 3) execution; and 4) analysis. The IntelliSTAR ™ architecture, depicted in Figure 1 consists of the following four intelligent processes: 1) planner; 2) controller; 3) subsystem specialist; and 4) analyst. The planner maintains and updates plans based on current objectives. The controller coordinates the creation of an agenda of tasks and supervises its execution of the specialists. Finally, the analyst assesses the degree of success or failure of an executed mission. These processes working in harmony comprise the IntelliSTAR ™ architecture and rely on proven artificial intelligence techniques involving real time control architectures, resource bounded problem solving, diagnosis, machine learning and planning.

ROCKWELL'S AUTONOMY LEGACY AND PROGRESS

Rockwell has evolved architecture and the appropriate tools to increase autonomous capability for space systems. In 1985 Intellistar, an overall architecture for on-board satellite autonomy was established. In 1988 the controller concept was further analyzed and refined. This included development of the initial controller. The primary function of the controller is to coordinate the generation of a schedule for executing selected missions of a satellite or a group of mission related payloads. The controller generates a schedule with the help of other intelligent processes within the various subsystems or payloads. During 1989 an enhanced controller scheduling prototype and the subsystem specialist design was developed. The operation of this prototype was introduced at the 1990 Guidance and Control Conference. In 1990 several spacecraft behavioral models, the surveillance controller and the spacecraft simulator were developed. Additionally, the iterative and reactive planner was enhanced this year to dynamically respond to mission objectives and spacecraft and environmental constraints. This planner is implemented in knowledge engineering environment using frame based techniques. This AI approach promotes rapid prototyping and a more responsive system than conventional methods. The behavior models developed include the autonomous navigation, electrical power, thermal control, attitude control, and communications. These models realistically simulate an operating spacecraft environment. Also, considerable systems and software progress concentrated on building PAVLOV, a tool to validate the architecture and its operation. This paper will focus on PAVLOV's contribution to validating the architecture.

ROCKWELL'S PAVLOV SPACECRAFT SIMULATOR

Rockwell built PAVLOV, a realistic and dynamic spacecraft testbed to evaluate and test the IntelliSTAR ™ architecture. This testbed minimizes the amount of engineering effort required to specify a system and provide a means to evaluate the architecture. This testbed uses subsystem definitions and a translator to produce C Language Integrated Production System (CLIPS) executable simulation code of a user prespecified system. The engineer specifies satellite systems using an input modeling language called Satellite Modeling Language (SML). The SML provides an adequate user interface that can be used to specify the space system simulation in terms of objects, operations, and environmental effects. For example a space system can be described as one system or several subsystems or units within each subsystem. These subsystem definitions include the electrical power (EPS), thermal control (TCS), attitude control (ACS), navigation (NAV), and communication (COM) system. Systems operations and environmental effects are described in terms of rules which define how the objects behave given certain environmental conditions. The specified input is then processed by a modeling language translator which produces the executable simulation code. Consistent format of the output code is automatically provided by the translator. A diagram of the PAVLOV simulation definition process is in Figure 2.

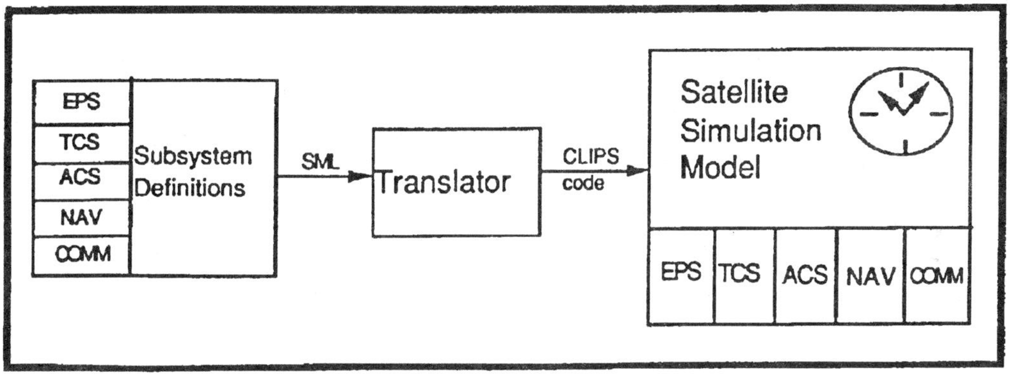

Figure 2 Pavlov Simulation Process

SUMMARY

The Rockwell IntelliSTAR TM architecture is being refined for use on state of the art space systems using Artificial Intelligence technology. When implemented on an actual space system this architecture will reduce future space operational costs and increase effectiveness in operating and controlling space systems. The initial objectives of the Rockwell project are to dynamically schedule on-board satellite resources and activities. In the process, Rockwell is developing techniques which can simplify manned and unmanned space operations and improve the productivity of ground controllers. The Rockwell approach is based on examining system operations, operational data and gaining feedback from satellite operators. This feedback and data was used to construct the prototype demonstrated in this conference on a personal computer. Additionally, the Rockwell Space Operations Company, a Rockwell Division which operates the Space Shuttle, has a extensive, PC based data base on spacecraft manned and robotic operations which can assist this effort.

ACKNOWLEDGMENTS

The authors wishes to acknowledge the Rockwell team responsible for developing the Satellite Controller concept and building the prototype. This team is in Rockwell's Software and Simulation Systems Department of Space Systems Division Seal Beach facility. Linas Raslavicius, the Project Coordinator, and Tom Gathmann, the lead engineer, are the original architects of the Intellistar TM concept. Other members involved in the design and evolution of the concept and prototype include Joyce Kubo and Barbara Pauls. The authors also thank the many people from the Rockwell Artificial Intelligence Center, Rome Laboratory, AF Space Command, AF Space Division, The Phillips Laboratory, NASA, DARPA and the Aerospace Corporation for their encouragement and comments throughout this project. Special thanks goes to Capt Rick Higgens from the USAF Phillips Laboratory for his constructive comments on the Independent Research & Development project upon which this paper is based.

BIBLIOGRAPHY

1. Barry, John M. "Applications of Expert Systems to Command and Control of Satellites," Aerospace Applications of Artificial Intelligence Conference, Oct 1988, Dayton, Oh.

2. Barry, John M., Sary, Charisse. "Increasing Autonomy through Satellite Expert System Scheduling" AIAA/SOLE Second Space Logistics Symposium, Sept 1988, Costa Mesa, Ca.

3. Barry, John M., Gathmann, Thomas P. "Expert System for On-Board Scheduling and Control." NASA Artificial Intelligence Conference, Nov 1988, Huntsville, Ala.

4. Barry, John M., Raslavicius, L. and Gathmann, Thomas P. "Spacecraft Autonomy Standards for Autonomous Operations." 1991 AIAA Science and Technology Exposition, Reno, Nevada. (Jan 1991)

5. Gathmann, T. Johnson, S., Pitts, J. *Artificial Intelligence (AI); Applications in Satellite Autonomy.* FY88 IR&D Final Report, IL No. 792-110-88-086

6. Gathmann, Thomas P. "A Dynamic Satellite Simulation Testbed Based on CLIPS and CLIPS Derived Tools" 1st CLIPS Users Group Conference, NASA Conference Publication 10049, NASA Johnson Space Center, Houston, Texas, (August, 1990)

7. Gathmann, Thomas P. and Halker, Donna. "Towards a Manageable Solution to the Iterative Development of Embedded Knowledge Based Systems. IEEE Conference on Managing Expert System Programs and Projects, Washington D.C. (September, 1990)

8. Gathmann, Thomas P. and Raslavicius L, "Systems Approach to the Satellite Operations Problem. 25th Intersociety Energy Conversion Engineering Conference (IECEC-90), Reno, Nevada (August, 1990)

9. Pauls, Barbara and Sherman, Mark. "Satellite Simulation Utilizing CLIPS" 1st CLIPS Users Group Conference, NASA Conference Publication 10049, NASA Johnson Space Center, Houston, Texas, (August, 1990)

10. Sary, C., Gathmann, T. Kass, J. *Satellite Controller Concept.* FY88 IR&D Final Report, IL No. 792-110-88-083, Sept. 1988.

11. Gathmann, T. Johnson, S., Pitts, J. *Artificial Intelligence (AI); Applications in Satellite Autonomy.* FY88 IR&D Final Report, IL No. 792-110-88-086

12. Gathmann, Thomas P., Kubo, J. *Satellite Autonomy Generic Expert System (SAGES)*, Rockwell International, Satellite & Space Electronics Division (S&SED) FY 1988 IR&D Final Report (Sept 1989).

13. Gathmann, Thomas P., Johnson, S., Butcher, L. Sease, J., *Satellite Controller (SATCON) Concept*, Rockwell International Satellite & Space Electronics Division FY 1989 Project Final Report.

14. Gathmann, Thomas P., Kubo, J. et al. *On-Board Satellite Autonomy Architectures*, Rockwell International Satellite & Space Electronics Division FY 1990 Project Report.

15. Gathmann, T., Kubo, J. *Satellite Autonomy Generic Expert Systems (SAGES)*. FY89 IR&D Final Report, September, 1989.

16. Gathmann, T., Kubo, J., et al. *Satellite Controller Concept (SATCON)*. FY89 Final Report, September, 1989.

DEMONSTRATION DISKETTE

Copies of the IntelliSTAR [TM] demonstration diskette may be obtained for government employees by writing the author at the following address:

John M. Barry, Space Systems Division, Rockwell International - M/S AE09, P.O. Box 7009, Downey, California 90241-7009

AUTHORS' BIOGRAPHIES

John Barry is the Program Development Manager for Research & Technology at Rockwell's Space Systems Division in Downey, Ca. Prior to joining Rockwell he was an Air Force Officer for 20 years serving the USAF Space Division and Rome Air Development Center in senior engineering and management positions on the Space Shuttle, KC-10 Aircraft, the Defense Meteorological Satellite Program, the Titan III program and the Strategic Defense Initiative. He has a B.S. Degree in Math and Physics, from LeMoyne College, an M.S. Systems Management from the University of Southern California, an M.S. in Logistics Management from the Air Force Institute of Technology and an M.S. in Computer Science from Northrop University.

Linas Raslavicius is the Manager for Software Research & Technology at SSD's Seal Beach facility and an consultant to the current project. He has a B.S. in Information Engineering and M.S. in Electrical Engineering from the University of Illinois. He has over 14 years experience in designing satellite software systems.

Thomas Gathmann is a Member of the Technical Staff and the current Technical Lead for the Autonomy IR&D project at SSD's Seal Beach facility. He has B.S. in Aerospace Engineering from Cal Poly, Pomona, and an M.S. in Aerospace Engineering from the University of Southern California. He has over 9 years experience in computer science with 7 years experience in Artificial Intelligence and expert systems at Rockwell. He is a primary instructor of the Rockwell 200 hour Expert Systems course developed by Teknowledge.

Section III
CHALLENGES AND APPROACHES TO SPACE-BORNE POINTING CONTROL

SESSION III

Chairperson:	Col. Roland H. Worrell, III SDIO/PTB
Co-Chairperson:	Robert Laskin Jet Propulsion Laboratory
Local Chairperson:	Arthur Throckmorton Martin Marietta Astronautics

The following paper numbers were not assigned:

AAS 91-057 to -059

AAS 91-050

THE CONTROLS-STRUCTURES INTERACTION GUEST INVESTIGATOR PROGRAM - AN OVERVIEW AND PHASE I EXPERIMENTAL RESULTS

Rudeen Smith-Taylor* and Sharon E. Tanner†

The integration of control systems with the structural dynamics of flexible spacecraft can significantly benefit performance. Because of this need to integrate both control systems and structural dynamics, a multidisciplinary approach is required for the design of the large, flexible spacecraft of the future. Such is the approach of the NASA Controls-Structures Interaction (CSI) Program whose objective is to develop and validate the technology needed for the structure and the control system to interact beneficially to meet the requirements of future space missions.

Through the CSI Guest Investigator (GI) Program, researchers from industry and academia use government testbeds to validate their advanced control techniques and integrated designs. The objective of the GI program is to support CSI technology advancement by 1) involving the most renown CSI researchers from academia and industry, 2) providing them with the most advanced CSI test facilities for their experimental validation, and 3) disseminating these experimental results to the research community.

Phase I of the GI program was initiated in 1988, with the selection of eight CSI researchers from industry and academia, to conduct research in two government test facilities. The 2-year activity will be discussed in this paper in the form of a program review, description of the test facilities, and a summary of the experimental results. Information on the next phase of the program will also be presented.

INTRODUCTION

The behavior of large flexible space structures of the future will necessitate that control system designers be cognizant of the structural dynamics of the spacecraft. The integration of high performance control systems and flexible structures must be designed so as to improve system performance and prevent any detriment to system stability. Because of the uncertainties involved in controlling flexible structures, the design of these advanced control systems cannot rely solely on analytical development but requires experimental validation on dynamically realistic and structurally

* GI Program Manager, CSI Office, MS 298, NASA Langley Research Center, Hampton, Virginia 23665-5225.

† Structural Dynamics Division, NASA Langley Research Center, Hampton, Virginia 23655-5225.

complex test facilities.[1] This integrated approach, referred to as controls-structures interaction technology, is the focus of the NASA Controls-Structures Interaction (CSI) Program. The CSI Program is managed by the Office of Aeronautics, Exploration, and Technology (OAET) and is an element of NASA's Civilian Space Technology Initiative. The Program is a multidisciplinary research activity whose objective is to develop and validate the technology needed for future spacecraft to meet increasingly demanding mission requirements. Three NASA Centers, the Langley Research Center (LaRC), the Marshall Space Flight Center (MSFC), and the Jet Propulsion Laboratory (JPL), are cooperatively developing this technology.

In Phase I of the CSI Guest Investigator (GI) program, researchers from industry and academia have participated in a 2-year research activity in which they used government testbeds to test controller design techniques. The objective of the GI program is to support CSI technology advancement by 1) involving the most renown CSI researchers from academia and industry, 2) providing them with the most advanced CSI test facilities for their experimental validation, and 3) disseminating these experimental results to the research community. Phase I of the GI Program is nearing completion, having produced valuable results with increased appreciation of the need for experimentally validating CSI research techniques and methodologies. This paper will briefly discuss the NASA CSI Program, followed by a discussion of the GI program with emphasis on the ground test facilities, research methodologies, and experimental results.

THE CSI PROGRAM

Future NASA space missions will require increased pointing precision, precise attitude control, and the use of multi-payload platforms with interacting control systems.[2] The mission requirements will be for control systems that are both highly integrated into and highly interactive with the large, flexible structures. To meet these technology goals, the CSI Program approach is to: 1) develop and validate integrated design/analysis methods; 2) develop and demonstrate ground test methods/techniques to establish on-orbit performance; 3) obtain in-space experimental data to validate design/analysis and ground test methods; and 4) establish design methods and criteria for qualification of spacecraft for future space missions.[2]

The five CSI program elements addressing these issues are: 1) configurations and concepts, 2) integrated analysis and design methods, 3) ground testing methods, 4) in-space flight experiments, and 5) a GI program. The three NASA centers have specific areas of expertise which support these elements. The JPL effort emphasizes development of design technology for optics-class applications and micro-precision controlled structures. LaRC emphasizes multiple payload platforms and global control of large antennas, in addition to providing management for the technical program and the GI program. MSFC is concerned with 1) flight qualification methods and flight experiments/tests, and 2) the Controls, Astrophysics, and Structures Experiments in Space (CASES) flight experiments.

GUEST INVESTIGATOR PROGRAM

The GI program objectives are to 1) solicit and support the most promising CSI research, 2) provide the most advanced testing facilities for experimental validation of this research, and 3) make the experimental results available to the CSI community. Because the advancement of CSI technology greatly benefits from the participation of renown researchers from academia and industry, a general solicitation for participation in the GI program is made to obtain the most promising research approaches. The submitted proposals are reviewed and evaluated by an intercenter selection team. The selections for award are based on technical merit, utilization of the current test facilities, and cost. The GI program manager is at LaRC with technical monitors located at each test facility. Additional government and contractor personnel provide the hardware, software, and operational support required to conduct the experiments.

Phase I began in February of 1988 with eight awards, five to universities and three to industry. The principal investigators and their primary research thrust are shown in Table 1.

Table 1

PHASE I GUEST INVESTIGATORS

University/Industry	Principal Investigator	Primary Thrust
Arizona State	Dr. Bong Wie	Collocated/Noncollocated Controllers
Boeing Aerospace	Dr. Michael Chapman	Nonlinear Math Modeling
Cal Tech	Dr. John Doyle	Noncollocated Controller Design
U Cincinnati	Dr. Randall Allemang	System ID and Controller Design
Dynamic Engineering	Wilmer Reed	Passive & Active Suspension Sys Dsgn
Harris Corporation	Dr. David Hyland	Noncollocated Controller Design
MIT	Dr. W. Vander Velde	Sys ID and Fault Detection & Isolation
Purdue	Dr. Robert Skelton	Noncollocated Controller Design

The research thrusts and approaches are as diverse as the organizations represented. Two of the activities (Boeing Aerospace and Dynamic Engineering) were completed in 1 year and dealt with hardware modeling and the design of suspension systems, respectively. The results of these two research activities are contained in references 7 and 8, and will not be discussed in this paper. The other six activities were of 2-year duration and concentrated on system identification and/or the development and validation of active control techniques. The research goals of these six included application of specific modern control theory, disturbance rejection, failure detection, and the effect of signal-to-noise ratio on the observed nonlinearities in the hardware. The results of this research will be discussed following the descriptions of the two test facilities used during Phase I: the Mini-MAST testbed at LaRC, and the Advance Control Evaluation for Structures (ACES) testbed at MSFC.

CSI GROUND TEST FACILITIES

Mini-MAST Test Facility

The Mini-MAST facility is a 20-meter long deployable/retractable truss, located at the NASA Langley Research Center.[3] Manufactured by The Astro Aerospace Corporation using flight-quality materials, Mini-MAST is a representative truss for use in future space missions. The Mini-MAST was designed and built as a laboratory model of the 60-meter MAST truss to be flown by the LaRC Control of Flexible Structures (COFS) Program; hence, the name Mini-MAST. The COFS flight program was subsequently cancelled. The Mini-MAST is one of the first realistic testbeds for CSI research.

The cantilevered truss is supported vertically from a rigid foundation. The truss construction has three graphite/epoxy member types: longerons that run parallel to the beam axis; battens that form the triangular cross-sections; and diagonals in the beam face planes. The longeron and diagonal members are hinged with pinned connections to titanium corner-body joints. The battens are rigidly connected to the corner bodies. The beam has 18 bays which are deployable and retractable, 2 bays at a time. The current configuration of Mini-MAST is as shown in Fig. 1.

Fig. 1 Mini-MAST Test Facility

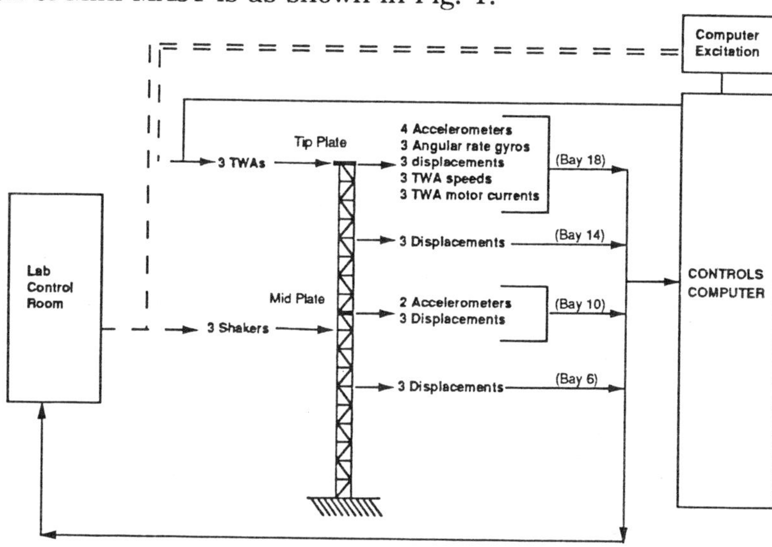

A stiff platform at the tip, bay 18, and one near the mid-point, bay 10, hold servo accelerometers for linear acceleration measurements and rate gyros for angular rate measurements. Noncontacting displacement sensors distributed along the beam axis are mounted alongside the truss for observing lateral displacement. These are connected to a mainframe real-time controller computer via fiber-optic cables for providing the feedback signal for the actuators. Three torque wheel actuators (TWAs), mounted on the tip platform and weighing approximately 85 pounds each, are the only actuators available for control. Disturbance input to the structure can be provided by the TWAs or by 3 Unholtz-Dickie 50 lb shakers attached at bay 9.

The Mini-MAST truss has 5 structural modes below 10 Hertz (Hz). The first pair of bending modes are at approximately 0.86 Hz, the first torsion mode is at 4.2 Hz and a second pair of bending modes at approximately 6.1 Hz. There are 113 vibrational modes (local modes of the diagonals) between the 2nd bending modes and the 2nd torsion mode, which is at approximately 22.9 Hz.[4]

A finite element model (FEM) is provided to each GI for simulation and analysis prior to facility testing. Because the TWAs can cause structural damage to Mini-MAST under certain conditions, each controller is first run through a series of simulations to verify system stability and determine maximum member loads exerted on the truss elements.

ACES Test Facility[5,6]

The ACES ground test facility is located at the NASA Marshall Space Flight Center, Huntsville, Alabama. The test article is a deployable beam approximately 14 meters in length, supporting an offset antenna. The beam was built by Astro Research as a flight backup for the Voyager magnetometer boom. This beam is extremely lightweight (about 5 lbs) and very flexible. It is symmetric and triangular in cross section with three continuous longerons forming the corners. The cross members divide the beam into 91 sections, each having equal length and mass and similar elastic properties. The beam exhibits a longitudinal twist of approximately 260 degrees when fully deployed. The ACES configuration is shown in Fig. 2.

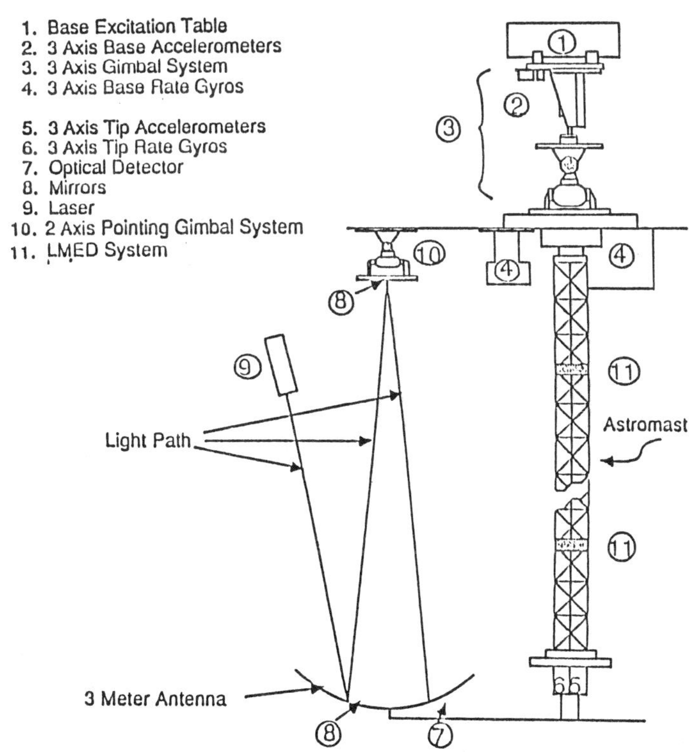

Fig. 2 ACES Ground Test Facility

An antenna and two counter balance legs are appended to the beam tip which, with the pointing gimbal arms at the base, form a configuration having modal characteristics of large space structures. The Aces configuration has 42 modes with frequencies under 9 Hz. A hydraulically-operated base excitation table provides disturbance inputs and a 3-axis gimbal system provides rotational control. Orthogonal pairs of linear momentum exchange devices (LMEDs) are located at two discrete locations along the length of the beam. The LMEDs are proof-mass actuators which provide translational control forces. Rate gyros and accelerometers are mounted at the base and tip of the beam. An optical system consisting of a fixed position laser, 2 mirrors, and a 2-axis detection plane provides a measure of control system performance. One of the mirrors of the optical system is mounted to a 2-axis advanced pointing gimbal system (AGS) located on an extension arm appended to the base of the beam. This gimbaled mirror is used in a closed-loop image motion compensation (IMC) system. The system is equipped with 9 control actuators and 22 sensors. The control and instrumentation system is supported by a computer system including an array processor that has a sample data rate of 50 Hz. A FEM of the beam and a nonlinear simulation model are made available to each GI for controller analysis prior to testing.

PHASE I RESEARCH EXPERIMENTS AND RESULTS

Experimental results of the six GI teams that performed research in the areas of system identification and controller development are discussed in this section in the order of the GI's initial visit to Mini-MAST. Each GI was required to perform validation testing at both government testbeds described above, concentrating on a single testbed for a year. Some experimental results from the second year are not yet available.

Arizona State University

Dr. Bong Wie demonstrated the simplicity and effectiveness of applying classical control designs to realistic CSI testbeds. Various single-input-single-output (SISO) controllers were simultaneously applied to bending and torsional motion. Nonminimum phase compensation was employed for periodic disturbance rejection. With Mini-MAST, sensor output decoupling was required since the individual displacement sensors are inherently coupled. Decoupling was not necessary with the ACES testbed.

Both collocated and noncollocated SISO controllers were demonstrated. Decoupled displacement measurements at bay 18 provided the Mini-MAST collocated feedback for the three torque wheel actuators at the same location. Similarly, the decoupled displacement measurements from Bay 10 supplied feedback to the same actuators at the tip for the noncollocated controllers. At ACES, Dr. Wie used four separate SISO loops which were identified by the Harris Corporation researchers. (see Fig. 8) None of the ACES feedback signals were collocated.

Nonminimum phase filtering was successfully employed to add damping to secondary modes. By not restricting filter zeros to the left-half s-plane, these filters were shown to increase closed-loop damping of flexible modes while tolerating significant model uncertainty. The root locus method, Bode plots, and iterative refinement were the primary means used in developing robust compensators.

In the periodic disturbance rejection demonstration, an internal model for the disturbance (with known frequencies, but unknown magnitudes and phases) was included as part of the compensator. The disturbance rejection filter is made of individual filters, each designed for a specific frequency. The full filter then has as many pole-zero combinations, or dipoles, as frequencies in the periodic disturbance.[9]

The success of applying the previously described classical control concepts to the Mini-MAST testbed can be seen in the results shown in Fig. 3.[9] A collocated Mini-MAST controller provided 20 percent active damping for the first bending mode, as compared to 2 percent inherent damping (Fig. 3(a)). Active damping was 15 percent for the first bending mode in the noncollocated controller shown in Fig. 3(b).

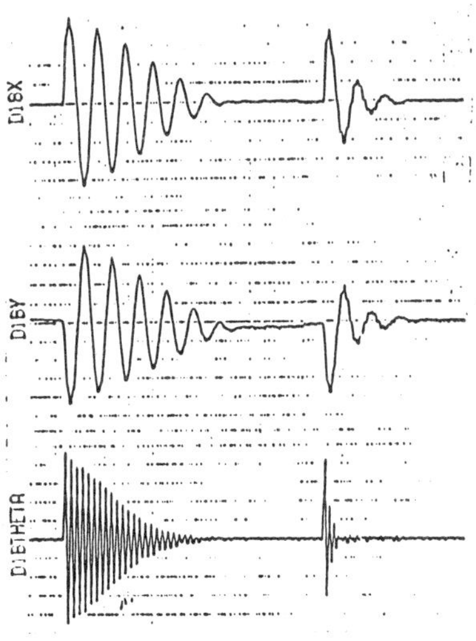

Open-loop Closed-loop

Fig. 3(a) Collocated Controller

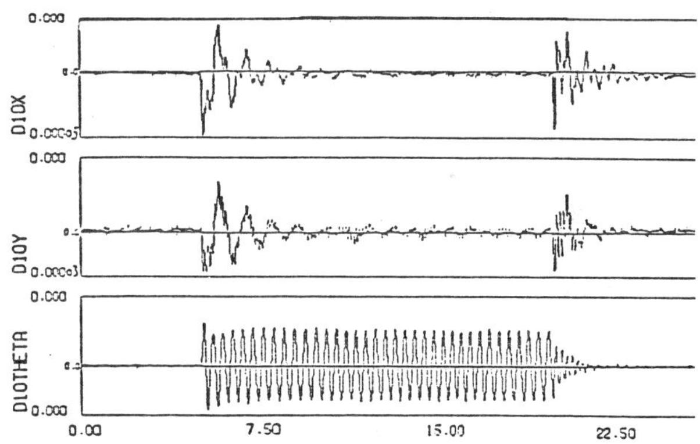

Fig.3(b) Noncollocated Controller

Data from the ACES testbed, AGS loop, is shown in Fig. 4. The open loop time history of the rate gyro at the base to a step disturbance is shown in Fig. 4(a). Active damping is provided by a classical SISO controller, Fig. 4(b). The inclusion of the periodic disturbance rejection filter improves the controller performance as shown in Fig. 4(c).[17]

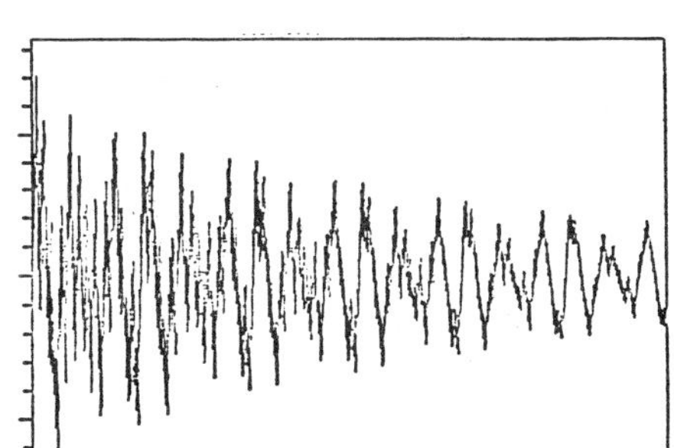

Fig. 4(a) Open-loop

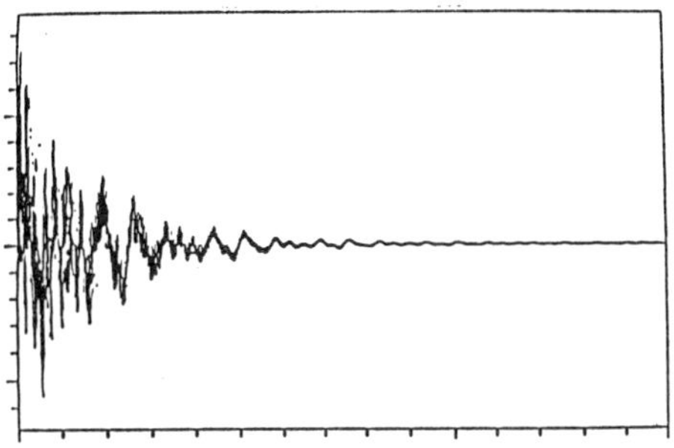

Fig. 4(b) Active Damping

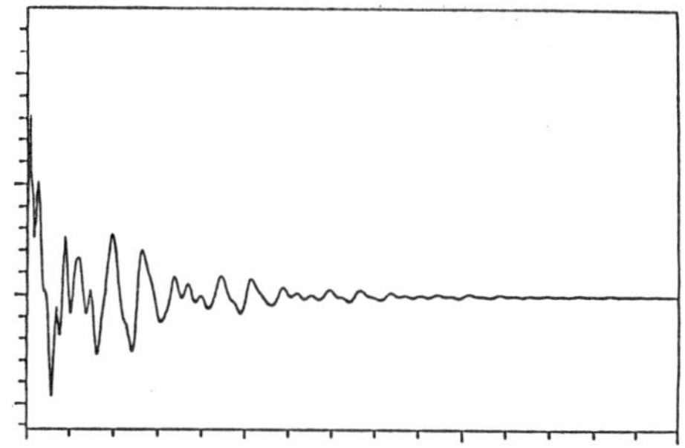

Fig. 4(c) With Rejection

Purdue University

Dr. Robert Skelton used Modal Cost Analysis (MCA) with self-tuning Output Variance Constraint (OVC) controller design to develop multiple-input-multiple-output (MIMO) controllers. An iterative procedure was applied, integrating both system identification and control law development. Model order reduction is required to make the complex dynamics associated with realistic large space structures more manageable. Controller performance is used as a criteria in the model reduction process, ensuring the appropriateness of the analytical model for a particular controller design. Hence, the controller is model-based and the model parameters are adjusted through controller performance, requiring an iterative process.[10]

MCA provided the basis for model order reduction. MCA includes a closed-form solution for the weighted quadratic cost functions associated with each chosen system output. Controller design was integrated through use of an output weighting matrix Q, obtained through application of the OVC control design algorithm. Using updated weights in Q as design parameters in the MCA produces a more appropriate reduced-order model for controller design. Design specifications and noise covariance are considered design parameters for OVC controllers. Specifications influence the weights assigned to modes and noise, thereby influencing weights assigned to estimator gains and input signals. The iterative process acts as a self-tuning mechanism, producing a series of controllers from low to high gain.

At both Mini-MAST and ACES, Dr. Skelton first used the FEM provided and then developed models by applying the Q-Markov Covariance (QMC) equivalent realization algorithm to experimental data. QMC models based on white noise excitation differed substantially from QMC models based on pulse excitation.

Because displacement sensors are not available in space, only platform-mounted accelerometers and rate gyro sensors were used for feedback on the Mini-MAST testbed. Displacement sensors were used only for evaluation of controller effectiveness. The low frequency first bending modes were weakly observable by the platform accelerometers. Consequently, less active damping was added to the first bending modes than to the second bending and torsion modes. These results are shown in Figs. 5(a) and 5(b)[16], where open and closed loop data are shown as dashed and solid lines, respectively. Simulation results showed using displacement feedback could significantly improve damping of the first bending mode.

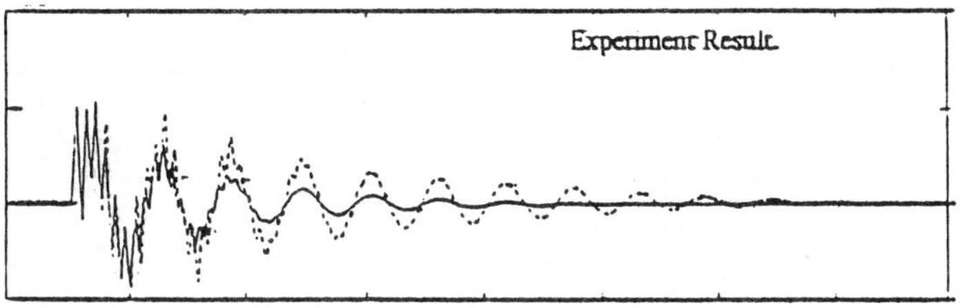

Fig. 5(a) 1st & 2nd Bending Modes

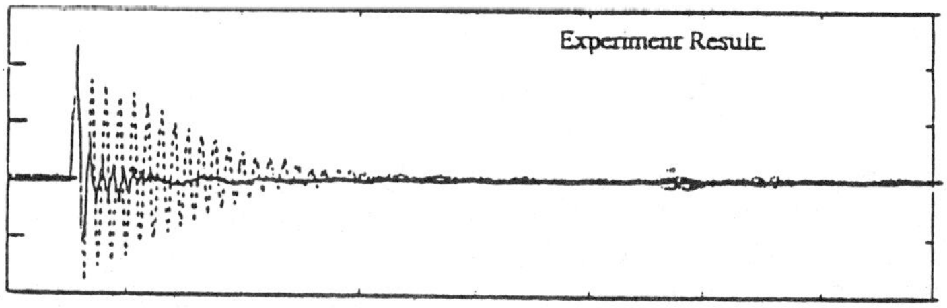

Fig. 5(b) Torsion Mode

Massachusetts Institute of Technology

Dr. Wallace Vander Velde demonstrated failure detection and isolation (FDI) techniques at both Mini-MAST and ACES. Generalized parity relations and failure detection filters were the two methods selected for investigation since both are applicable to sensors and actuators alike. Neither require hypotheses concerning possible failures thereby reducing the computation required. At Mini-MAST, FDI was also investigated during closed-loop testing, using the control law of Dr. Wie.

The FDI of Mini-MAST's three-tip displacement sensors, using the generalized parity relations method, produces a scalar residual r(t). This residual will be zero only when the input measurements are noise-free in an accurately modeled system with all sensors functioning perfectly. With non-ideal systems, the goal is to ensure that when a sensor fails the residual produces an identifiable signature distinctive from the background noise created by unmodeled dynamics and measurement noise.[11]

The parity equation can be derived from a state-space model of the system dynamics using an autoregressive technique. However, Dr. Vander Velde's research showed that parity equations derived directly from experimental input-output data performed more effectively than corresponding model-based equations by eliminating errors in modeling the dynamics of the structure. Single Sensor Parity Relations (SSPR) and Double Sensor Parity Relations (DSPR) were also compared. SSPR are simpler to create and implement. Isolation of the faulty sensor is trivial with SSPRs since only one sensor is involved. However, the magnitude of the failure signature with SSPRs can be inadequate if the sensor fails in the off mode. Consequently, DSPRs are favored in order to increase the reliability of the failure detection. DSPRs involve decision logic to isolate the failed sensor. For example, Fig. 6 shows typical residuals for failure of sensor #1, evident in both r_{12} and r_{13}, where r_{XY} is the residual from the DSPR of sensor X and sensor Y. The residual, r_{23}, would not present such a signature.[11]

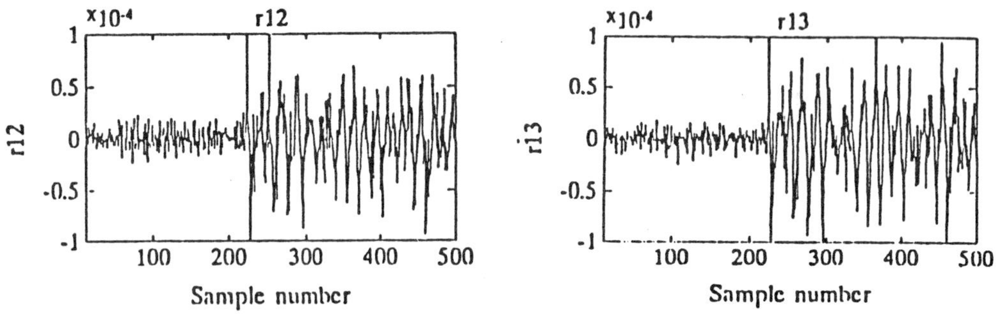

Fig 6. DSPR Residuals r_{12} and r_{13}

Sampling frequency and the number of time lags used in the equations also increase the magnitude of the failure signature of the residual, whether using SSPRs or DSPRs. Larger failure signatures were produced by longer sample periods, as illustrated in Figs. 7(a) and (b). In Fig. 7(a), the sample period was .015 seconds. In Fig. 7(b), the sample period was .030 seconds.[11] Sensor failure was simulated at the same point in the time history, even though the sample numbers differ due to the change in sample frequencies.

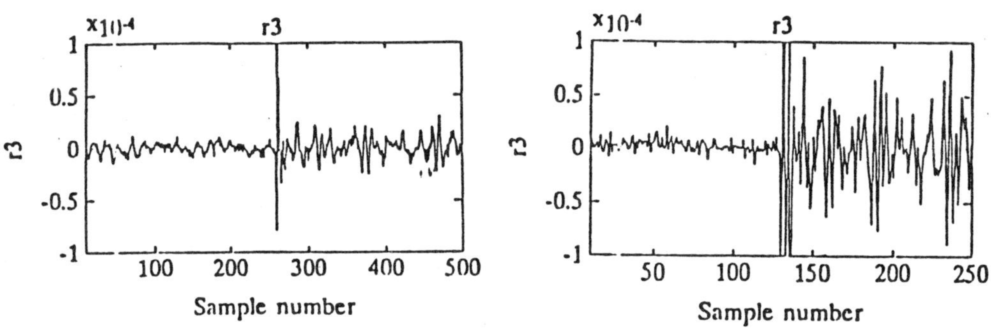

Fig. 7(a) & (b) Residual for Identified SSPR for Sensor Failure

Harris Corporation

Dr. David Hyland led the Harris research team in application of the Optimal Projection Approach for Uncertain Systems (OPUS) for controller design. Dr. Hyland, assisted by Dr. Emmanuel Collins, Douglas Phillips, and James King, applied Maximum Entropy/Optimal Projection (MEOP), an OPUS controller process, in the design of their robust, high performance controllers.

The MEOP process begins with low- to moderate-authority controllers to which robustness is added through application of the homotopy algorithm, creating a maximum entropy design. Maximum entropy designs increase gain stability, allow order reduction in the controller bandwidth, and improve the controller's tolerance of uncertainties in damping, frequency, and location of system zeros. Next in the MEOP process, the order of the controller is reduced through a balanced grammian reduction, to give a good approximation of the optimal projection equation. The homotopy algorithm is used again to increase controller authority, this time creating the optimal projection design.[12]

Like Dr. Skelton, the Harris researchers chose accelerometers and rate gyros as feedback sensors for Mini-MAST, eliminating displacements sensors which are not available in space. They emphasized controller simplification as essential in meeting stringent on-orbit processing power limitations. Selecting the design model was the first crucial step in the controller design process. At ACES, the researchers developed a state space model using the Eigensystem Realization Algorithm. The Mini-MAST FEM predicted system responses with sufficient accuracy to be used as the design model.

At each testbed, a decentralized controller architecture was developed, integrating several reduced-order controllers, each operating on an individual loop within the system. At ACES, four SISO loops dominating the system dynamics were identified, as shown in Fig. 8. These loops were subsequently used by other researchers.

Fig. 8 ACES SISO Loops

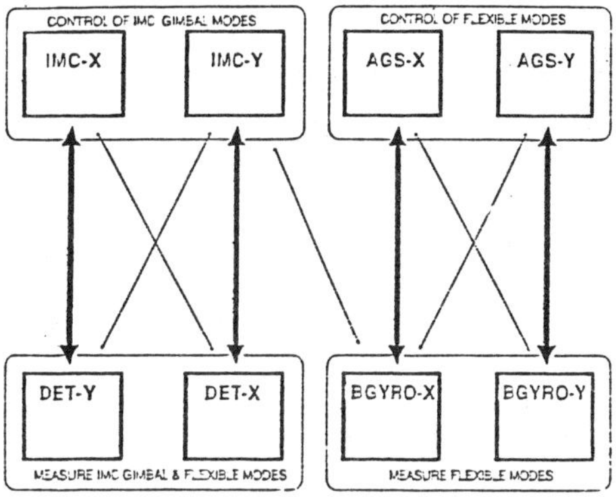

An additional LMED loop, using the four collocated accelerometers as feedback signals, was effective in adding damping to the higher frequency vibrations when integrated with the IMC and AGS loops. Stroke limitations with the LMED complicated this controller design. Ultimately, the controller included a high-pass filter cascaded with a low-pass filter, each discretized separately using bilinear transformation with frequency prewarping. Tests were performed with individual decentralized controllers and then with integrated controllers. Fig. 9 shows typical experimental results from ACES, where the integrated controller clearly yielded better performance than any of the three individual feedback controllers.[13]

At Mini-MAST, three loops were used: one SISO loop to control the first torsion mode, and two MIMO loops to control the first and second bending modes. Mini-MAST results are shown in Fig. 10. With a gain block of 10, the classical rate feedback provided more than 50 percent damping to the torsional mode (Fig. 10(a)). Bending modes were controlled by a series of MIMO controllers, coupled with the torsion output feedback, shown in Fig. 10 (b), (c), and (d) in order of increasing effectiveness. Results shown in these figures are from a baseline LQG controller; a reduced-order LQG controller with precompensator; and a MEOP controller with precompensator, which also damped the higher frequency second bending modes.

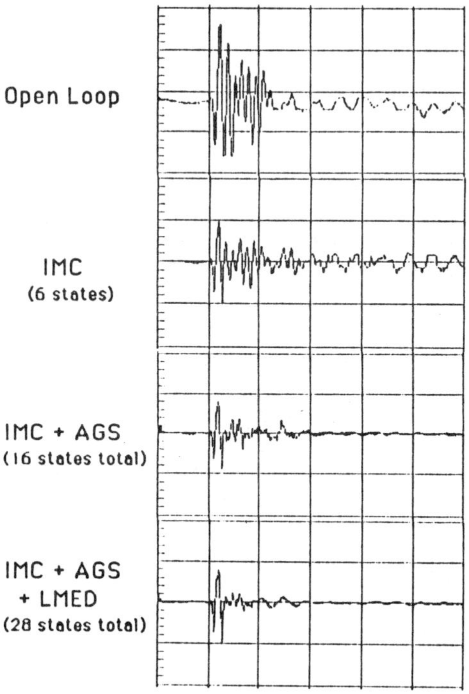

Fig. 9 ACES Results

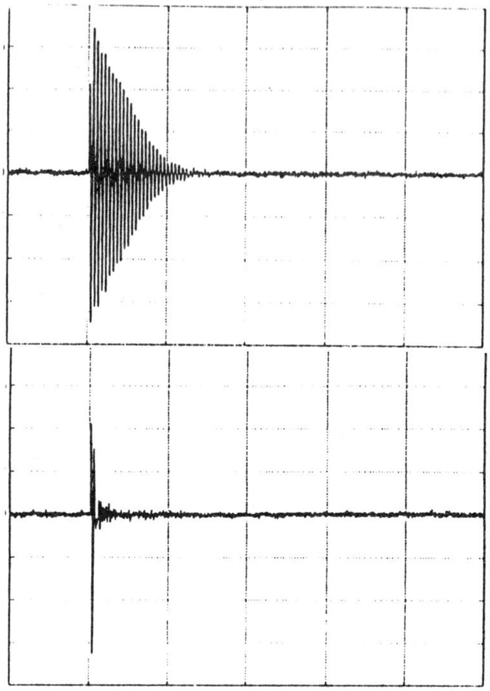

Fig. 10(a) Torsional Mode

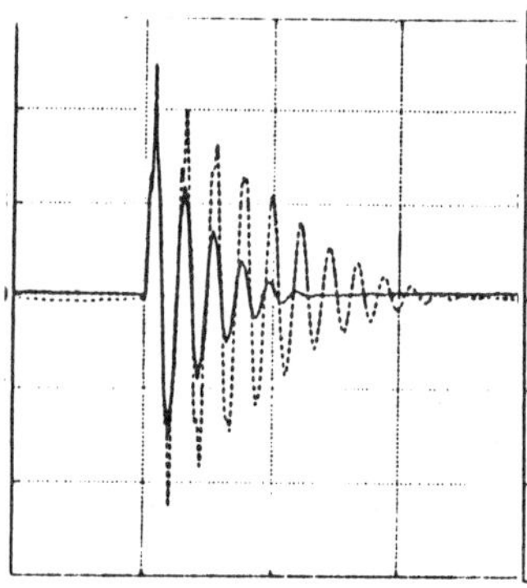

Fig.10(b) Baseline LQG

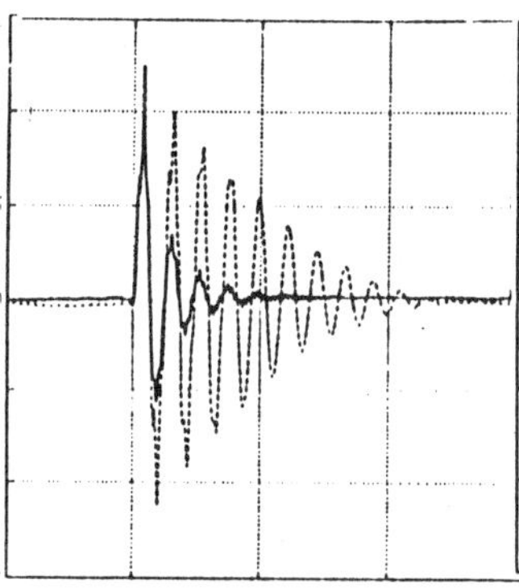

Fig.10(c) Precompensator

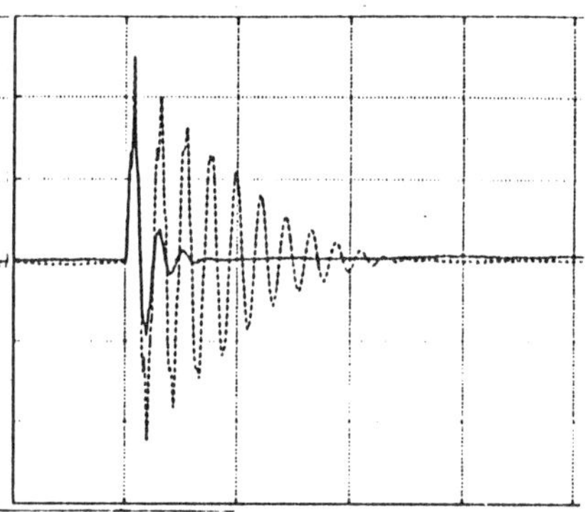

Fig.10(d) MEOP

California Institute of Technology

Dr. John Doyle and Dr. Gary Balas used μ-analysis and synthesis for their research effort. Control design using μ-synthesis is an iterative process which alternates between solving an H_∞ optimal control problem and a structure singular value (μ) analysis problem. μ-synthesis can directly account for known and unknown errors such as structural modes eliminated from the reduced order design models, unmodeled sensors and actuator dynamics, or inaccurate damping, frequencies, or mode shapes. Fig. 11 shows the system block diagram used for initial Mini-MAST control designs with both additive and multiplicative uncertainty descriptors. Subsequent designs added parametric uncertainty associated with variations in natural frequencies.

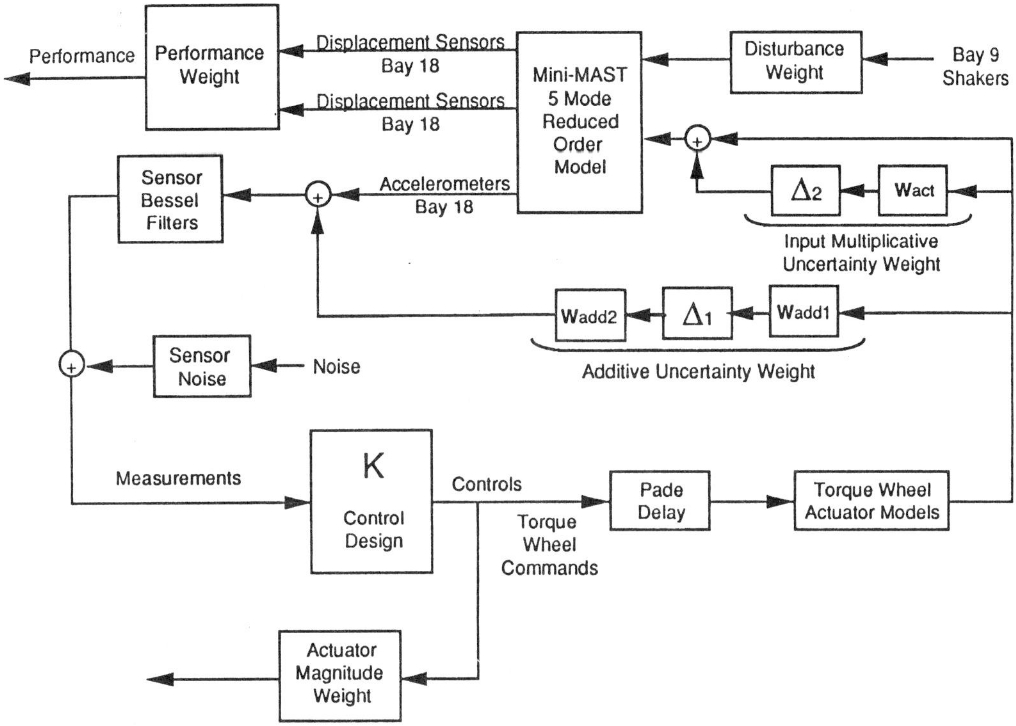

Fig. 11 System Block Diagram with Uncertainties

At the ACES testbed, modeling difficulties created the need for excessively large uncertainty values, which in turn severely penalized controller performance. This experience proved useful, however, in emphasizing the importance of accurate models.[18]

The researchers successfully applied the same techniques to the Mini-MAST facility. Analytical studies were conducted, assuming no uncertainty in the system, to select feedback sensors. Bay 18 accelerometers were found to be sufficient for observing and controlling the five modes below 10 Hz which dominate structural response. Attenuation of truss displacement at Bays 10 and 18 was selected as the performance objective for all controllers.

Eighteen controllers were designed, varying actuator magnitude weights and uncertainty descriptions. Experimental results closely correlated with simulations. Fig. 12 shows results of one of the most aggressive controllers. The truss first bending motion is attenuated in approximately 1 second as seen in the tip displacement sensors (Fig. 12(a)). The second bending motion is attenuated within 3 seconds, as seen in the Bay 10 displacement sensors (Fig. 12(b)). Open loop decay time is approximately 17 seconds for the same excitation.

Mini-MAST Results

Fig. 12(a) Bay 18

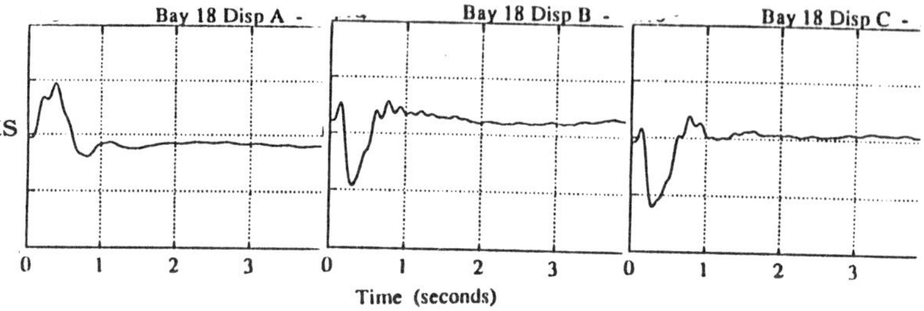

Fig. 12(b) Bay 10

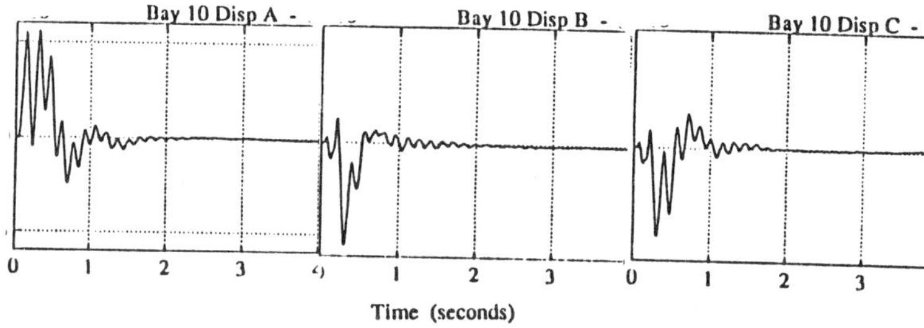

University of Cincinnati

The University of Cincinnati research effort was led by two principle investigators. Dr. Randall Allemang led the effort to develop reliable state space models which Dr. Gary Slater used for control law development. At ACES, scheduling delays resulted in Dr. Slater developing controllers based on FEMs provided by MSFC; but for Mini-Mast, Dr. Allemang's identified models were used.

In applying system identification techniques to actuator/structure/sensor systems at the ACES and Mini-Mast facilities, Dr. Allemang highlighted the complexity that time delays add to the model. Current modal parameter estimation techniques can be used to create accurate modal models of combined systems only by accounting for time delays present in the data. The time delays can then be treated as measurement delays in the controller design.

The system identification work at ACES addressed nonlinearities, variations in system dynamics over time, and the lack of anti-aliasing protection. Nonlinearities were defined to be small enough to still permit a linear model. However, significant variations in system behavior were noted in frequency response functions (FRF's) taken 3 months apart. Physical changes in the base gyros were suspected since only FRF's involving those instruments were affected. Dr. Allemang also predicted aliasing problems due to filter limitations in the control system hardware. Such problems later surfaced in Dr. Slater's control law applications.[14]

Autoregressive techniques were used to estimate system parameters for the modal model from the FRF data. Typical measured and synthesized FRF's from the ACES testbed are shown in Fig. 13. Residues for each system pole were calculated and used to estimate mode shapes and modal mass. These in turn were used to create the state-space model with modal displacements and velocities as the states.

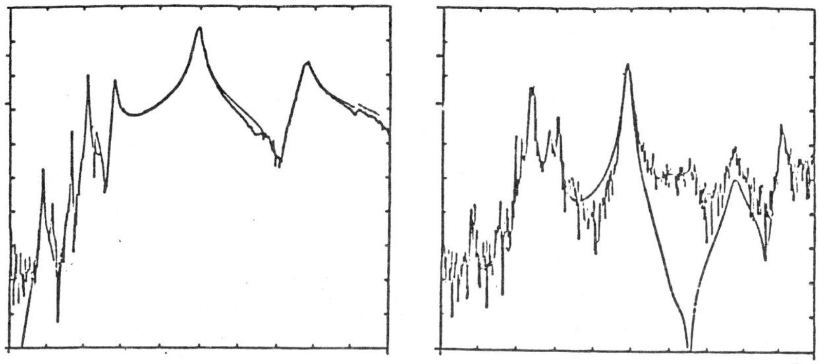

Fig. 13 Measured and Synthesized FRFs from ACES

The ACES state-space model would have benefited Dr. Slater since significant modeling errors in the FEM reduced the effectiveness of the positivity designed multivariable controllers. In addition, while the positivity approach guarantees stability if sensors and actuators are ideal, application of this technique to real hardware showed its limitations. The AGS loop provided ideally positive real transfer functions suitable for positive real controllers. However, application of the controllers resulted in instabilities which in turn led to the development of multivariable scaling and phase compensation. Results of a stable controller with compensator are shown in Fig. 14.[15]

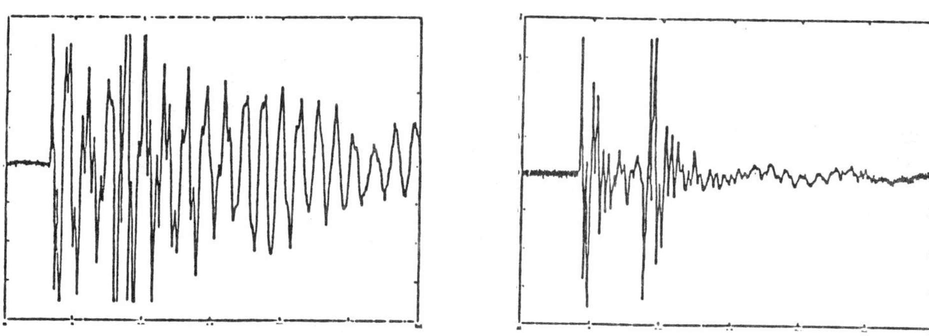

Fig. 14 AGS Results, Open Loop and Closed Loop

To maintain the positive definiteness of controller matrices when coupled with the compensators, comparable scale factors had to be applied to both the inputs and outputs of the system. Independent scale factors could be applied to all channels of the controller, thereby allowing on-line tuning of the system response. Such scaling holds promise for future large space structures. This procedure can tolerate a zero scale factor, thus compensating for a sensor or actuator failure without requiring a redesign of the controller.

SISO positivity controllers applied to the LMED loop at ACES were stable and produced a measurable improvement in damping. In addition, since one LMED was not functioning at that time, a zero scale factor was added to the compensator for a multivariable LMED controller which was stable even with one failed actuator.

FUTURE PHASES AND PLANS

The GI program was designed as a multi-phased research activity utilizing the government's most advanced test facilities. With Phase I coming to completion and with newly developed and enhanced CSI testbeds, the solicitation for Phase II of the GI Program was issued in June 1989. The Air Force Phillips Laboratory joined NASA for this phase, thereby strengthening the GI program. Phase II will provide three unique and challenging testbeds for validation of CSI technology. The number of proposals received in response to this solicitation indicated not only interest in the program but commitment to the advancement of CSI technology. Table 2 shows the Phase II selections.

Table 2

PHASE II GUEST INVESTIGATORS

University/Industry	Principal Investigator	Primary Thrust
Boeing Aerospace	David Warren	CMG/RCS Pointing & Slewing
Harris Corporation	Dr. David Hyland	Noncollocated Controller Design
Martin Marietta Corp.	Dr. Eric Schmitz	Smart Struts & Controller Design
MIT	Dr. Andreas von Flotow	Passive Damping &Controller Desgn
Texas A&M University	Dr. Srinivas Vadali	Controller Design

The following is a description of the three test facilities to be used by the GI's in Phase II.

At LaRC, the CSI Evolutionary Model (CEM) testbed will be used by the Phase II GI's and by future CSI researchers. As the name implies, the CEM will change over time in size, complexity, and experimental capabilities. The Phase O configuration consists of a 4-longeron truss, 16.8 meters in length and divided into .25-meter cubic bays. An 8-rib reflector, 4.9 meters in diameter, is mounted on one end with a laser detector system mounted atop a vertical truss at the other end. The model is suspended using two cables attached to support trusses as shown in Fig. 15.[1]

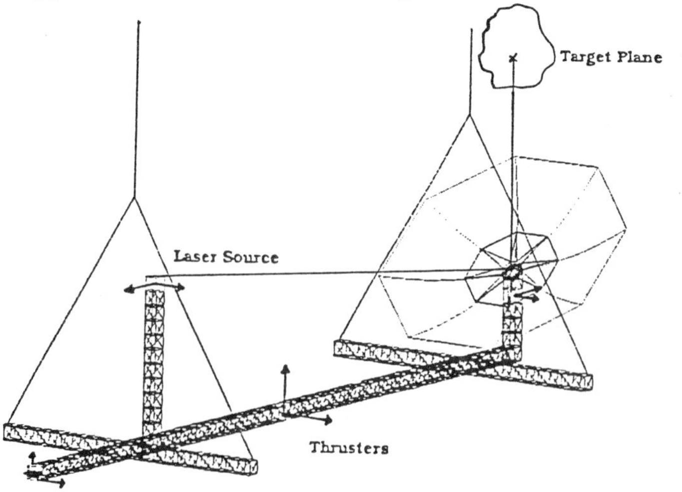

Fig. 15 CSI Evolutionary Model Schematic

At MSFC, the Control, Astrophysics, Structures Experiment in Space (CASES) ground test facility hardware is a 32-meter Solar Array Flight Experiment (SAFE) boom which was flown on the STS-41D Shuttle mission. The inverted boom, as shown in Fig. 16, is secured to a support structure which in turn is attached to an airbearing tripod system that translates in the horizontal plane and rotates about the longitudinal axis.[17]

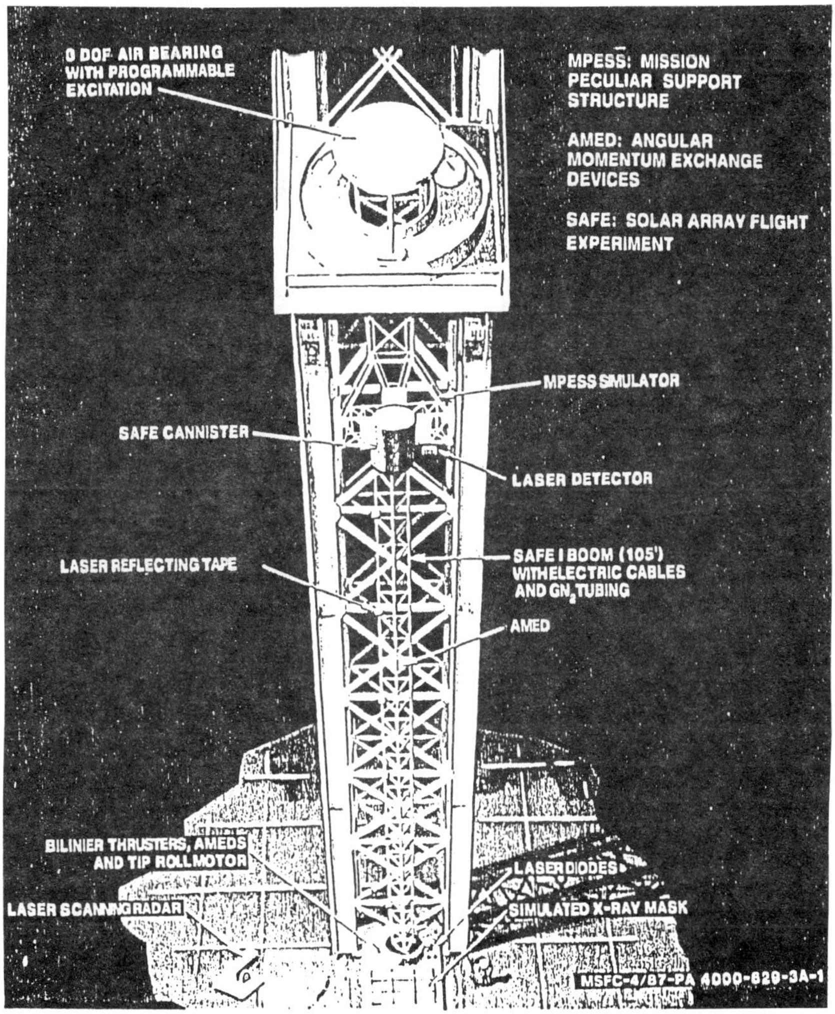

Fig. 16 CASES Ground Test Facility

The Air Force Advanced Space Structure Technology Research Experiment (ASTREX) ground test facility is located at the Phillips Laboratory, Edwards AFB, CA. The initial ASTREX experiment article is a 5-meter structural model of a 3-mirror space based laser beam expander supported by a spherical air bearing atop a pedestal. The air bearing mount provides the test article with three degrees of angular freedom as shown in Fig. 17. The article is housed in a constant temperature, air-tight enclosure to minimize disturbances.[17]

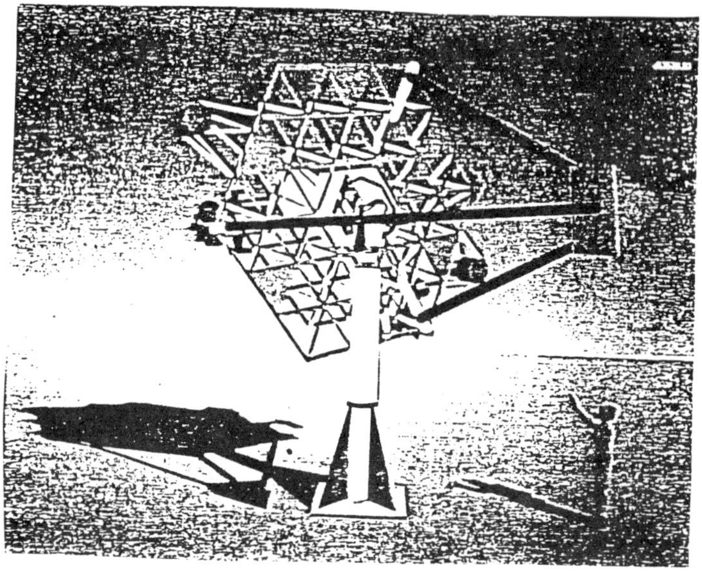

Fig. 17 ASTREX Schematic

CONCLUDING REMARKS

The objective of the Controls-Structures Interaction program is to integrate the design of the control system and the structural dynamics of large, flexible spacecraft to meet the control system performance requirements of 21st-century space missions. The three field centers cooperatively developing NASA CSI technology are Langley, Marshall, and JPL. Through a joint phase of the Guest Investigator Program, the Air Force Phillips Laboratory has recently become a participating member of the CSI activity.

After a 2-year activity, Phase I of the CSI GI program is coming to completion. During this time, principal investigators using ground test facilities at LaRC and MSFC, conducted experiments to validate CSI techniques in system identification and controls development. The objective of the GI program is to support CSI technology advancement by 1) involving the most renown CSI researchers from academia and industry, 2) providing them with the most advanced CSI test facilities for their experimental validation, and 3) disseminating these experimental results to the research community.

The primary goal of all the researchers involved in this advancement of CSI technology is to better understand the practical limitations of simplifying theoretical assumptions. The sharing of information and experiences toward a common goal was prevalent during the course of the 2 years.

The experimental results from two NASA testbeds has proven that advanced ground facilities, with flight-like hardware structures, are effective for the validation of CSI technology. Phase I results have shown that without knowledge of the structural dynamics to be controlled, many of the most accepted control techniques demonstrated only marginally improved performance results. When these same techniques were integrated with knowledge of the structural dynamics, the result was considerably better performance.

The Phase I GI program has been instrumental in advancing CSI technology, producing valuable experience to be shared with the research community. With these results at hand and the improved government test facilities, the research effort from Phase II and subsequent phases promises to result in even more impressive demonstrations of integrated design. The technology for an integrated control/structure design for future space structures is nearly at hand and the newly developed CSI testbeds promises to make the ground validation of this technology possible.

ACKNOWLEDGEMENTS

Our appreciation goes out to the Guest Investigators and the many people that have supported the NASA CSI GI program in so many ways.

At NASA Langley Research Center, we want to thank the members of the CSI Ground Test Team and the CSI Analysis and Design Methodology Team that served on the technical review committee, as test conductors, as technical monitors, as facility coordinators, and to all that gave hardware, software, and operational support for the Mini-MAST facility.

AT NASA Marshall Space Flight Center, our thanks to John Sharkey, Henry Waites, Alan Patterson, and the personnel of Control Dynamics Company for their hardware, software, and operational support of the ACES facility.

REFERENCES

1. W. H. Fleming: Future Directions in Control Theory, A Mathematical Perspective. SIAM Report of the Panel on Future Directions in Control Theory, 1988.

2. J. R. Newsom, W. E. Layman, H. B. Waites and R. J. Hayduk: The NASA Controls-Structures Interaction Technology Program. 41st Congress of the International Astronautical Federation, Dresden, GDR, Report No. IAF-90-290, October 1990.

3. L. R. Adams: Design, Development and Fabrication of a Deployable/Retractable Truss Beam Model for Large Space Structures Application. NASA CR-178287, June 1987.

4. S. E. Tanner and et al: Mini-MAST CSI Testbed Preliminary User's Guide. NASA TM 102630, 1991.

5. J. P. Sharkey: Active Control Evaluation for Spacecraft (ACES) Ground Test Facility User's Guide. March 1989.

6. R. D. Irwin, V. L. Jones, S. A. Rice, S. M. Seltzer and D. J. Tollison: Active Control Technique Evaluation for Spacecraft (ACES). Final Report to Flight Dynamics Lab of AF Wright Aeronautical Labs, Report No. AFWAL-TR-88-3038, June 1988.

7. J. M. Chapman: Nonlinear Modeling of Joint Dominated Structures. NASA Contractor Report 4324, September 1990.

8. R. R. Gold, I. P. Friedman, W. H. Reed III and W. L. Hallauer: Suspension Systems for Ground Testing Large Space Structures. NASA Contractor Report 4325, October 1990.

9. B. Wie, L. Horta and J. Sulla: Active Structural Control Design and Experiment for the Mini-MAST. Proceedings of the 1990 American Control Conference, San Diego, CA, May 1990, pp. 1428-1434.

10. C. Hsieh, J. Kim and R. E. Skelton: Closed Loop Lab Tests of NASA's Mini-MAST. Proceedings of the 1990 American Control Conference, San Diego, CA, May 1990, pp. 1441-1444.

11. W. E. Vander Velde and C. M. Van Schalkwyk: Failure Detection and Isolation Experiments with the Langley Mini-MAST. Proceedings of the 1990 American Control Conference, San Diego, CA, May 1990, pp. 1422-1427.

12. E. G. Collins, Jr., D. J. Phillips and D. C. Hyland: Design and Implementation of Robust Decentralized Control Laws for the ACES Structure at Marshall Space Flight Center. Proceedings of the 1990 American Control Conference, San Diego, CA, May 1990, pp. 1449-1454.

13. E. G. Collins, Jr., D. J. Phillips and D. C. Hyland: Design and Implementation of Robust Decentralized Control Laws for the ACES Structure at Marshall Space Flight Center. NASA Contractor Report 4310, July 1990.

14. R. J. Allemang, S. Shelley, D. L. Brown and Q. Zhang: Practical Experience with Identification of Large Flexible Structures. Proceedings of the 1990 American Control Conference, San Diego, CA, May 1990, pp. 1445-1448.

15. G. Slater, A. Bosse and Q. Zhang: Practical Experience with Multivariable Positivity Controllers. Proceedings of the 1990 American Control Conference, San Diego, CA, May 1990, pp. 1445-1448.

16. Phase I Guest Investigator Program, 2nd Mid-Year Review, Williamsburg, VA, July 1990.

17. NASA Research Announcement NRA-89-LaRC-3, Controls-Structures Interaction Experiments for Large Space Structures Guest Investigator Program, June 1989.

18. Phase I Guest Investigator Program, Year-End Review, NASA Langley Research Center, Hampton, VA, January 1990.

AAS 91-051

DEMONSTRATIONS OF LSS ACTIVE VIBRATION CONTROL TECHNOLOGY ON REPRESENTATIVE GROUND-BASED TESTBEDS*

David C. Hyland, Douglas J. Phillips and Emmanuel G. Collins, Jr.†

Many future space missions involving flexible structures may require active vibration control to satisfy mission objectives. Thus, it is important for active control of flexible structures to be practically demonstrated in ground based experiments. These experiments can validate (or invalidate) existing theories and technology and provide directions for future research. This paper discusses two experiments conducted by Harris which successfully demonstrate control of flexible structures. The first experiment involved control design and implementation for the ACES structure at NASA Marshall Space Flight Center while the second experiment was conducted in-house at Harris using the Multi-Hex Prototype structure. The paper concludes with some remarks on the lessons learned from conducting these experiments.

INTRODUCTION

Many large space system concepts will require active vibration control systems to satisfy critical performance requirements such as line-of-sight (LOS) accuracy and constraints on rms surface roughness. In order for these concepts to become operational it is imperative that active vibration control be practically demonstrated in ground based experiments.

In regard to theoretical developments, these experiments serve a two-fold purpose. First of all, they can validate (or even invalidate) existing theories proposed for the control of flexible structures. Secondly, the insights gained from actual hardware experiments elucidate important issues in practical control design and thus provide needed direction for future theoretical research. For example, based on our experience, it is clear that sampled-data control design and analysis are important (though somewhat neglected) areas of research. In the past, the integration of theory and experimentation has perhaps been lacking in the modern controls community. However, recent trends seem to show that this integration is increasing.

In addition to validating and guiding theory, experimentation also highlights the primary technology issues involved in implementing active control for flexible

* Sponsored in part by the National Aeronautics and Space Administration under contract NAS1-18872 and Air Force Office of Scientific Research under contract F49620-87-C-0108.

† Harris Corporation, Government Aerospace Systems Division, MS 22/4842, Melbourne, Florida 32902.

structures. For example, experiments tend to reveal the importance of sensor and actuator development tailored for flexible structures. Also, the number of design iterations required in the control design portions of these same experiments and the large amounts of data manipulation reveal the need for the continual evolution of efficient and ergonomically sound design environments such as MATLAB and MATRIX_X.

Over the past seven years, five major active vibration control experiments have been performed by Harris, using structures having a wide range of complexity and employing both in-house and Government supplied test facilities [1-5]. In this paper, we particularly focus on two experiments involving test articles of representative complexity that are traceable to two major classes of Large Space Systems.

The primary methodology chosen for control design in this experiment is the Maximum Entropy/Optimal Projection methodology [7-13], a subset of the Optimal Projection Approach for Uncertain Systems (OPUS) [14,15] which allows for the simultaneous trade-off of five fundamental issues in control design: actuator sizing, sensor accuracy, controller order, robustness and system performance. The Maximum Entropy/Optimal Projection approach was developed particularly to allow high performance, robust control law design for flexible structures. The design equations consist of four coupled matrix equations which specialize to the standard LQG Riccati equations when the plant is known perfectly and a full order controller is desired.

The first experiment discussed herein was conducted on the ACES test-bed [4,5] at NASA Marshall Space Flight Center under the auspices of the NASA Control/Structures Interaction Guest Investigator Program. The ACES system involves a configuration traceable to large RF system concepts with antenna reflector, metering truss and surrogate feed. Line-of-sight (LOS) errors are monitored using an optical measurement system and the control objective is maximal reduction of LOS jitter due to supporting base-plate disturbances. Although the system exhibits very complex dynamics, OPUS design technology was used to obtain a highly robust decentralized control system consisting of simple, easy-to-implement sub-controllers. Experimental results show that this controller achieves very substantial reductions in LOS error and metering truss vibration.

The second experiment considered here is based upon the Harris Multi-Hex Prototype Experiment (MHPE) apparatus [2,3] developed on Harris IR\&D. This is a large optical reflector structure comprising a seven panel array, supporting truss, and a secondary support structure which typifies a number of generic characteristics of optical system concepts. The MHPE was designed to study the dynamics of multi-segment optical structures, in particular the issue of primary mirror dephasing. To reduce the controller complexity high bandwidth precision actuators and sensors were developed [6]. Both decentralized and centralized controllers were tested. Results show more than an order of magnitude reduction in both LOS jitter and dephasing.

The paper is organized as follows. The next section discusses control design and implementation for the ACES structure at NASA Marshall Space Flight Center. The following section discusses the corresponding results for the Harris Multi-Hex Prototype Experiment (MHPE) apparatus.

The paper concludes in the final section with some remarks on the lessons learned from conducting these experiments.

THE ACES EXPERIMENT

The ACES experimental testbed [16] is located at NASA Marshall Space Flight Center (MSFC). The basic test article, a spare Voyager Astromast, is a deployable, lightweight (about 5 pounds), lightly damped beam, which is approximately 45 feet in length. The ACES configuration, shown in Figure 1 consists of an antenna and couterweight legs appended to the Astromast tip and the pointing gimbal arms at the Astromast base. Overall, the structure is very flexible and lightly damped. It contains many closely spaced, low frequency modes (more than 40 modes under 10 Hz.)

The goal of the control design is to position the laser beam in the center of the detector. The detector and pointing gimbals are each positioned on the end of a flexible appendage to increase the difficulty of the control problem. The lack of information about the appendage motion also adds complexity to the controller design (i.e., there is no accelerometer or gyro at the location of the gimbals or the detector).

The actuators and sensors available for controller implementation are listed in Figure 1. In our control design and implementation we used 8 control inputs and 8 measurement outputs. The inputs were the X and Y torques of the IMC gimbals, the the X and Y torques of the Advanced Gimbal System (AGS) and the X and Y forces of the two Linear Momentum Exchange Device (LMED) packages. The measurements consisted of the X and Y detector (DET) position outputs, the X and Y base gyro (BGYRO) rate outputs and the X and Y outputs of the LMED accelerometers. The disturbances were chosen to be position commands to the Base Excitation Table (BET). The BET motion is regulated by an analog controller which allows any type of BET movement within the frequency limitation of the hydraulic system.

Modeling procedure for the ACES structure

In this experiment a finite element model (FEM) was supplied by MSFC. We generated frequency response functions (FRF's) to evaluate the FEM. In general, the FEM predicted significantly different frequency responses than those obtained by actually testing the structure.

The FRF's of the IMC-Y to DET-X and IMC-X to DET-Y loops revealed that those loops are influenced very little by the flexible modes of the structure. It follows that the IMC gimbals are not capable of controlling flexible modes to improve LOS performance. Thus, if one considers the four actuator inputs (IMC-X, IMC-Y, AGS-X, AGS-Y) and the four sensor outputs (DET-X, DET-Y, BGYRO-X,

BGYRO-Y), it is not necessary to feed back the BGYRO outputs to the IMC gimbals since the BGYRO's primarily contain information about the behavior of the flexible modes which the IMC's cannot control. In addition, the DET outputs do not contain much (if any) useful information for the AGS gimbals that is not already provided by the BGYRO's. Thus, the achievable performance cannot be improved by feeding back the DET outputs to the AGS gimbals or the BGYRO outputs to the IMC gimbals. As illustrated by Figure 2, analysis of test data also revealed that within the decentralized structure described above there were four dominant loops: AGS-X to BGYRO-X, AGS-Y to BGYRO-Y, IMC-X to DET-Y, and IMC-Y to DET-X.

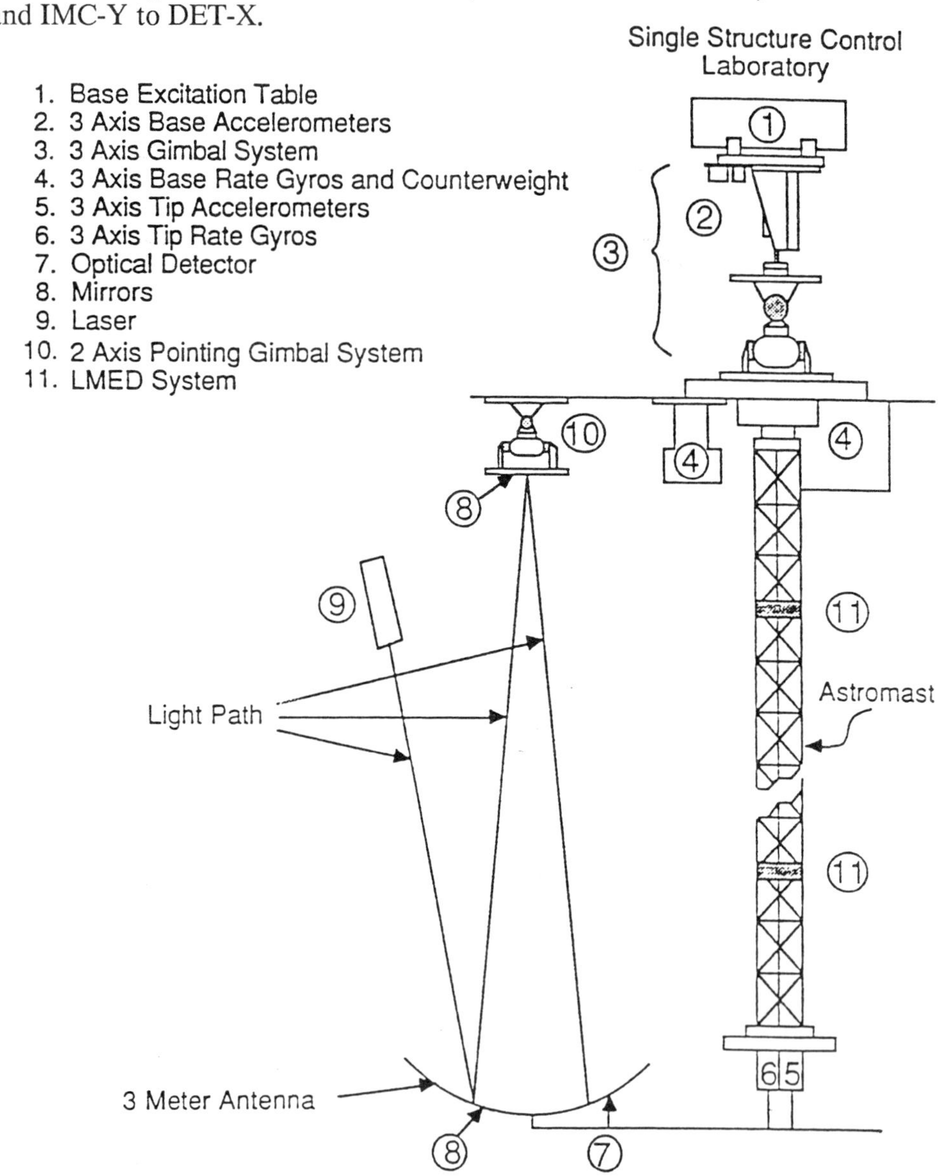

Figure 1 The ACES Experiment

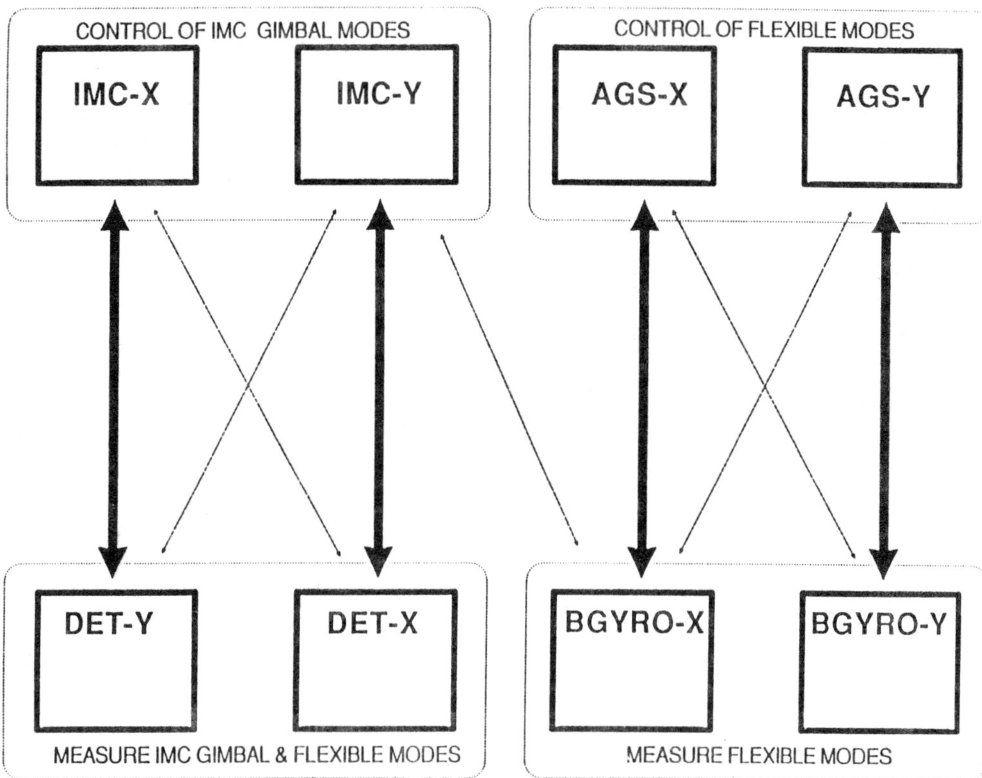

Figure 2 The pattern of strong and weak interaction for the ACES structure.

Due to time and budgetary constraints we decided to develop control design models by using the Eigensystem Realization Algorithm (ERA) [17,18] to develop state space models of the four dominant loops instead of modifying the FEM. It is important to note that since all test data were collected by using the control sensors and actuators, models generated from this data should account for the sensor and actuator dynamics.

Due to a change during the project in the internal configuration of the LMED's, we delayed control design activities involving these devices. Ultimately, it was decided to simply feed back the collocated LMED accelerometers to the corresponding LMED force axis and to determine simple dynamics for these controllers by using crude models of the loops, developed by information provided to us by MSFC and our knowledge of proof mass devices. These collocated LMED loops did provide significant performance improvement.

The control design models for the AGS-X to BGYRO-X and AGS-Y to BGYRO-Y loops were respectively 17th and 19th order continuous-time models. First-order all-pass filters were incorporated into these models to account for the computational delay. The control design models for the loops IMC-X to DET-Y and IMC-Y to DET-X were both 4th order discrete-time design models. Ideally, for digital implementation of a control algorithm it is better to design the controller directly using discrete-time representations of the system model. We chose to design the AGS to BGYRO feedback loops using continuous-time models due to the need for robust control and because of a greater maturity in the theory and software development for robust control in the continuous-time setting.

Performance of the integrated controller for the ACES Structure

The final (integrated) controller consisted of decentralized controllers for each of the three major feedback paths. Since the order of the controllers for the IMC gimbal to detector loops, the AGS gimbal to base gyro loops, and the collocated LMED loops were respectively 6th, 10th and 12th order, the integrated controller consisted of 28 states. The controller was evaluated for disturbance commands to BET-X and BET-Y.

Controller (Controller order)	Mean Error & Standard Deviation	dB Improvement	Test Result Time Histories for X LOS
Open Loop	mean = 0.0329	————	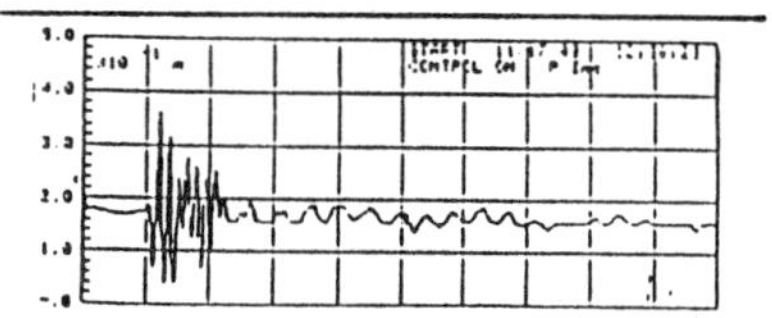
	S.D. = 0.0256	————	
IMC (6 states)	mean = 8.5×10^{-5}	51.8	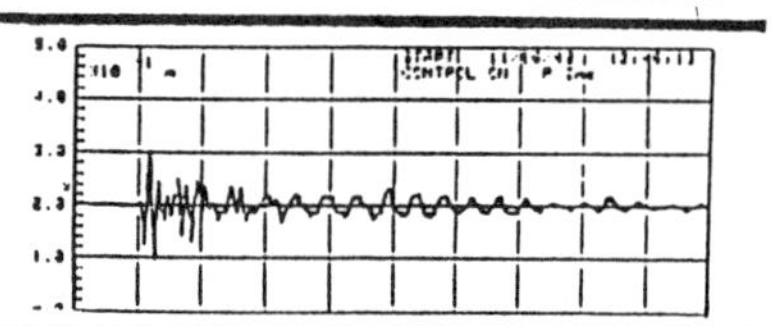
	S.D. = 0.01742	3.34	
IMC + AGS (16 states total)	mean = 8.4×10^{-5}	51.9	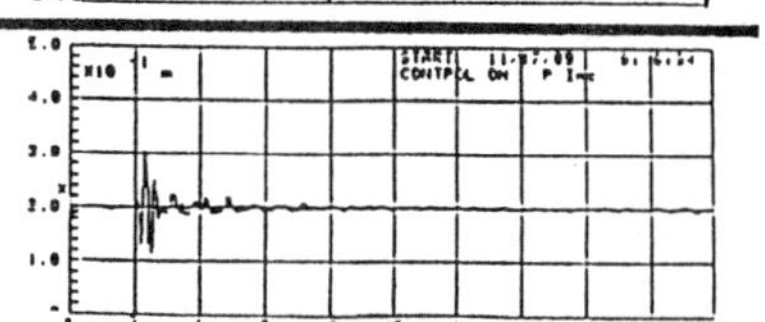
	S.D. = 0.0107	7.58	
IMC + AGS + LMED (28 states total)	mean=4.2×10^{-5}	57.9	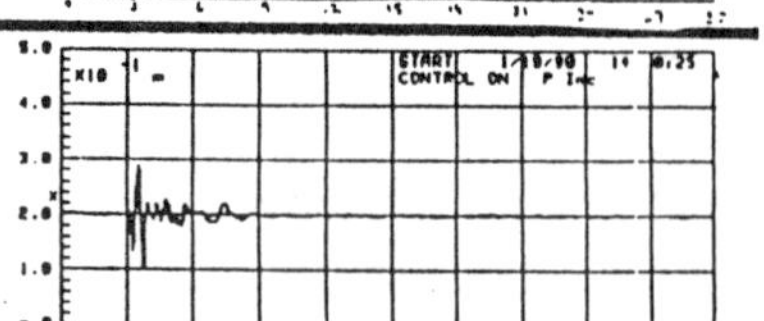
	S.D. =9.0×10^{-3}	9.0	

Figure 3 LOS-X performance for the ACES experiment as the individual controllers were combined

The two measures of performance that were computed to compare the open-loop and closed-loop performance are the mean ($\bar{q}$) and the standard deviation (σ), defined respectively as [16, p. 2--24]

$$\bar{q} \triangleq \frac{1}{N}\sum_{i=1}^{N}$$

$$\sigma \triangleq \sqrt{(\sum_{i=1}^{N} q_i^2 - N\bar{q}^2)/(N-1)}$$

where for a given sensor output N is the number of recorded samples over a specified time interval and $\{q_i\}_{i=1}^{N}$ is the sequence of sampled values. The performance improvement in dB for both the mean and standard deviation is defined as

$$\text{Performance Improvement} = 20\log_{10}\left(\frac{\text{OL performance}}{\text{CL performance}}\right).$$

For pulse commands to BET-X Figure 3 shows the resultant performance improvement respectively in the DET-X response as the controllers were combined. The relatively large improvements in mean are due to large open-loop detector biases which were effectively eliminated using feedback. In general, the mean values were dominated by the size of the initial biases.

Figure 4 shows a direct comparison of the open loop response of the X axis of the detector to the closed loop response with the integrated controller.

THE MULTI-HEX PROTOTYPE EXPERIMENT

The Multi-Hex Prototype Experiment [2] (MHPE) emulates the generic properties of large space structure concepts pertaining to high frequency RF or optical applications. The MHPE contains five major subsystems: the structure, vibration platform, actuators and sensors, the interface electronics, and the MCX-5 Data Acquisition and Control Computer.

The structure is composed of a Multi-Hex panel array mounted on top of a six-member truss which in turn is mounted on top of a vibration platform. As shown in Figure 5, the Multi-Hex panel array is a seven panel graphite/epoxy array wherein each panel is a hexagonal box truss measuring 4.3 feet across. The whole array weighs 177 pounds and is 12.8 feet across. On top of the Multi-Hex panel array a tripod tower supports the secondary reflector structure. The six-member truss consists of six aluminum tubes which attach to the array at three locations (every other corner of the center panel) and to the vibration platform. The structure also contains forty-two aluminum honeycomb facets which emulate a carbon-carbon mirror surface in weight and stiffness. A finite element model has been created for this MHPE structure. From this model fifty modal frequencies and modeshapes have been extracted. Typical modeshapes may be categorized into several groups. One group of modeshapes involves bending of the outer panels with little motion of the center panel. These modes depend on the stiffness of the joints that connect the outside panels to the center panel. Another group of modes involves significant bending of both the array panels and the six-member support truss. Finally there are numerous highly localized modes wherein a subset of the panels vibrate while the remaining panels are nearly quiescent.

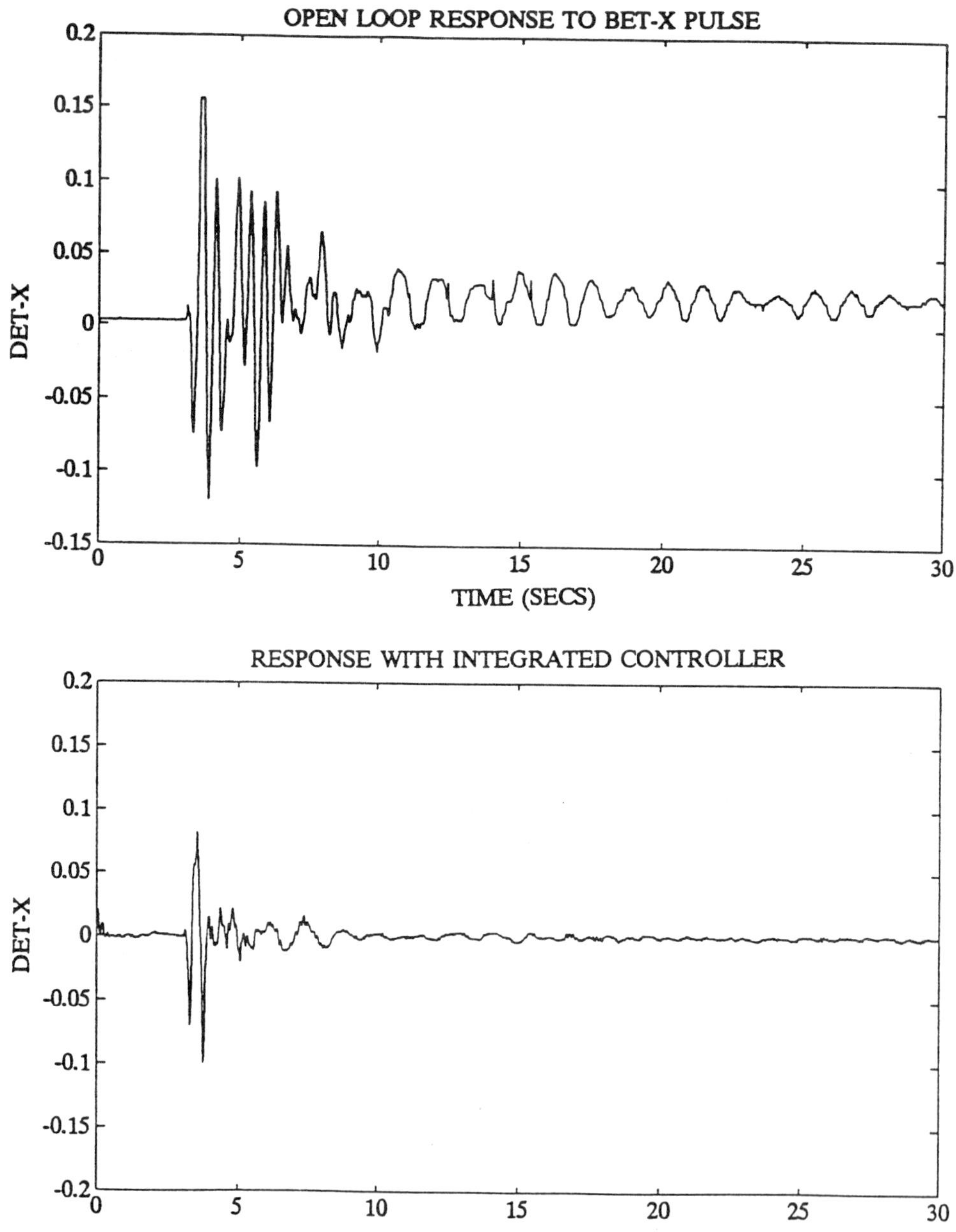

Figure 4 Open loop vs. closed loop LOS-X performance with the integrated controller for the ACES structure.

Figure 5 The Multi-Hex Prototype Experiment.

The vibration platform consists of a five foot diameter, one inch thick aluminum table to which three vertically directed disturbances can be applied (to simulate aft body disturbances from the spacecraft) by electrodynamic shakers. Mounted in tandem with each shaker is an air bag isolator with associated bracketry to support the weight of the entire structure and to isolate it from ground vibration.

The actuators and sensors consist of six Linear Precision Actuator (LPACT) units and twenty-four inertial grade accelerometers. The LPACT device, shown in Figure 6, is a bearingless, linear voice coil actuator designed, fabricated, and tested at Harris Corporation [6]. Each LPACT utilizes a proof mass mounted accelerometer to close a force control loop which serves to override nonlinearities and to reduce the effect of the restraint flexure mass resonance. Without the force compensation, a second order, very lightly damped resonance at 9 Hz is present in the transfer function from voltage input to force output. With the force compensation this actuator has a flat response from 5 to 500 Hz (see Figure 7). This device can provide a maximum force of five pounds and has a twenty micro-pound force resolution. With each LPACT there is an externally mounted collocated accelerometer for the implementation of a decentralized (using collocated actuator/sensor only) structural control loop and/or a centralized (utilizing all of the actuators and sensors) control loop. Each accelerometer has one micro-g resolution with a flat response from 0 to 300 Hz. LPACTs are situated in the center of each of the six outer hexagonal panels.

Figure 6 Linear Precision Actuators.

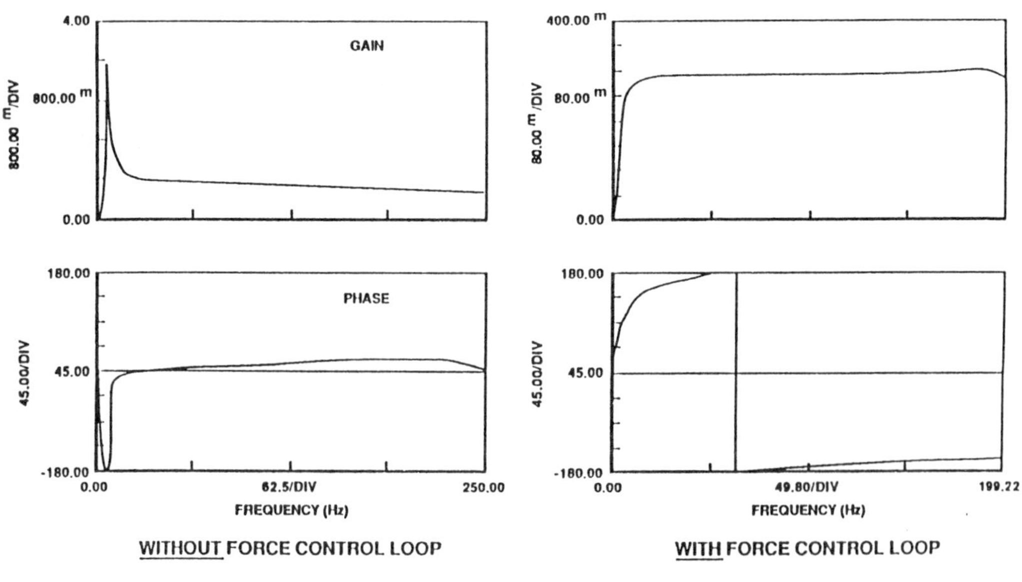

Figure 7 Dynamics of an LPACT with and without internal force control.

The interface electronics consist of the signal conditioning electronics, force loop compensation electronics, decentralized control electronics, LPACT interface electronics, and the servo amplifiers. The signal conditioning, force loop compensation and decentralized control electronics, and the LPACT interface electronics reside in the Structural Interface Assembly (SIA) and the servo amplifiers reside in the Servo Amp Assembly (SAA). The signal conditioning electronics scale the sensor outputs so that they are in the measurement range of the A/D converter of the Data Acquisition Control Computer; they also filter the signals to reduce aliasing effects. The force loop electronics takes the proof mass mounted accelerometer measurement and provides the electronic compensation required to generate a current command to close the force loop. The decentralized analog control electronics use the collocated accelerometer measurement input and provide up to eighth order compensation, which generates a command to close the decentralized control loop. The LPACT interface electronics sum the current commands from the force loop, decentralized loop and the command from the centralized control loop. The current command from the LPACT interface goes to the servo amplifier which drives the LPACT unit.

The MCX-5 computer performs the data acquisition and control required to close the centralized vibration suppression control loop during a controls experiment. This computer is used to correlate the test data with the modeling, analysis and simulation results.

The design goals of this experiment are to reduce the mirror panel dephasing and line-of-sight (LOS) error by 30 dB. The preliminary state space plant model contains the first twenty modes of the structure.

The first controllers successfully implemented were a set of six decentralized collocated analog controllers. To implement these controllers the Hybrid Accelerometer, a special type of sensor, was designed and fabricated. The Hybrid accelerometer was designed to take advantage of the high accuracy of the servo accelerometer (QA-700) and the high bandwidth of a piezoelectric accelerometer (PCB 303A). These two accelerometer measurements are electronically combined to provide an acceleration measurement which has a flat frequency response from DC to 5 KHz. With LPACT dynamics flat from 5 to 500 Hz and hybrid accelerometer dynamics which are flat from DC to 5KHz a relatively simple controller can be designed that is positive-real up to 1000 Hz. These loops add damping to the unmodeled high frequency modes that can potentially destabilize sample-data controllers.

This decentralized control system architecture provides a high degree of fault tolerance. That is, if any of the actuators and sensors fail, the remaining actuator/sensor pairs will provide reduced but stable vibration attenuation performance. Also, since the loops are independent of each other, synchronization during turn on is not required.

Since the addition of the tower, 3 additional LPACT units have been mounted on top of the tower in a delta configuration. Hybrid controllers have been designed consisting of a 6 input-6 output digital controller and simple analog loops. The digital portion of these controllers provide a high degree of attenuation for modes whose frequencies are less than 120 Hz while the analog loops provide attenuation of both low and high frequency modes. Figure 8 shows the performance of a hybrid controller under persistent disturbance excitation with a 35.8 Hz constant amplitude sinusoidal disturbance. Figure 8 also displays the fault tolerance of the hybrid controller by showing that stability is maintained and the performance degrades gracefully as the LPACTs are turned off one-by-one. To demonstrate the high bandwidth of the analog loops a constant 411 Hz disturbance was applied and the resulting accelerometer responses are shown in Figure 9.

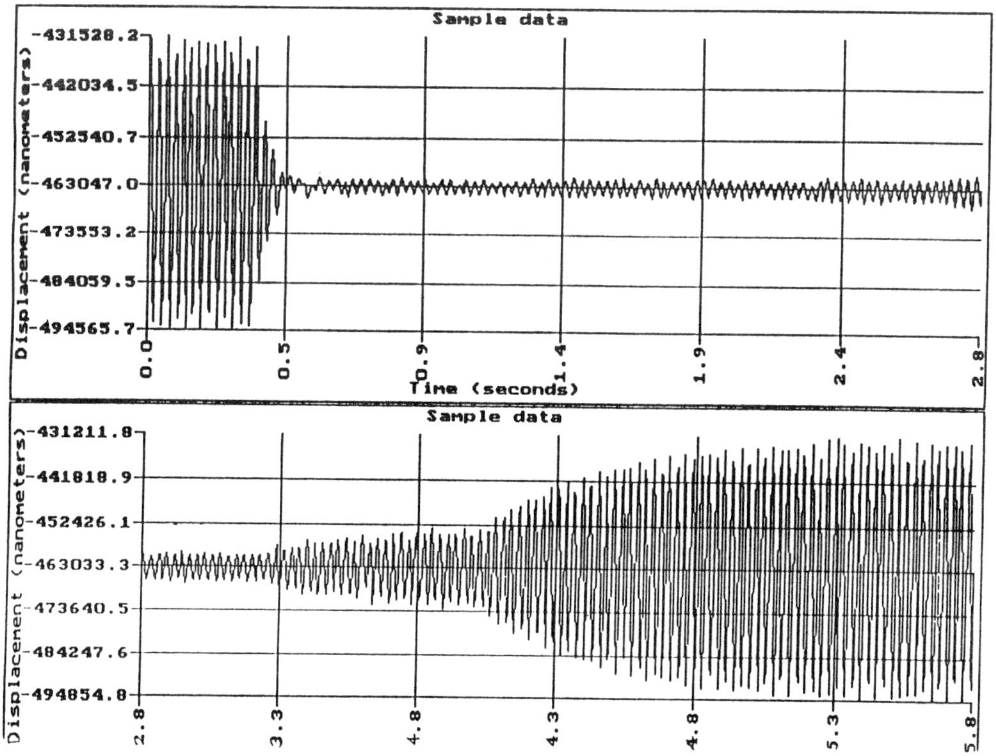

Figure 8 Performance of a hybrid controller for the MHPE as the LPACTs are turned off one at a time. (The controller was turned on after 0.4 sec).

CONCLUSIONS

These experiments have demonstrated successful control system design and implementation for flexible structures and have provided validation of the Maximum Entropy/Optimal Projection approach to controller design. Based on our experience with these experiments we make the following remarks and recommendations for future development in the theory and technology associated with active control for flexible structures.

(i) For complex structures finite element models will often not be sufficiently accurate for designing implementable controllers. Thus, identification tools (such as the Eigensystem Realization Algorithm) need to be further developed to increase their accuracy and efficiency in producing models of flexible structures.

(ii) Actuator and sensor dynamics can be a very influential factor in control design. It is very important to develop sensors and actuators with dynamic characteristics tailored to achieving high performance for a given structure (or class of structures).

(iii) Sampled-data issues can also be very influential factors in control design although these issues are somewhat neglected by modern control theories. Much more attention should be given to sampled-data issues by control theoreticians.

(iv) Design of implementable controllers requires numerous iterations and the manipulation of large amounts of data. This necessitates the continual evolution of efficient and ergonomically sound design environments such as MATLAB and MATRIXx.

(v) Relatively simple controllers (i.e., low order and/or decentralized) controllers can provide significant performance improvement.

(vi) A practical control design and implementation approach is to start with "simple" controllers and increase controller complexity as needed to increase performance.

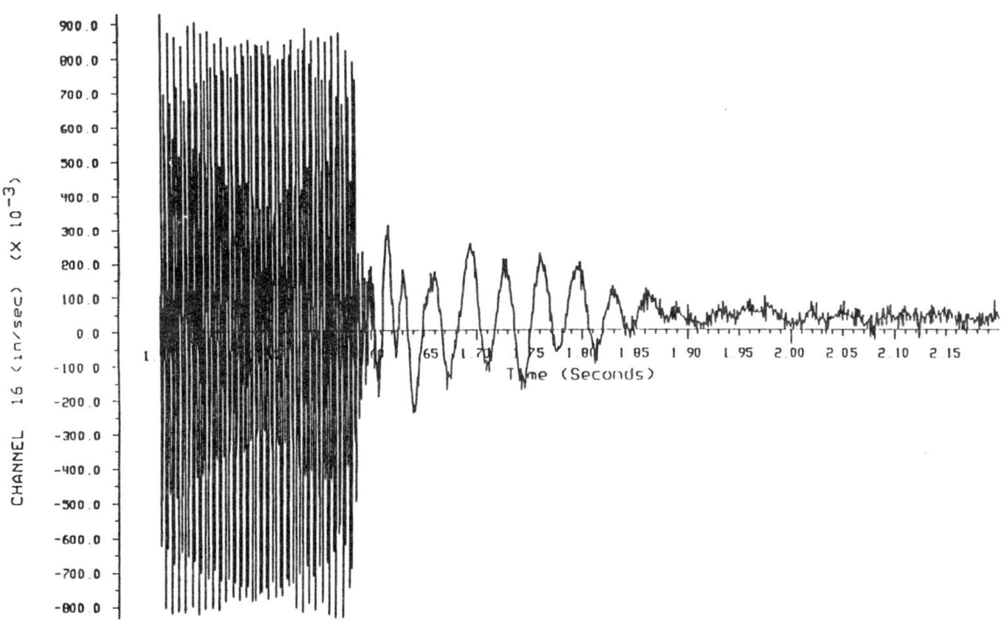

Figure 9 Performance of the MHPE controller for a 411 Hz disturbance. (The controller was turned on after 1.58 secs.)

REFERENCES

1. S. W. Greeley, D. J. Phillips, and D. C. Hyland, "Experimental Demonstration of Maximum Entropy, Optimal Projection Design Theory for Active Vibration Control," *Proc. Amer. Contr. Conf.*, Atlanta, GA, June 1988, pp. 1462--1467.

2. D. C. Hyland and E. G. Collins, Jr., "A Robust Control Experiment Using an Optical Structure Prototype," *Proc. Amer. Contr. Conf.*, Atlanta, GA, June 1988, pp. 2046--2049.

3. D. J. Phillips, E. G. Collins, Jr., and D. C. Hyland, "Experimental Demonstrations of Active Vibration Control of Flexible Structures," *Proc. IEEE Conf. Dec. Contr.*, pp. 2024--2029, Honolulu, HI, December 1990.

4. E. G. Collins, Jr., D. J. Phillips, and D. C. Hyland, "Design and Implementation of Robust Decentralized Control Laws for the ACES Structure at the Marshall Space Flight Center," *Proc. Amer. Contr. Conf.*, San Diego, CA, May 1990, pp. 1449--1454.

5. E. G. Collins, Jr., D. J. Phillips, and D. C. Hyland "Robust Decentralized Control Laws for the ACES Structure," *Contr. Sys. Mag.*, to appear.

6. D. C. Hyland and D. J. Phillips, "Development of the Linear Precision Actuator," *Proc. 11th Annual AAS Guid. Contr. Conf.*, Keystone, CO, January 1988.

7. D. C. Hyland, "Minimum Information Stochastic Modeling of Linear Systems with a Class of Parameter Uncertainties," *Proc. Amer. Contr. Conf.*, Arlington, VA, June 1982, pp. 620--627.

8. D. C. Hyland, "Maximum Entropy Stochastic Approach to Controller Design for Uncertain Structural Systems," *Proc. Amer. Contr. Conf.*, Arlington, VA, June 1982, pp. 680--688.

9. D. C. Hyland and D. S. Bernstein, "The Optimal Projection Equations for Fixed-Order Dynamic Compensation," *IEEE Trans. Autom. Contr.*, Vol. AC--29, 1984, pp. 1034--1037.

10. D. S. Bernstein and D. C. Hyland, "The Optimal Projection/Maximum Entropy Approach to Designing Low-Order, Robust Controllers for Flexible Structures," *Proc. IEEE Conf. Dec. Contr.*, Fort Lauderdale, FL. December 1985, pp. 745--752.

11. D. S. Bernstein and S. W. Greeley, "Robust Controller Synthesis Using the Maximum Entropy Design Equations," *IEEE Trans. Autom. Contr.*, Vol. AC--31, 1986, pp. 362--364.

12. S. Richter and E. G. Collins, Jr., "A Homotopy Algorithm for Reduced Order Compensator Design Using the Optimal Projection Equations," *Proc. 28th IEEE Conf. Decis. Contr.*, Tampa, FL, Dec. 1989, pp. 506--511.

13. E. G. Collins, Jr., and S. Richter, "A Homotopy Algorithm for Synthesizing Robust Controllers for Flexible Structures Via the Maximum Entropy Design Equations," *Third Air Force/NASA Symposium on Recent Advances in Multidisciplinary Analysis and Optimization*, San Francisco, CA, September 1990, pp. 324--333.

14. D. S. Bernstein and D. C. Hyland, "Optimal Projection for Uncertain System (OPUS): A Unified Theory of Reduced-Order, Robust Control Design," in *Large Space Structures: Dynamics and Control*, S. N. Atluri and A. K. Amos, eds., Springer-Verlag, New York, 1988.

15. D. S. Bernstein and D. C. Hyland, "The Optimal Projection Approach to Robust, Fixed-Structure Control Design," *Mechanics and Control of Space Structures*, J. L. Junkins, Ed., AIAA Publication, 1990, to appear.

16. R. D. Irwin, V. L. Jones, S. A. Rice, D. J. Tollison and S. M. Seltzer, *Active Control Technique Evaluation for Spacecraft (ACES)*, Final Report to Flight Dynamics Lab of AF Wright Aeronautical Labs, Report No. AFWAL-TR-88-3038, June 1988.

17. J. N. Juang and R. S. Pappa, "An Eigensystem Realization Algorithm for Modal Parameter Identification and Model Reduction," *J. Guid. Contr. Dyn.*, Vol. 8, 1985, pp. 620--627.

18. J. N. Juang and R. S. Pappa, "Effects of Noise on Modal Parameters Identified by the Eigensystem Realization Algorithm," *J. Guid. Contr. Dyn.*, Vol. 9, 1986, pp. 294--303.

AAS 91-052

EXPERIMENTAL VERIFICATION OF NANOMETER LEVEL OPTICAL PATHLENGTH CONTROL ON A FLEXIBLE STRUCTURE

Michael O'Neal*, Daniel Eldred†, Dankai Liu† and David Redding‡

This paper describes an experimental facility being developed at Jet Propulsion Laboratory for demonstration and validation of control concepts arising out of NASA's Control Structure Interaction program. The facility is meant to be a ground testbed with relevance to a broad class of future precision optical space systems. The objective of the experimental program is to investigate a multi-layer control approach to the maintenance of nanometer-level optical pathlength stability in the presence of external disturbances and multiple structural resonances. Ultimately, the facility will explore the effect of applying, separately and in combinations, structural vibration suppression, vibration isolation, and active optical articulation. This paper describes the testbed facility, the structure, optics, sensors, actuators, and real-time computer and program development environment. Initial optical articulation experimental results are presented.

INTRODUCTION

The development of future space and lunar-based astronomical instruments such as telescopes and optical interferometers will require major advances in the control, alignment, and pointing of distributed optical systems mounted on large flexible space structures. For example, a feasibility study performed at the Jet Propulsion Laboratory (JPL) of a space-based optical interferometer called the Focus Mission Interferometer (FMI) [1] showed that a combination of disturbance isolation, active and passive structural vibration suppression, and active optical pathlength compensation would be required in order for the interferometer to function at full effectiveness.

No single layer of this multi-layer architecture would be sufficient, by itself, to ensure success of a system where micron-level open loop response needs to be quieted to the nanometer regime to meet the strict optical tolerances. In order to investigate the proper blending and interactions between the layers in this architecture at the nanometer level, JPL

* Member of Technical Staff, Mechanical Systems Development Section, Jet Propulsion Laboratory, California Institute of Technology, 4800 Oak Grove Drive, Pasadena, California 91109.

† Member of Technical Staff, Guidance and Control Section, Jet Propulsion Laboratory, California Institute of Technology, 4800 Oak Grove Drive, Pasadena, California 91109.

‡ Site Manager, Space Programs Directorate, Charles Stark Draper Laboratory, 555 Technology Square, Cambridge, Massachusetts 02139. Member AAS.

has constructed an experimental ground testbed, the "Phase B Testbed," as part of its ongoing effort in the NASA-wide Control Structures Interaction (CSI) program . The Phase B Testbed is part of an evolutionary chain of testbeds at JPL. Earlier work [2] has demonstrated the capability to perform structural vibration suppression, which is only one of the layers of this control architecture.

The configuration of the Phase B Testbed consists of a scaled subset of the FMI, and it exhibits many of the same problems as the actual instrument but is much simpler and more manageable in a laboratory environment. This is an important consideration in order for the Testbed to be successfully constructed and brought into working order within the budgetary and manpower constraints imposed on the program. Components of the Phase B Testbed include: a structure with multiple resonances within the bandwidth of interest; an optical path which is correlated with structural motion; sensors and actuators for isolating the disturbances, suppressing structural vibrations, and compensating the optical pathlength; and real-time computers for implementation of the control algorithms and recording the results. Some of the significant features which would exist in a working interferometer but which were omitted on the Phase B Testbed include: an external metrology system to estimate overall system alignment (but that would require dozens of laser interferometers); siderostats and fast steering mirrors for steering incoming light beams (which create an important disturbance source as well as wavefront tilt errors); rigid body motion control via momentum wheels and thrusters; and a track for increasing the optical delay path by as much as 10 meters. On the other hand, gravity loading, which puts the structure in an initial strain condition and causes a degree of nonlinearity, exists in the laboratory but not in an orbiting system. Seismic, acoustic, and electrical noise and air induced variations in the index of refraction further complicate ground-based experiments.

An important aspect of the CSI program is the integrated development environment that is associated with the development of the CSI methodology. In particular, since one of the CSI objectives is to design control systems simultaneously with the structures they act upon, it was deemed important that the structural analysts and the control analysts work closely together so that engineers in one discipline have an appreciation of the demands of the other, if not the analytical skills. Towards this aim, the offices of the CSI design team at JPL were collocated and cross-disciplinary training has been strongly encouraged. In further support of this, nearly all analysts' computers have been linked together on a local network. As evidence of the success of this interdisciplinary cross-training, structural analysts have designed and implemented working controllers, and control analysts have modeled the structures together with their controllers and performed structural identification on the Phase B Testbed.

PHASE B TESTBED DESCRIPTION

The structure comprising the Phase B Testbed (Fig. 1) consists of a vertical tower standing 2.5 m (8 ft) high and two horizontal arms attached to the top of the tower and oriented at right angles to one another. The tower is rigidly mounted on a massive block of steel (1500 kg, 3300 lb) which provides rigidity and a measure of isolation of the structure from vibrations emanating from the ground. The block is supported above the ground by three jack-screws, the height of which are adjustable. This design permits precise leveling of the block and furthermore eliminates the possibility of the steel block rocking on an uneven floor.

Figure 1: The JPL Phase B Testbed

The basic unit for the assembly of the Phase B Testbed tower and arms is a "bay," which consists of a 40.6 X 40.6 X 28.7 cm (16 X 16 X 11.3 in) truss structure constructed from 1.27 cm (0.5 in) aluminum tubes or "struts," with a wall thickness of 1.5 mm (.058 in). These bays can also be combined easily to form different geometries for future experiments. The struts are attached to aluminum "nodes" manufactured by Starnet Structures Inc. which have threaded holes drilled in them at the correct angles. All of the nodes are identical. In all, the Testbed is comprised of 14 bays, 186 struts and 60 nodes. The attachment of the struts to the nodes is facilitated by custom hardware which allows any strut to be removed and replaced without any other disassembly or disturbance to the structure. One side of the attachment hardware remains permanently in the node and can be adjusted to compensate for variations in strut lengths. The other side remains attached to the strut. Once a strut is slipped into place, a "B" nut is tightened to obtain a rigid connection. The attachment hardware is also designed to be flexible relative to the stiffness of the aluminum tubes so the moments transmitted through the nodes to the struts are minimized.

An optical motion compensation system is encased in a flexure mounted frame called the "trolley" and attached to the end of the longer horizontal arm. The purpose of this optical system is to provide precise control actuation of the optical light pathlength at high bandwidths. While it is beyond the scope of this paper to describe the intricate workings of an optical interferometer, such a system is required in a stellar interferometer to control the optical pathlengths, and this trolley is identical to one currently being implemented as part of the delay line on a ground-based interferometer. Functioning interferometers need to control wavefront tilt as well as optical pathlength, but it has been determined [1] that pathlength control is the more challenging requirement to meet.

A photograph of the trolley hardware is shown in Fig. 2, and a schematic is shown in Fig. 3. The basic optical configuration, called "Optics Option 0", comprises a cat's-eye retroreflector and incorporates both a primary and secondary mirror to reflect light outward in a direction parallel to the incoming light beam, although translated in position. The trolley frame is constructed from Invar, a nickel-iron alloy which exhibits low thermal expansion, and aluminum to create a thermally stable separation between the primary and secondary mirrors. The trolley is suspended from an aluminum base by a flexure panel on each end. These panels are rigid and are attached to the top of the trolley and the inside of the base with Bendix pivot flexures. This arrangement provides a large measure of mechanical isolation of the trolley from dynamic motion of the Testbed structure. The base is bolted to the structure by means of a mounting bracket, the stiffness of which can be varied to tune the frequencies of the local modes so they do not interfere with control designs.

The Phase B Testbed has a two stage compensation system to control the optical pathlength. A third, very course stage, which involves moving the trolley along a track to extend the stroke to several meters, is included in the FMI, but omitted from the Testbed, because it does not pose the same level of technical challenge. The first, or course, stage involves moving the entire trolley relative to its base, which is done via a voice coil actuator which reacts between the trolley and the base with a maximum output of ± 62 N (14 lb). The total stroke is 6 mm (0.25 in). Due to the mass of the trolley and limited frequency response of the voice-coil actuator, this stage works best at lower frequencies, nominally from steady state to 100 Hz. The trolley weighs approximately 13 kg (28 lb), so motion in this stage will significantly excite the Testbed structure.

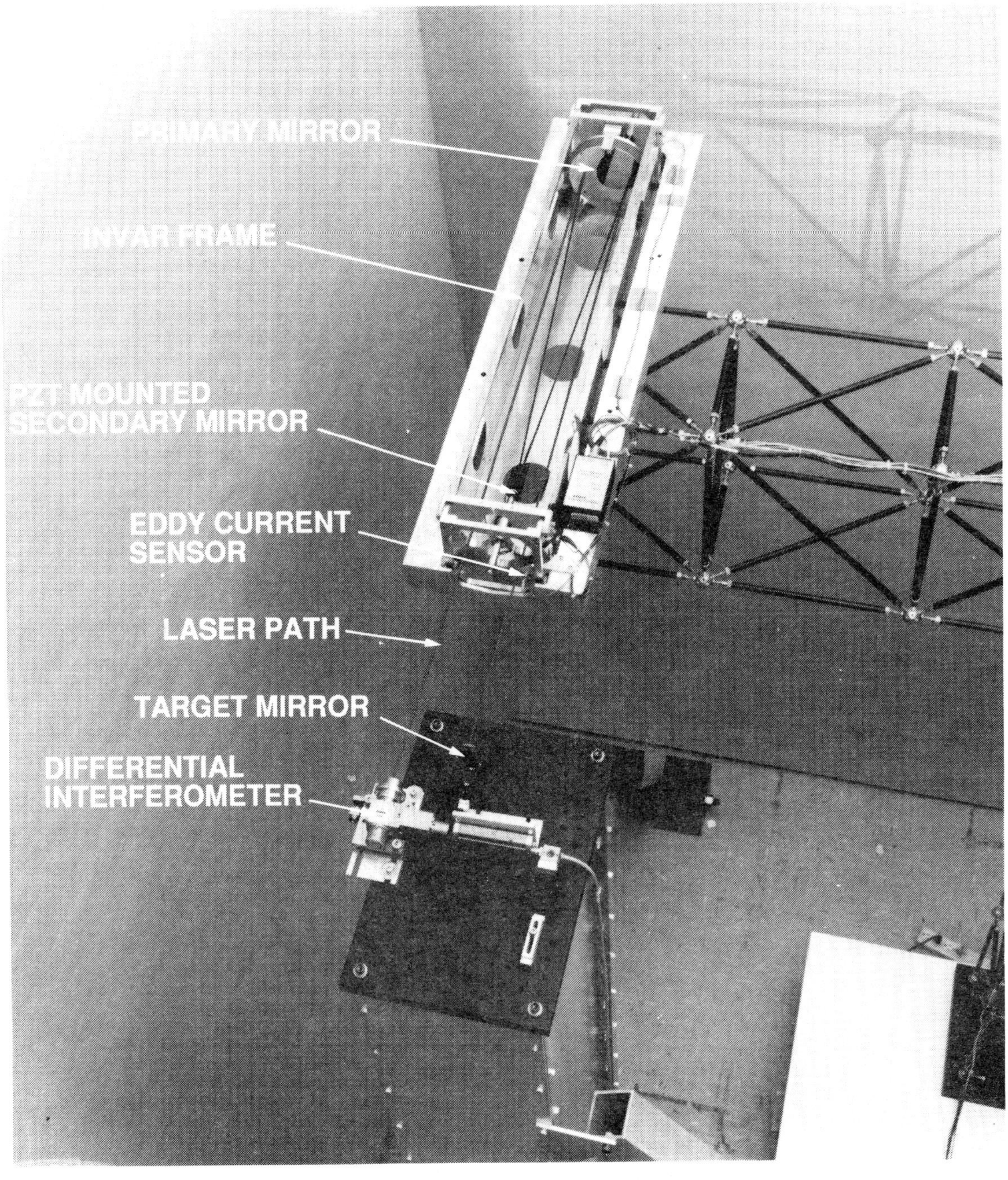

Figure 2: The Phase B Testbed Optical Motion Compensation System

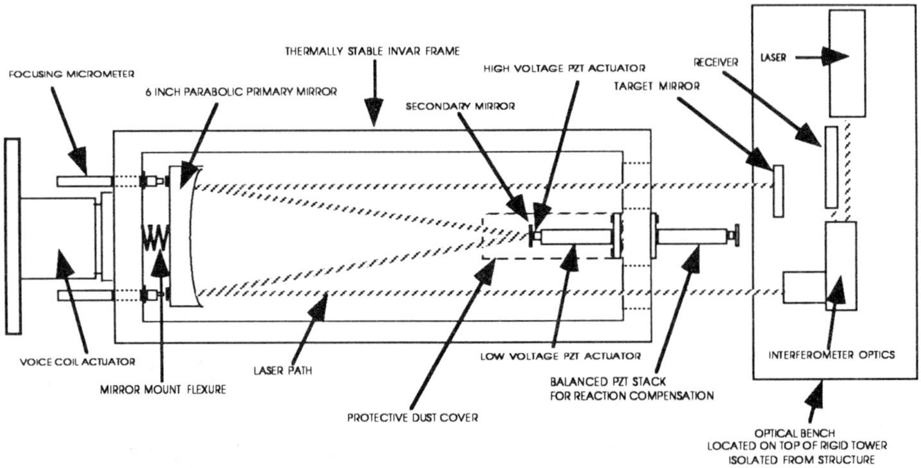

Figure 3: Schematic of Optical Motion Compensation System Showing Optics Option 0

The finer stage of motion compensation is accomplished by moving the secondary mirror relative to the trolley frame. The secondary mirror is mounted atop a two-stage piezoelectric stack (PZT); the coarse stage (low voltage PZT) having a stroke of 30 μm (1200 μin) and a resolution of 10 nm (0.4 μin), and the fine stage (high voltage PZT) having a stroke of 3 μm and a resolution of better than 1 nm. The bandwidth of the PZT actuators can be as high as 10 KHz. To prevent motion of the 9 g (0.02 lb) mirror from coupling with other dynamics, the PZT actuator was made to be reactionless by including an identical stack fed by the same command to move a counterweight in the opposite direction.

To increase the complexity of the experiment, additional optics can be added to the Testbed to couple structural motion into the pathlength more directly. In "Optics Option 1", shown in Fig. 4, motion of the trolley mount point is coupled directly to the pathlength because the target mirror is mounted on the trolley base which moves with the structure. "Optics Option 2", shown in Fig. 5, incorporates another structural point into the loop, so each mode will affect the pathlength differently. Additional complexity can be achieved by tuning the local modes of the trolley mounting bracket to be within the bandwidth of the control.

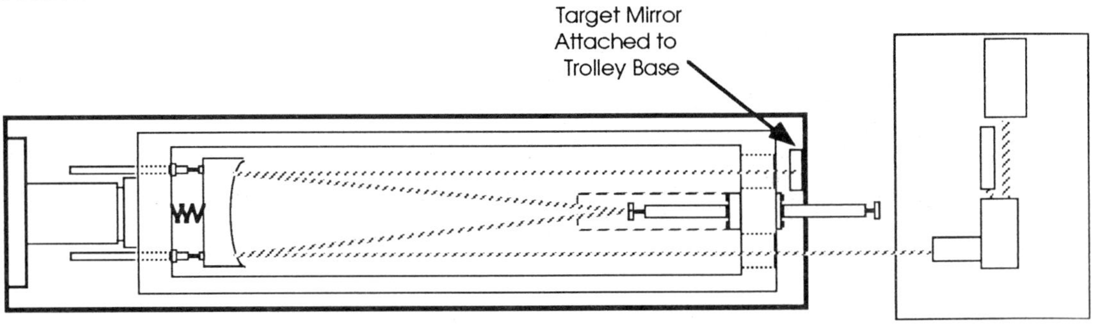

Figure 4: Optics Option 1

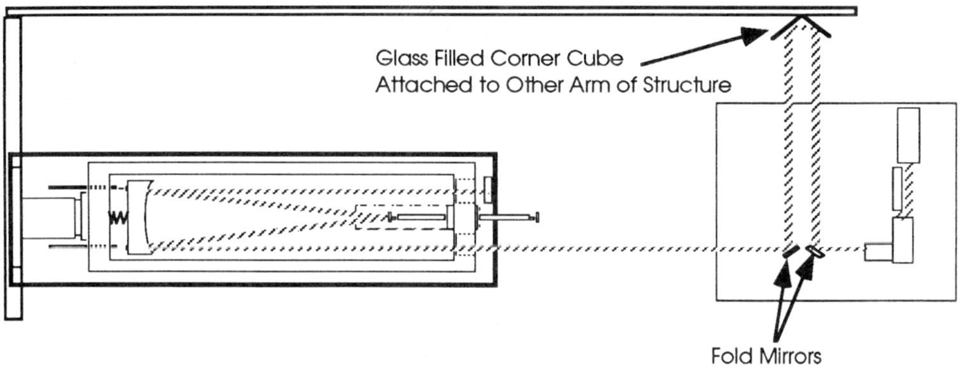

Figure 5: Optics Option 2

A Hewlett Packard heterodyne laser interferometer is used to measure pathlength to a resolution of 2.5 nm (.1 μin). Due to the geometry of the laser path on the Testbed, the distance the laser measures is actually four times what the optical path would be in an equivalent stellar interferometer, so the corresponding resolution of stellar pathlength is better than 1 nm. The target mirror for this interferometer is a flat mirror, so to first order the pathlength is insensitive to motions perpendicular to the optical path. The optics associated with this interferometer are mounted on an optical bench attached to the top of a 2.7 m (9 ft) high "laser tower". Since tower motions couple directly into pathlength, care was taken to make the tower as quiet as possible. The tower is an aluminum box-beam design which is intended to be as stiff as possible to avoid structural resonances and for maximum rejection of ambient disturbances from the laboratory space. Furthermore, the tower rests on a layer of energy absorbent neoprene rubber. Additional damping is achieved by placing sheets of neoprene between the panels and stringers of the box beam and under the optical bench. The residual tower motions are small and well within the control bandwidths, and thus are treated as an additional disturbance source. With the current configuration of the tower, the disturbance primarily consists of 0.75 μm (0.02 μin) peak-to-peak oscillations centered at 4 Hz and 70 nm (0.002 μin) at 17 Hz. Additional quieting can be achieved in the future by sensing the motion of the top of the tower with an accelerometer and either feeding the signal forward to the PZT mirror, or to a proof mass actuator located on the top of the tower.

A number of additional sensors are available and will be used in the future for monitoring structural motion for control implementation. Accelerometers can be located virtually anywhere on the trolley, structure, or laser tower. In general, the accelerometers are mounted to flats on the nodes using cyanoacrylate glue, and are removed by forcibly breaking the glue connection. In addition, a Kaman 2310 eddy current sensor is mounted on the trolley and outputs the relative displacement between the trolley and the structure. The PZT actuator is sufficiently stiff and devoid of dynamics so that its displacement is known directly from the input command voltage.

Though at present only the trolley actuators are being used, the Testbed is designed to incorporate a variety of actuators, which will be required to satisfy future control requirements. These actuators include: active struts, which employ piezoelectric elements to control the length of the struts and also employ eddy current sensors to monitor their elongation and load cells to monitor the applied load [3]; passive damping "D-Struts" manufactured by Honeywell, which employ a viscous fluid to achieve damping [4], and proof-mass actuators, which are currently being investigated.

Disturbance sources can be mounted virtually anywhere on the structure and can act in any direction. Currently, Mini-Shaker model 4810 shakers, manufactured by Bruel & Kjaer, Inc., are being used, which are capable of outputting 10 N (2.25 lb) peak force up to 18 kHz. Attachment of a shaker to the structure is achieved via "stingers," which consist of thin mechanical links between the shaker and the structure which are sized to avoid moment transmission and also to function as a mechanical fuse, breaking to protect the Testbed in case an attempt is made to apply too large of a disturbance source. In general, a force transducer is mounted in series with the stinger so that the magnitude of the disturbance can be measured. Currently, each shaker is suspended on a cable over 2 m (7 ft) long. With this arrangement, the shaker tends to sway about and thus the direction of the input force varies somewhat; to address this problem a fixture is currently being designed which allows a disturbance shaker to be rigidly affixed to any bay of the Testbed structure, and the force can be oriented in any arbitrary direction. This fixture can be modified to include another shaker so pure couples can be applied. These shakers can be used to emulate profiles from known space disturbance sources (reaction wheels, tape recorders, etc.). The fixture can also be modified to incorporate both hard and soft isolation actuators.

MODELING

A finite element model for the Testbed structure was developed using NASTRAN, employing beam elements for each of the struts. The model included 111 finite element nodes each with 6 degrees of freedom, for a total of 666 stiffness degrees of freedom. After Guyan reduction, the resulting eigenproblem included 237 degrees of freedom, and the standard NASTRAN eigenvalue routines were used to solve for the modal frequencies and mode shapes.

To verify the model for control system design, transfer functions between specific shaker locations and structural responses were determined experimentally. This was accomplished by attaching 45 accelerometers to the structure to measure key translational degrees of freedom. Eigenvalues and eigenvectors were determined with third party and in-house developed curve fitting algorithms. These results are compared to analytically generated transfer functions to identify any serious discrepancies and to experimentally determine the precise frequencies of the modal resonances. They are also used directly for control design.

A summary of the modes generated analytically and experimentally is shown in Table 1 below. The first structural mode consists primarily of torsion about the vertical axis, which couples directly to laser pathlength. Damping in the bare structure with the trolley replaced by an equivalent mass was found to be extremely low, identified as approximately 0.05% of critical damping in many modes. Addition of the trolley with its associated voice-coil actuator and cabling associated with accelerometers increases apparent damping to about 0.1-0.2% damping, depending on the mode and the details of the configuration. The modal frequencies and damping seen have a similar spectrum as those calculated for the FMI.

TABLE 1: MODAL SURVEY

Mode #	Modal Frequency (Hz) Analytical	Measured	Note
1	0.74	0.75	Trolley suspension mode
2	5.43	5.00	First structural mode
3	7.46	8.13	
4	11.68	11.88	
5	17.42	17.50	
6	20.84	23.13	
7	31.14	27.50	
8	40.47	40.63	
9	54.35	57.50	
10	65.70	64.38	
11	68.87	73.75	
12	80.84	82.50	
12a	*	96.88	No analytical mode found
13	106.41	110.00	
14	112.71	116.25	
15	125.83	125.00	

REAL-TIME COMPUTER

Control of the Testbed structure and optical pathlength is achieved using single board computers augmented by single-board array processors [5]. These computers are responsible for sampling the sensors, performing the necessary computations to compute the control gains, and sending these outputs to the various actuators to effect a change in the structure disturbance level. The computers, the array processors, additional memory, a board to read the laser interferometer, and the required analog-to-digital and digital-to-analog converters all reside in a VME bus chassis. This architecture is designed for maximum flexibility in the sense that additional computational power can be obtained simply by adding additional boards, and multiple control loops can be run independently or synchronized together as required.

Software for the real-time computers is designed to ease the analysts' task of implementing their controllers. Written primarily in C, the custom software includes routines for initializing and accessing the sensors, actuators and array processors and also includes a function for recording time histories of specified internal states. The software accepts designs written in discrete-time state-space format, which provides flexibility for a wide variety of linear feedback laws. The analysts generally synthesize their controllers using Pro-Matlab by The Mathworks, and accordingly the real-time software accepts parameter files written directly by Matlab.

The operating system for the real-time computers is VxWorks, by Wind Rivers. VxWorks executes a number of Unix-like commands, and includes the necessary functions for communicating between processors and synchronizing them as well as for debugging programs. Program development, execution and monitoring of data for the real-time computers is performed on Unix workstations which are linked to the real-time computers via Ethernet; in this case both Sun 3/60's and Sparcstation 1's are used. While code produced on the 3/60's is executable directly on the real-time computers, a cross compiler is required for code developed on the Sparcstations and has proved very successful.

The connectivity of the network allows equivalent access to the real-time computers from any workstation connected to the network. This makes it possible for analysts to develop code for the real-time computers from their desktop workstations and to debug and test it on the actual real-time hardware, all without leaving their offices. In fact, since the network is connected to the international Internet, any computer in the world which is also connected to this network is also capable of executing code in the real-time computers, and this capability has been demonstrated via remote login from across the country. Of course, necessary safeguards have been built in to protect the Testbed from unwanted intruders.

A schematic block diagram showing the interconnections among the components of the Testbed is shown below in Fig. 6. At the present time the hardware includes two real-time computers, a four-channel digital-to-analog converter, 16 Mbytes of memory, and the laser interface board. These components are resident in a VME chassis, which itself is mounted in a rack which also includes a monitor and assorted amplifiers and signal conditioning electronics for interfacing between the computers and the transducers. The V30XE and V3E real-time computers, manufactured by Heurikon, are based on the 68030 microprocessor and run with clock speeds of 20 and 25 mHz respectively. The array processor consists of a Quickcard, by CSPI, though it will be replaced shortly by a Supercard II. The digital-to-analog converter is manufactured by Data Translation, and uses 16-bit resolution. With this combination of processors, execution of a 40-state controller with 3 inputs and 3 outputs at a rate of 1 kHz has been demonstrated.

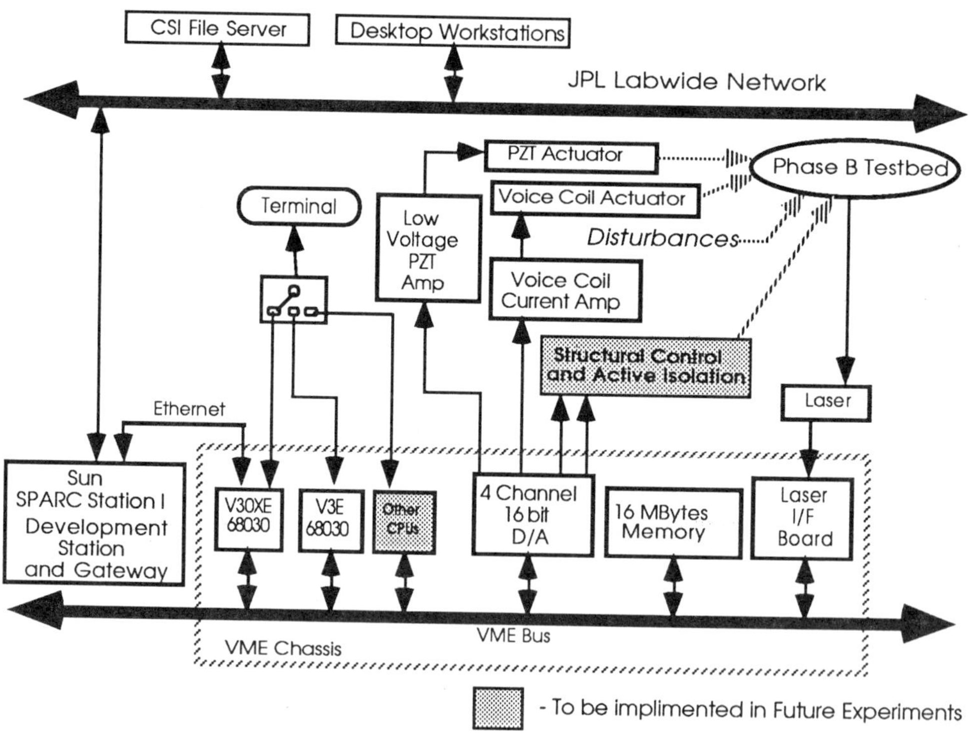

Fig 6: Block Diagram of the Real Time Computer and Testbed Interfaces

EXPERIMENT IMPLEMENTATION

Before experiments could commence, it was apparent that an assortment of problems had to be addressed. Most prevalent was the effect on the laser tower of seismic and acoustic noise present in the laboratory caused by nearby automobile traffic, the opening and closing of doors, movement and conversations of laboratory personnel, and even the 120 Hz hum of a transformer. As the tower was originally constructed, accelerometers mounted on its top detected as much as 100 μm motion in response to ambient noise. Addition of damping material to the tower reduced this level of excitation by more than two orders of magnitude within the frequency range that interferes with control design. The air turbulence of the air conditioning system also caused motion of the Testbed structure and trolley, so it is turned off during experiments with the side effect that the resultant temperature variations cause drift in the laser interferometer. It was discovered that electrical noise originating from the voice-coil amplifier caused mechanical noise in the trolley. These problems are typical of laboratory experiences.

A more serious problem was caused by the bandwidth of the piezoelectric actuator exceeding that of the laser interferometer. Large PZT commands result in pathlength changes occurring at a rate exceeding the Doppler limit of the interferometer, causing the interferometer to lose counts and therefore position. Large PZT commands also excite nonlinear effects in the PZT power amplifier, leading to large overshoot and ringing. These effects can be countered in the controller software by using a limiter on the PZT command, or by restricting the bandwidth of the PZT loop. A hardware fix has also been implemented in the form of an analog integrator with low-pass characteristics. Placed between the digital-to-analog converter and the PZT amplifier, it provides a limiting and smoothing function. Either approach is available to the investigators.

OPTICAL PATHLENGTH CONTROL

The objective of the control experiments is to reduce the jitter in the optical pathlength to a small fraction of a wavelength in the presence of representative disturbances. Analysis performed as part of the FMI study yielded a stellar optical pathlength jitter requirement of less than 5 nm RMS for proper functioning of a stellar interferometer. Ultimately a "layered" approach to control will be employed to achieve this goal, with disturbance and optics isolation, active and passive structural control, and optical pathlength compensation comprising the three control layers. For the present, only the optical compensation system is implemented, and the control results described below utilize only this type of control. Only Optics Option 0 was used. External disturbance sources were employed, and the environmental disturbances were evident (no additional active control of the laser tower was employed). Only the low voltage PZT and voice coil were utilized. The intention to date has been more to blaze a trail through the complexities of implementing any controller on the Testbed hardware, and less to obtain optimal performance. Nonetheless, performance of both controllers is quite good, as shown in the next section.

Two slightly different trolley controllers have been run to date. They differ more by implementation than by control strategy. Both are simple low-order compensators. "Controller 1" was designed using classical frequency response techniques. It is implemented in a single computer (the V30XE). It is a staged design, with the main control being supplied by a high-bandwidth PZT loop driven directly by the laser signal (Fig. 7). The voice coil is operated in a follow-up loop driven at low bandwidth by the PZT displacement. It is operated at a sample frequency of 2kHz. "Controller 2" demonstrates the splitting of control functions between multiple processors. As sketched on Fig. 8, the PZT loop is implemented in one computer (V3E) and the voice coil loop is implemented in another (V30XE). Both loops are driven by the laser output. By splitting functions, the PZT loop of Controller 2 can be run at a higher sample frequency of 7kHz.

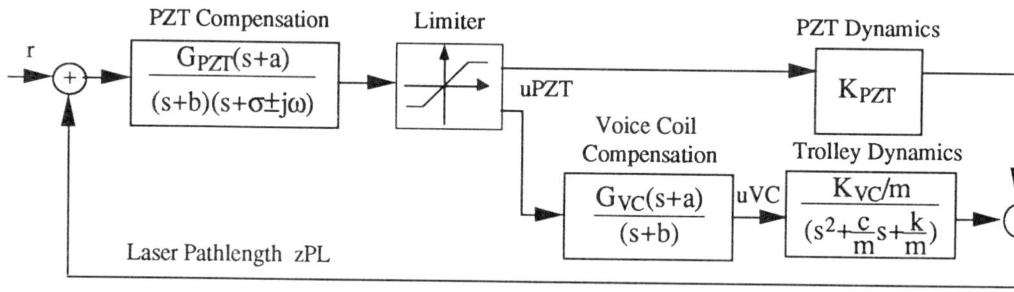

Figure 7: Block Diagram of Controller 1

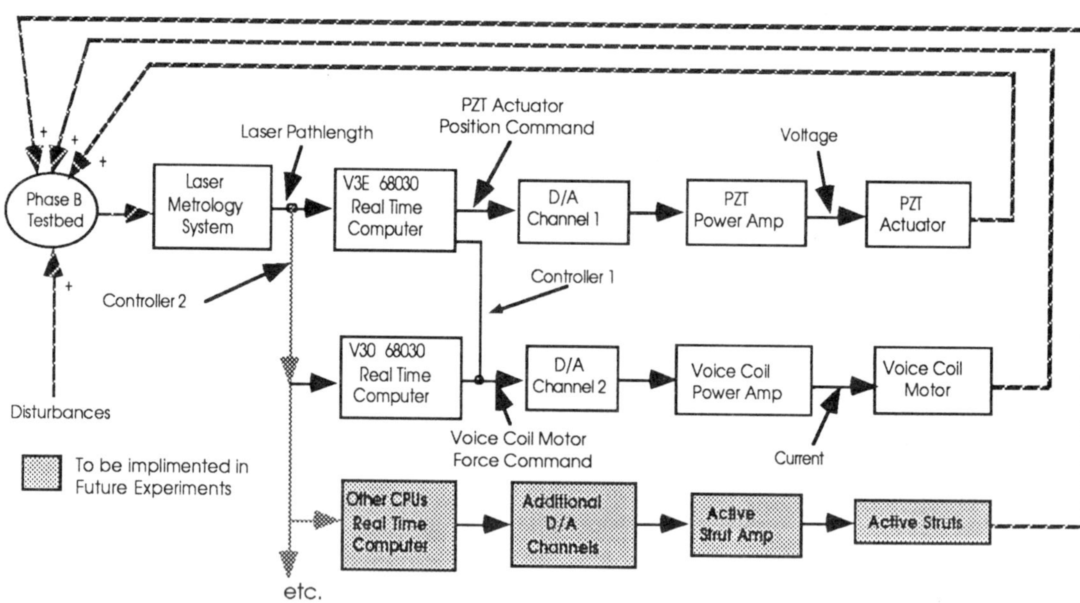

Figure 8: Block Diagram of Control Approaches

The trolley/PZT dynamical system set up in Optics Option 0, which uses the laser as primary sensor, is a collocated sensor/actuator control system. To first order, the structural dynamics provide a loosely coupled, out-of-the-loop disturbance source isolated by the trolley suspension. Inertially, the PZT and voice coil systems are very nearly uncoupled, due to the low PZT mass and the use of the counterweight. For control design purposes, it is not necessary to explicitly model the structure dynamics, nor is it required to model the small coupling between the PZT and trolley. Both controllers were designed using simple dynamics for the trolley and PZT: damped second-order oscillator dynamics for the trolley; and direct position actuation for the PZT.

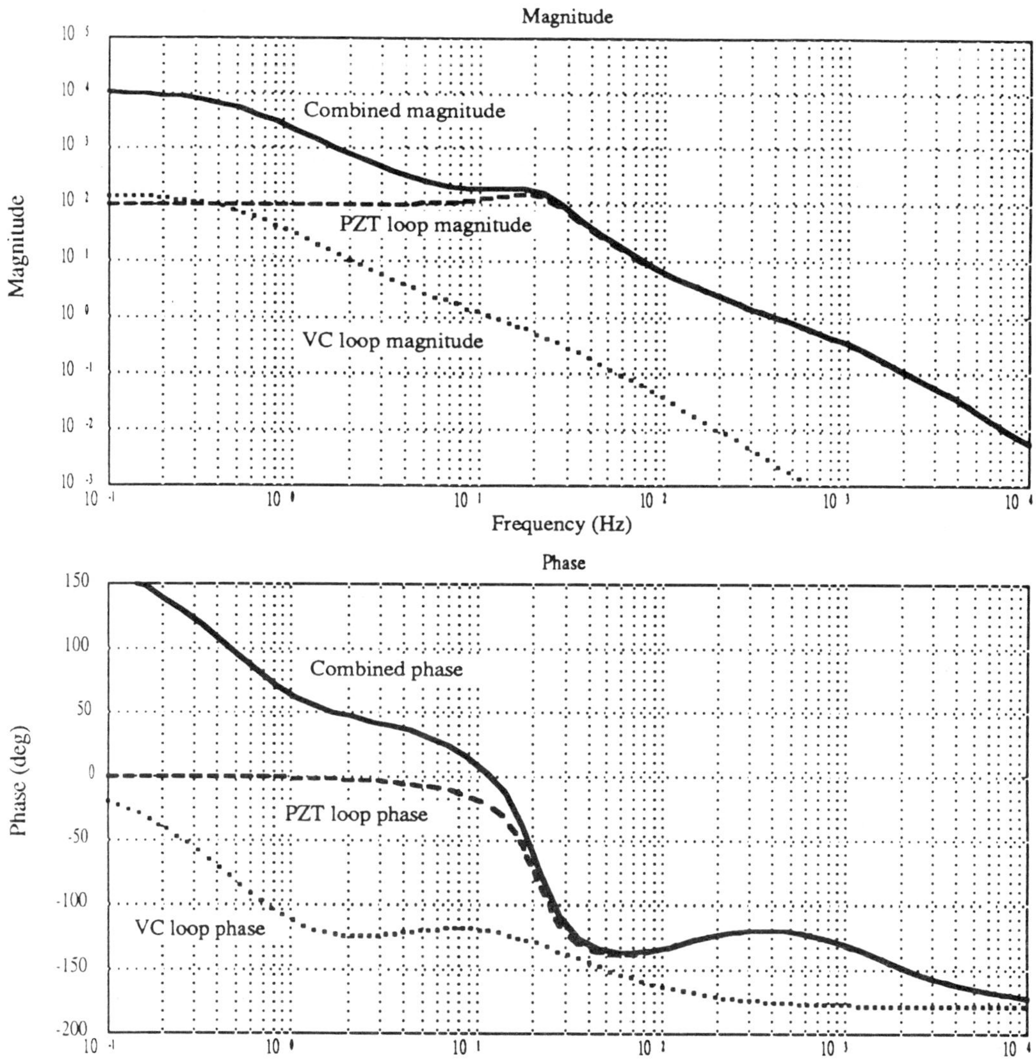

Figure 9: Frequency response of Controller 1

The functions of Controller 1 are sketched in the block diagram of Fig. 7, with parameter values as listed in Table 2. The PZT compensator uses a damped oscillator at 15 Hz, in conjunction with a lead filter, to obtain high gain at frequencies approaching the cut-off frequency of 350 Hz while still assuring good stability characteristics. Phase margin is greater than 50 degrees, as shown in Fig. 9. The voice coil compensator uses a lead filter to stabilize the trolley motion. The voice coil loop acts to desaturate the PZT and provide low frequency gain for error rejection. Combined system frequency response is shown on Fig. 9. As mentioned in the previous section, the limiter is required to prevent the PZT from outrunning the laser.

TABLE 2: PARAMETERS FOR CONTROLLER 1

Dynamics	PZT Compensator	VC Compensator
m=14.59 kg	a=725 rad/s	a=19.87 rad/s
k=144 N/m	b=8706 rad/s	b=198.7 rad/s
c=88 N-s/m	σ=94 rad/s	Gvc=-1.0e-3
Kpzt=-1.452		
Kvc=1.49e7	ω=94 rad/s	
	Gpzt=-1.5e-7	

Controller 2 was designed using a state-variable pole placement approach. The PZT loop, which is run at 7kHz in a dedicated computer, is a simple PI (proportional/integral) controller. The voice coil loop, running in parallel in a separate computer, consists of a second-order state estimator with closed loop poles critically damped at 25 Hz, and a PID (proportional/integral/derivative) controller with a 20 Hz bandwidth. It is run at 2kHz. A block diagram showing the design appears in Fig. 10, with parameters given in Table 3.

TABLE 3: PARAMETERS FOR CONTROLLER 2

Parameter	PZT Loop	Voice Coil Loop
Estimator frequency, f_k	-	25 Hz
Controller frequency, f_c	-	20 Hz
Controller damping ratio, ζ	0.707	0.707
Position feedback gain, CP	3500	$(2\pi f_c)^2 = 1.54 \times 10^4$
Velocity feedback gain, CV	-	$\zeta/(\pi f_c) = 1.10 \times 10^{-4}$
Integral feedback gain, CI	70	20
Estimator gain, K1	-	1000
Estimator gain, K2	-	$(2\pi f_k)^2 = 2.4 \times 10^4$
Actuated mass, M	negligible	13 Kg (28 lb)

No effort was made to optimize performance of these controllers. They were meant as vehicles with which to exercise the Testbed hardware and expose implementation problems. In that they succeeded quite well! The Testbed is now ready to handle more sophisticated control approaches.

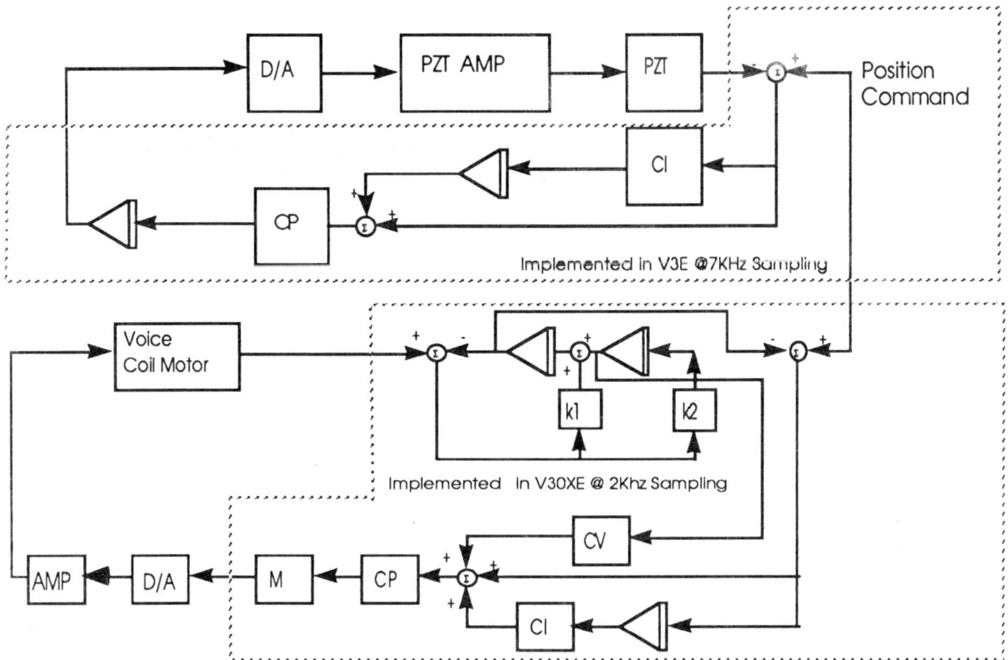

Fig. 10: Block Diagram for Controller 2

EXPERIMENTAL RESULTS

Controller 1 results are shown in Figs. 11-13. The experiment runs were five seconds in duration, with the controller being turned on at the one second point. Shown are time traces of the laser pathlength, a close-up of the laser pathlength period when the controller was on, a trace of the PZT command and a trace of the voice coil command. The first set of results (Fig. 11) were taken in quiescent laboratory conditions. The background disturbance in this case is mainly a 0.75 Hz, 80 micron oscillation due to air currents disturbing the trolley. This is easily handled by the controller. The closeup of the laser pathlength shows average peak-to-peak response under 100 nanometers, for an RMS response of roughly 30 nanometers. Expressed as equivalent stellar optical path, the RMS delay-line stability is roughly 8 nanometers.

The second set of results (Fig. 12) show performance with the structure being shaken at 5 Hz, near the first structural resonance. The disturbance shows up clearly in the laser signal before the control is turned on, as a 5 Hz oscillation of about 16 microns peak-to-peak amplitude. The controller is effective in eliminating the disturbance, with residuals of less than 100 nanometers peak-to-peak as in the no-disturbance case. With the controller on, the disturbance shows up in the PZT and voice coil commands.

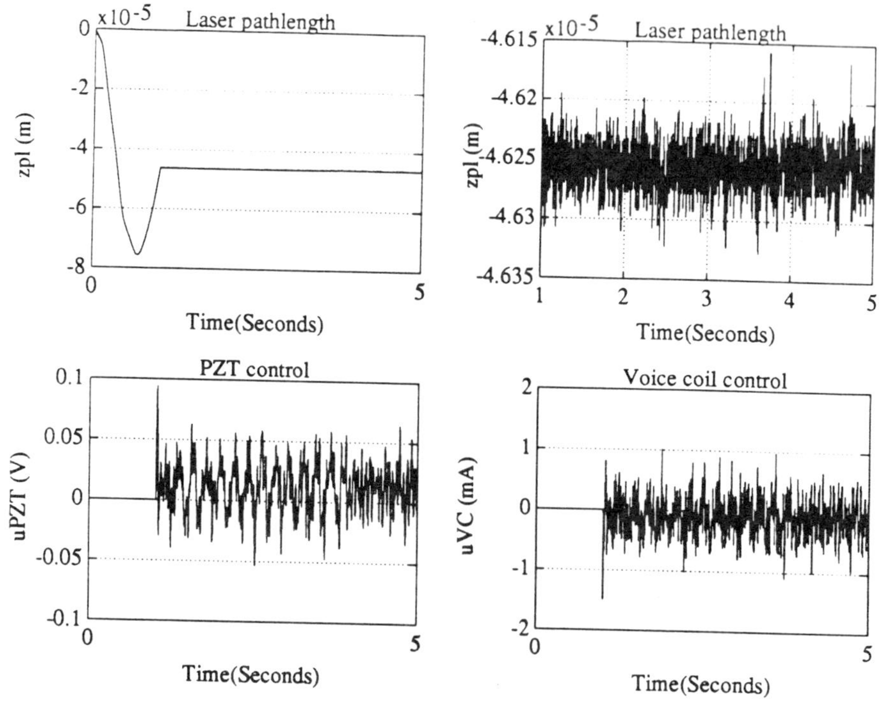

Figure 11: Results of Controller 1 on Quiet Structure

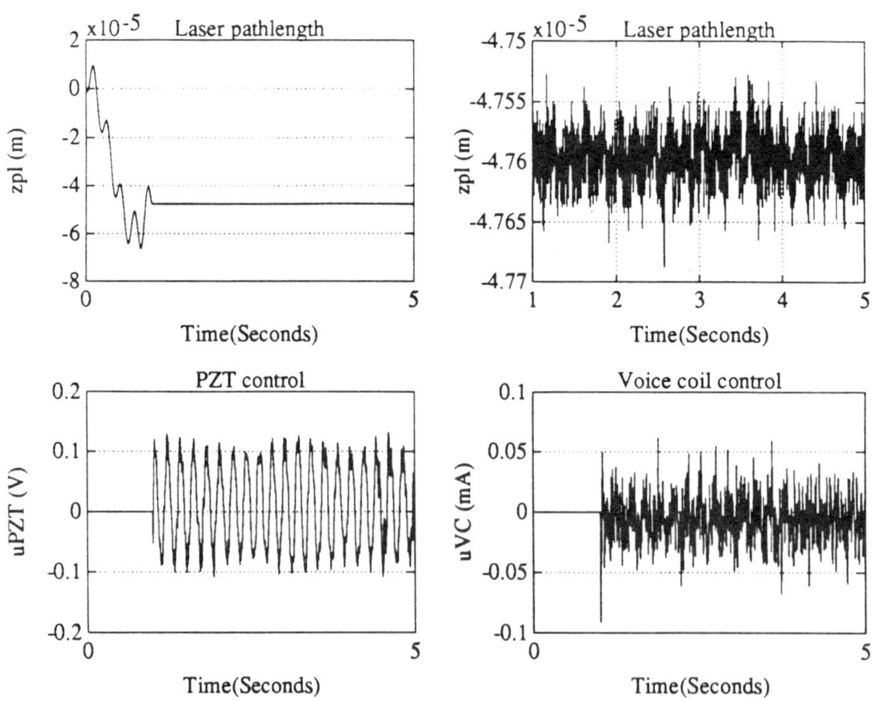

Figure 12: Results of Controller 1 with First Structural Mode Excited

The third set of results (Fig. 13) show performance with a higher-amplitude disturbance at 15 Hz, between structural resonance frequencies, to provide a more random input. During the uncontrolled period most of the motion shows up in the first structural mode. With the control on, various other modes can be seen in the control as well. Performance is not as good in this case, with a peak-to-peak laser pathlength amplitude of about 500 nanometers.

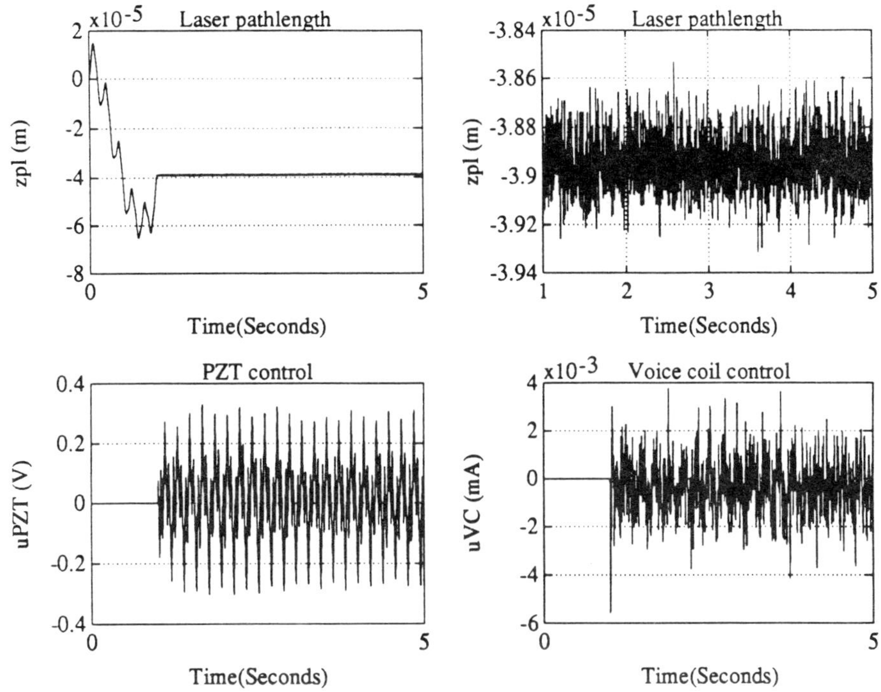

Figure 13: Results of Controller 1 with 15 Hz Disturbance

The results of Controller 2, shown in Figs. 14 and 15, are similar to those of Controller 1. These tests were 15 seconds in duration. The first 5 seconds show open loop response, the next 5 seconds show the response of the voice coil loop acting alone, and the final 5 seconds show both the PZT and voice coil loops working together. The first tests (Fig. 14) were performed with only the laboratory conditions applying a disturbance. The figure on the right shows a close up of the region between 8 and 12 seconds. Fig. 15 shows a similar run with a 5 Hz excitation applied to the structure. Again, the controller effectively reduces the disturbance.

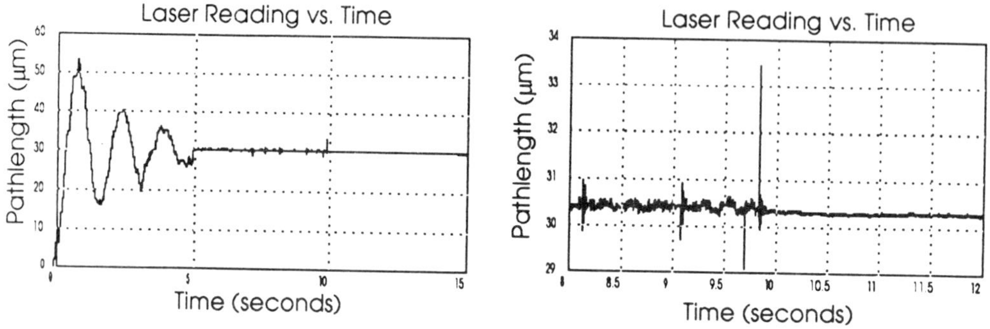

Fig. 14: Results of Controller 2 on Quiet Structure

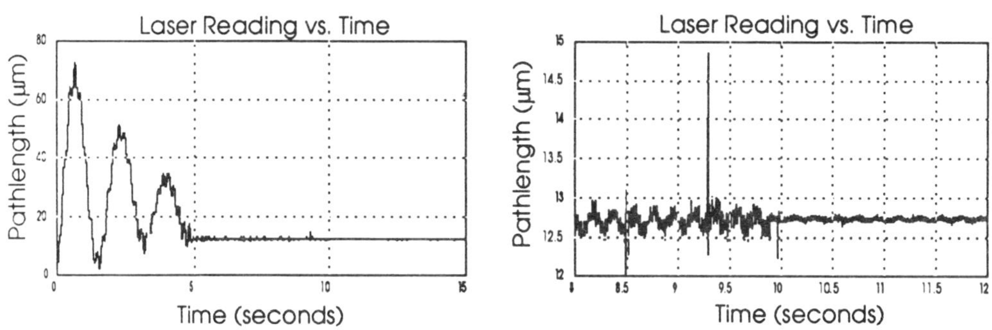

Fig. 15: Results of Controller 2 with First Structural Mode Excited

CONCLUSIONS AND FUTURE WORK

By pushing optical pathlength control into the nanometer regime, the first experiments on the Phase B Testbed have demonstrated the capability to explore and validate new Control Structure Interaction methodologies by implementing active and passive control using hardware similar to that planned for future space interferometers.

The Testbed will evolve into a more realistic simulation of the Focus Mission Interferometer by augmenting the optical configurations to couple more structural motion into the pathlength, adding larger (and more precisely regulated) disturbances, and introducing local modes due to optics mounts. This increased challenge will be met by advancing the technology of direct optical element control, as well as incorporating the other layers of the CSI control architecture: active and passive control of the structure, and active and passive isolation of optics and disturbances. The performance of the real-time computers will also be enhanced with the addition of faster single-board computers, faster array processors and a new release of the operating system.

Eventually, the structure will be made as quiet as possible in the current facility, but this may still be short of some of the requirements of the FMI and other missions due to the constraints of ground testing (acoustic, seismic, and electrical noise, the effect of gravity, etc.). The noise problems can be alleviated by moving the testbed to a facility where the environment can be controlled more carefully, and a much higher degree of seismic and acoustic isolation and electrical noise filtering is attainable. The gravity problem can be reduced somewhat by suspending the testbed with a zero spring rate suspension system. Both of these improvements are planned for the future.

ACKNOWLEDGEMENTS

The research described in this paper was performed by the Jet Propulsion Laboratory, California Institute of Technology, under contract with the National Aeronautics and Space Administration. The authors would also like to thank Bill Layman, Bob Laskin, Jim Fanson, Clark Briggs, John Spanos and the rest of the CSI team for all of their help with the Phase B Testbed experiments.

REFERENCES

1. Laskin, R.A. and M. San Martin, "Control/Structure System Design of a Spaceborne Optical Interferometer," AAS/AIAA Astrodynamics Specialist Conference, Stowe, VT, Aug. 1989.

2. Fanson, J.L., C.C. Chu, B.J. Lurie, and R.S. Smith, "Damping and Structural Control of the JPL Phase 0 Testbed Structure." Proceedings of the 1st U.S.-Japan Conference on Adaptive Structures, Maui, HI, Nov. 13-15, 1990.

3. Fanson, J.L., E.H. Anderson, D.M. Moore, and M.A. Ealey, "Development of an Active Truss Element for Control of Precision Structures," Optical Engineering, vol. 29, no. 11, Nov. 1990, pp. 1333-1341.

4. Wilson, J.F. and L.P. Davis, "Viscous Damped Space Structure for Reduced Jitter", 58th Shock and Vibration Symposium, Huntsville, AL, Aug. 1987

5. Fanson, J.L., H.C. Briggs, C.C. Chu, B.J. Lurie, R.S. Smith, D.B. Eldred, D. Liu, "JPL CSI Phase 0 Experiment Results and Real Time Control Computer", 4th NASA/DoD Control/Structures Interaction Technology Conference, Orlando, FL, Nov. 5-7, 1990

6. Spanos, J.T. "Control-Structure Interaction in Precision Pointing Servo Loops," Journal of Guidance, Control, and Dynamics, Volume 12, Number 2, Mar.-Apr. 1989, pp. 256-263.

AAS 91-053

ACQUISITION, TRACKING AND POINTING CONCEPT SOLUTIONS FOR AN INDEPENDENT COMPLEMENTARY SPACE EXPERIMENT (ICSE)

Sherman M. Seltzer*, Richard W. Schock*, Norman Fitz-Coy*, Ronald C. Francis*, Kent E. Weir*, Louis A. Morine†

The major objective of the Independent Complementary Space Experiment (ICSE) is to eject, acquire, track, and damage a test object by laser irradiation on orbit. This objective requires the transport of the experiment and test objects to orbit, the ejection of the test objects from the experiment spacecraft, and the acquisition, tracking, and pointing of the experiment with a sufficient accuracy and stability to allow the laser sufficient time to accomplish test object damage and/or destruction. This paper discusses the development, application and results of applying a unique approach of ejecting the test object in a direction normal to the spacecraft orbital plane as opposed to the traditional approach of ejecting within the spacecraft orbital plane. The target acquisition and tracking control, and the laser beam control optics were also transversely mounted to maximize control efficiency. The orbital dynamics of the relative motion between the spacecraft and the test object will be illustrated, and the flexible body effects will be assessed.

1.0 INTRODUCTION

The Independent Complementary Space Experiment (ICSE) is a concept that was designed to provide a low cost space experiment for risk reduction of the Zenith Star program and obtain data on the basic performance of a chemical laser in space. The resultant concept presents a challenge to the spacecraft control design.

The results of this experiment, when implemented, would provide data for establishing realistic design requirements for the Space Based Laser (SBL) functions. The results, scaled to the operational levels, would allow confident re-allocation of the performance error budgets as well as design criteria so that realistic subsystem requirements could be specified.

The primary objective of the ICSE was to "Evaluate the Operation of a Medium Power Chemical Laser and Demonstrate Its effectiveness in Space." A secondary objective was to collect ancillary data that would reduce the risk of future chemical laser experiments such as Zenith Star.

---

* Control Dynamics Company, 600 Boulevard South, Suite #304, Huntsville, Alabama 35802.

† W.J. Schafer Associates, 66 W. Springer Drive, Highlands Ranch, Colorado 80126.

The laser would be turned on remotely and its performance, such as power output, fluid flow, and resonator temperature, would be measured. Boresight alignment measurements would be achieved by observing the Medium Energy Laser (MEL) beam through the Acquisition, Track and Pointing (ATP) sensor and displaying this information on the ground. Boresight correction would be demonstrated by adjusting the steering mirrors to offset the measured boresight error and by confirming that the boresight correction was achieved.

The demonstration of the effectiveness of the laser against the Ballistic Missile Defense (BMD) threat would be accomplished by depressurizing (or destroying) a pressurized sphere or cylinder by laser irradiation within a relatively short time (such as 5 to 10 seconds).

The effectiveness of the laser against the midcourse threat can be achieved by irradiating inflatable balloons and light reentry vehicle (RV) replicas and collecting data consisting of the images of the inflatable balloon's observed destruction, and the thermal and velocity change signatures of the RV replicas resulting from the laser irradiation.

For the secondary objective, emphasis would be placed on the data that must be measured in space to anchor the ground test data and models to space conditions. Examples of this type of data are:

1. Exhaust effluent optics contamination
2. Exhaust effluent interference with the ATP sensing and the power on the target
3. Spacecraft vibration disturbances

1.1 Ground Rules

The intent of the ICSE was to design a system that will accomplish the stated objectives in a simple straight forward inexpensive way. There was no intention to define an optimum configuration.

The ground rules for this conceptual design were as follows:

1. The system must be compliant with the restrictive interpretation of the 1972 ABM Treaty.
2. The primary objectives are the only performance requirements. There are to be no extended corollary experiments such as vibration isolation or rapid retargeting. Phenomenology is to be limited to the capabilities of the ATP sensors with none added beyond what is needed for the primary objective.
3. Mission parameters are variables in the design process. That is, the range, test object size, etc. can be adjusted to make the system design and risk more favorable for the accomplishment of the primary objective.
4. The ATP system must be simple and low risk. To this purpose the ranges can be shortened and the test objects are augmented. This will allow the ATP system to use proven sensors and to make the ATP system easy to accomplish. In short, this is not a ATP experiment.

1.2 ICSE Overview

The overview of the ICSE Payload and Spacecraft are shown in Figure 1 The key features of the Payload include:

1. 100 kw HF linear laser with a 60 second total run time capability.
2. The optical system consists of a 1 meter primary mirror telescope with a unified optical bench and open loop off axis steering
3. Simple tailored Acquisition, Tracking, and Pointing system

The spacecraft is designed to utilize existing hardware and is configured to be launched on a Delta 7920 launch vehicle. The spacecraft carries four (4) test objects on board for ejection prior to the experiment operation. The Test Objects are ejected in a direction normal to the spacecraft orbital plane as opposed to the traditional approach of ejecting within the spacecraft's orbital plane. The target acquisition and tracking control, and the laser beam control optics were transversely mounted to maximize control efficiency. With this mounting configuration, the major torque perturbation caused by the laser exhaust force and the Gain Generator Assembly (GGA) exhaust was about the pitch axis. The pitch axis was parallel to the laser line of sight and therefore perturbations about this axis affected pointing to only a second order. Torque perturbations about the roll axis were very small in comparison to either pitch or yaw and therefore also second order.

The beam expander is not articulated relative to the spacecraft and there is no vibration isolation. The fine pointing of the Medium Energy Laser (MEL) beam is done by open-loop commanding of two steering mirrors (located just prior to the beam entrance through the primary mirror). There is no high frequency jitter control of the MEL beam.

Figure 2 shows the spacecraft layout as it would be within the payload compartment of the Delta Launch vehicle. Not shown are the 4 Test Objects which are mounted to the same interface as the Rocket Engine Module System (REMS) and are expelled aft.

The four test objects are a derivative of the GLOMR Satellite (Figure 3). They are passive except for the augmentation with beacons and/or retro-reflectors for tracking aids. These will be deployed from the spacecraft (one for each experiment scenario) as described above.

The operations center chosen for the ICSE was the RME operations center at Maui. The choice of ground station location was made in order to take maximum benefit of the existing equipment installations already in place for other experiments. Furthermore, there are other communication stations in the vicinity of Maui that increase the amount of time available for the experiment (especially with the chosen 2 day repeating orbit).

In addition, the telescopes at the Maui tracking station can be conveniently used for ground observations of the experiment as it is conducted. This will add additional important data to the experiment results and serve as a backup should anything interfere with the spacecraft on-board sensors.

1.3 Experiment Scenario

The geometry associated with the ICSE is shown in Figure 4. The ICSE spacecraft is in a 28 degree 340 KM altitude that provides low drag and a 2 day repeating orbit with four opportunities every two days (see Figure 5).

After all of the initial checkouts and calibrations, the system is ready to perform the planned experiments. At a point 90 degrees before a pass over Maui, the test object is ejected from the spacecraft in a direction perpendicular to the ICSE orbit. This creates the geometry shown in Figure 4.

The test object orbit is essentially flying in formation with the ICSE spacecraft in a crisscrossing manner. Aerodynamic drag on the test object causes it to stabilize in attitude and gradually increase the range relative to the spacecraft. Because of the crisscrossing nature of the orbits, the range and angular rates are periodic as shown in Figure 6. The range rate and angular rates are near zero when the test object is at its relative maximum over the Maui tracking stations.

As the spacecraft approaches Maui, the payload tracks the test object. The test object reaches its maximum distance and minimum relative velocity from the spacecraft (the ideal time for performing the experiment) just as it passes over the Maui command center. The payload then irradiates the test object for approximately 10 seconds while its ATP sensors record the event. The system is then shut down and reconditioned for the next experiment.

1.4 Control System

Figure 7 is a summary of the ATP concept for the ICSE and Figure 8 lists the key ATP attributes. This system provides simple but flexible operation with low risk. It consists of a visible camera with a zoom lens and an infrared sensor. The visible camera is used for both target acquisition and fine track. The infrared sensor is used to record the event and as a backup for tracking the target. Figure 9 shows the operation concept for vehicle control and precision pointing. Vehicle control is maintained by using the ATP sensors and inertial reference to command the vehicle thrusters. Precision pointing is accomplished by utilizing the same ATP sensors to command open loop control of the two steering mirrors containing precision position sensors.

The Hughes Dual Mode tracker processor is used to generate and make the control commands. This system provides flexible control operation and redundancy. The use of a zoom lens provides flexibility for uncertainties in the sensor performance.

The spacecraft will be stabilized in all three axes by an inertial reference, a simplified Acquisition, Track, and Pointing (ATP) system, and a simple hydrazine reaction control system. This will allow the system to be treaty compliant and allow a very simple ATP sensor suite to be employed.

The precision off axis steering is open loop, deriving its control from the ATP sensor, the spacecraft control system must minimize disturbances above the off axis control system bandwidth of 30 Hz. In addition, it must control the beam expander position to within the off axis pointing capability of +/- 5mrad. This requires careful control of the spacecraft and reaction control thrusters, is crucial to the performance of the program, and is the subject of this paper.

2.0 DESCRIPTION

Based on the flight experiment described in the Introduction, the subsequent paragraphs describe the development of the Acquisition, Tracking and Pointing (ATP) system for this flight experiment.

The relative position and velocities of the Test Object and the Spacecraft, after the inertial plane target ejection described in the Introduction, were assessed in detail to determine the acquisition and tracking rate requirements for design of the control system. This assessment and results are described in section 2.1.

In order to ensure that no coupling occurred between the flexible body modes of the experiment and the control system, a flexible body math model was developed. This model is described in Section 2.2, as is the results of the dynamic analysis. This model was then subsequently used in the control simulation model described in section 2.3. A control system was designed for the experiment and is also described in section 2.3. The results of the control simulation are described in section 3.0.

In order to insure additional irradiation opportunities, and considering the inertial plane target ejection causes the target to move perpendicular to the orbit plane and, on first pass, come within 250 meters of the spacecraft, a tracking simulation of that pass was conducted to determine if adequate control authority was available to track the target. This assessment is discussed in section 3.2.

Finally, the Conclusions and Recommendations are presented in section 4.0.

2.1 Spacecraft to Test Object Relative Motion Simulation

This section describes a simulation which was developed to study the relative dynamics between the ICSE spacecraft and the Test Object. A short discussion of the simulation models and verification is included. Results included in this section illustrate the relative motion for one particular scenario. The initial conditions associated with this scenario are also presented.

2.2.1 General Characteristics

The simulation integrates two sets of differential equations describing the motion of the CSE spacecraft and Test Object. It is assumed that the motion of each vehicle is governed by the effects of atmospheric drag and gravitational acceleration only. The equations of motion are initialized through input of vehicle state vectors. Other simulation parameters must also be defined within an input file, including: gravitational parameters, ballistic coefficients, final time, etc. The code is written in FORTRAN and is generic enough to operate on most all computers.

2.1.2 Models and Verification

The simulation incorporates both atmospheric and gravitational models for the Earth. The atmospheric model is based upon the 1962 Standard Atmosphere. It is assumed that the atmosphere rotates with the planet and no winds exist. Drag acceleration is computed as a function of atmospheric density, vehicle speed, and vehicle ballistic coefficient. The gravitational model is based upon an inverse squared force field modified using the J2, J3, and J4 harmonic terms. Equations of motion are integrated using a fourth order Runge-Kutta routine and a constant integration step size.

2.1.3 Results

A set of plots was produced to illustrate the relative dynamics between the ICSE Spacecraft and Test Object. The data represents motion of the spacecraft with respect to the Test Object through plots of relative position, range rate, and the line of sight angular rate. Time histories are presented either explicitly or through tic mark intervals of three (3) minutes.

Before deployment of the Test Object, the combined vehicle is assumed to be in a circular orbit about the Earth inclined at 28 degrees. The Earth's gravitational model includes the J2 term only. Deployment of the Test Object is in the direction of the spacecraft's orbital angular momentum vector. For this particular scenario, the initial conditions are presented in the following table.

Spacecraft:	
Altitude....................................	327.5 km
Radius....................................	6705.645 km
Speed.....................................	7.7099 km/sec
Ballistic Coefficient...............	100 kg/m**2
Test Object:	
Altitude....................................	327.5 km
Radius....................................	6705.645 km
Speed**..................................	7.7219 km/sec
Ballistic Coefficient..............	11.5 kg/m**2

** 12 m/sec delta-v imparted at deployment along spacecraft orbital angular momentum vector

The relative position data is represented with respect to a Radial, Tangential, and Normal (RTN) coordinate frame affixed to the Test Object. Therefore, the RTN frame rotates around the Earth at the Test Object's orbital rate. Figure 10 illustrates the relationship of this coordinate frame with respect to the Earth Centered Inertial (ECI) frame.

Relative position as a function of time is illustrated through plots of Radial position versus Tangential position (Figure 11) and radial position versus normal position (Figure 12). From the time history of the range (Figure 13) it appears that the Test Object and spacecraft apparently collide after approximately half a revolution around the Earth. In actuality, the vehicles are separated by a distance of approximately 250 meters. Initially, the mean of range rate is near zero (0.0) (Figure 14); however, it increases slightly with time. This deviation from a zero mean is attributed to differences in harmonic perturbations associated with each vehicle. Deviations may also be associated with variations in relative velocity vectors for differently inclined orbits. In other words, for rotating atmospheres with no wind, the direction of the resultant relative velocity vector is a function of the inclination of the orbit in which the Test Object and spacecraft orbital inclinations differ. The line-of-sight (LOS) angular rate (Figure 15) is the rotation of the line joining the Test Object and spacecraft. This plot shows the magnitude of the rate at which the Test Object would have to slew to keep a constant attitude with respect to the spacecraft. A three dimensional representation of the relative motion is also shown on Figure 16.

2.2 Finite Element Modeling

A finite element model (FEM) of the Independent Complementary Space Experiment (ICSE) was developed using a PC-version of the ANSYS FE program. The model consisted of 105 node points, 294 degrees of freedom (DOF), 28 beam elements, 32 plate elements, 45 concentrated masses, and 3 single DOF springs. The beam elements were used to model the spacecraft support structure and the solar array support structures, while the plate elements were used to model the solar arrays (16 plate elements per solar array). The 3 single DOF springs were used to model the Gain Generator Assembly (GGA) isolation system (i.e., isolation in the 3 orthogonal directions). All other components (referred to as payloads) were modeled as concentrated masses attached to the spacecraft support structure.

The model was dynamically tuned to satisfy three constraints. One constraint was a 15 Hz. cantilevered mode for the entire spacecraft (this is a launch constraint), another was a 55 Hz. cantilevered mode for the beam-expander (BX) assembly, and the third was a 4 Hz. mode for the GGA isolation system.

2.2.1 Solar Array Model

Each solar array was modeled as a 1.5 inch thick homogeneous material with 1.5x1.5x.25 inch aluminum angle beam as edge support as illustrated below. The center support structure was modeled as an aluminum tube 2.75 inches in diameter and 0.125 inches thick. Sixteen homogeneous plate elements were used to model each solar array, while the support structures for each array were modeled with beam elements. The properties of the homogeneous material used to model an array were derived from an aluminum honeycomb with a weight per unit area of 1.2 pounds/ft^2. The derived weight of the homogeneous material is 79.5 pounds. For a 1.5 inch thick plate, the density of the homogeneous plate was computed to be 0.3 slugs/ft^3. The elastic modulus of the plate material was maintained at 10×10^7 psi. (i.e., same EI as aluminum, but lighter.) The same elastic modulus was used for the array support structures.

Solar Array Configuration

2.2.2 *Spacecraft Model*

The spacecraft structure was modeled as an assemblage of beam elements. Each section of the spacecraft was modeled separately beginning with the beam-expander section and terminating with the target bay section. The properties of each section were derived based on a cylindrical shell of a given diameter; each section had a wall thickness of 0.5". (The taper of the target bay section was neglected.)

The elastic modulus of the BX section was adjusted until the first cantilevered mode of that section was at a frequency of 53 Hz. (desired frequency was 55 Hz.). The BX section was then attached to the remainder of the spacecraft and the elastic moduli of the remaining sections were adjusted until the first cantilevered mode for the entire spacecraft was at a frequency of 15 Hz. Stiffness "tuning" was accomplished with all payloads of concern appropriately attached to their respective sections of the spacecraft and the solar arrays in their stowed position.

The TREETOPS simulations were done for a free flying system and not for a constrained system. Therefore, once the two cantilevered constraints were used to develop the necessary stiffness for each section of the spacecraft, the launch vehicle constraint was removed, resulting in the desired free-free model of the spacecraft. All frequencies and mode shapes under 10Hz resulting from this model were then incorporated into the TREETOPS simulation model as illustrated in paragraph 2.3.

2.3 TREETOPS Simulation Model

A flexible body simulation was developed using TREETOPS to examine the pointing control capabilities of ICSE. The TREETOPS simulation used the following mass properties:

Mass	=	260	slugs		
I_{xx}	=	1374	sl - ft^2		
I_{yy}	=	3978	sl - ft^2		
I_{zz}	=	4564	sl - ft^2		
I_{xy}	=	-131	sl - ft^2		
I_{xz}	=	15.0	sl - ft^2		
I_{yz}	=	4.9	sl - ft^2		
C.G. location (x, y, z)			6.93	-.13	03 ft
Thruster locations (x, y, z)			2.0	2.8	2.8 ft
			2.0	2.8	-2.8 ft
			2.0	-2.8	2.8 ft
			2.0	-2.8	-2.8 ft
GGA Disturbance loc.(x, y, z)			10.97	0	0 ft
Tracker location (x, y, z)			12.85	2.36	2.83 ft

The origin of the coordinate system is at the rear of the vehicle as shown in Figure 2.3.1. Twelve thrusters were positioned as shown in Figure 17 and were used for attitude control. The attitude control sensors included a star tracker sensor for roll and yaw attitude pointing error and a rate gyro for pitch axis rolling control [the

LOS of the tracker and laser is out the pitch (y) axis]. The attitude pointing control is a discrete controller with a 0.02 second sampling period that uses a PD controller with deadband and hysteresis to determine whether or not to fire the thrusters. A diagram of the controller for each axis is shown in Figure 18. The position gains were 1.0 and the rate gains 0.77 sec. for each axis and the inner and outer deadbands were 0.5 and 1.0 milliradian, respectively. The thrusters were not throttleable, and each thruster had a thrust of 1.0 lb. The specific impulse of each thruster was 225 sec. Four thrusters were fired at one time for roll control, six for pitch control, and six for yaw control. Two disturbances were modeled in the rigid body simulation, the GGA disturbance force (in all three axes), and the laser exhaust force imbalance (in the z direction only).

All flexible modes under 10 Hz were used in the model. These modes are illustrated as follows:

11 Flexible Modes	Frequency (Hz)
SA SYMMETRIC FLAPPING MODE 1	1.06
SA ASYMMETRIC FLAPPING MODE 1	1.09
SA SYMMETRIC TWISTING MODE 1	2.33
SA ASYMMETRIC TWISING MODE 1	2.34
GGA ISOLATOR (AXIS 1)	4.13
GGA ISOLATOR (AXIS 2)	4.23
GGA ISOLATOR (AXIS 3)	4.24
SA SYMMETRIC FLAPPING MODE 2	6.64
SA ASYMMETRIC FLAPPING MODE 2	6.65
SA SYMMETRIC TWISTING MODE 2	8.84
SA ASYMMETRIC TWISTING MODE 2	8.84

3.0 SIMULATION RESULTS

3.1 TREETOPS Flexible Body Pointing Control Simulation Results

The flexible body pointing control capabilities of ICSE were examined for two cases of disturbance inputs. The disturbances were the GGA disturbance in three axes and the laser exhaust thrust imbalance in the z direction. Table 3.1.1 summarizes the disturbance inputs for the two cases. The PSD for the GGA disturbance with 2.9 lbs-RMS is shown in Figure 19. A time domain signal was generated from this PSD and used as the GGA disturbance input for cases 1 and 2. The thrust imbalance was a constant force with the magnitude shown in Table 3.1.1 for the two cases.

Table 3.1.1

FLEXIBLE BODY DISTURBANCE INPUTS

CASE NO.	GGA DISTURBANCE MAGNITUDE RMS-LB	THRUST IMBALANCE LB.
1	2.9	4.0
2	2.9	0.4

The LOS errors for the two cases are shown in Table 3.1.2.

Table 3.1.2
FLEXIBLE BODY POINTING ERROR SUMMARY

Case No.	Pointing Error - RMS	Milliradians (Degrees) Peak
1	.57 (.033)	.76 (.044)
2	.36 (.021)	.70 (.040)

Plots showing pointing error (yaw and roll), pointing error rate (yaw and roll), thruster force commands (yaw and roll), pitch angle, pitch rate, pitch thruster command, amount of propellant used, and spacecraft translation for the two cases are shown in Figures 20 through 35. The dominant disturbance is the laser exhaust thrust imbalance. For both cases the torque from the thrusters is greater than the torque from the exhaust thrust imbalance.

3.2 Tracking Simulation Results

The tracking of a target by ICSE was examined using TREETOPS. The mass properties and thruster configuration were the same as those of the flexible body model in sections 3.1. The field-of-view of the tracker was ± 8 milliradians and the target was 10 kilometers from the ICSE spacecraft. Figure 36 shows the two cases used to determine the maximum tracking rates in roll and yaw. For case 1 (roll rate case), the target enters the field-of-view along the z axis. For case 2 (yaw rate case), the target enters the field-of-view along the x axis. The maximum roll rate that can be tracked is 1.05 degrees/second and the maximum yaw rate is 0.65 degrees/second. The yaw inertia is greater than the roll inertia. Figures 37 through 42 show the tracking error, tracking error rate and thruster commands for both cases.

4.0 CONCLUSIONS AND RECOMMENDATIONS

Based on the results of this study, the following conclusions are drawn:

1. The inertial plane target ejection provides a well ordered experiment scenario with multiple test opportunities.
2. The ICSE space experiment was well controlled within the required plus or minus 2 milliradians.
3. The flexibility effects incorporated into the model were noted in the simulations but did not inhibit the ability of the control system to control within required specifications.
4. Ensure proper differences in spacecraft and target ballistic coefficient, to control intersection range.

SBL "Brassboard"

- Payload
  - 100 KW Linear HF Laser
  - 60 Second Run Time
  - Uncooled Optics
  - 1 M Primary Mirror
  - All Other Optics ~20 cm
  - Beam Expander Non Articulated Side Looking
  - Open Loop Off-Axis Steering With ATP
- Spacecraft
  - 3 Axis Stabilized
  - 2.7 MB/S Down Link To SGLS (DMSP Heritage)
  - Hot Fire End-To-End Test
  - 4 Augmented Test Object Spheres
- Launch
  - Delta 7920
  - Cape Canaveral
  - Orbit: 28 Deg Incl., 324 KM Altitude
  - 2 Day Repeating Orbit

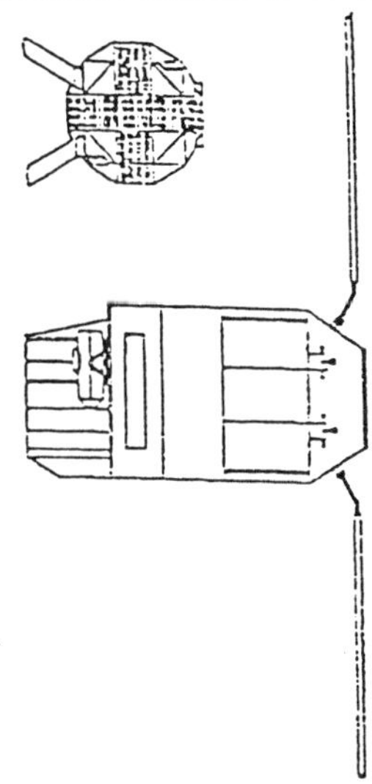

Figure 1. Overview of the ICSE

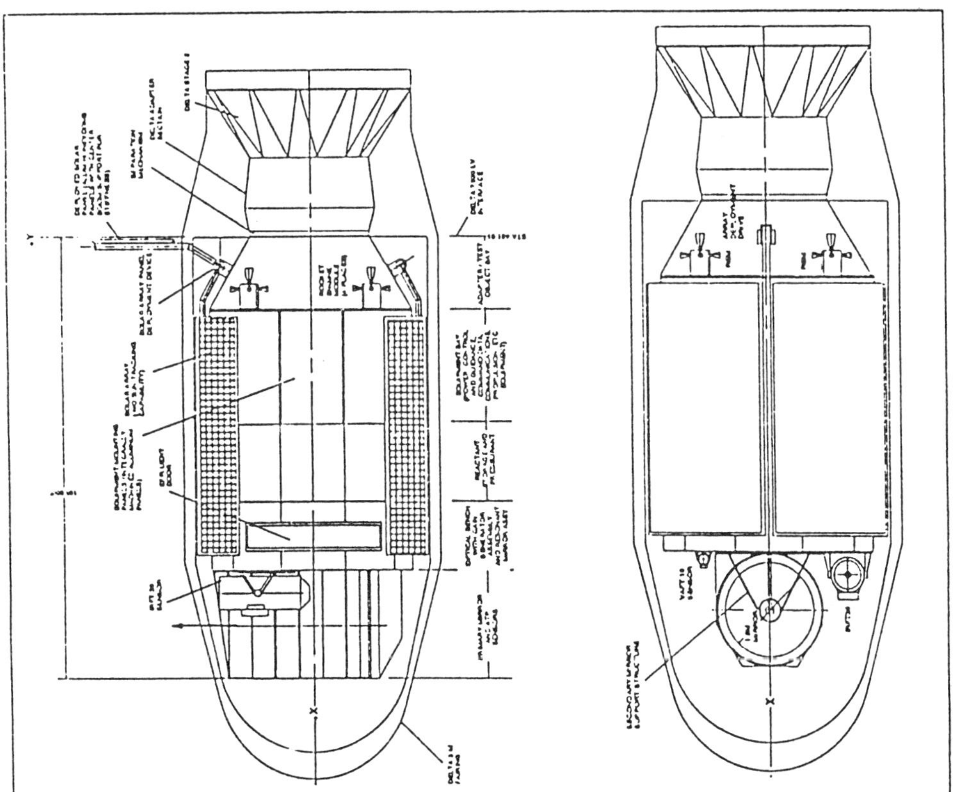

Figure 2. ICSE Layout Within the Launch Vehicle

- Original GLOMR
  - 2 Year Life
  - 324 Km Altitude
  - DSI Cost To DARPA $1M
  - No Tumble
- DSI Has Supplied ROMs For Test Object
- Gas Generator Or Spring Ejection
  - 10-20 m/s
- Modifications (3) Test Objects
  - Maintain Aerodynamics
  - Pressurized Kevlar Outer Hull
  - Internal Liner/Support
  - Internal Battery Power System Supplies Beacons
  - Programmable Timer/Communications
  - Current Weight Estimate 50 Lbs
  - No Tumble, No Spin
- Modifications For 1 Target Board Test Object
  - Add Sensors
  - Add Communications

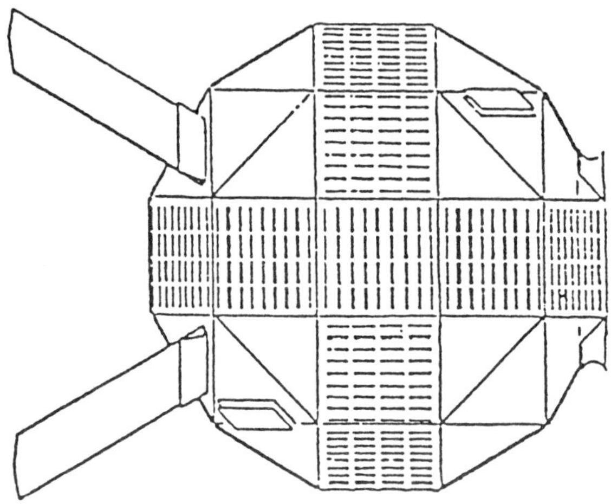

Figure 3. Test Object

Sequence Of Events For IC Event

- Eject Test Object At 1/4 Orbit Before Hawaii
- Align Spacecraft LOS To Test Object
- Capture Test Object
- Establish Fine Track
- Irradiate Test Object For 10 Seconds
- Observe Event
- If Event Or If 10 Seconds Occurs, Shut Down

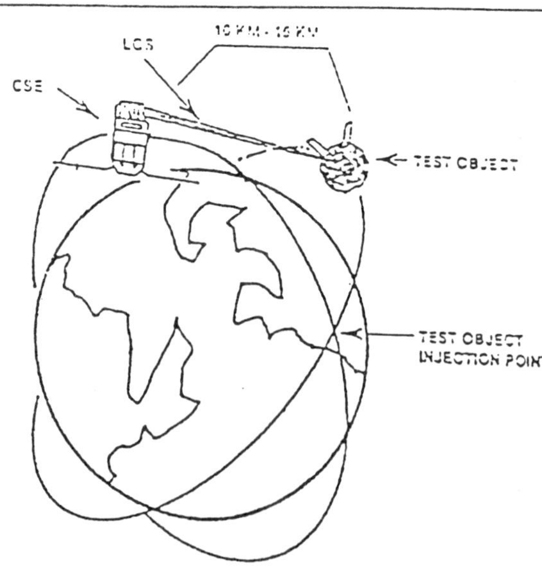

Figure 4. Experiment Geometry

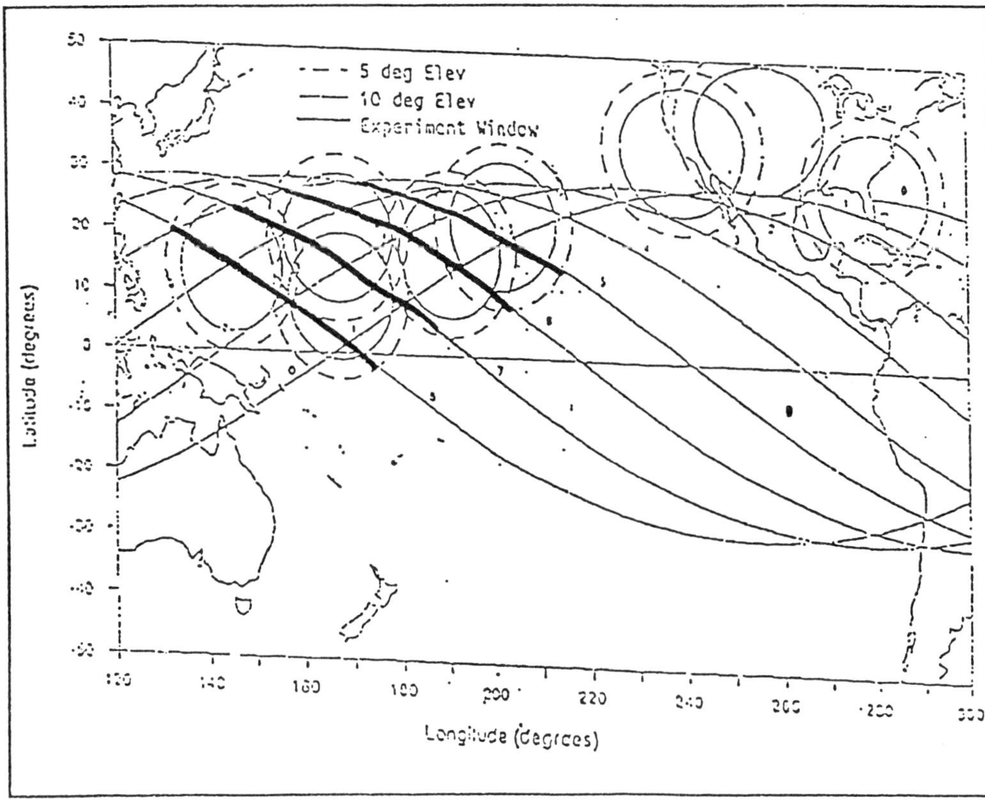

Figure 5. Successive Engagement Opportunities Over Hawaii

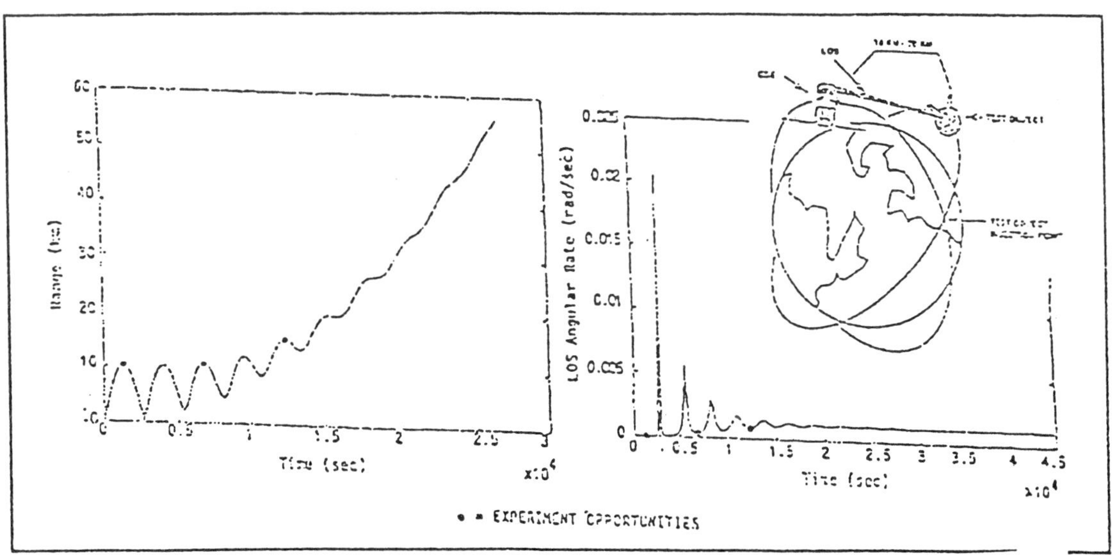

Figure 6. Experiment Scenario Opportunities

• Acquisition Sensor	• VCT 10 With Zoom (480 x 760 CCD Array) • Hardmounted To Optical Bench
• Fine Track Sensor	• VFT 10 With Zoom (480 x 760 CCD Array[a] • IRFT 30 (160 x 244 PtSi Array)
• Track Processor	• DMT
• Range Measurement	• Triangulation Of Target Beacons Using FPA's
• Beam Exp Articulation	• Non Articulated
- Beam Exp Actuators	• None
- Beam Exp Isolators	• None
• Off Axis Control	• Optical Steering
• Boresight	• Hardmount Sensors To Bes Optical Axis
• Data Acquisition (ATP)	Solid State Recorder

Figure 7. ATP Concept Summary

• Spacecraft Pointing	• ±2 mrad 2 Sigma Accuracy Required
• Targets	• Augmented With Beacons
• Capture Sensor	• 10 cm Visible Fine Track (VFT) With Zoom • Hardmounted To Optical Bench
• Fine Track Sensor	• VFT With Zoom • 30 cm Infrared Fine Tracker (IRFT)
• Track Processor	• Hughes Dual Mode Tracker (DMT)
• Range Management	• Triangulation Of Target Beacons Using Focal Plane Arrays
• Data Capture	• Solid State Recorders

Figure 8. Key ICSE ATP Payload Attributes

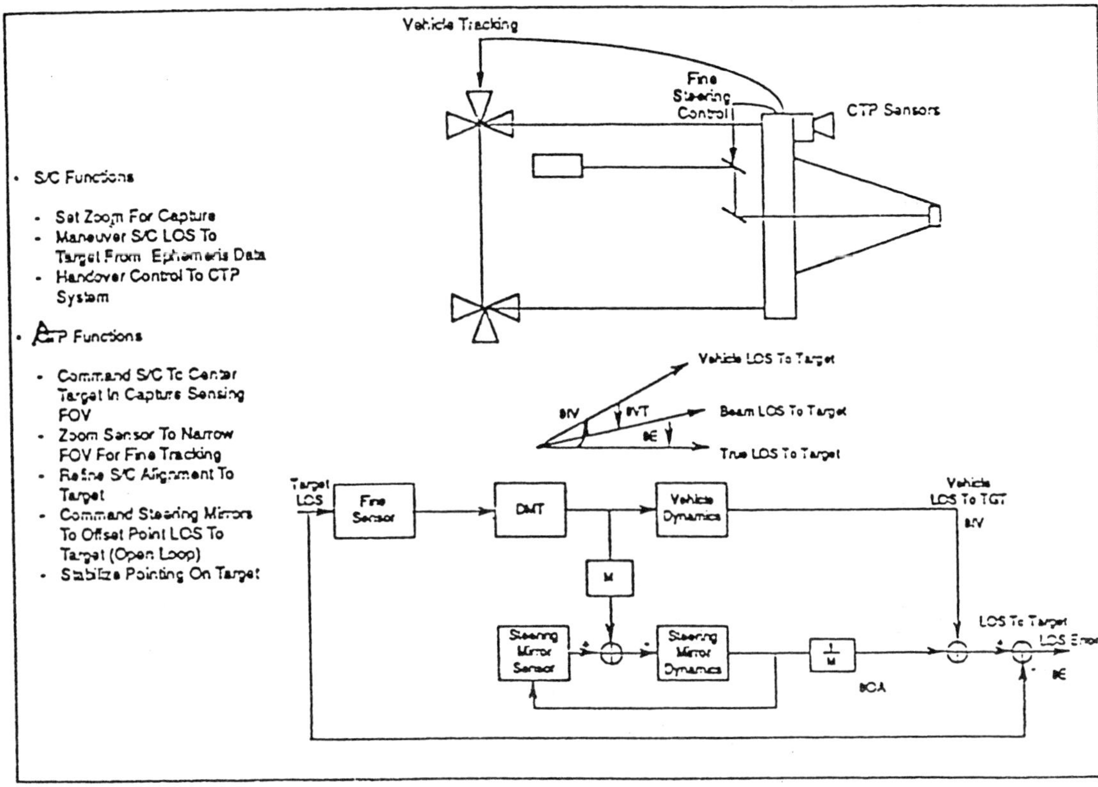

Figure 9. ICSE ATP Operation

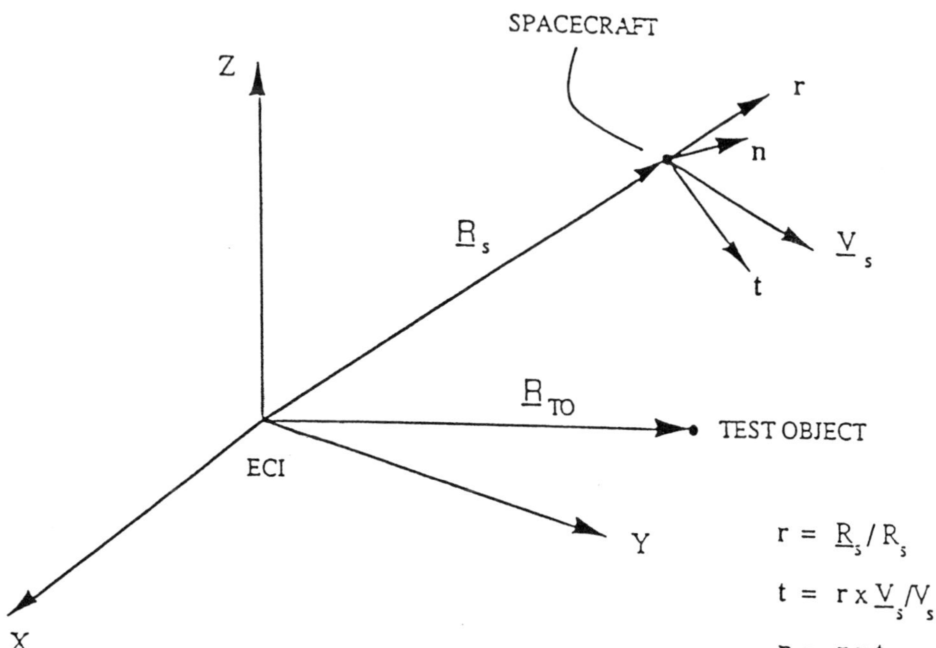

Figure 10. Coordinate System

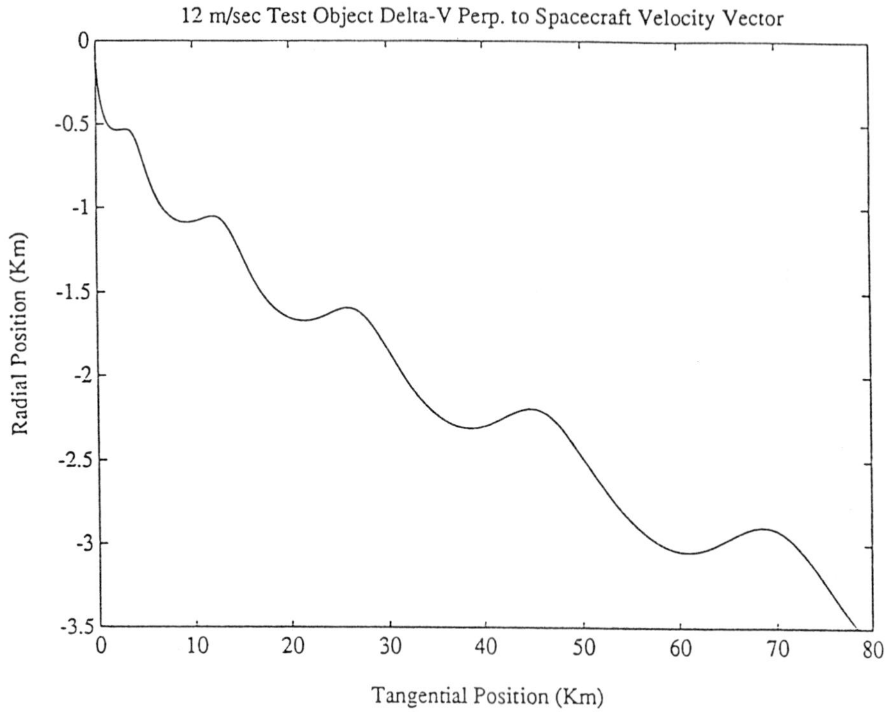

Figure 11. Radial Position vs Tangential Position

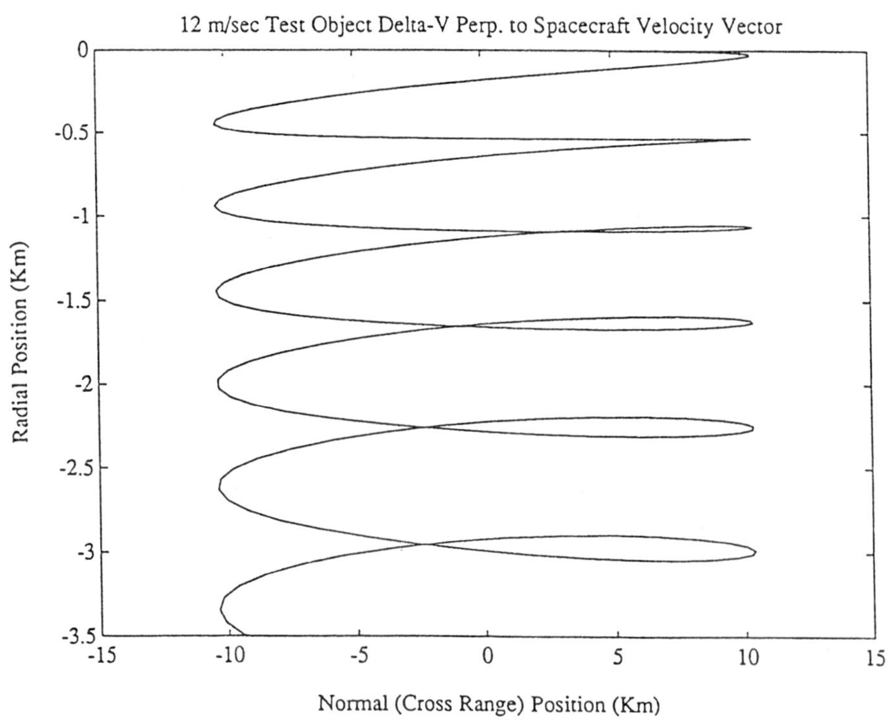

Figure 12. Radial Position vs Normal Position

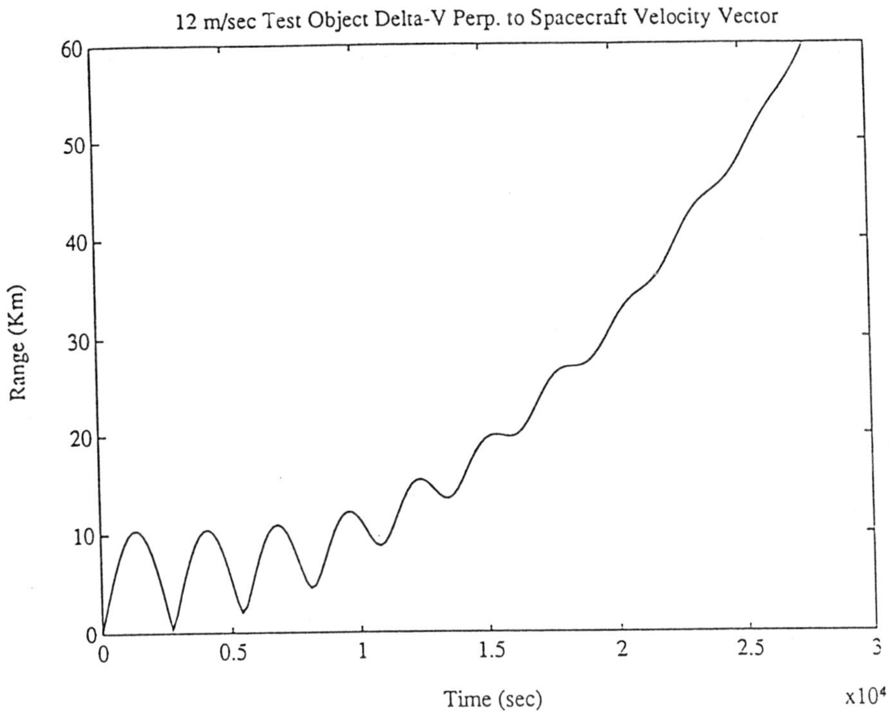

Figure 13. Range vs Time

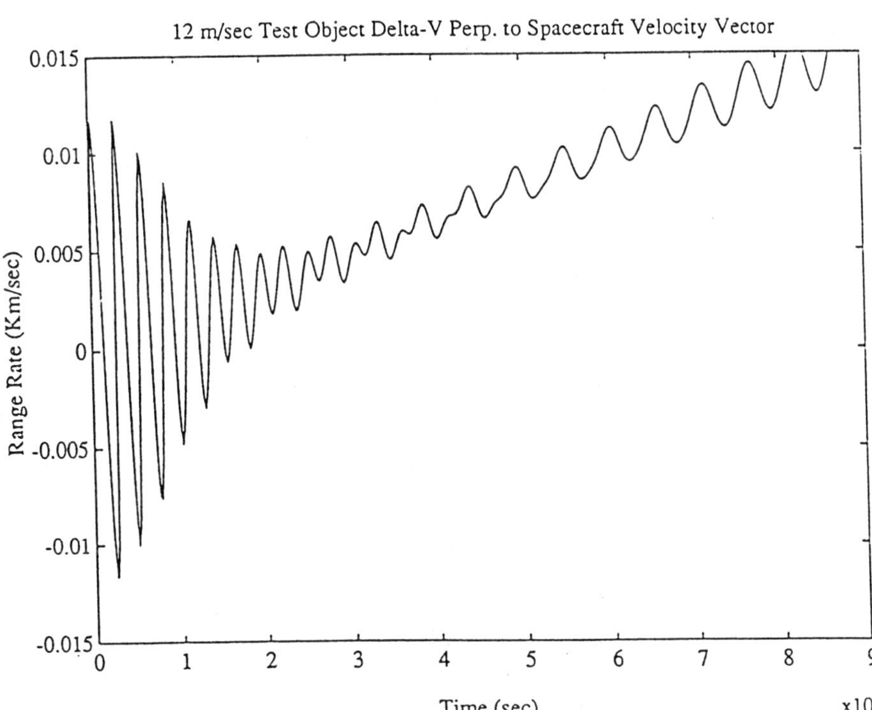

Figure 14. Range Rate vs Time

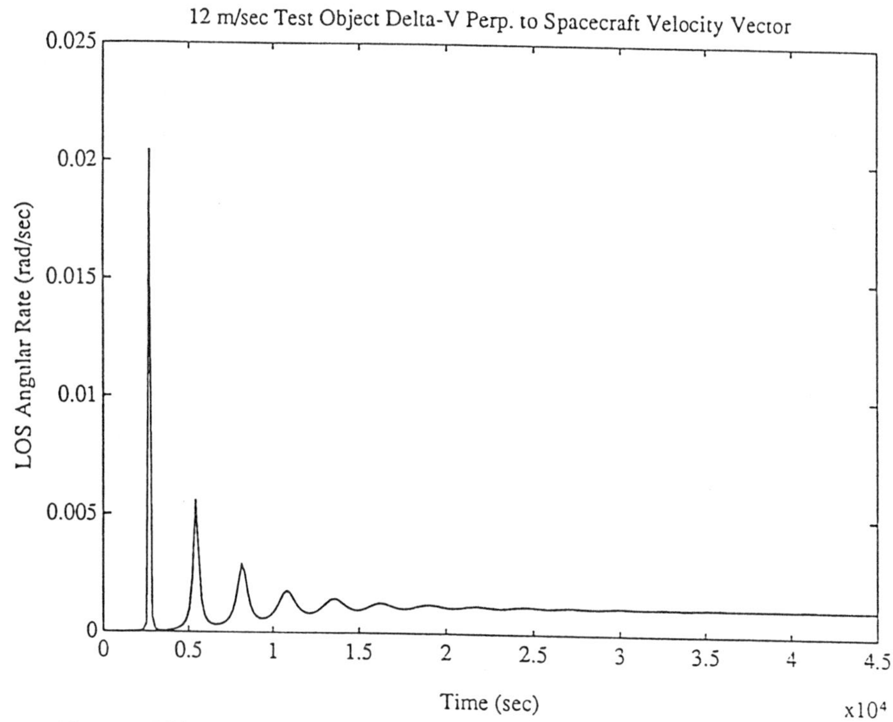

Figure 15. LOS Angular Rate vs Time

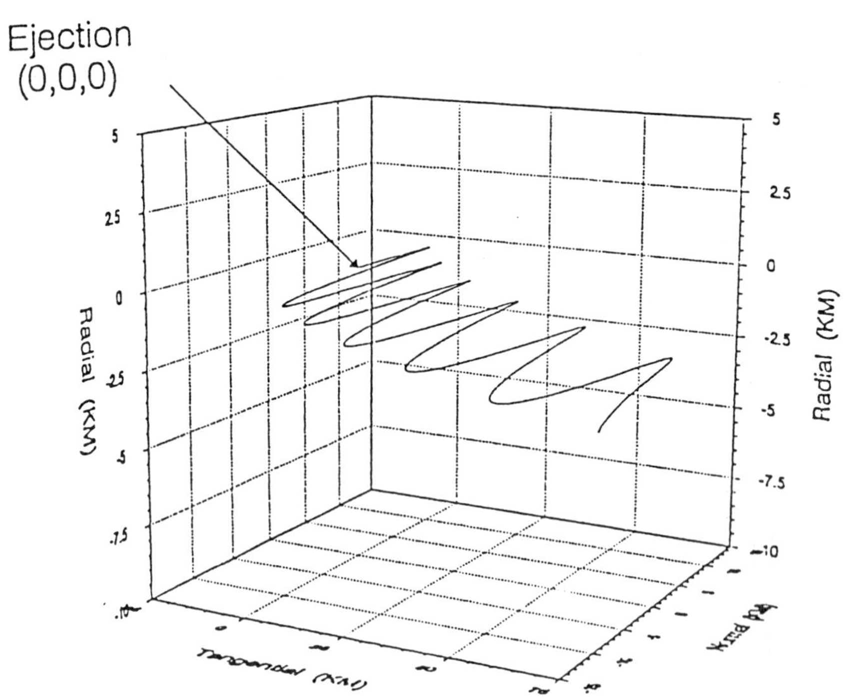

Figure 16. Relative Position of Test Object with Respect to Spacecraft for Out-of-Plane Ejection

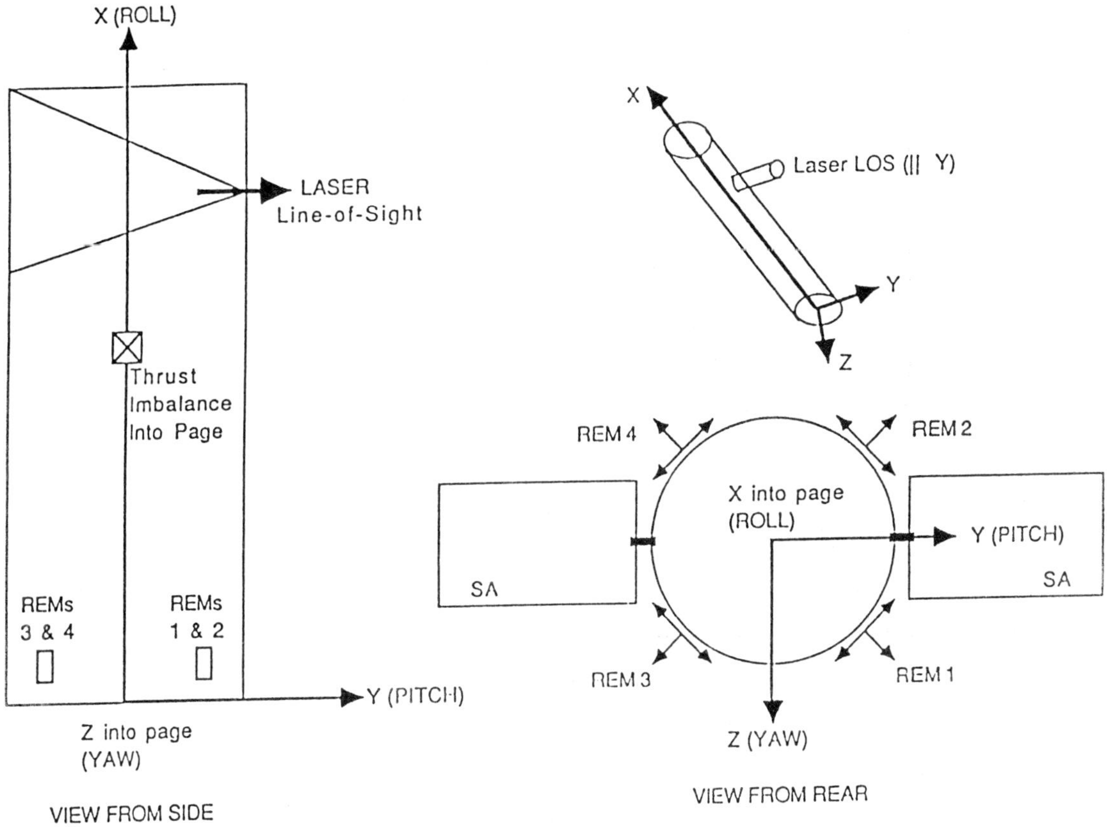

Figure 17. ICSE Spacecraft Coordinate System and Thruster Configuration for the Flexible Body Simulation Model

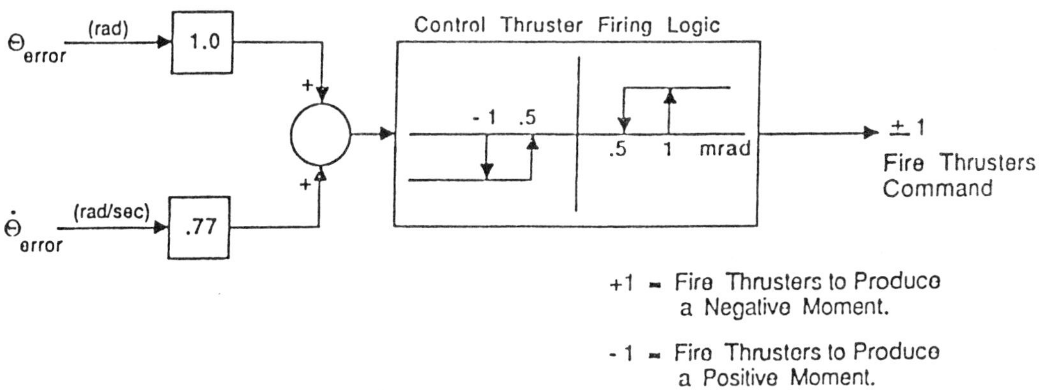

Figure 18. ICSE Attitude Controller for Roll, Pitch and YAW AXES

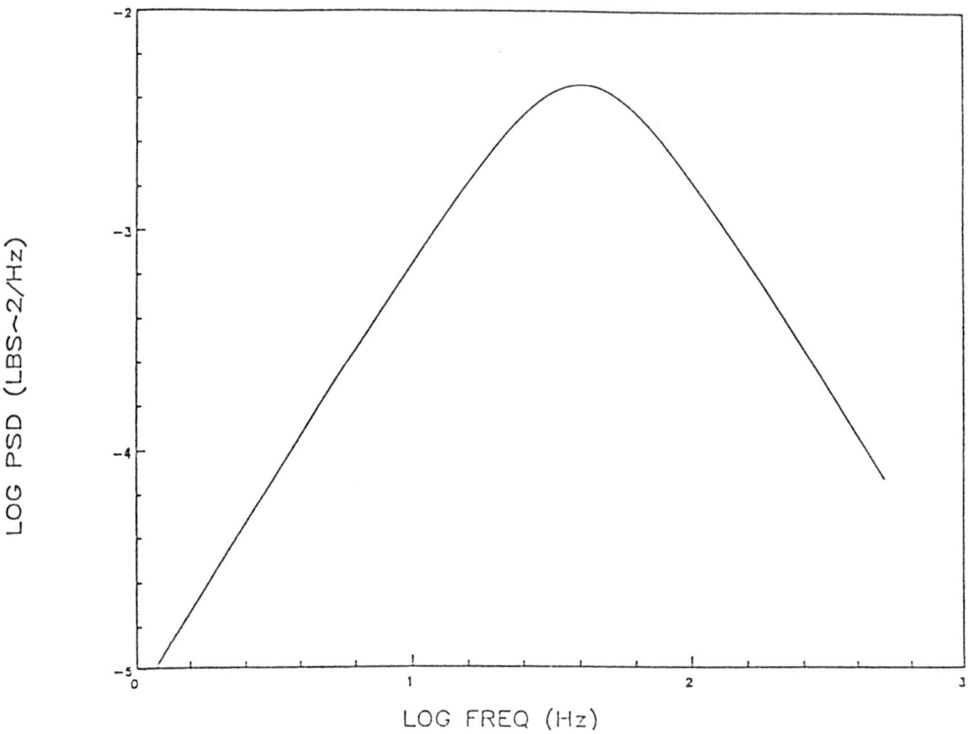

Figure 19. PSD of GGA Disturbance used for cases 1 and 2

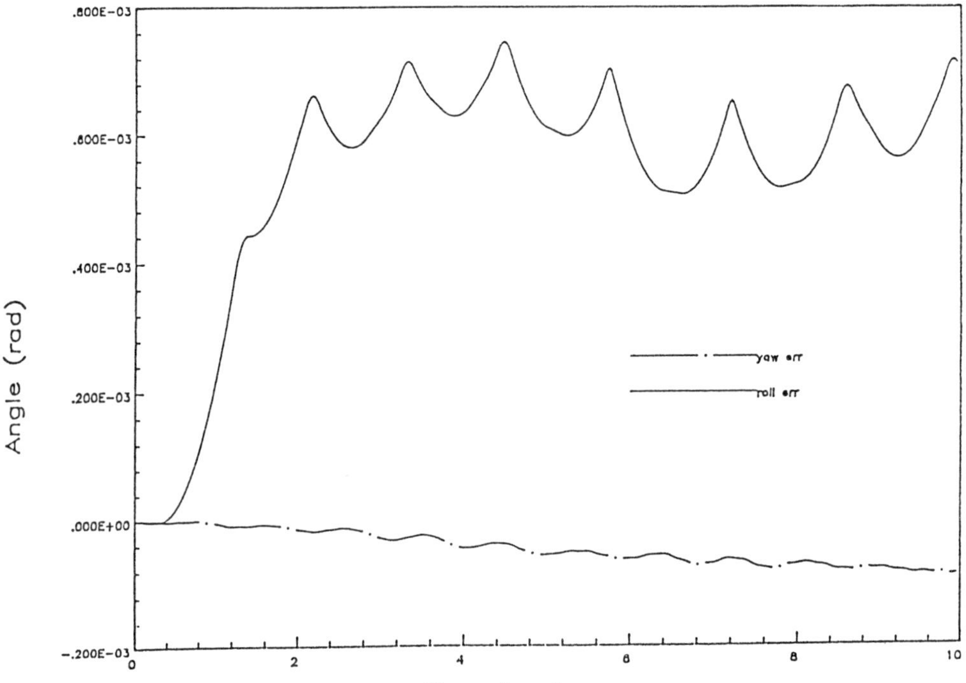

Figure 20. Roll and YAW Pointing Error for Case 1

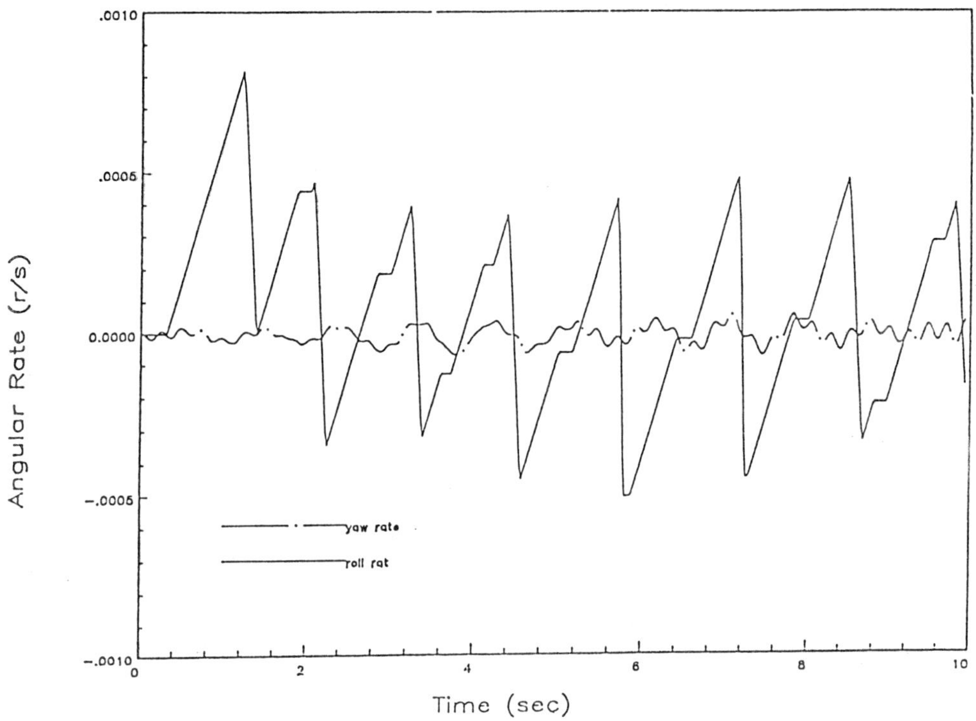

Figure 21. Roll and YAW Pointing Error Rate for Case 1

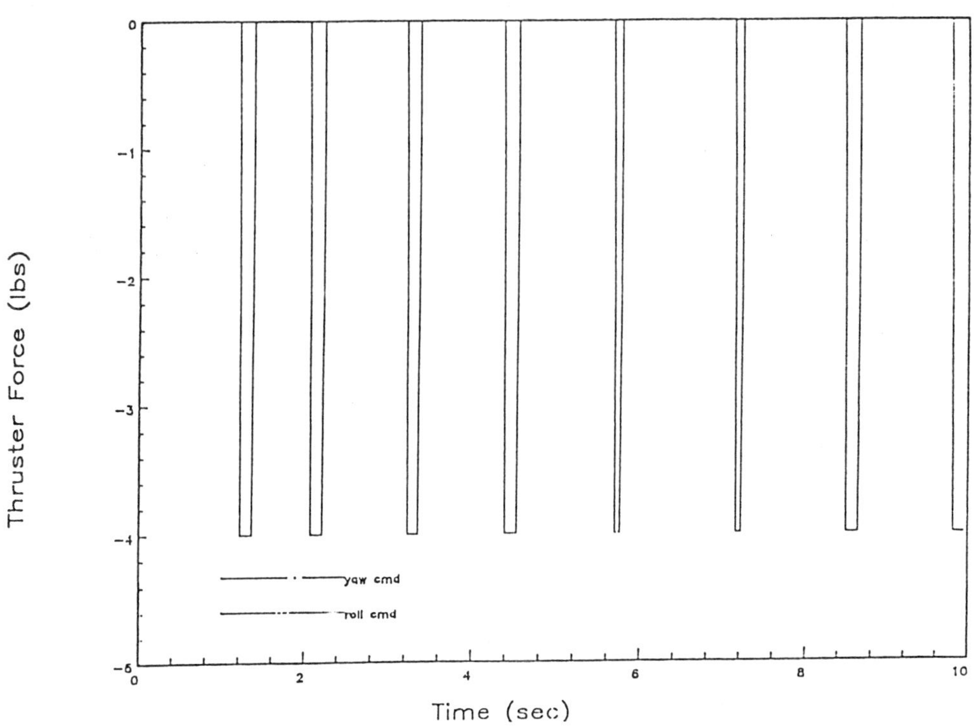

Figure 22. Roll and YAW Thruster Force Commands for Case 1

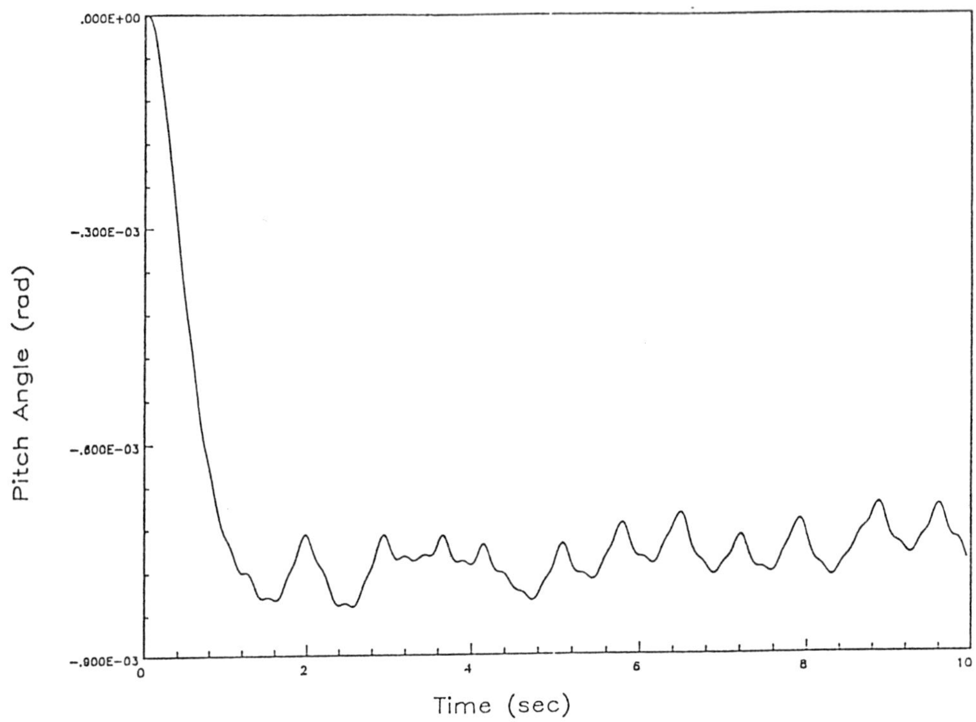

Figure 23. Pitch Angle for Case 1

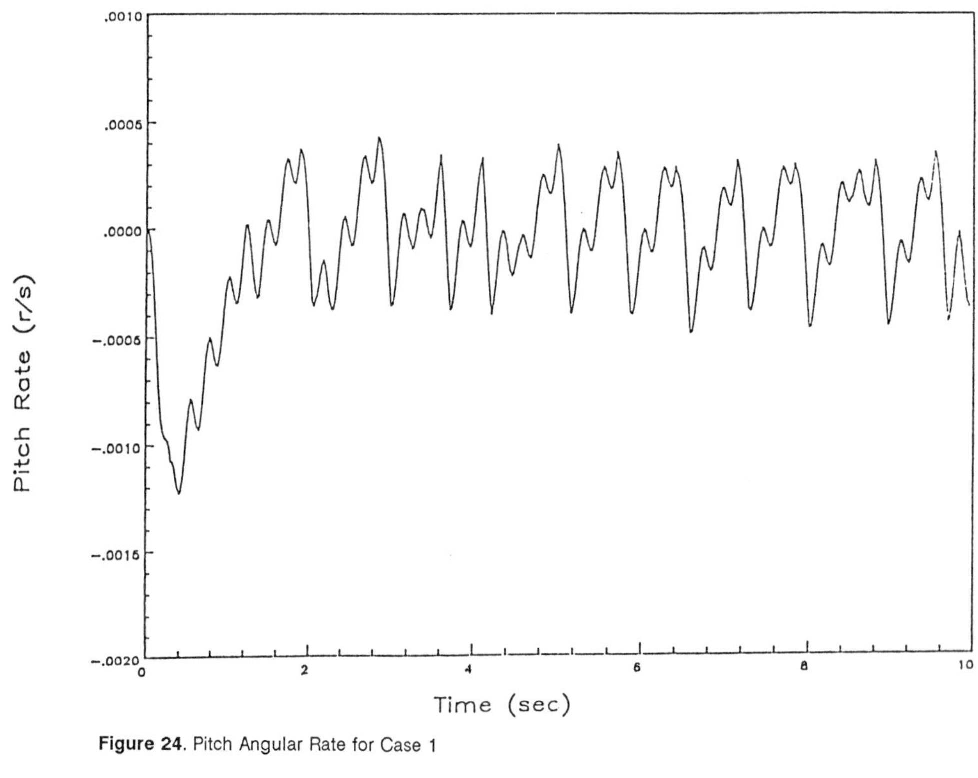

Figure 24. Pitch Angular Rate for Case 1

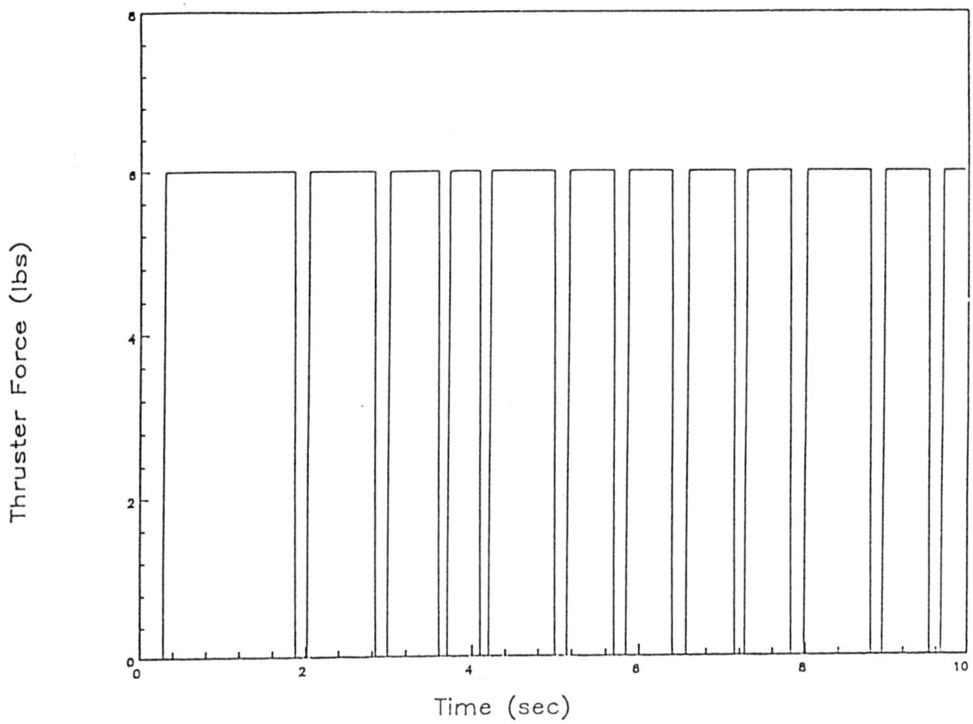

Figure 25. Pitch Thruster Force Commands for Case 1

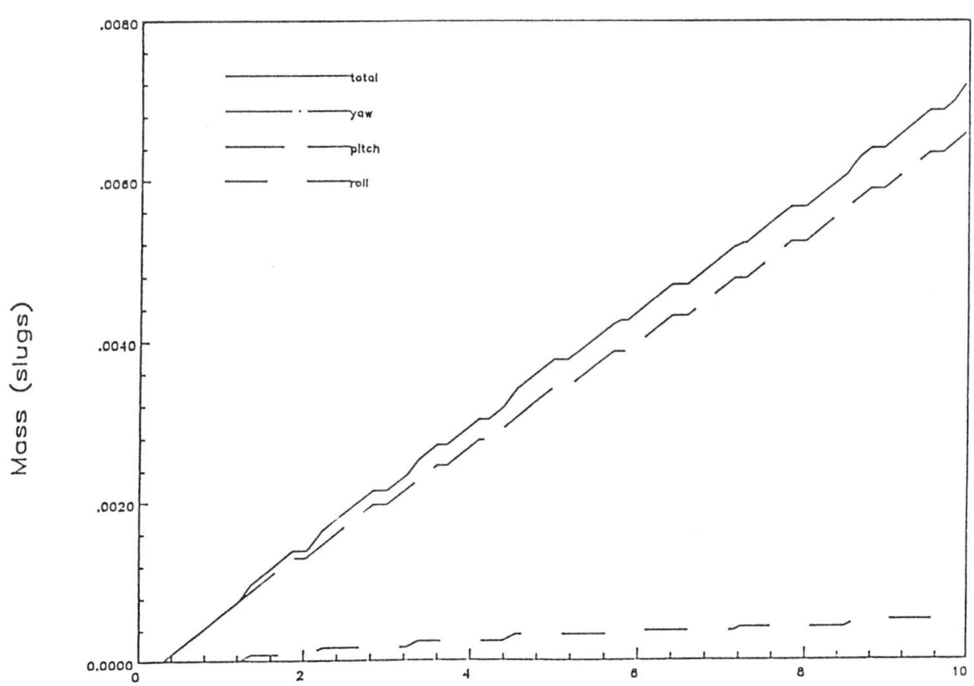

Figure 26. Amount of Propellant Used for Case 1

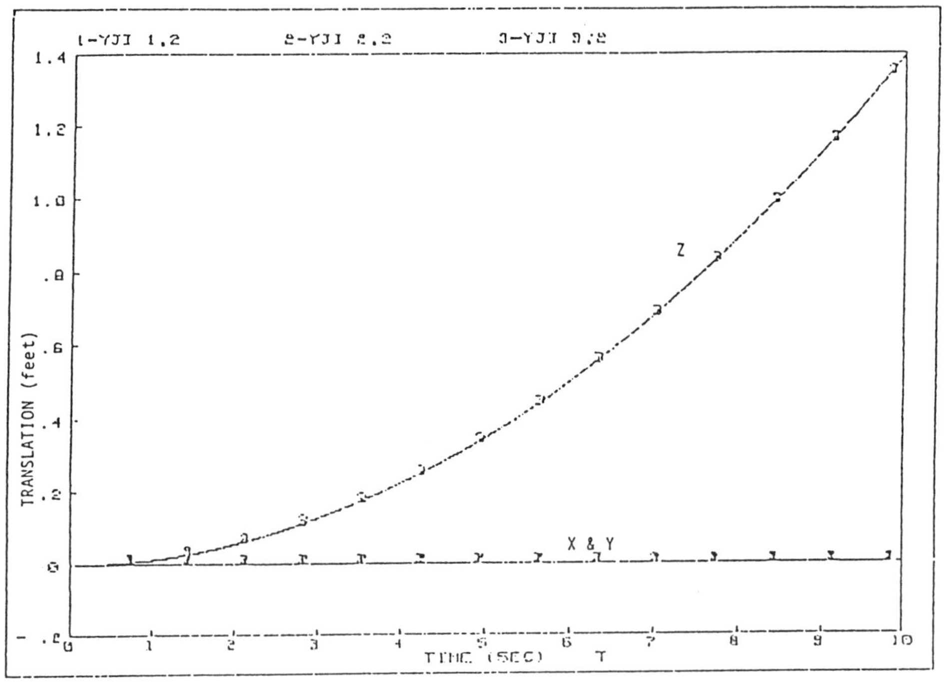

Figure 27. Spacecraft Translation for Case 1

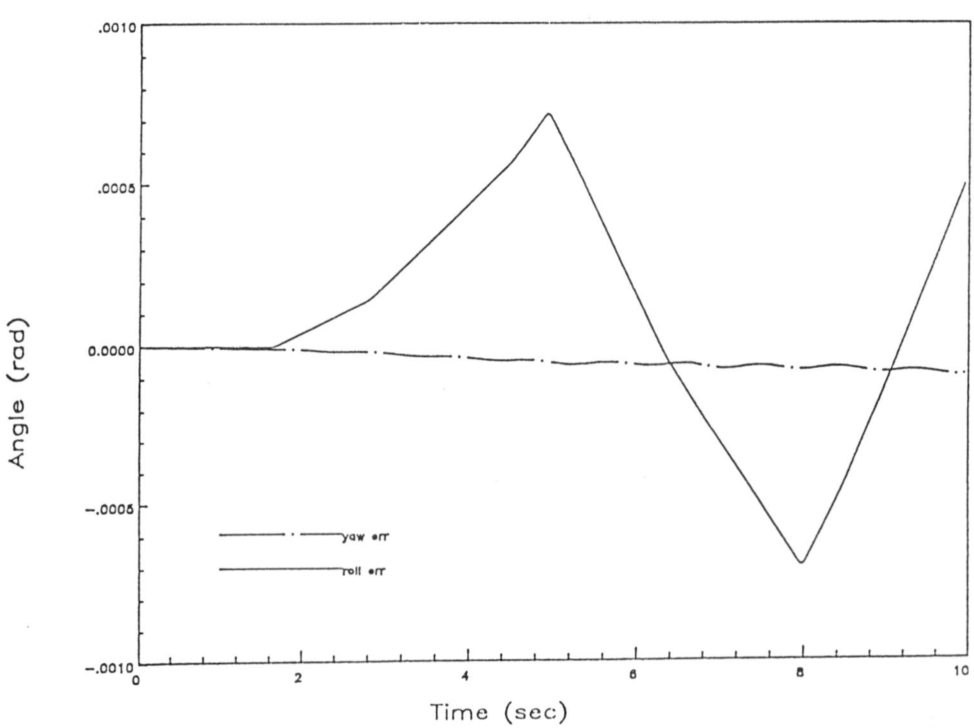

Figure 28. Roll and Yaw Pointing Error for Case 2

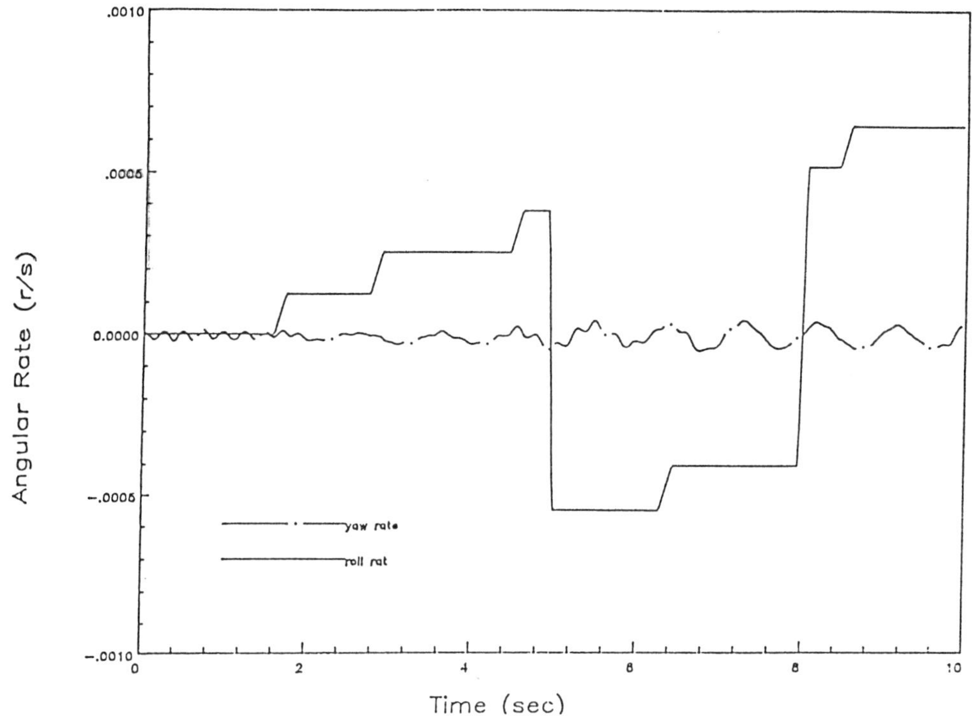

Figure 29. Roll and YAW Pointing Error Rate for Case 2

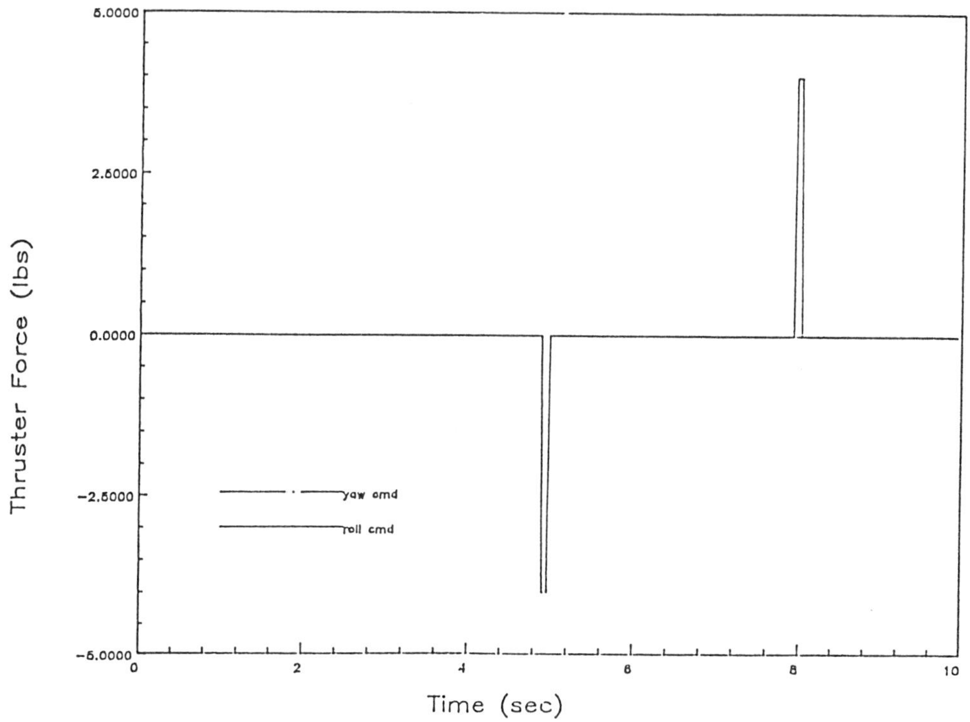

Figure 30. Roll and Yaw Thruster Force Commands for Case 2

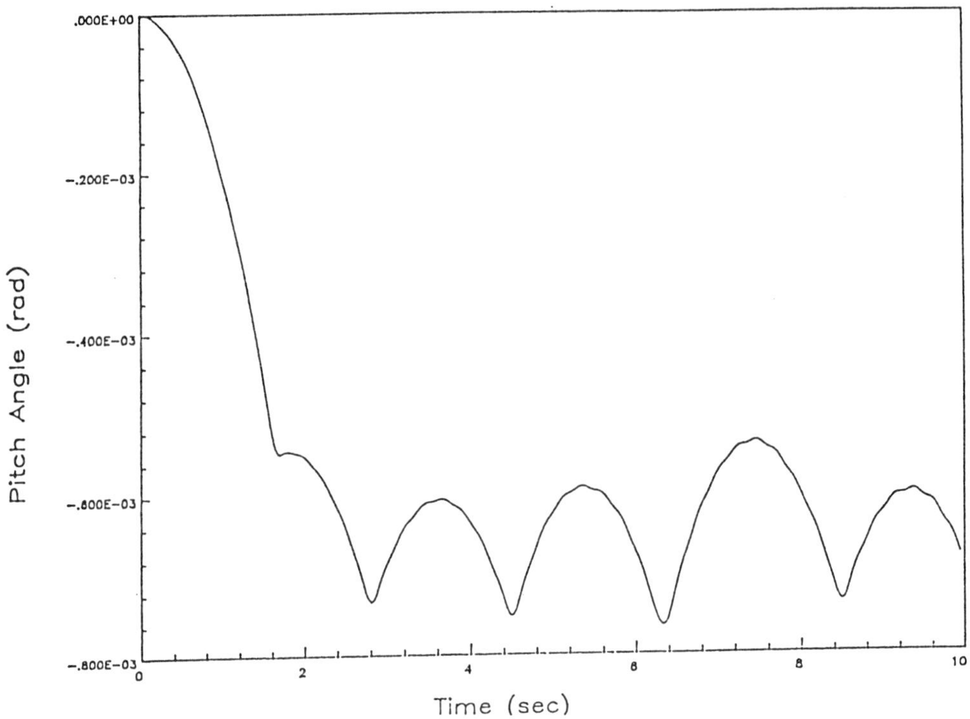

Figure 31. Pitch Angle for Case 2

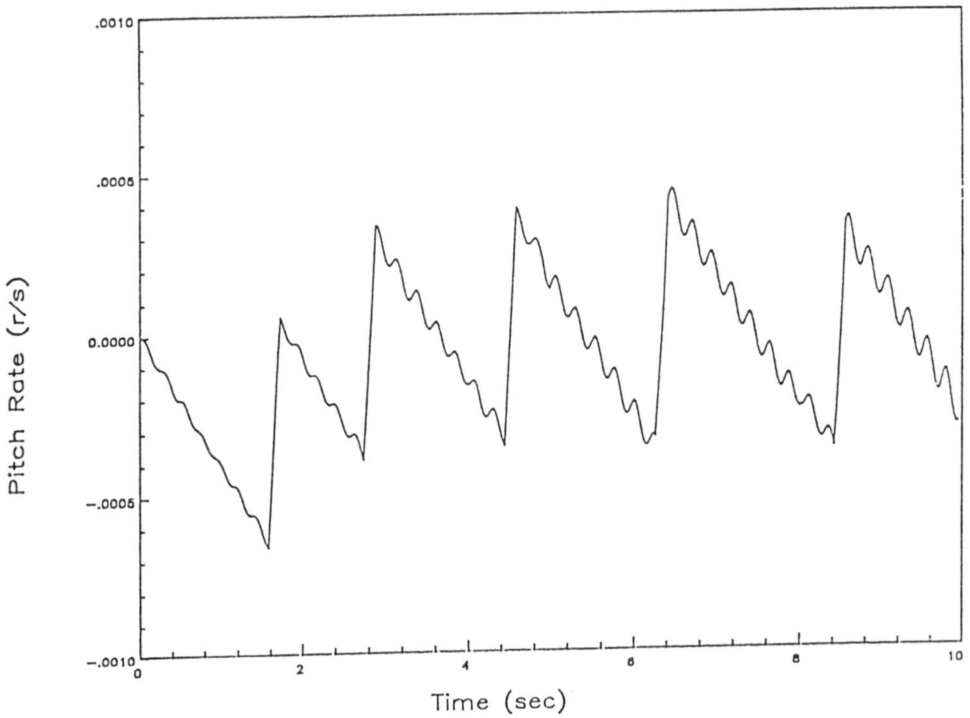

Figure 32. Pitch Angular Rate for Case 2

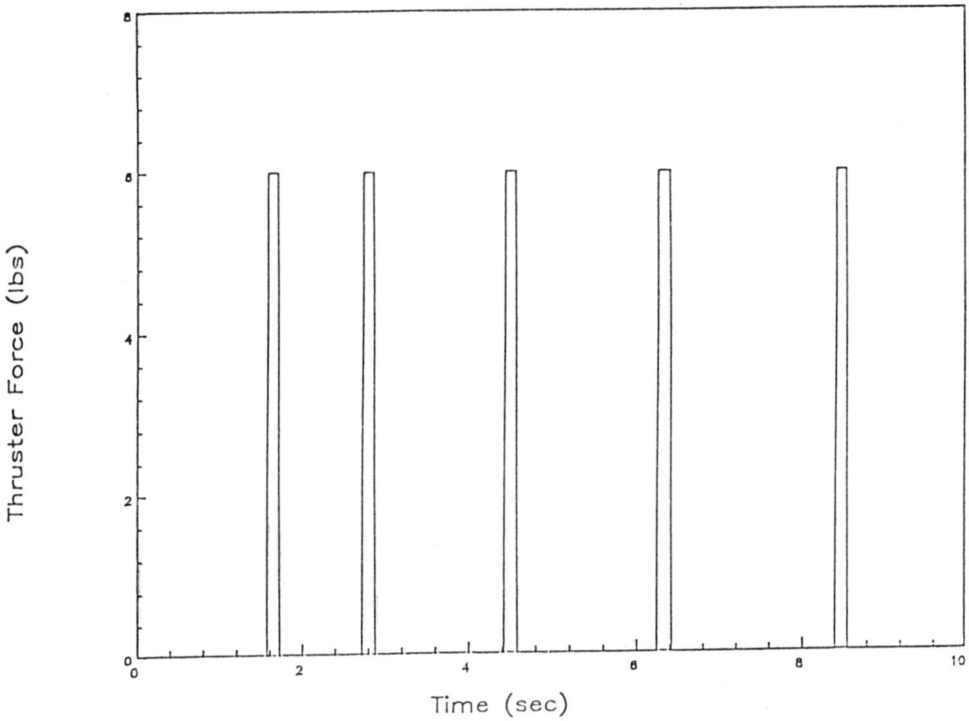

Figure 33. Pitch Thruster Force Commands for Case 2

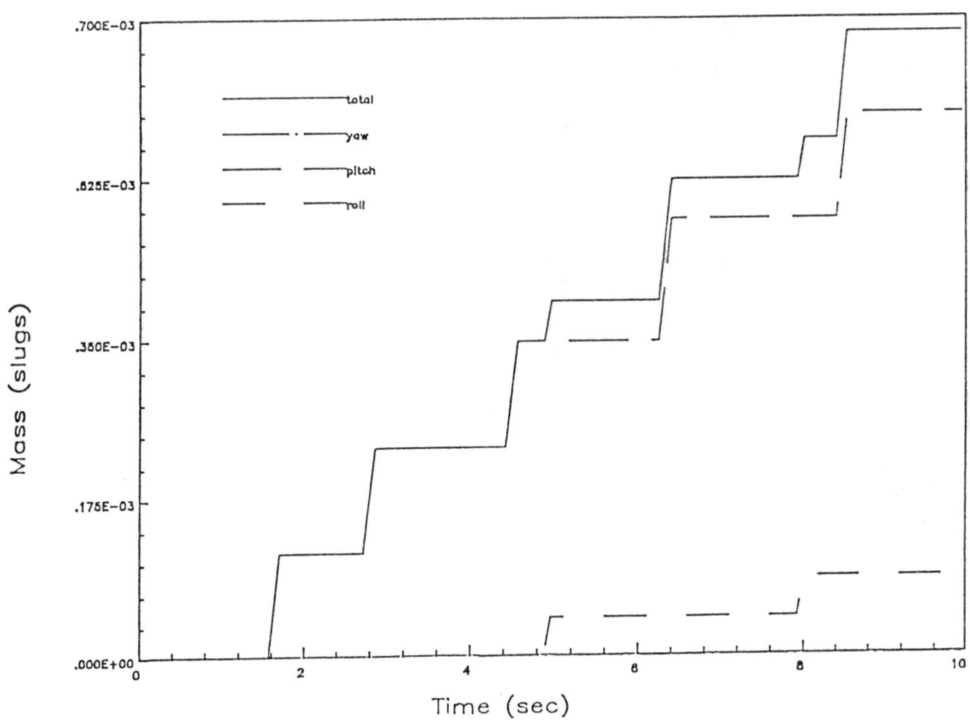

Figure 34. Amount of Propellant Used for Case 2

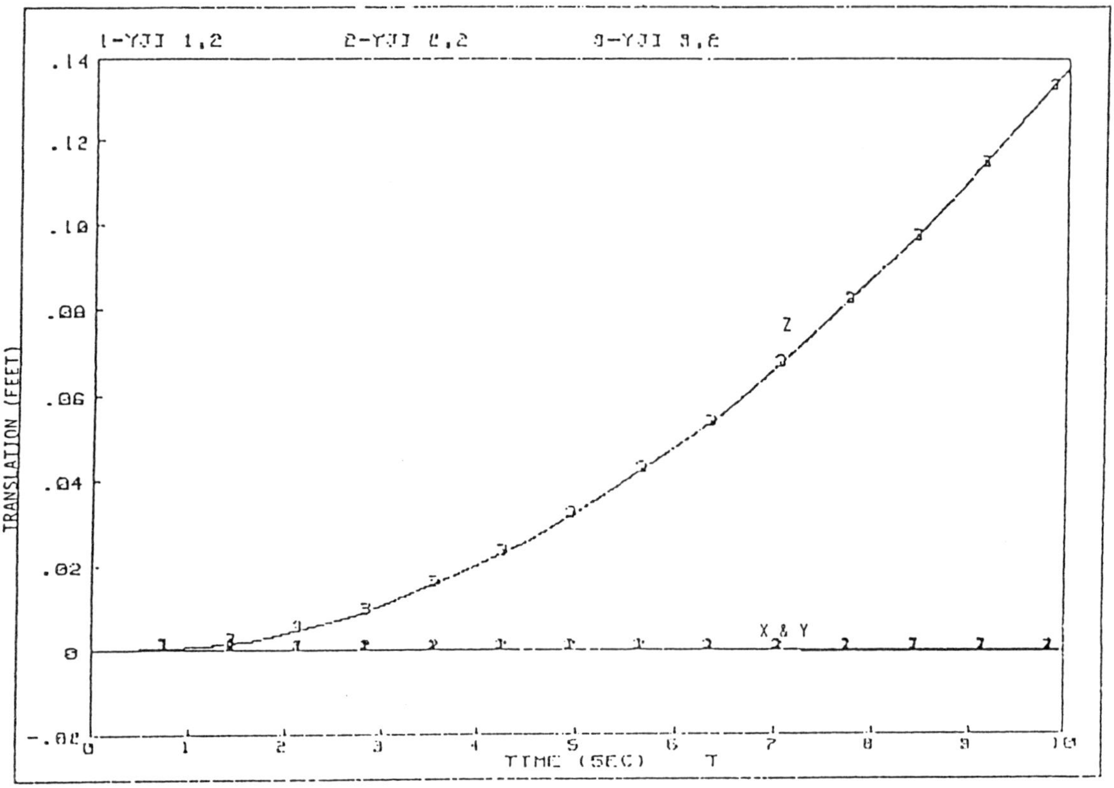

Figure 35. Spacecraft Translation for Case 2

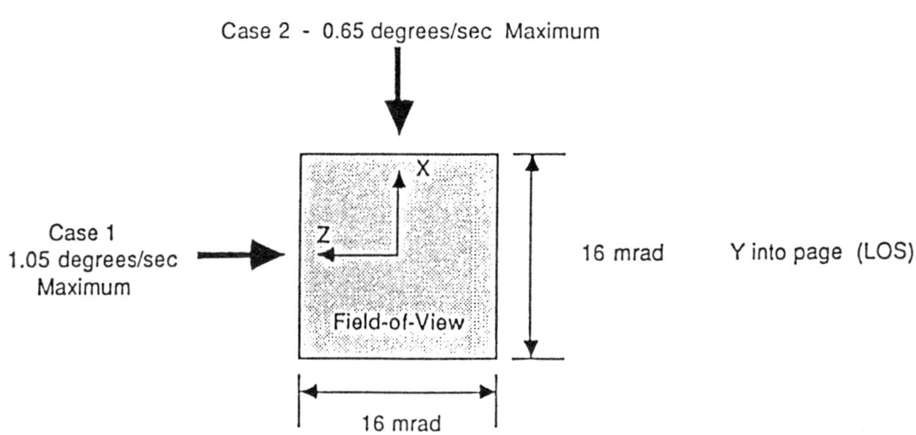

Figure 36. Tracking Results for ICSE

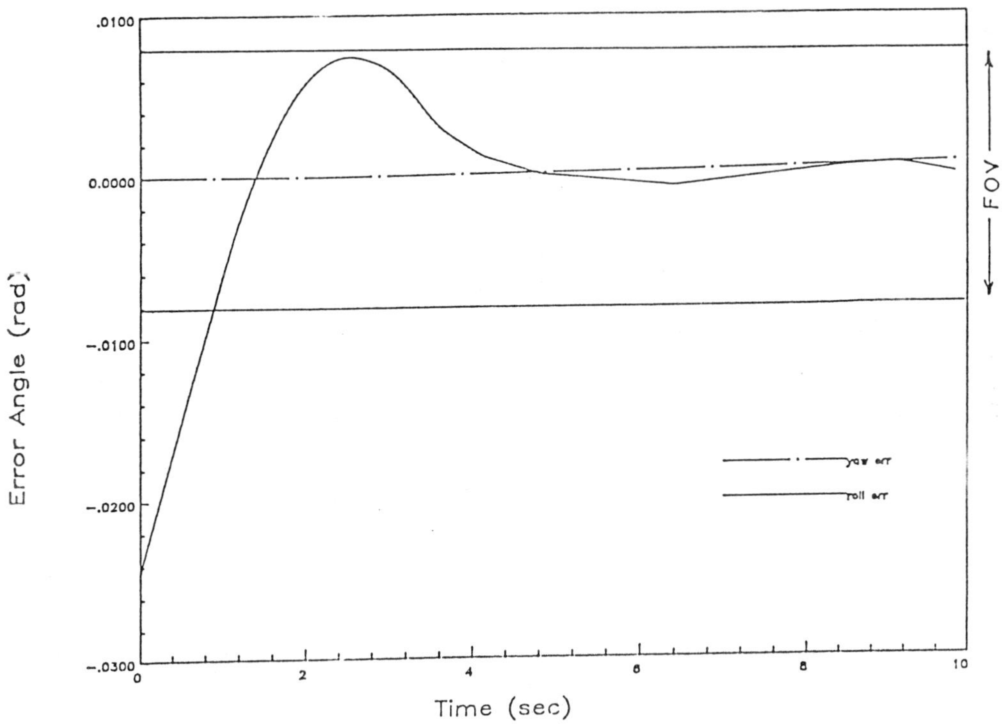

Figure 37. Tracking Error Angle for Roll Rate of 1.05 Deg/Sec

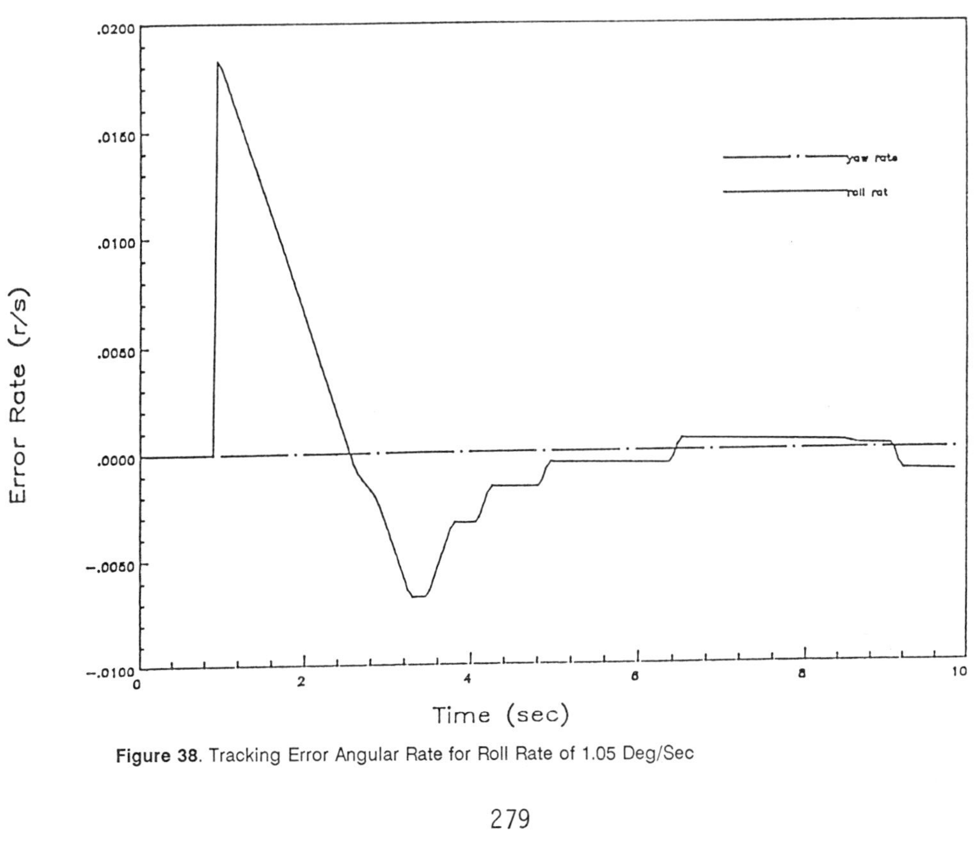

Figure 38. Tracking Error Angular Rate for Roll Rate of 1.05 Deg/Sec

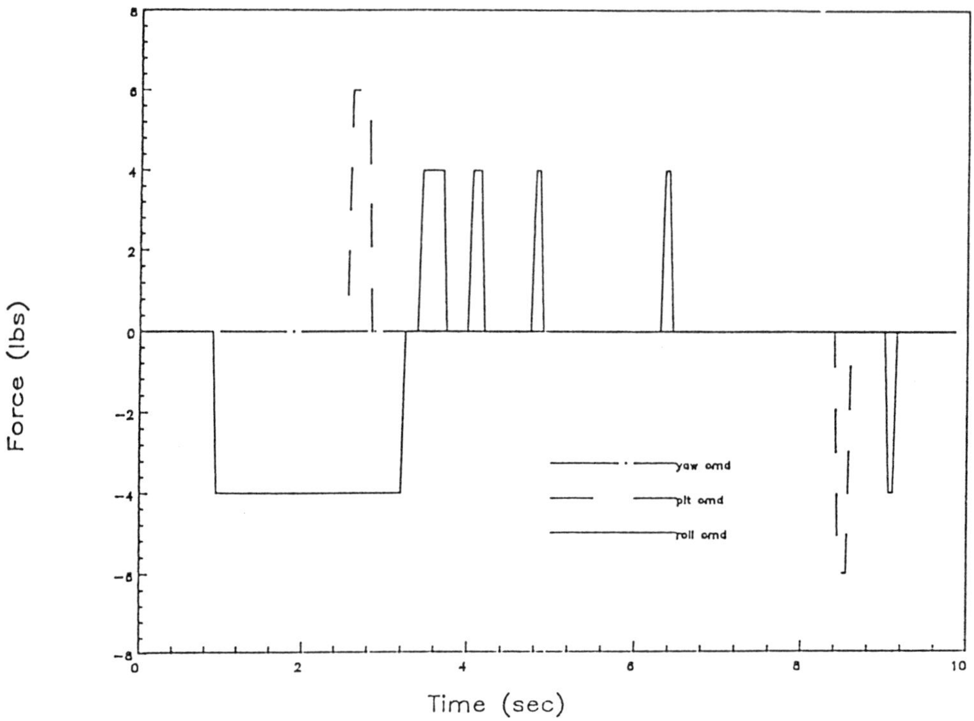

Figure 39. Thruster Commands for Tracking of Roll Rate of 1.05 Deg/Sec

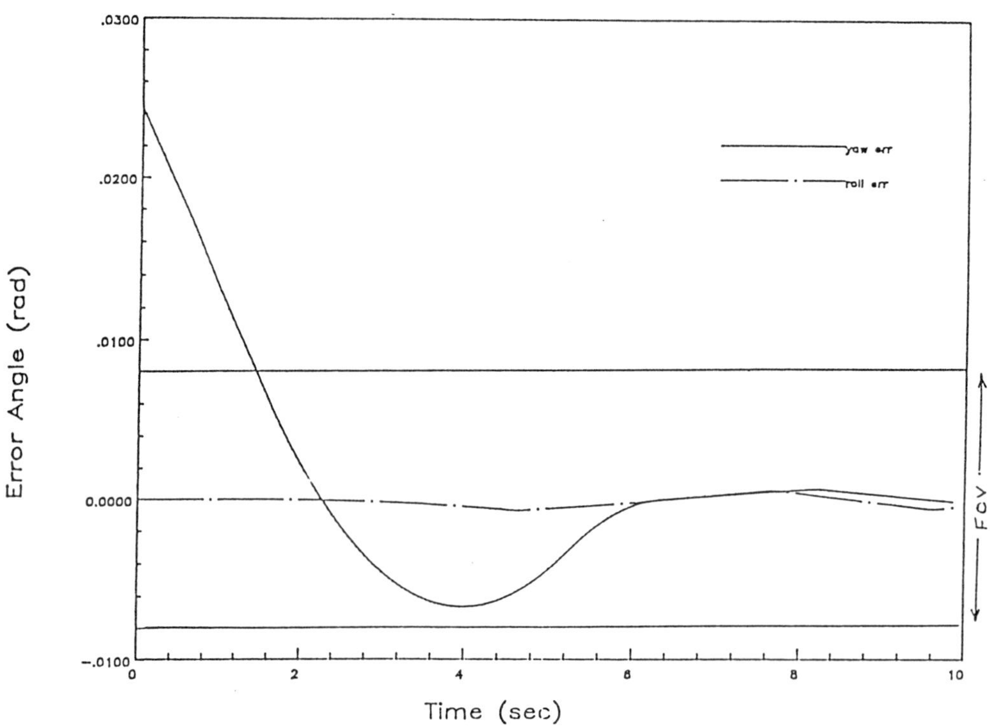

Figure 40. Tracking Error Angle for Yaw Rate of 0.65 Deg/Sec

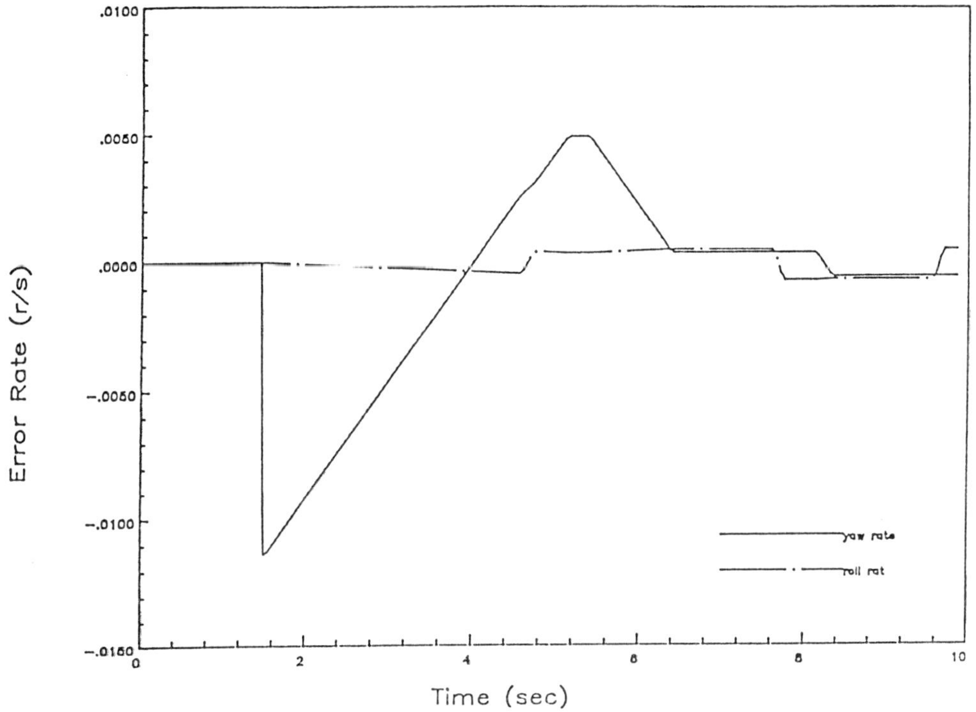

Figure 41.Tracking Error Angular Rate for a Yaw Rate of 0.65 Deg/Sec

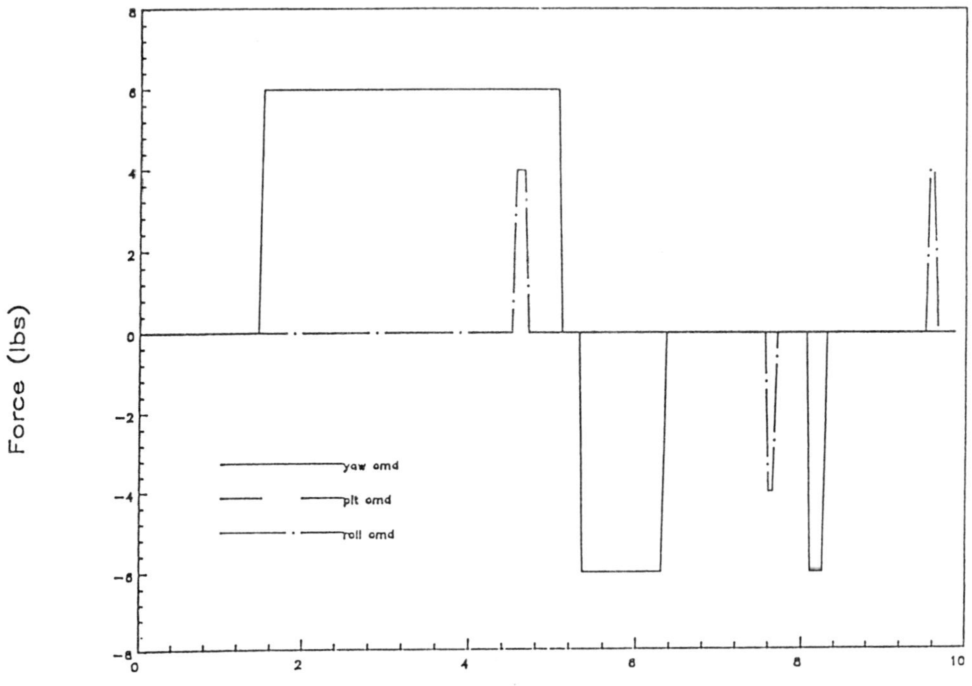

Figure 42. Thruster Commands for Tracking of Yaw Rate of 0.65 Deg/Sec

AAS 91-054

A WAVEFRONT COMPENSATION APPROACH TO SEGMENTED MIRROR FIGURE CONTROL

David Redding*, Bill Breckenridge†, George Sevaston† and Ken Lau†

Segmented large mirrors have several advantages over conventional monolithic optics, especially for use in space. To be used effectively for astronomy, however, they must be actively aligned and controlled to extreme levels of precision. We consider this "figure control" problem for a spaceborne sub-millimeter wave telescope, the Precision Segmented Reflector Project Focus Mission Telescope. We show that performance of any figure control system is subject to limits on the controllability and observability of the quality of the wavefront. We present a "Wavefront Compensation" method for the Focus Mission Telescope which uses mirror figure sensors and three-axis segment actuators to directly minimize wavefront errors due to segment position errors. This approach shows significantly better performance when compared with a "Panel State Compensation" approach.

INTRODUCTION

The NASA Precision Segmented Reflector (PSR) program is developing technologies to enable the flight of large spaceborne astronomical telescopes starting sometime around the turn of the century. Much of our current activity centers on the PSR Focus Moderate Mission Telescope (FMMT), a medium-sized (3.65 m aperture) space-based submillimeter-wave telescope which exemplifies many of the technical difficulties that must be overcome.

This paper addresses the problem of figure *maintenance* control for the FMMT primary mirror, which is formed of seven separate segments. Figure maintenance control is used when the ideal configuration of the panels with respect to each other is known. It uses sensors that directly sense motions of the panels to determine their errors relative to the ideal state. These errors are corrected by actuators that rotate and translate the panels. Figure *initialization* control is the process of determining the ideal configuration for the panels and moving them there. The FMMT initialization controller uses the science detector directly. By looking at known bright calibration stars, the panels are moved to maximize the image quality directly. FMMT figure initialization control will be described in Ref. 1. An overall block diagram of the FMMT figure control system is provided in Fig. 1.

Figure maintenance is used between initializations to hold the figure of the mirror steady while taking science data. It is intended to counter the long-term thermal drifts that the structure experiences. For the FMMT, these errors are expected to be 10-200 microns at the segment centers over the course of a single orbit. They occur on a relatively long time scale, typically tens of minutes. Consequently, the FMMT figure maintenance controller has a low bandwidth on the order of 0.1 Hz.

The FMMT also experiences vibrational disturbances due to the presence on the structure of vibrating machinery, such as reaction wheels and a secondary mirror chopping mechanism. These are low amplitude, high frequency effects, causing segment motions of 0.01-0.8 microns peak at frequencies ranging from 0.5-500 Hz. They are not affected by the figure maintenance controller, being beyond its effective bandwidth. They are passed through directly to the panels causing acceptable levels of image jitter.

* Los Angeles Area Site Manager, Charles Stark Draper Laboratory, Inc. † Jet Propulsion Laboratory, California Institute of Technology, 4800 Oak Grove Drive, Pasadena, California 91109.

The objective of the FMMT figure maintenance controller is to *maximize the optical quality of the entire telescope* with respect to primary mirror segment position errors. It does this by moving the primary mirror segments to cancel optical aberrations induced on the wavefront by rotational and translational disturbances of the segments. Our controller uses an accurate mathematical model of the FMMT optics to compute the control that *minimizes the wavefront error* of the telescope. This differs from previous work, which seeks to minimize segment position errors directly, without accounting for their effect on the ultimate imaging performance of the telescope.[2,3]

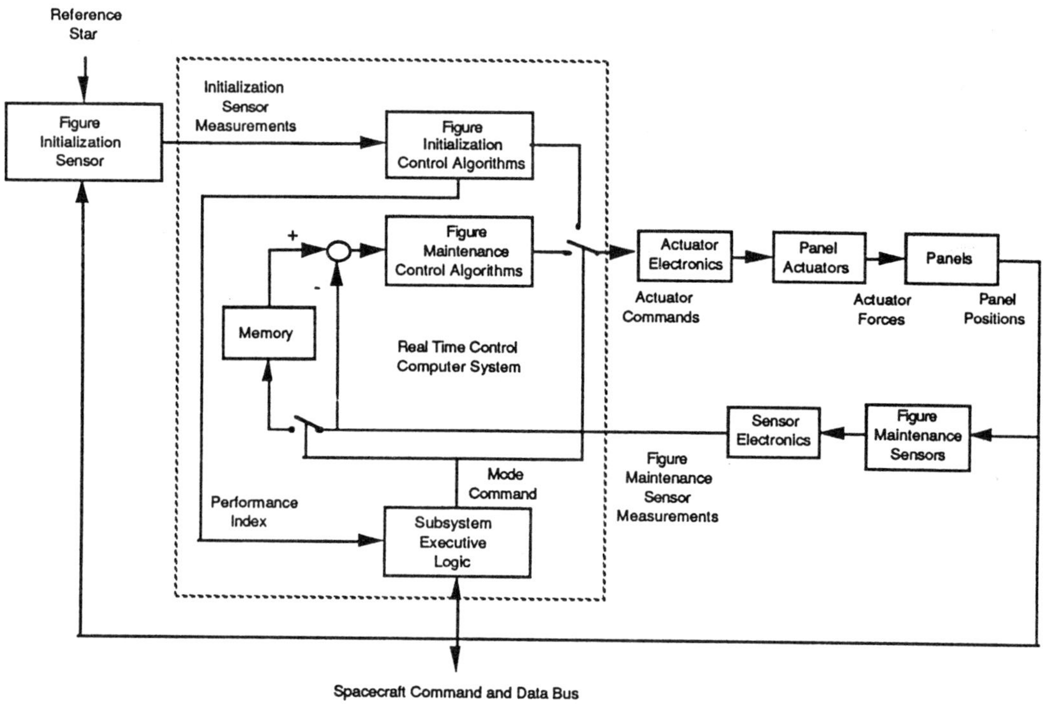

Figure 1. FMMT figure control system block diagram.

This paper presents a simplified analysis that addresses the fundamental questions of figure maintenance control. We treat the FMMT as a quasi-static system, which is appropriate given the long time-scale of the thermal disturbances as compared to the maintanence control bandwidth. We consider only the six outer segments, using the center segment as a reference for the rest. We neglect the effects of segment surface distortions and motions of other optical elements, though these can easily be included in in the analysis if desired. We concentrate on identifying the basic limiters of performance for figure control systems: controllability, observability, actuator and sensor noise. We address these and other issues in the context of the FMMT, but the methods we use, and the conclusions we draw, can be applied to any segmented-mirror telescope.

The paper proceeds as follows. The next section provides a summary description of the FMMT. We then discuss the problem of modelling the telescope so as to correctly quantify the effects of mirror segment errors on the wavefront. The fourth section derives a "Wavefront Compensation" (WFC) segment control approach which minimizes the wavefront effects of segment errors. We consider the issue of controllability, showing that there exist fundamental limits to the ability of any figure control system to compensate for segment errors. The fifth section considers several segment figure sensor configurations to determine their ability to observe the segment error contribution to wavefront error. We show that some configurations do not provide full observability of the wavefront. Finally, we put the estimator and controller together into an integrated control algorithm which is tested using covariance and Monte Carlo analysis. The results, which compare WFC to a direct "Panel State Compensation" (PSC) segment error control approach, show that WFC is significantly better.

DESCRIPTION OF THE FMMT

The FMMT main optics form an *f*/10 Cassegrain telescope, consisting of a very fast *f*/0.4 segmented parabolic primary mirror and a hyperbolic secondary mirror, as sketched in Fig. 2. The FMMT also uses 1:1 relay optics to reimage the light in a cryogenic dewar which contains the pixillated science detectors (not shown). Currently, we compute the pupil function of the system at the spherical reference surface (RS) located just past the secondary. This reference surface is aligned with the nominal spherical wavefront of the unperturbed system. One of the baselined science detectors is a 10×10 bolometer array which is sensitive in the 100-300μm wavelengths of light. This bolometer is used by the initialization controller in the initial phasing of the panels, as well as to collect astronomical science data.

The primary mirror is supported by a trusswork backplane structure constructed of high stiffness, low thermal coefficient-of-expansion members. The segments attach to the nodes of the truss and are influenced by motions of the nodes. The thermal and other slow motions of the optics are expected to be between 10-235 μm and 10-200 μrad over the maintenance cycle.

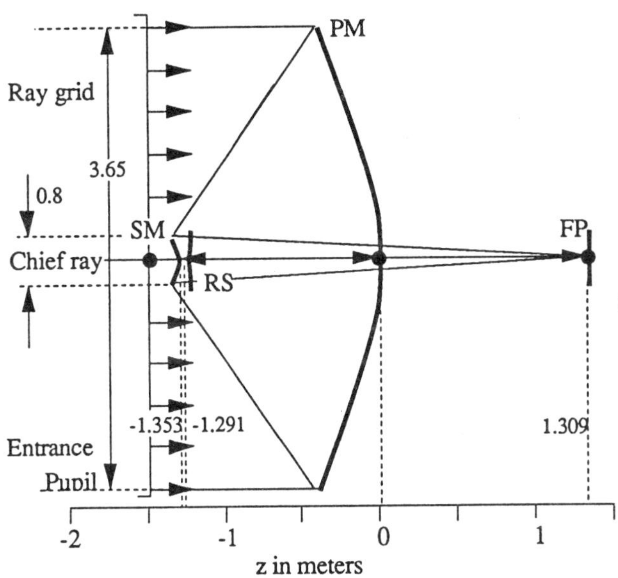

Figure 2. FMMT main optics.

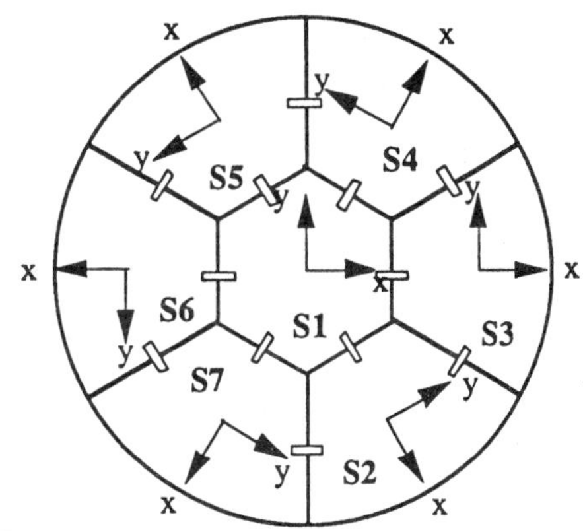

Figure 3. FMMT primary mirror showing segments, representative sensor locations, and segment actuator coordinates.

Each segment panel is separately actuated in three degrees of freedom. This is accomplished using three linear actuators acting in the segment z-axis at the three segment attach points. For this analysis we resolve the three available control degrees of freedom into tip (rotation about the local y-axis); tilt (rotation about x); and piston (translation along the local z-axis) commands acting at the center of the panels. The segment actuator coordinates are centered in each panel with the x-axis pointing radially away and tangent to the surface, the z-axis along the normal, and the y-axis forming a tangent in the circumferential direction (Figs. 3 and 4). One-sigma actuator noise levels are expected to result in approximately 350 nm error in piston and 350 nrad in tip and tilt. The three uncontrolled degrees of freedom per panel consist of x- and y-decenter (translation along x or y) and twist (rotation about z).

Also indicated on Fig. 3 are the locations of figure sensors for one candidate sensor configuration. Two types of figure sensors are used in this analysis. Edge-sensors use an array of four laser interferometers to directly measure relative piston (z-axis translation) and hinge angle (x-axis rotation) between two segments, as indicated on Fig 5. Gap sensors using interferometric, inductive or capacitive devices can be used to measure relative displacements in the y-axis direction. One-sigma measurement noise of about 1 nm in piston and gap was assumed; hinge angle noise of about 1 nrad was also assumed.

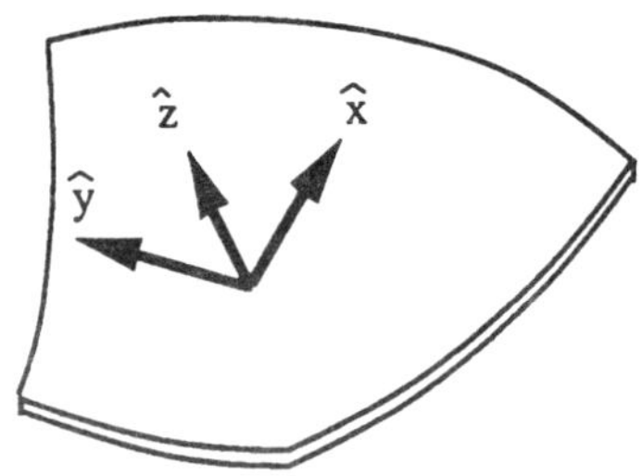

Figure 4. Sketch of segment coordinates

MODELLING

The first step in creating a controller of this type is to formulate the objective function in terms of the error states and the control variables. In our case, the objective function is wavefront error squared at the exit pupil, the error states are the 6DOF rotational and translational motions of the six outer segments, and the controls are the 3DOF (tilt, tip and piston) actuators of each segment. The second step is to take the derivatives of the objective function with respect to the error states and control variables. These derivatives provide a linear model of the influence of the errors and controls on the objective. The third step is to invert the linear model (in the least-squares sense) to obtain expressions for the control in terms of the errors, as weighted by the error contribution to the objective function. The result is the optimal control gain, i.e. the control that minimizes the objective function.

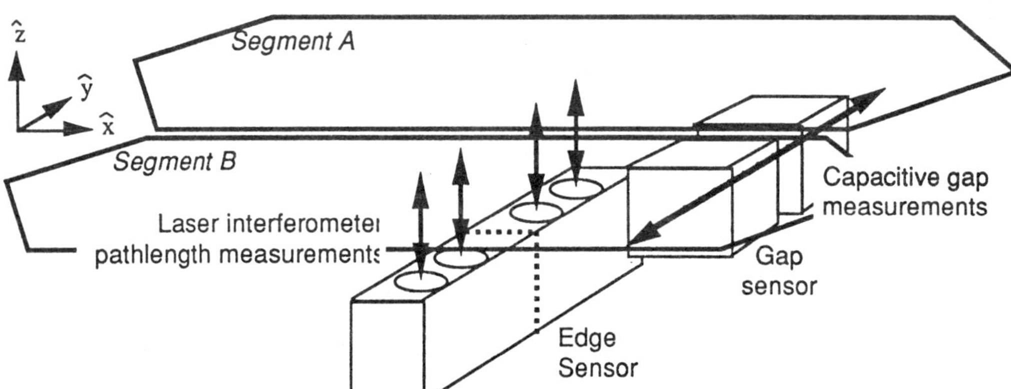

Figure 5. Figure sensor geometry.

A similar process is followed to create an estimator for processing segment figure sensor measurements into segment error state estimates. Here we start by expressing the sensor measurements as a linear function of the segment state errors using the kinematics of the FMMT structure. This measurement equation is linearized and inverted (using weighted least-squares) to obtain equations for estimator gains, giving segment state errors as a linear function of sensor measurements. An important question for the estimator is whether or not it provides full observability of those segment errors that affect wavefront error. This question also requires the use of the optics modelling and is addressed in the next section.

In most of this analysis we treat the FMMT as a quasi-static system. We consider the errors to be initial-condition errors and we neglect structural and thermal dynamics. In the case of the thermal disturbances, the time scale is so slow compared to segment servo dynamics that the controller is well approximated as a periodic single-stage controller without relevant dynamics. As for the structural vibrational disturbances, they occur so fast compared to the maintenance controller bandwidth that they pass through unimpeded.

This section of the paper summarizes the FMMT modelling process. The wavefront error is best modelled using ray-trace optics. Ray-tracing treats light beams as bundles of straight-line rays. Each ray is broken into segments, with direction changes occurring at reflective and refractive surfaces. For the FMMT, we trace a bundle of n (a large number) collimated rays starting at the entrance pupil. This represents the light from a point object at infinity (not a bad approximation to a distant star). We follow each ray as it hits one of the primary mirror segments, then the secondary mirror, and then finally the spherical reference surface, which has its center of curvature at the nominal focal point. In the nominal case, the pathlength of each ray is exactly the same. When the system is perturbed, the length of some of the rays will change. Differencing the perturbed and nominal pathlengths gives the ray OPDs (optical path differences) at the reference surface. The OPDs in turn determine the phase of the complex pupil function at the reference surface. The focal plane image is determined by propagating the wavefront determined by the pupil function to the focal plane using physical optics techniques. System image quality is indicated by the RMS wavefront error computed directly from the ray OPDs.

To compute the ray-trace models we follow the analysis of Refs 4-5 as realized in the Controlled Optics Modelling Package (COMP) computer program[6]. The COMP routines provide a full ray-trace (and differential ray-trace) analysis capability that can be run inline with other computer programs, or separately on a stand-alone basis. We use two COMP models of the FMT. One is an exact nonlinear ray-trace model and the other is a linearized model. The exact model is used to generate data for specific cases, as well as to compute the nonlinear Monte Carlo results presented below. The linear model, which is set up using COMP to directly calculate the partial derivatives of the rays at the reference surface with respect to the segment states and control actions, is used in the analysis to determine controllability and observability of the controller, and to perform covariance analyses. The mathematics underlying COMP are summarized briefly in the next few paragraphs; see Refs. 4-6 for details.

COMP represents the i-1th ray segment in terms of its starting point $\vec{p}_i$ with respect to the vertex of the ith surface, its direction $\hat{r}_{i-1}$, and its length L_{i-1} (see Fig. 6). Starting at $\vec{p}_i$, the ray proceeds in the direction $\hat{r}_{i-1}$ until it hits the ith surface. Conic surfaces are represented by a dyad $\mathbf{M}$ and a vector $\vec{\mathbf{N}}_0$, which are functions of the geometric focal length f, eccentricity e and principal axis direction $\hat{\psi}$

$$\mathbf{M} \equiv \left(\mathbf{I} - e^2\hat{\psi}\hat{\psi}\right) \tag{1}$$

$$\vec{\mathbf{N}}_0 = -f\hat{\psi} \tag{2}$$

The distance travelled to the surface is given by L_i, which is a solution of

$$\left(\hat{r}_{i-1} \cdot \mathbf{M} \cdot \hat{r}_{i-1}\right)L_i^2 + 2\hat{r}_{i-1} \cdot \left(\mathbf{M} \cdot \vec{p}_i + \vec{\mathbf{N}}_0\right)L_i + \vec{p}_i \cdot \left(\mathbf{M} \cdot \vec{p}_i + 2\vec{\mathbf{N}}_0\right) = 0 \tag{3}$$

The point of incidence on the surface is given by

$$\vec{\rho}_i = \vec{p}_i + L_i\hat{r}_{i-1} \tag{4}$$

$\vec{\rho}_i$ is the starting point for the next (i[th]) ray segment: the process repeats until the ray reaches the last surface. The direction of the ray reflected from the surface is given by

$$\hat{\mathbf{r}} = \mathbf{R} \cdot \hat{\mathbf{i}} \tag{5}$$

where **R** is the "reflection dyadic"

$$\mathbf{R} = \mathbf{I} - 2\widehat{\mathbf{N}}\widehat{\mathbf{N}} \tag{6}$$

and the normal at the point of incidence is

$$\widehat{\mathbf{N}} = \mathrm{unit}\left(\vec{\mathbf{N}}_0 + \mathbf{M}\vec{\mathbf{p}}_i\right) \tag{7}$$

The effect of perturbations either in the incoming ray or in the surface attitude and position is to change the outgoing ray state. COMP represents perturbed rays in terms of small ray direction perturbations $d\hat{r}$, transverse aberration or "beamwalk" $\vec{\gamma}$, and OPD dL (Fig. 7). Ref. 4 presents closed-form equations for the derivatives of ray states with respect to perturbations of the incident ray and of the surface perturbations $\vec{\theta}$ (rotation) and $\vec{\delta}$ (translation). These derivatives are chained to develop a linearized model of the complete optical system. For the FMMT we are concerned with the OPD at the exit pupil only. The COMP linear optical model takes the form

$$\vec{w}_0 = \mathbf{C}\,\vec{x}_0 \tag{8}$$

Here $\vec{w}_0$ is a vector of n ray OPDs, and $\vec{x}$ is the error state, composed of six 6-vectors $\vec{x}_{\mathbf{seg}i}$ (one for each of the outer segments). We take the $\vec{x}_{\mathbf{seg}i}$'s to be in segment coordinates, so that each is in the form

$$\vec{x}_{\mathbf{seg}i} = \left[\theta_x\ \theta_y\ \theta_z\ \delta_x\ \delta_y\ \delta_z\right]^T \tag{9}$$

C is thus an array of partial derivatives relating the ray OPDs to element perturbations

$$\mathbf{C} = \begin{bmatrix} \dfrac{\partial L_{\mathrm{ray1}}}{\partial \vec{\theta}_{\mathbf{seg2}}} & \cdots & \dfrac{\partial L_{\mathrm{ray1}}}{\partial \vec{\delta}_{\mathbf{seg7}}} \\ & \vdots & \\ \dfrac{\partial L_{\mathrm{rayn}}}{\partial \vec{\theta}_{\mathbf{seg2}}} & \cdots & \dfrac{\partial L_{\mathrm{rayn}}}{\partial \vec{\delta}_{\mathbf{seg7}}} \end{bmatrix} \tag{10}$$

Details of the derivation of C are available in Ref. 4.

Two other linear models are required in the derivation of the controller. The first is the control action matrix **A**, which scales actuator motions u directly into segment coordinates. **A** is a 36 by 18 matrix which simply projects the three controls (tip θ_x, tilt θ_y and piston δ_z) into segment coordinates. $\vec{u}$ is the control vector made up of the six individual segment controller vectors $\vec{u}_{\mathbf{seg}i}$, where each $\vec{u}_{\mathbf{seg}i}$ is

$$\vec{u}_{\mathbf{seg}i} = \begin{bmatrix} \theta_x \\ \theta_y \\ \delta_z \end{bmatrix} \tag{11}$$

Given that the controls act along the segment coordinates, the 6 by 3 **A** matrix for one segment is

$$\mathbf{A}_{\mathbf{seg}i} = \begin{bmatrix} \hat{x}_{\mathbf{seg}i} & \hat{y}_{\mathbf{seg}i} & \begin{matrix}0\\0\\0\end{matrix} \\ \begin{matrix}0\\0\\0\end{matrix} & \begin{matrix}0\\0\\0\end{matrix} & \hat{z}_{\mathbf{seg}i} \end{bmatrix} \tag{12}$$

The full **A** matrix is

$$\mathbf{A} = \begin{bmatrix} \mathbf{A}_{seg2} & 0 & 0 \\ 0 & \ddots & 0 \\ 0 & 0 & \mathbf{A}_{seg7} \end{bmatrix} \tag{13}$$

The effect of actuator motions in segment state coordinates is

$$\vec{x}_{act} = \mathbf{A}\,\vec{u} \tag{14}$$

Adding the effect of actuator motions to those of the segment errors, the total wavefront error is

$$\vec{w}_1 = \vec{w}_0 + \mathbf{CA}\,\vec{u} = \mathbf{C}\,\vec{x}_0 + \mathbf{CA}\,\vec{u} \tag{15}$$

The controller objective function is the total wavefront error squared:

$$J = \vec{w}_1^T\,\vec{w}_1 \tag{16}$$

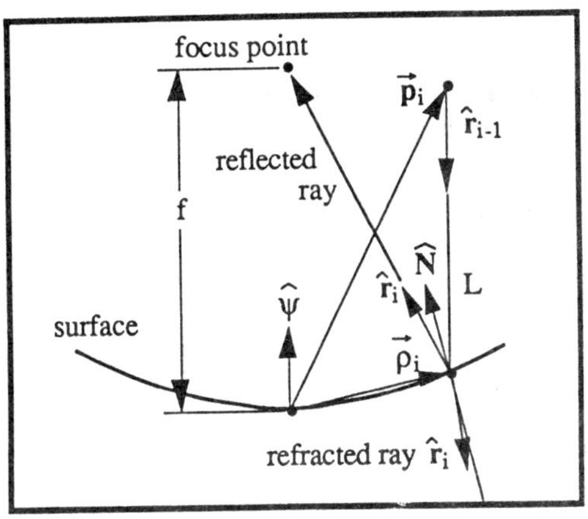

Figure 6. Exact Ray-Trace Geometry

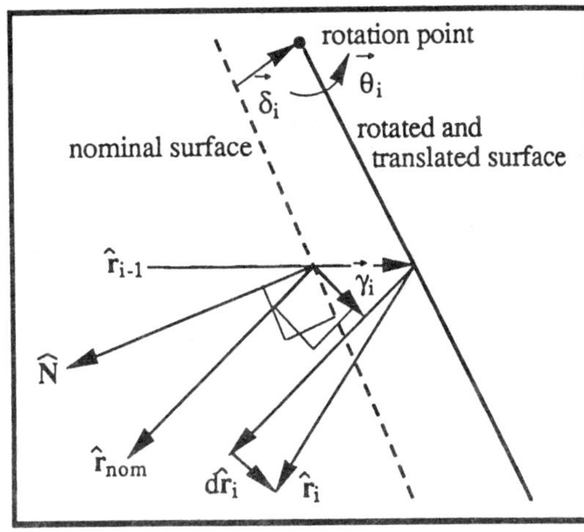

Figure 7. Perturbation Geometry

The last part of the modeling task is the derivation of the measurement equations. These depend on the particular sensor configuration, and we will consider several. All use the same basic elements: edge sensors and gap sensors located between adjacent panels. The edge sensors we consider use laser interferometers configured to measure two degrees of freedom between the two segments: relative piston (local z translation) and relative tilt (local x or y rotation), as sketched in Fig. 5. The gap sensors are capacitive devices that, when used, are mounted in pairs at the same locations as the edge sensors. They provide measurements of relative gap (local x or y translation) (Fig. 5). Both sensor systems can provide excellent performance, yielding accuracies in the nanometers and nanoradians.

Direct derivation of the linearized measurement equations is straight-forward. In general, the form of the measurement equation is

$$\vec{z} = \mathbf{H}\vec{x} \tag{17}$$

As an example, consider the edge and gap sensors between two panels i and j as sketched on Fig 8. The 3 measurements are formed into the measurement vector $\vec{z}_{ij}$, which is a function of the two segment states

$$\vec{z}_{ij} = \begin{bmatrix} \text{hinge angle} \\ \text{piston} \\ \text{gap} \end{bmatrix} = \begin{bmatrix} \dfrac{\partial \vec{z}_{ij}}{\partial \vec{x}_i} & \dfrac{\partial \vec{z}_{ij}}{\partial \vec{x}_j} \end{bmatrix} \begin{bmatrix} \vec{x}_i \\ \vec{x}_j \end{bmatrix} \tag{18}$$

The partials are derived by inspection as

$$\frac{\partial \vec{z}_{ij}}{\partial \vec{x}_i} = \begin{bmatrix} -\hat{x}_{ij} & 0\,0\,0 \\ \hat{z}_{ij}^T \vec{\rho}_i^{\times} & -\hat{z}_{ij}^T \\ \hat{y}_{ij}^T \vec{\rho}_i^{\times} & -\hat{y}_{ij}^T \end{bmatrix} \tag{19}$$

and

$$\frac{\partial \vec{z}_{ij}}{\partial \vec{x}_j} = \begin{bmatrix} \hat{x}_{ij} & 0\,0\,0 \\ -\hat{z}_{ij}^T \vec{\rho}_j^{\times} & \hat{z}_{ij}^T \\ -\hat{y}_{ij}^T \vec{\rho}_j^{\times} & \hat{y}_{ij}^T \end{bmatrix} \tag{20}$$

The full set of measurement equations for each of the configurations is an elaboration of this simple case, with the measurement vector $\vec{z}$ consisting of the $\vec{z}_{ij}$ from each sensor, and $\vec{x}_0$ being the full error state

$$\vec{z} = \mathbf{H}\vec{x}_0 + \vec{r}$$

Here $\vec{r}$ is sensor noise. The measurement matrix $\mathbf{H}$ is built up of the partials of the measurements as above

$$\mathbf{H} = \begin{bmatrix} \frac{\partial \vec{z}_{12}}{\partial \vec{x}_1} & \frac{\partial \vec{z}_{12}}{\partial \vec{x}_2} & & \\ & \ddots & & \\ & & \frac{\partial \vec{z}_{67}}{\partial \vec{x}_6} & \frac{\partial \vec{z}_{67}}{\partial \vec{x}_7} \end{bmatrix} \tag{21}$$

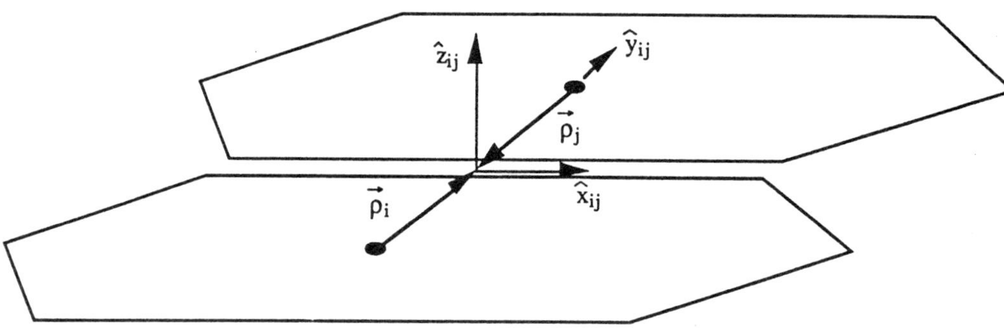

Figure 8. Figure sensor geometry.

WAVEFRONT CONTROLLABILITY

In keeping with our use of a quasi-static model of the FMMT, we treat the control problem as a single-stage optimal control problem. We derive a control that takes the objective function - the wavefront error squared - to its minimum possible value in a single instantaneous step. This is not a bad approximation for our system, as the time scale of the control is very short compared to the thermal disturbances. In the real system the control will be implemented using a dynamic controller that takes the panel and actuator dynamics into account. Our intention here is to address the underlying controllability question: Can the segments be controlled to completely eliminate the effects of segment errors on the wavefront? The answer is a function of the optics and the system geometry, and is the same for both the quasi-static and dynamic controllers.

The control problem is to minimize the objective function J over the transition from $\vec{w}_0$ to $\vec{w}_1$. We derive the optimal control by finding the value of $\vec{u}$ that renders J stationary, or (in other words) satisfies $dJ = 0$ for non-zero values of $d\vec{u}$:

$$dJ = d\vec{w}_1^T \vec{w}_1 = d\vec{u}^T A^T C^T \left(CA\vec{u} + C\vec{x}_0\right) = 0 \tag{22}$$

This condition is satisfied by the control

$$\vec{u}_{opt} = -\left[A^T C^T C A\right]^{-1} A^T C^T C\vec{x}_0 \tag{23}$$

The control as implemented includes actuator noise $\vec{q}$, and uses the estimated panel error state $\vec{x}_{est}$. Defining **G** as the control gain matrix, the control as implemented is

$$\vec{u} = -G\vec{x}_{est} + \vec{q} \tag{24}$$

where

$$G = \left[A^T C^T C A\right]^{-1} A^T C^T C \tag{25}$$

The same result can be derived by choosing **G** to minimize the variance of the residual wavefront error, or as a minimization with the transition equation adjoined to the objective function as an equality constraint, as discussed in Ref. 7.

Assume that we know the error state exactly, so that $\vec{x}_{est} = \vec{x}_0$. How good is the optimal control at correcting the wavefront effects of those errors? Inserting the control into the transition equation, the residual segment error is

$$\vec{x}_1 = \vec{x}_0 + A\vec{u} = \left(I - AG\right)\vec{x}_0 + A\vec{q} \tag{26}$$

The residual wavefront error due to control is

$$\vec{w}_1 = C\vec{x}_1 = C\left(I - AG\right)\vec{x}_0 + CA\vec{q} \tag{27}$$

There are two contributors to the residual wavefront error. The first is the contribution of actuator noise $\vec{q}$. This cannot be completely eliminated, so it must be brought to acceptable levels by careful design of the hardware and electronics. The actuator noise provides a "noise floor" independent of the size of the initial errors below which performance cannot be improved.

The second contributor is controllability error. The controllability condition for this controller is that, neglecting noise, the wavefront error can be taken to zero from an arbitrary initial state. For the wavefront error to go to zero, the product **C(I - AG)** must go to zero. There are two possible ways for this to be true: either the difference **(I - AG)** must be zero; or the non-zero parts of it must be orthogonal to the wavefront error model **C**. The first condition is satisfied when **AG = I**. This is possible only when there are as many control degrees of freedom as there are error states, so that **AG** is of full rank. This is not the case for the FMMT, nor is it the case for most other segment control systems, as the actuators provide three degrees of freedom per segment only (tip, tilt and piston) leaving another three without direct control (twist and x and y decentration).

The second way in which there can be no controllability errors occurs when the residual segment errors are orthogonal to wavefront error. There are three possible cases here. The first case occurs when errors in the uncontrollable error states simply do not result in any wavefront error. Then there are no controllability errors because all contributors to the wavefront error are directly controlled. This would be true for the FMMT if the primary mirror were flat, so that decenter and twist errors did not change the reflected wavefront.

The second case occurs when the uncontrollable error states couple into the wavefront, but in such a way that they can be exactly recovered by the controlled degrees of freedom. This would be true if the primary mirror were spherical. Then a decenter error, which is not directly controlled, could be exactly compensated for by the wavefront compensation controller, which would use the controllable degrees of freedom to piston and tilt the segment until the original figure and wavefront are exactly recovered. This can happen because the shape of any section of a sphere is identical to any other section of that sphere, except for its position and orientation.

The third case occurs for the FMMT and most other astronomical telescopes which have neither flat nor spherical primary mirrors (the FMMT has a paraboloidal primary, others commonly use a hyperboloidal primary). Aspherical mirrors do not have the property that any section is the same shape as any other section, and in general, a decenter or twist error cannot be fully recovered by tip, tilt and piston actions. It can be mitigated to a large extent by the Wavefront Compensation type of controller, which uses the knowledge of the optics built into the **G** matrix to minimize the residuals. For this most common case, however, there will always be a finite controllability error.

An example will serve to illustrate the point. Consider the FMMT with a segment error of 100 μm radial (x) decenter of Segment 3. This is an uncontrollable segment error with significant contribution to wavefront error. The uncorrected wavefront is as shown in the surface plot of Fig. 9; the RMS wavefront error is 5.93 μm. The optimal control is to tilt and piston the segment, with θ_y=-26.8 μrad and δ_z=-0.847μm. The residual wavefront error is 0.39 μm; the corrected wavefront is as shown on Fig. 9.

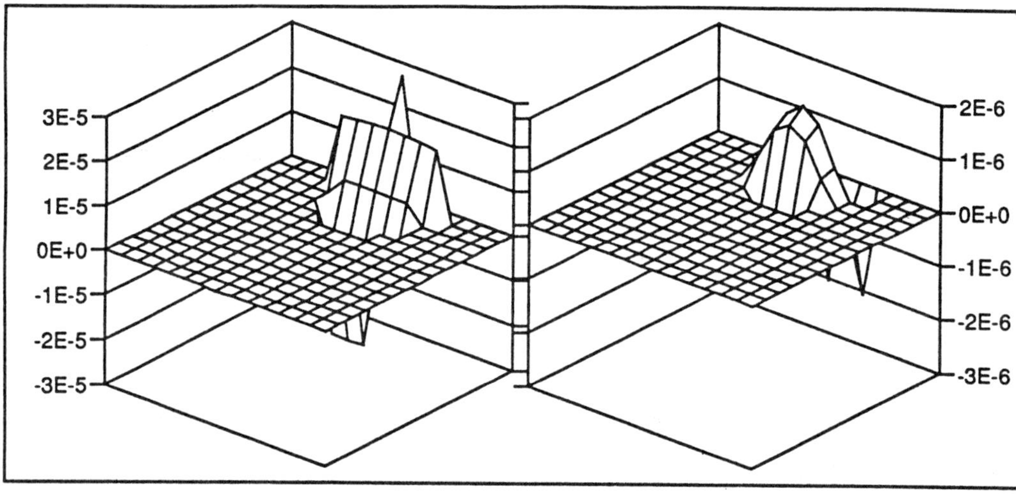

Figure 9. Uncompensated and compensated OPD plots for 100 μm segment 3 x-axis translation error.

How much of a problem the controllability errors are in general depends on several factors. How eccentric - how far from spherical - are the segment shapes? This depends on the *f* number of the optic, and the FMMT, with a very fast *f*/0.4 primary and relatively large segments, has serious problems with controllability. How accurate are the initial alignments of the telescope? The controllability error magnitude depends on the expected segment misalignments. If these are small in the uncontrollable directions, or can be adjusted periodically, then the controllability errors may not be larger than the noise floor. If they are large and are not adjustable, as is the case for the FMMT, than they are a concern.

More definitive results for the WFC are presented after a discussion of the panel error state estimation problem.

WAVEFRONT OBSERVABILITY

The WFC controller requires that an accurate estimate of the segment state errors is available. This estimate must observe not only the controllable parts of the panel error states (tip, tilt and piston), but also the uncontrollable parts (decenter and twist), so that it can compensate for their contributions to the wavefront error. In this section we derive the equations for a general panel state estimator, which converts measurements of segment positions and angles obtained using edge- and gap-sensors into estimates of segment error states. We then test the estimator with a few different configurations of figure sensors to determine observability of the contributors to wavefront error. It turns out that some configurations used by other workers[2,3] do not provide full observability of the segment errors that contribute to wavefront error, and therefore are not usable in the WFC approach. We identify and describe a couple of others that do provide full observability of all the important segment errors.

As we did for the controller discussion, we treat the estimation problem as a single-stage estimation problem, deriving a weighted least-squares estimator of the panel states (Ref. 7). We then test it for observability in each of several sensor configurations. The results we obtain for the single-stage estimator represent the bounds of performance of the multi-stage Kalman Filter estimator that would be implemented in the flight system.

Consider the general problem of estimating the segment error states. We start with some initial segment error state $\vec{x}_0$ with covariance X_0. The sensor measurement signals $\vec{z}$ are obtained via the measurement equation discussed above

$$\vec{z} = \mathbf{H}\vec{x}_0 + \vec{r} \tag{28}$$

where $\vec{r}$ is the measurement noise, characterized as zero-mean noise with covariance **R**. The particulars of **H** depend on which sensor configuration is used, as discussed previously. We assume that the previous best estimate of the segment state was zero. The objective function for the weighted least-squares estimate of $\vec{x}_0$, following Ref. 7, is

$$J = \frac{1}{2}\left[\vec{x}_0^T \mathbf{X}_0^{-1}\vec{x}_0 + (\vec{z} - \mathbf{H}\vec{x}_0)^T\mathbf{R}^{-1}(\vec{z} - \mathbf{H}\vec{x}_0)\right] \tag{29}$$

We pick our estimated state $\vec{x}_{est}$ to minimize this function. Differentiating, we determine $\vec{x}_{est}$ as the value of $\vec{x}_0$ which yields dJ=0

$$dJ = d\vec{x}_0^T\left[\mathbf{X}_0^{-1}\vec{x}_0 - \mathbf{H}^T\mathbf{R}^{-1}(\vec{z} - \mathbf{H}\vec{x}_0)\right] = 0 \tag{30}$$

The result is the well-known least-squares estimator,

$$\vec{x}_{est} = \mathbf{K}\vec{z} \tag{31}$$

where the gain **K** is

$$\mathbf{K} = \mathbf{P}^{-1}\mathbf{H}^T\mathbf{R}^{-1} \tag{32}$$

and

$$\mathbf{P}^{-1} = \mathbf{X}_0^{-1} - \mathbf{H}^T\mathbf{R}^{-1}\mathbf{H} \tag{33}$$

The full estimate including noise is

$$\vec{x}_{est} = \mathbf{KH}\vec{x}_0 + \mathbf{K}\vec{r} \tag{34}$$

so the error $\vec{e}_x$ in the estimate of the segment error state is

$$\vec{e}_x = \vec{x}_0 - \vec{x}_{est} = (\mathbf{I} - \mathbf{KH})\vec{x}_0 - \mathbf{K}\vec{r} \tag{35}$$

The covariance **E** of the figure estimation error $\vec{e}_x$ reduces to **P** (Ref. 7):

$$\mathbf{E} = (\mathbf{I} - \mathbf{KH})\mathbf{X}_0(\mathbf{I} - \mathbf{KH})^T - \mathbf{KRK}^T = (\mathbf{I} - \mathbf{KH})\mathbf{X}_0 = \mathbf{P} \tag{36}$$

The panel error state estimate can projected to the wavefront using the **C** matrix. The result is the estimated wavefront error $\vec{w}_{est}$:

$$\vec{w}_{est} = \mathbf{C}\vec{x}_{est} \tag{37}$$

The error in the estimate of the wavefront error $\vec{e}_w$ is

$$\vec{e}_w = \mathbf{C}(\vec{x}_0 - \vec{x}_{est}) = \mathbf{C}(\mathbf{I} - \mathbf{KH})\vec{x}_0 - \mathbf{CK}\vec{r} \tag{38}$$

The covariance of the error in the wavefront estimate is

$$\mathbf{E}_w = \mathbf{CEC}^T = \mathbf{CPC}^T \tag{39}$$

This equation shows that, like the control errors, the estimation errors have two sources. One source is the sensor noise $\vec{r}$. For the FMMT the sensor noise is expected to be very low, in the nanometer range. This is small enough that the sensor noise will not be very significant, as shown later.

The second source of error in the wavefront estimate is the observability error. Full observability requires that the estimation error be zero in the absence of noise. This is true only if $\mathbf{C}(\mathbf{I} - \mathbf{KH}) = 0$, either by the difference $(\mathbf{I} - \mathbf{KH})$ going to zero, which happens with full observability of the segment errors, or by the residuals of the segment error estimate $(\mathbf{I} - \mathbf{KH})$ being orthogonal to the wavefront, so that the product $\mathbf{C}(\mathbf{I} - \mathbf{KH})$ goes to zero.

We have tested the observability of several figure sensor configurations for figure and wavefront observability. The tested configurations are shown in Fig. 10. We first checked the rank of the figure estimate observability matrices **KH** for each configuration. It turns out that none of the configurations provide full observability of the segment errors. We then examined their ability to estimate wavefront error. To do this we computed the inverse figure estimate error covariance $\mathbf{E_w}$ and then decomposed it using singular value decomposition (SVD) techniques. We separated observable modes from unobservable modes to help in characterizing the shapes of the unobservable modes. We also computed overall performance as the RMS of the wavefront estimation error.

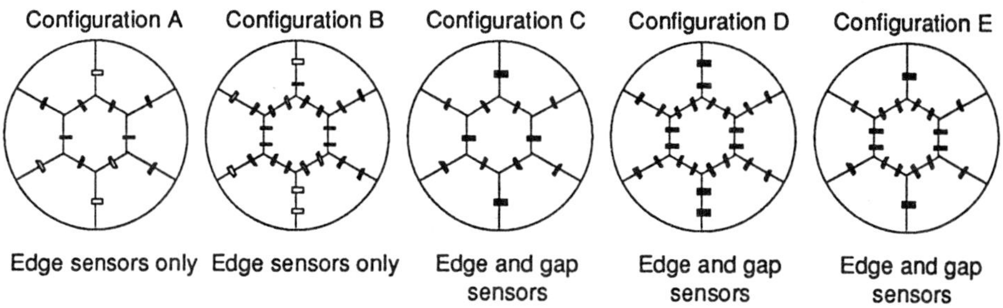

Figure 10. Sketch of tested sensor configurations.

Two configurations that rely on edge sensors alone (A and B). These configurations do not provide full observability of the wavefront errors. Neither did configuration C. Two configurations with gap sensors were found to provide good estimates of the wavefront error (D and E). Of these we prefer E as it requires fewer sensors. This configuration is used in computing the results of the next section.

There are two main types of unobservable modes, as sketched in Fig. 11. The first "spin" mode involves panels rotating about the optical axis. This mode is present in all of the sensor configurations, but it turns out not to produce any wavefront error, as it does not change the shape of the mirror, and so is orthogonal to C. The second "spread" mode occurs in configurations without gap sensors. This involves the segments moving together radially away from the optical axis with a slight amount of rotation, so that the total motion is orthogonal to all of the sensor axes. This motion couples quite strongly to the wavefront error, and as it is unobservable, causes the wavefront error to be unobservable.

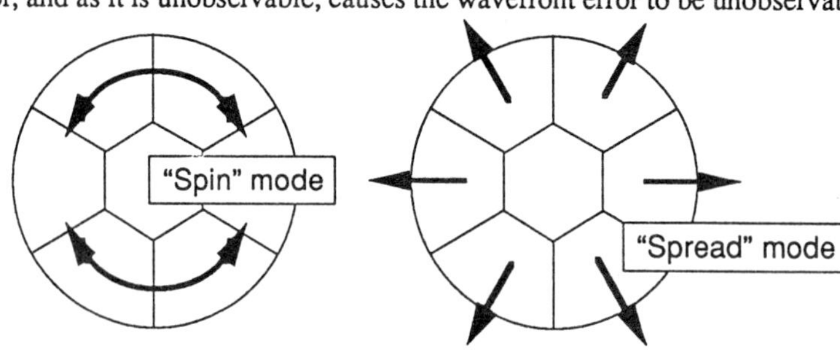

Figure 11. Sketch of unobservable modes.

PANEL STATE COMPENSATION

Previous work in this area[2,3] has used controllers that minimize "figure" error residuals, as opposed to our approach, which minimizes wavefront error residuals. In order to provide a point of comparison for our approach, we derive a "Panel State Compensation" (PSC) controller that minimizes figure error as represented by the segment error states. It must be noted that the segment error states may not be the best representation of "figure." Another approach[2] is to project the states onto three surface points (located at the linear segment actuators) and minimize the surface errors at those points. Alternatively, large numbers of sample points, or polynomial representations of the surface, can be used as better representations of the surface figure. These other approaches can be lumped under the general name of "Figure Compensation" (FC) controllers. The PSC approach, by minimizing state errors in segment coordinates, can be seen as minimizing 6DOF surface errors at the center of each segment.

The objective function for the PSC controller is the state error squared:

$$J = \frac{1}{2}\vec{x}_1^T\vec{x}_1 \tag{40}$$

where the transition from initial to final segment state is:

$$\vec{x}_1 = \vec{x}_0 + A\vec{u} \tag{41}$$

The control is the value of $\vec{u}$ that satisfies $dJ = 0$,

$$\vec{u} = -[A^TA]^{-1}A^T\vec{x}_0 \tag{42}$$

Note that the product A^TA is equal to the identity (see Eq. 11). As implemented, the PSC control has the same form as the WFC control,

$$\vec{u}_{PSC} = -G_{PSC}\vec{x}_0 + \vec{q} \tag{43}$$

The difference is in the gain matrix,

$$G_{PSC} = [A^TA]^{-1}A^T = A^T \tag{44}$$

This matrix is a transformation of the initial state error back into the controlled states, so that the effect of the PSC control is simply to null the controllable state errors, leaving the uncontrollable state errors untouched. The residual wavefront error is simply the projection of the uncontrollable state errors and the actuator noise to the wavefront:

$$\vec{w}_{PSC} = C(I - A^T)\vec{x}_0 + CA\vec{q} \tag{45}$$

In the example of Fig. 9, the PSC controller has no effect, as the segment error is entirely in an uncontrolled direction. The PSC leaves the initial 5.93 μm RMS wavefront error untouched, as compared to the WFC, which reduces the error by a factor of 15.

RESULTS

We can evaluate the performance of the WFC and PSC controllers over a broad range of initial error conditions most easily using covariance analysis. This is, of course, a linear technique and therefore represents some level of approximation. To back up these results we have also run two Monte Carlo simulations, each representing 100 specific cases computed using the exact nonlinear ray-trace optical model. The results agree very closely with the covariance analysis.

The covariance analysis results include the effects on the wavefront of both the control and estimation errors. The total residual wavefront error for the WFC controller is

$$\begin{aligned}\vec{w}_1 &= C\vec{x}_1 = C(\vec{x}_0 + A\vec{u}) = C(\vec{x}_0 - AG\vec{x}_{est} + A\vec{q}) \\ &= C(I - AGKH)\vec{x}_0 - CAGK\vec{r} + CA\vec{q}\end{aligned} \tag{46}$$

The covariance of the total residual wavefront error is

$$W_1 = C(I - AGKH)X_0(I - AGKH)^T + CAGK\,R\,K^TG^TA^TC^T + CA\,Q\,A^TC^T \tag{47}$$

The expected value of the wavefront error is

$$e_{WFC} = \sqrt{\mathrm{trace}(W_1)/n_{ray}} \tag{48}$$

This is a function of the initial error covariance and the sensor and actuator noise covariances.

Figure 12 shows the expected wavefront error versus initial segment error for fixed values of sensor and actuator noise. The sensor noise was assumed uncorrellated with an expected value of 1 nm in gap and piston and 1 nrad in hinge angle for each sensor. Actuator noise levels were assumed uncorrellated at 350 nm in piston and 350 nrad in tip and tilt for each segment. The contributions of sensor and actuator errors are shown on Fig. 12. The segment errors, also assumed uncorrellated, were varied together in the ratio 1.175:1 between rotational and translational errors.

These results show that the WFC controller is quite effective at reducing the effects of segment errors, providing over two orders of magnitude improvement in the wavefront error for larger errors, and providing actuator noise-limited performance for smaller errors.

The total residual wavefront error for the PSC controller, which we assume uses the same estimator as the WFC controller, is

$$\vec{w}_{PSC} = C\left(I - AA^TKH\right)\vec{x}_0 - CAGK\vec{r} + CA\vec{q} \tag{49}$$

The covariance of the PSC-controlled wavefront is

$$W_{PSC} = C\left(I - AA^TKH\right)X_0\left(I - AA^TKH\right)^T + CAGK\,R\,K^TG^TA^TC^T + CA\,Q\,A^TC^T \tag{50}$$

The expected value of the PSC wavefront error is

$$e_{PSC} = \sqrt{\mathrm{trace}\left(W_{PSC}\right)/n_{ray}}$$

The PSC wavefront error is also plotted on Fig. 12 for the same conditions as the WFC controller. As shown, the PSC controller is uniformly 15 times worse than the WFC until the WFC reaches the actuator noise floor.

Two sets of Monte Carlo runs were also made. These use the exact nonlinear model of the FMMT to compute RMS wavefront error. Initial conditions for each run of the Monte Carlo simulator was initialized using numbers generated by a Gaussian random number generator, using the same stastical parameters that went into the covariance cases. Both controllers were run for each set of initial conditions. The results are shown on Fig. 12; they agree quite closely to the covariance analysis results.

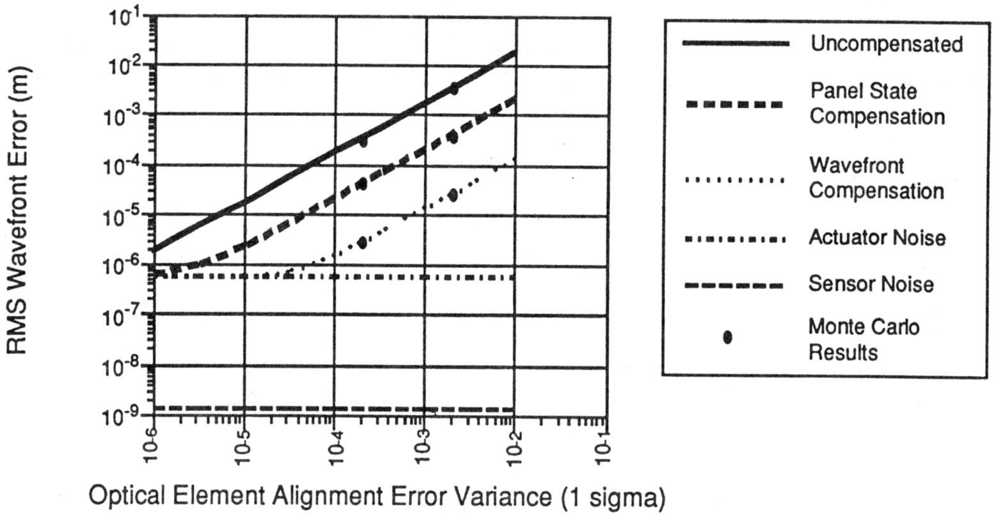

Figure 12. Control results.

DISCUSSION

The results show that, with respect to RMS wavefront error, the WFC approach does very well, and in fact does a lot better than the PSC approach. This is not surprising, since the WFC control is optimal with respect to RMS wavefront error. The improvement comes because the WFC approach is able to both observe and compensate for the wavefront effects of uncontrollable segment position errors. This ability is due to the incorporation of an accurate model of the optical effects of the segment errors into the WFC controller.

By minimizing the wavefront error, the WFC controller maximizes the overall optical quality of the telescope with respect to the modelled degrees of freedom. Other error sources, such as motions of the secondary mirror, relay optics and focal plane, and thermal deformations of the primary mirror segments, also contribute to the overall system performance. The WFC framework can easily be expanded to include all these effects, provided appropriate sensors and actuators are added as well. To realize this, the state vector, which in this paper contained only the six rigid body states of each of six primary reflector panels, would be augmented to include the six secondary mirror rigid body states, the six focal plane rigid body states, and as many primary segment deformation states (e.g., thermal mode shapes) as are deemed important. Similarly, the control vector would be augmented with secondary mirror tip, tilt, x-decenter, y-decenter and focus, and panel deformation control inputs. The measurement vector would be modified to include more samples of the primary segment surfaces, secondary positions and orientations, and focal plane positions and orientations. The basic mathematics of the problem, which yield an optimal control solution for the modelled degrees of freedom, are unchanged from the results presented in this paper.

The WFC controller should outperform any of the FC approaches mentioned above with respect to wavefront error. The differences need not be as great as those seen between the WFC and PSC aprroaches, depending on the same observability and controllabilty considerations addressed here. Telescope optical parameters, such as focal ratio and relative segment size, will significantly affect the performance of any controller.

With respect to the FMMT, the current design is changed somewhat from that described in this paper. The statistics assumed here for actuator and sensor errors were chosen to reflect what were considered reasonable levels of performance. The actual error budget for FMMT components will be discussed in Ref. 8. The sensor configurations discussed in the paper were investigated in an effort to identify a practical edge sensor based option that yielded full observability of all degrees of freedom that effect performance. The configuration that is actually baselined for the FMMT is an optical truss, and contains neither edge nor gap sensors. It uses laser interferometers with beam launching elements supported on an optical bench located behind the secondary which illuminate retro-reflectors mounted on the primary panel front surfaces. The optical truss yields full observability of the segment position modes, and additionally has the potential for observing deformation modes. The FMMT baseline figure control concept, as well as the rationale for its selection, will be described in Ref. 8

CONCLUSION

We have examined the problem of segmented mirror figure control to determine the fundamental limits to performance. We presented a method of modeling the optical effects of the segments that enabled us to derive a controller that directly maximizes optical quality of the entire telescope with respect to segment position errors. Eccentric mirror surface curvatures and limited control degrees of freedom, in addition to actuator noise, were shown to limit the ultimate controllability of any figure controller. Several figure sensor configurations were examined to determine their ability to observe those segment errors that couple to wavefront error. Results for the FMTT show that the wavefront errors due to segment errors can be reduced by over two orders of magnitude.

ACKNOWLEDGEMENT

The research described in this paper was carried out at the Jet Propulsion Laboratory, California Institute of Technology, under contract with the National Aeronautics and Astronautics Administration.

REFERENCES

1. D. Redding, W. Breckenridge, K. Lau, and G. Sevaston, "Segmented Mirror Figure Control for a Space-Based Far-IR Astronomical Telescope," to appear at the 1991 SPIE Aerospace Sensing Conference.

2. T.S. Mast and J.E. Nelson, "The Figure Control of Segmented Telescope Mirrors," Ten Meter Telescope Report No. 80, March, 1983.

3. J.-N. Aubrun, K.R. Lorell,.T.S. Mast and J.E. Nelson, "Dynamic Analysis of the Actively Controlled Segmented Mirror of the W.M. Keck Ten-Meter Telescope," IEEE Control Systems Magazine, December, 1987.

4. D. Redding and W. Breckenridge, "Linearized Ray-Trace Analysis," Paper No. ?, International Lens Design Conference, June 1990.

5. D. Redding and W. Breckenridge, "Optical Modeling for Dynamics and Control Analysis," Memo DCR-90-30, August 22, 1990.

6. D. Redding, "Controlled Optics Modelling Package User Document," Memo DCR-90-50, September 17, 1990.

7. Y.C. Ho. and A.E. Bryson, *Applied Optimal Control*, Hemisphere Publishing, 1975.

8. G. E. Sevaston, D. Redding, H. Kadogawa, K. Lau, B. Levine, N. Nerheim and W. Breckinridge, "The Precision Segmented Reflectors Moderate Mission Figure Control Subsystem," to appear.

AAS 91-055

A CLASS OF NEURAL NETWORKS AND THEIR APPLICATION

Noureddine Kermiche and Renjeng Su*

In recent years dynamic neural networks have been demonstrated as a potential approach to Content Addressable Memory. The main idea there is to devise a learning algorithm for a dynamic neural network so that a set of desirable states may be made into attractors, which represent memory of information. The goal of this paper is to present a particular architecture of neural network for which the learning of the desirable attractors can be synthesized directly. This architecture and the learning scheme has an important property that they can be used for modeling systems with unknown dynamics.

INTRODUCTION

In recent years dynamic neural networks have been demonstrated as a potential approach to Content Addressable Memory , see, for example, Ref 7. The main idea there is to devise a learning algorithm for a dynamic neural network so that a set of desirable states may be made into attractors, which represent memory of information. The goal of this paper is to present a particular architecture of neural network for which the learning of the desirable attractors can be synthesized directly. This architecture and the learning scheme has an important property that they can be used for modeling systems with unknown dynamics.

A NEURAL NETWORK ARCHITECTURE

The architecture of the neural networks studied in this paper is illustrated in **Fig.** 1.

---

* Department of Electrical and Computer Engineering, University of Colorado, Campus Box 429, Boulder, Colorado 80309-0429. Research supported in part by Center for Space Construction and a research grant from NSF, DDM-8454192.

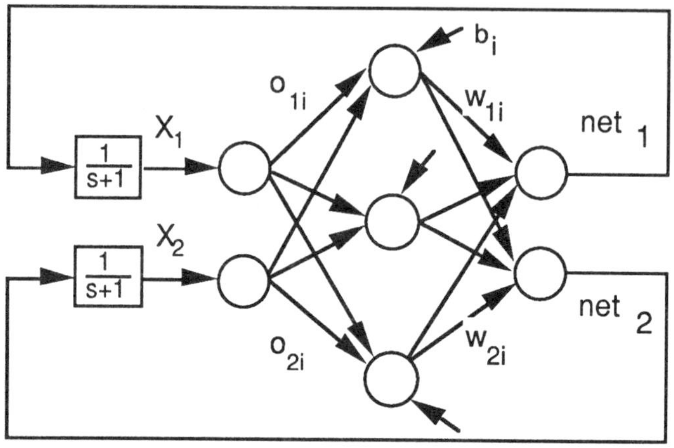

Fig. 1

The main idea of this architecture is connecting feedback paths of simple dynamics around a feedforward neural network. The dynamics in the feedback paths used here are simple first-order lags. The resulting neural networks are a very rich set of nonlinear dynamical systems. By inspection, the systems in **Fig** 1 can be modeled by a set of differential equations

$$\frac{dx_1}{dt} = - x_1 + net_1(x_1, x_2)$$
$$\frac{dx_2}{dt} = - x_2 + net_2(x_1, x_2) \tag{1}$$

with net(x_1, x_2) representing the feedforward neural nets. Our feedforward neural nets are three-layer nets with a layer of input nodes, a layer of output nodes and a layer of hidden nodes. The nodes on the input layer and output layer are simply registers which register the input and output values with no further processing. The hidden nodes receive the weighted sum of the input values and pass the result through a smooth sigmoid-like function. The function used in this paper is the logistic function $s(\xi) = (1+e^{-\xi})^{-1}$.The values at the output nodes are weighted sums of the results from the sigmoidal function. Precisely, the feedforward neural nets are defined by the following equations

$$\begin{aligned} net_1(x_1, x_2) &= \sum_{i=1}^{N} w_{1i}\, s(o_{1i} x_1 + o_{2i} x_2 + b_i) \\ net_2(x_1, x_2) &= \sum_{i=1}^{N} w_{2i}\, s(o_{1i} x_1 + o_{2i} x_2 + b_i) \end{aligned} \tag{2}$$

where w's, and o's are the weighting constants, called interconnects. It has been shown, for example, Ref. 3 and Ref. 5, that such feedforward nets can approximate any continuous functions. This is a significant analytical result for it implies that the neural nets (1) can approximate any dynamic systems which are defined by continuous vector fields. A natural application of the neural nets (1) is therefore in the area of modeling and simulation. An important feature of the dynamic neural nets is their structural uniformity. For neural nets representing different dynamics, they only differ in the number of nodes and particular values of the interconnects; the structure remains invariant. With this feature they provide an important alternative for modeling nonlinear dynamical systems.

ATTRACTORS AND CONTENT-ADDRESSABLE-MEMORY

Attractors are stable states of a dynamical system. At an initial state near a neighborhood of an attractor, the state trajectory will converge to the attractor. The concept of attractors has been used to produce memory devices. A prerequisite of this application is to be able to efficiently establish attractors at desirable locations, that is, to write information into the memory. This is a synthesis problem: given a state space and a set of states as desirable location of attractors, how can a dynamic system be established in the state space such that the specified states are the only attractors? The main goal of the paper is to show that this synthesis problem can be solved by the capability of our neural nets in dynamics modeling.

The idea is illustrated as follows. Consider the dynamic neural net (1) and a point (x^*_1, x^*_2) in the state space. Let Ω be a neighborhood around (x^*_1, x^*_2). Suppose

$$\begin{aligned} net_1(x_1, x_2) &= x_1^* \\ net_2(x_1, x_2) &= x_2^* \end{aligned} \tag{3}$$

over Ω, then the dynamical system (1) takes the form

$$\frac{dx_1}{dt} = - x_1 + x_1^*$$
$$\frac{dx_2}{dt} = - x_2 + x_2^* \tag{4}$$

Clearly the point (x*1, x*2) is an attractor in Ω (**Fig**.2).

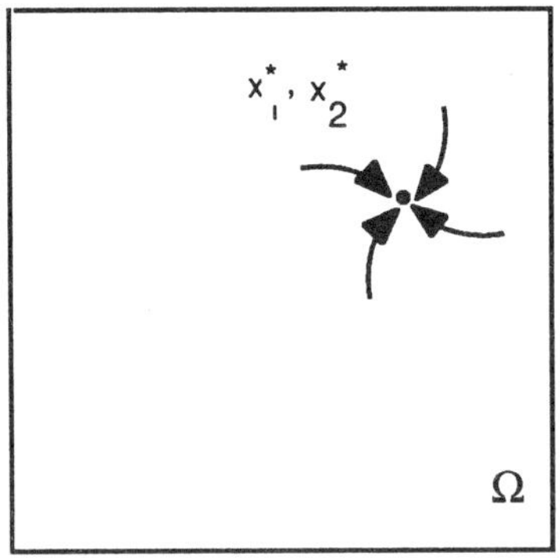

Fig.2 Feedback network with one attractor.

How can a dynamic neural network be trained, i.e. its interconnects be set, so that a certain point become an attractor? There are two approaches to this training problem. First, since system (4) has an attractor at (x^*_1, x^*_2), we may simply take the feedforward nets (2) and train them with different points (x_1, x_2) so that Eq. (3) are satisfied (usually in an approximate sense). This is a problem of neural network learning of static functions. A second approach to setting attractors is a dynamic one. It is based on the idea of training the dynamic neural net (1) so that it approximates the dynamic system (4) in the sense that the trajectories of the neural network system close to those of the reference system (1). The dynamic approach is a more powerful one of the two, because it can be generalized to the case when the neural nets are used to model systems with unknown attractors. The rest of the paper will detail the dynamic approach.

ATTRACTORS LEARNING AS AN OPTIMIZATION PROBLEM

Consider a given stable linear systems

$$\frac{dx_1^d}{dt} = - x_1^d + x_1^k$$
$$\frac{dx_2^d}{dt} = - x_2^d + x_2^k \tag{5}$$

$(x_1{}^k, x_2{}^k)$ with $k = 1,...,P$ being a set of constant states which are the attractors of the system. Let Ω^k be a neighborhood of the attractor $(x_1{}^k, x_2{}^k)$. The sets Ω^k are assumed to be disjoint (**Fig.** 3).

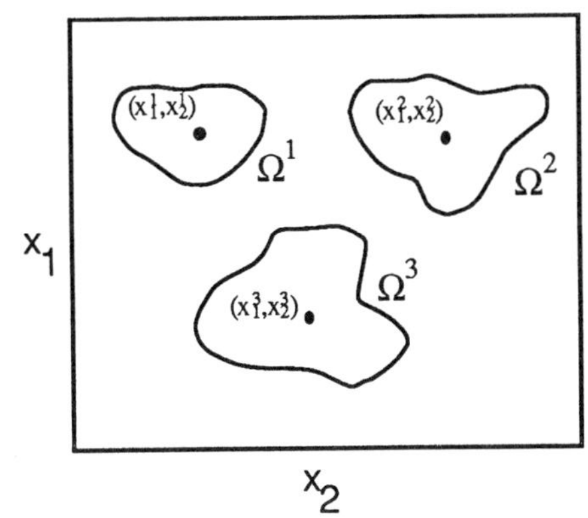

Fig. 3. Attractors with respective neighborhoods.

Consider a neural net system defined by Eq. (1). To set the interconnects we want to find a set of (w,o,b) such that the index function

$$E = \int_0^T (x-x^d)^{tr} Q (x-x^d)\, dt \tag{6}$$

is minimized for some fixed time $T > 0$, when $x(0) = x^d(0)$. There Q is a positive definite matrix and tr denotes matrix transpose.

The solution of the neural net system (1) is a function of the form

$$x\,(t) = x\,(\,t;\, x\,(0),\, w,\, o,\, b) \tag{7}$$

With a fixed T, the performance index is a function of the quadruple (x(0), w, o, b).

$$E = \int_0^T (x-x^d)^{tr}\, Q\, (x-x^d)\, dt = \Phi\, (x(0),\, w,\, o,\, b) \tag{8}$$

To further eliminate the dependency of E on the initial state, we seek a value of (w, o, b) which minimizes E over a set of randomly generated initial states. The objective function then becomes

$$J(w,\, o,\, b) = E_{x(0)} \left[\Phi\, (x(0),\, w,\, o,\, b)\right] \tag{9}$$

where $E_{x(0)}$ refers to the expected value over a set of x(0).The problem is then to find a set of connection weights to minimize (9). Uniform probability density function is chosen to generate the initial states in Ω^k, k = 1,...,P.

A LEARNING ALGORITHM

Let us define

$$\mathbf{x} = \begin{bmatrix} x_1 \\ x_2 \end{bmatrix}, \mathbf{x}^{\mathbf{d}} = \begin{bmatrix} x_1^d \\ x_2^d \end{bmatrix}, \mathbf{net} = \begin{bmatrix} net_1 \\ net_2 \end{bmatrix} \text{ and } \mathbf{e} = \begin{bmatrix} e_1 \\ e_2 \end{bmatrix} = \mathbf{x} - \mathbf{x}^{\mathbf{d}}$$

From (1) and (5) we can write

$$\frac{d\mathbf{e}}{dt} = - \mathbf{e} + \mathbf{net}(\mathbf{x}) - \mathbf{x}^k,\ \mathbf{x}(0) = \mathbf{x}^d(0),\ \mathbf{x}^d(0) \in \Omega^k \tag{10}$$

The condition $\mathbf{e}(0) = \mathbf{0}$ and the Lipschitz continuity of the right hand side of (10) give the unique solution

$$\mathbf{e}(t) = \int_0^t \exp^{-(t-\tau)} \left[\mathbf{net}(\mathbf{x}(\tau)) - \mathbf{x}^k\right] d\tau \tag{11}$$

The problem of learning this set of P attractors becomes one of minimizing

$$E = \int_0^T \mathbf{e}(t)^{tr}\, \mathbf{e}(t)\, dt \tag{12}$$

Using Schwarts inequality, we can write

$$E \le T\, E_1 \tag{13}$$

where

$$E_1 = \int_0^T \left[\mathbf{net}(\mathbf{x}(t)) - \mathbf{x}^k\right]^{tr} \left[\mathbf{net}(\mathbf{x}(t)) - \mathbf{x}^k\right] dt \tag{14}$$

Gradient-descent method can be applied to both (11) and (14). In (11) the error signal used is computed from the state trajectories of the neural network and the reference dynamical system. In (14), the error signal used is related directly to the difference between the nonlinearity of the neural network and the desired linear function of the desired dynamical system. Gradient descent on (14) involves less computation.

An update rule based on the functional given in (14) can be derived as follows.

$$\begin{aligned}
w_{1i}^{j+1} &= w_{1i}^{j} + \varepsilon \int_0^T v_1^j(t)\, s(\xi^{i}(t))\, dt \\
w_{2i}^{j+1} &= w_{2i}^{j} + \varepsilon \int_0^T v_2^j(t)\, s(\xi^{i}(t))\, dt \\
o_{1i}^{j+1} &= o_{1i}^{j} + \varepsilon \int_0^T \left[w_{1i}^j\, v_1^j(t) + w_{2i}^j\, v_2^j(t)\right] x_1^j(t) \frac{d}{d\xi_i^j(t)}\, s(\xi_i^j(t))\, dt \\
o_{2i}^{j+1} &= o_{2i}^{j} + \varepsilon \int_0^T \left[w_{1i}^j\, v_1^j(t) + w_{2i}^j\, v_2^j(t)\right] x_2^j(t) \frac{d}{d\xi_i^j(t)}\, s(\xi_i^j(t))\, dt \\
b_{i}^{j+1} &= b_{i}^{j} + \varepsilon \int_0^T \left[w_{1i}^j\, v_1^j(t) + w_{2i}^j\, v_2^j(t)\right] \frac{d}{d\xi_i^j(t)}\, s(\xi_i^j(t))\, dt
\end{aligned} \tag{15}$$

where

$$\xi_i^j(t) = o_{1i}^j\, x_1^j(t) + o_{2i}^j\, x_2^j(t) + b_i^j$$

$$v_1^j(t) = net_1(\, x_1^j(t),\, x_2^j(t)) - x_1^k$$

$$v_2^j(t) = net_2(\, x_1^j(t),\, x_2^j(t)) - x_2^k$$

The parameter ε is called the learning rate. The index j denotes the iteration index and k the index of a specific attractor. The learning iteration starts with arbitrary initial values for (w,o,b). A random initial condition x(0) in one of the Ω^k sets is generated. The error signal between the linear and nonlinear responses is obtained for a finite period of time [0, T]. With a proper choice of the learning rate ε, all the terms in the update rule are computed to find the new values for (w,o,b). A new random initial condition in one of the Ω^k sets is then produced and the process is repeated until (w,o,b) converges to a certain value (w*,o*,b*). Due to the stochactic nature of the process, the interconnects (w,o,b) can only be expected to converge to (w*,o*,b*) with small fluctuations. The problem of scanning the sets Ω^k is discused next.

CONTINUOUS LEARNING OF ATTRACTORS

The method presented in this proposal is based on gradient descent method which has the following interesting result. Consider the case when an attractor is learned, and the learning procedure is repeated to learn about a second attractor.

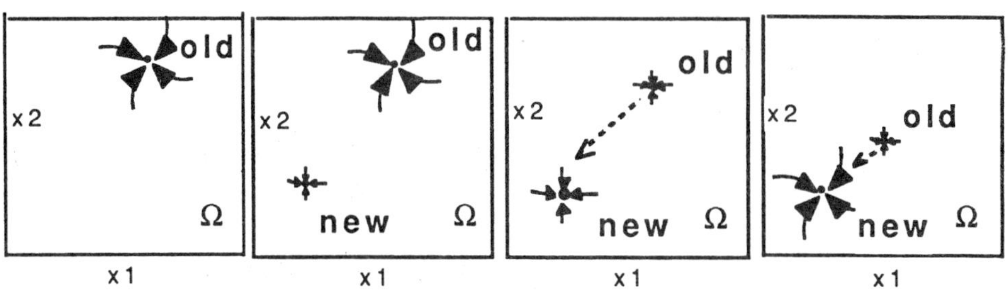

Fig. 4 " Forgetting" in feedback networks.

If the learning iterations of the second attractor is conducted without revisiting the first attractor, the basin of attraction of the first attractor will shrink, and the location of the first attractor will move toward the second attractor. It will finally fuse with the new attractor and disappear. In order to maintain the attractors previously learned, they must be revisited intermittently, while the new attractors are being learned. This interesting phenomenon may serve as a way to remove the information from the memory.

EXAMPLES

Two examples are shown here to demonstrate the idea discussed in this paper.

Example 1.

In this example we use a network with 20 nodes in the hidden layer to learn a dynamics of four attractors at.(.5,.5) (-.5,.5) (.5,-.5) (-.5,-.5). The result is shown in **Fig.** 5. There we use the vector field to illustrate the outcome from the learning.

Example 2.

This example shows how the concept of attractor-learning may be applied to devising a memory for pattern restoration. Using 10 nodes in the hidden layer, we consider a system with 100 input nodes representing a map of 10 X 10 pixels. The states can take any gray value between 0 and 1. The network learns to distinguish patterns given in **Fig.** 6 and **Fig.** 7. After learning, when a noisy pattern is given to the network, it converges to the closest pattern to the given initial condition. **Fig.** 8 and **Fig.** 9 illustrates the results.

Fig. 5 4-attractor dynamical neural network.

Fig. 6 Pattern1

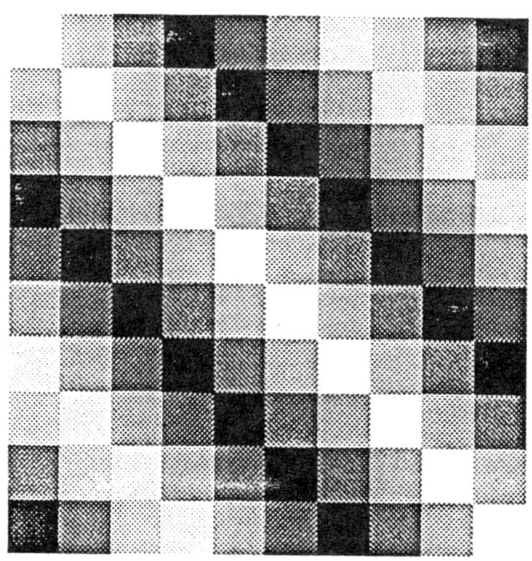

Fig. 7 Pattern2

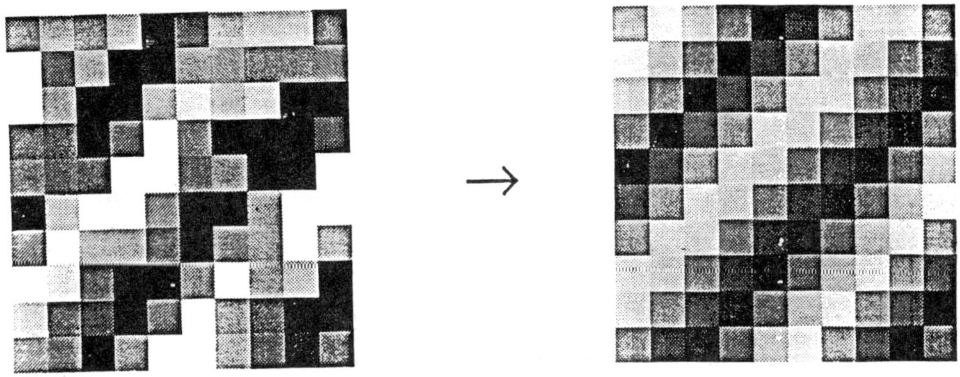

Fig. 8 Convergence when the initial condition is closer to pattern1

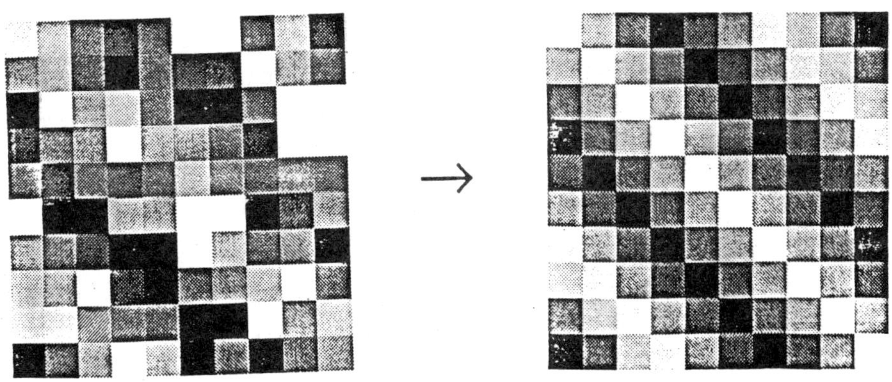

Fig. 9 Convergence when the initial condition is closer to pattern2

REFERENCES

1. M.A. Aizerman, E.M. Braverman and L.I. Rozonoer, " Theoritical fundations of he potential function method in the problem of training automata to classify input situations, " *Automat. Remote Contr.* (USSR), vol. 25, no. 6, 1964.
2. M.A. Aizerman, E.M. Braverman and L.I. Rozonoer, " The potential function method in the problem of estimating the characteristics of the functional transformer by randomly observed points, " *Automat. Remote Contr.*
3. G. Cybenco, " Approximation by superpositions of a sigmoidal function, " *Mathematics of Contrl, Signals and Systems*, vol. 2, pp. 303-314, 1989

4. J. E. Dennis, Jr., " Non-linear least squares and equations," In *The State of the Art in Numerical Analysis*. New york: Academic Press, 1977.
5. K. Funuhashi, " On the approximate realization of continuous mappings by neural networks, " *Neural Networks,* vol. 2, pp. 183-192, 1989
6. Hopfield, J. J. 1982. Neural networks and physical systems with emergent collective computational abilities. *Proceedings of the National Academy of Science* 79:2554-58.
7. Hopfield, J. J. 1984. Neurons with graded response have collective computational properties like those of two-state neurons. *Proceedings of the National Academy of Science* 81:3088-92.
8. Hopfield, J. J., and Tank, D. W. 1985. Neural computation of decisions in optimization problems. *Biological Cybernetics* 52:141-52.
9. K. Hornik, "Multilayer feedforward networks are universal approximators," *Neural Networks*, Vol 2, ppl 359-366, 1989.
10. Narendra, K.S., and Pasthasarathy, K. " Indentification and control of dynamical systems using neural networks,"*IEEE Transactions on Neural Networks,* vol. 1, pp. 4-27, 1990.
11. Pearlmutter, B.A. "learning state space trajectories in recurrent neural networks, "*Neural Computation*: Vol. 1, pp. 263-269, 1989. pp. 4-27, 1990
12. Pineda, F.J. "Generalization of backpropagation to reccurent and higher order networks", In: Proceedings of IEEE Conference on Neural Information 611, 1987.
13. D. E. Rumelhart, D. E. Hinton and R. J. Williams, *Parallel Distributed Processing: Exploration in the Microstructure of Cognition*. Cambridge, MA: MIT Press, 1986
14. Ya. Z Tsypkin, *Adaptation and Learning in Automatic Systems*. New york: Academic Press, 1971.

AAS 91-056

THERMAL ELASTIC SHOCK AND ITS EFFECT ON TOPEX SPACECRAFT ATTITUDE CONTROL

Darrell F. Zimbelman*

Thermal Elastic Shock (TES) is a twice per orbit impulsive disturbance torque experienced by low-Earth orbiting spacecraft. The disturbance is attributed to vehicle passage in and out of the Earth's umbra, during which, large flexible appendages undergo thermally induced bending caused by rapidly changing thermal conditions. Flexible structures, in particular solar arrays, experience rapid cooling during entrance and rapid heating at exit. The fundamental equations used to model the TES disturbance torque for typical spacecraft appendages (e.g. solar arrays and antenna booms) are derived in detail. In particular, the attitude pointing performance of the TOPEX spacecraft, when subjected to the TES disturbance, is analyzed using a three-axis, non-linear, time-domain simulation. Results indicate that the TOPEX spacecraft will exceed its roll axis attitude control requirement during penumbral transitions, and remain in violation for approximately 150 *sec* each orbit until the umbra collapses. A localized active control system is proposed as a solution to minimize and/or eliminate the degrading effects of the TES disturbance.

INTRODUCTION

In the early 1980's, an unexpected perturbation was exhibited by the LANDSAT 4 and 5 satellites. The full-up on-orbit LANDSAT satellite, a three-axis stabilized Earth-pointing vehicle, is shown in Fig. 1. An anomalously large, twice per orbit, thermally induced spacecraft perturbation was observed in the flight data.

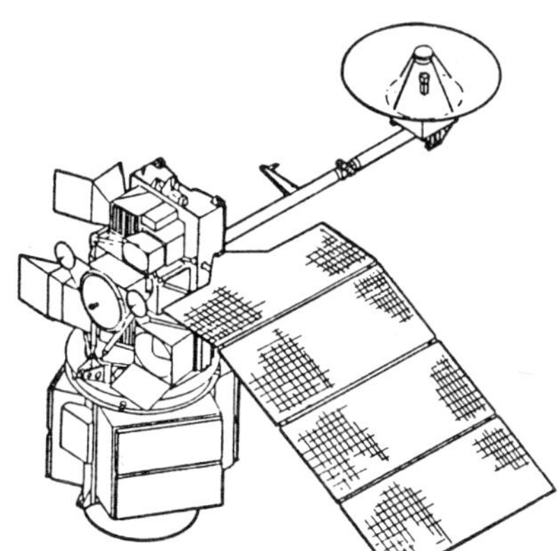

Fig. 1 Deployed LANDSAT-4 Satellite Configuration

* Engineer, GN&C Department, Fairchild Space & Defense Corporation, MS G43, 20301 Century Blvd., Germantown, Maryland 20874-1181.

Fig. 2 shows the LANDSAT-4 on-axis roll derived rate telemetry data over one orbital period. This disturbance was correlated with the spacecraft's entrance and exit to the eclipsed region of the orbit plane where it was determined that the single, large solar array experienced the phenomena of TES[1]. TES is defined as the motion of large, flexible appendages caused by rapidly changing thermal conditions. For the case of LANDSAT, the vehicles experienced a rapid cooling effect during umbra entrance (sunset) and rapid heating during exit (sunrise). The rapid change in

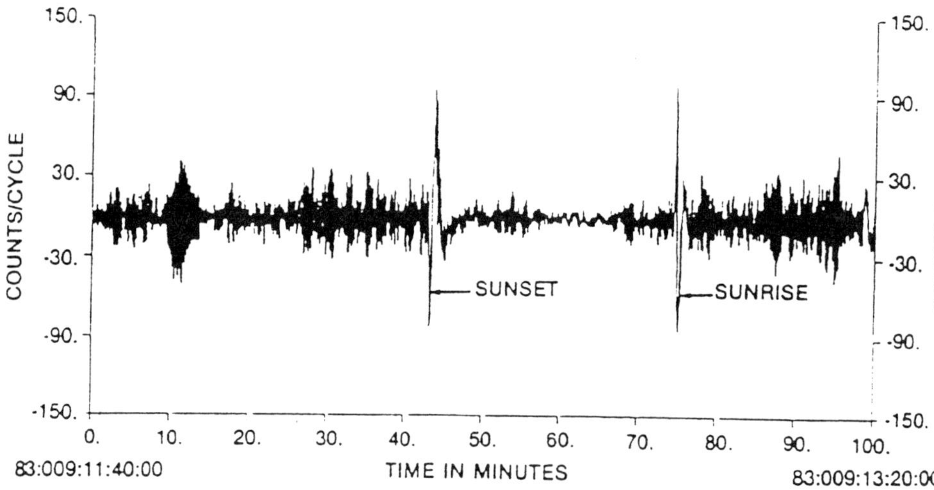

*Where: 1 count = 0.05 Arcseconds and, 1 cycle = 0.512 Seconds

Fig. 2 LANDSAT-4 Roll Axis Derived Rate Telemetry Data Over One Orbit

temperature caused the hot surface of the array to bend, either from thermal compression (sunset) or thermal expansion (sunrise), resulting in a disturbance torque acting on the spacecraft during penumbral transitions.

The dual-array Communications Technology Satellite (CTS) also experienced a TES perturbation, but to a much lesser degree. Flight performance for the CTS three-axis stabilized mode showed that the boom housing the solar blankets experienced a small thermal shock when entering and exiting the Earth's umbra[2]. The difference in perturbation magnitude between LANDSAT and CTS was a result of the differences between the two spacecraft designs. The TES disturbance is most pronounced for asymmetric satellite configurations such as the single-wing LANDSAT vehicles. Satellites possessing a symmetrical dual-wing array design, such as CTS, are significantly less affected by TES since the motion of both arrays tends to be self-compensating. However, a single-wing array configuration is sometimes required to provide an unobstructed radiator view of cold space for proper thermal control.

The Hubble Space Telescope (HST) is the most recent spacecraft to experience a disturbance associated with differential heating. The HST has been troubled by a thermally induced disturbance correlated with the vehicle's night and day transitions. Configuration of the HST is quite similar to the CTS design in that the HST contains two large solar arrays oriented symmetrically about the main body of the spacecraft. Like the CTS, the HST TES disturbance is attributed to the thermal deformation of the bi-stem booms which are used to secure the solar array blankets. Several analyses of the problem indicate that each 12.1 m long boom-panel assembly is subjected to a thermal gradient of $10^{\circ}C$ which requires approximately 300 sec to reach a thermal equilibrium state, while further analyses show that the solar arrays oscillate for periods up to 6 min with tip deflections on the order of 25 cm^3.

A future spacecraft highly susceptible to the TES disturbance is the proposed TOPEX spacecraft scheduled for launch in June of 1992. This assumption is based on the TOPEX vehicle design. TOPEX is a three-axis stabilized, nadir-pointing scientific satellite traveling in a near circular orbit at an altitude of 1335.5 km and an inclination of 65°. The fully deployed on-orbit vehicle configuration is illustrated in Fig. 3. A comparison of Figs. 1 and 3 reveals the same general configuration, including the presence of a single, large solar panel. Recall that it was the LANDSAT

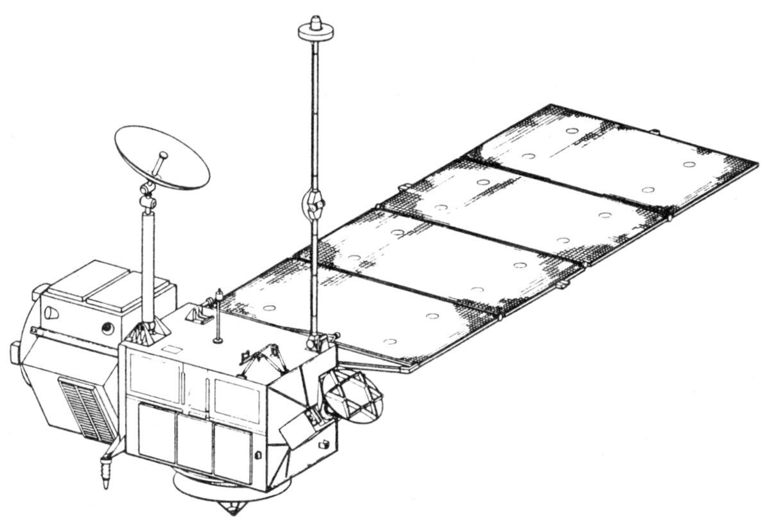

Fig. 3 Deployed TOPEX Satellite Configuration

solar array, undergoing thermal deflections during the transition to and from the umbra, which degraded the attitude pointing performance. TOPEX will be used to remotely monitor the world's oceans over a three year period as part of the joint TOPEX/POSEIDON program sponsored by the National Aeronautics and Space Administration (NASA) and the Centre National d'Etudes Spatiales (CNES). The primary objective of the TOPEX satellite will be to measure sea surface height using a dual frequency radar altimeter. This particular application, referred to as satellite altimetry, is part of an ongoing effort to study global ocean-atmosphere interaction. For the TOPEX/POSEIDON mission, current accuracy estimates for the sea surface height measurement stand at approximately $\pm 14\ cm$[4]. Although this accuracy estimate represents an increase over TOPEX's predecessors, e.g. GEOS-3, SEASAT and GEOSAT, further reduction is still necessary in order to assess the role of key oceanographic processes occurring at or below the 10 cm level. Meeting the desired mission objective requires the placement of somewhat stringent attitude pointing requirements on the TOPEX satellite. The TOPEX 1σ attitude control requirements are 144 *arcsec* (0.04°) for roll and pitch, and 504 *arcsec* (0.14°) for yaw. However, during Normal Mission Mode (NMM) (i.e. the primary science data collection mode) the roll axis is expected to violate its attitude control requirement due to TES. Exceeding the baseline attitude requirements can result in the degradation or loss of scientific data. In the case of the radar altimeter, exceeding the attitude control requirements can result in seriously degraded performance, even temporary loss of lock. Table 1 lists the range of antenna boresight pointing error versus expected radar altimeter performance. It is clear from Table 1, that pointing excursions beyond 1800 *arcsec* (0.5°) will cause the instrument to break lock, resulting in temporary data loss. This condition is highly undesirable and should be avoided. However, as was evidenced by LANDSAT, the magnitude of the TES disturbance could cause attitude perturbations large enough to exceed altimeter pointing requirements and corrupt the generated data set. Thus the need for accurate disturbance models is motivated by the precision attitude requirements for the TOPEX/POSEIDON mission. In particular, a detailed model describing the TES perturbation is necessary in order to predict the overall attitude response of the TOPEX vehicle.

The focus of this research is to investigate the phenomena of TES with an emphasis on Earth orbiting space vehicles. The intent is to shed light onto the mathematical formulation of the disturbance torque created as the spacecraft passes in and out of the Earth's shadow cone. It will be shown that the TES disturbance is a function of the time rate of change of the thermal gradient across a flexible appendage as well as the mass and material properties of the structure. A detailed model to predict the thermal gradient across a solar array will be developed and applied to the TOPEX satellite. With respect to the TOPEX satellite, the magnitude of the predicted TES disturbance torque will be determined. The estimated TES disturbance torque will then be used to study the effect of this disturbance on the TOPEX Attitude Control System (ACS) using a three-axis, non-linear, time-domain simulation. Finally, an active control system will be introduced to accommodate the TES disturbance. The design is centered on the concept of localized control with the goal of reducing the disturbance torque before it interacts with the vehicle base-body. The control system is independent of the vehicle core-body ACS and will be adapted to a TOPEX satellite design in order to numerically demonstrate the feasibility of the concept.

TES DISTURBANCE MODEL

The TES disturbance torque model uses the concept of linear thermal expansion to describe the angle through which a structure will bend under the influence of thermal heating. Utilizing this physical principle, the appendage thickness can be expressed as a function of the material properties (α_{cte}), the thermal gradient (ΔT) and the radius of curvature (ρ_b):

$$d = \rho_b \alpha_{cte} \Delta T \tag{1}$$

Rearranging and solving for the radius of curvature gives:

$$\rho_b = \frac{d}{\alpha_{cte} \Delta T} \tag{2}$$

The above relation holds true for both the top and bottom surfaces (i.e. $\rho_b \approx \rho_{b_{top}} \approx \rho_{b_{bottom}}$) under the assumption that the separation distance is small, $d << \rho_b$. Now assume that the structure can be represented as a slender rod with length L and mass M. The mass distribution about the fixed end of the rod is given by:

$$I = \frac{1}{12} L^2 M + \frac{1}{4} L^2 M \tag{3}$$

If the rod length and mass are divided into n pieces then a discrete representation for the rod inertia becomes:

$$I = \sum_{i=1}^{n} \left(\frac{1}{12}\left(\frac{L}{n}\right)^2 m_i + \left[\frac{L}{n}\left(i - \frac{1}{2}\right)\right]^2 m_i\right) \tag{4}$$

where the ith mass element is expressed as:

$$m_i = \frac{M}{n} \tag{5}$$

By factoring out the mass term, Eq. (4) can be rewritten as an equivalent inertia distribution (lumped mass) given as:

$$I = \sum_{i=1}^{n} l_i^2 m_i \tag{6}$$

where l_i is of the form:

$$l_i = \sqrt{\frac{1}{12}\left(\frac{L}{n}\right)^2 + \left[\frac{L}{n}\left(i - \frac{1}{2}\right)\right]^2} \tag{7}$$

Next, let I_b be an equivalent inertial displacement defined as the rod inertia, Eq. (6), multiplied by the angle through which the structure rotates relative to the constrained end. The structure is assumed to deform in a circular arc of radius ρ_b under the influence of a uniform temperature gradient applied over the entire length. Then for the ith inertia element and the bend angle of the ith mass element:

$$I_b = \sum_{i=1}^{n} l_i^2 m_i \theta_i \tag{8}$$

The arc length each element of mass travels is given by:

$$s_i = l_i \theta_i \tag{9}$$

Rearranging and substituting Eq. (9) into Eq. (8) yields:

$$I_b = \sum_{i=1}^{n} l_i m_i s_i \tag{10}$$

The change in radius of curvature associated with each l_i increment, depicted in Fig. 4, is expressed as follows:

$$\Delta\rho_b = \rho_b - \rho_b \cos(\phi_i) \tag{11}$$

where the curvature angle, ϕ_i, associated with each l_i is given by:

$$\tan\phi_i = \frac{l_i}{\rho_b} \tag{12}$$

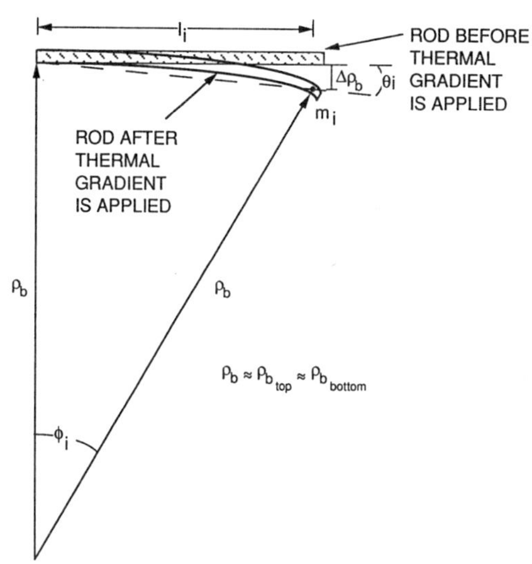

Fig. 4 Change in Radius of Curvature for Each Inertia Length Increment

If the assumption is made that each inertia length increment is small compared to the radius of curvature, such that the ratio $\frac{l_i}{\rho_b} < 0.176$, then the curvature angle, ϕ_i, is less than 10°. Under this approximation, the curvature angle for the ith element can be simplified as:

$$\phi_i \approx \frac{l_i}{\rho_b} \tag{13}$$

If it is further assumed that each mass element only travels through a small angle, i.e. $\theta_i < 10°$, then the arc length, s_i, is essentially linear and equal to the change in radius of curvature given by Eq. (11). This approximation can be stated as:

$$s_i \approx \Delta\rho_b \tag{14}$$

Thus from Eqs. (11), (13) and (14),

$$s_i \approx \rho_b - \rho_b \cos(\frac{l_i}{\rho_b}) \tag{15}$$

Substituting Eq. (15) into the derived expression for I_b, Eq. (10), yields:

$$I_b \approx \sum_{i=1}^{n} l_i m_i \rho_b [1 - \cos(\frac{l_i}{\rho_b})] \tag{16}$$

The development of the TES disturbance model through Eq. (16) is similar to the derivation given by Ref. 1. Eq. (16) defines a general analytical expression to describe the dynamic motion of a slender rod of finite thickness given an applied thermal gradient. This equation can also be used to represent the movement of thin beams and thin flat plates since each geometric figure shares a common expression for the inertia about one end.

Intuitively, the first and second time derivatives of Eq. (16) would yield the momentum and torque time histories describing the TES disturbance. An inspection of the LANDSAT-4 derived rate telemetry data (Fig. 2) reveals that the controlled response of the vehicle exhibits an impulsive velocity in one direction immediately followed by a similar motion in the opposite direction, and finally a decaying step of the initial sign. This responsive rate motion indicates that the angular attitude deviation of the vehicle is represented as an impulsive error in one direction followed by an exponentially decaying step error of opposite sign. Since the LANDSAT satellite ACS was designed to accommodate both good command following performance and low-frequency disturbance torque rejection capabilities, it is assumed that TES disturbance torque is of high frequency and furthermore, that the TES disturbance torque possesses the same two-phase function characteristic, i.e. an impulse followed by an exponentially decaying step of opposite sign. However, the resulting torque function calculated using Eq. (16) only captures the latter part (i.e. the exponentially decaying step) of the overall disturbance. This is because I_b is a multivalued function at $t = \tau$ which yields a step discontinuity at $t = \tau$ for the first time derivative of I_b. Neglecting the step discontinuity in the second time derivative eliminates the impulsive term present in the expected structure of the disturbance torque. Thus, in order to capture the disturbance torque function which completely characterizes the observed on-orbit attitude response, fundamental continuous theoretical functions are derived to explain the underlying mathematics behind the TES phenomena. In the following development a continuous function for I_b, analogous to the discrete I_b expression given in Eq. (16), will be developed.

Let I_b be defined as a continuous function of the form:

$$I_b = a[u(t+\infty) - u(t-\tau)] + f(t-\tau)u(t-\tau) \tag{17}$$

where

$$\begin{aligned} a &\equiv \text{constant} \\ f(t-\tau) &\equiv \text{given function of time} \end{aligned}$$

and $u(t-\tau)$ is a unit *step* function defined as:

$$u(t-\tau) = \begin{cases} 0 & \text{if t} < \tau \\ 1 & \text{if t} \geq \tau \end{cases} \tag{18}$$

In order for I_b to be continous at $t = \tau$, the following matching condition must hold true:

$$f(t-\tau)|_{t=\tau} = a \tag{19}$$

Momentum is defined as the first derivative of I_b with respect to time, i.e. $\frac{dI_b}{dt}$. Computing the first time derivative of Eq. (17) gives the following expression for momentum:

$$H_b = a[\dot{u}(t+\infty) - \dot{u}(t-\tau)] + f(t-\tau)\dot{u}(t-\tau) + \dot{f}(t-\tau)u(t-\tau) \tag{20}$$

where $\dot{u}(t-\tau)$ is a *Dirac* function described as:

$$\dot{u}(t-\tau) = \begin{cases} \infty & \text{if t} = \tau \\ 0 & \text{otherwise} \end{cases} \tag{21}$$

The second derivative of Eq. (17) with respect to time, $\frac{d^2 I_b}{dt^2}$, yields the equation for the torque:

$$\begin{aligned} T_b &= a[\ddot{u}(t+\infty) - \ddot{u}(t-\tau)] + f(t-\tau)\ddot{u}(t-\tau) \\ &\quad +2\dot{f}(t-\tau)\dot{u}(t-\tau) + \ddot{f}(t-\tau)u(t-\tau) \end{aligned} \tag{22}$$

It is assumed that all terms containing the derivatives of the unit *step* function $u(t+\infty)$, i.e. $\dot{u}(t+\infty)$ and $\ddot{u}(t+\infty)$, have no contribution to the dynamics at $t \geq 0$. Thus Eqs. (20) and (22) reduce to the following form:

$$H_b = -a\dot{u}(t-\tau) + f(t-\tau)\dot{u}(t-\tau) + \dot{f}(t-\tau)u(t-\tau) \tag{23}$$

$$\begin{aligned} T_b &= -a\ddot{u}(t-\tau) + f(t-\tau)\ddot{u}(t-\tau) \\ &\quad +2\dot{f}(t-\tau)\dot{u}(t-\tau) + \ddot{f}(t-\tau)u(t-\tau) \end{aligned} \tag{24}$$

The *doublet*, $\ddot{u}(t-\tau)$, in Eq. (24) which is multiplied by the function $f(t-\tau)$ can be expressed as two terms given as[5]:

$$f(t-\tau)\ddot{u}(t-\tau) = f(0)\ddot{u}(t-\tau) - \dot{f}(0)\dot{u}(t-\tau) \tag{25}$$

Finally, if the matching condition, Eq. (19), requiring I_b to be continuous at $t = \tau$ and the *doublet* equation, Eq. (25), are applied to both Eqs. (23) and (24), these expressions further reduce to the following:

$$H_b = \dot{f}(t-\tau)u(t-\tau) \tag{26}$$

$$T_b = \dot{f}(t-\tau)\dot{u}(t-\tau) + \ddot{f}(t-\tau)u(t-\tau) \tag{27}$$

The fundamental relationships, given by Eqs. (26) and (27), can be used to compute the TES disturbance momentum and torque using the time derivatives of the known function $f(t-\tau)$. An inspection of Eq. (26) indicates that the momentum is comprised of the first time derivative of the function $f(t-\tau)$ multiplied by a *unit* step function. The presence of this step discontinuity is consistent with the expected form of the first derivative of Eq. (16) as previously discussed.

In order to determine the functional form of $f(t-\tau)$, a time dependent analytical relationship for the thermal gradient is assumed. A relatively simple exponential function has been choosen to represent the thermal gradient for umbra entrance and is expressed as:

$$\Delta T_{sunset} = \begin{cases} q & \text{if } \mathrm{t} < \tau_{\text{sunset}} \\ ce^{\frac{-(t-\tau_{sunset})}{k}} + b & \text{if } \tau_{\text{sunset}} \leq \mathrm{t} < \tau_{\text{sunrise}} \end{cases} \tag{28}$$

where

$$\begin{aligned} k &\equiv \text{decay constant} \\ b &\equiv \text{temperature bias} \\ c &\equiv \text{temperature scaling} \\ q &\equiv \text{temperature constant} \\ \tau_{sunset} &\equiv \text{umbra entrance time constant} \end{aligned}$$

A similar expression is used for sunrise and is given by:

$$\Delta T_{sunrise} = \begin{cases} b & \text{if } \tau_{\text{sunset}} < \mathrm{t} < \tau_{\text{sunrise}} \\ c(1 - e^{\frac{-(t-\tau_{sunrise})}{k}}) + b & \text{if } \mathrm{t} \geq \tau_{\text{sunrise}} \end{cases} \tag{29}$$

with $\tau_{sunrise}$ being the umbra exit time constant. The parameters used to calculate ΔT_{sunset} and $\Delta T_{sunrise}$ can be prescribed to match the predicted thermal response of the structure. Substituting Eqs. (2) and (28) into Eq. (16) yields the following expression for I_b at sunset:

$$I_{b_{sunset}} \approx \sum_{i=1}^{n} [\frac{l_i m_i d}{\alpha_{cte} q}(1 - \cos[\frac{l_i \alpha_{cte} q}{d}])] + \sum_{i=1}^{n} [\frac{l_i m_i d}{\alpha_{cte}[ce^{\frac{-(t-\tau_{sunset})}{k}} + b]} (1 - \cos[\frac{l_i \alpha_{cte}[ce^{\frac{-(t-\tau_{sunset})}{k}} + b]}{d}])] \tag{30}$$

Substituting Eqs. (2) and (29) into Eq. (16) yields the following expression for I_b at sunrise:

$$I_{b_{sunrise}} \approx \sum_{i=1}^{n}[\frac{l_i m_i d}{\alpha_{cte} b}(1-\cos[\frac{l_i \alpha_{cte} b}{d}])] + \sum_{i=1}^{n}[\frac{l_i m_i d}{\alpha_{cte}[c(1-e^{\frac{-(t-\tau_{sunrise})}{k}})+b]} (1-\cos[\frac{l_i \alpha_{cte}[c(1-e^{\frac{-(t-\tau_{sunrise})}{k}})+b]}{d}])] \quad (31)$$

It can be seen that Eq. (30) is equivalent to Eq. (17) with:

$$a_{sunset} = \sum_{i=1}^{n}[\frac{l_i m_i d}{\alpha_{cte} q}(1-\cos[\frac{l_i \alpha_{cte} q}{d}])] \quad (32)$$

and

$$f(t-\tau_{sunset}) = \sum_{i=1}^{n}[\frac{l_i m_i d}{\alpha_{cte}[ce^{\frac{-(t-\tau_{sunset})}{k}}+b]}(1-\cos[\frac{l_i \alpha_{cte}[ce^{\frac{-(t-\tau_{sunset})}{k}}+b]}{d}])] \quad (33)$$

Likewise, it is evident that Eq. (31) is equivalent to Eq. (17) with:

$$a_{sunrise} = \sum_{i=1}^{n}[\frac{l_i m_i d}{\alpha_{cte} b}(1-\cos[\frac{l_i \alpha_{cte} b}{d}])] \quad (34)$$

and

$$f(t-\tau_{sunrise}) = \sum_{i=1}^{n}[\frac{l_i m_i d}{\alpha_{cte}[c(1-e^{\frac{-(t-\tau_{sunrise})}{k}})+b]} (1-\cos[\frac{l_i \alpha_{cte}[c(1-e^{\frac{-(t-\tau_{sunrise})}{k}})+b]}{d}])] \quad (35)$$

If the thermal gradient is represented as ΔT without any subscript, then a general expression for the function $f(t-\tau)$ can be formulated as:

$$f(t-\tau) = \sum_{i=1}^{n}[\frac{l_i m_i d}{\alpha_{cte}\Delta T}(1-\cos[\frac{l_i \alpha_{cte}\Delta T}{d}])] \quad (36)$$

It is evident that Eq. (36) is equivalent to Eq. (16) with ρ_b replaced by Eq. (2). This result makes sense as the physical relationships used to derive Eq. (16) must be contained in the mathematics. Examination of Eq. (27) shows that the second derivative of $f(t-\tau)$ only yields one term of the general TES torque expression (i.e. the exponentially decaying step). The step discontinuity present in Eq. (26) is needed to capture the impulsive term in Eq. (27). This result confirms the previous discussion with respect to the first and second derivatives of Eq. (16). Thus, the momentum and torque equations can be computed using Eqs. (26) and (27), along with Eq. (36). They are as follows:

$$H_b = \begin{cases} 0 & \text{if t} < \tau \\ \sum_{i=1}^{n} -\frac{l_i m_i d\Delta T}{\alpha_{cte}\Delta T^2}\left(1 - \cos\left[\frac{l_i \alpha_{cte}\Delta T}{d}\right]\right. & \\ \left. -\frac{l_i \alpha_{cte}\Delta T}{d}\sin\left[\frac{l_i \alpha_{cte}\Delta T}{d}\right]\right) & \text{if t} \geq \tau \end{cases} \tag{37}$$

$$T_b = \begin{cases} 0 & \text{if t} < \tau \\ T_\varsigma + \frac{\Delta T H_b}{\Delta T} - \frac{2\Delta T H_b}{\Delta T} & \\ + \sum_{i=1}^{n} \frac{l_i^3 m_i \alpha_{cte}\Delta T^2}{d\Delta T}\cos\left[\frac{l_i \alpha_{cte}\Delta T}{d}\right] & \text{if t} \geq \tau \end{cases} \tag{38}$$

with

$$T_\varsigma = \begin{cases} H_b & \text{if t} = \tau \\ 0 & \text{if t} > \tau \end{cases} \tag{39}$$

where $\dot{\Delta T}$ and $\ddot{\Delta T}$ are the first and second time derivatives of the thermal gradient. The TES disturbance momentum and torque can be evaluated using Eqs. (37), (38) and (39) for any one of three geometric shapes. These include slender rods, thin beams and thin flat plates. The momentum and torque expressions are entirely dependent on the applied thermal gradient and its sucessive derivatives along with the mass and material properties of the selected geometric object. The TES disturbance torque, described by Eqs. (38) and (39), possesses the two-part characteristic exhibited in the controlled response of the LANDSAT-4 satellite.

THERMAL GRADIENT PREDICTION

It was shown in the previous section that the TES disturbance torque is dependent, in part, on the succesive derivatives of the thermal gradient across a flexible structure. In order to predict the temperature difference between any two adjacent body surfaces, a comprehensive understanding of the heat fluxes acting on the spacecraft is necessary. Earth-orbiting satellites are subjected to three sources of incoming radiation: emitted solar, Earth reflected solar (albedo) and Earth emitted infrared (EIR). Nadir-pointing satellites, such as LANDSAT and TOPEX, experience distributed heating on the vehicle base-body as the orientation changes throughout the orbit. The solar array, on the other hand, experiences extreme differential heating, as one surface is always oriented nearly perpendicular to the sun. The cell or hot side of the solar array receives contributions from incoming solar, albedo and EIR radiation. Alternatively, the back or cold side of the panel experiences heating from only two sources: albedo and EIR. Furthermore, during passage through the umbra, the back side of the structure receives little or no incoming heat flux and thus will primarly radiate energy to space. Once the appropriate heat fluxes are determined, the thermal response of the solar array can be predicted using numerical modeling techniques.

A numerical model of a thin rectangular flat plate is constructed and used to predict the surface temperatures in the presence of an applied heat flux. The plate is assumed to be infinitely-long and infinitely-wide and possess a small finite thickness (i.e. $d << L$). The slab is also assumed to be made of a single, homogenous material with even heating ($Q_{hot}(t) \neq Q_{cold}(t)$) applied across each face of the solar panel (isothermal surfaces) and no internal heat generation. Under these assumptions, no heat flow exists along the length or across the width of the array (i.e. $\frac{\delta^2 T(x,y,z,t)}{\delta y^2} = \frac{\delta^2 T(x,y,z,t)}{\delta z^2} = 0$). Thus heat will only flow through the thickness of the array from the sun-pointing to the anti-sun pointing surface. In general, the three-dimensional heat equation is utilized to mathematically solve transient heat flow problems. However, for this model the three-dimensional heat equation reduces to a one-dimensional representation described as:

$$\rho c_p \frac{\delta T(x,t)}{\delta t} = K_o \frac{\delta^2 T(x,t)}{\delta x^2} \tag{40}$$

where

$$\begin{aligned} t &\equiv \text{time} \\ \rho &\equiv \text{density of material} \\ c_p &\equiv \text{specific heat of material} \\ K_o &\equiv \text{conductivity of material} \\ T(x,t) &\equiv \text{temperature at point x and time t} \\ x &\equiv \text{spatial location through thickness of plate} \end{aligned}$$

In order to solve this parabolic partial differential heat equation, an initial condition must be specified at $t = 0$. The initial condition for this model is given by:

$$T(x,t)|_{t=0} = T_o \tag{41}$$

where T_o is a constant initial temperature. This condition is valid for the region:

$$0 \leq x \leq d \tag{42}$$

where d is the thickness of the slab. In addition to the initial condition, sufficient boundary conditions must be applied to both faces of the plate. The surface boundary conditions can be derived from the general form of the radiation heat exchange boundary condition. For this condition, the transfer rate of heat flow across a boundary is a function of the emitted surface energy and the absorbed incident radiant thermal energy, both emitted and reflected. The radiation exchange between the plate and the surrounding medium, e.g. the vacuum of space, is acknowledged by emittance coupling. The boundary condition is expressed as:

$$-K_o \frac{\delta T}{\delta n_\perp} = \sigma \epsilon_s (T_s^4 - T_{medium}^4) - \alpha_s Q_i \tag{43}$$

where

$$\begin{aligned}
\sigma &\equiv \text{Boltzmann's constant} \\
T_s &\equiv \text{surface temperature} \\
T_{medium} &\equiv \text{temperature of surrounding medium} \\
\alpha_s &\equiv \text{absorptivity of the surface} \\
\epsilon_s &\equiv \text{emissivity of the surface} \\
n_\perp &\equiv \text{surface outward normal direction} \\
Q_i &\equiv \text{incident radiant heat flux}
\end{aligned}$$

For the hot surface of the plate ($x = 0$) the applied boundary condition, from Eq. (43), is stated as follows:

$$K_o \frac{\delta T(x,t)}{\delta x}|_{x=0} = \epsilon_{hot}\sigma(T^4(x,t)|_{x=0} - T^4_{medium}) - Q_{hot}(t) \tag{44}$$

with ϵ_{hot} representing the emissivity of hot surface. The heat flux, $Q_{hot}(t)$, is given by:

$$Q_{hot}(t) = \alpha_{hot}(Q_s(t) + Q_{albedo}(t)) + \epsilon_{hot}Q_{eir}(t) \tag{45}$$

where

$$\begin{aligned}
Q_s(t) &\equiv \text{incoming solar radiation} \\
Q_{eir}(t) &\equiv \text{Earth emitted infrared radiation} \\
Q_{albedo}(t) &\equiv \text{Earth reflected solar radiation} \\
\alpha_{hot} &\equiv \text{absorptivity of the hot surface}
\end{aligned}$$

The cold surface boundary condition ($x = d$) is similar to the hot condition and is given as:

$$-K_o \frac{\delta T(x,t)}{\delta x}|_{x=d} = \epsilon_{cold}\sigma(T^4(x,t)|_{x=d} - T^4_{medium}) - Q_{cold}(t) \tag{46}$$

with $Q_{cold}(t)$ expressed as follows:

$$Q_{cold}(t) = \alpha_{cold}Q_{albedo}(t) + \epsilon_{cold}Q_{eir}(t) \tag{47}$$

where α_{cold} and ϵ_{cold} are the absorptivity and emissivity of the cold surface. Eqs. (40), (44) and (46) can be transformed into explicit finite difference equations using the central, forward and backward finite difference formulas, respectively, in order to numerically solve the transient heat flow problem. First-order time, $O(\Delta t)$, and second-order space, $O(\Delta x^2)$, finite difference expressions for the one-dimensional heat flow equation and the surface boundary conditions are represented as:

$$T_i^{t+1} = \frac{\alpha_d \Delta t}{(\Delta x)^2}[T_{i+1}^t - 2T_i^t + T_{i-1}^t] + T_i^t \tag{48}$$

$$T_1^t = \frac{2\Delta x}{3K_o}(Q_{hot}^t - \epsilon_{hot}\sigma[(T_1^{t-1})^4 - (T_{medium}^{t-1})^4]) + \frac{4}{3}T_2^t - \frac{1}{3}T_3^t \tag{49}$$

$$T_N^t = \frac{2\Delta x}{3K_o}(Q_{cold}^t - \epsilon_{cold}\sigma[(T_N^{t-1})^4 - (T_{medium}^{t-1})^4]) + \frac{4}{3}T_{N-1}^t - \frac{1}{3}T_{N-2}^t \tag{50}$$

where

$i \equiv i^{th}$ node

$\Delta t \equiv$ integration time step

$N \equiv$ number of nodes across plate thickness

$\Delta x \equiv$ spatial distance between adjacent nodes

$\alpha_d \equiv$ thermal diffusivity of the material ($\alpha_d = \frac{K_o}{\rho c_p}$)

The notation, T_b^a, is used to denote the ath time and the bth node across the array thickness. Since the finite difference approximations are consistent, numerical stability is a necessary and sufficient condition to insure convergence. Applying a stability analysis to Eq. (48) yields a conditionally stable situation governed by the following stability criteria:

$$\frac{\alpha_d \Delta t}{\Delta x^2} \leq \frac{1}{2} \tag{51}$$

The spatial (temporal) increment is computed using Eq. (51) given a specified time (space) increment. The number of nodes across the array thickness is determined using the following equation:

$$N = \frac{d}{\Delta x} + 1 \tag{52}$$

Once the material properties of the flat plate are selected and the proper heat fluxes acting on the plate are determined, Eqs. (48), (49) and (50) can be utilized to predict the thermal response of the solar array.

The TOPEX solar array is composed of aluminum honeycomb sandwiched between two aluminum facesheets. The mass and material properties of the honeycomb, along with the surface absorptivities and emissivities are listed in Table 2. For the TOPEX numerical model, the array is assumed to be comprised of a solid slab of aluminum honeycomb. Since the array is assumed to be a single, homogenous surface, the material properties (i.e. K_o, ρ and c_p) are held constant across the array thickness. This assumption remains valid for the density, specific heat and thermal conductivity since the aluminum honeycomb contributes over 95% of the total array thickness. However, in order to account for the different layered materials of the true array, the thermal diffusivity was selected to be a linear function of the spatial location across the array thickness, i.e. $\alpha_d = \alpha_d(x)$. This type of function is necessary to accurately predict array surface temperatures since they are highly sensitive to the thermal diffusivity. On the other hand, the temperature difference between the two surfaces, the primary interest of this research, is driven by the selection of the thermal conductivity.

For the TOPEX vehicle, the longest umbra (2080 *sec*), and consequently the shortest penumbra (10 *sec*), occurs when the angle between the sun and the orbit plane, β', is zero. This situation corresponds to the worst case thermal condition as the satellite will reach its minimum thermal state. Therefore, array heat flux estimates were calculated for $\beta' = 0°$ using the Thermal Radiation Analysis SYStem (TRASYS) software package. These estimates were then used to generate the predicted thermal response of the TOPEX solar array using the detailed thermal model. Fig. 5 illustrates the array thermal gradient response over one orbital period, beginning at spacecraft 6 am. Spacecraft local 6 am time is defined to be

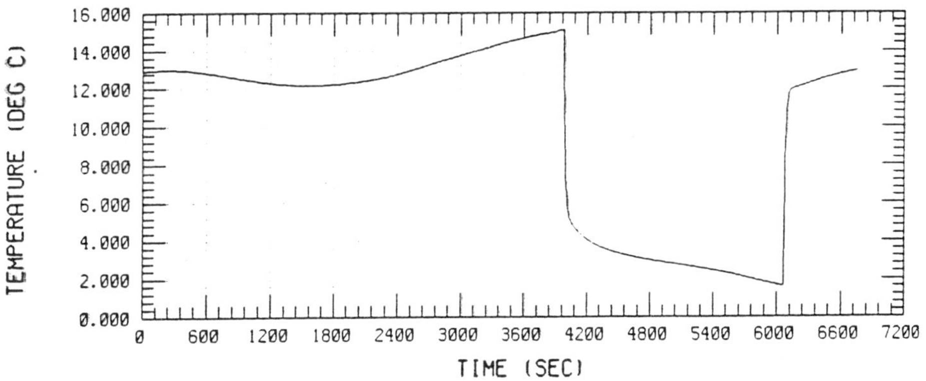

Fig. 5 TOPEX Solar Array Thermal Gradient for $\beta' = 0°$

90° in a clockwise direction from the Earth-sun vector projection onto the orbit plane. The thermal gradient exhibits an exponential decrease at umbra entrance from $15°C$ to $4°C$ in approximately 100 *sec*. Similarily, the gradient increases in an exponential fashion from $1.5°C$ to $12.5°C$ over a span of 100 *sec* at exit. These rapid changes in the thermal gradient act to produce the TES disturbance torque. The exponential character of the thermal gradient justifies the selection of the exponential analytical thermal gradient function used in the derivation of the TES disturbance torque model. For $|\beta'|$ angles greater than 56°, the umbra collapses and the thermal gradient remains essentially constant resulting in no TES disturbance. Thus for the TOPEX spacecraft, the TES disturbance will only be of concern for $|\beta'|$ angles less than 56°. Unfortunately, this condition represents approximately 80% of the proposed three year TOPEX/POSEIDON mission.

TOPEX TES DISTURBANCE TORQUE

During NMM operation, the TOPEX spacecraft will employ two different yaw control laws, one for local vertical flight and one for sinusoidal yaw steering. The sinusoidal yaw steering is required to achieve near perfect array sun-pointing as $|\beta'|$ increases since the solar array drive unit only encompasses one degree of rotational freedom. The yaw steering mode enables the vehicle to provide the additional degree of freedom. The control laws are governed by the following equation:

$$\psi_c = \begin{cases} 90^\circ + (90^\circ - \beta')\cos\Omega & \text{if } \beta' \geq 15^\circ \\ 0^\circ & \text{if } 0^\circ \leq \beta' < 15^\circ \\ 180^\circ & \text{if } -15^\circ < \beta' < 0^\circ \\ -90^\circ - (90^\circ + \beta')\cos\Omega & \text{if } \beta' \leq -15^\circ \end{cases} \tag{53}$$

where

$$\psi_c \equiv \text{commanded yaw attitude angle}$$
$$\Omega \equiv \text{orbit angle relative to local spacecraft 6 am}$$

The $\pm 15^\circ$ limit is selected to distinguish between the zones of operation due to power considerations. The reversal in fixed yaw orientation from 0° to 180° as the sun passes through the orbit plane, $\beta' = 0^\circ$, insures proper thermal control of sensitive spacecraft instrumentation. Consequently, the different yaw control during NMM operation requires the examination of the TES disturbance for both local vertical flight and the yaw steering mode.

The magnitude of the TOPEX solar array TES disturbance can be estimated using Eqs. (28), (29), (37), (38) and (39). The parameters used to define the thermal gradient analytical functions, Eqs. (28) and (29), can be evaluated using the predicted thermal gradient response. These parameters remain valid for $|\beta'|$ values less than 56°. In order to distribute the TES disturbance into the spacecraft axes, the array orientation as the vehicles enters and exits the Earth's shadow cone is needed. The array outward normal, with respect to the vehicle roll axis (+x axis), is always 215° at entrance and 325° at exit. Since the array is co-aligned with the spacecraft pitch axis, the TES disturbance will only be distributed into the roll and yaw axes. During sunset, the impulsive component of the roll TES disturbance torque, for $\beta' = 0^\circ$, has a magnitude of -1.71 $N-m$ while the exponentially decaying step reaches a peak of 0.051 $N-m$. At sunrise, the roll disturbance signal reverses sign. The yaw disturbance impulse is -2.45 $N-m$ followed by a step of 0.074 $N-m$ and remains consistent during both penumbral transitions. A frequency analysis of the roll and yaw disturbance signals reveals that each frequency up to the Nyquist limit contributes to the construction of the disturbance signals. For the case when the vehicle -x axis is in the direction of the velocity vector, $\beta' = -0^\circ$, the roll disturbance remains equivalent to the roll signal for $\beta' = 0^\circ$, while the yaw disturbance shows a sign reversal. The yaw steering mode roll axis TES disturbance torque for $\beta' = \pm 15^\circ$ is identical to the local vertical case, however, the structure of the yaw curve now shows a sign change between sunset and sunrise for this mode.

TOPEX ATTITUDE POINTING PERFORMANCE

During NMM operations, the Attitude Determination and Control Subsystem (ADCS) must provide precise pointing knowledge and control of the radar altimeter antenna boresight axis relative to the nadir direction. Given estimates of the TES disturbance, as well as other environmental perturbations such as aerodynamic drag, gravity-gradient and solar radiation pressure, the attitude pointing performance of the TOPEX satellite can be evaluated using a three-axis, non-linear, time-domain simulation.

An analysis block diagram of the TOPEX roll-axis NMM digital controller is depicted in Fig. 6. The form for the pitch and yaw axes controllers is identical

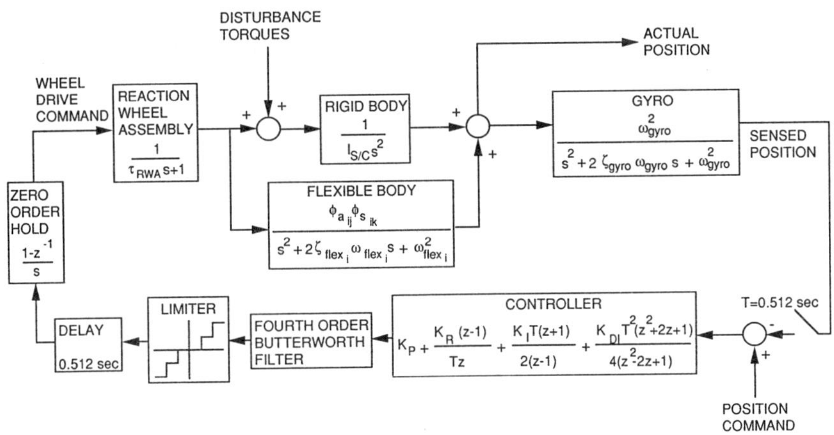

Fig. 6 TOPEX NMM Roll Axis Analysis Block Diagram

to Fig. 6. The baseline controller $-3\ dB$ closed-loop bandwidth is selected to be $0.01\ Hz\ (0.063 \frac{rad}{sec})$ for all three axes. A detailed Controls/Structure Interaction analysis has been performed using a NASTRAN Finite Element Model (FEM) of the fully-deployed TOPEX vehicle[6]. In order to attenuate the degrading effects of structural vibration on system stability, a fourth-order *Butterworth* digital filter is implemented into the design of the NMM controller. However, the *Butterworth* filter will introduce a significant lag in the NMM ACS response, causing the disturbance torques to have a greater impact on pointing performance. The NMM controller shown in Fig. 6 provides four distinct control actions: Proportional control, Derivative control, Single-Integral control and Double-Integral control. This form of control logic results in an ACS which will not only zero-out steady-state errors due to constant disturbance torques but also function to decrease attitude errors due to time-varying disturbance torques. The TOPEX NMM controller design is also advantageous in that, on a per-axis basis, the control law gains are simple functions of only three parameters: the spacecraft inertia ($I_{S/C}$), the desired controller -3 dB closed-loop bandwidth and a single controller break frequency[6]. The NMM digital controller is implemented in a three-axis, non-linear, time-domain computer simulation in order to evaluate the roll, pitch and yaw pointing performance degradation caused by the environmental perturbations and in particular, the TES disturbance torque. The attitude simulation includes models of the ADCS NMM sensor/actuator hardware, the On-Board Computer attitude control algorithms and the TOPEX spacecraft three-axis flexible-body dynamics.

The principal TOPEX ADCS NMM sensor hardware package consists of three integrating rate gyros, one dedicated to each spacecraft axis, while the actuator package contains four Reaction Wheel Assemblies (RWA's), one aligned along each vehicle axis with the fourth skewed equiangular from the roll, pitch and yaw axes. Each RWA nominally operates between $\pm 27\ N-m-sec$ with a control torque authority of $\pm 0.14\ N-m$. The skew wheel is dynamically used to bias the three orthogonal axes reaction wheels in order to prevent zero-speed crossings. Momentum unloading of the wheels is provided via three magnetic torquer bars using a cross-product law with the Earth's magnetic field. The spacecraft rigid body dynamics and flexible characteristics are modeled as:

$$W_{S/C}(s) = \frac{1}{I_{S/C}s^2} + W_{flex_i}(s) \tag{54}$$

where the $W_{flex_i}(s)$ terms are computed using the second order transfer function shown in Fig. 6. The ith flexible mode is represented as ω_{flex_i} and the damping associated with the ith mode is ζ_{flex_i}. The numerator is formulated as the product of the jth actuator and k^{th} sensor modal gain matrix elements for the ith mode, $\phi_{a_{ij}}\phi_{s_{ik}}$ respectively. Two composite NASTRAN FEM's were developed to predict the resonant modes of the deployed spacecraft. The first FEM assumes the array is positioned in the spacecraft pitch-yaw plane, while the second model is for an array oriented in the roll-pitch plane. For this analysis, the first FEM was utilized to provide the TOPEX flexible data. Each sensor/actuator pair of modal gain elements was subjected to a modal significance analysis[6]. Seven predominant modes were retained and used to simulate the flexible characteristics of the TOPEX satellite. Five of the seven modes are associated with the bending and tortional vibration of the solar array. The damping ratio for all modes is assumed to be 0.001.

Simulation results for the TOPEX spacecraft attitude pointing performance during NMM local vertical flight are described in this section. Only the roll axis will be shown since this axis was expected to violate the NMM pointing requirment. Fig. 7 shows the roll axis attitude error beginning at local 6 am for $\beta' = 0^\circ$ over one orbit. This error represents the controlled response of the vehicle coupled with the environmental disturbance torques acting on the spacecraft. The initial transient is due to simulation initialization and should be ignored. The TES disturbance is clearly seen in Fig. 7. The roll axis reaches a peak error of -411.10 *arcsec* during sunset ($t = 4000\ sec$) and 405.5 *arcsec* at sunrise ($t = 6100\ sec$). From Fig. 7, the roll axis attitude error violates the 144 *arcsec* attitude control requirement for a period of about 153.6 *sec*. The pitch and yaw attitude signals remain below their respective error limits throughout the orbit. The altimeter antenna boresight pointing error is estimated by performing a Root Sum Squared (RSS) of the roll and pitch axes errors. The boresight error reaches a maximum value of 418.46 *arcsec* at

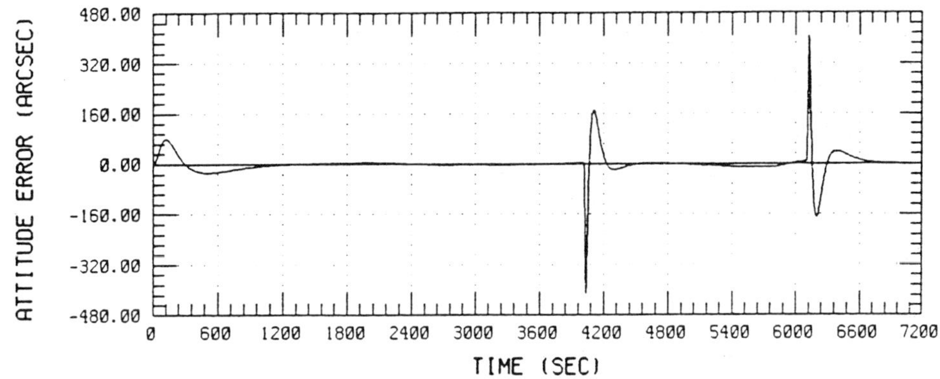

Fig. 7 TOPEX Roll Axis Attitude Error for $\beta' = 0°$

sunset and 420.76 *arcsec* at sunrise and never exceeds the nominal operating range of 0 to 504 *arcsec* (see Table 1). Derived rate error for the roll axis is presented in Fig. 8. Comparison of Figs. 2 and 8 reveals a identical structural response. The vehicle roll, pitch and yaw attitude errors, after the 180° yaw turn around maneuver ($\beta' = -0°$) are quite similar, both in magnitude and structure, to the case when $\beta' = 0°$ except the yaw axis signal reverses sign.

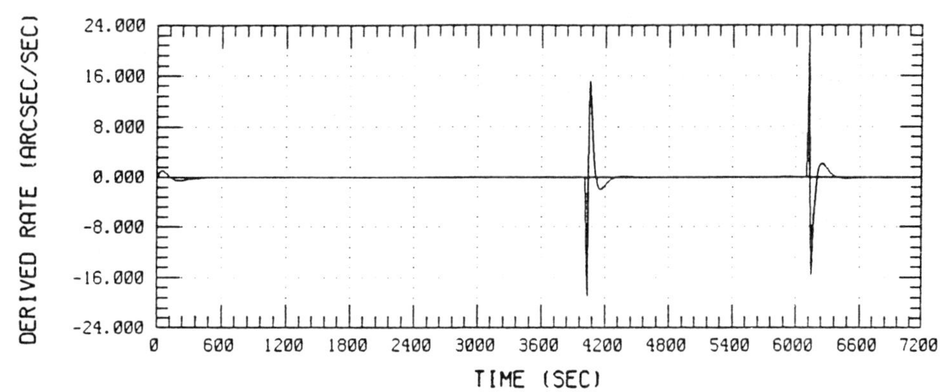

Fig. 8 TOPEX Roll Axis Derived Rate Error for $\beta' = 0°$

The TOPEX roll attitude error during yaw steering mode, for $\beta' = 15°$, is shown in Fig. 9. As in the local vertical mode, this figure is for one orbit period and begins at local spacecraft 6 am. The characteristic of the roll reponse at sunrise and sunset remains consistent with the roll attitude error presented in Fig. 7. However, a small amplitude sinusoidal variation is present during non-TES portions of the orbit which is attributed to coupling with the yaw axis. During this control phase, the roll axis again temporarily exceeds the NMM attitude control requirement for

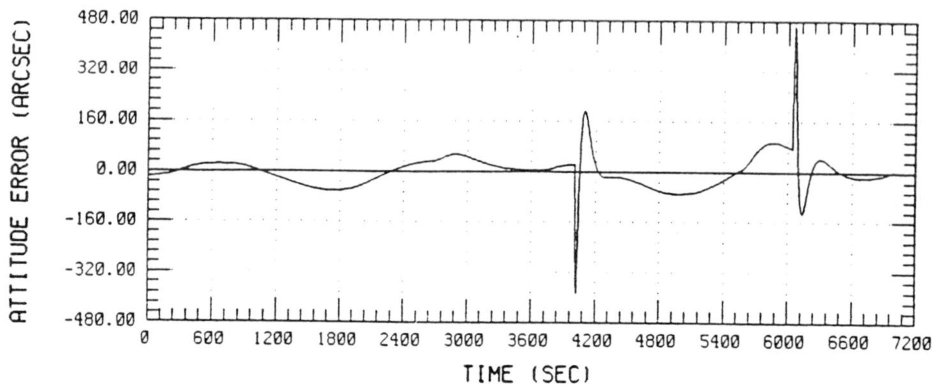

Fig. 9 TOPEX Roll Axis Attitude Error for $\beta' = 15^\circ$

about 127.0 *sec*. The radar altimeter antenna boresight pointing error reaches a peak of 390.94 *arcsec* and 475.57 *arcsec* at sunset and sunrise respectively. This pointing error remains within the nominal operating region of 0 to 504 *arcsec* for the entire orbital period. For the case when $\beta' = -15^\circ$, the attitude errors are similar to the errors for $\beta' = 15^\circ$ except the sinusoidal variation is 180° out of phase.

For both modes, local vertical and yaw steering, the TOPEX satellite responds in a well damped manner when subjected to the TES disturbance. The vehicle takes approximately 500 *sec* to completely recover from the applied TES perturbation. The short duration and large magnitude of the impulsive disturbance dominate the response time of the ACS and the torque authority present in the RWA's. However, for regions of the orbit where the TES disturbance is absent, the TOPEX ADCS is able to compensate for the other environmental disturbances without introducing attitude errors which exceed the NMM attitude control requirements. Furthermore, the combined roll and pitch axes attitude deviations are small enough to prevent the radar altimeter from experiencing any degraded performance.

TES CONTROL SYSTEM

To counter the TES disturbance torque, a localized active control system (i.e actuator, controller and sensors) will be mounted on or near the disturbing member. Fig. 10 illustrates this concept by showing a typical satellite solar array structure. The sensors are rigidly attached to the panel surfaces while the controller and actuator are placed in the yoke of the array. This design is advantageous in that the flexibility problems associated with non-colocated sensors and actuators is avoided. The proposed control system is a combination of hardware and software. The hardware components consist of a distributed thermal sensor network, a micro-computer and a small RWA. The thermal sensor network will be strategically mounted on each face of the solar panel as shown in Fig. 10. The flexibility built into the software design will allow on-orbit adjustments to accommodate the true nature of the TES

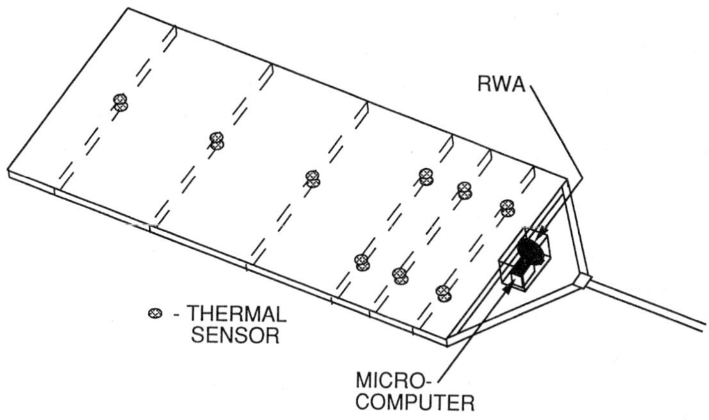

Fig. 10 Solar Array with TES Control System

disturbance torque. The complete system block diagram is depicted in Fig. 11. An inspection of Fig. 11 shows that the system uses the thermal sensor network to provide temperature measurements at various locations on the solar array. The measurement data are then fed into an optimal estimation routine based on the Gauss-Markov theorem[7]. An optimal estimate of the thermal gradient that exists across the panel at a desired location is then used to predict the magnitude of TES

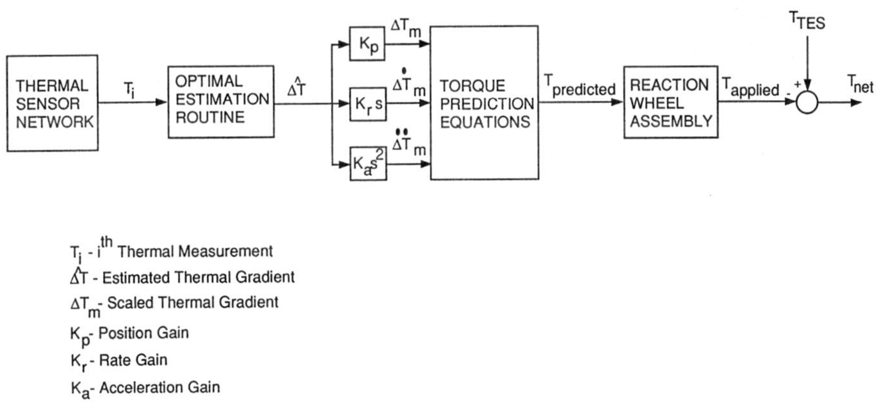

Fig. 11 TES Control System Block Diagram

disturbance torque using the detailed analytical model. This control scheme is not an "open-loop" prediction technique due to the fact that thermal sensor feedback is employed. The generated torque prediction is used to drive the small RWA in response to the TES disturbance. Temperature gains, K_p, K_r and K_a, have been included to tailor the system response and the digital implementation will allow for the aforementioned on-orbit adjustments.

The localized control system is demonstrated using the TOPEX NMM three-axis, non-linear, time-domain simulation. The optimal estimation routine is not implemented in the simulation. Instead, a thermal gradient estimate is assumed to already have been provided by the routine. The TES disturbance torque equations are used to compute a torque command to the array RWA. The array RWA is assumed to operate with a momentum range of $\pm 20\ N-m-sec$ and provide a torque capability of $\pm 0.2\ N-m$. The time constant for the thermal sensors is assumed negligible and the execution rate of the localized controller is 0.064 sec (15.625 Hz). Only the local vertical mode for $\beta' = 0°$ will be used to show the theoretical feasability of this control system concept. In particular, the roll axis attitude error will be examined. Initially, the software gains K_p and K_r are set to unity. If the gain K_a is also initialized to one, the attitude error response of the TOPEX spacecraft is worse than if no controller was used. The magnitude of the roll and pitch attitude deviations cause the altimeter pointing error to enter the degraded region of operation. This is because the solar array control system is not able to remove the impulsive portion of the TES disturbance, but can entirely accommodate the decaying step characteristic. Thus the vehicle base-body ACS must provide all the torque authority to return the vehicle to a zero error state. If on the other hand the gain K_a is set to -1.0, the control system will act to minimize the impulsive disturbance and magnify the exponentially decaying step disturbance. This approach takes advantage of the step disturbance to help pull the vehicle back to zero error. Similarly, if the acceleration gain is further decreased, the peak attitude error response attributed to the impulsive disturbance continues to decrease, but the error induced by the amplified decaying step begins to grow and will eventually exceed the antenna pointing requirement. Thus the gain can be optimized to provide the smallest NMM attitude control requirement violation period for the roll axis. Fig. 12 shows a plot of the roll axis control requirement violation period, for sunset, as

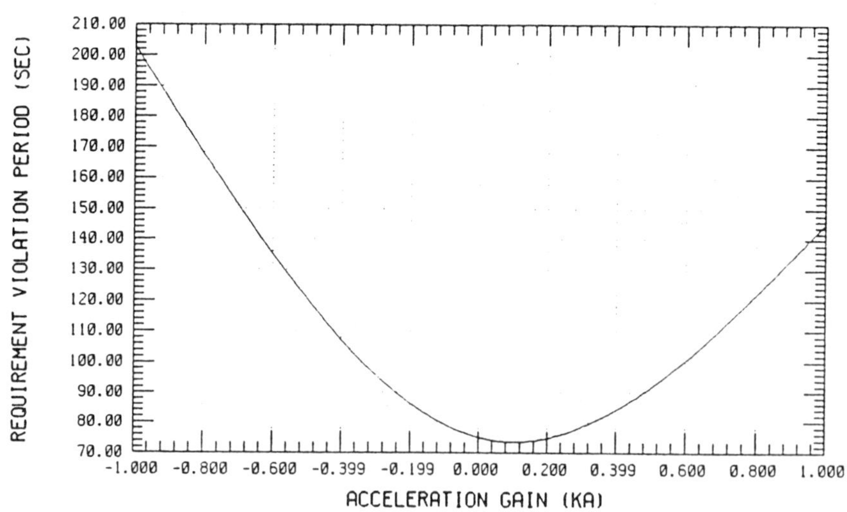

Fig. 12 Roll Axis Attitude Violation Period Versus K_a for Sunset

a function of K_a. A value of 0.07 corresponds to the shortest violation period for the roll axis attitude error and is therefore selected as the acceleration gain. Using this gain, the peak roll axis attitude errors for sunset and sunrise are -390.05 *arcsec* and 384.30 *arcsec* respectively. This represents a decrease of roughly 20.0 *arcsec* in comparison with Fig. 7. The antenna pointing error now reaches a peak error of 397.11 *arcsec* at sunset and 398.85 *arcsec* at sunrise. The magnitude of these peak pointing errors indicates that the altimeter will not experience any degraded performance. Unfortunately, the localized control system does not seem to provide any additional benefit, for this example, other than a slight reduction in the roll peak attitude error.

Now suppose that the torque capability of the solar array reaction wheel is increased by an order of magnitude to 2.0 $N-m$. Although this torque capability is not currently available in any RWA, it is a good theoretical example to demonstrate the full potential of the local control system. Since the RWA torque authority is not equivalent to the magnitude of the TES impulsive torque, the acceleration gain, K_a, is still needed to aid in the vehicle recovery from the TES disturbance. In a similar fashion to the previous case, the acceleration gain is selected to be 0.5. For this situation, the roll axis attitude error never exceeds ± 120.00 *arcsec*, as shown in Fig. 13, and thus stays within the NMM attitude control requirement of 144

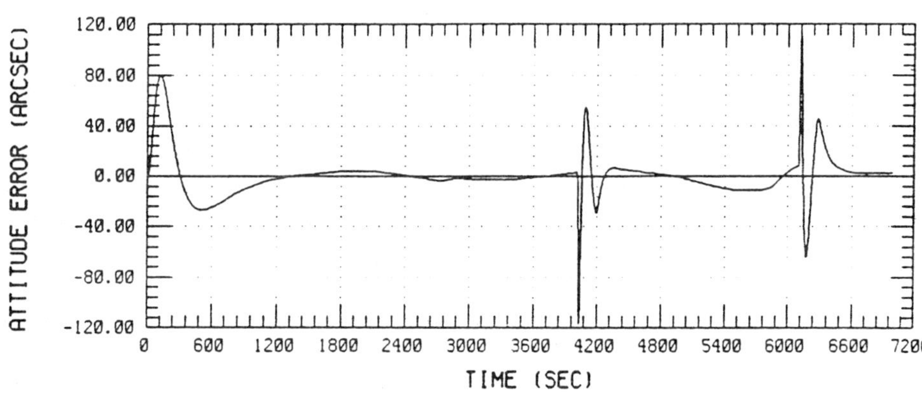

Fig. 13 TOPEX Roll Axis Attitude Error for $\beta' = 0°$ ($K_a = 0.5$)

arcsec. Furthermore, the peak altimeter pointing error decreases to 120 *arcsec*. Thus the proposed control system has the ability to suppress the adverse effect of the TES disturbance torque for a TOPEX spacecraft given a sufficient amount of torque capability. In a more general sense, the control system could minimize the effect of the TES disturbance for any spacecraft given the proper sizing of the torque actuator.

CONCLUSION

Comparison of the simulated TOPEX roll axis derived rate error for local vertical mode $\beta' = 0°$ with the LANDSAT-4 roll axis derived rate telemetry data shows excellent agreement in structure but differs in magnitude by about 10 *arcsec*. The magnitude offset can be attributed to specific differences between the TOPEX and LANDSAT vehicles. For instance, the TOPEX spacecraft roll and yaw axes inertias are approximately three times larger than the LANDSAT-4 roll and yaw inertias. Furthermore, the TOPEX solar array is 35% thicker with an inertia that is six times larger than the LANDSAT-4 array inertia. Since these dissimilarities could easily account for the magnitude deviations, it is concluded that the TES disturbance torque model presents a valid mathematical representation of the physics behind the TES phenomena.

Numerical simulations predicting the attitude pointing performance of the TOPEX spacecraft indicate that the vehicle will exceed its roll axis NMM attitude control requirement during penumbral transitions when the TES disturbance is applied. The roll axis is expected to violate the acceptable attitude error limit of 144 *arcsec* for a maximum period of approximately 150 *sec*. Fortunately, the TOPEX satellite ACS responds in a well-damped manner to null the TES disturbance and does not introduce attitude deviations that exceed the 504 *arcsec* limit which would cause the altimeter to experience degraded performance. The other environmental disturbances produce a negligible effect on the vehicle pointing performance.

Based on the numerical results for the TOPEX attitude pointing performance with the inclusion of the localized array controller, it can be concluded that the proposed control system presents a viable solution for accommodating the TES disturbance. Simulations showed that the roll axis control requirement violation period was not significantly altered when the torque authority of the RWA was 0.2 $N - m$. If the torque capability of the RWA is increased by an order of magnitude, theoretical results indicate that the vehicle axes, in particular the roll axis, will never exceed their attitude control requirements when the spacecraft is subjected to the TES disturbance. However, using a 2.0 $N - m$ RWA is not a practical solution since the torque capability of the largest available RWA is only 0.8 $N - m$. If the 2.0 $N - m$ RWA is assumed to represent an upper bound, using a reaction wheel with a torque authority of about 0.6 $N - m$ could appreciably reduce the impact of the TES disturbance for a spacecraft similar to the TOPEX vehicle.

ACKNOWLEDGMENT

A portion of the work described in this paper was performed for the Jet Propulsion Laboratory, California Institute of Technology, sponsored by the National Aeronautics and Space Administration.

REFERENCES

1. Jasper, P.E. and S. Neste, "UARS Solar Array Snap," *General Electric Space Division-Inter Office Memo*, U-1K21-UARS-481, July 9, 1986.

2. Vigneron, F.R. and R.A. Miller, "Flight Performance of the Stabilization System of the Communications Technology Satellite," *Journal of Guidance and Control*, Vol. 1, No. 6, pp. 404-412, 1978.

3. Asker, J.R., "Hubble Beset by Radiation Effects, Solar Array Vibrations, But Controllers Expect Early Resolution of Both Problems," *Aviation Week and Space Technology*, Vol. 132, No. 25, pp. 26, June 18, 1990.

4. Stewart, R., L.-L. Fu and M. Lefebvre, *Science Opportunities from the TOPEX /POSEIDON Mission*, Jet Propulsion Lab (JPL) Publication 86-18, JPL, Pasadena, California, 1986.

5. Kaplan, W., *Operational Methods for Linear Systems*, Addison-Wesley Publishing Company, Massachusetts, 1962.

6. Dennehy, C.J., B.B. Lee and R.V. Welch, "Controls/Structures Interaction Analyses and Simulation for the TOPEX/POSEIDON Satellite Normal Mission Mode Digital Controller," *Dynamics of Flexible Structures in Space*, Cranfield Institute of Technology, Cranfield, Bedford, U.K., May 15-18, 1990.

7. Zimbelman, D.F., R.V. Welch and G.H. Born, "Optimal Temperature Estimation for Modeling the Thermal Elastic Shock Disturbance Torque," *Submitted to Journal of Spacecraft and Rockets.*

Table 1
EXPECTED RADAR ALTIMETER PERFORMANCE AS A FUNCTION OF ANTENNA POINTING ERROR

Antenna Pointing Error ($arcsec$)	Expected Radar Altimeter Performance
$0.0 \rightarrow 504.0$ ($0.0^\circ \rightarrow 0.14^\circ$)	Normal operation
$504.0 \rightarrow 1080.0$ ($0.14^\circ \rightarrow 0.3^\circ$)	Some performance degradation
$1080.0 \rightarrow 1800.0$ ($0.3^\circ \rightarrow 0.5^\circ$)	Serious performance degradation for up to 100 second spans of data
> 1800.0 ($> 0.5^\circ$)	Probable loss of altimeter lock. Definite data loss or systematic altitude errors for up to 100 second spans

Table 2
TOPEX SOLAR ARRAY PROPERTIES

Definition	Symbol	Value
Length	L	7.7216 (m)
Width	W	3.302 (m)
Thickness	d	0.034925 (m)
Mass	M	215.55 (kg)
Density	ρ	2702.0 ($\frac{kg}{m^3}$)
Thermal Conductivity	K_o	1.3056 ($\frac{W}{m \cdot K}$)
Specific Heat Capacity	c_p	896.0 ($\frac{J}{kg \cdot K}$)
Coefficient of Thermal Expansion	α_{cte}	23.4×10^{-6} (K^{-1})
Solar Absorptivity	α_s	Front 0.79
		Back 0.15
Emissivity	ϵ_s	Front 0.81
		Back 0.86

Section IV
INTERNATIONAL SPACE PROGRAMS

SESSION IV

Chairperson:	Jacques Busseuil Aerospatial/Strategic and Space Systems Division
Co-Chairperson:	Lyle Jenkins NASA Johnson Space Center
Local Chairperson:	Jim Taylor Martin Marietta Astronautics

The following paper numbers were not assigned:

AAS 91-067 to -069

AAS 91-060

EVALUATION OF FLIGHT RESULTS OF THE ROSAT ATTITUDE MEASUREMENT AND CONTROL SYSTEM

W. Schrempp, A. Böinghoff, E. Brüderle, N. Gradmann, H. Herbst, L. Kaffer and W. Oesterlin*

ROSAT, launched on 31th of May 1990 has opened a new era of x-ray astronomy. Up to now it has performed an all-sky survey on x-ray sources and will additionally perform detailed investigations of some thousand targets by pointing its telescopes on them. The attitude measurement and control system was designed to work autonomously to a large extent with a measurement accuracy of a few arcsec and a control accuracy of a few arcmin. The paper describes the first eight successful months of operation in space.

SURVEY OF THE ROSAT ATTITUDE MEASUREMENT AND CONTROL SYSTEM AND ITS DEVELOPMENT

The ROSAT Mission

ROSAT is a large astronomical satellite designed to observe X-Ray sources. It was built as part of a cooperative U.S.-German-U.K. program with primary responsibility for manufacture of the spacecraft falling to German industry contracting to the German Research and Testing Institute for Aeronautics and Astronautics (DLR). It was financed by the German Ministry of Research and Technology and NASA who provided the launch vehicle. The spacecraft carries German, American and British payloads, the main telescope consisting of a Wolter mirror built under the scientific direction of the Max-Planck Institute. Responsibility for the Attitude Measurement and Control Subsystem (AMCS) was with Messerschmitt-Bölkow-Blohm, who carried out design, development and test of the subsystem as a whole, and also manufactured or procured the measurement and control hardware. Prime contractor for the entire spacecraft was Dornier System Company.

The scientific mission consists of 2 different operational modes which also define the essential tasks of the attitude measurement and control system.

o Scan mode (sky survey)

  In this mode one axis of the S/C is oriented to the Sun and the S/C rotates around the Sun line with a scan rate equal to the orbital rate. The telescopes mounted perpendicular to the rotation axis scan the sky along a great circle. Due to the apparent motion of the Sun of 1°/day with respect to the stars, a complete survey of the celestial sphere is performed within half a year (180 days).

* Messerschmitt-Bölkow-Blohm GmbH., Space Communications and Propulsion Systems Division, W-D-8012 Ottobrunn, Germany. Mailing Address: Postfach 801169, W-D-8000 München 80, Germany.

o Pointing mode

After the sky survey long-term observation of selected targets will be performed with high pointing accuracy and at the end of each observation period wide-angle slew maneuvers are performed to the next target.

The low circular orbit of 580 km altitude and the use of a single ground station result in ground contact times of a few minutes separated by periods which can last up to many hours, so that nearly all operations must be performed autonomously. In addition, provision must be made for the storage of large numbers of telecommands with time reference (time-tags) and the position and magnitude data for several hundred stars.

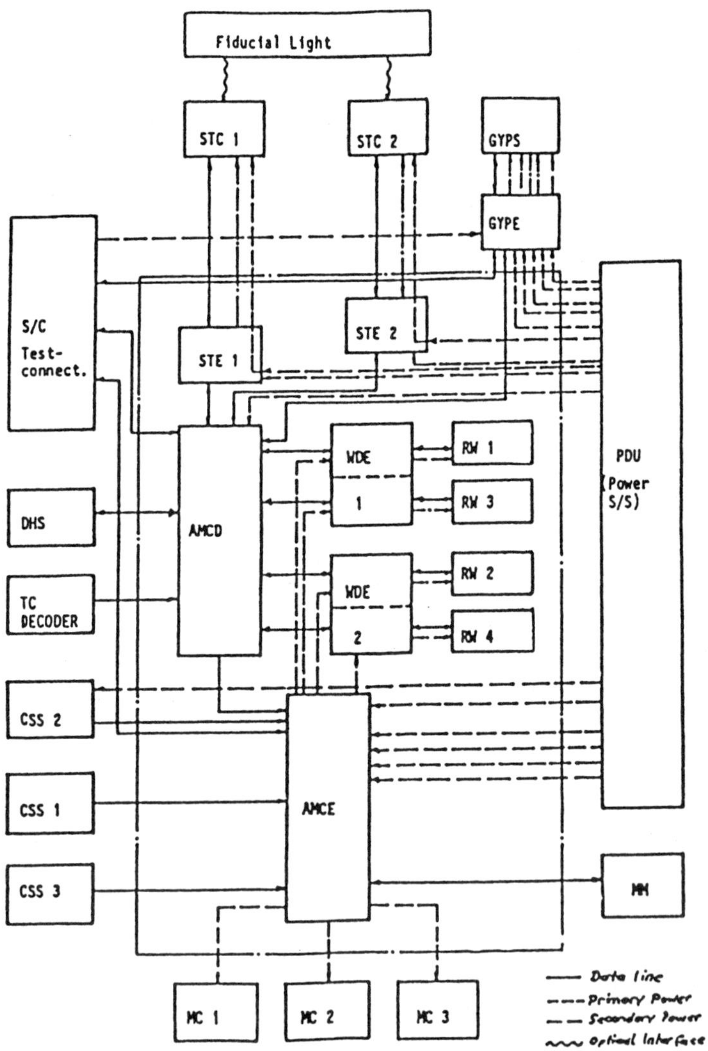

Figure 1 AMCS block diagram: (ST = Star Tracker, GYP = Gyro Package, RW = Reaction Wheel, WDE = Wheel Drive Electronics, AMCD = Attitude Measurement and Control Data System, AMCE = Attitude Measurement and Control Electronics, CSS = Coarse Sun Sensor, MM = Magnetometer, MC = Magnetic Coil).

System Description

The attitude measurement and control system block diagram is shown in Figure 1.

Basis of the attitude measurement system is the gyro package and the star sensors. Three coarse Sun sensors are mainly for acquisition purposes. A magnetometer measures the Earth magnetic field in case computation is not possible. The alignment between the telescope and each star tracker is measured with high accuracy by an alignment control system which projects artificial stars (fiducial light stars) from the telescope focal plane into the star trackers.

Due to the high moments of inertia and the large amount of slew maneuvers, reaction wheels are used as torquers. They are desaturated by magnetic coils which react with the Earth magnetic field.

The connection between the different equipment is made by the AMCD and the AMCE. The AMCD is the microprocessor based onboard computer (OBC) which contains the AMCS S/W in PROMS. A large RAM area is used for computation and also provides means for additional programs which can be commanded from ground to supplement the onboard S/W where necessary. The AMCE contains the non-digital ancillary functions of the system (amplifiers, switching circuitry etc.). The technical parameters of star tracker and gyro which are mainly responsible for the measurement and control accuracy are as follows:

o Star sensors:

  The two star sensors (manufacturer MBB/SIRA) have a field of view of 5.9° x 4.4° each and a magnitude range of $m_V = +6.5$ to 0. Both star sensors are aligned $\pm 3^{\circ}$ apart from the telescope axis so that their fields of view are overlapping in a small region. Each star sensor is able to search and track 3 real stars and 1 fiducial light star simultaneously, the fiducial light stars serving as a means for in-flight measurement of misalignments between telescope and star sensor axes. The measurement accuracy is characterized by the following figures:

  Pointing mode:

  Systematic error - $\leq$ 1.7 arcsec

  Noise equivalent angle - $\leq$ 0.5 arcsec

  Scan mode:

  Systematic error - $\leq$ 10 arcsec

  Noise equivalent angle - $\leq$ 5 arcsec

  Resolution - $\leq$ 0.35 arcsec

  Magnitude measurement accuracy - $\leq$ 0.25

o Gyro package:

  The gyro package consists of four high-precision single axes rate integrating gyros (manufacturer FERRANTI), where three gyros are aligned along the S/C axis and the fourth one in the diagonal output with a resolution of 4/3 arcsec/pulse, a fine resolution between $\pm$ 4/3 arcsec is obtained by a 12 bit A/D conversion of an analog output.

Further important figures are 0.2% scale factor accuracy, 1°/h untrimmed maximum drift rate, 0.005^{o}/h in 1 hour and 0.01^{o}/h in 24 hours random drift. In normal operation all four gyros are operating and the output of three gyros is selected automatically for control purposes.

Operational Modes and Failure Detection, Isolation and Recovery

o Safe mode

The safe mode is designed to protect the spacecraft by automatically acquiring the Sun and maintaining the solar array normal to the Sun line so that power and thermal requirements for the spacecraft are met.

o Standby mode

The standby mode provides the transition from safe mode to pointing and scan mode, which are the normal operational modes of the system and used for experiment data collection.

o Scan mode

This is considered to be the nominal operational mode during the first six months of the satellite lifetime. The S/C rotates at a scan rate of about 4 arcmin/sec around its x-axis which is oriented towards the Sun. Thus, after half a year the traces of the experiment cover the complete celestial sphere.

o Pointing mode

When the satellite has scanned the complete sphere, after about six months it will be operated in the pointing mode for the remaining nominal lifetime of about one year. In this mode the satellite has to acquire selected orientations and maintain them within a pointing accuracy of some arcmin.

o Degraded mode

Due to the fact that all the S/W is stored in PROM in the AMCD no modification of this S/W is possible during the mission. Therefore one additional mode is implemented whose S/W can be loaded from the ground station into the RAM in order to be flexible enough to modify the S/W due to unforeseen circumstances. At the end of the development phase a nominal degraded mode was already defined being the pointing mode with a triangular motion in x- and y-axis with commandable amplitude and period (wobble mode).

o Automatic Failure Detection, Isolation and Recovery

Due to the fact that the satellite is operated from one ground station only, the maximum contact time per orbit is only about 8 minutes with a maximum possible period of 16 hours without contact. Automatisms are therefore implemented for the AMCS to be able to survive failure cases.

AMCS Development

After several pre-studies, the final concept of the AMCS was established and detailed during Phase B2 in the second half of 1983. The high requirements of the system concerning measurement accuracies and operational aspects necessitated a large amount of development work in hardware, software and system testing.

Most of the equipment was already available or could be adapted to the ROSAT requirements. The star tracker, however, had to be newly developed since no equipment was available on the European or U.S. market for the application foreseen. After a competition phase, MBB together with Sira was entrusted with the development and manufacture of the ROSAT Star Tracker at the end of 1983. Together with the star tracker a high-precision test bench for sub-arcsec-testing had to be developed and set up at MBB. A fully functional EM of the star tracker was available at the end of 1985 and the qualified flight models of the sensor were delivered in spring 1988. Seeing the success and excellent in-orbit performance it can be said that the star tracker is one of the highlights of the ROSAT program.

An important contributor to the successful AMCS performance was also the comprehensive testing on subsystem level of the hardware and software. Subsystem testing was performed on 2 levels: static and dynamic testing. The static testing was performed partly closed loop with a special subsystem test equipment. This subsystem test equipment and the test algorithms developed and used in subsystem testing were later also employed in the test on system level. Another important part of the subsystem testing was done on a special dynamic bench test facility. Here the sensors were mounted on a 3-axis servo table and optically stimulated by an artificial Sun and artificial stars which allowed a subsystem operation close to the actual orbital and operational conditions. The automatic test sequences of static and dynamic testing formed also a useful basis for the establishment of the operational flight plan.

Static and dynamic testing of the EM took place in August 1986 and the FM testing with the complete flight hard- and software at the end of 1987.

ROSAT LAUNCH AND PREOPERATIONAL PHASE

The ROSAT groundstation operations are performed from Oberpfaffenhofen, Germany, by the German Space Operation Center (GSOC). The prelaunch training for the operations took place with a real time ROSAT simulator with a H/W-OBC in the loop bringing all participants and interested spectators the feeling of a careful preparation. On June 1, 1990, about 10 minutes before midnight, ROSAT was launched with a DELTA-II rocket from Cape Canaveral. The picture-perfect launch brought the spacecraft into its 580 km altitude, 53° inclination orbit. For the first weeks of operation telemetry data from the following groundstations could be made available at GSOC when the S/C was in their visibility range:

- Madrid/Weilheim
- Canberra
- Goldstone

with contact durations for a groundstation being about 8 minutes and contacts every 30 to 90 minutes for periods of about 8 hours per day followed by 16 hours without any groundstation coverage.

Prior to and during ascent GSOC had contact to the S/C via Cape Canaveral, 18 minutes after launch, Weilheim had the first direct contact when the S/C was still connected with the last rocket stage, 20 minutes later the Canberra contact showed the S/C in the expected safe mode with the proper orientation of the -x-axis to the Sun. The attitude control was in steady state with:

- o Attitude errors about y- and z-axis shown by the Sun sensor being in the order of some hundredth of a degree
- o Commanded wheel torques around zero
- o Low wheel rates
- o Mode rate coil operation for the wheel unloading

Of course this S/C-status led to a relief of the tension for all experts. ROSAT was in a safe situation and further operations could be envisaged without pressure.

Two weeks had been foreseen to place in operation all ROSAT control modes for the various mission phases. The first contact period in the night of the launch already brought a series of surprises and excitations but also successes. As a first maneuver a switch to the standby mode was commanded where the star sensors should be put into operation. The following maneuvers were as foreseen in the sequence of events:

- o Automatic star acquisition with transition to the pointing mode
- o Identification of the acquired stars by the groundstation
- o Command of 5 catalog stars and inertial attitude.

The first successful operations, were, however, followed by disappointment: After some control motion, all acquired stars were lost. For analysis purposes standby mode was again commanded. Additionally, the gyro drift biases which had been determined meanwhile and the eclipse surveillance were activated. An automatic switch to the safe mode after about an hour left the groundstation again with a lot of problems at the end of the contact period. The reason for the star loss after acquisition was explained by misalignment between the two star sensors which had not yet been determined. A corrective bias command was foreseen but not given. It was therefore decided to make the acquisition and operate in the pointing mode with only one star sensor. No explanation could be immediately found for the switch to safe mode and some OBC memory words were selected for the telemetry to ease the further search. Some days passed without visible progress. It was not possible to come to the pointing mode with inertial reference derived from stars. Investigations revealed the following intermediate results:

- o Automatic switch to safe mode: The reason was twofold: The Sun sensor delivered only 8.1 Volts as Sun presence signal instead of the specified 8.4 V as a minimum. In the onboard-S/W erroneously a limit of 8.2 V for Sun presence was programmed instead of the specified 7.0V.
- o Unsuccessful star acquisition: Improperly compensated gyro drifts were identified as the main reason.

In the contact period of the 7th of June, a great success was achieved:

- The gyro drift bias was determined again and the respective commands were sent. It turned out that with bias commands for x, y and z gyros of 0.062°/h, -0.062°/h, and 0.05°/h the remaining drift was smaller than 0.005°/h revealing the excellent performance of the gyros.

- With the proper gyro bias commands the star acquisition was now successful. A star catalog and inertial reference was commanded. The S/C was for more than one orbit period in the pointing mode deriving inertial reference from measured and identified stars.

- A scan catalog was commanded and the scan mode was entered from the pointing mode. It operated properly from the very beginning.

This enormous step forward during one night was still increased on the next day: After determination of the misalignment between the star sensors and respective correction via commands, the second star sensor was also activated for the control. The AMCS was now fully operable and the S/C was left now for several days in the scan mode.

No real setback in the ROSAT control took place in the following days. On the 12th of June the scan mode was left and pointing mode was entered again. The wobble mode was loaded and commanded. It worked properly. At first 50° slew was performed successfully. From further slews the gyro scale factors were determined and compensated.

Thus, the GSOC status report for ROSAT stated on 21st of June 1990, that the attitude measurement and control subsystem had an excellent behavior and that even big slew maneuvers of 50 and 180 degrees all went extremely well.

ANALYSIS OF TELEMETRY DATA

For an assessment of the attitude control of the nominal operational modes telemetry data for three typical modes of operation have been analyzed:

- A large angle slewing maneuver followed by steady state pointing
- A steady state scan operation
- A steady state wobble pointing.

Measurement Accuracy

For the slewing maneuvers but also for steady state periods without star reference especially in the scan mode the attitude accuracy is dominated by the gyro drift and scale factor values. With the commanded bias values mentioned in the previous section, a compensation better than 0.005°/h could be achieved, which is by far sufficient from the accuracy point of view. Even better values are only limited by the resolution of the commandable values. From large-angle slewing maneuvers, a scale factor 0.9979 was determined and compensated for the x gyro. With this value the slewing maneuvers can also be performed with a higher accuracy than required for a proper star acquisition.

The measurement accuracy in pointing and scan mode is mainly determined by the noise equivalent angle of the star sensors. Figures 2 and 3 show the noise history of a star measurement in the two axes for the pointing mode. In this figure the differences of two successive measurements are plotted which are representative then for the difference of two independent noise values. The following variances are then obtained for the sensor noise in pointing mode: σ_x = 0.45 arcsec, σ_y = 0.50 arcsec, which confirmed the expectation that the accuracy is higher than specified. Figures 4 and 5 give the noise for the scan mode where the same procedure as for the pointing mode is applied with additional consideration of the scan motion. For the scan mode the variances σ_x = 0.8 arcsec, σ_y = 0.9 arcsec are obtained which are also much lower than specified. The spikes in Figures 4 and 5 represent sensor blemishes (known imperfections in the star tracker CCD) which are marked in the measurement output not to be taken for attitude reference purposes.

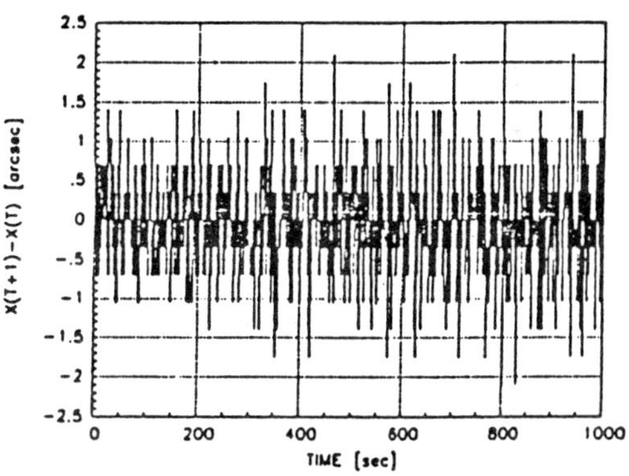

Figure 2 x-axis star sensor measurement noise in pointing

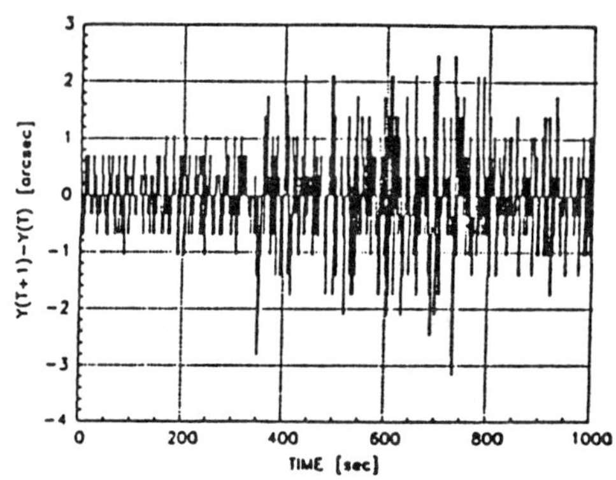

Figure 3 y-axis star sensor measurement noise in pointing

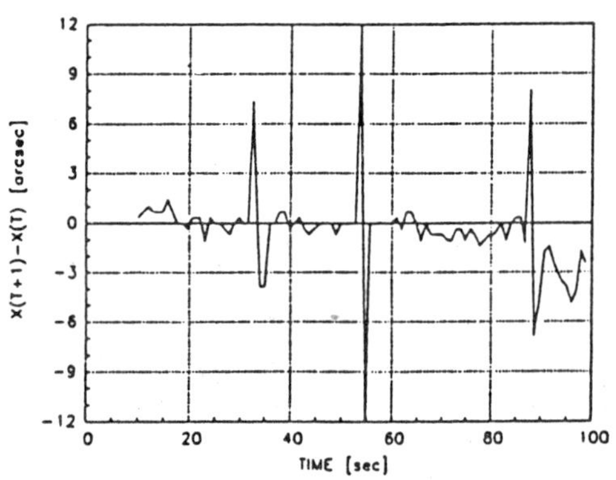

Figure 4 x-axis star sensor measurement noise in scan

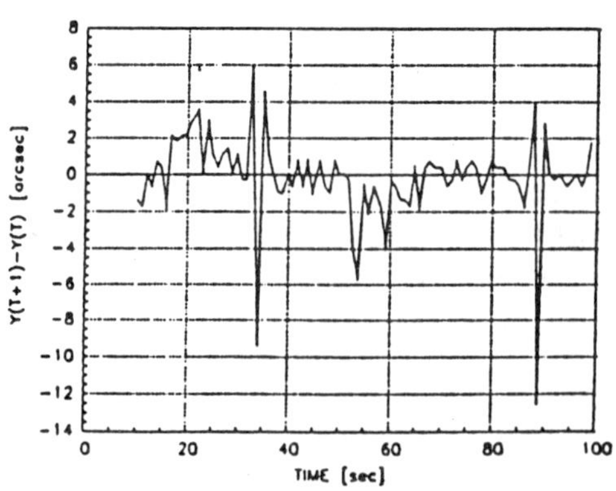

Figure 5 y-axis star sensor measurement noise in scan

Star Sensor Magnitude Measurement

From SKYMAP star catalog information (as spectral class and visual magnitude) an instrumental magnitude related to the ROSAT star sensor measurement has to be calculated, which has to be commanded together with the star positions of the operational pointing and scan mode catalogs. For such predicted star magnitudes a comparison has been made which actually measured magnitudes. Such measurements were obtained from the standby mode with a commanded scan rate. The measured stars were identified and the obtained magnitudes allowed a comparison with the predicted ones. It turned out the the prediction formula was fairly accurate. The maximum determined deviation was about $\Delta m = 0.56$ where the measurement resolution for the magnitude is given by $\Delta m = 0.25$. In general the deviation was only in the order of one quantization step, which is considered as an excellent result. Nevertheless, from the obtained measurements the formula for the prediction of the star magnitudes for the operational star catalogs was slightly adapted.

Control Accuracy

In Figures 6-9 the attitude and rate errors are given for the selected pointing mode and a scan mode steady state period. In pointing mode the attitude error does not exceed 0.6 arcmin in all axes, which is very small compared with the allowed 6 arcmin in x- and y-axis (3 σ values). These attitude errors include the influence of wheel stiction torques which can easily be identified from the wheel rate zero crossings in Figure 6. The S/C rates are always less than 1 arcsec/sec in the sensitive axes x and y and less than 3 arcsec/sec in z-axis.

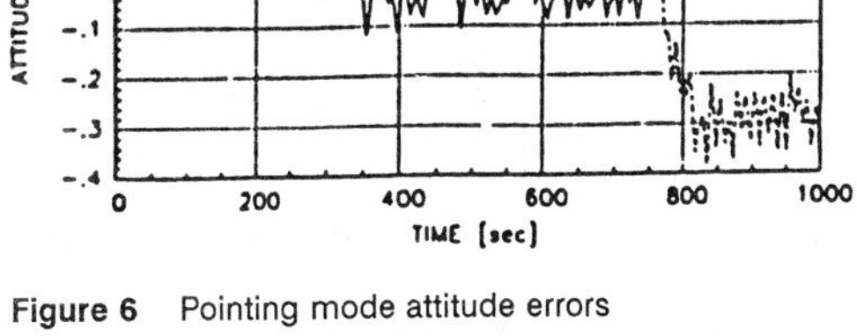

Figure 6 Pointing mode attitude errors

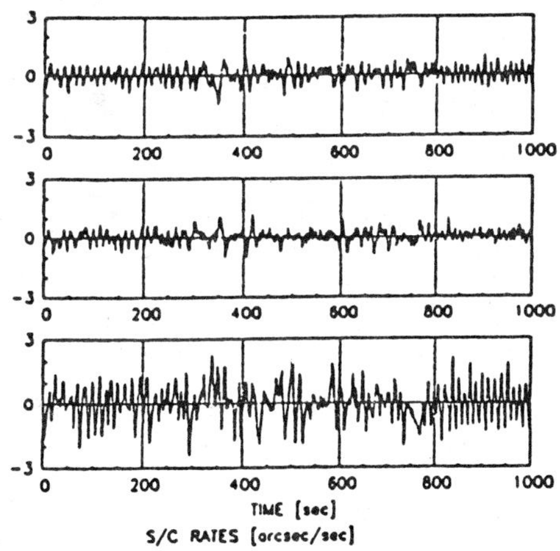

Figuro 7 Pointing mode rate errors

In the scan mode the attitude errors are less than 1.5 arcmin, which is also very small compared with the specified values of 15.8 arcmin for x- and y-min and 1.6 deg for z-axis (3 σ values). The rate errors in x- and y-axis are less than 3 arcsec/sec for x- and y-axis and less than 10 arcsec/sec for the z-axis which can be considered as a demonstration of the excellent performance.

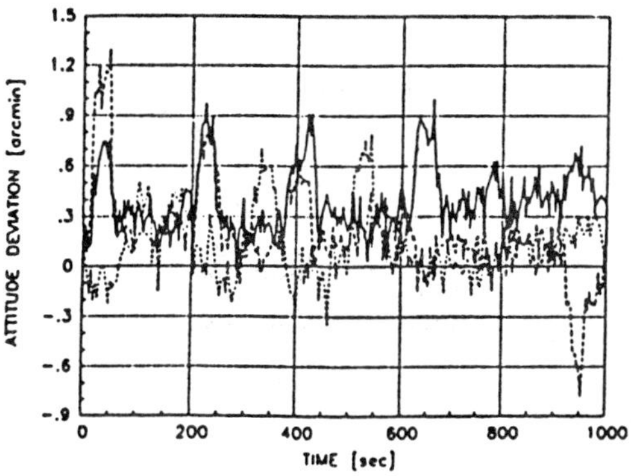

Figure 8 Scan mode attitude errors

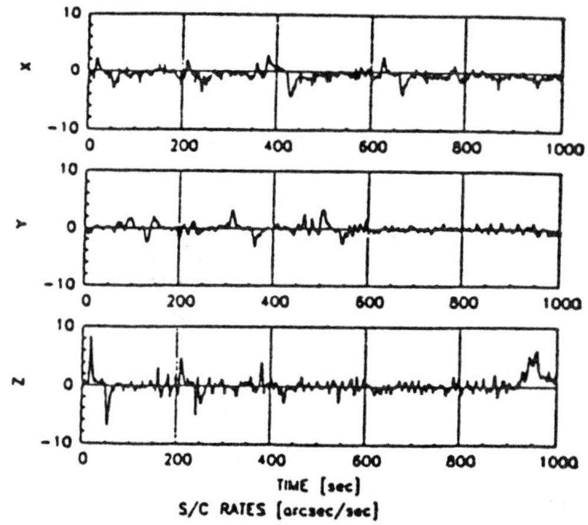

Figure 9 Scan mode rate errors

In Figure 10 NS gives for the scan mode case the number of stars which are available for attitude reference generation. It can be seen that for most of the time two stars are available, but there are also short periods where only one star or even none is available. SSC shows the steady state condition which is always given. EC indicates that no Earth cone violation has occurred.

In Figure 11 the S/C attitude with respect to a commanded pointing direction is shown for a wobble mode pointing. The commanded triangular wobble motion for x- and y-axis with an amplitude of 2 arcmin and a period of about 400 sec can easily be identified. The attitude error with respect to the commanded orientation is also by far within the accuracy requirement.

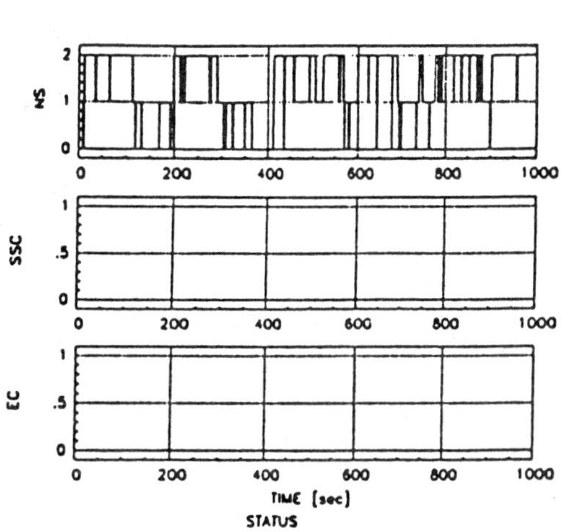

Figure 10 Scan mode: Number of reference stars and status signals

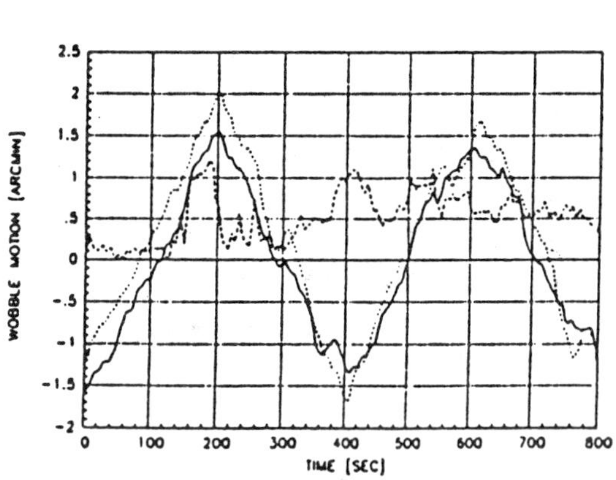

Figure 11 Wobble mode attitude with respect to reference target

Magnetic Unloading of the Reaction Wheels

Figures 12 and 13 show the reaction wheel rates for the selected periods of pointing and scan operation. In both cases, the magnetic field vector is computed by means of an Earth magnetic field model implemented in the OBC-S/W. The coil operations for the two cases to unload the accumulated angular momentum can be seen in Figures 14 and 15. In both cases the angular momentum of the entire S/C is very low and therefore only moderate coil operations can be observed. For the scan case the wheel rates are much higher than for the pointing case. It has, however, to be considered, that in scan operation the wheel rates nominally compensate the commanded S/C x-rate. The accumulated angular momentum of the S/C including the wheels is also very low for this case.

Due to the fact that the total angular momentum is controlled to zero, the scan mode nominal wheel angular momentum has to compensate for the body scan angular momentum (Figure 13), whereas for an inertial pointing the nominal wheel speed is zero (Figure 12). In this figure the wheel speed measurement resolution (quantization) can be recognized, which is about $\Delta\omega = 0.26$ rad/sec.

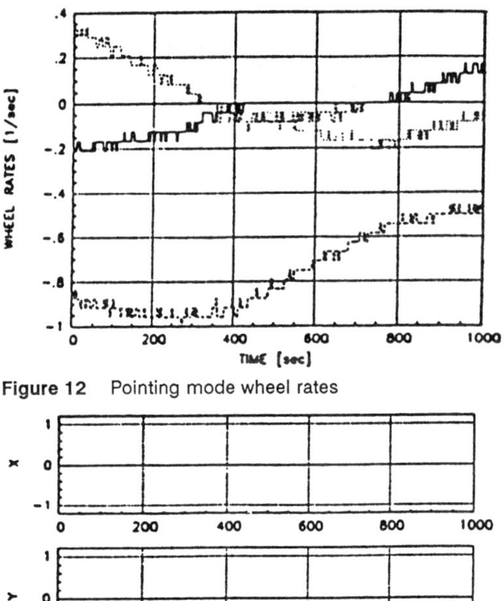

Figure 12 Pointing mode wheel rates

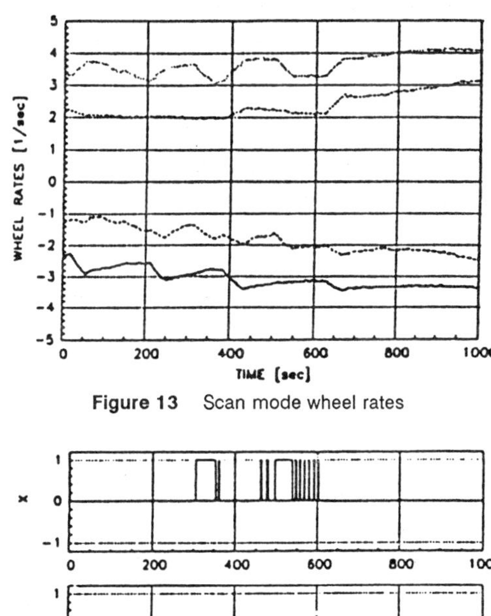

Figure 13 Scan mode wheel rates

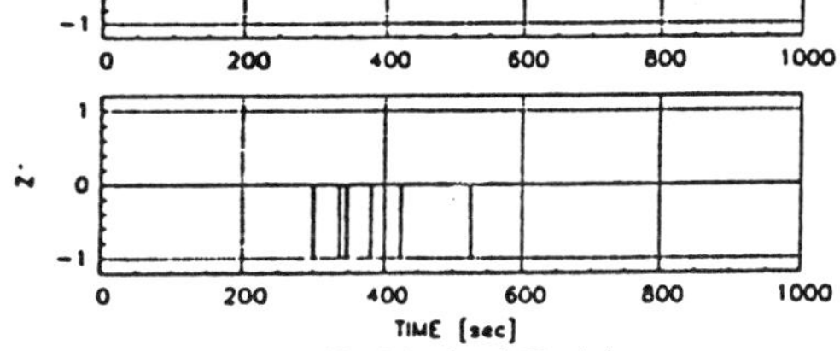

Figure 14 Magnetic unloading in pointing mode

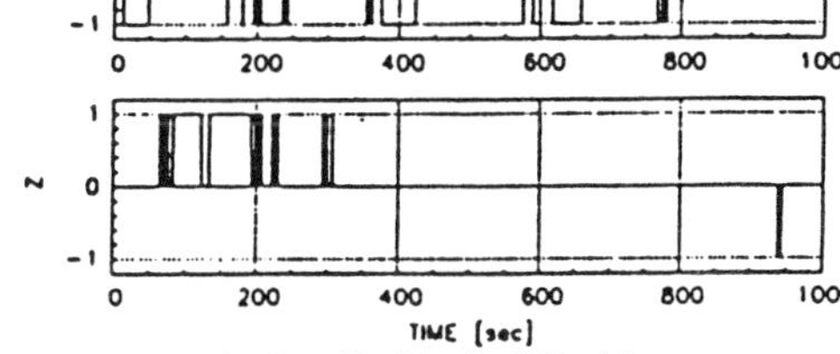

Figure 15 Magnetic unloading in scan mode

Slew Maneuvers

Figures 16-18 show a large angle slew maneuver typical for the reorientation of the S/C before a new pointing period. Figure 16 gives the S/C angular rates, Figure 17 the control torques and Figure 18 the reaction wheel rates. In Figures 18 and 19 the acceleration period can easily be seen. Due to power restrictions it is not allowed to operate all wheels simultaneously with maximum torque. Therefore the acceleration to the constant slew rate of 0.3°/sec is performed with only about 0.1 Nm for each wheel. The deceleration phase shows clearly three steps which have been chosen to avoid an oscillation transient to steady state conditions.

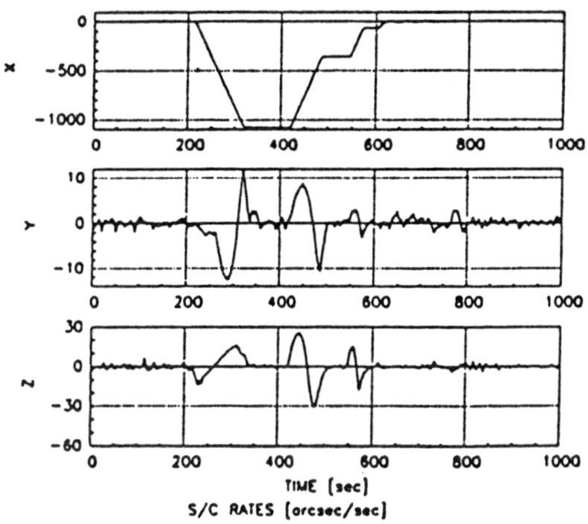

Figure 16 S/C rates for a slewing maneuver

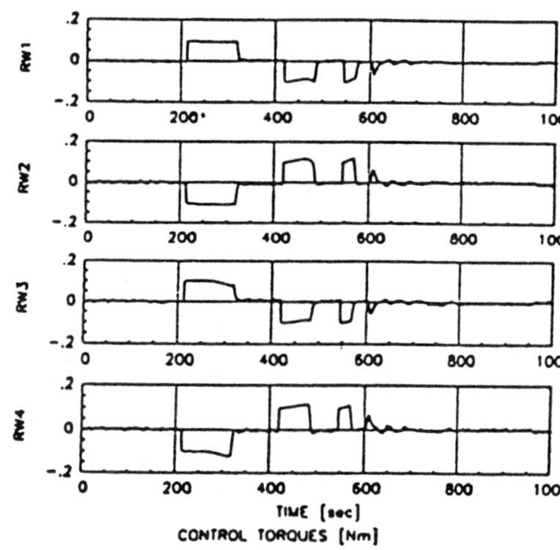

Figure 17 Commanded control torques for a slewing maneuver

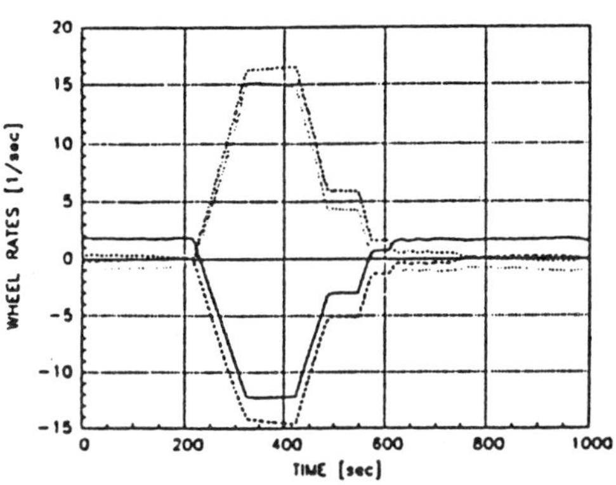

Figure 18 Wheel rates during a slewing maneuver

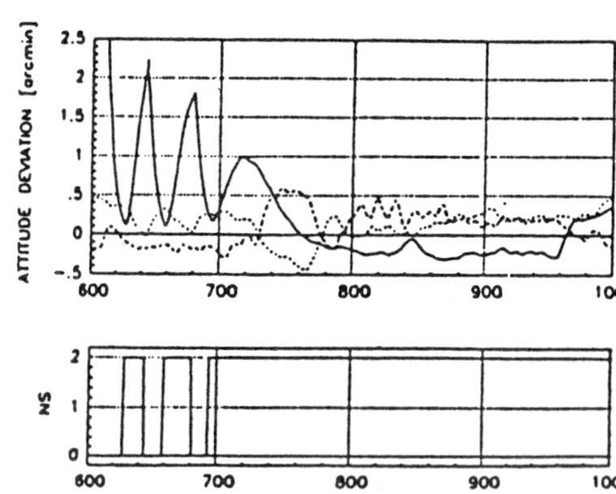

Figure 19 Transition to steady state after a slewing maneuver

Figure 19 shows the transient from slewing conditions to steady state pointing. When, after a slewing maneuver, steady state conditions are achieved, the control-S/W generates star sensor commands in order to measure the expected stars of the star catalog given with the pointing direction. NS indicates the number of stars which are identified w.r.t. the star catalog and which can be used for attitude generation purposes. It can be seen that due to the star sensor measurements the attitude error seen by the S/C increases. This is due to the fact that the attitude error built up during the slewing (uncompensated gyro drift) is now detected by the star sensors. If steady state conditions are reached, a filtered attitude update by the star sensor measurements take place. For larger attitude updates (larger than about 6 arcmin), the update angular rate exceeds the admissible S/C rate for steady state conditions, which interrupts the update procedure until steady state conditions are achieved again.

The total amount of x-axis error after the slewing is more than 6 arcmin which is corrected as can be seen in three steps in about 100 seconds. It has to be noted, that for this slewing maneuver the x-axis scaling factor was not yet compensated.

Star Identification

In the pointing mode at least two stars have to be measured and identified from a star catalog containing up to 6 stars. No problem could be detected for a large number of pointing directions with small- and large-angle slew maneuvers. In all cases, three axes reference could be obtained from measured and identified stars with respect to the commanded star catalog.

In the scan mode the star catalog contained 500 stars. For some days, once or twice per orbit, a wrong star identification was observed within a two-week scan period. The consequence was a short time loss of steady state but a self-remedy of the control system. On the 17th of July, however, after another 4 days scan period, a wrong star identification led to a loss of attitude with the consequence that no further stars could be measured and identified, leading to the necessity of a reacquisition maneuver. The reason was found in a star preceding a catalog star of the expected trace in a sensor FOV by only a small angle. The groundstation S/W, which produces the scan catalog, has therefore been modified in order to cancel those stars from the star catalog leading to the identification difficulties mentioned above.

ROSAT OPERATIONAL PHASE

After about 8 weeks of preoperational phase the normal all-sky survey observation program started on July 30 with the scan mode. An extract of the ROSAT AMCS diary for the 6-month scan operation is given subsequently.

The first event occurred on September 9 with an anomaly of star sensor 1. The sensor no longer delivered proper measurements. Error bits in the sensor housekeeping data were set only after some hours of malfunction. With switch-off and conservation switch-on the sensor recovery was successful. The anomaly is considered as a single-event upset.

The second anomaly occurred on September 25 with the star sensor 2. It no longer delivered valid measurements and the housekeeping data reported an "internal failure". Again the switch-off/switch-on procedure was applied. However, this time it was successful only for short periods and then the anomaly recurred. In the following weeks, periods with proper measurements and malfunction were alternating but the periods of malfunction were increasing and at the end of November the efforts for the sensor recovery were interrupted until the beginning of the pointing phase in February 1991.

Evaluation of available data indicated that the problem lies in the area of the signal processing electronics, not in the area of the optical sensing (CDD and directly associated electronics). Currently, it can be said that the failure of a single component would not explain the failure, and the most likely cause of the failure is a gradual degradation of a part of the RAM area in the conventional signal processing unit.

Due to the fact that there was always one star sensor in operation, the ROSAT mission was not interrupted by the malfunction of the other sensor. On the contrary, all trouble shooting for sensor 2 during September to November could be performed without interruption of the scientific mission.

The first interruption of the normal scan measurements occurred during the weekend before Christmas on December 21. After the regular reversal of the scan rate the AMCS was no longer able to identify stars. In the course of the reacquisition procedure on December 22 an increase of the x-gyro drift by a factor of 10 was detected, compared with the last drift determination in June. After compensation of the new drift the scan mode was again entered. However, during the slewing maneuver into the steady state scan, the surveillance of the gyros indicated a gyro error. This algorithm checks the relation between the 4 gyro outputs:

$$|\omega_S - \omega_X - \omega_Y - \omega_Z| = \omega_\varepsilon \leq \varepsilon$$

where the tolerable value ε had been set to 15 arcsec/sec. On 5 consecutive violations of the above relation the AMCS performed an automatic gyro test on the basis of conservation of the angular momentum. The S/C was therefore switched into the standby mode until December 27.

It was then concluded that the gyro test was initiated due to an additional degradation of the x-gyro scalefactor. The threshold ε was set to 30 arcsec/sec and the recovery procedure reestablished the normal scan operation still on the same day.

A second interruption of the normal scan operation occurred on the 5th of January 1991. Star sensor 1 failed and stopped delivering measurements. Switch-off/switch-on put the sensor back into operation. The reacquisition procedure revealed that the x-gyro drift was again about a factor of 10 higher than determined on December 27, being now about 5 arcsec/sec. The AMCS was therefore commanded to take the x-gyro no longer as a nominal measurement source and on January 7 normal scan operation was reestablished.

For both star sensor 1 failures on September 9 and January 5, it was found that they occurred during extreme solar activities, which supports the assumption that both occurrences were single event upsets.

The most severe incident, however, took place during the night of January 25/26 1991. The CPU 1 of the AMCD ceased operation and the automatic switch over to CPU 2 did not take place. The attitude was lost, and a very heavy drain of the battery was the consequence leading to a switch-off of gyros and wheels. After several hours of uncontrolled motion the groundstation was able to switch on CPU 2 gyros and wheels by direct ground commands and normal operational conditions could be reestablished. The reasons for this unfortunate sequence of events are under study but not yet known today. Again, a very heavy solar activity was observed during that night but it is not yet clear whether any correlation exists to the difficulties encountered. There is however one happy consequence: The switch-off and -on of the gyros brought the x-gyro back to normal drift performance of about 0.1 arcsec/sec drift (uncalibrated). The most probable explanation would be a loss of synchronism of the x-gyroscope which has been reestablished by the switching operation.

Despite the problems experienced, it must be pointed out that the loss of effective mission time is very small and that the scientific mission time line is still on schedule.

CONCLUSION

The results and the experience gained during the first weeks of ROSAT operation can be summarized as follows:

- o All AMCS modes can be operated as foreseen. They are easy to handle by the single ground station of GSOC, which is evidenced by the fact that after 6 weeks a quasi-routine operation could be established.
- o All essential equipment works very satisfactorily and the essential parameters measured in orbit are better than specified. Unfortunately the star tracker redundancy seems to be definitely lost.
- o A few improvements in the ground software for star selection were made after launch:
    - Adaptation of the star tracker calibration curve which transforms visible magnitude of the star catalog into instrument magnitude.
    - Suppression of stars which have a preceding star in scan direction in the operation star catalog of the scan mode. This resulted in a long-term undisturbed scan operation.
- o One small on-board software error has been found so far: wrong threshold of the Sun presence signal for eclipse surveillance. This is a very minor function and the problem has been corrected by an S/W patch which can be loaded in the free RAM area of the degraded mode.
- o The first scientific objective of the mission, i.e. the complete sky survey, is currently reaching its conclusion within the schedule foreseen.

In view of the high complexity of the AMCS subsystem, this is considered to be an excellent result, which gives great satisfaction to all people concerned with design, manufacture, operation and scientific utilization of the subsystem.

REFERENCE

1. Design and test of the ROSAT-AMCS, W. Schrempp, R. Alexander, A. Böinghoff, E. Brüderle, H. Herbst, L. Kaffer, B. Möckl, W. Oesterlin, K. Stinshoff., IFAC 10th World Congress, Munich, FRG, 1987. Vol. 6. pp. 13-21.

AAS 91-061

IMAGE NAVIGATION AND REGISTRATION OF SECOND GENERATION METEOSAT

Bernard Blancke*, James L. Carr† and Patrick Mauté*

The second generation geostationary european weather satellite, METEOSAT, will fly much more sophisticated instruments (imager and sounder) than the presently spinning METEOSAT. The feasibility of a three-axis stabilized METEOSAT spacecraft has been demonstrated by AEROSPATIALE for EUMETSAT under ESA contract. The design described here has mainly been optimized for image quality. It features a configuration with fixed solar arrays, attitude control with reaction wheels, an earth sensor, and a polaris sensor. The imager instrument features a thermally optimized design with dual object plane scan mirrors which minimizes disturbance torques and provides certain advantages for image navigation.

This paper presents the three-axis design for METEOSAT as well as analysis, key parameter trade-offs, and computer simulation results which demonstrate the expected image navigation and registration performance. The results presented in this paper are also of interest to the design of the GOES-N series of satellites which will provide the U.S. with its geostationary meteorological sensing capability in the next century.

INTRODUCTION

METEOSAT is the generic name of a series of geostationary meteorological satellites which has been operated since 1976 in the frame of the World Weather Watch, providing meteorologists with visible, infrared and water vapour imagery of the part of the world centered on the Greenwich meridian. The Second Generation METEOSAT will replace the first one within the end of the century, and will feature enhanced

* Aérospatiale Cannes, 100 Bd du midi, F-06322 Cannes la Bocca, France.

† Swales & Associates, Inc., 5050 Powder Mill Road, Beltsville, Maryland 20705.

capabilities (i.e. more channels, better resolution, sounding capability); to provide these, EUMETSAT customers have envisaged to change from spinned to three axis stabilized satellite design.

Figure 1 shows the layout of the Second Generation METEOSAT which carries three payload instruments: an Enhanced Visible and InfraRed Imager (EVIRI, 14 channels, 2.5 km sampling distance, 24 mn repeat cycle), a High Resolution VISible imager (HRVIS, 0.5 km sampling distance, 5 mn repeat time over Europe), and an Enhanced InfraRed Sounder (EIRS, 17 channels, 16 km sampling distance, 1.5 h repeat cycle over Europe).

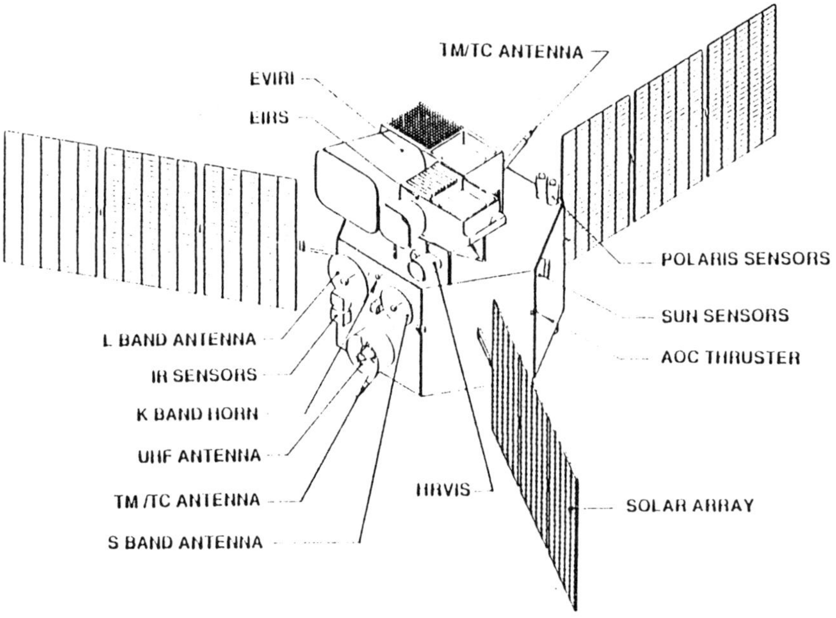

Fig.1 Three Axis Stabilized 2nd Generation Meteosat

The three axis stabilization ensures that the payload instruments are always viewing the earth, but has the disadvantage of inducing a diurnally varying thermal environment; this could cause instrument pointing and internal misalignment errors which would corrupt image navigation (knowledge of pixel location) and registration (stability of pixel location). The solution adopted to solve this problem is to optimize the design w.r.t. thermal deformations, and compensate for them during the rectification process using an Image Navigation and Registration (INR) process similar to that of GOES-I (ref. 1).

The following sections discuss the INR requirements, INR rationale , and sources of error; then simulations are presented, showing that the requirements are met in all cases.

INR REQUIREMENTS

The image quality requirements are summarized in Table 1, which shows the allowed one sigma errors expressed in kilometers at subsatellite point; relative errors refer to registration either between pixels within a given image (requirements over durations up to 2.4 mn) or between the same pixel in two successive images (requirements over 24 mn); absolute errors refer to image navigation, i.e. the process of determining the exact location on earth of a given pixel.

	MISSION A	MISSION B		
	EVIRI channel 14	EVIRI channels 1 to 13	HRVIS	EIRS
sampling distance	1 km	2.5 km	0.5 km	16 km
raw image distortion	0.5 km relative over 3.3 seconds	0.5 km relative over 3.3 seconds	0.4 km 2 km relative over 0.13 s 11 s	4.8 km relative over 3 mn
level 1.5A prediction error	12.5 km absolute 2.5 km relative over 2.4 mn	12.5 km absolute 2.5 km relative over 2.4 mn	5 km absolute 0.5 km relative over 7 seconds	T.B.D.
level 1.5B correction error	N.A.	2.5 km relative over 24 mn 0.625 km relative over 6.5 seconds	N.A.	N.A.
level 1.5C correction error	N.A.	2.5 km absolute 1 km relative over 24 mn	N.A.	8 km absolute

Table 1 INR Requirements Summary

There are four different levels of specification defined as:

1. raw image: this is the data directly output from the instruments, including full radiometric correction

2. level 1.5A: data at this level has been preprocessed using a predicted deformation model; this is based upon information about past and predicted performance

3. level 1.5B: data at this level has been preprocessed using predicted and real time information from satellite

4. level 1.5C: data at this level has been preprocessed using comparisons with a reference image

There are two types of missions envisaged: mission A with only the EVIRI imager augmented with a high resolution visible channel, and mission B with separate high resolution imager (HRVIS) and sounder (EIRS) in addition to the EVIRI.

INR RATIONALE

As can be seen from Table 1, stringent requirements on the raw image only concern short term registration; it is possible to cope with these by minimizing the high frequencies perturbations by design, without active compensation at instrument level. This approach firstly concerns the attitude control, which generates the platform short term stability; the key features of the attitude control design are the following:

* three reaction wheels are used as primary actuators, so that gyroscopic torques are negligible, and vibrations induced are at rather low frequencies, thus avoiding potential resonance problems (e.g. with the earth sensor scanner).

* magnetic bearings for the wheels minimize friction and wobble; this, associated with optical high resolution encoders ensure very neat wheel speed control (ref. 2).

* the sensors associated to these actuators are the Lockheed type 2 earth sensor for roll and pitch (which is to be flown on GOES-I), and the Sodern pole star sensor for roll and yaw (which has half the noise of the earth sensor).

* magnetic torquers are used as secondary actuators, allowing a momentum dump free of perturbation in roll and yaw; for the pitch axis, dumping is performed by thrusters during station keeping maneuvers every month (this is feasible thanks to the solar torque balance in pitch of this configuration).

Two other design features optimize the short term registration:

* the dual mirror optical path of the instruments (see figure 2) minimizes the rotating inertia (no gimbal) and thus the scanning perturbations.

* the three solar panels are fixed (no perturbation induced by solar array rotation).

Once short term requirements are met, image preprocessing will deal with absolute distortions such as biases and alignments, and long term varying distortions; these latter have two origins: actual position of the satellite, and thermal deformations (which mainly occur during +/- 4 hours around midnight as the solar rays enter the instruments field of view); both can be predicted by using measurements as inputs to prediction models:

* rangings from two ground stations are input into an orbit prediction model to predict the satellite position; it is only required from the second station to echo back the ranging tone sent by the first one.

* stars are viewed by the instruments as references in the vicinity of the earth, and these detections are input to a thermal deformation model to predict the thermal deformations as well as the constant biases and alignments; this process will perform well as long as the time constant of the thermal deformation is much higher than the period between star availability; moreover, this identification is facilitated by the optical design of the instruments which features decoupled scanning axes.

* as an option to stars, landmarks can be used as earth based references, but the accuracy achieved is not as good; this is shown in the simulation cases section.

SYSTEM ERROR SOURCES

To derive an INR performance, it is necessary to go through all the error sources, that is to identify the error path, between light emission from the earth scene and light reception by the instruments detectors; this path is the following:

* from earth scene to satellite local orbital frame (orbital errors)

* from satellite local frame to satellite reference frame (attitude errors)

* from satellite reference frame to detectors axes (alignments, thermal deformations and servo errors)

These errors are detailed in the following paragraphs

Orbit

Deviations from a perfectly circular equatorial orbit will cause image errors; the eccentricity will cause the longitude to oscillate at a daily period, while the inclination will cause both a latitude and a longitude (second order effect) oscillation; the mean deviation from geostationnary altitude will cause a mean longitude drift.

As these oscillations and drift are very slow with respect to the raw image jitter time constant specified, their impact on raw image quality will be very small, and compatible with the requirements without any correction.

For what concerns processed levels, the precision of orbit determination out of transients (station keeping maneuvres) is good (satellite positionning accuracy of about 0.3 km at 1σ), leading to residual errors well within the specification; just after a maneuvre, the orbit prediction is degraded until the algorithm has converged (about 2 days delay); this latter effect is minimized by chosing small correction amplitudes (for example, 0.2 degrees inclination deadband controlled with an accuracy of 5% leads to an absolute error of about 1 km at sub satellite point).

Attitude

The attitude errors are driven by the performances of the attitude control loops in response to internal and external torques; internal torques are:

* sloshing motion torques - due to the three axis stabilization of the satellite, the amplitude of these torques will be negligible out of transients; it may happen that a transition mode will be necessary to reach the required platform stability after a maneuvre.

* wheel wobble - this vibration induced by the misalignment (about 1 arcminute) between the wheel inertia axis and its rotation axis is also negligible due to the high inertia of the satellite.

* mirror motion torques - due to the optical design of the instruments, the moving mirrors have a minimum inertia (no gimbals). The only disturbance of some importance is the effect of the EVIRI (imager) blackbody calibration during the imaging of the other instruments, if they are not synchronized; but even this effect is within the raw image specification without active compensation; for processed levels, it can be easily taken into account as it is deterministic.

These internal torques are in fact not detected by the attitude control subsystem as either their amplitude or their time constant are too small.

The only torques effectively controlled are the external torques induced by the solar pressure; although these torques are very low and varying very slowly, the actuators (wheels) and sensors (earth and pole star) used in the control loops are sources of errors by themselves for the attitude control.

There are two types of attitude control errors which effect the image navigation and registration. The first is the jitter of the attitude control system. In the MSG-3X concept the roll and yaw axes are controlled using a polaris sensor (ref. 3). The pitch axis is controlled with an IR earth sensor which is very noisy (about 100 μradians (1-σ)); it is possible by optimizing the control filters to reduce this noise down to 5 μradians (1-σ) at the platform level. The earth sensor is also susceptible to drifts induced by variations in the earth's IR radiance due to clouds and radiance gradients. This is the second source of error. In the MSG-3X image navigation system slow drifts will be seen in the navigation data and corrections would be applied during the next day. The residual error due to earth sensor drifts in the navigated imagery would then be the day-to-day change in pitch drifts.

Thermal Distortions

The EVIRI instrument utilizes a pair of one degree-of-freedom scan mirrors instead of a single two degree-of-freedom gimbaled mirror as is used on GOES-I. The effect of thermal distortions on the two configurations is different, with the two mirror EVIRI system requiring one fewer parameter to characterize its optical misalignments. This section describes the effects of the EVIRI thermal distortions and how they can be compensated for in the image navigation process.

Figure 2 shows a schematic representation of the EVIRI optical system. Each of the three elements - slow-scan mirror (NS scan), fast-scan mirror (EW scan), and telescope - may be subject to tilts and rotations due to thermal distortions and manufacturing defects. They may also be subject to de-spacing and de-centering; however, the de-spacing and de-centering of optical elements (excluding those internal to the telescope) has no effect on the line-of-sight. For this reason only tilts and rotations of the mirrors and telescope are considered. The optical elements within the telescope and aft-optics are not treated individually here, instead they are considered as contributors to the telescope misalignment angles.

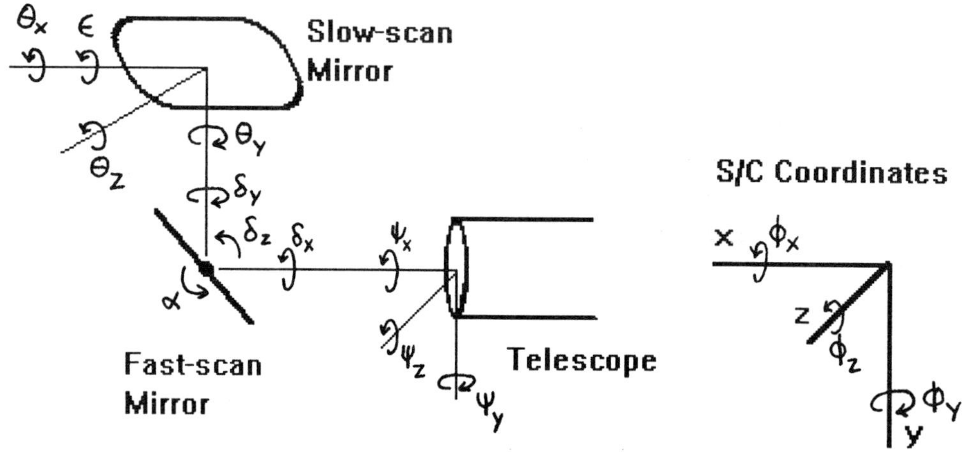

Fig.2 EVIRI Optical System

The line-of-sight may be represented by the X and Y components of a unit vector in the spacecraft coordinate system (figure 2). The computation of the net line-of-sight vector need only be performed to linear order in the small tilt and rotation angles:

$$\begin{bmatrix} X \\ Y \end{bmatrix} = \begin{bmatrix} S_\alpha \\ -C_\alpha S_\varepsilon \end{bmatrix} + \begin{bmatrix} 0 & C_\alpha C_\varepsilon & C_\alpha S_\varepsilon & 0 \\ -C_\alpha C_\varepsilon & 0 & S_\alpha & -S_\alpha C_\varepsilon \end{bmatrix} \begin{bmatrix} \phi_x' \\ \phi_y' \\ \phi_z' \\ \theta_y' \end{bmatrix} \quad \longleftarrow \text{Model}$$

$$+ \begin{bmatrix} 0 & 2C_\alpha(1-C_\varepsilon) \\ -C_\varepsilon(1-C_\alpha) & 2S_\alpha S_\varepsilon \end{bmatrix} \begin{bmatrix} \delta_y' \\ \delta_z' \end{bmatrix} \quad \longleftarrow \text{Tolerance} \qquad (1)$$

$$S_\alpha = \text{Sin } 2\alpha \qquad C_\alpha = \text{Cos } 2\alpha$$
$$S_\varepsilon = \text{Sin } 2\varepsilon \qquad C_\varepsilon = \text{Cos } 2\varepsilon$$

where α and ε are mechanical scan angles referenced to zero at nadir, and ϕ_x', ϕ_y', ϕ_z', θ_y', δ_y', δ_z' are linear combinations of ϕ_x, ϕ_y, ϕ_z, θ_x, θ_y, θ_z, δ_x, δ_y, δ_z, ψ_x, ψ_y, ψ_z; the primed angles have been defined so as to eliminate linear dependencies and to group effects in order of importance (equation 2).

$$\begin{bmatrix} \phi_x' \\ \phi_y' \\ \phi_z' \\ \delta_x' \\ \delta_y' \\ \delta_z' \\ \theta_x' \\ \theta_y' \\ \theta_z' \\ \psi_x' \\ \psi_y' \\ \psi_z' \end{bmatrix} = \begin{bmatrix} 1 & 0 & 0 & -1 & 1 & 0 & 2 & 0 & 0 & 0 & -1 & 0 \\ 0 & 1 & 0 & 0 & 0 & 2 & 0 & 1 & -1 & 0 & 0 & -1 \\ 0 & 0 & 1 & 0 & 0 & 0 & 0 & 0 & 1 & 0 & 0 & 0 \\ 0 & 0 & 0 & 1 & 0 & 0 & 0 & 0 & 0 & 0 & 0 & 0 \\ 0 & 0 & 0 & 0 & 1 & 0 & 0 & 0 & 0 & 0 & -1 & 0 \\ 0 & 0 & 0 & 0 & 0 & 1 & 0 & 0 & -.5 & 0 & 0 & -.5 \\ 0 & 0 & 0 & 0 & 0 & 0 & 1 & 0 & 0 & 0 & 0 & 0 \\ 0 & 0 & 0 & 0 & -1 & 0 & 0 & 1 & 0 & 0 & 0 & 0 \\ 0 & 0 & 0 & 0 & 0 & 0 & 0 & 0 & 1 & 0 & 0 & 0 \\ 0 & 0 & 0 & 0 & 0 & 0 & 0 & 0 & 0 & 1 & 0 & 0 \\ 0 & 0 & 0 & 0 & 0 & 0 & 0 & 0 & 0 & 0 & 1 & 0 \\ 0 & 0 & 0 & 0 & 0 & 0 & 0 & 0 & 1 & 0 & 0 & 1 \end{bmatrix} \begin{bmatrix} \phi_x \\ \phi_y \\ \phi_z \\ \delta_x \\ \delta_y \\ \delta_z \\ \theta_x \\ \theta_y \\ \theta_z \\ \psi_x \\ \psi_y \\ \psi_z \end{bmatrix} \quad (2)$$

It should be remembered that the angles 2α and 2ε never exceed 8.7° (the apparent earth radius) and so it makes sense to group the misalignment effects in powers of α and ε using the Taylor series. Those of low order in the scan angles will be relatively more important than those of high order. The recommended approach to image navigation is to solve for those misalignments of zeroth or first order and to minimize by tolerancing those misalignments of quadratic or higher order. The misalignment angles have been grouped according to this recommendation in equation (1). The set requiring modeling includes three angles ϕ_x', ϕ_y', and ϕ_z' which describe the attitude of the line-of-sight when the mirrors are in their home positions. The misalignment included in the modeled set is $\theta_y'=\theta_y-\delta_y$ which has an effect approximately proportional to the EW scan angle α and independent of the NS scan angle ε which causes a displacement of the line-of-sight in the NS direction.

Instrument Servo Errors

Pointing of the instrument line of sight is accomplished with two separate one degree of freedom scan mirrors under the control of servos; the servos sense the mirror scan angles using optical encoders; the fast scan mirror (EW) is mounted on flexural pivots and controlled in quasi open loop by a D.C. motor, while the slow scan mirror (NS) rotates about ball bearings and is controlled in closed loop by a D.C. motor.

Instrument servo errors may be classified as either dynamic or static; their minimization and compensation is done within the instruments and is transparent to the INR system. The net residual servo pointing errors may have both angle dependent and angle independent components; as parts age towards the end of life there may be an increase in the angle dependent component of servo pointing error.

INR SIMULATION STUDIES

The INRSIM computer code developed for analysis of the GOES-I INR system was modified for use in analyzing the MSG-3X system. The main modification was the replacement of the GOES-I optical sensitivity matrix by that in equation (1). The set of solution parameters is consequently smaller than for GOES-I since the EVIRI only requires solution for a single misalignment parameter, whereas two are required for GOES-I. Other modifications to the INRSIM code were relatively minor.

No thermal distortion profiles were available for the MSG-3X that could have been used as inputs for INRSIM, therefore, a set of GOES-I thermal distortion profiles were used. The GOES EW misalignment was taken to represent the modeled EVIRI misalignment and the GOES NS misalignment was taken to represent both of the un-modeled EVIRI misalignments. The effects of the un-modeled misalignments were incorporated as per equation (1). Figure 3 shows these assumed thermal distortions.

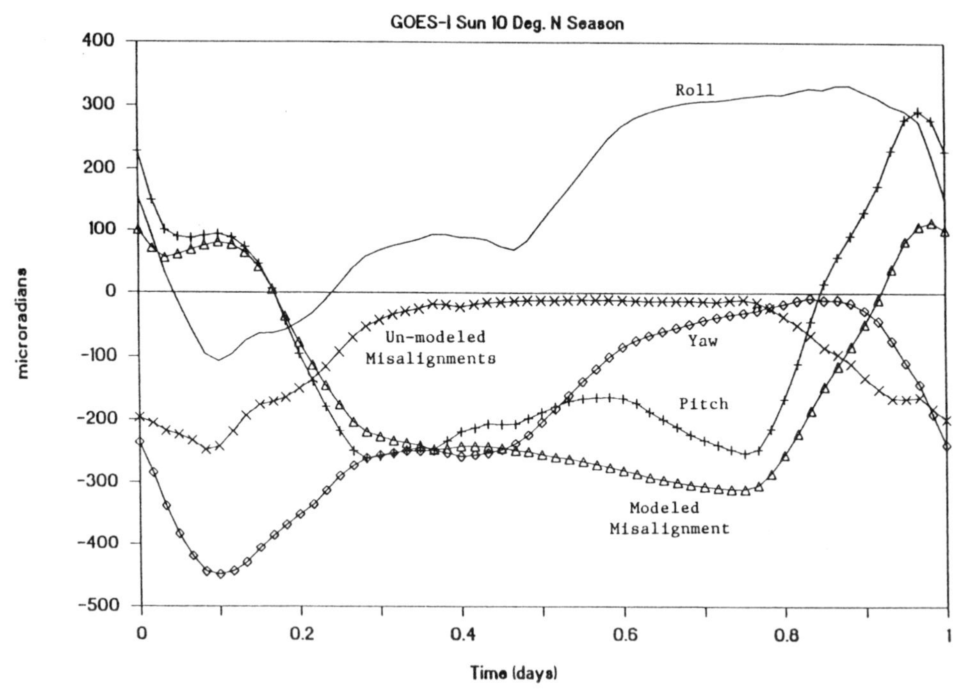

Fig.3 Thermal Distortion Profile

Simulation Cases

Three cases studied with the modified INRSIM computer code are presented here:

1. Quiescent case with landmark and range navigation with 15μradians (3σ) allocated for attitude control jitter.

2. Quiescent case with star and range navigation with 15 μradians (3σ) allocated for attitude control jitter.

3. Radiance gradient case with star and range navigation with 15 μradians (3σ) allocated for attitude control jitter.

The quiescent cases are those for which the earth sensor is not subject to drifts due to earth radiance variations caused by clouds and radiance gradients. The radiance gradient case involves earth sensor drifts which are considered worst case for the LMSC GOES Type-2 earth sensor. The radiance gradient drift profile shown in figure 4 has been developed for the GOES-I program using real atmospheric data and a model of the earth sensor. The effect is only applied in pitch for the MSG-3X configuration since the polaris sensor will be used in the roll control loop. Two ranging stations are utilized in the simulations with ranging pairs hourly. The station locations are at Kourou and Darmstadt. The use of two ranging stations for MSG-3X allows the navigation solution to be developed with stars and no landmarks.

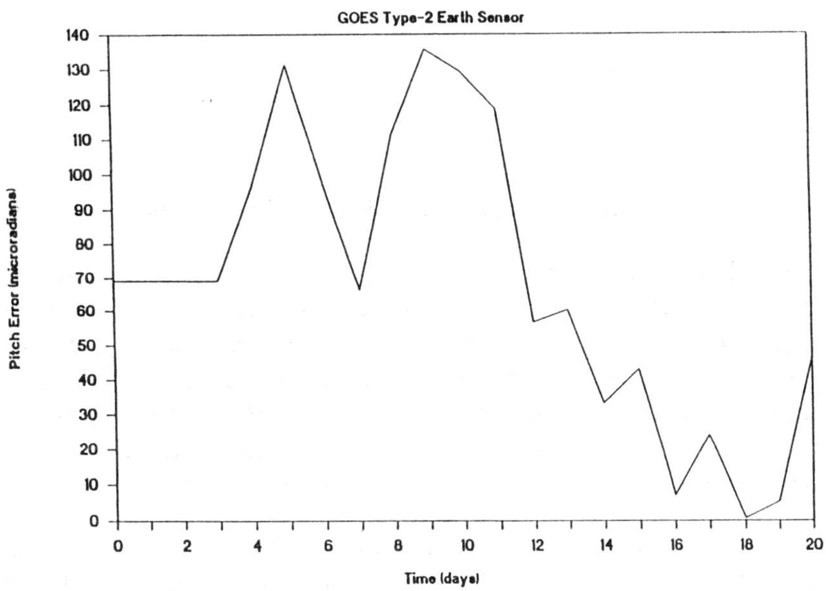

Fig.4 ES Radiance Gradient Disturbance

Table 2 summarizes the noise inputs for the baseline 15 μradian pitch control system. An allocation of 15 μradians (3σ) is made for the pitch axis jitter. The roll and yaw axis jitters are presumed superior due to the use of the polaris sensor. The random measurement errors for stars and landmarks contain components associated with the attitude control jitter, encoder errors, and measurement precision. The allocations for measurement precision have been set at one-half pixel (3σ). In addition to the random measurement errors, the systematic error for the landmarks has been set at one-half of a visible pixel. Random ranging errors of 15m (3σ) are assumed and a bias of 5m is added to each station's one-way range.

	Pitch - Axis			Roll - Axis		
Pointing Jitter (3-σ)						
Attitude Control Jitter		15μR			10μR	
Encoder Error		10μR			10μR	
Net Jitter (RSS)		18μR			14μR	
Measurement Errors (3-σ)	STAR	VIS.LMK	IR LMK	STAR	VIS.LMK	IR LMK
Attitude Control	15μR	15μR	15μR	10μR	10μR	10μR
Measurement Precision	15μR	15μR	35μR	15μR	15μR	35μR
Encoder Error	10μR	10μR	10μR	10μR	10μR	10μR
Net Random (RSS)	23μR	23μR	39μR	21μR	21μR	38μR
Systematic		15μR	15μR		15μR	15μR

Ranging Errors	
Random (3-σ)	15m
Bias	5m

Table 2 Simulation Error Summary

The modified INRSIM program has been configured to solve for an attitude and misalignment model with an eighth order Fourier series representing roll and pitch and a fourth order Fourier series representing yaw and the modeled misalignment. In addition there is a linear ramp function included with each functional series. This is for the purpose of absorbing the difference in end point values that might exist in an otherwise diurnal model due to aperiodic effects such as earth sensor drifts.

Table 3 summarizes the input configuration for the navigation simulations. The first set of results are shown in figures 5 and 6 which contain the absolute and 24 minute relative navigation errors at the NE 60°

Earth Central Angle (ECA) point for the quiescent case with landmark and range observations. In both figures the MSG specifications are met, however, there is less margin for the relative navigation specification. Figures 7 and 8 are the corresponding results for the second quiescent case where star observations are used in place of landmarks. Comparisons between the two sets of results shows that the star case's performance is slightly superior to the landmark case's particularly in the NS absolute navigation. This is for three reasons; first, the star measurement random errors are equal to the visible landmark errors and smaller than the IR landmark errors which must be used near orbit midnight. The second reason is that there are insignificant systematic errors associated with the stars (catalog errors should be at the 2 μradian level (3σ)). The third reason is that since the stars appear further away from nadir, their geometry is more favorable for the determination of the yaw and the modeled misalignment. Since the modeled misalignment displaces the line-of-sight in the NS direction this is likely the reason for the improvement in NS absolute navigation seen when stars are used.

- Ranging Stations at Darmstadt and Kourou
- GOES-I Thermal Distortion Model for Sun 10°N Season with Modeled and Un-modeled Misalignments
- Roll and Pitch Model: 8th-Order Fourier & Ramp
- Yaw and Misalignment Model: 4th-Order Fourier & Ramp
- 4 Landmarks Observed Every 1/2 Hour (Landmark Cases)
  1 Ranging Pair Every Hour (All Cases)
  4 Stars Observed Every 1/2 Hour between images (Star Case Only)
- 0.2° Inclination Orbit at 0° Station
- MSG Catalog for Landmarks

Table 3 Simulations Inputs Configuration

Figures 9 and 10 show the absolute and relative navigation errors for the radiance gradient case with star and range navigation. In all of the radiance gradient cases the EW absolute navigation errors have degraded relative to the corresponding quiescent case, however, the MSG specifications continue to be met.

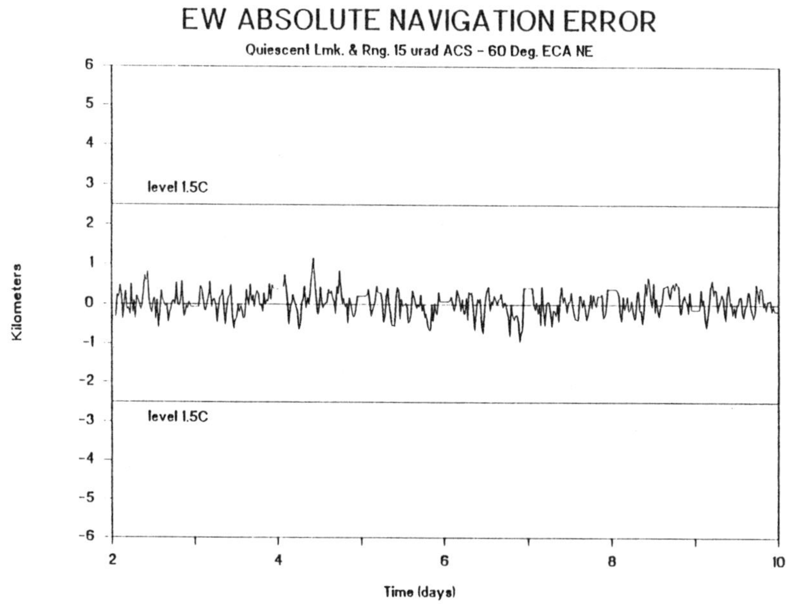

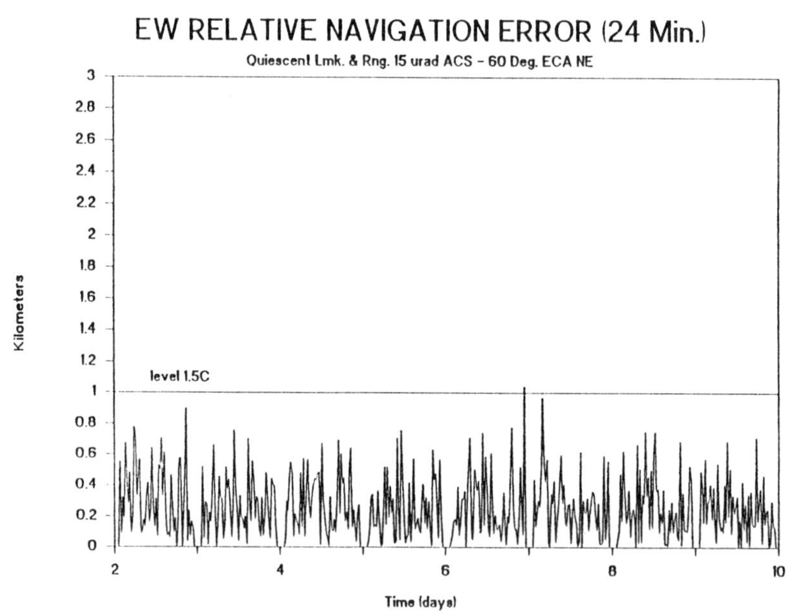

Fig. 5 EW Navigation Performance for the Quiescent Case with Landmark and Ranging Navigation

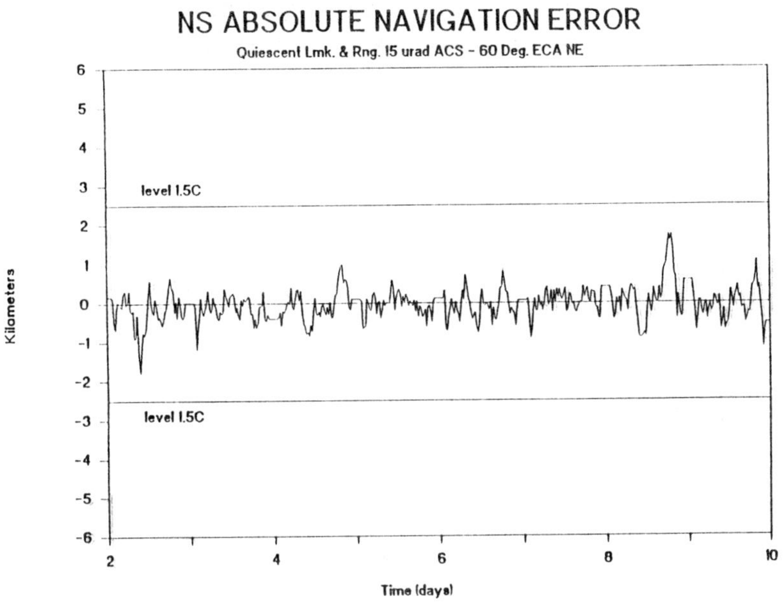

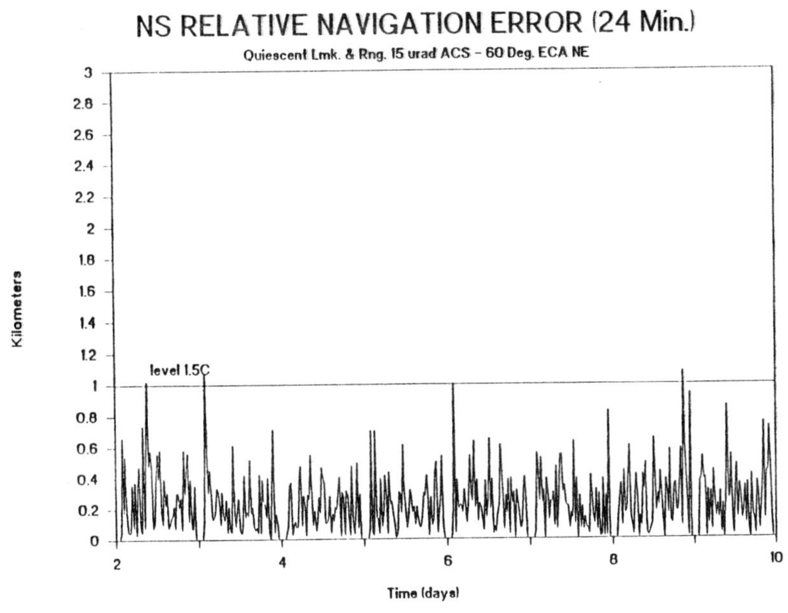

Fig. 6 NS Navigation Performance for the Quiescent Case with Landmark and Ranging Navigation

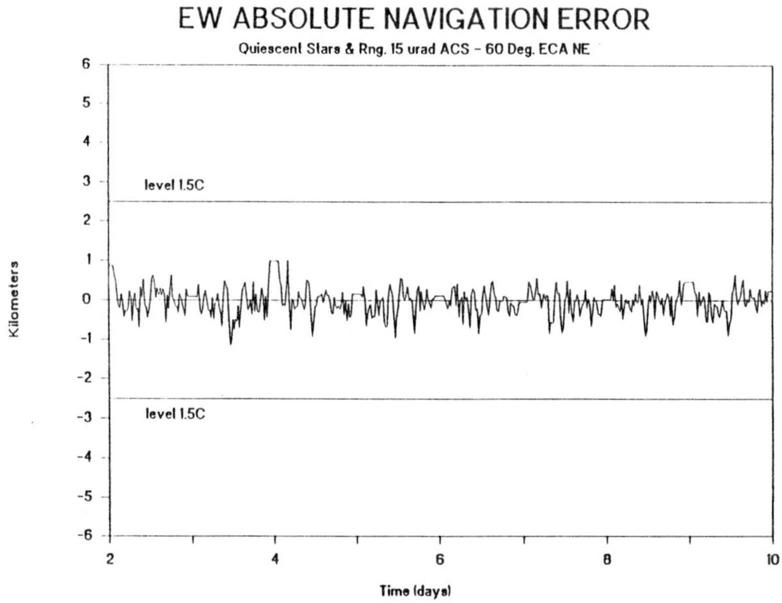

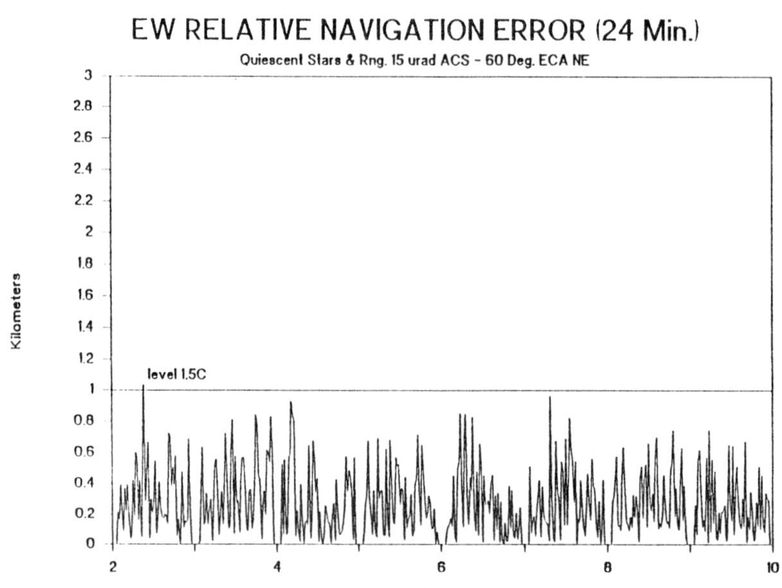

Fig. 7 EW Navigation Performance for the Quiescent Case with Star and Ranging Navigation

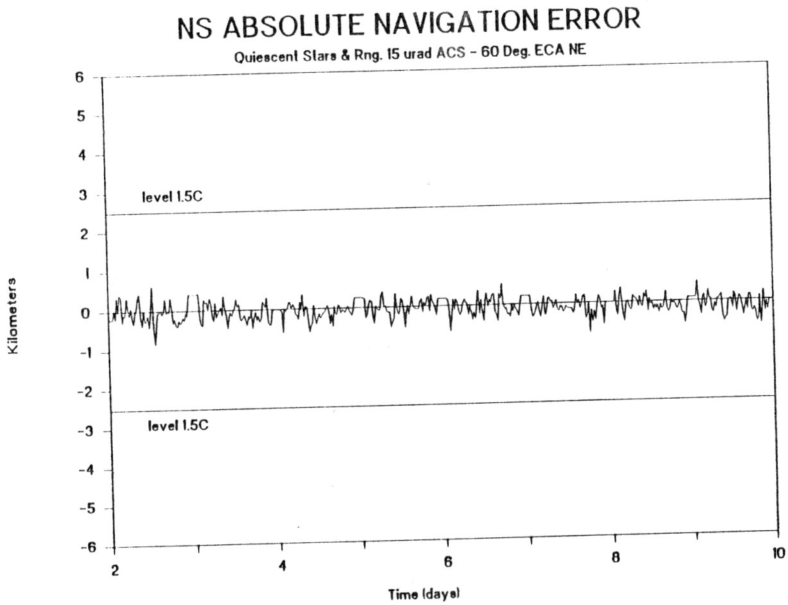

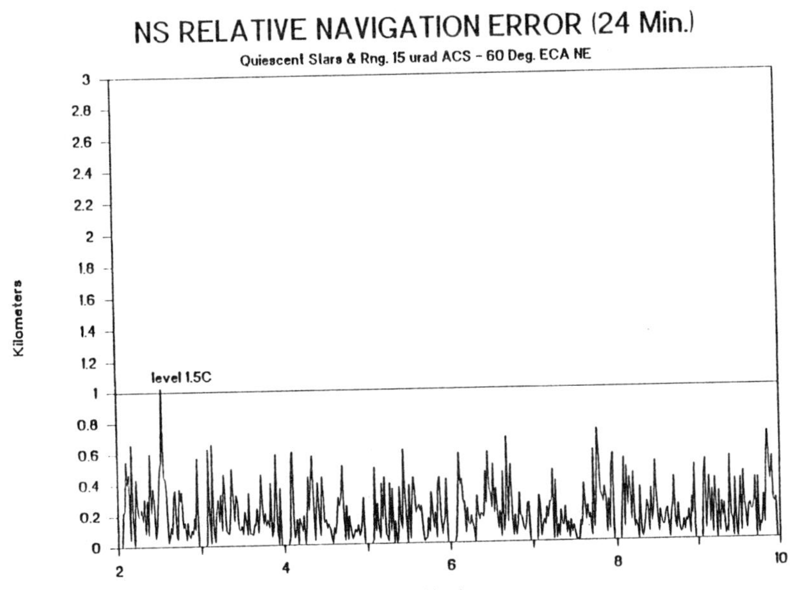

Fig. 8 NS Navigation Performance for the Quiescent Case with Star and Ranging Navigation

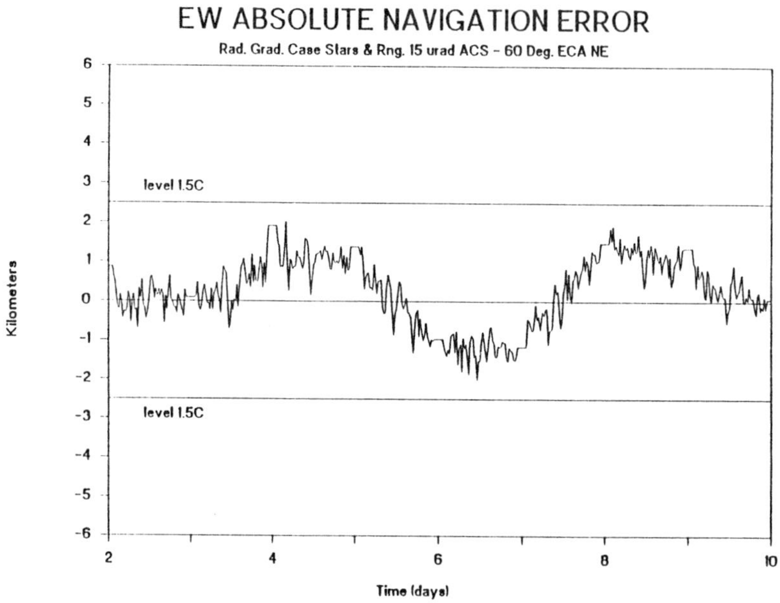

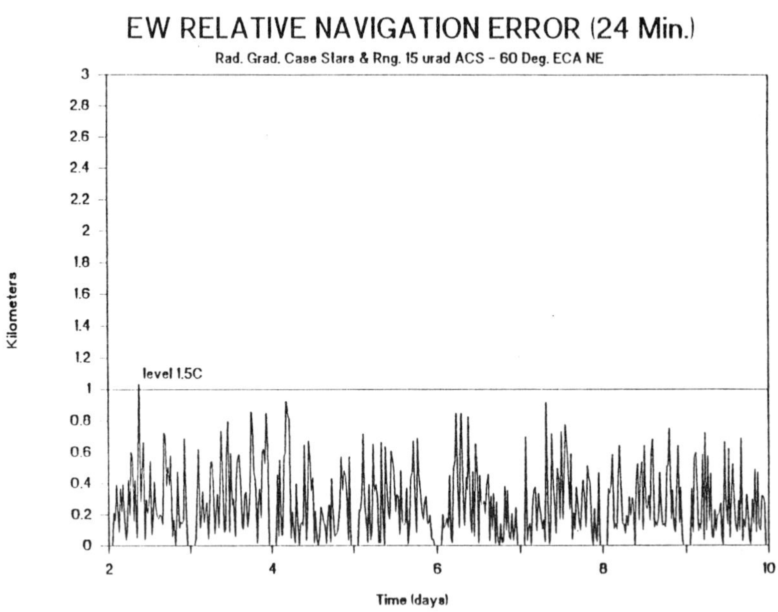

Fig. 9 EW Navigation Performance for the Radiance Gradient Case with Star and Ranging Navigation

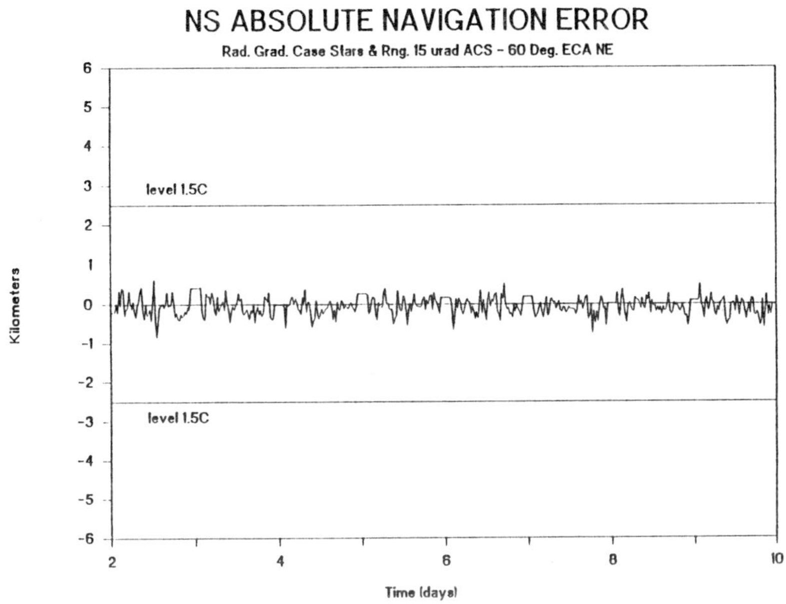

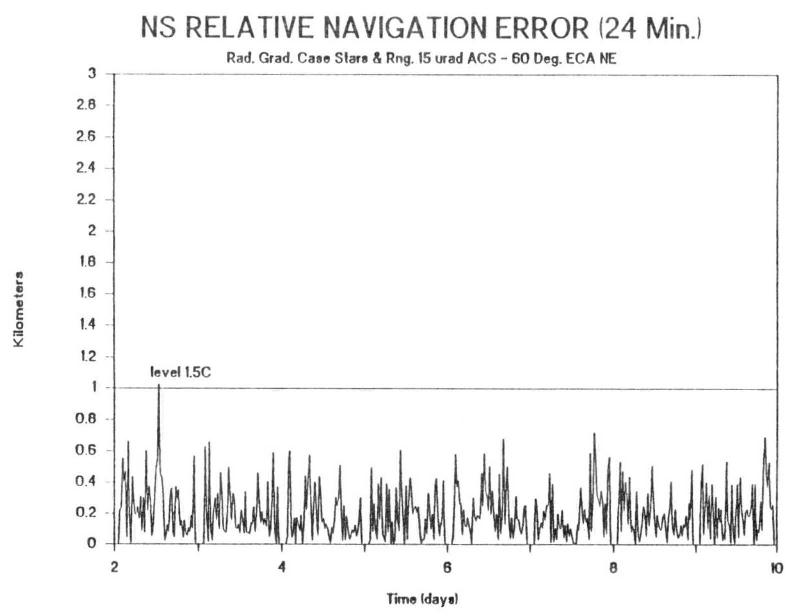

Fig. 10 NS Navigation Performance for the Radiance Gradient Case with Star and Ranging Navigation

AAS 91-062

CONTROL SYSTEM CHALLENGES ASSOCIATED WITH THE MOBILE SERVICING SYSTEM FOR SPACE STATION FREEDOM

Michael E. Stieber* **and Patrick T. K. Fung**†

Canada is participating in the international Space Station "Freedom" Program through the development and operation of the Mobile Servicing System (MSS) which will play a critical role in the assembly, maintenance and servicing of the station. The MSS is a robotic system featuring a large manipulator for massive payloads and a smaller manipulator for dexterous tasks. The MSS manipulators can be operated either manually by astronauts or in automatic modes. Advanced control system features and stringent requirements on performance and safety present technical challenges for the design and development of the MSS control systems.

INTRODUCTION

When the international Space Station "Freedom" (SSF) is assembled and operated in orbit later this decade, robot manipulators will play an essential role which will continue throughout the 30-year lifetime of the program. Canada has established a strong capability in space robotics through the development of the Remote Manipulator System (RMS) for the U.S. Space Shuttle. As a partner nation in the Space Station program, Canada is contributing a sophisticated robotic "Mobile Servicing System" (MSS) to the Space Station infrastructure, and Canada will also participate in the operation and utilization of the Space Station.

---

* Manager, Control Systems and Simulation, Space Station Program, Canadian Space Agency, R105 Montreal Road, Ottawa, Ontario, Canada.

† Manager, Control Systems Engineering, Advanced Technology Sysstems Group, Spar Aerospace Ltd., Canada.

This paper provides an overview of MSS, with focus on its manipulator systems, and a description of the MSS control systems, highlighting some of the technical challenges to be met during the development of the MSS.

OVERVIEW OF THE MOBILE SERVICING SYSTEM

Role and Configuration of the Mobile Servicing System (MSS)

The MSS has been assigned a predominant role in satisfying the following functions on the Space Station [1]:

- Servicing of external payloads
- Assembly of the Space Station
- External maintenance of the Space Station
- Transportation on the Space Station
- Deployment and retrieval of free flying vehicles
- Support of Extra-Vehicular Activities (EVA).

The broad scope of this role requires a versatile system with the capability to manipulate a wide range of objects at a number of different operating locations on the station.

The space segment of the MSS is shown conceptually in Figure 1. It consists of three elements, the Mobile Servicing Centre (MSC), the Special Purpose Dexterous Manipulator (SPDM) and the MSS Maintenance Depot (MMD). A set of MSS Control Equipment will be integrated with the Space Station Data Management System (DMS) to provide for operation of the MSS elements.

The core on-orbit element of the MSS is the MSC. The mobility of the MSC is provided by the Mobile Transporter (MT) developed by the United States. The Mobile Remote Servicer (MRS), consisting of the MSC Base System (MBS) and the Space Station Remote Manipulator System (SSRMS) is attached to the top of the MT. The SSRMS and the SPDM manipulator systems are the key elements of the MSS in terms of control systems and their role and physical configuration are described below in more detail.

This overview reflects the MSS design as baselined at the MSS Interim Design Review in August 1990 and reviewed at the Space Station Freedom Integrated Systems Preliminary Design Review in December 1990. At the time of this writing, the Space Station Freedom program is undergoing a restructuring mandated by the U.S. Congress which may affect the MSS program baseline.

The Space Station Remote Manipulator System (SSRMS)

The SSRMS is the larger of the robotic devices of the MSS, with a length of 17.6 metres. Its configuration is shown in Figure 2. Each end of the SSRMS is tipped with a Latching End Effector (LEE) incorporating a force/moment sensor. The LEEs are designed to attach to payloads and to serve as a base for the SSRMS when attached to a Power Data Grapple Fixture (PDGF).

The SSRMS has seven joints arranged anthropomorphically in clusters of three at each end forming "wrist" and "shoulder", and one central joint forming the "elbow". All joints have a range of travel of $\pm$ 270 degrees. The kinematic redundancy provided by the 7 joints increases the operational flexibility and helps to avoid kinematic singularities.

The symmetric arrangement of the joints and end effectors enables either end of the SSRMS to function as the base or tip and permits the SSRMS to relocate by stepping from one PDGF to another. PDGFs have been placed strategically on the station allowing the SSRMS access to areas which are beyond its reach from the MT.

Examples for the tasks performed by the SSRMS are the assembly (berthing) of the pressurized modules, the positioning of EVA astronauts, and the berthing of the Shuttle Orbiter when it visits the station. This task drives the payload capability of the SSRMS to 116,000 kg. The maximum tip velocity of the SSRMS is 0.37 m/s and in automatic mode its tip positioning accuracy is $\pm$ 4.5 cm with respect to the shoulder. This accuracy can be improved by using visual feedback. In addition to performing tasks alone the SSRMS will be used to transport and position the Special Purpose Dexterous Manipulator and the U.S. provided Flight Telerobotic Servicer (FTS). Power, data and video will be passed through the SSRMS for the operation of these devices.

The Special Purpose Dexterous Manipulator (SPDM)

Many of the tasks assigned to the MSS require the manipulation and precise positioning of small objects such as Orbital Replaceable Units (ORUs). Functions requiring work of this nature will be performed by the SPDM [2]. The SPDM will improve the efficiency of the SSF crew and reduce the need for EVA by providing a means to accomplish EVA tasks robotically.

The SPDM consists of three major assemblies, a base, an articulated body and two dexterous manipulator arms. This design concept is illustrated in Figure 3. The manipulator arms are approximately 2 metres long, each having seven rotary joints. Each arm is equipped with a Tool Changeout Mechanism (TCM) which also provides gripper functions. A video camera and light are built into the TCM to provide boresight views of work piece interfaces which must be grasped, and to assist in inspection tasks.

The wrist of each manipulator incorporates a force-torque sensor. The manipulators are attached to the top end of the body by a common roll joint.

The body is articulated, consisting of two folding segments and four joints. The body joints, combined with a roll joint in the base provide five rotary degrees of freedom for positioning the base of the dexterous manipulator arms relative to the worksite.

Accommodations are provided on the body for tools and temporary storage of small ORUs during operations. The SPDM base is configured as an extension to the SSRMS. It has a PDGF at one end and a LEE including a boresight camera at the other. The PDGF enables the SPDM to be picked up, transported and operated on the end of the SSRMS. The LEE can be used to grapple large payloads and serves to attach the SPDM to PDGFs on the MSC or elsewhere on the station for stand-alone operation.

The combined reach of the SPDM body base and arms measured to the wrist pitch joint of the arms exceeds 5 metres. The SPDM manipulators can each exert in excess of 110 N force and torques of 55 Nm. The payload capacity is 600 kg. When unloaded the SPDM manipulators can be positioned to within an accuracy of 0.6 cm without using the vision system and 0.125 cm using vision feedback.

MSS CONTROL SYSTEMS

MSS Control System Concepts

The MSS control system and human interface philosophy is based on the proven and successful concepts of the Shuttle RMS [3]. The motions of the SSRMS and the SPDM can be commanded on a joint by joint basis or as coordinated actions of multiple joints to effect the motion of an operator-defined point-of-resolution (POR).

Both human-in-the-loop and automatic modes are available. In the human-in-the-loop the operator can command translational and rotational velocities of the POR using a translational and a rotational hand controller with 3 degrees of freedom each, similar to the hand controllers of the RMS on the Shuttle Orbiter. Automated motions can be commanded via operator specified motion of a POR, pre-stored POR auto-sequences, or pre-stored joint position auto-sequences, or artificial vision feedback.

The following control features can be invoked by the operator:

- force-moment accommodation
- storage and replay of POR and joint trajectories
- simultaneous coordinated control of SSRMS and SPDM manipulators
- position/orientation hold on one or more degrees of freedom
- simultaneous coordinated control of the SPDM manipulators.

The SSRMS control system concept is shown in Figure 5. The functions of the control system are to operate the SSRMS arm and the LEE which acts as the tip of the arm. In order to achieve the required failure tolerance, there are two independent strings for all command and control functions on the SSRMS.

In human-in-the-loop operation, the operator specifies rate commands for the POR through the hand controllers at a control station located in a cupola on the station or in the aft flight deck of a Shuttle Orbiter berthed to the station. To reduce time delay in the control loop, the command signals from the hand controllers are sent directly to the SSRMS Arm Computer Unit (ACU). Other inputs are sent from the control station via the MSS Control and Computing Facility (MCCF) to the ACU. Based on the feedback from camera views and/or direct views, the operator corrects the arm trajectory by modifying the hand controller inputs. The POR position, orientation and velocities, joint positions and rates, force-torque sensor measurements can be displayed selectably.

In automatic trajectory operation, the operator specifies a desired destination in real time or selects an autosequence previously stored in the MCCF. The desired destination or the selected autosequence is then downloaded to the ACU for execution. The operator monitors the arm motions through camera views and/or direct views. He/she can vary the execution rate of the automatic motion.

Both in manual and automatic mode the POR commands received by the ACU computer are resolved into joint rate commands or joint position commands, or both, based on the inverse kinematics of the redundant manipulator. The joint commands are then sent to each of the joint servos to drive the SSRMS. In constrained motion situations the force-moment accommodation control feature can be invoked. It uses a 6-axis force/moment sensor at the tip of the SSRMS to automatically limit the SSRMS tip forces and moments in and about selected axes.

Artificial vision feedback is an other mode of arm control. In this mode, the target position and orientation obtained by the artificial vision unit are fed back to the arm computer for controlling the arm to track the target. The operator can vary the relative position and orientation between the target and the arm. The LEE has six basic functions: snare, de-snare, rigidize, de-rigidize, latch, de-latch. Each of the functions can be operated manually, or pre-programmed to operate in sequence.

The control organization of the SPDM is shown in Figure 6. The basic control concepts of each arm and the latching end effector are similar to those of the SSRMS. Unlike the SSRMS, the SPDM is manipulator system with tree topology, with a total of 19 individual joints. The body and arms can be operated in a coordinated manner. The TCM at each end of the arm can be controlled to hold special tools and ORUs. The programming of automatic sequences permits the automation of tasks such as the exchange of an ORU.

MSS Control System Challenges

Some of the system requirements with respect to the functionality, performance and safety of the MSS, together with constraints on weight and power, result in technical challenges to be met in the design and development of the MSS. We will now highlight some of the critical areas for control systems development.

Control/Structure Interactions: The very large payload capacity of the SSRMS, which exceeds its own mass by two orders of magnitude, can lead to very low fundamental frequencies of oscillation of the SSRMS/payload system. The SSRMS servomechanisms must be designed to cope with these oscillations which vary with payload size and manipulator configuration. Also, dynamic interactions with the Space Station attitude control system are of concern when moving large payloads with the SSRMS.

Force/Moment Control: The force/moment accommodation feature described earlier is a promising new development in space robotics which poses challenges in terms of sensors which can cover the full range of on-orbit operations, as well as in terms of the closed-loop control system implementation which is constrained by relatively low sampling rates. It is difficult to simultaneously achieve stability robustness and high performance in force/moment control of a large and structurally flexible manipulator like the SSRMS interacting with an environment with variable stiffness.

Coordinated Motion of Multiple Manipulators: Coordinated control of the two SPDM arms is required when they jointly manipulate the same object. Coordination of control between the SSRMS and the SPDM is required when the SPDM is operating from the end of the SSRMS and needs to reposition its body while its arm(s) is attached to a fixed object. The coordinated control strategies employed in these cases must avoid the buildup of forces in the closed kinematic chain and must achieve good synchronization of the motion of the different manipulators. Coordinated motion trajectories must be chosen such as to avoid reach limits or singularity zones of each one of the manipulator arms. When the SSRMS and the SPDM execute a coordinated motion (chain topology) the order of the inverse kinematic problem increases which can quickly become a computational burden.

Automatic Control using Artificial Vision System: Canada has gained considerable experience in vision systems for space applications through the tests carried out on the STS 41-G Shuttle mission in 1984 [4] and the development of a Space Vision System (SVS) to be demonstrated during the CANEX-2 experiments on a future Shuttle mission. However, the use of an artificial vision system for automatic control of an operational space manipulator poses additional challenges in terms of controlling the end point of a flexible structure with non-colocated sensors and actuators, and in terms of devising guidance algorithms which will minimize the probability of loosing the target from a relatively narrow field of view.

Prevention of Uncommanded Motions/Operations: The very stringent safety requirements on the MSS provide challenges in designing the MSS control and computer architecture such that health monitoring functions achieve a very wide failure coverage, and that failures will not cause uncommanded motion of the MSS or uncommanded release of a payload. Either one of these possibilities may cause damage to the station.

Machine Autonomy: The current state-of-the-art in robotics is not sufficiently mature for incorporating truly autonomous capabilities into the MSS. However, it is a goal of the MSS program to move towards an autonomous on-orbit capability during the 30-year operational life of the MSS. In addition to the automation features described earlier, the MSS baseline control systems provide "hooks and scars" for future incorporation of advanced high-level task planning systems. The development of these capabilities provides a challenge to the Canadian robotics research and development community. The automation technologies under development for the MSS are described in [5,6].

CONCLUSIONS

The MSS fulfils a critical role in the assembly, maintenance and servicing of Space Station Freedom. The MSS control systems build on the proven concepts of the successful Shuttle RMS in an evolutionary manner. Some of the advanced control system features such as force/moment control and automatic guidance of manipulators by artificial vision, as well as the large payload capacity and the stringent safety requirements provide new challenges for control design and development. These challenges are met by the ongoing MSS development program.

ACKNOWLEDGEMENTS

The authors wish to thank their colleagues at the Canadian Space Agency and at Spar Aerospace Ltd. who have contributed to the development of the MSS Control System.

REFERENCES

1. H.L. Werstiuk, D.M. Gossain, "The Role of the Mobile Servicing System on Space Station", IEEE Int. Conf. on Robotics and Automation, 1987

2. R.C. Hughes, D.G. Hunter, "The Special Purpose Dexterous Manipulator (SPDM): A Focus for Automation and Robotics on the Space Station", AIAA/NASA Symposium on Space Automation and Robotics, 1988.

3. R. Ravindran, K. H. Doetsch, "Design Aspects of the Shuttle Remote Manipulator Control," AIAA Guidance and Control Conference, 1982.

4. B.V. Tryggvason, H.F.L. Pinkney, S.G. MacLean, M. Garneau, C.I. Perrat, B.A. Aikenhead, "Mission 41-G Tests Supporting the Development of a Space Vision System", Canadian Aeronautics and Space Journal, Vol. 31, No. 3, 1985.

5. D.G. Hunter, "Automation of Mobile Servicing System Operations", International Symposium on Artificial Intelligence, Robotics and Automation in Space, 1990.

6. R. Ravindran, S.S. Sachdev, D.G. Hunter, "Automation and Robotics Technologies for the Mobile Servicing System", AIAA Space Programs and Technologies Conference, 1990.

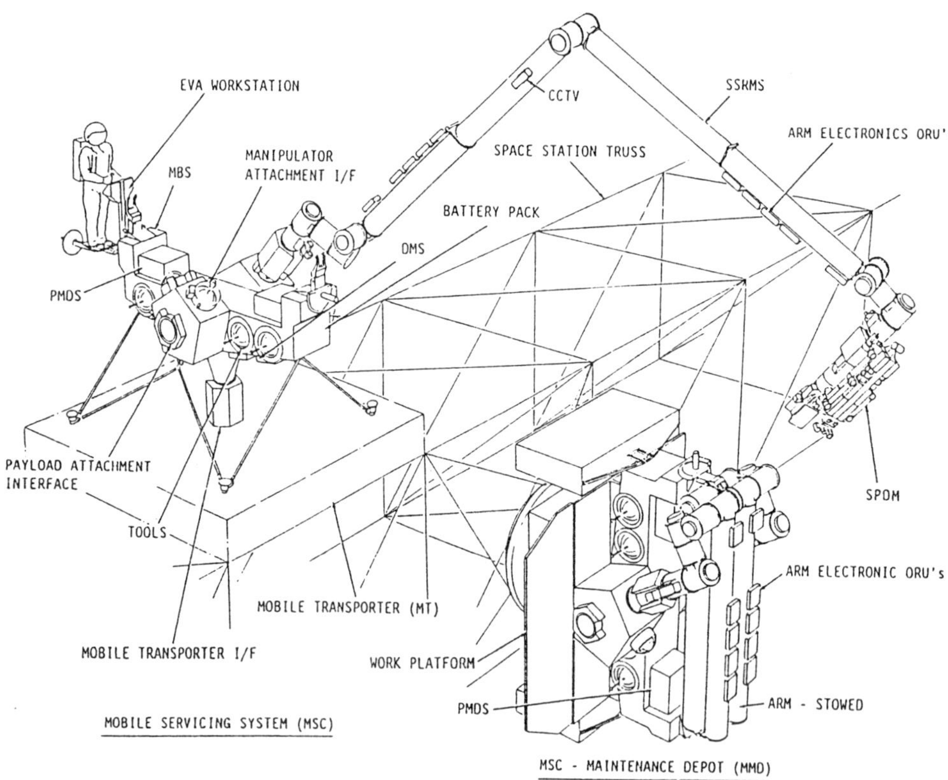

Figure 1: Mobile Servicing System on Space Station

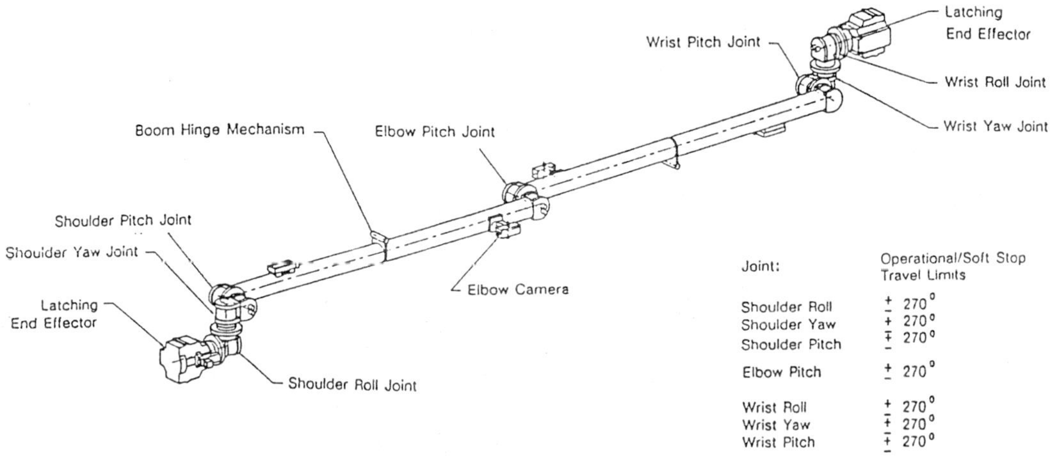

Figure 2: Space Station Remote Manipulator System (SSRMS)

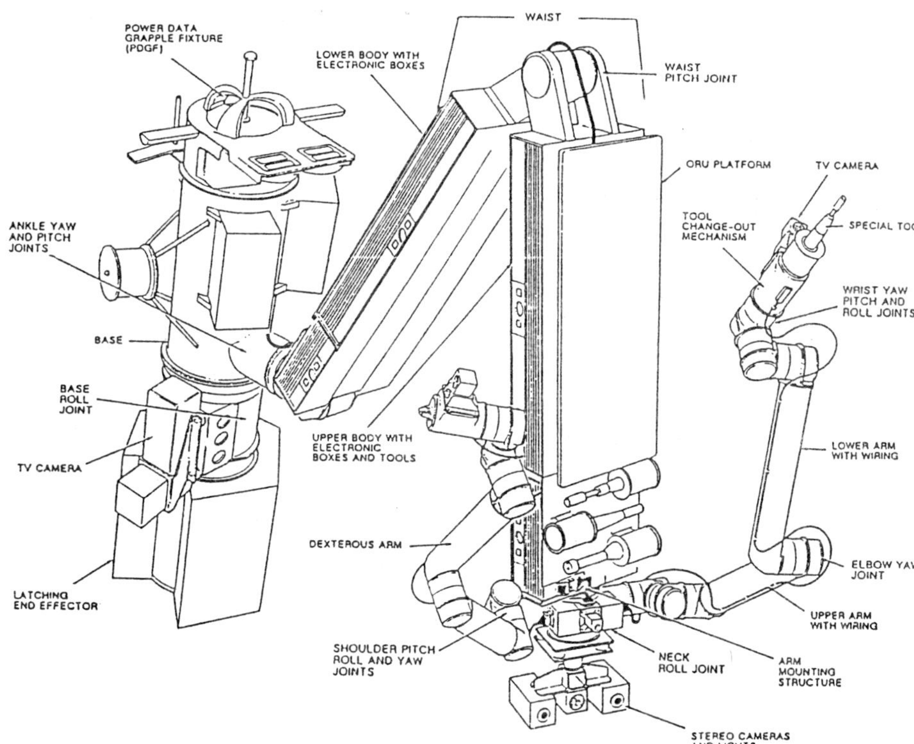

Figure 3: Special Purpose Dexterous Manipulator (SPDM)

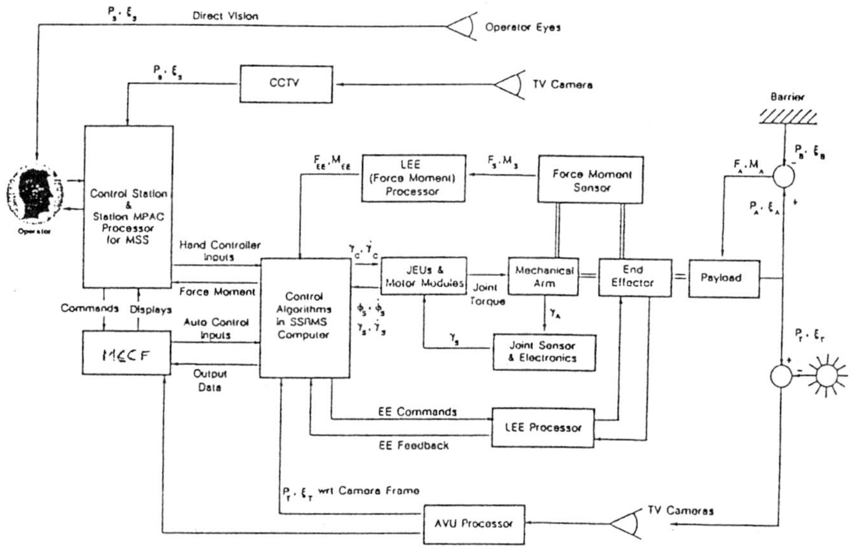

Figure 4: SSRMS Control Systems

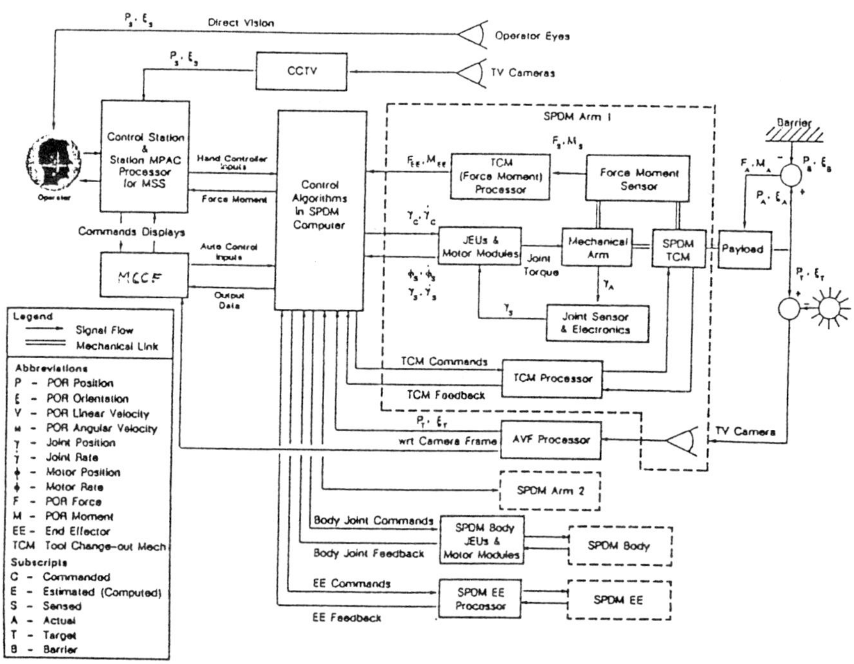

Figure 5: SPDM Control Systems

AAS 91-063

A GUIDANCE, NAVIGATION AND CONTROL SYSTEM FOR RENDEZVOUS AND PROXIMITY OPERATIONS BETWEEN HERMES, THE COLUMBUS FREE FLYER AND THE SPACE STATION FREEDOM

Wigbert Fehse*, Alberto Tobias*,
Calixte Champetier†, Jean-Michel Pairot† and Christian Pauvert†

A Rendez-Vous & Docking Proof-of-Concept Programme has been recently initiated by ESA, covering RV and proximity operations between Hermes, the Columbus Free Flyer and the Space Station Freedom. A complete RV System has been designed on the basis of the commonalities between the Hermes and the Columbus projects. Requirements have been derived, and software and hardware elements have been developed (or adapted) and verified in closed loop for each item of the RV System (docking/berthing system, RV sensors, onboard software, GNC processor, human operator intervention.

The present paper is focussed on GNC. The basic elements and principles (trajectories, equipments used) are presented, followed by some major RVD GNC technical features such as navigation with relative GPS, homing guidance, and the last meters approach prior to docking and berthing. Performances analysis results are provided. Finally, the software tool developed to validate the overall approach is described.

1 INTRODUCTION

The Guidance, Navigation and Control activity is only one of those addressed by the RVD-Preliminary Development Programme (RVD-PDP) described in section 2.2. Indeed, RVD-PDP deals also with Mission and Vehicle Management, Man-in-the-loop intervention, Docking Dynamics, Safety Analyses, and overall verification philosophy for RVD systems. What is presented here on GNC is then a partial result of a work in which GNC is considered within a much broader frame. In the present paper, the stress is essentially put on the automatic rendez-vous approach.

The RVD-PDP is oriented to the analysis of missions involving Hermes, the Columbus Free-Flyer (MTFF), and the Freedom Space Station (FSS). The mission analysed in priority in this paper are:

- Hermes docking with MTFF, where the MTFF is either earth- or inertially pointing
- MTFF berthing with FSS

which are the most promising missions in a near future.

In order to keep the possibility in the future for missions between Hermes and FSS, and to avoid then too high a cost of modifications on Hermes, the Rendez-vous system concept designed and tested during the RVD-PDP takes also into account potential missions between Hermes and FSS, considering both berthing and docking scenarios.

* European Space Agency, 8/10 Rue Mario Nikis, F-75738 Paris Cedex 15, France. † MATRA Espace, Rue des Cosmonautes, ZI du Palays, F-31077 Toulouse, France.

2 LOGIC FOR A RVD SYSTEM PRE-DEVELOPMENT PROGRAMME

2.1 Background and Scope of the RVD Proof-Of-Concept Programme

The European Space Agency started early in the 1980ties with a technology development programme for a Rendezvous and Docking capability, anticipated to be required in future European missions. After the development of strategies and system concepts, of technological items such as rendezvous sensors, latches and connectors for mating, GNC processor and prototype rendezvous control software and of analysis tools, such as guidance/control and docking simulations programs, it was evident that with such items alone one would not have yet a functioning RVD system. A proof that systems concept and elements design is correct in a real case in orbit could be performed, has still to be established. Such a proof can be provided only through integration and testing of a RVD system in closed loop in dynamic environment under realistic environment conditions.

In the second half of the 1980ties the Columbus and Hermes programmes emerged as the major European scenario requiring a rendezvous and docking capability. The Columbus Free Flyer (MTFF) was planned to act both as a chaser when rendezvousing with the Space Station Freedom, and as a target when being served by the European Spaceplane Hermes. Hermes would always have the role of a chaser in rendezvous operations with the Columbus Free Flyer or potentially with other Space Stations, such as Freedom or Mir.

Comparing the Hermes vs MTFF and the MTFF vs Freedom scenarios it was found that the approach scenarios and rendezvous strategies are very similar and that the basic requirements for the automatic rendezvous control systems aboard MTFF and Hermes are comparable. Human operator interaction with the automatic system would be required for the pilot in case of Hermes and for operators in the target (Freedom) and on ground for the unmanned MTFF. In both cases interaction of the human operator with the automatic system would be on relatively high level.

Also, it has been found that RVD is a new technique in Europe, in which too little experience had been established so far to start the flight item development phase without uncontrollable risk. It has been, therefore, concluded that a joint pre-development and performance verification of a basic automatic RVD onboard system by the Hermes and the Columbus projects would reduce development risks and cost significantly. In order to be representative for a real case this pre-development programme would have to include all systems and onboard operations implementation aspects concerning GNC, mission and vehicle management, human operator interaction, docking and berthing, and would also have to include the development of an adequate verification philosophy and of the necessary verification tools and facilities.

Ideally, at the finalisation of such a programme a complete RVD-GNC system (excl. thrusters) would be tested in closed loop under realistic conditions of motion and illumination conditions on a kinematic motion test bed and the docking/berthing system would be tested in real size and subjected to the real velocities, forces and torques on a dynamic motion test bed to demonstrate full compliance with the mission requirements.

However, due to schedule and funding limitations, the RVD Proof-Of-Concept Programme presently conducted in Europe, had to implement a first step, in which only:

- the sensors can be tested under realistic kinematic motion and illumination conditions,

- the RV control software in all its modes will be developed as non real-time software and will run in closed loop against an environment simulation program on a mainframe computer,

- one control mode will run in real time in a GNC onboard processor in closed loop with a real time environment simulator,

- a reduced size docking/berthing mechanism can be tested under realistic conditions of scaled down velocity and force/torque conditions

In a second step the development of the kinematic and dynamic motion facilities will have to be finalised and their readiness for RVD verifications will have to be evidenced. The validation with representative model test items will be at the same time the final step of the proof-of-concept for the RVD system.

2.2 Main Activity Lines in the RVD System Pre-Development Programme

Concepts for Rendezvous and Docking systems, strategies and operations have to consider two main spacecraft systems: the RV Control System, to which the Guidance, Navigation and Control (GNC) system belongs and the Docking/Berthing System (DBS). The RV Control System is in charge of Mission and Vehicle Management (MVM) tasks and Guidance, Navigation and Control functions by means of the GNC System.

Within the RV Control System the key innovative element are the MVM and the RV GNC algorithms, the RV relative navigation sensors and the GNC computers as the GNC algorithms are expected to size the performance of those computers.

The role of the various operators involved, on the chaser, on the target or on ground in GNC and MVM tasks of highly automated RV Control Systems is a key issue to be considered early in the definition of suitable concepts for the systems, RV strategies and operations.

GNC system and DBS can not be considered in isolation (1). The DBS takes over from the GNC the responsibility to guide the two spacecraft until final mating. Establishment of performance requirements must be the result of a process of fair distribution of errors allowed to the two systems.

Five major lines were identified: the MVM and GNC onboard algorithms, the real-time performance of those algorithms on a real GNC computer, the role of the operators in MVM and GNC tasks, the RV sensors and the Docking/Berthing System.

The RVD System Pre-Development programme (RVD-PDP) was structured in two phases. In a first phase, requirements of the Hermes and Columbus projects were identified and assessed. Baseline concepts were analysed and suitable "strawman" concepts were defined. The strawman concepts for all the above activity lines has to envelop the Hermes and Columbus baseline concepts to allow their proof and at the same time provide margin to assess their robustness. RVD strategies were defined for each Hermes and Columbus scenario for nominal and contingency conditions. Preliminary concepts for RV control systems, including among other concepts for MVM and GNC systems, were defined.

Safety, performance and operations analyses were carried out, as shown in section 4, and were iterated with the concepts for RVD strategies and RV Control System allowing to derive detailed concepts. RVD strategies were split in phases and these in modes with identification of transition criteria as shown in section 3. GNC structure and GNC algorithms were established.

A verification approach intended to provide proof of the concepts was defined - as shown in section 5 for the GNC system - and development requirements for models and verification tools were identified.

Phase 2 saw the split of the RVD-PDP in five main lines according to the above:

a RV Control Software

MVM and GNC algorithms were developed by MBB/ERNO (Germany) and MATRA (France) and coded. The algorithms had to be tested in closed loop with a high fidelity software environment simulator. A Rendezvous Verification Simulator (RVV) presented in section 5 allowed to implement the Hermes and Columbus scenarios, check the performances established in the analyses of phase 1 and provide proof of the MVM and GNC concepts.

b Restricted Real-Time Test with Real onboard Computer

Final proof of a concept should include the demonstration that all MVM and GNC algorithms run real time on the real computer in a simulated environment. The RVD-PDP focused on the last mode of the RVD mission, the final approach of Hermes to the MTFF. This is the most demanding mode in terms of RVD related issues, computing power, elements handled by the GNC computer and safety of the mission. The appropriate subset of the RV Control Software was coded and implemented by MATRA (France), GMV (Spain) and SAAB (Sweden) on an Advanced Guidance and Control processor (AGCP) developed by ESA for RVD missions and run against an environment simulator.

c Man-In-The-Loop

Candidate concepts for crew role identified in phase 1 were developed in phase 2. Additional GNC algorithms were designed and coded, Man Machine Interfaces were defined and implemented. Simulators for intervention in MVM and GNC tasks were developed with MBB/ERNO (Germany) and AEROSPATIALE (France)

d RV Relative Navigation Sensors

Sensors for relative navigation have been developed in Europe in recent years under funding of ESA or National Space Agencies. In RVD-PDP sensors from MBB (Germany), MATRA (2) (France) and SAAB (Sweden) were further developed for integration and test on the European Proximity Operations Simulator (EPOS) in Oberpfaffenhofen (Germany). EPOS is developed in parallel with the RVD-PDP by the DLR (Germany) in the frame of the RVD Proof-Of-Concept Programme and will allow to test current concepts for RV sensors under realistic motion and illumination conditions.

e Docking/Berthing System and Dynamics

The DBS performance and the concept itself are intimately related with the performances of the GNC systems of the chaser and target and the strategy for proximity operations, as detailed in section 4.3. Proof of the concept for RV Control System and RVD strategies shall go in parallel with the proof of the DBS concept. In RVD-PDP a model a Front End of the DBS was built by SENER (3) (Spain) and tested on a Docking Dynamics Test Facility at MATRA (4) (France). Simulations with a software simulation tool were run in parallel by AERITALIA (Italy) and results of tests and simulations were cross checked.

3 GNC CONCEPT FOR RENDEZVOUS

Trajectories

A set of trajectories were considered in the RVD-PDP for operations between the various involved spacecraft. Figure 3.a illustrates the operations of Hermes with the MTFF. Figure 3.b illustrates the operations of both Hermes and the MTFF with the Space Station Freedom.

Equipments

All the chaser equipments are connected to the GNC Real-Time Bus (GNC/RTB) for exchanges of information with the RV Control Software.

Sensors

The Global Positioning System (GPS) is utilized by the chaser spacecraft for the estimation of its absolute position during the whole orbital flight, up to a point where switching to relative navigation using another sensor is performed (see below). During the RV and proximity operations, the chaser GPS receiver works in relative mode (called RGPS) for the elaboration of the target-to-chaser relative position at long and medium ranges (typically up to 100-200 metres) from information delivered by the chaser GPS receiver, and by the GPS receiver mounted on the target spacecraft. A local link (Ku-band) between the target and the chaser spacecraft is used to allow the exchange of relevant information for the setup of the relative position. The processing of the chaser and target GPS measurements (RGPS) is performed within the GNC Application software, as described in section 4.1.

The Inertial Measurement Units (IMU), used during all the RV and proximity operations for attitude prediction (use of gyro information) are composed of three gyros and three acceleros for Hermes, and one 2-DOF gyro plus one accelerometer for MTFF. The accelero information shall be used for failure detection .

The Absolute Attitude Measurement Sensors:

- star sensor, providing inertial attitude measurement for both Hermes and the MTFF
- sun sensor, providing inertial attitude measurement (restricted to the MTFF)
- earth sensor, providing earth-related attitude measurement (restricted to MTFF)

Absolute attitude measurement sensors provide information for attitude update during all modes where the chaser is nominally earth- or inertially pointing.

The Rendez-Vous Sensor (RVS) is used for:

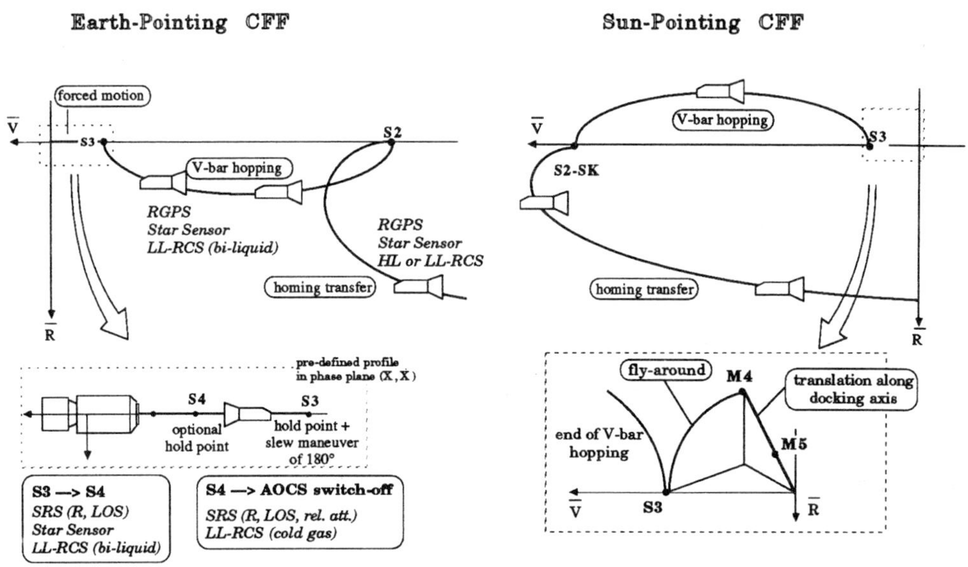

Figure 3.a : RV and Proximity Operations of Hermes with the Columbus Free Flyer

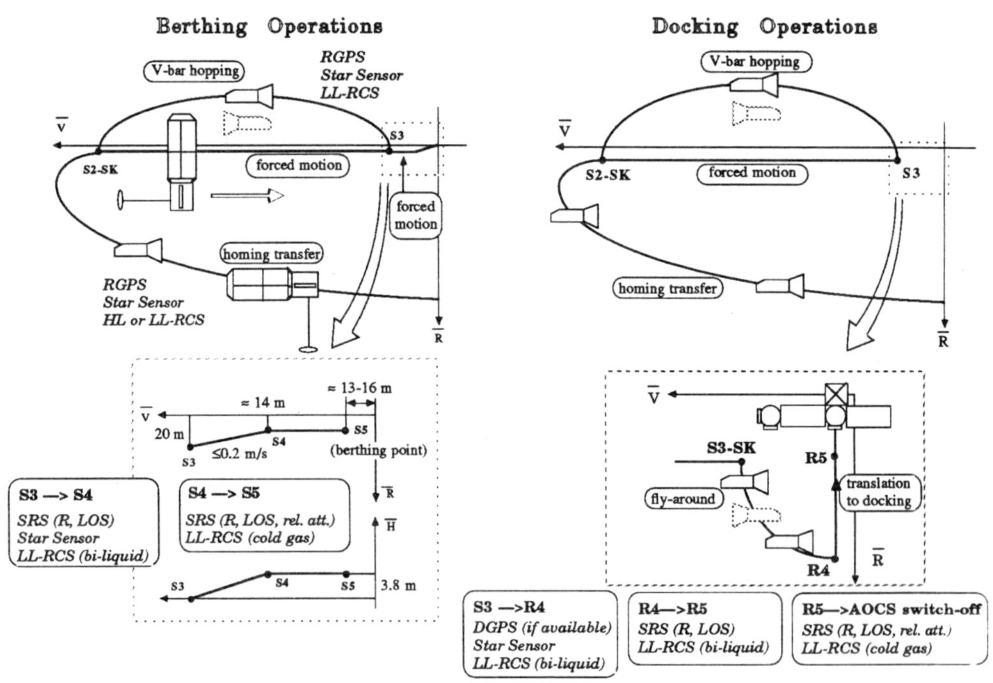

Figure 3.b : RV and Proximity Operations of Hermes and MTFF with Freedom

- the final translation to docking for operations of Hermes with the MTFF (along V-bar and along an inertially-pointing axis) and the SSF (along R-bar)

- the final translation to berthing for operations of both Hermes and the MTFF with the SSF

The RVS function is to deliver a measurement of the relative position and additionally of relative measurement in target close vicinity.

Two concepts of RVS were nominally considered within the PDP: a Short Range Sensor (SRS) and a Proximity Sensor System (PSS). An alternative concept for RVS is considered within the PDP, namely the Medium Range Sensor (MRS). The MRS may be assumed to be the back-up solution to RGPS for some operations.

Sensor type	Working range	Measurements
SRS (camera type)	200 m —> 0.5 m	above 20m: relative range measurement measurement of the relative LOS additionally a measurement of the relative spacecraft attitude for ranges below 20 metres.
PSS	15 m —> 0.1 m	relative atitude
MRS	1000 m —> 100 m	relative position

Actuators

The Intelligent Thruster Control Unit (ITCU) in charge of transforming the commanded forces and torques established with the GNC software, into information related to each individual unit of the considered thruster cluster. The ITCU additionally transforms the information commanded to each individual thruster, into force and torque vectors expressed in the chaser body frame (use of thruster arrangement matrix). The exchange of information between the GNC Computer and the ITCU are performed via the Real-Time Bus.

Thrusters

During the RV and proximity operations, the actuation of both Hermes and the MTFF, acting as chaser spacecraft, shall be achieved by means of thrusters as described below:

free drift and homing transfer	bi-liquid thrusters
closing and final approach	low level bi-liquid thrusters
close target vicinity : - 40 meters for Hermes when rendezvousing with the MTFF - 30 meters for Hermes when rendezvousing with the SSF - 30 meters for the MTFF when rendezvousing with the SSF	low level cold gas thrusters
withdrawal operations	cold gas or bi-liquid thrusters (as for the approach).

GNC Real-Time Bus (GNC/RTB)

The GNC/RTB is a local link between the equipments (sensors, actuators) and the GNC Computer.

GNC Computer

Within the RV Control System concept established for the PDP, two separate computers shall be considered, on which the RV control on-board software shall run:

- the System Computer (SYSC)
- the GNC Computer (GNCC)

The GNC computer includes the following functions:

- the GNC application software, ensuring the estimation of the chaser attitude and position, the guidance and the setup of forces and torques for spacecraft control
- the low level FDIR (functional FDIR for equipments)
- the management of the data (I/O management)
- the algorithms management
- an interface with the system computer

GNC Software

The Navigation function performs the following sub-functions:

sub-function	processing of the raw measurements	prediction	updating
input from	sensors	- on-board inertial reference units - target spacecraft via the local link (relative attitude prediction only) - ITCU reports	attitude: - inertial attitude measurement - RVS's position - target and chaser spacecraft (GPS), - RVS's

The Guidance function performs the following computations:

- reference strategies or reference trajectories
- corrections to be applied to the theoretical profile in order to take into account the realworld perturbations.

and delivers:

- linear/angular guidance acceleration
- the nominal linear/angular rates
- in some cases the nominal attitude/position

The guidance function is also in charge of calculating the acceleration needed to compensate for the chaser motion in an earth-pointing frame.

The Control function performs the relevant actions in order to calculate the control velocity or the impulsions or the mean angular/linear accelerations to perform the attitude/position control. It is in charge of controlling the linear and angular position and rates of the chaser wrt the nominal state elaborated by the guidance function.

4 DESCRIPTION AND PERFORMANCES OF GNC SYSTEM MAIN FEATURES

4.1 Navigation with Relative GPS

Rendezvous Operations using Relative GPS

The chaser navigation during the long and medium range rendezvous operations is carried out by means of the Relative GPS using the satellites of the NAVSTAR constellation. RGPS is assumed to be initiated during the chaser phasing operations along its free drifting orbit below the target orbit, so as to ensure the convergence of the on-board filters. Relative GPS navigation is used up to the point where a direct measurement of the relative position between the two involved spacecraft is made possible, i.e. a few hundreds of meters. Relative GPS is thus used during the following rendezvous operations:

- the homing phase, during which the chaser spacecraft is transfered from its free drifting phasing orbit up to a point on or close to the target orbit (a few hundreds of metres, mission-dependent);

- the closing phase, during which the chaser spacecraft is transfered up to a point still on/close to the target orbit, but at a shorter distance;

- the circumflight maneuver if required, allowing to acquire the target mating axis when not on the approach axis (e.g. the target inertially-pointing axis for the docking operations of Hermes with the sun-pointing Columbus Free Flyer, or the earth-pointing axis for docking operations of Hermes with the Space Station Freedom).

GPS Measurements and Absolute Navigation

The GPS absolute navigation filter, running before the rendezvous and proximity operations are initiated between both spacecraft, is propagated all along the mission in parallel with the relative GPS filter, so as to keep an inertial reference which may turn out to be necessary when escape maneuvers must be initiated during the approach operations to the target (contingency operations).

The NAVSTAR GPS is based on a constellation of 18 satellites (minimum number) placed on circular orbits at altitudes of 20,000 km. The information delivered by each GPS satellite (GPSS) consists in the current satellite ephemeris and of the GPS date. Comparing the delivered GPS date with the spacecraft on-board time yields the so-called pseudo-range information:

$$\| \underline{r}_H - \underline{r}_i \| . \delta t$$

where $\underline{r}_H$ is the spacecraft absolute position, $\underline{r}_i$ is the GPSS absolute position, and δt is the desynchronisation time between the spacecraft time and the GPSS time (all GPSS's are assumed to be synchronized). A range rate measurement may also be performed using the Doppler effect on the GPS signal carrier, and leads to an integrated doppler measurement. Proceeding by triangulation from the three measurements would lead to the desired absolute position if $\delta t = 0$. Since the spacecraft and GPS clocks are not synchronized, a fourth measurement is needed to recover a reliable position estimation. The triangulation accuracy crucially depends on the geometric space distribution of the visible GPS satellites. If $\underline{u}_i$ is the unit vector defining the line of sight (LOS) between the spacecraft and the GPSS number (i), and [H] a 4x4 matrix of the selected LOS, the parameter:

$$tr(H^T H)^{-1} \qquad \text{with } H = \begin{pmatrix} \underline{u}_1 & \underline{u}_2 & \underline{u}_3 & \underline{u}_4 \\ 1 & 1 & 1 & 1 \end{pmatrix}$$

named Global Dilution Of Position (GDOP) characterizes the achievable positioning performances. The smaller, the better the localization. With more than four visible GPSS's, a selection which minimizes the GDOP is performed on-board.

The main GPS measurement errors come from the Selective Availability (SA) concept attached to the NAVSTAR procedures; as a matter of fact, for civil applications, a low frequency coded degradation is intentionally superimposed to the original signal. The following other perturbations are also to be considered:

- the GPS receiver noise;
- the GPSS ephemeris errors;
- the ionosphere propagation;
- the time divergence between the spacecraft and the GPSS clocks.

Sources of errors	error (1σ)
SA + GPSS ephemeris + satellite clock	30 m
user receiver	10 m
ionosphere	15 m
Total	33 m

<u>Relative GPS Navigation Filter</u>

For all considered rendezvous and proximity operations, the target spacecraft is assumed to be equipped with GPS receivers, and performs a processing on raw measurements similar to that carried out aboard the active chaser spacecraft. The relative navigation uses the GPS measurements carried out by the receivers placed on both vehicles, and is performed aboard the approaching chaser spacecraft. The target spacecraft transmits its own raw GPS measurements to the approaching vehicle by means of a dedicated radio link. Each GPS measurement (target and chaser) consists in:

- the identity of the visible GPS satellites, on which a measurement has been performed;
- the GPS measurements proper (either pseudo-range or integrated doppler);
- the GPS receivers observation times;
- the ephemeris of the GPS satellites (those seen by the chaser only).

In case four or less GPS satellites are commonly seen by both the target and the chaser receivers, the corresponding measurements shall be used in the relative GPS navigation filter. In case more than four common GPS satellites are seen by both spacecraft (in fact never more than eight in the best case), the selection of the four "better" satellites is first to be carried out, which minimizes the GDOP. The information on these four selected satellites will be used for the Relative GPS navigation filter update.

The structure of Relative GPS navigation filter depends on the nature of the delivered measurements.

In casc of pseudo range measurements, the propagated state vector consists of:

$$\underline{Y} = \left(\underline{X}\,,\, \dot{\underline{X}}\,,\, \Delta T_R^R\,,\, E_{clock}^R\right)^T$$

In case of integrated doppler measurements, the propagated state vector consists of:

$$\underline{Y} = \left(\underline{X}\,,\, \dot{\underline{X}}\,,\, \Delta T_R^R\,,\, E_{clock}^R\,,\, I_R^1\,,\, \dots\,,\, I_R^{NR}\right)^T$$

where:
- X and $\dot{X}$ are the relative position and velocity vectors expressed in the target LVLH frame
- ΔT_R^R is the differential clock error
- E_{clock}^R is the differential frequency error
- I_R^i is the integrated doppler constant for measurement number (i)

The prediction of the position and velocity terms of the state vector is carried out by means of a first order Clohessy-Wiltshire model, fed with the propulsion information delivered by the on-board propulsion unit when any. In parallel with the Relative GPS navigation filter state vector, the error covariance matrix $\hat{P}_R$ of the Relative GPS navigation filter is propagated as follows:

$$\hat{P}_R = \Phi_R^T\, \hat{P}_R^-\, \Phi_R + Q_R$$

where :
- Φ_R is the transition matrix (depends on measurement type, pseudo-range or integrated doppler)
- Q_R is the state noise covariance matrix (depends on measurement type)

The update of the Relative GPS navigation filter is carried out on the basis of the GPS raw information delivered by both spacecraft receivers, when any. The information fed into the Relative GPS navigation filter is expressed in the form of an innovation $\underline{I}_R$, setup from:

- the current Relative GPS navigation filter state vector $\underline{Y}$;
- the current Absolute GPS navigation filter, propagated in parallel with the relative filter, and delivering an estimate of the chaser absolute position and velocity vectors wrt an inertial reference frame
- the estimated position and velocity of the relevant GPS satellites on a reference frame

In addition to this innovation, an observation matrix H_R associated to the selected GPS satellites is calculated. The update of the relative state vector $\underline{Y}$ is thus:

$$\underline{Y}^{+} = \underline{Y} + K_R \cdot \underline{I}_R$$

where K_R is the gain matrix of the Relative GPS navigation filter, calculated from:

- the estimated covariance matrix $\widehat{P}_R$
- the observation matrix H_R
- the measurement noise covariance matrix R_R

$$K_R = \widehat{P}_R \cdot H_R^T \cdot \left[H_R \cdot \widehat{P}_R \cdot H_R^T + R_R \right]^{-1}$$

Figure 4.a illustrates the GPS measurement processing performed aboard the chaser spacecraft. Figure 4.b illustrates the principle considered for the relative navigation using GPS information.

4.2 Guidance during the Homing Transfer

The homing phase is an orbital maneuver consisting in transfering the chaser from a free drifting orbit located below the target orbit to an orbit the altitude of which depends on the considered mission. The aimed orbit is either the target orbit, or located a few meters below. The position of the chaser wrt the target on the final orbit, also depends on the considered mission.

The homing trajectories are defined according to the principle of Hohmann transfer which performs the transfer thanks to horizontal thrusts. The constraints on the final altitude of the transfer determine mainly the forces to be commanded to the chaser, and the constraints on the final position determines mainly the moment when the transfer has to begin. Two classes of homing transfers may be considered:

1 Impulsive transfer

Two horizontal impulses are applied to the chaser (a first impulse ΔV_{X1} at the beginning of the transfer, a second impulse ΔV_{X2} at the end of the transfer). The transfer duration is half an orbital period. The impulse amplitudes are:

$$\begin{cases} \Delta V_{X1} = \dfrac{\omega (Z_0 - Z_F)}{4} \\ \Delta V_{X2} = \dfrac{\omega (Z_0 - Z_F)}{4} - \dfrac{3\, \omega Z_F}{2} \end{cases}$$

where :
- Z_0 is the chaser coordinate along the Z-axis of the target LVLH frame at homing transfer start
- Z_F is the aimed chaser coordinate along the Z-axis at homing transfer end
- ω is the target orbital rate

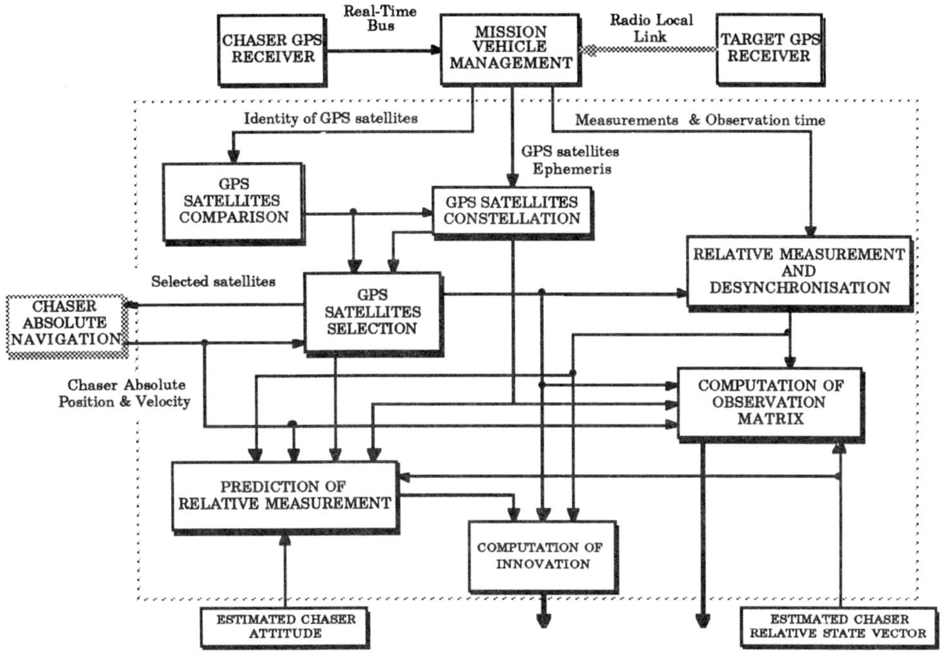

Figure 4.a : GPS Measurement Processing

DATA PROVIDED TO GNC FOR EACH SELECTED GPS S/L :
- Pseudo-Range measured by Target receiver and observation time
- Pseudo-Range measured by Chaser receiver and observation time
- GPS satellite Ephemeris

RELATIVE NAVIGATION FILTER STATE VECTOR :
- Chaser relative Position in Target local orbital frame
- Chaser relative Velocity in Target local orbital frame
- Differential clock error
- Differential clock frequency error

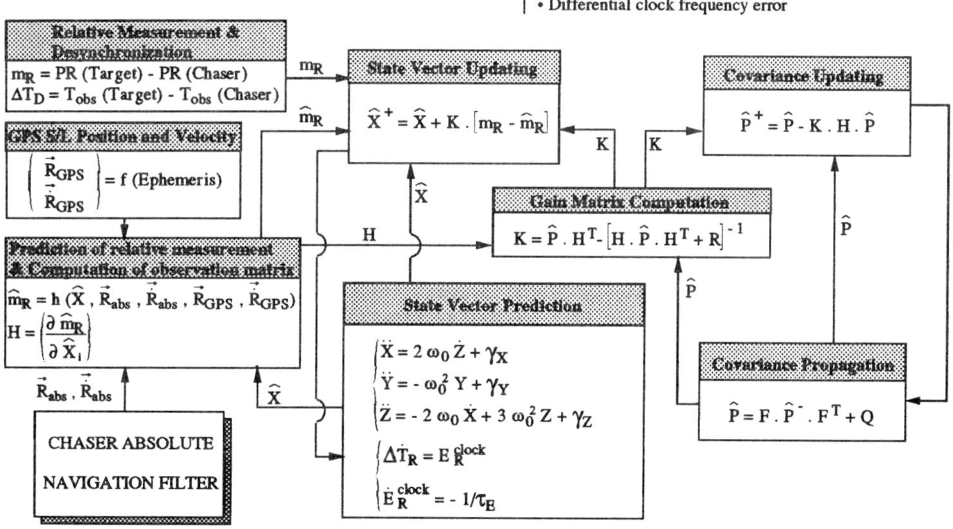

Figure 4.b : Principle of Relative Navigation using GPS

2 Continuous transfer

A constant acceleration γ_X is continuously applied to the chaser during the whole transfer. The transfer duration is one orbital period. The acceleration to be applied is

$$\gamma_X = \frac{\omega^2 (Z_0 - Z_F)}{4\pi}$$

In fact the actual guidance cannot be performed in such an ideal manner, because neither pure impulses nor a constant acceleration can be produced by actual on-board thrusters.

Quasi-impulsive transfer

The two-impulse guidance mentioned above is in fact replaced by a two-boost guidance profile, the amplitude of which γ_{max} is within the spacecraft acceleration capabilities, and with a variable duration (τ) depending on the required impulse.

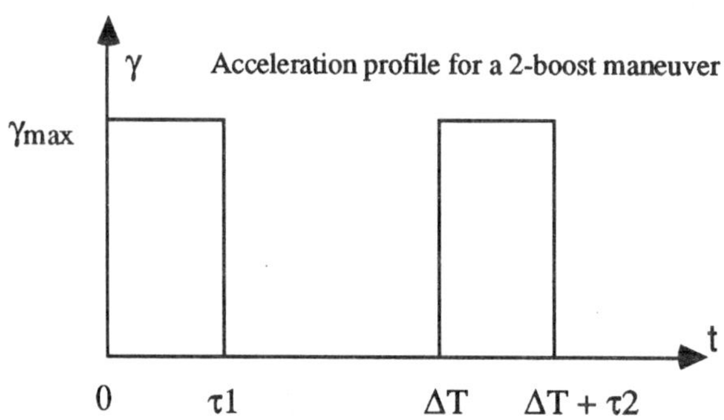

$$\Delta T = \frac{\pi}{\omega}$$

$$\tau 1 = \frac{\omega(Z_0 - Z_F)}{4\gamma_{max}}$$

$$\tau 2 = \frac{\omega(Z_0 - 7Z_F)}{4\gamma_{max}}$$

Quasi-continuous homing transfer

The proposed approach consists in spliting the continuous thrust into successive boosts. The total velocity increment is provided by performing (N) boosts of variable duration. The interval of time ΔT between two successive boosts is constant. To get closer to the ideal continuous strategy, (N) will have to be large enough.

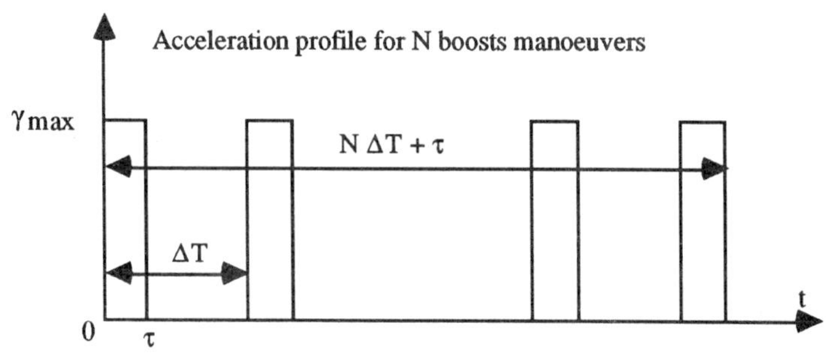

$$\Delta T = \frac{2\pi}{N\omega}$$

$$\tau = \frac{\omega(Z_0 - Z_F)}{2N\gamma_{max}}$$

Terminal-point guidance and control principle

During the homing transfer, a particular emphasis is put on the final aimed point in terms of position and of velocity. A terminal-point guidance has been selected to meet this constraint. The terminal-point guidance principle is to calculate, at each guidance sample time, a new admissible trajectory going though both the (estimated) chaser position and the desired final point, with the defined final velocity. The terminal-point guidance delivers the amplitude of the horizontal boost to be applied (parallel to the target LVLH X-axis), and the nominal chaser velocity at current time. The chaser estimated velocity is thus controlled so as to fit to the nominal one delivered by the guidance.

On-trajectory guidance and control principle

Since the algorithms used for terminal point guidance and control may nevertheless become singular in the vicinity of the aimed point, a change in guidance principle has to be made at the end of the homing transfer. It consists in controlling both the position and the velocity of the chaser such as to maintain the chaser position on a pre-defined guidance trajectory, which is the last estimated trajectory during the terminal-point guidance. As far as fuel consumption is concerned, this concept is less interesting than the first one, but offers better performances and robustness.

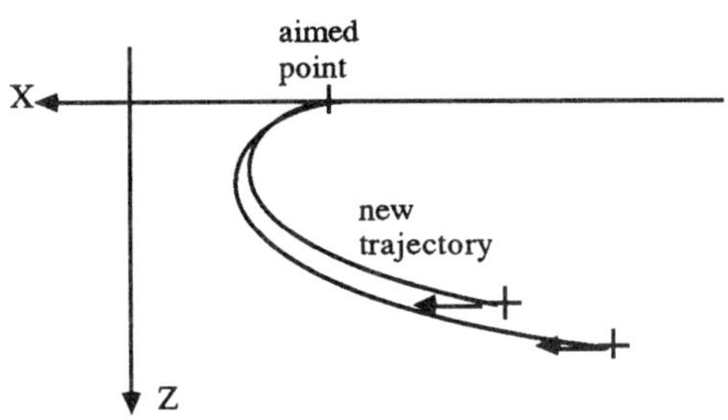

Terminal-point guidance

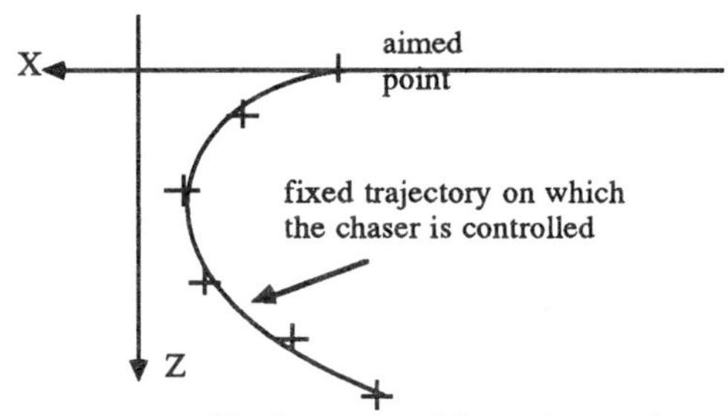

Trajectory guidance/control

Performances

The following performances can be expected during the homing transfer when using the combined terminal-point guidance + trajectory control scheme. The following sources of errors are accounted for:

- navigation errors (Relative GPS navigation errors)
- model errors (2nd order Clohessy-Wiltshire and J_2)
- differential air drag error : 10^{-6} m/s^2
- thrust errors (2% in module; 0.25 ° for bias on thrust direction ; 0.05 ° for noise on thrust direction)

		$N_{boosts} = 60$	$N_{boosts} = 100$	$N_{boosts} = 150$
X (m)	(3 σ)	11.0	6.6	4.8
Z (m)	(3 σ)	17.0	10.5	15.1
$\dot{X}$ (cm/s)	(3 σ)	4.1	2.4	2.0
$\dot{Z}$ (cm/s)	(3 σ)	2.1	1.4	1.5
fuel consumption (m/s)	(mean)	6.99	6.96	6.77

Switching from terminal-point guidance to trajectory control is performed 10 boosts before reaching the aimed point for this particular example.

4.3 Navigation, Guidance and Control during Final Operations to docking

The final operations to docking start when the chaser spacecraft is roughly aligned in position and attitude with the target docking axis, or with an axis parallel to it. They are often preceded by station-keeping operations, allowing e.g. the inspection of the docking port prior to the start of the final operations. Safety is a critical item during these operations. Collision has to be avoided, and a particular emphasis has to be put on the definition of the approach scheme along the docking axis. Docking operations may be performed:

- along the X-axis of the target LVLH frame (docking of Hermes with earth-pointing Columbus Free Flyer);
- along the Z-axis of the target LVLH frame (docking of Hermes with the Freedom Space Station);
- along one target axis which is maintained inertially-pointing (docking of Hermes with sun-pointing Columbus Free Flyer)

The safety profile is (rather) easy to setup in the case of operations with earth-pointing target spacecraft, but is almost impossible to establish in the case of operations with sun-pointing spacecraft.

Navigation

The navigation during the final operations to docking is based on the utilisation of optical sensors operating with a set of retro-reflectors located on the target spacecraft, which delivers:

- a measurement of the range between the target retro-reflectors and the sensor reference frame;
- a measurement of the line-of-sight between the target retro-reflectors and the sensor reference frame, in the form of e.g. 2 bearing angles;
- a measurement of the relative attitude between the target retro-reflectors and the sensor reference frame, if the shape of the target retro-reflectors allows for it. The relative attitude measurement is usually possible only when both spacecraft are in close vicinity (typically 10 to 20 meters).

At the start of the final operations to docking, the chaser position and velocity are estimated and controlled wrt a frame independent from the actual target attitude. This frame is the LVLH frame in case of operations with an earth-pointing target, and the target inertially-pointing frame in case of operations with a sun-pointing target. In parallel, the chaser attitude is estimated and controlled wrt a frame independent from the actual target attitude (chaser LVLH frame or inertial reference frame respectively).

At the end of the final operations to docking, the chaser position and attitude needs to be estimated (and thus controlled) wrt a frame related to the target spacecraft, since drastic tolerances are usually required at docking. The filter structure is thus different, and the coupling between position and attitude must be accounted for the filter setup. The relative position, velocity, attitude and attitude rate between vehicles docking ports are thus estimated and controlled.

The estimation filter is of (α,β) type; an example is given below for the lateral position/ velocity (Z):

$$\begin{cases} \dot{\hat{Z}} = \hat{\dot{Z}}_C + L_C \Omega_C^{meas} + \left(\hat{X} + L_T\right) \hat{\Omega}_T + k\left(Z^{meas} - \hat{Z}\right) \\ \dot{\hat{Z}}_C = \hat{\gamma}_C^Z + \hat{\dot{X}}\, \hat{\Omega}_T + l\left(Z^{meas} - \hat{Z}\right) \\ \hat{\dot{Z}} = \hat{\dot{Z}}_C + L_C \Omega_C^{meas} + \left(\hat{\dot{X}} + L_T\right) \hat{\Omega}_T \end{cases}$$

where :
- L_C is the distance between the chaser COM and the chaser docking port
- L_T is the distance between the target COM and the target docking port
- Ω_C^{meas} is the chaser gyro measurement
- Z^{meas} is the (Z) measurement reconstituted from the RV sensor measurements
- $\hat{\gamma}_C^Z$ is the estimated chaser acceleration in the reference frame
- (k) and (l) are the navigation gains

Guidance

The guidance profile along the approach direction must ensure, as far as feasible, a collision-free trajectory allowing to apply a passive safety concept. Starting from a hold point (station keeping operations), the trajectory allowing the chaser to dock with the target spacecraft must include at least:

- an acceleration phase;
- a deceleration phase;
- a constant velocity phase (docking must be performed with a constant approach velocity)

A constant velocity phase may be added between the acceleration and deceleration phases if necessary. The acceleration and deceleration phases may additionally be split into various phases allowing to meet the safety. Deeper analyses have been carried out in the case of the final approach of Hermes to an earth-pointing Columbus Free Flyer. In that particular case, an example of guidance profile shown on figure 4.c has been established. It consists in:

- a constant acceleration phase (compatible with Hermes propulsion capabilities)
- a constant velocity phase
- a "proportional braking phase", corresponding to a passive safety parabola
- an exponential braking, ensuring also a safe approach at a closer relative distance
- a constant velocity phase performed at 5 mm/s, ensuring soft docking operations

The overall duration of the final operations to docking is in the order of 39 minutes, i.e. lower than one orbital day (58 minutes).

<u>Control</u>

A control of the chaser velocity along the approach axis is performed all along the final approach to docking. The control box is fed with information delivered by both the navigation box, and the guidance box, as illustrated below. A proportional control on the difference between the nominal velocity delivered by the guidance, and the estimated one has been considered.

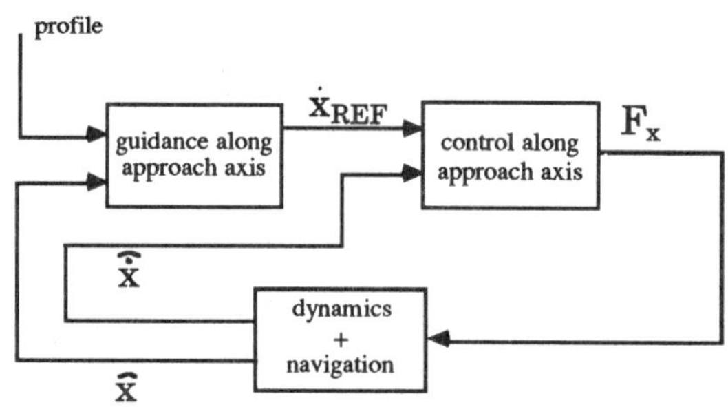

The lateral position and velocity are to be controlled to zero all along the final operations, so as to ensure performances at docking compatible with the docking/berthing system requirements. A proportional/ derivative control has been selected for that purpose. Attitude and attitude rate are controlled to zero all along the final operations, first wrt the chaser reference frame (LVLH or inertially-pointing frame), then wrt the target docking frame in the last part of the approach. A proportional/derivative control has also been selected for that purpose.

During the last metres of the final operations, a strong coupling exists between the position control and the attitude control, which cannot be achieved separately. Several concepts may be envisaged to counteract this effect. The selected concept consists in first transfering the chaser center of mass to the docking axis by applying a force, then to correctly orient the chaser wrt the target docking axis by applying both a force and a torque, as illustrated below.

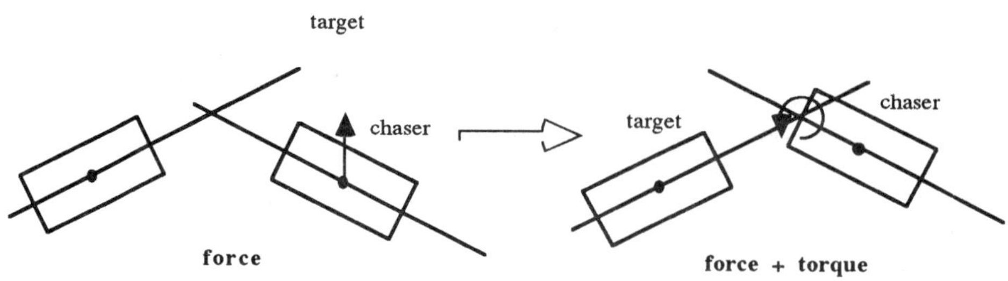

Figure 4.d illustrates the complete GNC scheme immediately prior to docking, when there exists a coupling between the relative attitude and the lateral relative position coordinates.

Performances

Performances of the complete GNC scheme proposed here for the final approach prior to docking have been assessed in the case of operations between Hermes and an earth-pointing Columbus Free Flyer. Considering a typical rendezvous sensor with the following errors:

range error	0.5 % x Range
relative attitude error	0.5 °
line of sight error	0.05 °

an analytical evaluation of the noise and bias transmission through the filter and an optimization of the error budget has yielded typical bandwidths for the navigation filter (for a measurement frequency of 1Hz). The guidance/control gains have been computed so as to minimize the transmission of the disturbances (sensor noise and bias, target angular motion) through the whole GNC filter.

bandwidths	Navigation	Control
relative attitude estimation	0.01 Hz	0.01 Hz
lateral position estimation	0.03 Hz	0.05 Hz
axial velocity	0.04 Hz	0.05 Hz

Figure 4.e shows the relative lateral position between spacecraft docking ports during the last part of the final approach, up to docking. The actual position coordinates and the estimated ones are represented.

Figure 4.f shows the relative attitude between the spacecraft docking ports during the last part of the final approach, up to docking. The actual relative attitude and the estimated one are represented.

4.4 Final Translation to Berthing

The berthing operations are performed between the Columbus Free Flyer or Hermes as chaser spacecraft, and the Freedom Space Station. The chaser is transfered by homing and closing maneuvers from its free drifting orbit to a point in front of the target, below the velocity axis (X-axis of the LVLH frame), at a few tens of meters from the target center of mass (typically 100 to 130 meters). The objective of this final translation to berthing is to transfer the chaser spacecraft in target vicinity, so as it is captured by the FSS manipulator, which then autonomously performs the berthing of the chaser to a dedicated station port. As for the final operations prior to docking, the final approach prior to capture must be safe, so as to avoid collision with the target spacecraft.

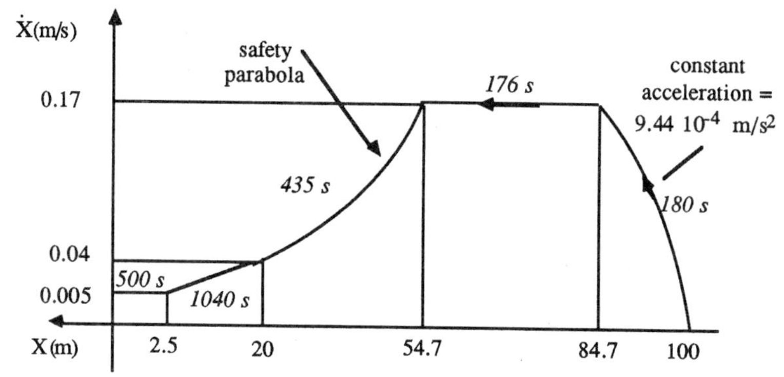

Figure 4.c : Example of guidance profile during Final Operations to Docking (Hermes/earth-pointing MTFF)

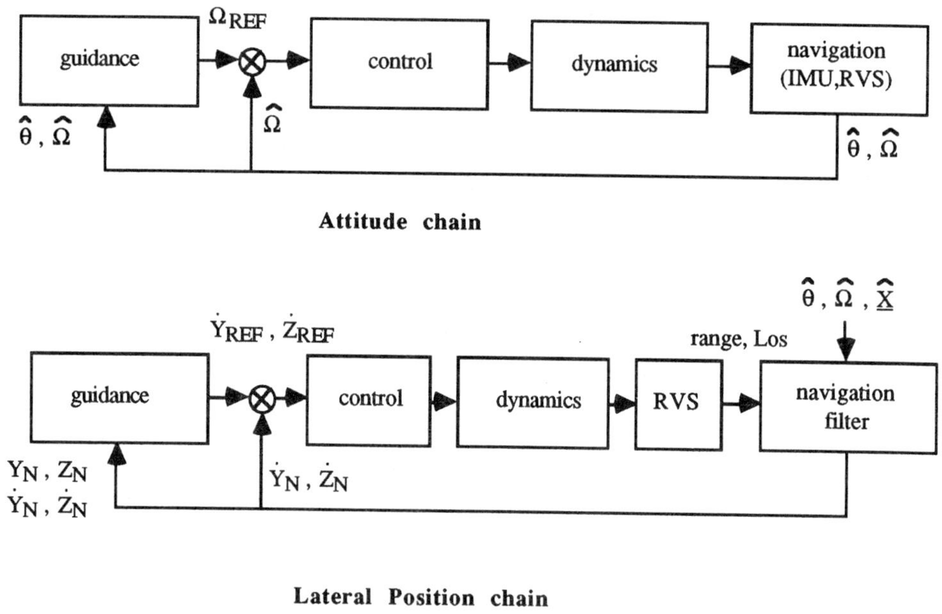

Figure 4.d : Attitude and Lateral Position GNC in the last meters before docking

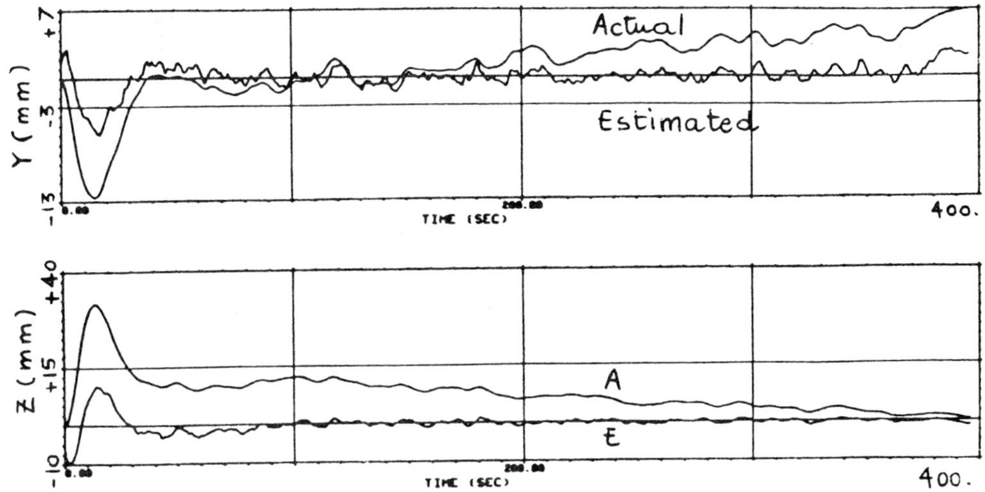

Figure 4.e : Relative lateral position between spacecraft docking ports during the final approach to docking

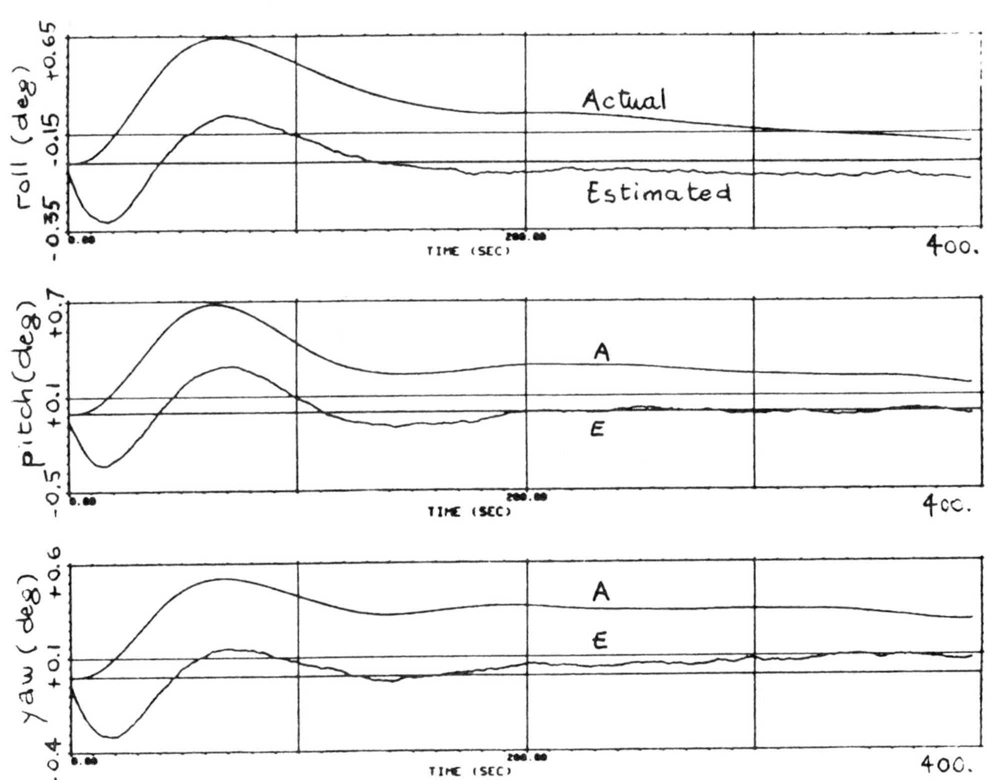

Figure 4.e : Relative attitude between S/C docking ports during final approach to docking

Performing the final approach below the velocity axis from a point in front of the target naturally prevents from failures of the propulsion system (at short term), since the resulting free drifting trajectory naturally recedes from the target, as shown below.

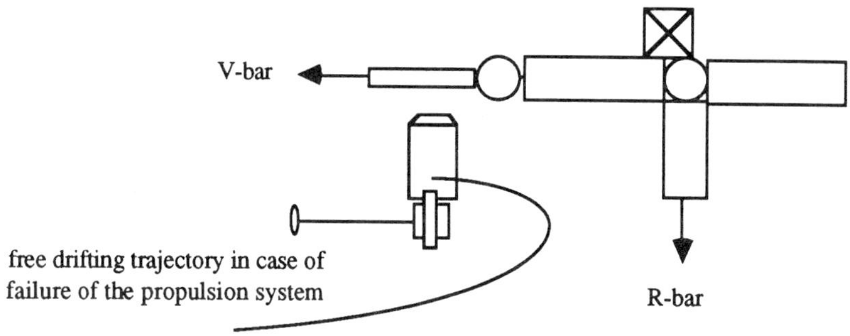

The final translation is assumed to be carried out with a cooperative target spacecraft, which delivers relevant information on its own angular position and rate. As for the final approach to docking, the final approach to berthing is split into two phases:

- a first phase, where the chaser position and attitude are estimated and controlled wrt ideal frames, i.e. regardless the actual target attitude about its nominal reference

- a second phase, where the chaser position and velocity are estimated and controlled wrt a real target frame. Nevertheless unlike the approach to docking, analyses have demonstrated that the chaser attitude does not need to be estimated and controlled wrt the target proper, since the FSS manipulator will ultimately compensate for relative attitude errors in closed-loop once the target is captured and grappled.

The chaser is guided so as to perform the transfer from a point P1 (end of closing) to a point P2 (where the manipulator performs capture and grappling). The starting point P1 is in the orbital plane, whereas the aimed point P2 is not. It is recommended to perform the final part of the approach along a straight line parallel to the V-bar axis, and the guidance profile shown below has been adopted for the final approach to berthing.

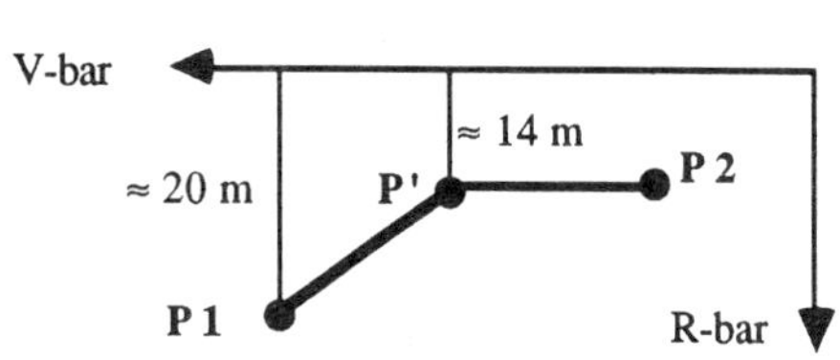

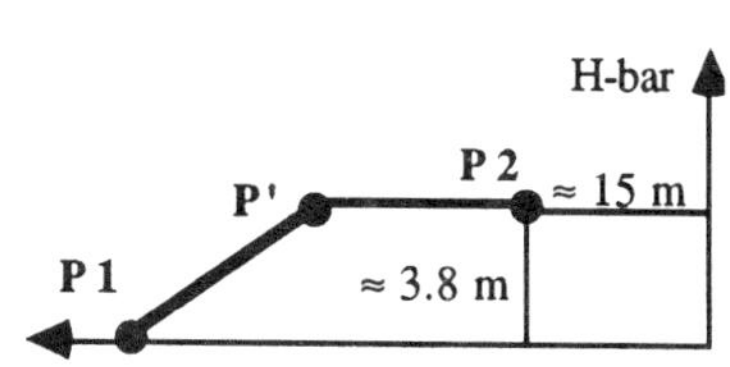

5 VALIDATION OF THE GNC SYSTEM FOR RENDEZVOUS AND PROXIMITY OPERATIONS

The GNC System defined for the rendezvous and proximity operations between Hermes, the Columbus Free Flyer and the Freedom Space Station is a part of an overall rendez control software (RVC) to be implemented aboard the chaser spacecraft. In addition to the pure GNC tasks to be perfomed aboard, the rest of the RVC, hereafter called the Mission and Vehicle configuration Management (MVM) software realizes the following set of tasks:

- the management of the various mission phases and modes
- the configuration of the vehicle resources according to the current mission phases or modes
- the detection, the isolation and the recovery from the various failures occuring in the rendezvous control system (FDIR)
- the management of the GNC and FDIR algorithms implemented aboard according to the current mission phase or mode
- the acquisition of the sensor measurements and the delivery of control orders to the propulsion unit through a real-time bus

The considered RV control system assumes a separation of the on-board software over two computers, one dedicated to all GNC tasks and to a part of the MVM tasks (algorithms management, partial FDIR, data acquisition/delivery on the real-time bus), and a second one dedicated to the rest of the MVM tasks.

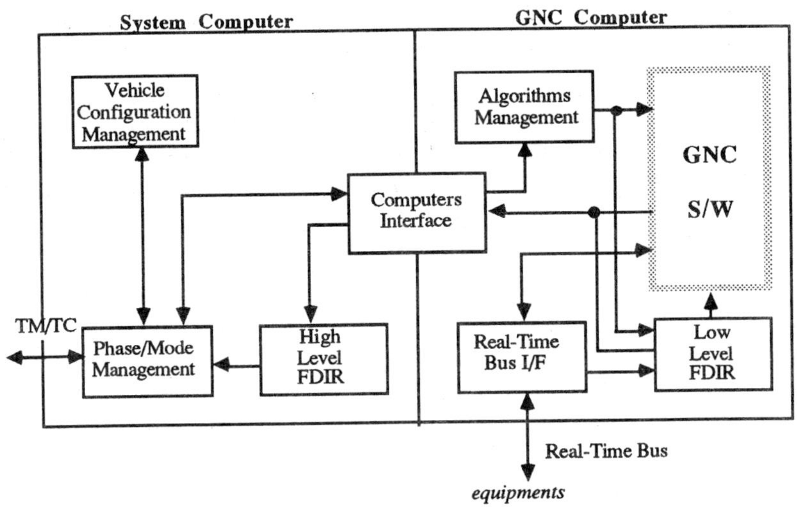

Figure 5.a : Simplified architecture for the overall rendezvous control s/w

A simulation tool for the validation of the GNC concept setup for rendezvous and proximity operations has been designed and developed. This tool includes models for the rendezvous control software (both GNC and MVM tasks), and additionally includes the simulation of a complete RVC environment, called the rendezvous verification software (RVV), including:

- a set of models for the simulation of the equipments aboard the chaser spacecraft (sensors, actuators) and aboard the target spacecraft;
- models of the spacecraft orbital motion and attitude dynamics, including the main external perturbations (absolute and differential air drag, plume impingement by thrusters,...);
- models of the interfaces between the on-board rendezvous contol system and the operators candidate to the mission control, i.e. the Hermes pilot, the target operator (FSS only) and ground operators;
- the simulation control system, including TM/TC handling and a set of facilities for data control and archiving.

The simulation tool, thus including both the RVC and the RVV, has been designed so as to be executed in real-time if needed. Additionally, the structure of the tool is such that the implementation of the software over separate computers is possible:

- one GNC computer, including all GNC and a part MVM tasks;
- one System computer, including the remaining MVM tasks;
- one environment computer, including the complete RVV

Figure 5.b illustrates the design of the RVV/RVC simulation tool, and in particular the separation of the software over several computers.

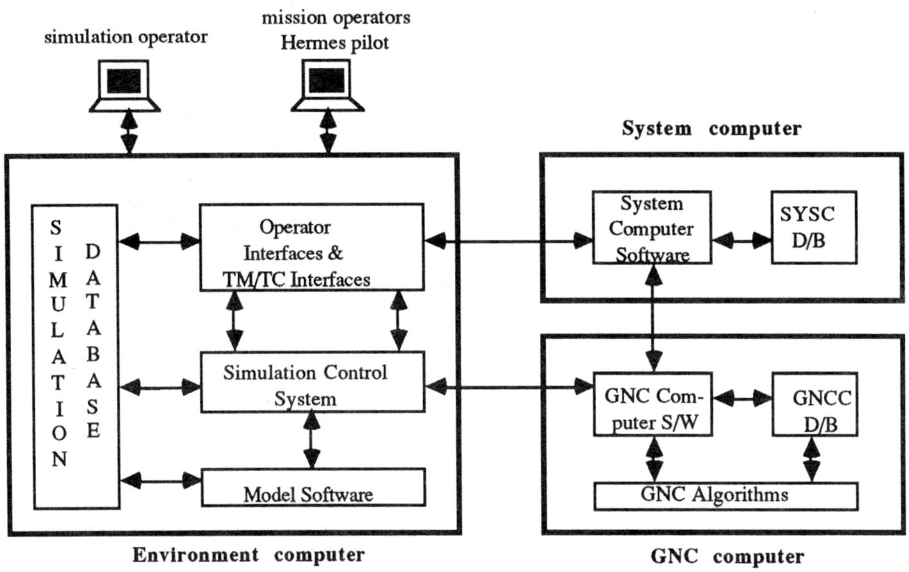

Figure 5.b : Design and architecture of the rendezvous control software verification tool

In order to be independent from the target machine, a Virtual Operating System (VOS) has been developed, acting as a small software layer between the application software and the target machine operating system.

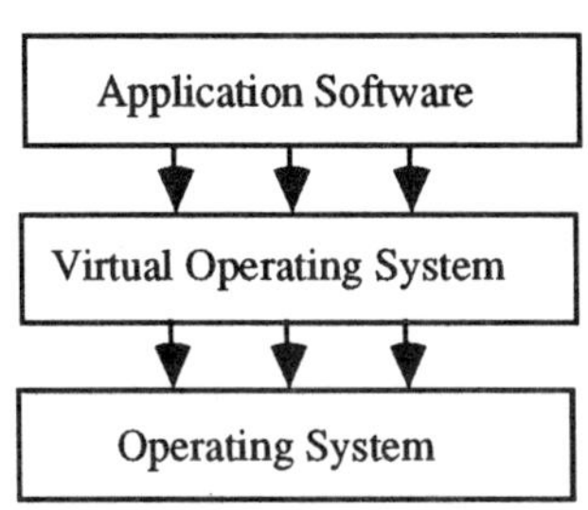

Non real-time software tests are performed with the developed software tool. The target machine for this purpose is be a SUN workstation operating with Unix. With this configuration, several Unix tasks have been defined to run the complete RVV/RVC software:

- one task for the functions running on the simulated GNC computer
- one task for the functions running on the simulated System computer
- several tasks for the functions running on the simulated environment computer

6 CONCLUSIONS

The RVD-PDP, when completed, shall provide both the Hermes and the Columbus Programmes with much valuable information as regards rendezvous and proximity operations. It shall significantly decrease the development risk of a RV system at project level, and thus shall avoid potential schedule delay and associated cost increase in case of system misdesign.

As concerns the GNC in particular, the RVD-PDP shall allow to:

- assess the suitability of the selected equipments (sensors, actuators),
- derive requirements on the GNC processor (necessary memory, CPU load,...)
- provide navigation and control performances during the approach and retreat phases
- evaluate the amount of code necessary to perform GNC tasks

A simulation tool highly representative of both the onboard software and of its environment will be put at projects disposal, allowing to account for new mission requirements coming from the programmes. This software tool shall allow to assess the suitability of new RV concepts, and thus will allow to validate a complete RV system in closed loop at relatively low cost compared to a real in-orbit demonstration.

ACKNOWLEDGMENT

The homing transfer strategy and results presented in section 4.2 have been established in the frame of a study performed by MATRA for the French Space Agency (CNES). The other results have been established in the frame of the ESA RVD-PDP contract.

REFERENCES

(1) C. Pauvert and al. - "Docking Mechanism Design : Analysis of Front End Requirements and Verification Tools " - Autonomous Rendezvous and Docking Conference - Houston, Texas - 15-16 August 1990

(2) C. Pauvert and al. - Description and Performances of the MATRA CCD Camera Sensor Autonomous Rendezvous and Docking Conference - Houston, Texas - 15-16 August 1990

(3) J.J. Gonzalez-Vallejo and al. - Docking Mechanisms : Some European Development Programmes - Houston, Texas - 15-16 August 1990

(4) G. Brondino and al. - A dynamics motion simulator for future European Docking Systems" - 24th Aerospace Mechanisms Synposium, Kennedy Space Center - 1990

AAS 91-064

HIGH PRECISION CONTROL SYSTEMS FOR TELESCOPES AND INTERFEROMETERS

H. Jörck*

Near diffraction limited observations with astronomical telescopes and long baselines interferometers require line of sight (LOS) stabilization in the milli arcsec regime and optical pathlength differences (OPD) in the nm regime, respectively. A typical example for a very high precision control system is given by the 1 m Planetary Telescope (PT) pointing and target tracking system, which has to achieve line of sight stabilities < 0.01 arcsec via onboard image processing. The investigated control system is based on active secondary mirror control, which is based on optical sensors only in order to overcome verification problems of high precision gyros. The required technologies should be verified by ground based laboratory and field tests. A good opportunity to gain experience with very high precision control systems is given by the 60 m optical delay line of the Very Large Telescope Interferometer (VLTI) of the European Southern Observatory (ESO), which is currently under study. Critical problem areas in the field of high precision space telescope Pointing & Target Tracking (P&TT) systems were investigated via nonlinear simulations.

1. Introduction

Ground based astronomical observations suffer from athmospheric effects, which cause wavelength limitations (athmospheric window) and image blur by athmospheric turbulences. Whereas the wavelength constraints can only be overcome by space borne telescopes, much effort has been spent on the development of suitable techniques to increase the image quality both of ground and space based telescopes. Three basic technologies are here under development:

- control of large mirrors via active optics (ESO–Very Large Telescope, FIRST)
- compensation of athmospheric turbulences via adaptive optics and
- long baseline interferometry (ESO–VLT–Interferometer, MARK III)

Active optics and long baseline interferometry are valuable techniques to increase the resolution of space telescopes, too.

* Messerschmitt-Bölkow-Blohm GmbH., Space Communications and Propoulsion Systems Division, W-D-8012 Ottobrunn, Germany. Mailing address: Postfach 801168, W-D-8000 München 80, Germany.

Sub milli arcsec resolution can be obtained from Fizeau (< 20 m baseline) and Michelson Interferometers[1]. The central problem of interferometers is the stabilization of the optical path length difference (OPD) between the subapertures. A Fizeau interferometer with a full uv plane coverage has the same properties as a telescope[1], and thus active optics can be regarded as the basic technique for correction of all types of wavefront errors.

A typical application of active optics is the *Hubble Space Telescope (HST)*, where low frequency spatial wavefront errors can be corrected by an active *primary mirror (M1)* and *secondary mirror (M2)*. *Line of sight stabilization (LOS)* means minimization of wavefront tilt errors, which can be achieved by two different control system philosophies:

- The HST approach, where active mirrors are used mainly for corrections of thermal misalignments of the mirrors, whereas the LOS is aligned by a conventional *rigid body* level 0 control system and
- the SIRTF/*Planetary Telescope (PT)* level 1 design, where the M2 is used additionally for high precision LOS stabilization, i.e. *image motion control (IMC)*.

These basic approaches are outlined in section 2.

Ideally active optics should be used to control the whole light path from the scientific target to the scientific instrument (S.I.). This approach overcomes all problems related to

- unprecisely known position of reference objects (stars),
- velocity aberration,
- thermal deformations of the telescope,
- misalignments between reference sensor and S.I. and
- structural vibrations of telescope and bus etc.

One of the basic tasks of any telescope control system is to reduce the LOS–Jitter due to structural vibrations, which cause image blur or impose observation constraints. Image motions can be induced by

- *rigid body* motions of the entire telescope due to excited flexible appendages – (sun baffle, solar panels etc.) and
- vibrations of the telescope elements (M1, M2, focal plane) relative to each other.

The evaluation of the LOS–jitter for the PT is given in section 3.

For a compensation of all these effects including thermal deformations, the reference signal should be consequently obtained from the S.I. itself in order to close the light path (or measurement loop) from the target to the S.I. This approach is realized in the PT–*Pointing and Target Tracking (P&TT)* system design, where the scientific *Photone Counting Camera (PCC)* provides reference signals obtained from the scientific target directly by onboard image processing.

The basic idea of the PT control concept is to use image motion control via coma free M2 rotations and onboard image processing to solve most of the problems related to the isolation of mechanical vibrations. The reference data obtained from focal plane sensors are fused with measured actuator outputs in a suitable designed model based measurement filter. It is one of the major advantages of the concept, that the fine pointing loop can be verified by an optical test bed without gyro stimulation, because gyros are only foreseen for acquisition purposes. The reference system is described in section 4.

The feasibility of the proposed control concept was evaluated via nonlinear overall system simulations. Results are given in section 5.

Most of the identified problem areas can be investigated with ground based telescopes and interferometers. Of special interest is a verification of optical structures with high dimensional stabilities in combination with milli arcsec or nm resolution secondary mirror control loops. As finally discussed in sect. 6, an optical delay line includes most of the elements of such control systems and is currently under study for the 60 m optical delay line for the *Very Large Telescope Interferometer* (VLTI).

2. Control Concepts

The discussion of space telescope control concepts below relies on the proposed 1 m Planetary Telescope *Pointing and Target Tracking (P&TT)* system described in more detail. The 1 m Planetary Telescope will combine high resolution, diffraction limited imaging quality and long integration times (> 10h) in order to investigate time-variable phenomena in the solar system and beyond. The characteristics of the mission are listed in table 2-1.

- Telescope
    * Aperture : 1 m
    * FOV : 10 arc min
    * Resolution : diffraction limited at 115 nm.
    * focal length : 30 m
- Mission
    * Minimum solar elongation angle < 25°
    * Orbit: elliptical 24 h
      $H_P * H_A = 1000 * 70726$ km, i = 28°
    * Nominal life time > 2y, operational > 5y

Table 2-1: Telescope and mission data

The focal plane instrument payload comprises three imaging spectrophotometers, which cover wavelengths from 110 nm to 8μm and imaging photon counting and CCD cameras from 110 nm to 1μm. The P&TT specifications are given in table 2-2.

Relative pointing and tracking error (RPE) (over > 10 h)
< 0.02 arc sec rms

Short term relative pointing error (RPE/ST, LOS jitter)
< 0.01 arc sec

Target tracking:	* offset tracking via guide stars and precomputed trajectories
	* direct object tracking via on-board image processing
Reference objects:	* point sources (stars)
	* extended sources
	* surface features (e.g. Red Spot of Jupiter)

Table 2-2: P&TT requirements

The usual way for performance verification of a telescope attitude control system is to calculate the RPE and absolute pointing error w.r.t. a reference frame fixed e.g. in the Spacecraft (S/C) center of mass (COM), that means, the payload (telescope) is assumed to be rigid and only the impact of flexible appendages on LOS movements is taken into account. In the case of high resolution imaging wavefront errors of the overall telescope system have to be calculated.

Typical error sources of the telescopes can be split basically into 3 categories:

- optical (telescope) errors (MTF_{opt}), (MTF: modulation transfer function)
- sensor errors ($MTF_{S.I.}$) and
- stabilization errors (MTF_{stab})

The overall system performance can be calculated from

$$MTF_{SYS} = MTF_{OPT} * MTF_{S.I.} * MTF_{STAB.}$$

Most of the errors can be compensated by an active optics system as shown in fig. 2-1.

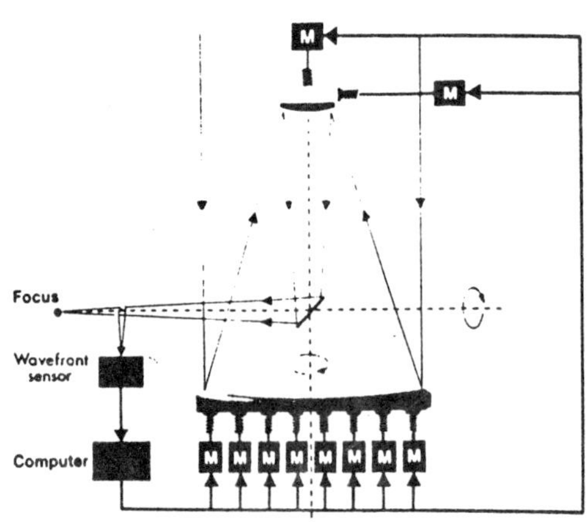

Fig. 2-1: Active Optics

Telescope control systems utilizing active optics can be classified as

- quasi static systems, where LOS alignment is achieved by slewing the *rigid* telescope and active optics are only used for calibration of thermal and launch errors, and
- image motion control systems based on high resolution M2 actuators.

Defocus and tilt errors are normally corrected by screw or piezo driven M2 actuators, whereas in case of infrared telescopes like SIRTF and FIRST (ground based: VLT) chopping of the mirror may be additionally required to remove the infrared backgound.

The HST control concept relies on a quasi static active optics system, a conventional S/C AOCS based on reaction wheels unloaded by magnetic torquers (see e.g. ROSAT) and a strapdown system [3] based on focal plane reference sensor. The major difference between PT and SIRTF[4] control concepts on one hand and the HST level 0 system on the other hand is, that in the former approaches the M2 is used to control low frequency image motions induced by internal disturbance sources, which are listed in table 2-1. The integrated PT control system architecture is shown in fig. 2-2.

Source	HST	PT
Long term thermal/mechanical errors	–	–
tape recorder	–	0
antenna manoeuvre/tracking	–	0
S.I. movable parts	–	0
rate gyro assembly	–	0
reaction wheels	--	--

Table 2–1: Internal disturbance sources

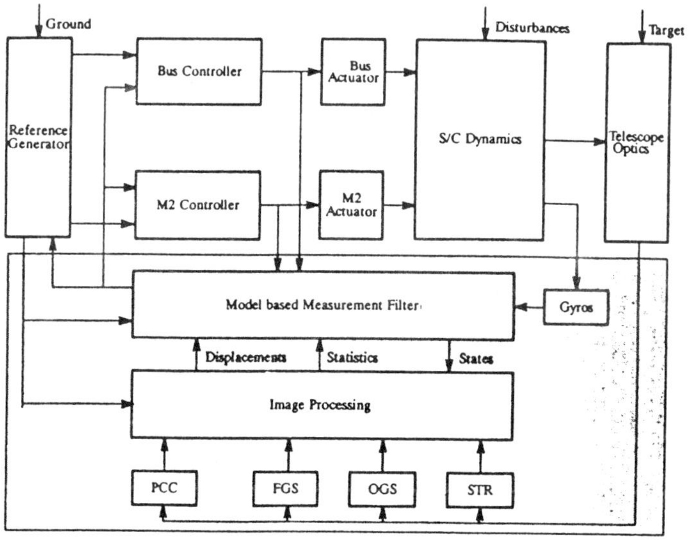

Fig 2-2: Integrated PT control system architecture

The main internal noise sources are the bus actuators, which have to compensate external disturbances like solar drag, gravity gradient torques, etc. The induced structural vibrations cause image blur and thus the implementation of low noise actuators is necessary. Residual image motions (wavefront tilts) are compensated by the M2 control loop. Both loops use reference estimates obtained from the same measurement system, which is based on sensor fusion principles as explained in more detail in sect. 4. The control system bandwidth is determined by the behaviour of the bus and telescope structure and thus the requirements for the control system should be minimized by load path optimizations.

3. Actuator Trade-Off

FE-Models

The LOS-jitter measured in the focal plane by e.g. the fine guidance sensor (FGS) establishes in case of the PT the major selection criteria for all P&TT related trade offs. The LOS-Jitter calculation is based on the optical transfer function of the telescope (including the FGS relay optics), which was combined with a *Finite Element* (FE) model of the entire S/C (fig. 3-1). The structure was excited by models of measured actuator noise spectrums.

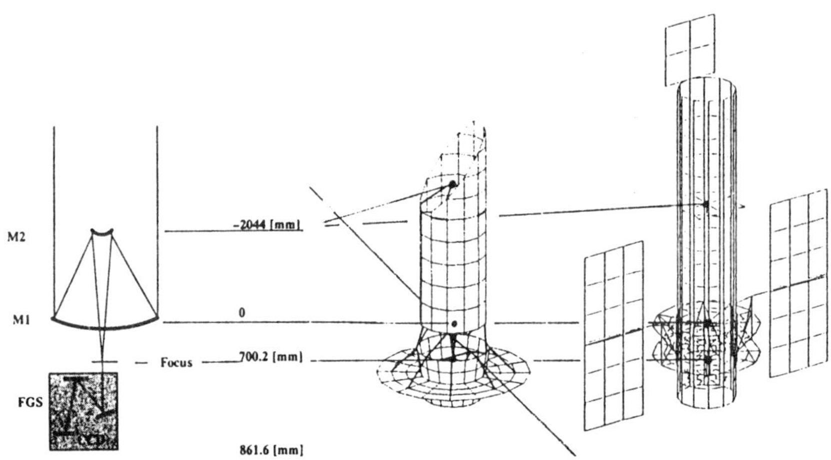

Fig. 3-1: Telescope / FGS combination with Delta *and Ariane*-FE-Models

The LOS-displacements were evaluated as functions of movements and tilts of M1, M2 and FGS. Point spread function distortions due to mirror misalignments were not considered in the LOS-jitter computations. Actuators were modelled as being hard

mounted on the bus. The S/C mass properties are summarized below:

S/C Mass	1300 [kg]
Inertias	diag (1370,1810,850) [kgm^2]

The external disturbance torques considered in the simulation were gravity gradient, solar drag and magnetic disturbance torques, which can be compensated either by a cold gas reaction system (e.g. Hipparcos 20 mN thruster) or reaction wheels (e.g. Teldix 0.2 Nm RSR 14). The wheels may be unloaded by cold gas thrusters or magnetic torquers (< 600 Am^2). A trade-off between the two options has to take into account the induced LOS-jitter (without M2 controller) and the basic features listed in table 3-1.

<u>**Cold gas thrusters**</u>

+ impulse excitation of LOS-jitter only
- limited slew rates
- limited life time
- contamination of optics possible
- limit cycle > 1 arcsec
- no redundancy after M2 failure (limit cycle > 1 arcsec)

<u>**Reaction wheels**</u>

- resonance excitation
+ high slew rates feasible
+ no fuel consumption (mag. torq.)
+ clean actuators (mag. torques)
+ high torque resolution possible
+ LOS stability < 0.1" without M2 feasible

Table 3-1: Basic features of investigated bus actuators

The noise models used were based on TELDIX actuator measurements shown in fig. 3-2. The LOS-jitter was calculated by nonlinear simulations for the thrusters and by frequency domain techniques for the reaction wheels. The simulation results for a pure cold gas reaction system are given in fig. 3-3. The large (high frequency) peaks measured by the FGS are due to the valve action of the thrusters. Although the one sided limit cycle pulses induce jitter peaks in 0.02 arcsec regime, long term (> 10 sec) rms levels are sufficiently small and can meet the overall LOS stability goal. However, due to the wide dead band of > 1 arcsec hard reliability requirements are imposed on the M2 loop.

In case of wheels unbalance effects of the rotor, bearing noise etc. may cause resonance excitations of structural modes. Since worst case resonance peaks are difficult to find with simulation techniques, the overall system transfer function was decomposed by the Complex Singular Value Decomposition (CSVD) method to calculate the impacts of force and torque disturbances of all 4 wheels on the FGS-LOS jitter (see fig. 3-4). These calculations have shown, that the effects of wheel disturbance torques and forces are nearly uncorrelated, because structural modes are well separated w.r.t. disturbance inputs and wheel disturbances are not synchronized.

As indicated, the jitter without the M2 loop is below 0.02 arcsec for the dominant solar panel modes (to be compared with limit cycles > 1 arcsec for thrusters). The maximum high frequency noise level based on hard mounted Teldix RSR 14 is expected to be < 1 milli arcsec.

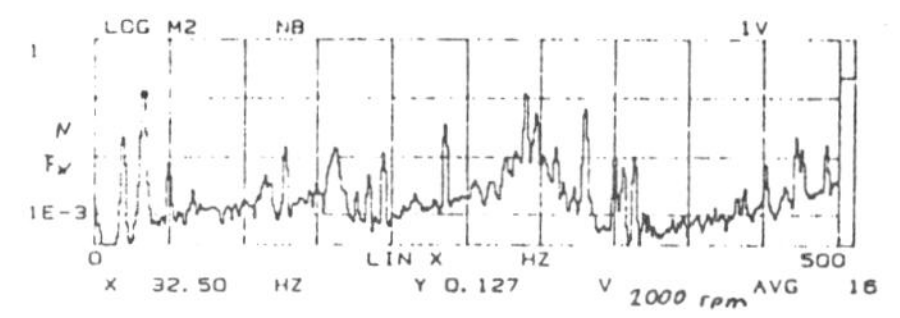

Hard mounted RSR 14

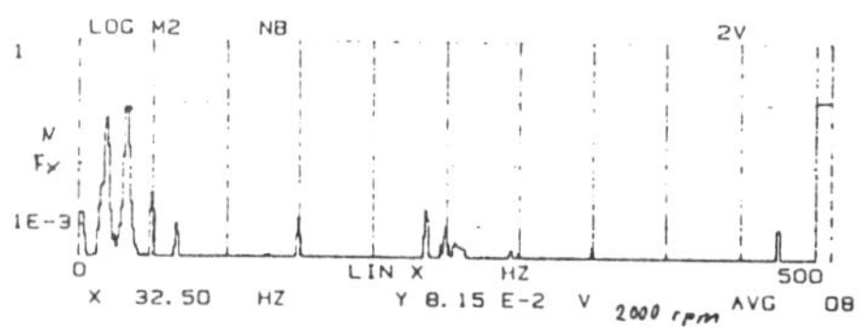

Soft mounted RSR 14

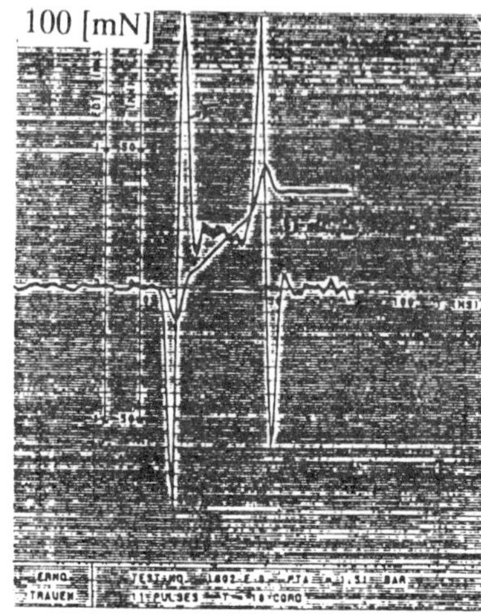

Fig. 3-2: RSR 14 noise spectrum at 2000 rpm and measured thruster forces

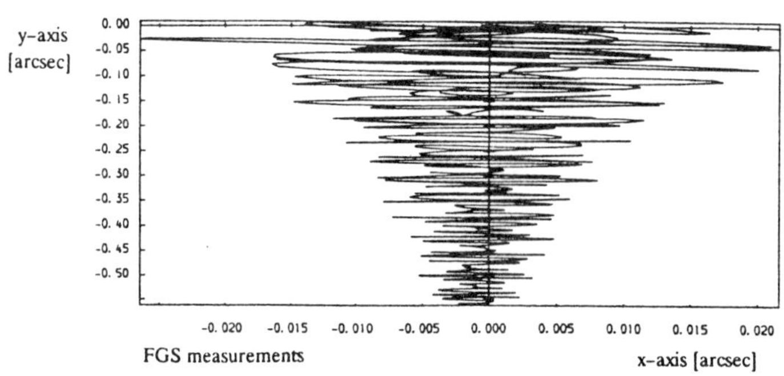

Fig. 3-3: Induced LOS-jitter on the FGS by a thruster pulse

These investigations have established the control system architecture and bandwidth: to handle the worst case jitter induced by solar panel motions, the M2 loop bandwidth should be in the range of 3–5 Hz. However, one has to keep in mind that the validity of FE–models is limited to 80–100 Hz and thus the behaviour of the structure in the acoustic frequency regime up to 500–1000 Hz is not known. These problems can be handled by vibration isolation mounts (spring/mass systems). Since the IMC loop compensates all low frequency motions induced by wheels, these mounts can be designed flexible about all axes, whereas the HST wheel damping units are designed to be stiff about the wheel spin axis[5].

Due to the advantages listed in table 3–1 and the low noise characteristics of present reaction wheel designs wheels unloaded by magnetic torques in combination with the active M2 were chosen as baseline.

M2 actuator

As discussed, an active M2 can be used to remove defocus, tilt and decentering coma errors. Furthermore, the M2 may be used for chopping purposes to remove the IR background noise. In case of the PT only corrections of defocus errors and tilt errors

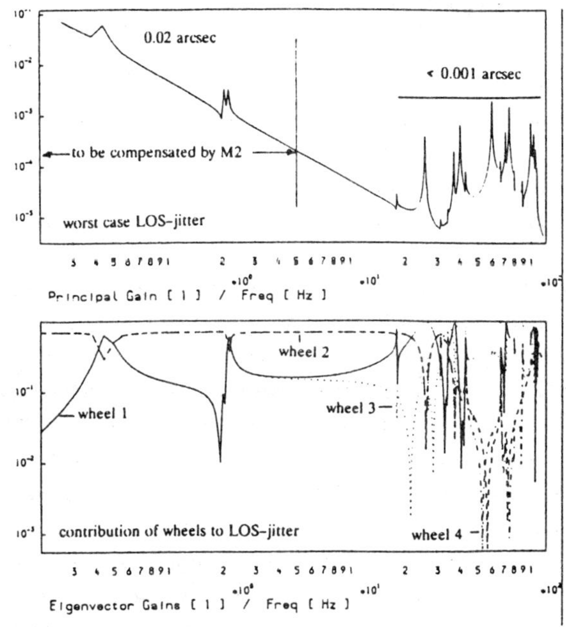

Fig. 3-4: SVD of the FGS-LOS jitter induced by reaction wheels

are foreseen, where the latter corrections have to be achieved by coma free mirror rotations as shown in fig. 3-5.

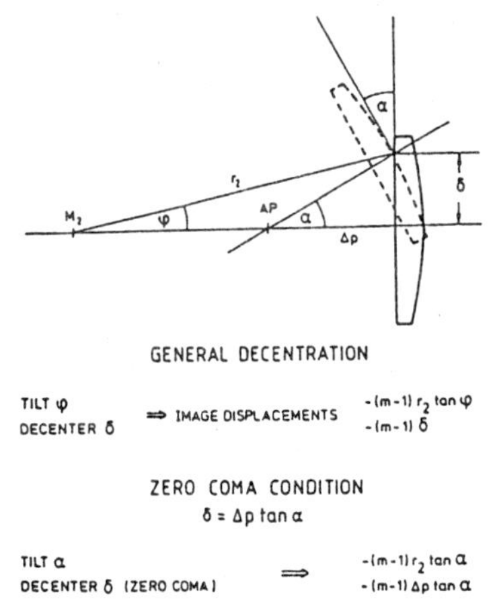

Fig. 3-5: Correction of wavefront tilt errors (IMC) via comafree M2 rotations

The present actuator design utilizes a local (low authority) control loop based on a PZT actuator and inductive differential position sensors to achieve sub m arcsec resolutions at a control loop bandwidth > 60 Hz. The resolution of the actuator is mainly limited by the sensor noise, where a position measurement error of 2nm corresponds in the present design to 1 milli arcsec tilt error at sky.

4. Reference System

Sensor Configuration

The reference system has to provide attitude information both for the 'coarse' bus controller (< 0.1 arcsec accuracy) and the IMC loop used for fine pointing in the < 0.01 arcsec regime. Conventional S/C attitude measurement systems in the arcsec regime are based on external *star trackers (STR)*, whereas systems in the subarcsec accuracy

regime utilize *focal plane (FP) fine guidance sensors (FGS)* to eliminate problems due to thermal misalignments between S.I. and reference sensor. The application of a FP–FGS is sometimes limited by practical constraints, e.g. especially when the wavelength bands, which can be used for reference purposes, differ from those of scientific interest (see e.g. FIRST). In this case sensors outside the optical path of the telescope may be required, which can be coaligned with the S.I. via *attitude transfer systems (ATS)*. Such systems can be based e.g. on fiducial light systems (ROSAT), Roddier sensors for coalignment and cophasing of space interferometer mirrors[6] etc. However, this approach suffers often from kinematical constraints, i.e. not all kinds of misalignments may be measured by such systems and consequently, whenever feasible, focal plane sensors should be implemented. In case of the HST the reference is obtained from a strap down system utilizing high precision gyros (0.25 marcsec resolution) and a FP–FGS employing a Koesters prism interferometer[7].

Besides thermal misalignments structural vibrations of the telescope have to be taken into account in a very high precision environment. The simplified dynamical model of the PT is sketched without modal and optical gains in fig. 4–1. Obviously *rate gyro assembly (RGA)* and FGS are not collocated and thus it has to be carefully evaluated, which vibrational modes can be detected by a strapdown system.

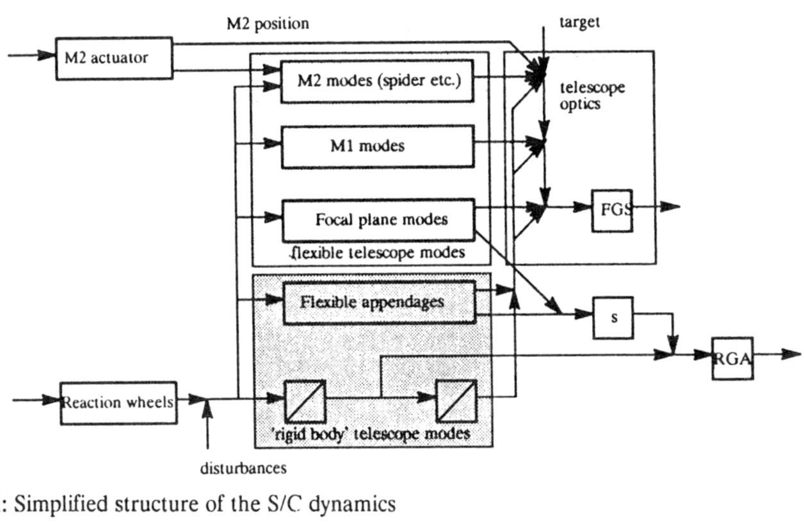

Fig. 4–1: Simplified structure of the S/C dynamics

Image Motion Control

Other impacts on the reference configuration result from the need of image motion control. If only LOS motions due to vibrations of flexible appendages have to be considered, the M2 can be steered by a strapdown system. Otherwise, if telescope flexiblities have to be measured too, the M2 control has to rely on focal plane camera signals. Two image motion control implementations are applicable as shown in fig. 4–2.

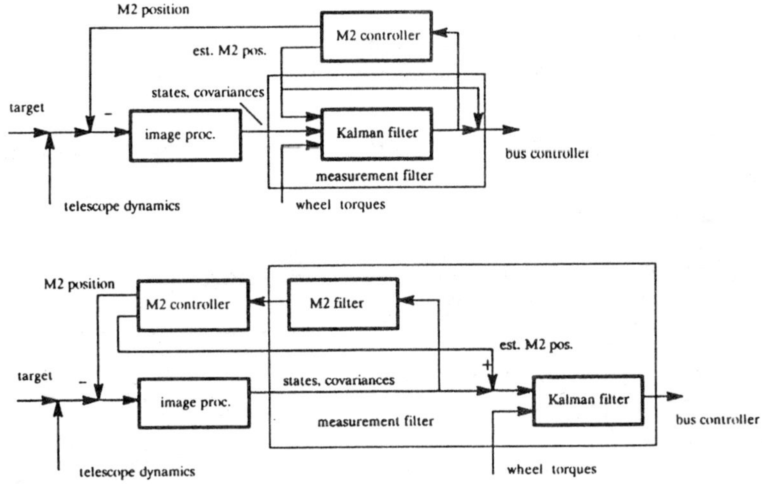

Fig. 4–2: IMC implementations

IMC based on estimated states requires exact knowledge of the S/C and telescope dynamics whereas in the second approach a Kalman filter is only used to generate rate and attitude estimates for the bus controller. The first approach is preferable, if the S/N ratio of the measurements can be improved by a dedicated Kalman filter.

The architecture of the filter is determined by kinematic parameters to be estimated. Some of the parameters are essential for LOS–control (S/C attitudes, rates), whereas others (relative S/C/target motion; target rotation) may be implemented for further improvements of the estimates.

The first structural analysis shown in fig. 3–4 indicates that high frequency vibrations induced by the reaction wheel bearing noise are sufficiently small and thus a strapdown approach would meet the PT–P&TT requirements. Nevertheless a pure optical reference system without gyros for fine pointing was chosen as baseline to overcome the verification problems w.r.t. high precision gyros[7]. The drawback of this approach is the decreased S/N ratio of the measurements due to the required high FGS sampling rate. Although some improvements w.r.t. conventional STR designs are possible e.g. with special CCD (CID) readout techniques, increased telescope apertures are required to solve these problems. A pure optical reference system is meaningful, if stringent pointing and target tracking requirements allow a sharing of the telescope light beam between S.I. and reference camera or if reference signals can be obtained directly from the S.I. as e.g. in case of the PT–photone counting camera (PCC).

Sensor Fusion

Another approach for enhencement of measurement accuracies is given by the implementation of sensor fusion principles. The purpose of multi sensor fusion for telescope applications is twofold:

- extension of FOV of high precision focal plane sensors for acquisition purposes and
- improvement of the measurement accuracy and bandwidth.

A typical information fusion scheme for high precision space telescopes is shown in fig. 4–3, where apriori informations, actuator output estimates and reference measurements are fused in a suitable designed measurement filter. This filter has to provide both time series and spatial filtering for image processing. Spatial filtering is required for the direct object tracking capabilities of the PT.

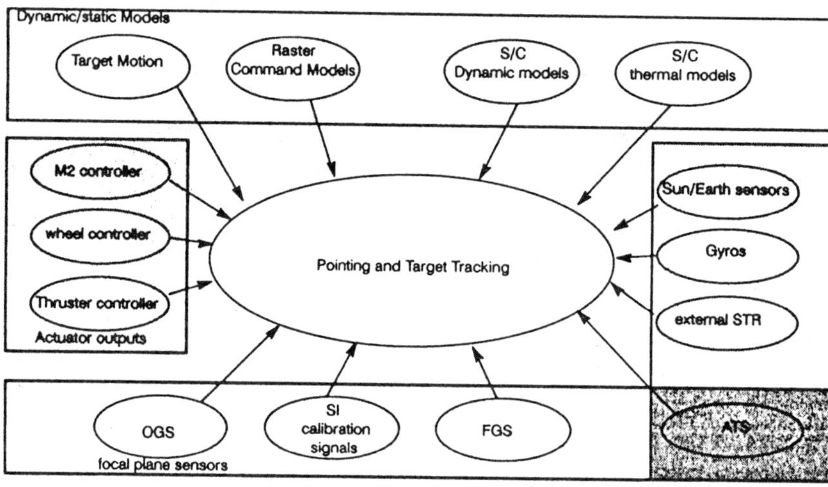

Fig. 4–3: Sensor and information fusion scheme

Direct Object Tracking

The scientific targets of the PT especially in the solar system differ not only w.r.t. their image size on the focal plane, but also w.r.t. their photon statistics and relative motion. Relative motion conventionally means the movement of the target center of mass (COM) w.r.t. the spacecraft. As shown in fig. 4–4 ,the PT–P&TT system has especially in case of surface tracking, e.g. tracking on the Red Spot of Jupiter, to deal with the target rotation, too.

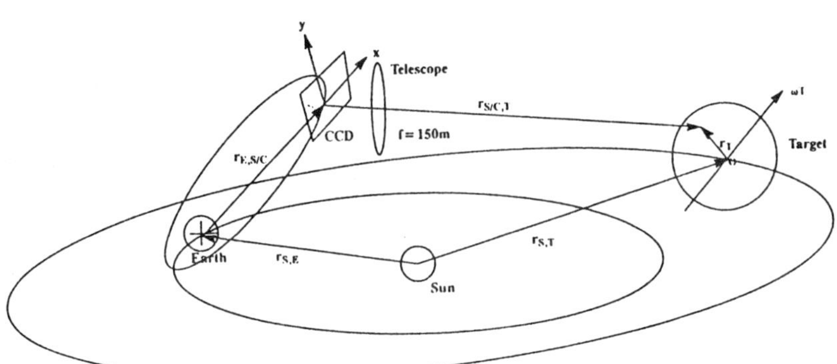

Fig. 4–4: Target tracking kinematics

In more detail, for observations in the solar system, the P&TT has to track disks, edges and surfaces of planets, asteroids etc. and photon clouds in case of faint objects. Investi gations have shown[8], that the tracking capabilities of state of the art guide star offset tracking systems like the HST control system are limited in case of solar system objects. For this reason and to maintain the required LOS stability even in case of long term observations (> 10 h), direct object tracking of surface features was chosen as the 'normal mode' of the PT, although onboard image processing needs powerful computers.

The inertial or relative position of the telescope LOS w.r.t. the target is determined by the focal plane reference sensors. The target surface is projected on the image plane of the sensor as shown in fig. 4-4. The relative motion between telescope and object causes motions of features in the image plane (x, y). The velocity field of this motion is called optical flow (x(x,y), y(x,y)[9], which is given by the telescope-target kinematics. The related kinematic parameters can be calculated in the frequency domain[10] or in the object space domain using the raw FGS data[11].For the latter approach two image processing methods were investigated in more detail, the direct comparison of image areas by cross-correlation and a pixel-recursive gradient method[10]. Dedicated methods were investigated for tracking of edges, where image segmentation algorithms have to be used if the target surface exhibits no sufficient contrast features. For tracking of very faint objects usual offset tracking schemes via guide stars may be used at reduced accuracy levels.

Sensor Arrangement

To cover target acquisition and tracking modes, the sensor arrangement shown in fig. 4-5 was chosen .

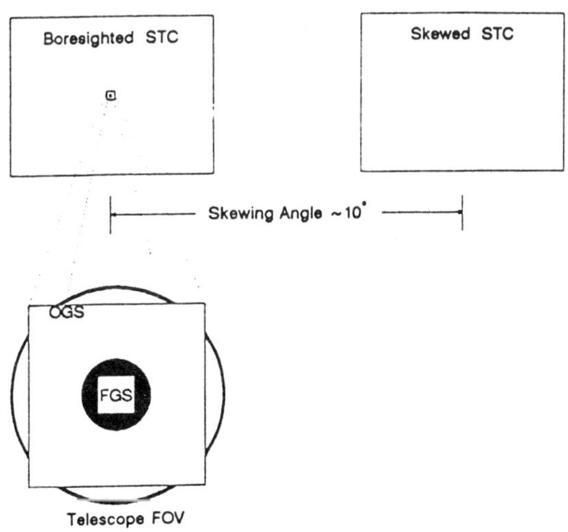

Fig. 4-5: Sensor arrangement

For fine guidance purposes, the scientific PCC or alternatively a dedicated CCD FGS can be used. For offset guidance purposes an *Offset Guidance Sensor (OGS)* was implemented, which covers the whole unvignetted telescope field of view (FOV). Target acquisition will be performed with a strapdown system based on coarse acquisition gyros and external ROSAT type star trackers (STR). The sensor data are listed in table 4–1. Furthermore, both the OGS and STR can be used for roll axis stabilization of the telescope.

SENSOR	CCD Pixels	Focal Length	Field of VIEW	IFOV	Sensitivity	Frame Rate
STAR TRACKER	385 x 288	f = 82mm	4.4 x 5.9 deg	55 x 55 arcsec	Mv = +6.5	1Hz
FINE GUIDANCE SENSOR (1)	1242 x (576 x 2)	f = 30m	192 x 90 arcsec	0.15 x 0.15 arcsec	Mv = +14.5	25Hz
(2)		f = 150m	38 x 18 arcsec	0.03 x 0.03 arcsec	Mv = +12	
OFFSET GUIDANCE SENSOR	770 x (576 x 2)	f = 6m	10 x 7.5 arcmin	0.75 x 0.75 arcsec	Mv = +14.5	25 Hz

Table 4–1: Data of optical sensors

Actuator Output Measurements

The S/N ratio of the inertial sensors can be improved by a measurement filter, which makes use of deterministic actuator inputs. In a high precision environment actuator outputs instead of commands should be used. Since both types of actuators, reaction wheels and M2 actuator utilize internal control loops to follow the commands exactly, the internal actuator measurements can be used directly by the measurement filter.

Digital wheel control systems under development are based on model following torque control utilizing an internal speed control loop to overcome torque irregularities due to stiction, friction etc.[12] (Torque resolutions > 14bit are feasible). However one has to keep in mind, that internal speed measurements provide torque informations only about the spin axis and not about disturbance forces due to unbalance effects.

Since the M2 position is measured w.r.t. the spider, bias values due to thermal misalignments have to be detected by the Kalman filter. Furthermore, if the induced telescope vibrations are not sufficiently small, an internal fiducial light system has to be implemented to detect the M2 position w.r.t the focal plane reference frame.

Dynamical Models

The implementation of dynamical models describing the short term LOS motion due to actuator induced jitter is the most problematic issue of an IMC filter design. If solar panel modes are excited by wheel unbalances, jitter amplitudes of > 1 pixel/sec may be observed without using the M2 for corrections. Improvements of the S/N ratio of the processed FGS measurements require the exact knowledge of the S/C dynamics within the Kalman filter unless a strapdown algorithm based on high precision rate gyros would be used. Since in practice eigenfrequencies and damping ratios even of low frequency structural modes are only poorly known, this approach would require noise and parameter adaptive measurement filters. For sake of simplicity, the alternate version of the IMC loop, where M2 controller is connected directly to the image processing outputs, was implemented in the nonlinear overall simulation of the PT–P&TT.

5. PT Simulation Results

The feasibility of the proposed control concept was checked via nonlinear simulations, which were based on the kinematics shown in fig. 4–4 and the optical and dynamical models as previously discussed. The simulated target motion using Mars pictures is shown in fig. 5–1. More details concerning especially the image processing algorithms are given in Refs. [10, 11] and [2].

The results were obtained from a first basic controller design with the following features: A state of the art PID reaction wheel controller based on the estimated quaternions and rates combined with magnetic wheel unloading via the crossproduct law[13] was used for bus control. The estimates were calculated by a low order Kalman filter, whereas the proportional M2 controller is based directly on the processed FGS data filtered by a second order structural filter to avoid M2 spider mode excitations.

Fig. 5–2 shows a simulation run with measurement noise characteristics obtained for tracking on the Mars surface. The achieved accuracies are listed in table 5–1. The relative long term LOS stability error of the entire system w.r.t. the reference frame is < 0.012 arcsec rms for surface tracking. The sensitivity of the investigated measurement system is sufficient for most of the solar system targets even in the direct object tracking mode.

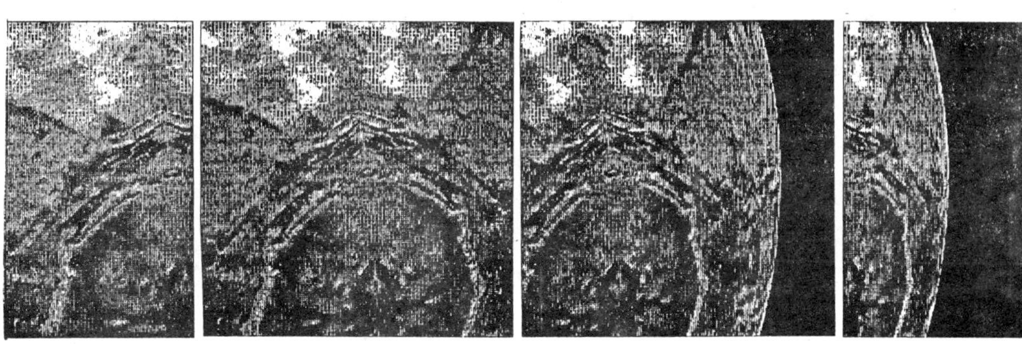

Fig. 5–1: Simulated target motion

Although the chosen P&TT approach has already met the requirements, a lot of work remains still to be done:

- the computer load has to be reduced by improvement of image processing and filtering algorithms (e.g. adaptive filters)
- the interaction of bus actuator disturbances and the M2 actuator with the telescope structure has to be checked by ground based laboratory and field tests

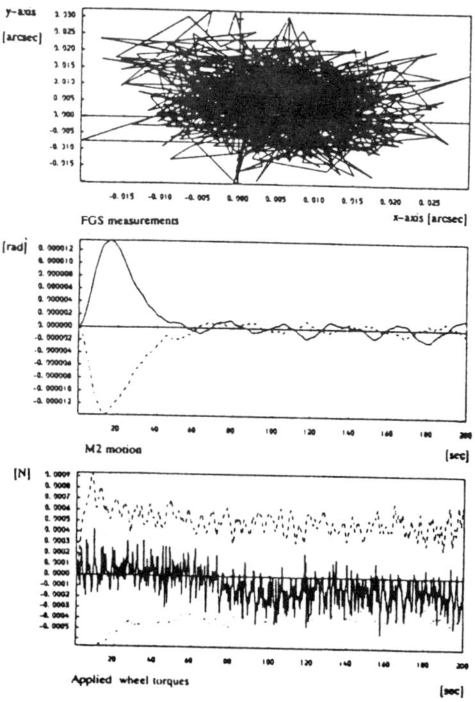

Achieved accuracies :

Measurement accuracies for surface tracking:
(25 Hz sampling rate)

Gradient method :	< 0.004 arcsec rms
Correlation method:	< 0.002 arcsec rms

Measurement accuracy for star tracking :

< 0.001 arcsec rms

f = 30 : Mv = 14
f = 150: Mv = 12

Achieved P&TT accuracies w.r.t the FGS fixed reference frame:

Bias* :	< 0.012 arcsec
LOS-jitter	< 0.007 arcsec rms

* Remark : to be removed by more sophisticated control laws

Fig. 5–2: 200 sec closed loop simulation for surface tracking

- the M2 actuator has to achieve sub milli arcsec or sub nm resolution respectively, and is therefore one of the critical elements of a multi stage control concept.

Most of these critical issues can be investigated with ground based optical long baseline interferometers such as the Very Large Telescope Interferometer (VLTI). For this reason MBB is currently investigating together with Carl Zeiss Jena and the Technical University of Braunschweig the 60 m optical delay line of the VLTI. A brief overview about the current status of the related investigations is given below.

6. Delay Line Control

The VLTI is a 120 m long baseline interferometer comprising the four 8 m telesocpes and a number of 2 m class auxiliary telescopes to achieve a full u, v plane coverage and time resolution[14]. The interferometer will be used both as a closed loop phased interferometer and as open loop coherent interferometer for tracking on faint objects, where no fringe tracker error signals are available. The photon starved open loop mode requires a high dimensional stability of the overall systems and especially of the 60 m optical delay line, which is based on a 0.8 m diameter cat's eye retroreflector. Besides the OPD stability < 7 nm at 100 Hz (< 30 nm/20 Hz) a high dynamic range, both in distance (from 20 μm to 60 m) and velocity (< 0.5 μm/sec to > 5 mm/sec) is required for OPD control.

Basically a control system architecture similar to the PT–P&TT (fig. 2–2) is applicable, where the OPD control is achieved by a *coarse* bus (i.e. cat's eye carriage) controller based on a segmented long stator synchronous linear motor and a sub nm resolution M2 for fine control (see e.g. [15]). The measurement system has to rely on sensor fusion principles, where image processing for fringe tracking, delay line metrology based on a laser interferometer, auxiliary sensor like e.g. accelerometers and sensors for lateral motion control have to be combined in a suitable, noise adaptive measurement filter. In case of the photon starved open loop tracking the M2 cannot be used for OPD fine control, because the complete optical path cannot be measured by laser interferometer due to straylight reasons. However, first straylight calculations have shown, that a corner cube can be placed on the rear of the M2 actuator (fig. 6–1).

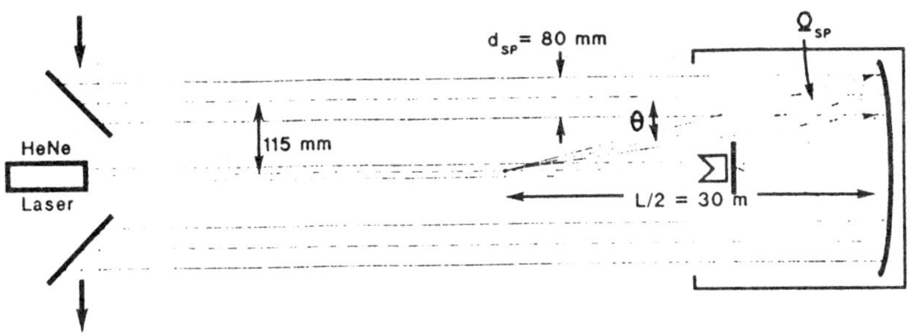

Fig. 6–1: Air-mirror scatter geometry of the delay line

This measurement geometry requires a high dimensional stability of the cat's eye assembly and low noise levels of the cat's eye suspension and environment.

Three types of low friction, low noise suspensions were assessed,

- a mechanical system based on large low friction wheels,
- a state of the art airbearing and
- a magnetic suspension derived from magnetic levitation vehicles as shown in fig. 6–2[16].

Due to the large alignment effort for the track and supporting structure the mechanical system was skipped. Although a magnetic bearing requires much more development effort than an airbearing, it has some interesting features:

- The magnetic suspension provides active vibration isolation limited only by the sensor noise (accelerometers, air gap, coil current).
- In combination with a laser based track reference system, which *commands* the flight height above the track, only a coarse alignment of the supporting structure and the foundation is required.

To reduce the heat dissipation of the configuration shown in fig. 6–3, a controlled permanent magnet[17] was developed as shown in fig. 6–4, where a current bias can be removed by adaptation of the air gap of the magnetic shunt.

First simulations have indicated, that the suspension can isolate the optics from disturbances induced by an imperfect track and environmental noise. If the present results

Fig. 6-2: Magnetic levitation systems for transportation vehicles

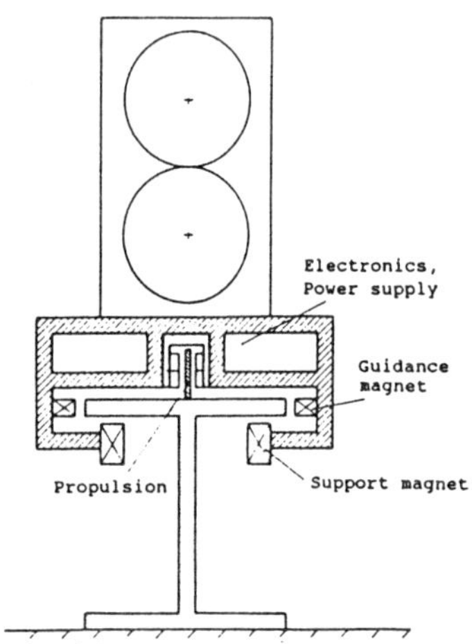

Fig. 6-3: Magnetic suspension

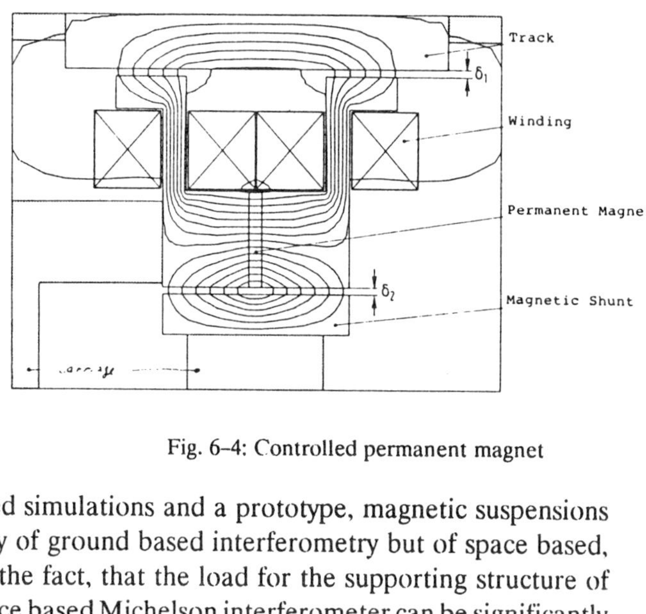

Fig. 6-4: Controlled permanent magnet

can be verified by more detailed simulations and a prototype, magnetic suspensions cannot only improve the quality of ground based interferometry but of space based, too. Of special interest is here the fact, that the load for the supporting structure of delay lines or telescopes of a space based Michelson interferometer can be significantly reduced by laser beam controlled magnetic suspensions.

7. Conclusion

High precision line of sight control requires a careful harmonization of optics, structural and control dynamics. The feasibility of the overall system verification is a major design driver for the control system layout. In order to eliminate e.g. problems related to the verification of a high precision gyro based strapdown system, a reference system based only on optical focal plane sensors was investigated for the 1 m Planetary Telescope. Image motion based on estimated states instead of processed measurements requires the precise knowledge of the S/C dynamics, which is usually only poorly known. In the context of present study, an image motion control loop based directly on processed camera data was implemented. With this first approach a LOS–jitter < 0.007 arcsec was obtained for tracking on the Mars surface. Further improvements based on adaptive measurement filters are possible. These nonlinear filters have to be optimized w.r.t. measurement performance, stability margins and overall computer load.

Most of these technologies have to be applied for ground telescopes and especially long baseline interferometers, too. Special developments in this area such as magnetic suspensions can reduce the accuracy requiremens for the supporting structure of ground and space based systems significantly.

8. Acknowledgments

This work was funded by the German BMFT/DLR as part of a Phase 1 study conducted together with DORNIER and CARL ZEISS being responsible for the spacecraft and telescope, respectively. Wheel noise measurements data were kindly provided by TELDIX. The MBB team members G. Lange, K.–H. Thörmer, D. Wolf, H. Zinner and the author also whish to express their thanks to the scientific group of the Planeten–Teleskop for many helpful discussions.

Acronyms

ESO	European Southern Observatory
FIRST	Far InfraRed Space Telescope
FE	finite element
FGS	fine guidance sensor
FOV	field of view
FP	focal plane
HST	Hubble Space Telescope
IMC	image motion control
LOS	line of sight
M1	primary mirror
M2	secondary mirror
OGS	offset guidance sensor
OPD	optical pathlength difference
P&TT	pointing and target tracking (system)
PCC	photon counting camera
PT	Planetary Telescope
RGA	rate gyro assembly
RPE	relative pointing error
S/C	spacecraft

S.I.	scientific instrument
SIRTF	Space InfraRed Telescope Facility
STR	star tracker
SVD	singular value decomposition
VLT	Very Large Telesope

References

1.J.E. Noordam et. al.
A proposed Medium–Term Strategy for Optical Interferometry in Space,
ESA SP–1135, Aug. 1990

2.H.Jörck, G.Lange, D.Wolf, H.Zinner,
High precision LOS stabilization for a 1 m space telescope with onboard image processing,
SPIE Vol. 1306, 1990

3.J.J. Rodden et.al.
Flight Software Operation of the Hubble Space Telescope Fine Guidance Sensor,
IFAC 10th World Congress AC, Vol. 6, pp. 73ff, Munich, July 1987

4.A.J. Pue et.al.
Configuration Trade–offs for the Space Infrared Telescope Facility Pointing Control System.
J. Guidance Vol. 10, No.2, March–April 1987

5.J.J. Rodden et.al.
Line of Sight Performance Improvement with Reaction Wheel Isolation,
AAS 86–005, pp. 71ff

6.Y.B. Bely et.al.
HARDI: A high angular resolution deployable interferometer for space,
SPIE Vol. 1130, 1989

7.H. Dougherty et.al.
Noise Characteristics and Minimization of a Precision Gyroscopic Rate Sensor,
IFAC Conf. AC in Space, Nordwijkerhoot, The Netherlands, 1982

8.H.A. Weaver et.al.
Planetary Science Update,
Space Telescope Science Inst. Newsletter, Febr. 1988, pp. 3

9.H. Zinner,
Determining the kinematic parameters of a moving imaging sensor by processing

spatial and temporal intensity changes,
J. Opt. Soc. Am. A3(9), pp. 1512 ff, 1986

10.P.S. Maybeck et.al.
Adaptive Tracking of Multiple Hot Spot Target IR Images,
IEEE Trans.AC. Vol. AC–28, No. 10, pp. 937ff, Oct. 1983

11.H.Zinner et.al.
Navigation of autonomous air vehicles by passive imaging sensors,
Proc. AGARD:Unmanned Air Vehicles, San Francisco, pp. 34.1ff, 1988

12.U. Bichler,
A Microprocessor Controlled Reaction Wheel,
ESA Workshopt on Advanced Technologies for
S/C Attitude Control, Noordwijk, The Netherlands, Oct. 1989

13.H. Dougherty et.al.
Magnetic Control systems for Space Telescope,
J.Astr.Sc. Vol. XXX, No. 3, pp. 229ff, July 1982

14.J.M. Beckers et.al.
The VLT Interferometer, I. Proposed Implementation
SPIE, Vol. 1236

15.M. Shao et.al.
The Mark III stellar interferometer
Astronomy and Astrophysics 193, 357 ff, 1988

16.E. Gottzein
Das "Magnetische Rad" ,
VDI Fortschritts–Bericht R8, Nr. 68

17.H. Weh, H. May
Fast acting magnets for transportation purposes,
Int. Conf. on Maglev Transport 85, pp. 155 / 163, 1985, Tokyo, Japan

AAS 91-065

ADVANCED GNC CONCEPTS AND TECHNIQUES FOR AN INTERPLANETARY MISSION

Calixte Champetier*, Pascal Régnier* and Jean de Lafontaine†

Among the challenging interplanetary missions foreseen at the beginning of the third millennium, the ROSETTA Comet Nucleus Sample Return mission, has as primary objective to land on a comet nucleus and to bring back to the Earth a scientifically meaningful sample of its subsurface.

The objective of this paper is to present preliminary results of on-going studies performed in the frame of the ESA Technological Research Programme supporting the ROSETTA project. After the review of the ROSETTA mission requirements and the baseline scenario, new Guidance, Navigation and Control (GNC) concepts applicable to the most critical phases of the mission are investigated : autonomous navigation during the orbiting phase, detection and avoidance of debris, autonomous GNC during the descent and landing phase, control and monitoring of the touch down and the drilling operations.

Quantitative results concerning the debris avoidance and the descent and landing are presented. Further investigation areas are pointed out : implementation of an UD extended Kalman filter for the autonomous navigation during the orbiting phase, refined performance assessment for the descent and landing phase, simulation of the contact dynamics at touch down.

INTRODUCTION

Among the challenging interplanetary missions foreseen at the beginning of the third millennium, the ROSETTA Comet Nucleus Sample Return mission has as primary objective to land on a comet nucleus and to bring back to the Earth a scientifically meaningfull sample of its subsurface.

The ROSETTA mission is one of the four "Cornerstones" proposed by ESA, as a collaboration project with NASA. It was defined in the ESA's long term plan "Horizon 2000" established in 1984. Since April 1986, studies have been carried out by ESA and the European industry. MATRA as a prime contractor performed the Mission Definition Study and the System Definition Study[1] which ended in October 1990.

The mission has been studied on the basis of a joint NASA/ESA contribution, the spacecraft being composed of three modules ; the carrier spacecraft contributed by NASA as a Mariner Mark II (MMII) derived unit from the CASSINI/CRAF spacecraft, a lander and an entry vehicle, the last modules being developed by ESA. Launch is assumed with a TITAN/CENTAUR vehicle provided by NASA and mission control is performed through the NASA's Deep Space Network (DSN).

* MATRA Espace, 31 rue des Cosmonautes, F-31077 Toulouse Cedex, France.

† ESA, European Space Research and Technology Centre (ESTEC), P.O. Box 299, NL-2200 AG Noordwijk ZH, The Netherlands.

Close cooperation on mission and spacecraft design has been initiated between the ESA engineers, the JPL team assigned by NASA to the carrier spacecraft, and the European industry.

In support to the system studies, an intensive Technological Research Programme is conducted by ESA to master the required new technologies. In this frame, MATRA is involved, as prime contractor in studies concerning the definition, the development and the validation of advanced guidance, navigation and control concepts and techniques applicable to the ROSETTA mission[5,6].

This paper will present the current status of the analyses in that field, focussing on the most critical points of the mission. After the review of the mission scenario and the requirements related to the GNC system, the proposed GNC strategies and concepts will be presented for the near comet phases of the mission, including orbiting phase, descent and landing phase, touch down and on-comet operations. Advanced concepts like autonomous navigation during orbiting phase, autonomous detection and avoidance of debris, active control of the touch down are discussed and further investigation areas are pointed out.

REFERENCE MISSION

Mission Requirements

A detailed presentation of the mission requirements and scenario can be found in Ref. 1-2. The main mission requirements are related to the comet characterisation, the landing and the sampling operations.

Table 1
COMET CHARACTERISTICS

Properties	*Minimum*	*Typical*	*Maximum*
Albedo	*0.01*	*0.05*	*0.5*
Radius (km)	*1*	*2.5*	*10*
Density (g/cm³)	*0.05*	*1*	*1.5*
Mass (kg)	8.10^{11}	7.10^{13}	6.10^{15}
Spin period (h)	*3*	*10*	∞

Comet characteristics (cf Table 1) are important parameters which drive the definition of the scenario, the design of the spacecraft and that of its GNC system. An essential requirement is that a safe landing must be ensured despite a poor a priori knowledge of these parameters (environment, kinematics, gravity field, shape, surface characteristics ...).

Other high level constraints must be taken into account in the operation scenario and spacecraft design :

- the proposed design must support one failure.
- the communications must be kept available as much as possible at all time during the mission.
- mission timing must be compliant with a return flight to the Earth : the near-comet operations must not take more than 200 days.

Mission Scenario

In order to avoid the hazardous environment of an active comet around the perihelion, the comet encounter is scheduled at the aphelion. The earliest candidate comet is Hartley 2 with a possible launch date on 2002 with TITAN/CENTAUR launcher. Comparable launch opportunities appear in the years 2003 and 2007, but with improved TITAN or Shuttle C launch.

The Rosetta mission can be decomposed into the following phases. An Earth-to-Earth phase is foreseen for gravity assist. The spacecraft is initially injected on an Earth escape hyperbola trajectory (V_{∞} = 5.09 km/s, d = 3.4 deg). About one year later, a retargetting maneuver (943 m/s) will put the spacecraft into an Earth fly-by trajectory. Due to radio isotopic elements on board, a safe Earth approach and fly-by is performed (down to 300 km). After a 41 month cruise-to-comet phase during which only small orbit control maneuvers are performed, the near-comet operations are initialized. The first critical operation is the detection of the comet. Depending on the nucleus size, albedo and Sun illumination angle, the detection by means of the S/C imaging system takes place from 6 10^5 km to 3 10^6 km from the comet. In the worst case, a minimum time of 3 days is left to the ground segment to process the sky images for comet identification and to optimize the rendezvous maneuvers.

The comet approach phase from 300 000 km (first rendezvous maneuver) to close vicinity of the nucleus (130/35 km) lasts 33 days. Three or four maneuvers are foreseen to reduce the relative velocity down to 25 m/s at 5000 km and 2 m/s near the nucleus. The approach trajectory is designed to optimize the observability of the relative position/velocity of the spacecraft (bias approach) and to minimize the damage due to impacts of debris (low velocity).

Once the spacecraft is close to the nucleus, a comet global mapping phase (21 days) is initiated. The spacecraft is injected into a comet polar orbit compatible with a good illumination angle. During this phase, an overall mapping of the illuminated part of the comet will be performed in order to determine a preliminary topography of its surface and to recognize candidate landing area of scientific interest. The orbit altitude for global mapping takes into account the maximum number of images for a complete coverage, the comet spin rate, the synchronization of the spacecraft with the comet areas to be mapped and the ground station visibility.

After the global mapping phase, potential landing sites are identified from ground image processing. The spacecraft is then injected into a specific orbit for detailed mapping of the candidate landing area. A period of 18 to 37 days is allocated to this phase (transfer and detailed mapping). Elliptical trajectories with typical periapsis from 5 km to 10 km altitude over the landing sites are selected. Apoapsis altitude may vary from 10 km to 190 km depending on the comet gravity field. Periapsis altitude is optimized according to the performance of the on-board instruments utilized for scientific and topographic data collection.

The descent phase (cf Fig.1) will be initiated from an elliptical orbit close to the detailed mapping one, but with a periapsis altitude of typically 3 km. This orbit is used to achieve the relative phasing of the spacecraft with the landing site while intensive ground tracking accurately reconstructs the dynamical state of the spacecraft. After one or two revolutions on this orbit, the deorbiting maneuver command is uplinked to the spacecraft and executed at the pre-selected time. The resulting homing orbit brings the spacecraft to a position and velocity that allows the use of the on-board instruments to synchronize the spacecraft motion with that of the landing site and initiate the autonomous final descent. At about 1 km altitude, a (typically) 1 m/s vertical descent is initiated.

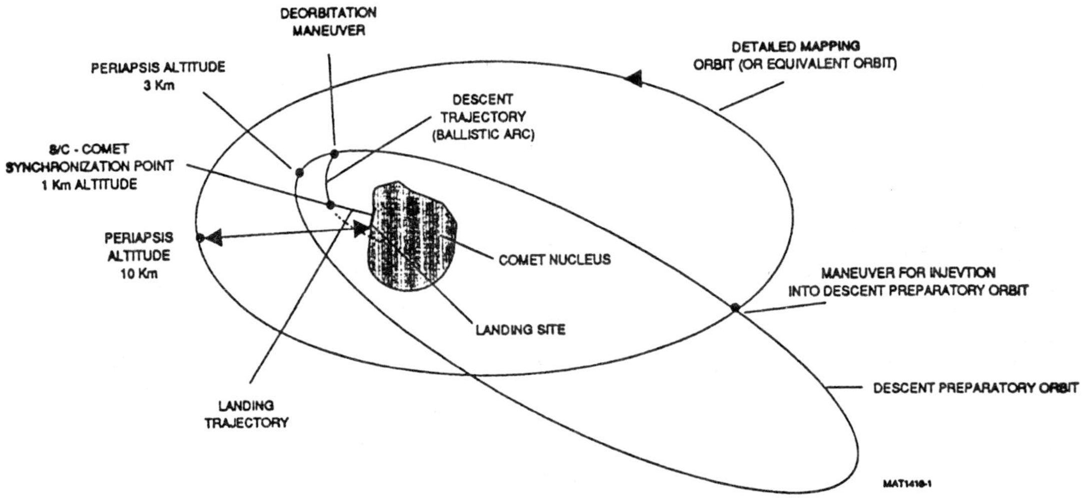

Fig. 1 : The ROSETTA Descent Scenario

The vertical velocity at touch down is reduced to 10 cm/s and the residual kinetic energy is dissipated into shock absorbers. Due to the weak gravity field (typically 70 μg), the low spacecraft stability margins (static stability angle less than 20 deg), and the poor knowledge of the comet surface characteristics, the GNC system will ensure an active stability at touch down to prevent from tipping over in case of local surface irregularities and sliding in case of icy slope. After being completely stabilized, anchoring is performed as soon as possible and the ground communication is established for monitoring the on-comet operations.

Spacecraft design and GNC instruments

The major driver for spacecraft design is the reuse of the Mariner Mark II (MMII) module developed for CASSINI and CRAF missions under NASA/JPL responsability. An optimum concept leading to minimum modification of the baseline on the MMII is shown on Fig.2. The spacecraft consists of three main modules :

- the cruiser and its scan platform (based on MMII) utilized during all the mission except reentry ;
- the lander carrying the sampling devices and left on the comet after departure ;
- the aerocapsule attached on the side of the MMII and carrying the sample container brought back to Earth.

Table 2 identifies the instruments utilized for both scientific and GNC purposes. All the instruments are mounted on the scan platform. The narrow angle CCD camera (NAC) is used for the detection of the comet and for the mapping. A pointing stability of half a pixel (3 microrad) over the exposure time (from 0.1s to 1s) is required for the comet detection and constitutes the major driver for the scan platform control design. The star and target tracker ASTROS II or the wide angle camera (WAC) is used for the updating of the gyro measurements and as a navigation aid in the vicinity of the comet. It provides the ground with images from which the main characteristics of the comet (shape, gravity field, rotation) are identified. The inertial lines-of-sight to natural landmarks located on the comet surface constitute the main information used for this purpose.

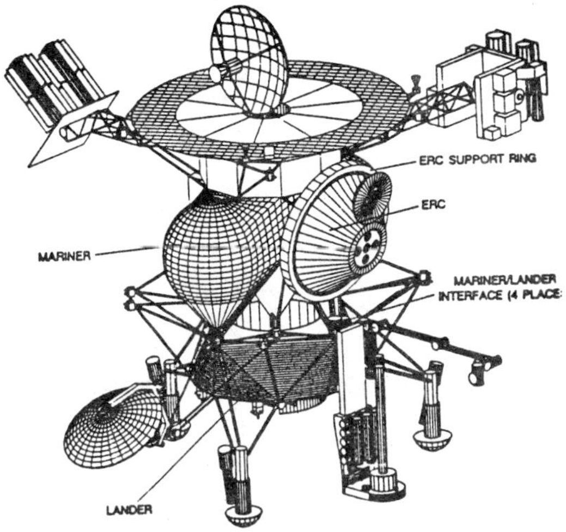

Fig. 2 : The ROSETTA Spacecraft Configuration with Deployed Appendices

The measurement of the spacecraft position and velocity relative to the comet surface is performed with a radar system and a laser mapper. These instruments also detect the spacecraft orientation relative to the surface in order to ensure an upright attitude at touch down. The radar system combines both range and Doppler measurements. The half-meter antenna has four beams, one directed along the boresight and the three others squinted at 2 deg. At large distance from the comet (up to 200 km), the boresight beam provides range measurements. At a lower range (below 2 or 3 km), the three offset beams provide range and range rate measurements in three independent directions. With the three range measurements, the normal to the comet surface is derived, thus providing a two-axis relative attitude measurement. From the range rate measurements along the three beams, the relative spacecraft velocity in three axes is computed.

The position of the spacecraft in the horizontal plane related to the landing site is estimated from the laser instrument by using elevation-map correlation techniques. During the global and detailed mapping phase of the comet, topographic maps of the landing site (500m x 500m) will be developed on ground using all the available instruments. This map will be stored in the on-board computer before descent initiation. During the guided phase of the descent, the laser system will scan a sub-area of about 30m radius, perform the correction with attitude and range variations, and correlate them with the stored map in order to derive the relative horizontal position. In the preliminary design[10], the field-of-view of the laser beam is ±4 deg, which permits to utilize the laser above 500 m altitude. A duration of typically 200 s is required for obtaining the sub-area scanning.

AUTONOMOUS NAVIGATION CONCEPTS FOR THE ORBITING PHASE

In the current nominal status of the ROSETTA mission concerning the observation phase, ground-based navigation using DSN tracking measurements and on-board line-of-sight measurements will ensure an accurate reconstitution of the spacecraft motion relative to the comet[7,8,9] (a few tens of meters in position and a few mm/s in velocity). At the same time, identification of the comet kinematic and gravity field will be performed to improve the poor a priori knowledge of the parameters affecting the prediction accuracy. The results of this ground activity will be uplinked to the spacecraft before autonomous descent and landing maneuvres are initiated. The on-board GNC system will then propagate its estimated dynamical state, using simplified model of the environment and measurements.

Table 2
SPACECRAFT NAVIGATION INSTRUMENTS

Narrow angle camera (MMII)		*Radar system (Doppler)*	
Field of view (deg)	*0.35x0.35*	*squint angle (deg)*	*2*
Pixel size (microrad)	*6*	*operating range (m)*	*20-2000*
Focal length (mm)	*2000*	*accuracy (mm/s)*	*1*
Wide Angle Camera (MMII)		*Laser system*	
Field of view (deg)	*2.8x2.8*	*field of view (deg)*	*4*
Pixel size (microrad)	*48*	*operating range (m)*	*20-5000*
Focal length (mm)	*250*	*accuracy (cm)*	*60*
Star and Target Tracker (MMII)		*update frequency (Hz)*	*1*
Field of View (deg)	*11.5x11.5*	*mapping accuracy (m)*	*10*
Pixel size (microrad)	*6*	*mapping update period*	*200s*
Focal length (mm)	*2000*		
Radar system (range)			
beamwidth (deg)	*0.4*		
operating range (km)	*0-200*		
accuracy	*5m bias + 1%*		
frequency (Hz)	*1*		

During the mapping phase, autonomous navigation is necessary to ensure spacecraft safety in the case of non nominal events leading to back-up modes. The ability to achieve this autonomy is one of the key feature of the current ESA contract on "Autonomous and Advanced Navigation Techniques Study"[6]. Navigation techniques have been developed to solve this problem by using the on-board cameras, the star and target tracker and the radar which provide as informations: the line-of-sight to the optical centre (barycentre of the illumitated part of the comet in the focal plane of the camera), the apparent diameter, the limb, the line-of-sight to natural landmarks, the range to surface along the pointing direction.

To efficiently process these measurements, on-board models for the comet shape and rotational dynamics are necessary. Furthermore, a model of the comet gravity field is required for the prediction of the spacecraft motion when orbiting. The initial parameters appearing in these models will be uplinked to the spacecraft and their estimation propagated by the on-board navigation filter, along with the spacecraft relative position and velocity vectors.

The objective of the on-board GNC system will be to acquire and maintain safe trajectories (back-up elliptic orbits, quasi-inertial or hovering points) in case of contingency, so as to minimize the impact on the mission schedule. The major non nominal events which may occur during this phase are the loss of communications with ground, the unavailability of the ground segment processing, the loss of the nominal trajectory due to debris avoidance, if such a system is proved to be necessary and feasible by the current study. The last point will be discussed later on.

To permit analytical computations in the measurement processing, the proposed on-board comet shape model is an ellipsoid. This model can be refined without increasing too much the complexity by adding an offset between the center of mass and the geometric center, and principal inertia axes different from geometric axes. The comet rotational dynamics is modeled as a torque free body rotating motion expressed with Euler angles in the principal inertia axes frame. The comet gravitational field is modeled as a second-order potential of an homogeneous ellipsoidal body expressed in the principal inertia axes. With this model, only two harmonic coefficients, directly related to the comet mass and the three principal moments of inertia, are involved. The spacecraft relative dynamics equations are derived from the comet models described above and a simple solar pressure model which is the main source of perturbing accelerations at the considered ranges to the comet (see Fig.3). Due to high nonlinearities of measurement and prediction models, and weak conditioning of the problem, an extended Kalman filter with UD propagation has been designed to propagate the estimation of the relative dynamical state of the spacecraft.

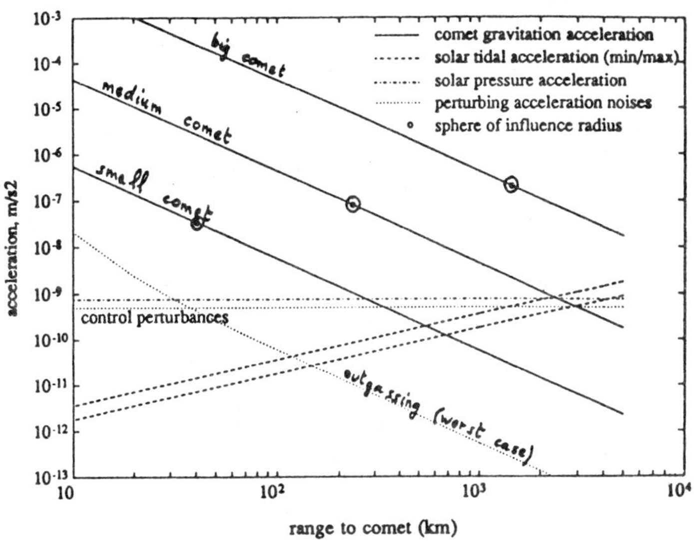

Fig. 3 : main perturbing accelerations acting on the spacecraft during mapping phases

Measurement models are designed to predict the measurements provided by the on-board instruments, cameras and radar, in order to produce innovations for use in the navigation filter. Considering the expected irregular shape of the comet, the necessary simplicity of the on-board models, and the fact that the comet surface as viewed by the camera is partly illuminated during the mapping phase, optical measurements like the apparent diameter or the limb are expected to provide poorly reliable informations. So, the effort has been put on the modelization of the line-of-sight to the optical centre which provides a smoother information. A center finding process has been developed using the on-board comet ellipsoidal shape model combined with the estimation of the comet attitude and the camera pointing direction. This process consists in computing the equations of the ellipses which delimit the illuminated area of the comet nucleus model in the camera focal plane, and then deriving the barycenter of this area by a dilatation technique (see Fig.4).

The modelization of the range to surface measurement presents no major difficulty. The line-of-sight to landmarks is a well-adapted concept for relative navigation since this measurement is rather precise and contains non ambiguous informations on the comet rotation and the distance to comet (if several landmarks are tracked). The on-board implementation of the large amount of image processing required by landmark tracking remains to be proven, even with a good initialization by the ground. However, this concept has been kept for further comparative analyses.

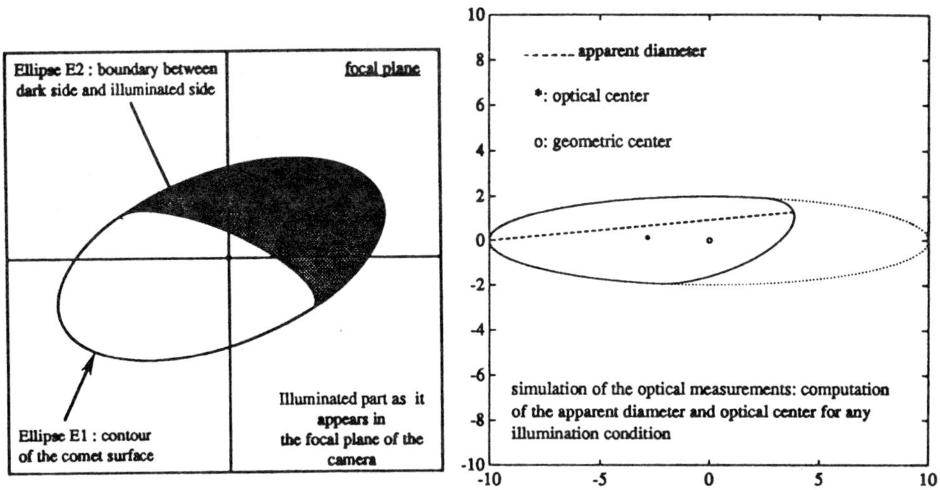

Fig.4 : Analytical calculation of the barycentre of the illuminated part for onboard measurement models

The detailed selected concepts are summarized in Table 3. For each subphase, the navigation filter will process the listed measurements, and the identifiable comet parameters will be estimated in the filter state vector along with the spacecraft relative position and velocity. These comet parameters will be updated and initialized by ground estimated values.

AUTONOMOUS DETECTION AND AVOIDANCE OF DEBRIS

The investigation of this problem is also one of the main issues of the on-going study, which has already produced instructive analyses on the need and feasibility of such a system.

Two kinds of debris are likely to be found in the vicinity of a comet near aphelion :

- debris which are co-orbiting the Sun outside the comet sphere of influence which may be considered as quasi-fixed in a comet centered Sun pointing frame, at typical ranges from 40 / 1000 km to 100,000 km to the comet. Detection and avoidance conditions are the following:
  - spacecraft relative velocity from 100 m/s to 5 m/s
  - debris straight ahead the spacecraft are dangerous
  - very good visibility conditions

Table 3
IN-ORBIT AUTONOMOUS NAVIGATION CONCEPTS

SUBPHASE	MEASUREMENTS	COMET MODELS	IDENTIFIABLE PARAMETERS	APPLICABILITY	COMMENTS
Low Drift 5000 kms —> 500 kms	• LOS to GC • apparent diameter • LOS to landmarks	• attitude • mass • ellipsoidal shape • offset COM-GC • landmarks catalog	• Euler angles + p,q,r • mass • offset COM-GC • 2 inertia ratios • landmarks coordinates	• fully illuminated nucleus • outside the comet sphere of influence	The 2 inertia ratios appear in the Euler equations of p,q,r
Inertial points 500 / 100 kms	• LOS to GC • radar altimeter (towards GC) • LOS to landmarks	same as above	• Euler angles + p,q,r • offset COM-GC • 2 inertia ratios • landmarks coordinates	• fully illuminated nucleus • known gravitational field	The gravitational field can not be estimated because of the numerous ΔV implemented
Hover points 100 / 10 kms	• LOS to optical center • radar altimeter (towards GC) • LOS to landmarks	• same as above • gravitation harmonics	same as above	• partially illuminated nucleus • known gravitational field	The optical center finding process is used to predict the Line Of Sight
Orbits 500 kms -> 10 kms	same as above	same as above	• Euler angles + p,q,r • mass • offset COM-GC • 3 moments of inertia • landmarks coordinates	• partially illuminated nucleus • inside the comet sphere of influence	Gravitation harmonics coefficients are expressed wrt the 3 principal moments of inertia
Orbits < 10 kms	• radar altimeter (towards landmarks) • LOS to landmarks	• attitude • mass • gravitation harmonics • landmarks catalog	• Euler angles + p,q,r • mass • 3 moments of inertia • landmarks coordinates	• partially illuminated nucleus • at least one landmark visible	The optical center measurement is prohibited because the nucleus does not fit in the camera FOV

GC = geometric center, COM = center of mass

- debris which are orbiting the comet inside the sphere of influence, at altitudes from ε to 40 / 1000 km depending on the comet size, with any orbit eccentricity and inclination. Detection and avoidance conditions are the following:
  - spacecraft relative velocity from ε to 5 m/s
  - dangerous debris must be searched in all directions
  - visibility conditions depend on the angle between the LOS to the debris and the Sun direction (from 0° to 180°)
  - the search can be limited to a cone centered around the spacecraft velocity vector because debris in head-on collision orbits are the most dangerous.

A preliminary collision probability study[12] using IRAS observations of the Tempel 2 comet applied to the ROSETTA nominal approach and global mapping phases has produced the results shown on Fig.5 (probable numbers of collision with given size particles), assuming that debris are of the same size and uniformly distributed from 0 to 100,000 km to the comet. The two preceding assumptions are very pessimistic but are very simple to formulate for preliminary analyses. More realistic conditions would involve an exponential geographic distribution of debris with respect to the range to comet and equally distributed in size. These worst case conditions are to be kept in mind when examining the results produced by this preliminary analysis.

The main conclusion is that the worst case probability to encounter dangerous debris which would damage the spacecraft is 1% during the approach phase segment at 100 m/s velocity and 0.001% during the global mapping phase.

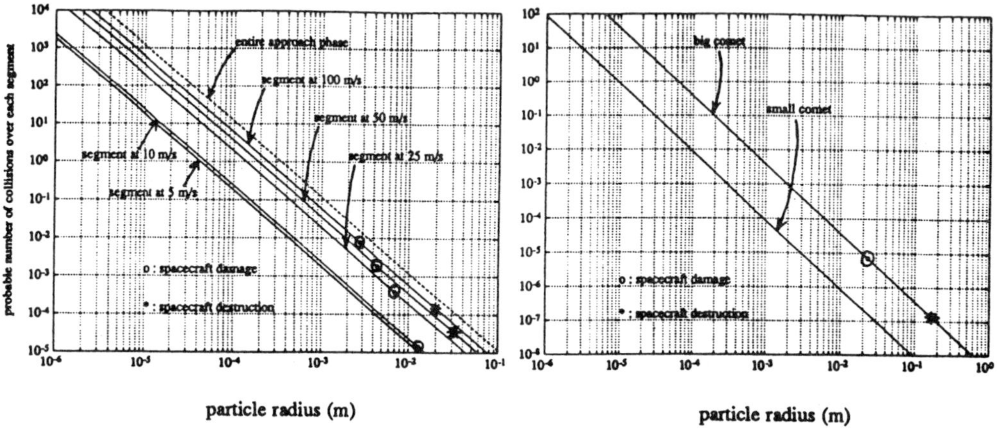

Fig.5 : Collision Probabilities (approach and global mapping phases)

Two informations are necessary to ensure the observability of the debris trajectory: the line-of-sight provided by an optical sensor and the range to the debris provided by a microwave sensor. The optical sensor sensitivity in visual magnitude is the main restriction for the debris detectability. It depends on the debris minimum dimension to be detected and on the illumination conditions (see Fig.6).

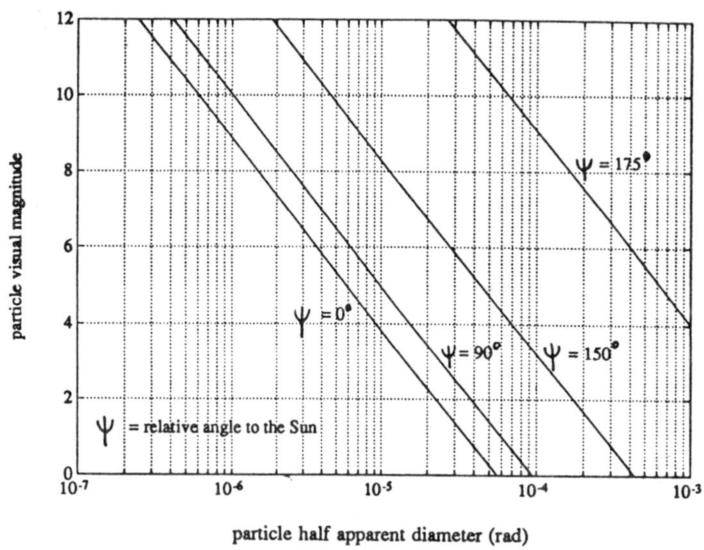

Fig.6 : Visual Magnitude of debris vs illumination angle and size

Assuming that the detection and tracking of debris can be performed, the problem addressed now is the capability of the GNC system to predict a collision hazard with a sufficient time margin to perform an avoidance maneuver. The following assumptions have been considered :

- the debris detection and tracking is assumed in a cone centered around the spacecraft velocity vector, with LOS and range measurements
- the problem reduces to a fixed spacecraft with approaching debris on straight lines with all velocities
- the analysis is two-dimensional (symmetric problem).

The following issues must be analyzed :

- tracking sensor performances in LOS and range measurements bias and noise and sensitivity in visual magnitude
- estimation of the debris relative motion
- collision hazard evaluation
- avoidance maneuver duration and efficiency
- GNC system performances in detection delays and false / missed alarms probabilities.

A discrete Kalman filter is used to estimate the debris relative motion. The collision hazard is evaluated by a test on the filter outputs: debris position and velocity vectors, to know if the predicted trajectory encounters a spherical safety domain of radius 10m around the spacecraft center of mass. A certain number of successive filter outputs are tested in order to decide an alarm set off or an alarm cancellation, taking into account the predicted time to go before collision and the time required to implement an efficient avoidance maneuver. The direction of the avoidance boost is optimized with respect to the predicted debris trajectory. The 8 sets of 3x10N hydrazine thrusters are used, providing a 40N thrust level in all directions without any attitude maneuver. With the above mentioned conditions, the minimum avoidance maneuver duration is 35s.

The GNC system has been simulated and tested in four different environments of debris trajectory and two levels of sensor performances, presented in the tables below.

Table 4

REAL WORLD ENVIRONMENT

ENVIRONMENT #	1	2	3	4
debris trajectory incidence angle (deg)	0	0	0	45
debris relative velocity (m/s)	100	25	5	5
detection distance (km)	5	2.5	1	1

SENSOR PERFORMANCES

SENSOR #	1	2
range noise	1%	5%
range bias	1 m	1%
LOS noise (deg)	0.1	1
LOS bias (deg)	0.1	1

The remaining parameter to be chosen is the minimum distance D_{MIN} from the debris trajectory to the spacecraft safety domain, the sign of which is an indication if there is collision or not. The performance criteria to be evaluated are the following :

- debris position and velocity estimation errors
- false and missed alarms probabilities :
  - D_{MIN} is varied such that debris encounter the safety domain. If no collision is detected or if an alarm is set off too late, the result is a missed alarm
  - D_{MIN} is varied such that debris travel outside the spacecraft safety domain. If an avoidance maneuver is ordered, the result is a false alarm
- detection delay performances : time passed from the detection by the sensor to the alarm set off by the GNC system in case of collision.

The last two results are derived from Monte-Carlo analyses on 200 time simulations. Estimation errors are shown on Fig.7, false / missed alarms probabilities on Fig.8 and detection delay performances in Table 5.

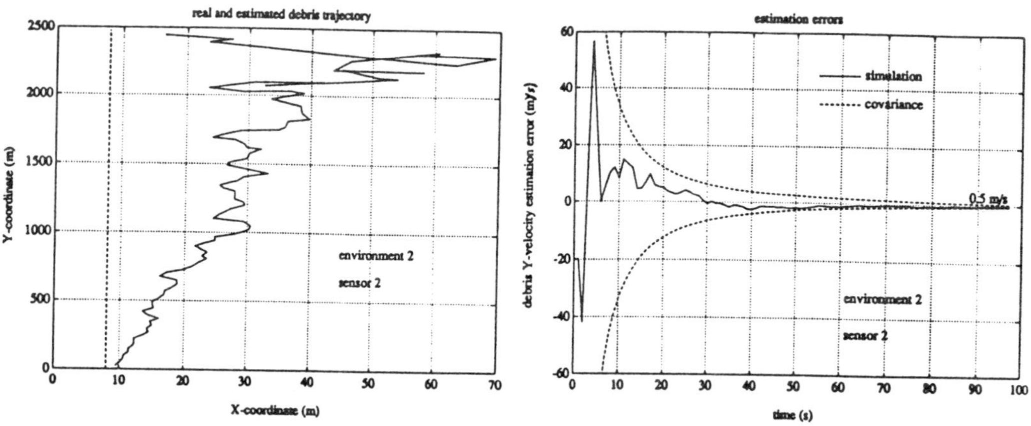

Fig.7 : Estimation errors

The use of the sensor # 2 degrades significantly the estimation errors in position and velocity by a factor of 5, whereas the velocity estimation errors with sensor # 1 are less than 1% of the debris absolute velocity, which enables to perform an efficient collision hazard evaluation.

The false / missed alarms probabilities and the detection delay performances allow to validate the GNC system in environments # 2, 3 and 4 with sensor # 1 performances. Bad results obtained in environment # 1 show that the GNC system has no time enough to react to a dangerous debris in only 50s.

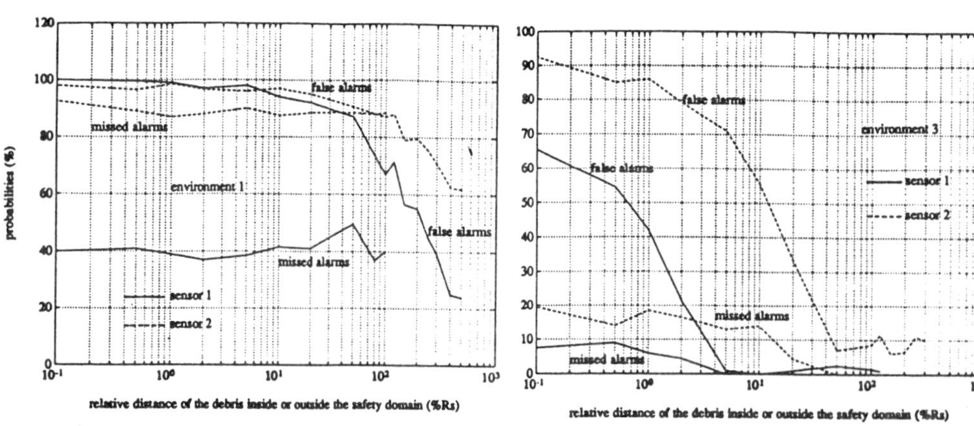

Fig.8 : False / missed alarms probabilities

Table 5
DETECTION DELAYS

		CENTER IMPACTING DEBRIS		TANGENTIAL DEBRIS	
		mean	std	mean	std
ENVIRONMENT 1	SENSOR 1	12.9 s	3.4 s	14.5 s	5.3 s
	SENSOR 2	23.4 s	7 s	25.8 s	7.98 s
ENVIRONMENT 2	SENSOR 1	25.3 s	5.2 s	47.6 s	18.8 s
	SENSOR 2	53.7 s	7.7 s	56.1 s	10.9 s
ENVIRONMENT 3	SENSOR 1	25.3 s	5.1 s	111 s	41 s
	SENSOR 2	70.4 s	21 s	118 s	46 s
ENVIRONMENT 4	SENSOR 1	29.5 s	5.8 s	229 s	60 s
	SENSOR 2	90.6 s	28.9 s	198.3 s	63.8 s

Two main conclusions can be derived from this study.

Concerning the collision probabilities, it is clear that more realistic assumptions on the debris distribution in location and size will significantly decrease the alarming 1% probability to encounter dangerous debris during the approach phase segment at 100 m/s velocity. On the other hand, the corresponding probability during orbiting phases could be increased due to a larger density of debris closer to the comet. Analyses performed up to now do not allow to conclude for or against the need of a debris avoidance GNC system and further investigations should focus on comet scientific observations allowing to derive precise informations on the debris distribution in location and size.

Concerning the feasibility of a debris avoidance GNC system, the study has produced the following results :

- a delay of 55 to 75s is necessary for the GNC system with sensor # 1 specifications to predict a collision and implement an assured avoidance maneuver, 35s of which must be dedicated to the boost duration. This total reaction time does not significantly depend on the spacecraft relative velocity and is lower when the velocity is higher (55s for 100 m/s velocity impacting debris)
- given this performance and with the current optical and microwave sensors designs, it seems to be not possible to detect and avoid impacting debris at 100 m/s velocity. The feasibility is constrained by the optical sensor sensitivity in visual magnitude which must be about 10 Mv (from Fig. 6) and by the microwave sensor ability to measure the range to 3mm radius debris up to 5500m.
- under good conditions (environments # 2, 3, 4 and sensor # 1), the GNC system proved to be strongly performant, with no missed alarm for debris approaching closer than 8 meters to the spacecraft center of mass. Therefore, the feasibility of a debris avoidance GNC system for use during the orbiting phase and the low drift approach where the spacecraft relative velocity is less than 25 m/s seems to be demonstrated by this preliminary study, but is limited by the following restrictions:
  - the tracking sensor main specifications are:

    LOS noise: 0.1°　　range noise: 1%　　sensitivity: 6 Mv

    detection distance of 7mm radius debris by a microwave sensor: 1750 m
  - the specification on the sensor field of view depends on the collision probability with dangerous debris coming from other directions than the opposite to the spacecraft velocity vector, which has not been investigated
  - with the above mentioned optical sensor sensitivity of 6 Mv, the search for dangerous debris will be limited by worst case illumination conditions

corresponding to a Sun incidence angle of approximately 155° (from Fig. 6). In other words, the cone centered around the direction to the Sun in which dangerous debris can not be detected because of bad illumination conditions has an half angle of 25°.

– the distinguishability of debris from the star background in the optical sensor field of view has not yet been demonstrated.

GNC CONCEPTS FOR THE DESCENT PHASE

This critical part of the mission has been extensively studied and preliminary results have been presented in earlier papers[3,4]. A detailed description of the concepts, strategies and performance assessment for the descent phase can be found in Ref. 3. Only the main results are presented here. As previously described (cf Fig.1), the descent scenario involves three main phases :

- the descent preparation ;
- the deorbitation maneuver and the homing ;
- the synchronisation maneuver and the final descent.

The first two phases are nominally commanded from ground, while the third one is entirely autonomous. The objective of the above mentioned analyses was to determine the performance of the GNC system in terms of dispersion in position and velocity at touch down. These performances are quite dependent on the design of the descent preparation orbit and that of the homing orbit which are not yet frozen. This design presently under investigation[11] depends on the requirements of continuous visibility to the Earth for communication, an orbital period compatible with the convergency of the orbit determination, a safe periapsis and visibility to the landing site.

Crucial parameters to be taken into account are the position and velocity estimation errors at the deorbitation point which depend on the preparation orbit and on the ground navigation performances along this orbit. Ground navigation have been extensively studied[7,8,9]. Typical performances (from Ref. 7) are indicated in the table below:

Table 6
S/C INITIAL STATE ERRORS
(1 σ at deorbiting maneuver)

Coordinates	*Position*	*Velocity*
radial	*70 m*	*0.92 mm/s*
transverse	*70 m*	*0.92 mm/s*
normal	*14 m*	*0.31 mm/s*

In Ref. 3, two homing strategies have been proposed for performance assessment :

- "slow" descent (with laser mapper),
- "fast" descent (without laser),

The first strategy takes into account the constraints imposed by the laser mapping. The second one is an alternate strategy to the situation where the laser mapper has failed in an earlier phase of the mission, or when no significant features on the comet surface allow its unambiguous operation.

Dispersions of the spacecraft state have been propagated around the selected homing trajectories for different types of comet taking into account :

- initial dispersion errors (Table 6)
- dipolar gravity field (0.1% 1σ on the comet gravitational parameter)
- non gravitational accelerations (comet outgassing)
- attitude errors at firing and thruster misalignments (0.5 deg bias, 0.5 deg noise 1σ)
- thruster magnitude error (1%).

Thc analyses have shown that for all comet types, the performances in the slow descent case are compatible with the laser mapper operating range.

After the synchronization maneuver, the final part of the descent is under the control of the GNC system which drives the spacecraft in closed-loop along the desired trajectory. A vertical descent trajectory has been selected aiming at a 10cm/s vertical descent at touch down. When the laser system is used, the horizontal position with respect to the landing site frame is estimated every 200s up to a 500m altitude by elevation map correlation. Performance analyses using Monte Carlo simulations have been carried out[3] for three different scenarios (the landing site being located on the equator) :

A : Slow descent using the laser mapper, with an equatorial homing orbit ;
B : Same as Case A, polar homing orbit ;
C : Fast descent from 500 m without laser mapper, equatorial homing orbit.

These simulations have been performed by using a three degrees-of-freedom simulation tool developed by MATRA and INISEL under ESA contract[5], and improved by ESTEC team. Monte Carlo analysis was conducted to evaluate the obtainable landing accuracy in the three cases. As initial conditions for the final descent, the estimated state at the end of the homing orbit is assumed. In addition to these random initial conditions, the following set of parameters is made varying from one simulation to the other : the comet kinematics, the comet gravitational field, the landing site coordinates, the instruments errors (sensors and actuators). Accuracy of the radar and laser instruments are those given in Table 2. The results of these simulations are presented in Table 7. In Case A, the dispersions are lower than in Case B, due to a lower thrusting effort required. Laser-assisted navigation provides much better performance than unassisted navigation, as expected. Nevertheless, the landing errors are still below 100 m. In addition to accuracy, of major interest is the magnitude of the horizontal and vertical velocity dispersions at touch down. For a zero commanded horizontal velocity and a 10 cm/s commanded vertical velocity, Table 8 shows the RMS touch down velocity for Case A and demonstrates a performance better than 3 cm/s along each axis.

Table 7
LANDING ACCURACIES

	Case A	Case B	Case C
	bias + noise RMS Errors (m)		
East	*14.8*	*21.1*	*73.2*
North	*14.5*	*16.2*	*20.7*

Table 8
LANDING VELOCITY

	RMS Errors (mm/s)
East	*21.9*
North	*8.4*
Vertical	*129.7*

Refinements to the GNC and environment simulators have been recently initiated under ESA contract[6] to MATRA, GMV and INISEL. Major improvements are going to be implemented concerning the environment simulator, the instrument modelization, the spacecraft dynamics, the GNC system :

- environment : tumbling movement, improved non gravitational forces models, realistic fractal shape for the landing site area, complex gravity field ;
- instruments : modelization of the attitude and pointing sensors and actuators, refinements in the modelization of the radar and the laser instruments ;
- spacecraft dynamics : modelization of all the degrees of freedom in translation and rotation of the central body, the scan platform, the high gain antenna, modelization of the platform boom flexibility and the liquid sloshing ; simulation of the contact forces and touch down dynamics ;
- GNC system : implementation of candidate attitude and articulation guidance and control laws ; implementation of new GNC concepts for the landing phase (laser and radar assisted navigation, unassisted navigation) ; implementation of control strategies at touch down.

All these improvements are expected to provide quite realistic assessment of the GNC system performances and will be used for comparative analyses of candidate GNC strategies and specification of the baseline GNC system.

TOUCH DOWN CONTROL

The objective of the analyses carried out for the touchdown phase was to propose some concepts compliant with different lander kinematic and attitude conditions at landing and comet environmental factors. Investigations performed on a passive touchdown - i.e. without any control of the spacecraft attitude and velocity - have shown very low stability margin and great risks of tipping-over for the spacecraft under low comet gravitational acceleration (10^{-4} m/s^2) and/or high comet surface friction coefficient (0.9) (see Fig.9). Hypotheses on the geometry and the inertia of the spacecraft are given in the table below.

Table 9
SPACECRAFT CHARACTERISTICS

Mass (kg)	*2600*
Inertia (kg/m²)	*3200*
Vertical distance COM to footpads (m)	*2.35*
Horizontal distance COM to footpads (m)	*1.70*
Horizontal distance COM to drilling unit (m)	*1.80*

As a consequence, active control at touch down appears mandatory. Two concepts for an active touchdown have been developed. Both require the cold gas thrusters to perform a vertical thrust within the spacecraft triangular base which stabilizes the lander by pressing it towards the comet surface. In the first approach, the signals issued from contact sensors are used to identify the lander pads in contact with the comet surface and to drive the touchdown according to the following sequence: when a pad P1 is touching the comet, a control torque is generated around the direction parallel to the other pads P2-P3. When two pads rest on the surface (e.g. P1 and P2), a control torque around the line joining these two pads is generated. The control torques are opposite and proportional to the angular rate around the corresponding comet-fixed direction. This relative rate is not measured, but can be estimated from the gyroscope measurements and the estimated comet angular rate. It is worth noting that the error on the estimated angular rate (a few µrad/s for typical gyros and a 10 h comet rotation period) is negligible with respect to the minimum angular rate destabilizing the spacecraft from its stable attitude (5 10^{-4} rad/s for a worst case gravity field).

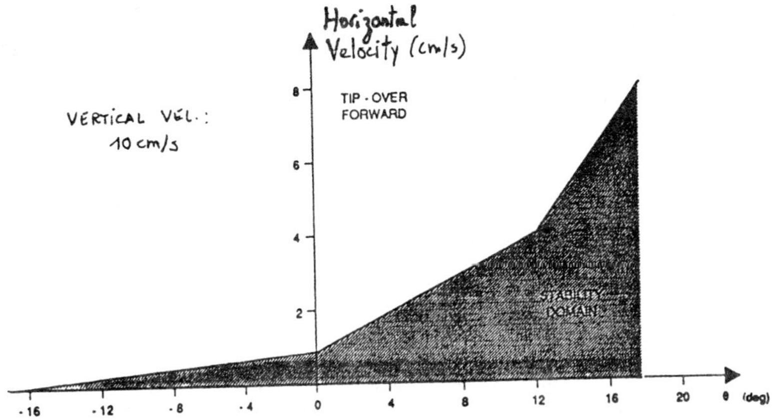

Fig.9 : Stability region for passive touchdown

The second control concept does not rely on any measurement of the spacecraft rotation. The contact of the lander with the comet is detected by the accelerometers measurements which are affected proportionaly to the ground reaction force. Then, a constant force along the spacecraft vertical axis is created towards the comet. Its primary effect is to create an artificial gravity field which increases the static stability margin. This concept certainly induces a smaller stability margin than the previous one. However, it is much robust from a system point of view since no information needs to be uplinked from the contact sensors mounted on the lander footpads to the on-board computer. Moreover, this control approach is not sensitive to the destabilizing effect of the flexible modes of the platform boom which may pollute the gyro measurements at touch down.

Before proceeding to the anchoring phase, the spacecraft must be stabilized on the comet. This implies to control any undesirable rotation and sliding with respect to the surface. Tipping over is monitored and controlled using the gyro measurements, as for the first touch down control strategy. No measurement of the relative velocity of the spacecraft with respect to the comet is available from the radar Doppler below an altitude of 20 m. Therefore, the sliding is controlled in open-loop by implementing a vertical thrust to increase the friction force. Assuming a landing on an icy slope of less than 30 deg with a residual lateral velocity of 3 cm/s, the required time to stop is given on Fig.10 as a function of the thrust magnitude.

The 8 sets of 3 x 30 N cold gas thrusters used to control the landing operation give two possibilities to perform a pure vertical force: 60N or 120N. Based on a propellant consumption criterion, the optimal selection depends on the landing site slope α and the surface friction coefficient f:

$\mathrm{tg}\,\alpha < f \Rightarrow 60\,\mathrm{N}$ selected;

$\mathrm{tg}\,\alpha > f \Rightarrow 120\,\mathrm{N}$ selected.

The worst admissible case, characterized by $\alpha = 30°$ and $f = 0.1$ under high local gravity ($4.2\ 10^{-3}$ m/s) corresponds to a consumption of 2.2 kg.

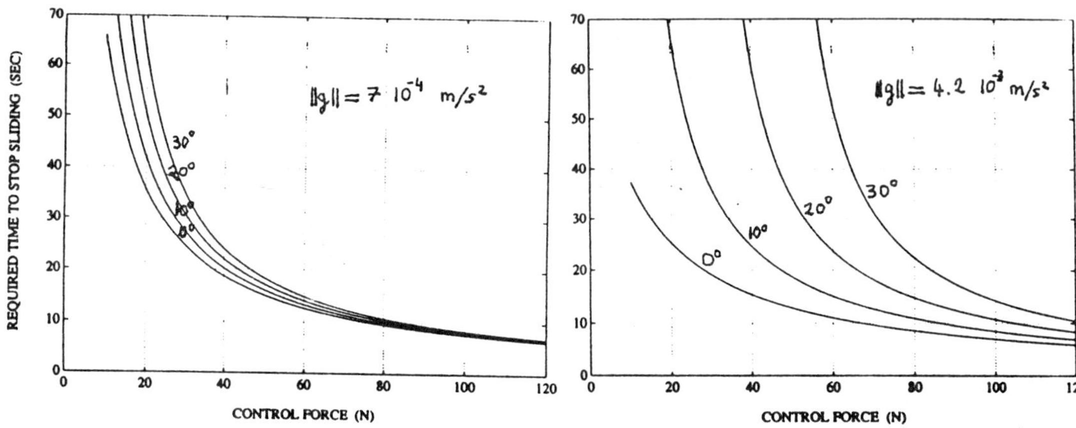

Fig. 10 : Minimum time to stop sliding vs slope

Monitoring of the sampling operations

Once the spacecraft is anchored on the comet, the drilling operations are initiated. In the nominal mode, the anchoring system is compensating the drilling resulting forces (100 N) and torques (10 Nm) as shown in Fig.11. These operations will be supervised by the accelerometers. The failure of one of the anchors causes a step of acceleration along the spacecraft normal of about 0.03 m/s^2. Two 70N biliquid thrusters vertically mounted above the footpads on each side of the drilling unit are sufficient to stabilize the spacecraft if anchoring no longer operates (see Fig.12). Their actuation prevents the lander from taking off and sliding due to the drilling resulting forces and torques: pads P1 and P2 are quite not constrained, while P3 is pressed into the ground. Then, the ratio between the tangential and the normal components of the ground reaction force is less than 0.08 (to be compared with the surface friction coefficient for ice f = 0.1) and sliding is prevented.

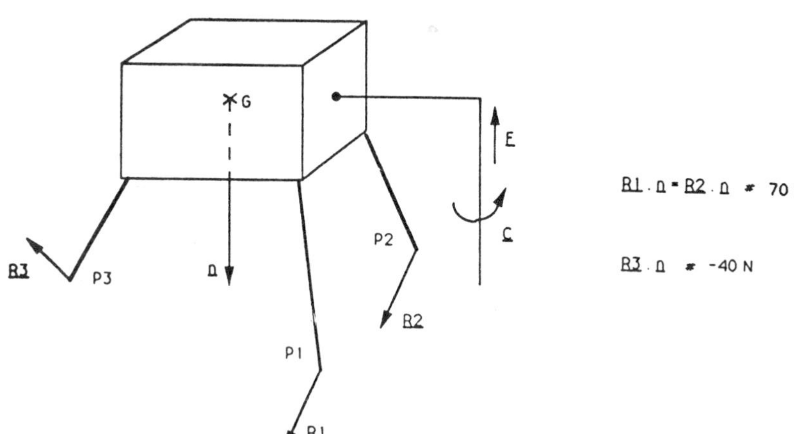

Fig.11 : Nominal sampling mode (with anchoring)

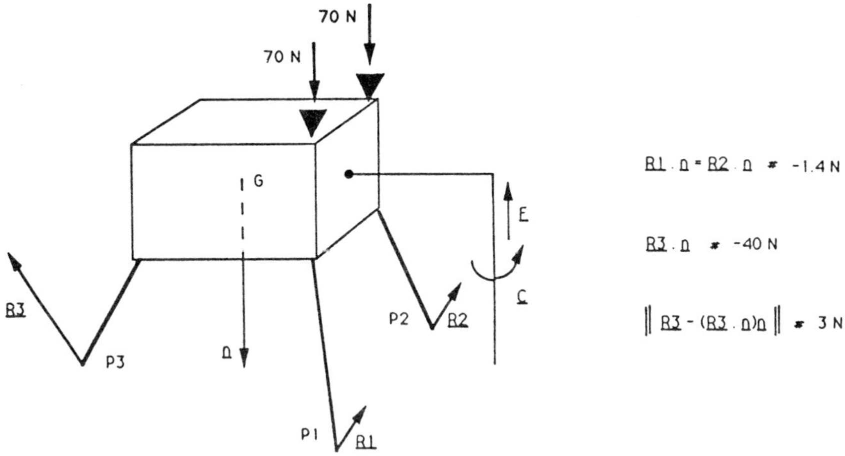

Fig.12 : Back-up sampling mode (without anchoring)

In order to detect any destabilization of the spacecraft, the drilling operations will be actively monitored. The angle between the lander vertical axis and the estimated comet gravitational acceleration vector is calculated from the accelero measurement at a given frequency. If it exceeds a pre-defined threshold (depending on the static stability angle of the spacecraft), a safe mode is activated.

CONCLUSION AND FURTHER INVESTIGATIONS

This paper has described the current status of the GNC analyses carried out in parallel to the ROSETTA System Definition Study in the frame of the ESA Technological Research Programme supporting the ROSETTA project.

After the description of the mission requirements and the mission scenario, investigations and analyses of autonomous GNC concepts for the near comet phase of the mission has been performed. Strategies for autonomous navigation during the orbiting phase have been proposed. Performance assessment of such concepts will be performed through the simulation tool currently in development.

Debris avoidance concepts for the observation phase has been analyzed. It has been shown that the detection of the debris must occur typically 60s before the impact, taking into account standard performances for optical and microwave detection instruments. This means that during the high velocity approach (100m/s), debris avoidance is not realistic, while during the orbiting phase, such a concept could be implemented. The necessity of its implementation on-board crucially depends on the repartition of the debris around the comet. More informations must be gained through intensive observation campaigns before taking such a decision.

Strategies for the descent and landing phase and associated GNC performances have been presented as results of previous analyses[3,7,8,9]. The dispersions at touch down are typically 20 m (1σ) with the laser system. In the case of a laser system failure, the landing accuracy that can be achieved with radar measurements is of the order of 75 m (1σ). The touch down horizontal velocity is below 3 cm/s. These results have been obtained by using a three degrees-of-freedom simulation tool. Extension of this tool to take into account all the degrees-of-freedom of the spacecraft and its articulated devices, and more realistic environment and instruments models is being performed and will be used for further analyses.

As concerns the touch down, preliminary analyses using a simplified 2D model have shown quite low stability margins for a passive touch down. Active control and monitoring concepts have been proposed in this paper. The above mentioned simulation tool will include the simulation of the dynamics at contact and will be used for validating the proposed concepts.

ACKNOWLEDGMENT

We would like to acknowledge the MATRA team working on the ROSETTA mission and the autonomous navigation techniques for its contribution to the elaboration of this paper.

REFERENCES

1. ROSETTA/CNSR System Definition Study, Draft Final Report, *ESA Contract* 8228/89/NL/SK, MATRA ESPACE, Oct. 1990

2. Present Definition of the ROSETTA/CNSR Mission, C. Etienne, IAF-90-434, *IAF Congress*, Oct. 6-12, 1990, Dresden, GDR

3. Autonomous Spacecraft Navigation and Control for Comet Observation and Landing, J. de Lafontaine, AIAA-90-2911, *Astrodynamics Conference*, Aug. 20-22, 1990, Portland, Oregon

4. Navigation in the Vicinity of a Cometary Nucleus, V. Guillaud, J. de Lafontaine, AAS-89-207-1, *AAS/GSFC International Symposium on Orbital Mechanics and Mission Design*, April 24-27, 1989, Goddard Space Flight Center, Greenbelt, Maryland

5. Autonomous and Advanced Navigation Techniques Study, Final Report, *ESA Contract* 7562/88/NL/MAC, MATRA ESPACE, April 1989

6. Autonomous and Advanced Navigation Techniques Study, Phase 1, *ESA Contract* 8808/90/NL/PM(SC), MATRA ESPACE

7. On the ROSETTA Comet Approach Phase Orbit Strategy and Navigation, M. Bello-Mora, MAS Working Paper 309, European Space Operations Centre, June 1990

8. Navigation of the Landing of a Spacecraft on a Comet Nucleus, J.K. Miller, CNSR/JPL TechNote 23, Jan. 1990

9. Orbit Determination of the Comet Rendezvous/Asteroid Flyby Mission : Post Rendezvous Phases, J.K. Miller, C.J. Weeks, L.J. Wood, AIAA-89-0348, *27th Aerospace Sciences Meeting*, Jan. 9-12, 1989, Reno, Nevada

10. Study and Design of a Comet Approach and Landing System, Draft Final Report, *ESA Contract* 7988/88/NL/PB(SC), Officine Galileo, June 1990

11. Orbit Strategies and Navigation Near a Comet, *ESA contract* 8947/90/D/MD, British Aerospace

12. Approach Strategy for the Rendezvous with Comet Temple 2, R.Y Kakuda, R. Ionasescu, L.E. Dickey, AIAA 86-2116, Astrodynamics Conference, Williamsburg, VA, Aug. 18-20, 1986

AAS 91-066

TELECOMMUNICATIONS, NAVIGATION AND INFORMATION MANAGEMENT CONCEPT OVERVIEW FOR THE SPACE EXPLORATION INITIATIVE PROGRAM

Jerome A. Bell*, Elaine Stephens† and Gregg Barton‡

Current manned-mission operations involving telecommunications, information systems, and navigation (TISN) functions are highly operator-intensive. There are large classes of near real-time operational decisions, for both TISN and general operations, which must be made in-situ at Mars due to the long delay time to Earth-based mission decision elements and return.

The Space Exploration Initiative (SEI) program will incorporate mission and infrastructure elements on the surfaces of the Moon, Mars, and Earth; in orbit around the Moon, Mars and Earth; and on vehicles in transit between surface and orbit locations. The TISN must be integrated into an end-to-end data system that will facilitate information flow from element to element throughout the SEI program. Generally, the mission TISN interfaces include: (a) flight vehicles and the Earth mission control center; (b) flight vehicles and flight vehicle; (c) flight vehicles and the outpost; (d) flight vehicles and extravehicular personnel; (e) flight vehicles and the Space Station Freedom; (f) the outpost and the Earth mission control center; (g) the outpost and surface elements within the outpost zone; (h) outpost and surface elements beyond the outpost zone; and (i) the Earth control center and surface elements beyond the outpost zone. However, specific interfaces and infrastructure will be a function of the specified mission architecture and the technology and research/advanced development objectives that are ultimately realized.

This paper will describe: (a) a general overview of the concepts for TISN (b) key engineering, operations and mission issues that have been identified, and (c) the results from NASA's FY90 architecture option definition activity.

---

* NASA Johnson Space Center, Lunar and Mars Exploration Program Office, Houston, Texas 77058.

† NASA Johnson Space Center, Tracking and Communications Division, Houston, Texas 77058.

‡ Charles Stark Draper Laboratory, 555 Technology Square, Cambridge, Massachusetts 02139.

INTRODUCTION

Current manned-mission operations involving telecommunications, information systems, and navigation (TISN) functions are highly operator-intensive. The expected complexity of planned lunar and Mars local mission operations, if forced to operate using attended monitor and control techniques, could be nearly intractable, or at least unaffordable. There are large classes of near real-time operational decisions, for both TISN and general operations, which must be made in-situ at Mars due to the long delay time to Earth-based mission decision elements and return. Not only must local decision making be software assisted, but the links to local terminals must also distribute this information rapidly and primarily operate autonomously.

The Space Exploration Initiative (SEI) program will incorporate mission and infrastructure elements on the surfaces of the Moon, Mars, and the Earth; in orbit around the Moon, Mars, and Earth; and on vehicles in transit between surface and orbit locations. The distributed SEI mission elements will be linked and tracked by communication and navigation systems, and it will require an equally distributed information system. The TISN must be integrated into an end-to-end data system that will facilitate information flow from element to element throughout the SEI program.

This paper will describe: (a) a general overview of the concepts for TISN; (b) key engineering, operations, and mission issues that have been identified; and (c) the results from NASA's FY 90 architecture option definition activity. This paper summarizes the work performed by a consolidated NASA and contractor team through FY 90 for various SEI mission options (see acknowledgements). The reference mission is derived from work performed on the 90-day study (Ref. 1).

SEI PROGRAM DATA SYSTEMS OVERVIEW

The SEI data system includes all data systems within the mission elements, mission operations control systems, as well as the TISN support infrastructure. The support infrastructure systems are intended to provide transparent service to the distributed science and mission elements. In order to achieve this goal, the systems internal and external to the support infrastructure must operate, interface, and evolve in a manner that assures desirable operations. Although the specific SEI program data system elements may be designed, assessed, and implemented separately, all elements will be required to operate together to meet the operational needs of the SEI. Therefore, operation of the SEI program data system as a whole must be considered at every phase of system design and implementation.

The end-to-end data functions may be divided into two domains: the user functions and the TISN functions. User functions include unique data processing on both the transmitting and receiving ends. The user must, however, be compatible with the TISN infrastructure. The infrastructure functions include the determination of vehicle state vectors, the storage and transmission of data, and the necessary associated data processing. Generally, the mission TISN interfaces include:

a. Flight vehicles and an Earth mission control center
b. Flight vehicles and flight vehicle
c. Flight vehicles and the outpost

d. Flight vehicles and extravehicular personnel
e. Flight vehicles and the Space Station Freedom
f. The outpost and an Earth mission control center
g. The outpost and surface elements within the outpost zone
h. The outpost and surface elements beyond the outpost zone
i. An Earth control center and surface elements beyond the outpost zone

However, specific interfaces and infrastructure will be a function of the specified mission architecture and the technology and research/advanced development objectives that are ultimately realized. One prime example of this is mission crew size. While the maximum crew size is set by the level of surface activity, the criteria for minimum crew size is not easily separable. Certainly, one component will be the required crew activity during transit. During missions to Mars, the lag time of up to 40 minutes round trip precludes dependence upon the Earth for time-critical responses, although not ruling out support from the Earth for other situations. The need for an integrated approach for employing expert systems, crew availability, and training concepts for vehicle and systems control is without question.

REFERENCE TISN CONCEPT OVERVIEWS

The SEI TISN architecture will be a safe, effective, and economical architecture comprised of a complementary combination of Earth-based, vehicle-based, and lunar/Mars local-based systems. The architecture will support three basic levels of service — nominal, degraded, and mission critical. The TISN will evolve over time to meet the increasing requirements as the program evolves from the precursor to the Mars utilization phases. The systems will be designed to allow for advances in technology. The general theory for evolution will nominally be to upgrade during regularly scheduled maintenance, refurbishment, or replacement, and as new satellites or systems are emplaced.

Telecommunications

The reference telecommunications infrastructure includes the Deep Space Network (DSN), the Advanced Tracking and Data Relay Satellite System (ATDRSS), the outpost communications center, the Mars Relay Satellites (MRS), and the vehicle communications systems (external).

The reference communications support for the robotic missions to Mars would evolve as a series of communications orbiters are emplaced. Initial in-situ communications would be direct to the Earth or by means of each mission's corresponding carrier spacecraft, which would serve as a simple communications relay to the Earth. The communications infrastructure for the Mars robotic missions would evolve to a satellite network consisting of two communications orbiters in Mars stationary orbit. The DSN would initially support the SEI program with shared use of one of its subnets; later, in the precursor mission, an SEI dedicated subnet would be used.

As the lunar manned mission progresses, unattended operations are planned to evolve. First, an unattended telecommunications operations testbed would be deployed on the Moon during the emplacement phase, which would lead to unattended operations demonstrations during the later consolidation phase. The end product would be a system design and operational techniques for use in all telecommunications elements for the lunar utilization phase, at Mars for signal acquisition and data transfer, and for remote mission operations. Also during the lunar mission, the DSN would add two more dedicated subnets to provide support to the program.

The MRS would be placed in orbit around Mars 1 year before the first manned mission arrives. The MRS would replace earlier communications orbiters used during the robotic missions. The DSN would evolve further to support the manned Mars missions by adding two additional 34m dedicated subnets with Ka-band capability.

Navigation

Navigation in general is the determination of the current and future translational and rotational vehicle state vector (position, velocity, attitude, and attitude rate) for space flight, atmospheric flight, and mobile surface systems. This state vector may be inertial, relative to another spacecraft or celestial body, or relative to a surface-centered coordinate frame. The navigation system will support a multitude of guidance and control functions (e.g., planetary transfers, orbital operations, descent and ascent flight phases, instrument pointing and tracking, surface roving, etc.) as well as mission planning, scientific experiments, and post flight reconstruction and data analysis.

Previous lunar and planetary missions have relied heavily on ground-Earth-based tracking to provide state vector updates. There are several reasons why this approach will not be adequate for the Mars exploration program. First, accuracies necessary for several critical parts of the reference mission, such as Mars aerocapture and precision landing on the surface, would require local (planet relative) updates. Second, the long round-trip light time to Mars of up to 40 minutes and the limited observation geometry from Earth-based tracking (DSN) would preclude rapid and accurate state vector updates. Third, for routine planetary operations such as surveys from orbit, a local autonomous scheme may be more cost effective than long-term use of ground tracking alone. The reference navigation infrastructure would utilize the combined attributes of the DSN and onboard systems.

Infrastructure and Evolution. The reference navigation infrastructure consisted of three components: the Earth-based infrastructure, the spacecraft and mobile surface elements, and the lunar and Mars local infrastructure. The supporting Earth radiometric tracking stations, user vehicle navaids, and local infrastructure (surface beacons and navsats) are shown in Fig. 1. These systems will be arranged in a hierarchy (primary, secondary, and tertiary) for each mission phase. The primary level would support nominal mission operations complemented by the secondary level whose function would be to improve the performance or replace the primary whenever it was unavailable. The tertiary mode would provide the minimum support required for safe crew return when neither primary nor secondary were available.

The Earth-based navigation system is supported by the ground network (GN); the ATDRSS; the DSN; the global positioning system (GPS); and a mission navigation center on Earth which processes radiometric data from the ATDRSS, the GN, and the DSN tracking stations. In contrast, the radiometric data from the GPS is processed real time onboard the spacecraft. The optical sensors onboard can be processed either on ground or real time onboard.

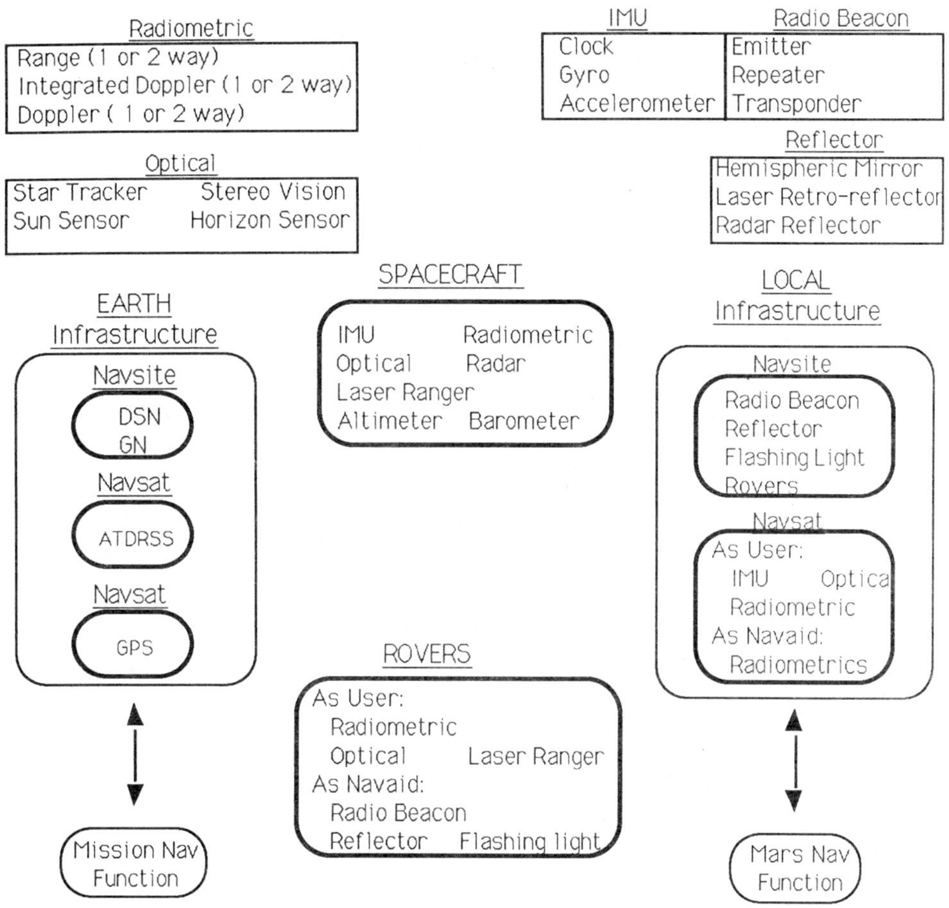

Fig. 1 Navigation Infrastructure Concepts

The ATDRSS, the GPS, and the GN can only support the vehicle while it is in Earth orbit. Earth-based navigation constitutes the baseline from which the mission navigation system will evolve and provides capabilities for testing, calibration, and back-up. The Earth-based system will be an upgraded version of the current planetary exploration navigation system where the required accuracies will be achieved utilizing ranging and Doppler, supplemented by very long baseline Interferometry (VLBI) tracking techniques.

All spacecraft would have an onboard state-of-the-art inertial measurement unit (IMU). The IMU is essential during real-time critical phases and could provide a secondary/tertiary role for trans-lunar and trans-Earth navigation. However, IMU data alone could not always meet the critical accuracy requirements. Other onboard navigational sources, such as radiometric, optical, or radar, etc., would be required to satisfy the need for real-time navigation. The spacecraft on-board navigation would be developed such that it had the appropriate level of performance, autonomy, fault detection and isolation, redundancy, and graceful performance degradation under failures. The spacecraft onboard navigation systems would be as common as appropriate between the lunar and Mars vehicles.

Local-based navigation would be supported by an infrastructure of navaids, navsites, and navsats. Radio beacons such as emitters, repeaters, or transponders are typical active navaids mounted on navsats or navsites. Reflectors or mirrors are typical passive navaids used on navsites. Navsats are orbital platforms providing systems to support radiometric navaids with potential global coverage depending on orbital mechanics. Navsats provide dynamic geometry for good observability but require ephemeris maintenance support from either the Earth-based infrastructure or the navsites. Navaids may be mounted on communication or imaging satellites or on dedicated navigation satellites. Navsites are surface platforms that are either stationary or mounted on a rover also providing systems to support radiometric or reflector navigation. Navsites require emplacement and survey and, except for the rover platform, would only be required to be surveyed once. Navsites could be surveyed either by VLBI or by the navsats.

Lunar mission navigation. The Earth-based infrastructure would provide the primary navigation for Earth-to-orbit, low-Earth-orbit operations (orbit maintenance, return capabilities, etc); trans-lunar injection and flight; and trans-Earth injection and flight. Earth-based navigation would also provide the initial spacecraft state following lunar orbit insertion and assist in the task of surveying navigation beacons on the lunar surface or on rovers. The Earth-based navigation infrastructure would also provide secondary and tertiary navigation in support of lunar orbit maintenance, descent, ascent, and rendezvous. As the infrastructure became capable of autonomous operations, the Moon would be used as a testbed for the Mars infrastructure and operations. In order to implement this testbed, the lunar and Mars infrastructures would be as common as possible where cost efficient, feasible, and appropriate.

The vehicle navigation systems could include an IMU, radiometric, radar, laser ranger, altimeter, and a barometer. The radiometric measurement sources with the IMU and the lunar-based infrastructure would be the primary navigation support for orbit maintenance, de-orbit, descent through landing, ascent through insertion, and early rendezvous targeting. Optical data would support line-of-sight tracking and IMU alignment and can also provide secondary and/or tertiary support for lunar orbital insertion, orbit maintenance, rendezvous, and Earth return pre-aerocapture. Lasers would provide primary support navigation for proximity rendezvous targeting and can also support surface ranging during landing. In addition, the terminal descent landing phase can be supported by radar.

Mars mission navigation. The Earth-based infrastructure would provide the primary navigation support for Earth-to-orbit, low-Earth-orbit operations (orbit maintenance, return capture, etc.); trans-Mars injection; and interplanetary cruise. Earth-based navigation would also be required for supporting initialization of Mars planet-relative navigation from 2 days out (or closer) from entry interface. In the secondary and/or tertiary roles, the Earth-based component can provide support for Mars pre-aerocapture, orbit maintenance, and trans-Earth injection. The VLBI is being considered as a possible source for surveying in planetary surface beacons or rovers. The feasibility of this role would be based on the trade-offs on the beacon power and antenna size requirements.

The Mars-based local navigation infrastructure would evolve during the precursor missions, the lunar tests and engineering demonstrations, and the later manned missions as new requirements and capabilities are identified. The first manned mission to Mars would have at least a primitive navsat and navsite network in place. Earth-based navigation has been the current practice; however, near real-time planet local navigation support would be considered essential for operations during critical phases of the mission such as aerocapture, landing, and rendezvous. It would also be considered necessary, even during temporary failure of the communication links, that a basic level of near real-time navigation be available since round-trip light time from Earth to Mars can be as long as 40 minutes and navigation requires long tracking arcs. Near real-time local navigation would provide faster and more accurate navigation due to better observation geometry. The navsats and navsites, in combination with the onboard (vehicle) sensor suite, would provide the primary navigation role in pre-aerocapture, orbit maintenance, descent, landing, surface operations (habitat and rovers), ascent and rendezvous, and trans-Earth injection navigation.

The radiometric measurement sources with the IMU and the Mars-based infrastructure would be the primary navigation support for pre-aerocapture, orbit maintenance, de-orbit through landing entry, surface operations (habitat and rovers), ascent through insertion, and long-range rendezvous. Optical data would support line-of-sight tracking and IMU alignment and could also provide secondary and/or tertiary support for pre-aerocapture navigation, orbit maintenance, rendezvous, and Earth return pre-aerocapture. Lasers would provide primary support navigation for short-range rendezvous and can also support surface ranging during landing. In addition, the terminal descent landing phase could be supported by radar or baro altimeters.

Precursor mission navigation. The precursor missions would be used to emplace the initial planetary-based navigation systems. Another objective of the robotic precursor missions would be to gain knowledge and demonstrate critical flight phase capabilities; including aerocapture, precision landing, and autonomous rendezvous and docking before the first manned mission.

Mars pre-aerocapture/aerocapture have been identified as a critical operation due to extremely tight navigation knowledge requirements. The precursor missions would be required to obtain more knowledge of the Mars atmosphere. Then the precursor missions would demonstrate the use of the onboard system, in combination with the Earth and Mars-based infrastructures, to ensure that they were adequate and capable of meeting aerocapture requirements for manned missions.

Navigation during a highly autonomous precision landing would be influenced by many factors such as vehicle performance characteristics, atmospheric conditions, wind speeds, and planetary surface conditions. The precursor missions would be required to return environmental and surface knowledge data; survey the landing sites; perform a hazard-free ground truth verification, and emplace navaids to support manned precision landings. The precursor missions would be used to demonstrate the operation of the onboard navigation system during execution of a precision landing.

The technology of autonomous rendezvous and docking could be demonstrated and verified by precursor missions in low Earth orbit or low lunar orbit. However, for complete verification of the Mars-based infrastructure, precursor missions would be required to demonstrate the capability to perform a rendezvous and docking sequence before manned vehicles would be committed to flight.

Information Systems

The SEI information infrastructure is represented in terms of nodes, interfaces, and data flows. Each of the nodes would contain a portion of the distributed system. Some nodes, such as those on the vehicles, at the outposts, and at ground control, would provide centralized support to their area with powerful computer systems to provide the information services, while other nodes, such as the relay satellites, would contain embedded information system functions appropriate for their own needs. Local information control systems would monitor and control local nodes based on the network schedule requirements. The system would have the ability to trace failures and perform fault isolation. Standards, such as communications formats, data structures, data types, coding, access keys, and priority designations for data transmission and storage throughout the system would be implemented for use by all mission and support elements.

Mission elements would require onboard storage capabilities for command loads and return telemetry when communications support is not available. Mars mission elements would require more capable onboard information systems than the lunar mission elements due to the long communication time delays to Mars.

Earth-based systems would plan and schedule the tracking and data acquisition networks. A centralized information control system would be located on the Earth to monitor the network and provide backup control. This centralized control may be co-located and closely integrated with the telecommunications network control. The Earth-based information system infrastructure used for the robotic missions would also support lunar manned missions. However, the elements would require an upgrade to accommodate additional service requirements. An unattended operations testbed would be deployed on the Moon during the emplacement phase and would be used for automated operations demonstrations during the consolidation phase. As automated operations capabilities became available, prime system control and monitoring functions might be distributed, with the central, Earth-based information systems (IS) control systems operating in a backup mode. The end product would be a system design and operational techniques for use in information systems on the Moon and Mars for data transfer and remote mission operations.

Earth-based IS, used for lunar manned missions, would support Mars manned missions where feasible. Some upgrades would, however, be required to accommodate the vastly different processing requirements of Mars manned mission support. In some instances, the upgrades would require whole new systems and facilities. Mars manned missions would require substantial in-situ IS capability for management, control, and data transfer operations. The largely unattended systems developed during lunar manned missions would be essential to facilitate local Mars real-time operations.

Onboard IS would be required on the relay satellites and the mission elements to manage the data flow and provide temporary storage as necessary. Initial relay satellites would be operated in either a tape record and playback mode or a real-time bent-pipe transmission configuration. Later replacement satellites may have more sophisticated switching modes.

ALTERNATE ARCHITECTURE TISN SUPPORT OPTIONS

Subsequent to the NASA 90-day study, additional TISN analyses were conducted during 1990 to support a NASA internal study on additional alternate mission architecture options. The mission options that were considered and the TISN support are described below.

Option 1

The goal of Option 1 architecture was to establish the capability of extended manned stay times (1 to 1.5 years) on Mars. The sites on Mars chosen for these outpost spanned widely different environments and represent areas of high research potential in the area of Martian geological, atmospheric and biological processes. Existing technology was used for initial Mars missions while high-leverage technologies were to be developed at a lunar testbed site and as part of the initial Mars missions. Fig. 2 depicts the evolution of TISN support.

Option 2

The goal of Option 2 architecture was to establish at least one permanent human settlement on the lunar and martian surfaces. The intent was not just to survive but thrive independently of any support from Earth. Large crew sizes, up to 30, required the highest level of communications of any of the architectures. Fig. 3 depicts the evolution of TISN support.

Option 3

The goal of Option 3 architecture was to enable a suite of relatively austere Mars missions including a short-duration (90-day), manned mission to the martian surface. A precursor lunar outpost was used almost exclusively to develop technologies and operation techniques for use in later Mars missions. Only limited "suitcase" science was assumed. A minimum of Mars robotic precursor activity verified a safe site for human landing. Subsequent to successful completion of the first manned mission and analysis of the data, a decision would be required about whether or not to continue manned missions to Mars. Because of the emphasis, this architecture resulted in the least extensive TISN support and a minimum amount of new technology. Fig. 4 depicts the evolution of TISN support.

Option 4

The goal of the Option 4 architecture was to accomplish systematic explorations of diverse locations on both the Moon and Mars. Globally distributed robotic explorations of both bodies led the way to characterize the environment. Later human landings at diverse sites coordinated continued robotic activity in an effort to gain as much knowledge and experience about the two bodies. Based on this newly acquired knowledge, mission planners would decide where and what kind of permanent outpost and facilities to set up on the planetary surfaces. Fig. 5 depicts the evolution of TISN support.

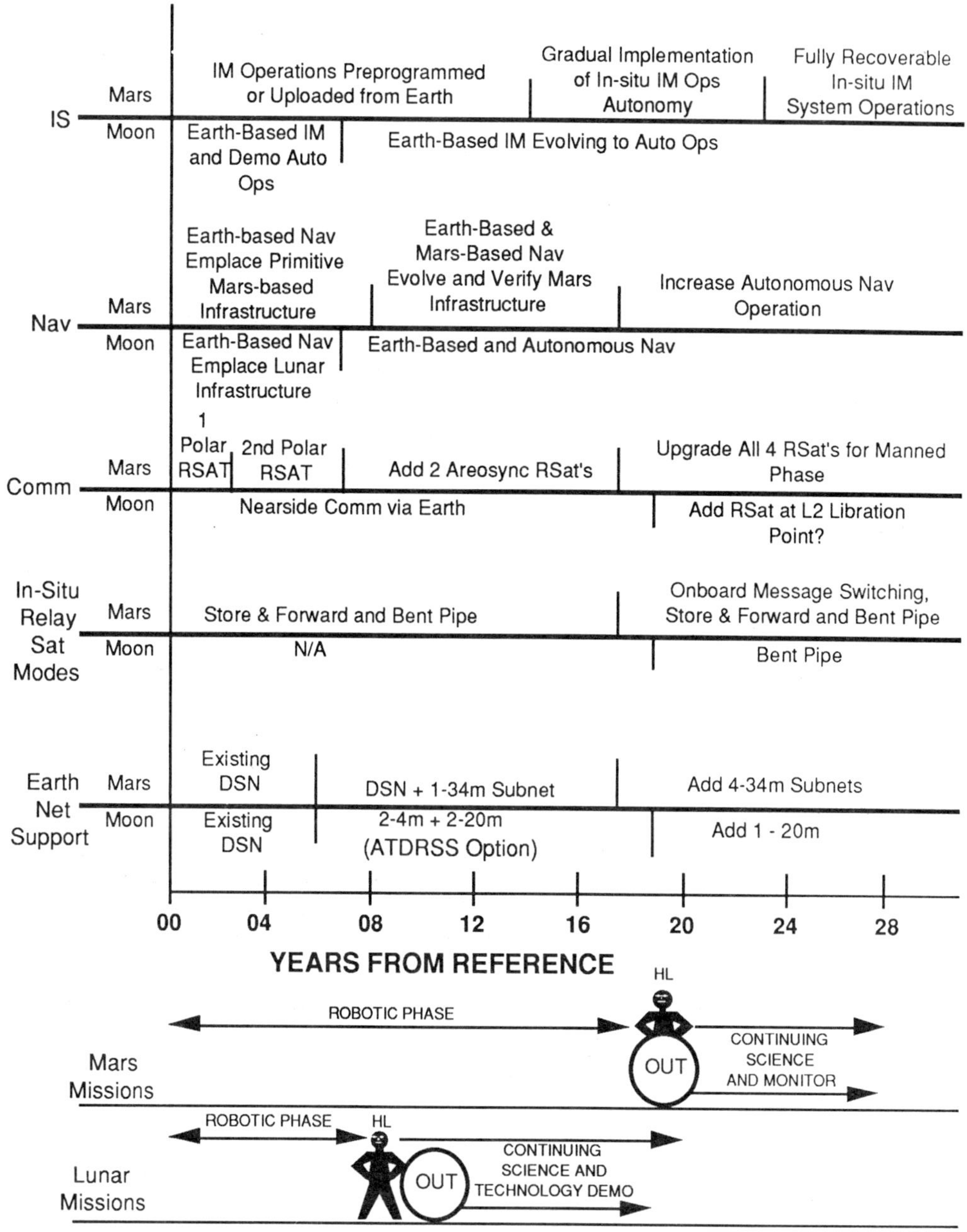

LEGEND HL - Human Landing
OUT - Outpost

Fig. 2 TISN Support Estimate Option 1

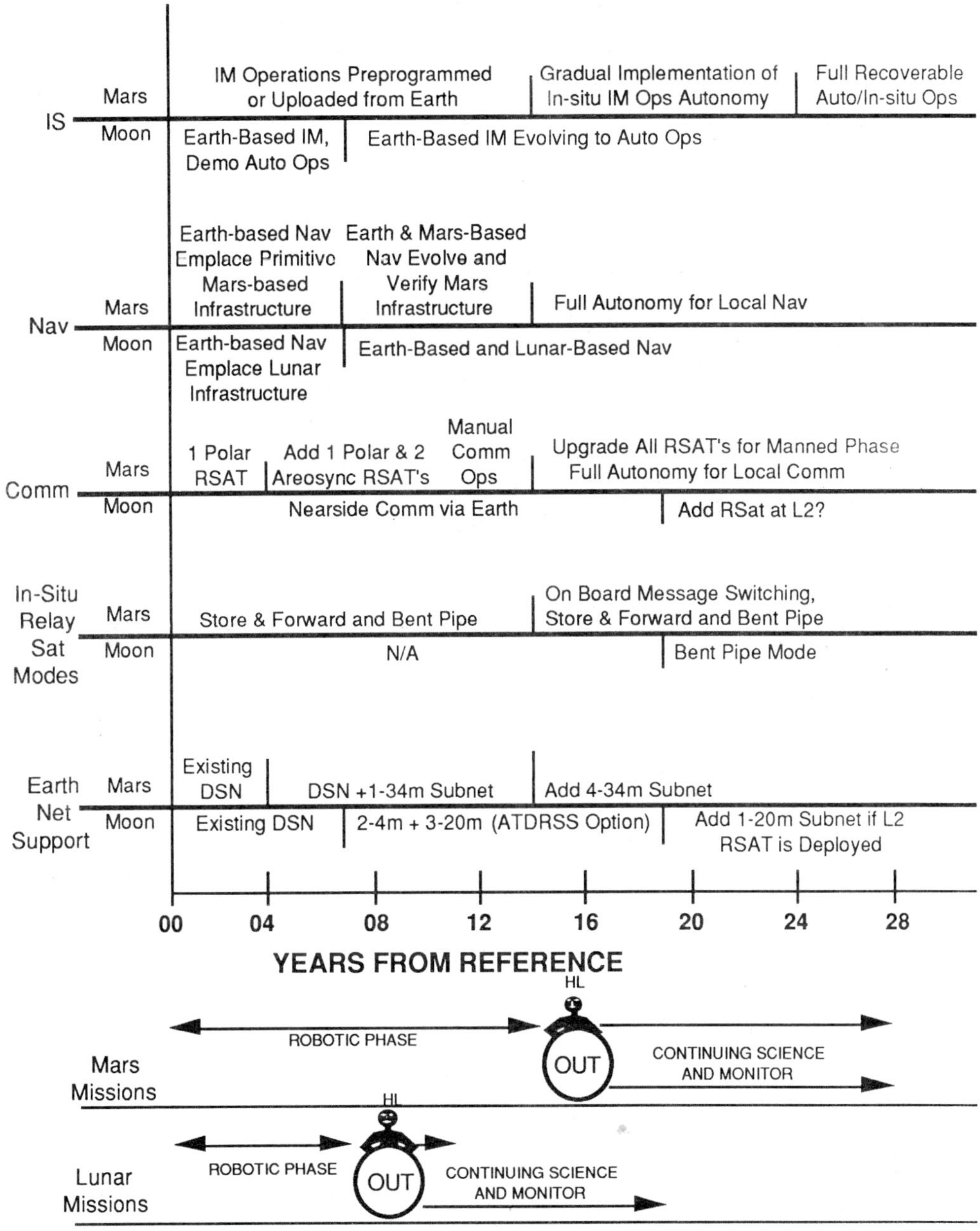

LEGEND HL - Human Landing
OUT - Outpost

Fig. 3 TISN Support Estimate Option 2

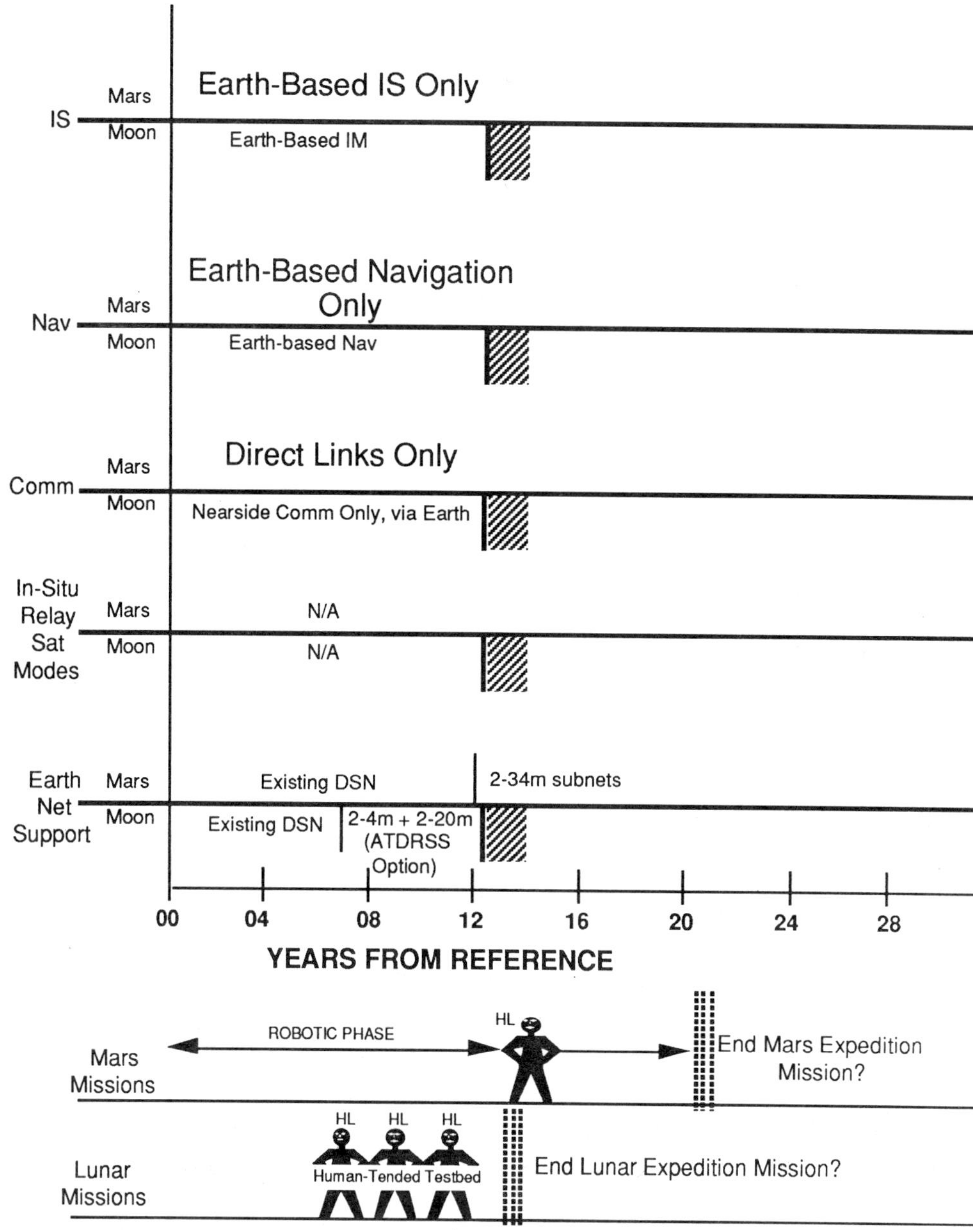

LEGEND HL - Human Landing
OUT - Outpost

Fig. 4 TISN Support Estimate Option 3

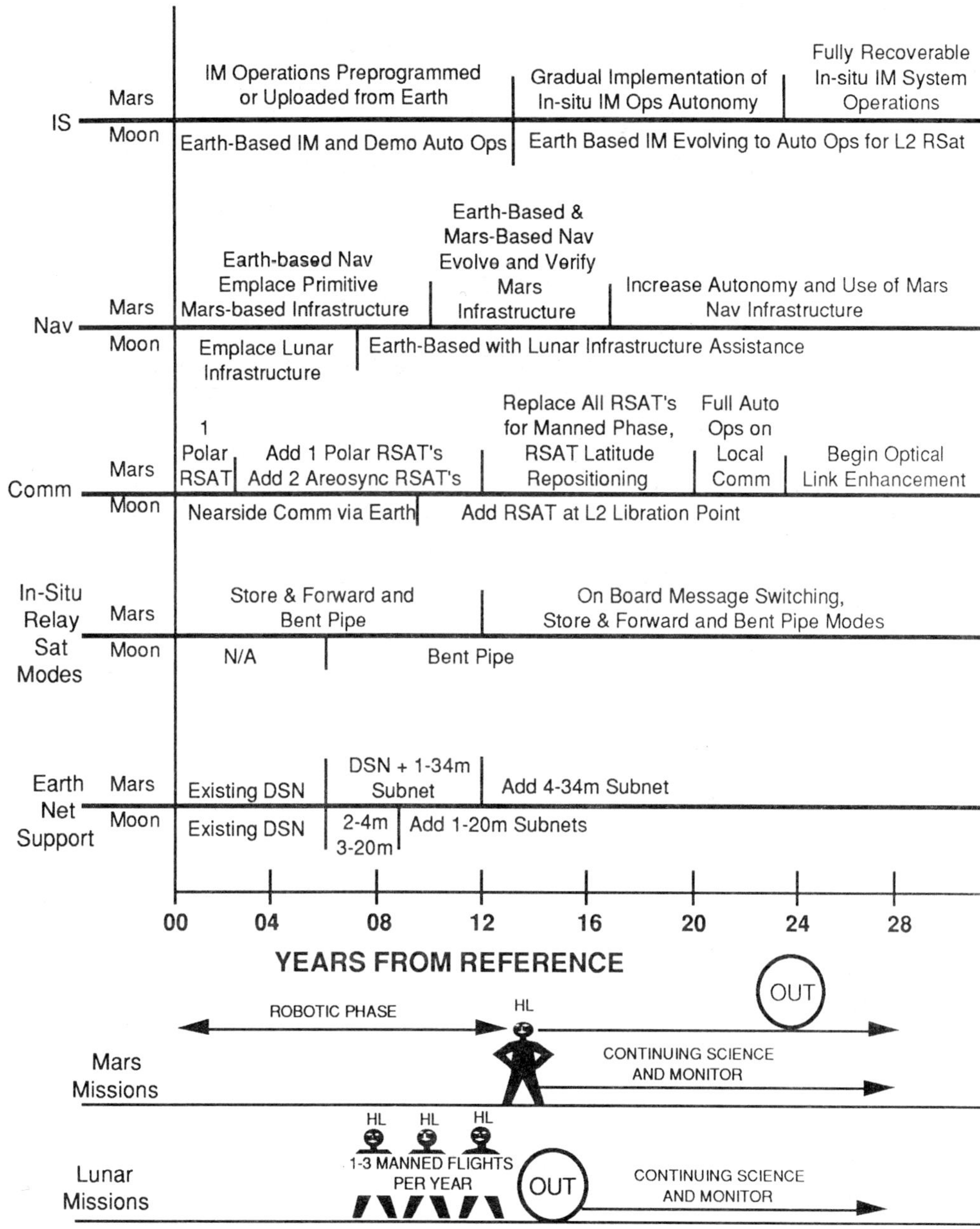

LEGEND HL - Human Landing
OUT - Outpost

Fig. 5 TISN Support Estimate Option 4

CONCLUDING REMARKS

While the TISN work to date has been studied in context with specified mission architectures, a number of technical capabilities, issues, and concerns have been identified that are relatively architecture independent. The specific architectures tend to drive TISN from the standpoint of technology availability, mission schedules, and support capacity. In this context, the SEI program will be reevaluating the TISN architecture based on the soon-to-be-released mission architectures from the synthesis committee.

Applicable flight experience and detailed analysis, to date, is limited with respect to the requirements that will be imposed by manned missions to Mars, and to an extent, to the Moon. Key capabilities will be derived from considerations of crew operations and safety, mission abort strategies, multiple landings to a given planet location, flight time constraints, external support responsiveness, etc. Mars missions flown to date (and those studied such as the Mars Rover Sample Return) have never had the need for all of the above requirements/constraints levied simultaneously. Likewise, although there is a knowledge base from the Apollo program, that knowledge is tied to lunar orbit rendezvous mission approach, and the knowledge transfer by inference to a direct lunar descent/ascent is subject to question. As shown in Fig. 6, the required robustness of any space system will need to increase to compensate for reduced flight and system operating experience. (Ref. 2) The aforementioned concerns give rise to the need for an operational implementation approach as shown in Table 1. (Ref. 3.)

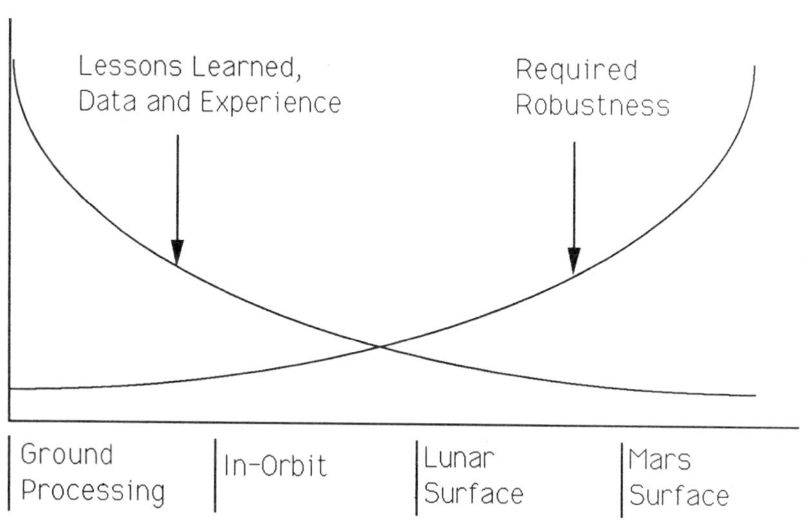

Fig. 6 Extent of our Current Lessons Learned

Table 1
EVOLUTION STRATEGY FOR OPERATION MANAGEMENT

	Earth-based	Moon	Mars
Mission Management (Ops & Science Policy)	Central focus and final authority Diverse participants		
Operations Direction	Earth orbit activity Initial lunar ops	Testbed for Mars	Mars tactical operations
Systems Monitoring and Control	Earth orbit activity Initial lunar ops	Some Mars prototyping	Real time and near real time systems control
Systems Support	Earth based	→	→
Planning	Planning data base Initial 24-48 hour plans	Look ahead planning	→
Science/Exploration Direction	Earth based IOC	Lunar surface based at FOC	Mars based
Science Support	Earth based	→	→

Key Issues

System Definition and Buildup. The development of an architecture should consider the complementary benefits of the Earth-based infrastructure, while recognizing the long-term benefits of autonomous operations. System level trades will be required to define the infrastructure elements (number, type, and location) and ensure compatibility between elements. System buildup must consider the progression of mission requirements (short and long term), the opportunities for deployment/emplacement of system elements, and the expected operational lifetime of these elements.

Reliability and Performability. Long-lifetime operational issues, critical operational performability, and overall system performability in the face of failures will be a critical factor in the system and element design. This may involve knowledge-based systems and will involve some fault tolerant capability at the system level or within the individual elements to ensure acceptable availability and performance over the mission timeline.

Emplacement. Many factors will influence the deployment of the TISN infrastructure, such as the progression of requirements, the deployment opportunities, the location of each element (i.e., in orbit or on the surface), and the implications of a failure in one element. For orbiting systems elements, the parameters of the orbit and maintenance of that orbit must be considered. For surface elements, issues such as the importance of controlling precisely the emplacement location on the surface, inertial frame, or with respect to surface features must be evaluated. These will have an impact on the complexity of the emplacement and surveying process.

Hardware Issues. Some of the hardware issues that must be addressed include the weight/power/volume required by both the users and infrastructure elements, the best type of measurement (ranging, Doppler, one way vs. two way, user of interferometric or synthetic aperture techniques), trades between power and accuracy, continuous versus intermittent operation, antenna requirements, etc. The number of elements in the infrastructures and the integration of these into an architecture will also be important. The impact of the operating environment, management/political environment, and problems presented by a long operational lifetime must also be addressed in selecting appropriate hardware.

Demonstration. Early design phases should address the need for a demonstration of elements of the system at the Earth and/or the Moon. This demonstration should consider the degree of risk associated with the design and a thorough understanding of the costs and benefits of a demonstration. The use of surface beacons on Earth to support the ephemeris updating of an orbiting satellite is one example. It should be noted, however, that the greatest risk may result from emplacement failure rather than failure of the electronics.

Navigation required accuracy. Navigation accuracy requirements for all flight phases have not been defined. Related requirements include convergence speed, coverage, and the payoff resulting from improved accuracy. Examples of this payoff include reduced vehicle L/D for aerocapture, improved orbital mapping, precision (pinpoint) landing, reduced operating range requirements for vehicle-to-vehicle relative navigation prior to rendezvous, fuel savings resulting from more accurate translational maneuvers in all phases, and enhanced scientific return from the mission.

Planned/Current Activities

Current plans are to continue the telecommunications activities, to begin to develop the architecture for the integrated mission information system, and to further analyze the requirements and systems options for the integrated navigation architecture.

The telecommunications work will need to focus on the data types, data rates, and connectivity needs for each user to improve the traffic models that have been used to date. The information systems work will initially focus on the functions and user requirements that the information systems must support. The navigation activity will begin to address the options shown in Table 2.

Table 2
PLANNED NAVIGATION ARCHITECTURE STUDY AREAS

Mission Phase	Option
Planetary Capture (Aerocapture, propulsive)	1. *Earth-based radiometric 2. Earth-processed optical 3. Onboard radiometric using local beacons and navsats 4. Onboard optical using Mars moons
Orbital Operations	1. Earth-based radiometric 2. Onboard radiometric using local beacons and navsats 3. Landmark tracking 4. Mars-based network navigation (like TDRSS) 5. Onboard optical using Mars and its moon
Descent (with and without surface aids)	1. Onboard inertial aided by altimetry 2. Radiometric aiding with local navsats 3. Radiometric aiding with surface beacons 4. Passive and active optical image correlation with surface features 5. Synthetic Aperture Radar (SAR, low priority alternative)
Surface Survey (landing site or beacon)	1. Earth-based radiometric 2. Local radiometric using "survey satellite", horizon tracker, bootstrap from another beacon 3. Processing of descent images (post-landing analysis) 4. Optical line-of-site from orbiting satellite
Mobile Surface Operations (rover)	1. Earth-based radiometric 2. Onboard optical using surface features 3. Onboard inertial navigation (IMU, odometer, etc.) 4. Onboard radiometric using navsats (and surface beacons at short range)
Rendezvous - High Energy Elliptic hp = 250 km ha = 33809 km	Sensor options 1. Ranging 2. Doppler 3. Passive optical 4. Laser 5. Radar
Rendezvous - Low Energy Circular hp = ha = 500 km	
Rendezvous - Hyperbolic V• = 6-12 km/sec	
Interplanetary Coast	1. Earth-based radiometric 2. Onboard passive optical to planets
Heliocentric Transfer (electric propulsion)	1. Earth-based radiometric 2. Onboard IMU combined with one-way range or Doppler

* DSN (Doppler, range, Delta VLBI, etc.)
hp = periapse altitude ha = apoaps altitude16 V• = Mars approach velocity

ACKNOWLEDGEMENTS AND REFERENCES

The following works were used as reference material for this paper:

1. E. M. Stephens, *Telecommunication, Navigation, and Information Management Infrastructure and Operations (Option 5 Reference)*, NASA/JSC 24573(U), August 22, 1990.

2. MacDonnel Douglas Space Systems Co., *SEI Integrated Logistics Working Group Presentation*, November 6 - 8, 1990

3. T. W. Eggleston, *SEI Operations Concept Briefing to Level I*, Mission Operations Directorate, NASA/JSC briefing, January 17, 1991.

The authors would like to thank the following people for their contributions: Angie Armstead, Ray Hartenstein, Jackee Todd, and Phil Shapiro of Computer Sciences Corp.; John Bessant, IBM; Antha Adkins, Lockheed Engineering Sciences Co; Thomas Stuart, Science Applications International Corp.; Dave Bell, Rolf Hastrup, and J.R. Hall, Jet Propulsion Laboratory; Don Hei, NASA Goddard Space Flight Center; and Dave Long, NASA Johnson Space Center.

Section V
RECENT EXPERIENCES

SESSION V

Chairperson:	Doug Broome NASA Headquarters
Co-Chairperson:	Bernard Kaufman Naval Research Laboratory
Local Chairperson:	Hugh Davis Ball Aerospace Systems Group

The following paper numbers were not assigned:

AAS 91-077 to -079

AAS 91-070

ATTITUDE CONTROL EXPERIENCE WITH THE COMBINED RELEASE AND RADIATION EFFECTS SATELLITE

William Frazier*, Kevin Saylors†, Fred Patton† and Kjell Stakkestad‡

The Combined Release and Radiation Effects Satellite (CRRES) was launched on July 25, 1990. The attitude control is based on a spin-stabilized, open-loop system design. Attitude determination is performed and control maneuvers are planned by the Air Force Consolidated Space Test Center (AF/CSTC). The performance of the spacecraft and ground segment attitude determination and control subsystem is shown to meet all requirements. An overview of the vehicle dynamics is also given, as both a rigid and flexible-body, following deployment of the four 50-meter wire booms. It is seen that the spacecraft dynamics are approximately as predicted by analysis. However, ambiguities inherent in using digital Sun sensor telemetry to verify acceptable spacecraft dynamics are also discussed.

INTRODUCTION

Mission Overview

The Combined Release and Radiation Effects Satellite (CRRES) program is a joint undertaking, with funding for the spacecraft and Department of Defense (DOD) science instruments from DOD, and managed by Air Force Space Systems Division Space Test and Transportation Programs Office. Overall program management is by NASA Marshall Space Flight Center, with funding for the respective science from NASA headquarters. The spacecraft bus was designed, built, and tested, and science experiments were integrated by Ball Space Systems Division. The launch was by the first commercial Atlas I on July 25, 1990, with procurement and launch facilities provided by NASA and AF. All spacecraft health and status monitoring and control are performed by Air Force

* Ball Space Systems Division, P.O. Box 1062, Boulder, Colorado 80306-1960.

† Scitor Corporation.

‡ Lockheed Technical Operations Company, 1309 Moffett Park Drive, Sunnyvale, California 94088.

Consolidated Space Test Center (AF/CSTC). In particular, the open-loop attitude determination and control system (ADCS) relies heavily on ground-based planning and evaluation, and is the subject of this paper.

The (contingency) target orbit for the actual launch date was a 350-km perigee altitude, 18.1-deg inclination, 200-deg argument of perigee, and 35,786-km apogee altitude [1,2,3]. All of these target elements were achieved satisfactorily except the last, which was approximately 2,200 km low. The science objectives to be accomplished from this orbit fall into three broad categories: 1) NASA performance of active chemical release experiments in the ionosphere and magnetosphere, 2) DOD studies of the natural radiation environment and its effects upon numerous microelectronic components as CRRES travels through the radiation belts, and 3) DOD low-altitude studies of ionospheric irregularities. The CRRES configuration with the six rigid and four 50-meter-long wire antenna booms deployed is shown in Fig. 1. The passive ring nutation damper can also be seen on the top deck, beneath the solar arrays.

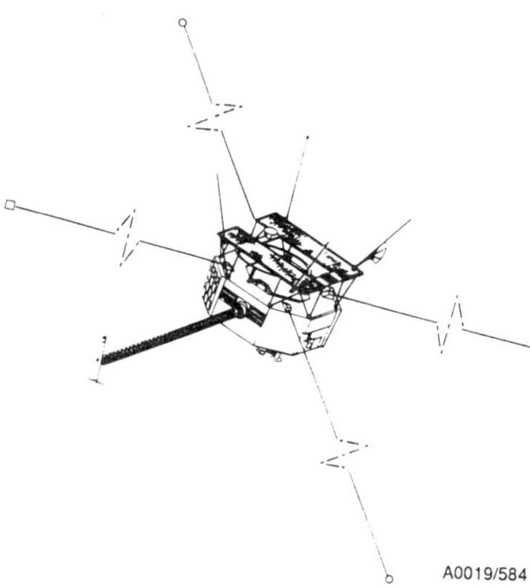

Fig. 1 CRRES configuration with booms deployed

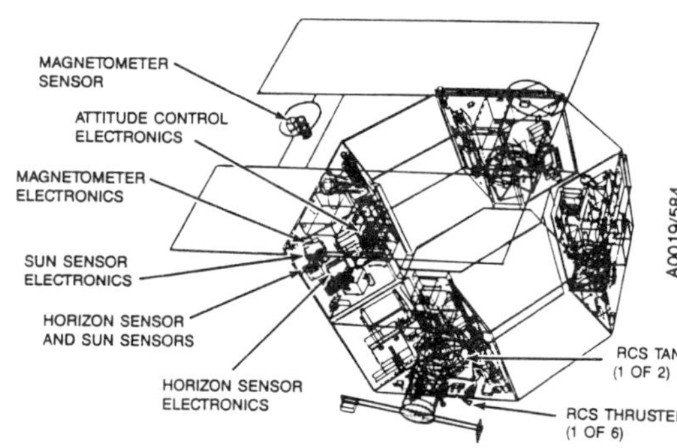

Fig. 2 ADCS component locations on CRRES

Attitude Determination and Control System

The CRRES ADCS was described previously in Ref. 4, and the component locations are indicated in Fig. 2. In summary, the primary method of attitude control is the angular momentum generated by spinning about the primary inertia axis. Spin-axis attitude and phase are measured by redundant digital Sun sensors, a fixed-horizon sensor, and a three-axis magnetometer. Magnetometer signals are conditioned and zero-crossings of one component are time-tagged, to generate a data set analogous to the horizon-sensor rotation angles. This attitude data is telemetered to the ground from both real-time measurements and tape recorder playback, and inertial attitude determination is then performed off-line, every day or two. The attitude is predicted by the ground software for times of interest in between updates.

Since the solar arrays are pointed in the direction of the momentum vector, it must be precessed occasionally to follow the apparent motion of the Sun, and to counter various disturbance torques, primarily due to gravity gradient. This is accomplished by the Reaction Control System (RCS), which employs 30 kg of hydrazine in each of two tanks in a blow-down design. The RCS includes three pairs of thrusters; precession/ΔV, spin-up, and spin-down. The latter two act as coupled pairs to control the spin rate, while the former has two modes; precession and ΔV. For a summary of ADCS mission events, the RCS fuel budget is given in Table 1. Based on events during spacecraft initialization, actual fuel use on-orbit has been within a few percent of the budget, to the extent that the actual maneuvers have matched their descriptions in the budget analysis. Therefore, the derived requirement to stay within the fuel budget appears to be satisfied up to the present time.

Table 1
RCS FUEL BUDGET FOR REVISED MISSION

1) Spin-up for wire boom deployment	1.6
2) Spin-down after boom deployment	1.8
3) Orbit adjusts (TBD)	15.0
4) Attitude maintenance (3 yr)	8.8
5) Spin rate maintenance	0.2
6) Ullage and modelling error	6.2
Sum:	33.6
Total available:	60.0
Margin:	26.4

The Attitude Control Electronics (ACE) box is responsible for time-tagging each of the attitude sensor measurements listed above, and for computing and telemetering a running average of the spin period. It also controls the three pairs of RCS thrusters, according to two different types of programs that are loaded into random access memory (RAM) by serial-digital commands from the ground. "Continuous Thrust" commands specify 1) a burn duration, and 2) either the spin-up, the spin-down, or the ΔV/precession thruster pair. "Precession Program" commands specify each of 1) the number of pulses to be applied, 2) the duration of each pulse, and 3) the delay off of the Sun crossing time to fire the first pulse. The ΔV/precession thrusters are fired alternately each one-half spin period. Since the spin period used for this timing is the average prior to the maneuver, the thruster pulse times are in effect predetermined, and the resulting path is approximately a great circle.

When either a Continuous Thrust or Precession command is received by the ACE, it is temporarily echoed in telemetry and then held in RAM until an "execute" command is received. This provides operational time to verify that the correct command is loaded prior to execution, and is another level of safety past the required thruster "enable" commands.

The ACE also provides an initialization/contingency mode of seeking the Sun in a rhumb-line precession maneuver, which cuts off at a 14.5 deg Sun aspect angle. However, the command invoking this function is disabled shortly after launch, for reasons described in the next section.

Spacecraft nutation induced by thruster firings is passively damped by a partially filled ring nutation damper[5].

Flight Support Software

The CRRES open-loop control system relies on operations performed by AF/CSTC under the Data System Modernization (DSM) environment. The flight support software used for CRRES includes access to large amounts of generic "core" DSM software, in addition to almost 20 thousand lines of Mission Unique Software (MUS) code. The functional flows between the MUS and the core DSM programs are diagrammed in Fig. 3. The duties of these software packages can be inferred from the list of acronyms included in the figure.

Orbit Adjusts

As mentioned above, the spacecraft ADCS and ground software designs support selection and performance of ΔVs for a moderate amount of orbit maintenance or modification. However, at the time of writing this paper, the requirements for such orbit adjusts are still under consideration, and none have been performed as yet. Consequently, this ADCS and software capability is not considered within the present scope.

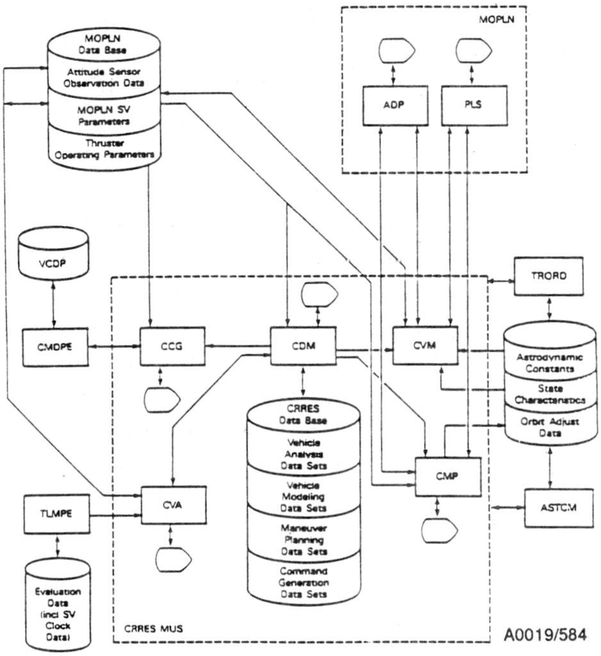

Core DSM :
ASTCM - Astrodynamics Common
ADP - Attitude Determination Processing
CMDPE - Command Planning and Evaluation
MOPLN - Mission and Orbit Planning
PLS - (maneuver) Planning Support
TLMPE - Telemetry Processing and Evaluation
TRORD - Tracking and Orbit Determination
VCDP - Vehicle Command Data Pool

CRRES MUS:
CCG - CRRES Command Generation
CVA - CRRES Vehicle Analysis
CDM - CRRES Data Base Manager
CMP - CRRES Maneuver Planning
CVM - CRRES Vehicle Modeling

Fig. 3 CRRES MUS functional flow diagram

SYSTEM REQUIREMENTS AND DESIGN

Flexible Versus Rigid Body

The ADCS requirements have evolved somewhat since described in Ref. 4. This is due primarily to the Challenger disaster, the subsequent switch to an expendable launch vehicle, and deletion of the Low-Earth Orbit (LEO) mission phase. During the two-to-three months in LEO, the CRRES was to have performed the chemical releases while a rigid body spinning at roughly 20 RPM. This mission phase, and its initialization by release from the space shuttle, was the source for the auto-Sun acquisition and multiple-pulse precession maneuvers described above. After pointing adjustments by the ADCS, a CRRES-internal Solid Rocket Motor (SRM) was to have carried the vehicle into Geosynchronous Transfer Orbit (GTO). The four 50-meter-long wire booms were then to be deployed, and the vehicle spun-down to the operational spin rate of 2 RPM in GTO.

With the switch to the Atlas I, the SRM and half of the chemicals were off-loaded, and the spacecraft was inserted directly into a GTO, from which the remaining chemical releases are to be performed mostly near perigee. The CRRES requirement to perform a perigee-raise ΔV as part of initialization and the concomitant large spacecraft point maneuver were also deleted.

The Centaur upper stage was successful in providing a 12-deg off-Sun angle and 2.2-RPM initial spin rate. During initialization, the spacecraft then spent only seven days as a rigid body, prior to wire boom deployment. As will be seen in the next section, the spacecraft main body nutation is highly coupled with the wire boom motions. Due to uncertainties in the various natural frequencies of this flexible body, only single-pulse precession maneuvers are allowed after wire boom deployment. This means that the auto-Sun acquisition and multiple-pulse large precession maneuver capabilities of the ADCS, intended for the rigid-body LEO mode, must not be used in the present GTO-only mission, and hence, are not discussed further.

ADCS Performance Requirements

The ADCS performance requirements salient in the GTO-only mission in progress are listed in Table 2.

Table 2
ADCS PERFORMANCE REQUIREMENTS FOR REVISED MISSION

Function	Performance Requirement
Attitude-sensor data	Allow determination of the spin-axis attitude and phase to within $\pm$ 2 deg (1σ)
Inertial spin-axis drift	Not exceed 2 deg per orbit revolution (1σ)
Rigid-body nutation damping	Less than 5 hours to half-amplitude after precession maneuvers, at 2 RPM
Flexible-body nutation and spin-rate damping times	Dominated by the wire booms; expected to be several hours

The last requirement in Table 2 is deliberately vague, because the wire booms are Government-furnished equipment (GFE). Their inertias about the spacecraft spin axis are roughly 70 percent that of the main body, and so spacecraft nutations are highly coupled with wire boom up/down motion, and spin-rate changes are highly coupled with the boom's side-to-side motion. Neither of these coupled modes are quickly damped by the spacecraft itself at 2 RPM. The wire booms contain some self-damping mechanisms whose time constants were not well known prior to launch, so the above requirement implicitly including this GFE performance was left somewhat soft.

ATTITUDE DETERMINATION AND CONTROL PERFORMANCE

Each of the ADCS performance requirements listed in Table 2 are now examined in sequence, and the predicted versus actual performance is discussed.

Attitude Determination Accuracy

Predicted Performance. The accuracy of the CRRES attitude determination was analyzed and is presented in Ref. 4. To abstract that reference, a Monte-Carlo-type analysis was done, numerically testing the attitude error resulting from variations in all independent variables. This analysis included thorough modeling of all attitude sensor errors, and hence, shows the sensitivity of attitude to such errors for various geometrical conditions. Important assumptions for the GTO case included: 1) attitude determinations based on real-time collection of 30 observation sets, 2) a spacecraft wobble angle of 0.25 deg (1σ), and 3) off-Sun angles ranging from 10 through 60 deg. In practice, each of these assumptions is conservative. Typical attitude errors are estimated by averaging the errors found by allowing the Sun angle to vary as in assumption (3), and the clock angle around the Sun to go through 360 deg. Typical 1σ attitude determination errors are then estimated to be approximately 0.5 deg.

On-Orbit Performance. Actual attitude determination accuracy is difficult to quantify, due to lack of independent methods of calibration. Residuals from the CRRES weighted-least-squares attitude fits can be used only as a relative measure, for example, to show the quality of fits using the magnetometer compared to those using the horizon sensor. Sensitivities to Sun-vehicle-Earth geometries can also be explored in this manner, although neither of these studies has been explicitly done. What has been done operationally is to compare the results of attitude fits based on neighboring observation sets. This always yields attitudes that agree to within 0.5 deg (a necessary but not sufficient condition for meeting the expected performance).

Attitude Drift

Predicted Performance. The inertial attitude drift has been estimated to be less than 0.4 deg/day. This is due primarily to gravity gradient torques, and to a lesser extent to aerodynamic and magnetic torques as well. Solar radiation pressure torque is negligible.

On-Orbit Performance. Actual attitude drift in declination, and that predicted by the MUS for a typical five-day span, are shown in Fig. 4 (for this case, drift in right ascension is much less than in declination, and so is not shown.) The observed drift is indicated by the solid line segments connecting the open squares, which represent actual attitude determinations. A linear fit to the observed drift indicates an average rate of approximately 0.16 deg/day, or less than half the expected worst-case performance.

Attitude Prediction Software Performance. One of the functions of the CVM MUS module shown in Fig. 3 is to provide vehicle-specific torque models. The total torque vector is the sum of the calculated aerodynamic, gravity gradient, and magnetic torques. These vectors are calculated at a series of points in the vehicle's predicted orbit to produce a time-ordered list of vehicle attitudes.

Representative differences between the observed and predicted attitude are visible in Fig. 4. The jagged line formed by the string of solid diamonds indicates the attitude predicted from the time of the first attitude determination on 19 November. The near-discontinuities in this line are due to the torque impulse on the vehicle at perigee crossings, every 10 hours. The error rate in declination is virtually zero, but is approximately 0.03 deg/day in right ascension, or roughly 19 percent of the drift magnitude, which is about as expected. This performance could be improved by manually adjusting the assumed vehicle characterisitics, such as drag coefficient, projected areas, residual magnetic moment, or mass properties. However, improved attitude prediction accuracy is not required for CRRES, since the attitude is re-determined frequently anyway, and then adjusted approximately every 10 days to follow the Sun.

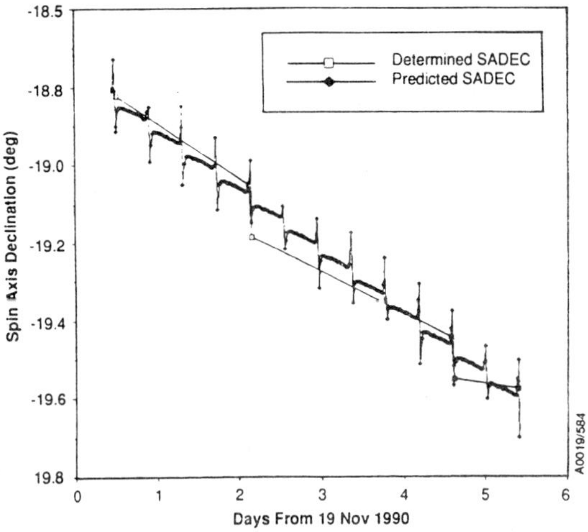

Fig. 4 Spin axis declination comparisons

Rigid-Body Nutation Damping

Predicted Performance. Analysis extended from that shown previously in Ref. 5 has predicted a half-amplitude nutation damping time of approximately 3.2 hr. At the 2 RPM GTO spin rate, the oil in the ring damper oscillates back and forth in laminar flow, termed the "spin synchronous" mode. This time constant neglects any nutation damping resulting from fuel slosh, and so should be expected to be slightly conservative.

On-Orbit Performance. The largest precession maneuver performed with CRRES as a rigid body was approximately 9 deg. At that time, five of the six rigid booms had been deployed. The maneuver was done using the ACE precession program function, with a string of 10 thruster pulses, 400 ms long each. At the end of this maneuver, CRRES had roughly 1.25 deg of nutation visible in Sun sensor telemetry (i.e., 2.5-deg difference between max and min Sun angles). A plot of the Sun sensor data, taken from tape recorder playback, for the first 80 minutes after this maneuver is given in Fig. 5.

Inferring the CRRES nutation on-orbit from Sun sensor data presents several difficulties, discussed to varying degrees in Ref. 6: 1) asymmetries in the spacecraft transverse inertias mean that the actual nutation will be greater than that observed, 2) the actual Sun angle varies continuously at approximately the inertial nutation rate, whereas the observed Sun angles vary discretely at roughly the body nutation rate, and 3) the Sun sensor least-significant bit (LSB) of 0.5 deg adds further uncertainty to the measurements.

Difficulty (1) above may be addressed by the methods presented in Ref. 6. In short, it notes that by conservation of angular momentum, the nutation amplitude in the direction of any given body-fixed axis will vary inversely with the inertia about that axis. Thus, for example, if the Sun sensor is pointed normal to the minimum (transverse) inertia axis, it has the possibility of detecting the maximum nutation angle. In the case of CRRES at this time in the mission, the Sun sensor looks 56.6 deg away from the minimum inertia axis, and as a result, it is capable of observing only a portion of the maximum excursions of the figure

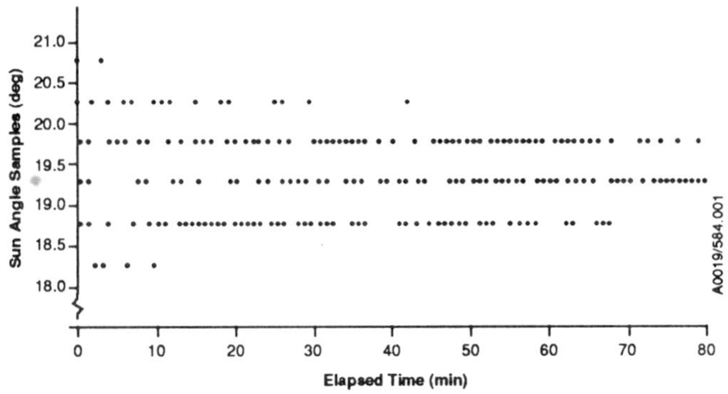

Fig. 5 Rigid-body nutation damping @ 2.57 RPM

axis from the momentum vector. Specifically, the maximum nutation is estimated as

$$\theta_{max} = .714\ \Delta\beta_m$$

where $\Delta\beta_m$ is the maximum observed difference in Sun angles.

Difficulty (2) refers to the fact that infrequent sampling of the nutation angle by the Sun sensor means that actual observation of the maximum $\Delta\beta_m$ is unlikely, and can only be expected after a large number of data samples have been collected. Unfortunately, during this time the nutation may damp considerably. Difficulty (3) regarding the LSB may be handled for worst-case analysis simply by adding (or subtracting) one-half of an LSB to any Sun sensor data of interest.

Fortunately for the issue of nutation *damping*, the Sun sensor telemetry difficulties are largely irrelevant because it is only the *relative* change in nutation (or Sun angle observations) that is important. For a first approximation, then, the envelope in Fig. 5 may be used to see that the nutation has damped to half-amplitude in approximately one hour. One adjustment that must be made to this damping time, however, is to note that it is inversely proportional to the square of the spin rate[5]. Thus, to compare with the required performance at 2 RPM, the observed time constant must be scaled by $(2.57/2.0)^2$, for an expected time of about 1.64 hours at 2 RPM. The rigid-body nutation damping thus exceeds the 5-hr requirement by at least a factor of three. This is due in part to the neglect of the fuel slosh viscous effects.

The difficulties with using the Sun sensor to observe spacecraft nutation are quite relevant when the magnitude of the nutation is of concern, as it is during maneuvers after extension of the wire booms.

Flexible-Body Dynamics: Active Spin-Down Procedures

Two types of maneuvers are critical once the wire booms are deployed; spin rate changes and precession maneuvers. The primary goal in spin-rate control is to get from the higher spin rate used for wire boom deployment, down to the operational rate of 2 RPM. Due to their large inertia about the spin axis, the deployment of the wires passively spins-down the vehicle by a factor of 0.59. The spin rate at the start of deployment was planned to be 20 RPM, but due to a last minute change was actually 15.5 RPM. Thus, the initial spin rate for CRRES as a flexible body was 9.1 RPM, from which the operational 2 RPM was sought.

Neither the spin-down nor the precession maneuvers have explicit performance requirements, but both imply some serious derived requirements. In particular, they must be feasible in a timely and, especially, a safe manner. In practice, it turned out that both of these goals were difficult, but were accomplished successfully. The following paragraphs deal with verifying the predicted flexible-body dynamics. For both types of maneuvers, the nominal magnitudes were carefully approached after on-orbit calibration of the vehicle dynamics models.

Predicted Performance. Analysis of the flexible modes of the spacecraft/wire boom system in response to spin-down torques has indicated that the mode of primary interest is simply a torsional effect, where all four booms swing side-to-side in unison, and the spacecraft spin rate oscillates. However, a secondary mode also exists in which the booms move toward and away from each other in a scissors manner, without coupling into the main body. Beating between these two modes is visible in the top two plots in Fig. 6, which represent simulation of the side-to-side motion of one sample of each of the two pairs of booms. Coupling into up/down motion and spacecraft nutation is predicted to be negligible, and so is not shown. The third plot is the predicted instantaneous spin rate of the main body, and the fourth is the quantized spin period to be expected from the Sun sensor data. It is significant to note that although the boom motion is what needs to be controlled (to less than 20 deg), in real life it is only the spacecraft spin period measured at discrete points that is observable in telemetry.

The strategy chosen to perform the spin-down procedure is to use a series of spin-down pulse doublets. Since the periods of the two boom pair's side-to-side primary mode are nearly equal, a second pulse may be fired one-half boom period later, to cancel the apparent motion of the booms with respect to the main body. (This does not effect the secondary scissors mode, which fortunately has a negligible amplitude.) In this manner, a doublet is to be performed every spacecraft contact (about every three hours), until the operational rate is achieved. This nominal sequence is expected to require about 10 days of intense operational activity.

The ratio of the spacecraft spin period to wire boom period is estimated to be a constant value of 0.223. Boom stiffness increases with the square of the spin rate, and the boom period goes with one over the square-root of the boom stiffness, so the ideal pulse spacing (i.e., half the boom period) increases linearly with spin rate. Similarly, the allowed length of each thruster pulse decreases with spin rate, as the boom stiffness also decreases. The pulses for each doublet are sized to cause less than 10 deg of boom motion each, so under the worst case of the second pulse firing at exactly the wrong time, boom motion would be tolerable. Using these relationships, the pulse spacing and duration were precomputed as function of spin rate for a nominal spin-down sequence, and tabulated for convenient reference.

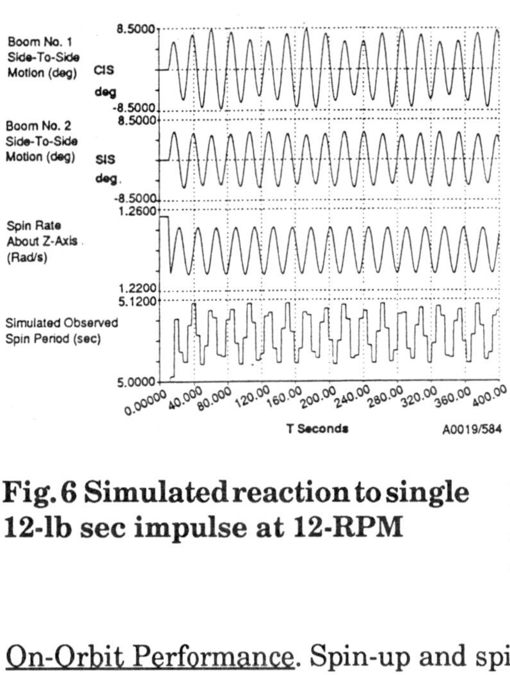

Fig. 6 Simulated reaction to single 12-lb sec impulse at 12-RPM

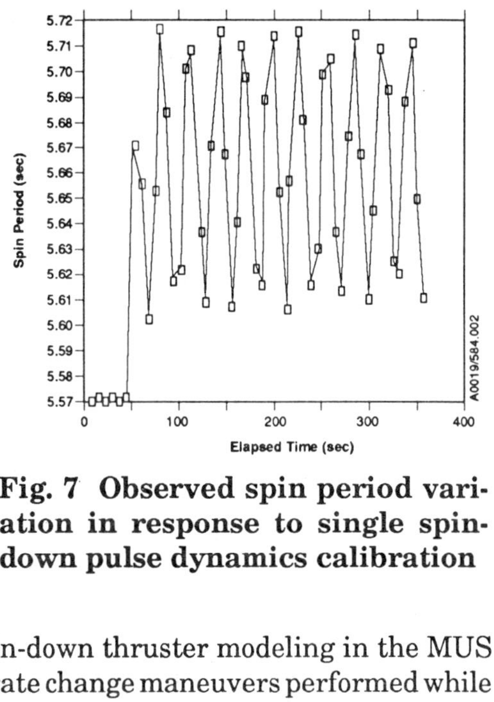

Fig. 7 Observed spin period variation in response to single spin-down pulse dynamics calibration

<u>On-Orbit Performance</u>. Spin-up and spin-down thruster modeling in the MUS was calibrated using data from the spin rate change maneuvers performed while as a rigid body. A thruster efficiency factor of 0.889 was derived empirically after one iteration. Use of this scale factor resulted in maneuver errors of less than 0.5 percent, so it was used for all subsequent spin-rate change planning.

The first step in implementing the spin-down procedure was to verify the analytic prediction that the (initial) spin-to-boom period ratio was indeed 0.223, since this is what determines the timing of the second pulse of each doublet. To evaluate this on-orbit, a 1.2-sec single spin-down thruster pulse was fired, to generate dynamics similar to that predicted in Fig. 6. Telemetered Sun-crossing times were used by the MUS to calculate "instantaneous" spin-period data, that is, averaged over each revolution of the satellite. This created the data set shown in Fig. 7, which was then processed to first extract the constant component, and then to perform a weighted-least-squares fit to the sinusoidal spin-period oscillations. Since these oscillations are due to the swinging of the booms, their period may be inferred from this fit. In this manner, it was found that the ratio of the spin period prior to the pulse was 0.224 times the boom period, or within 1 percent of that expected.

Despite the close agreement with the predicted boom period, it was more convenient to replace the precomputed table of nominal spin-down pulses with the equivalent algorithm, so that it was not necessary to stay in step with the anticipated spin rates. This also allows selection of the pulse spacing as either one-half or three-halves of the boom period. The latter was necessary at the higher spin rates, in order to allow uplink and verification of the commands necessary in between the two pulses.

In order to ensure spacecraft safety during the aggressive spin-down operations (i.e., every contact, around the clock), it was necessary to institute clear pass/fail criteria for each doublet prior to loading the commands for the next maneuver. To support this, it was required to collect at least 10 minutes of real-time Sun sensor data immediately after each doublet. This was processed by the MUS, to list the instantaneous spin periods, which were manually screened for maxima and minima. The difference between these extremes was required to be less than 0.005 times the initial spin period. Since it was scaled by the current spin rate, this single criteria was valid throughout the spin-down. It detects residual oscillations due to pulse-timing errors equivalent to roughly 17 deg of boom motion, or about twice as bad as the performance demonstrated in the first few doublets. (It should be noted that this criteria worked well to detect residual boom motion due to pulse *phasing* errors, but that it was impossible to actually observe boom *amplitude* during the motion prior to the second pulse.)

Failure of the doublet screening criteria above was first threatened when unscrupulous rounding of the uplinked CSM time-tags was allowed to cause the worst-case time delay error of near 1 sec. After this was diagnosed, an automated method of selected CSM time-tags was installed in the ground software, to keep the time delay error within 0.5 sec, and the problem did not recur. A second failure of the residual-to-initial spin period ratio criteria occurred when the spin rate became low enough (at 3.3 RPM) that the three-half boom period interval mentioned above became too sensitive to small errors in its calibration. Consequently, the one-half boom period pulse spacing was implemented, which allowed successful completion of the spin-down sequence.

In summary, the flexible-body spin-down was performed as safely as possible, in a controlled, though adaptive, environment, on-time in about 10 days.

Flexible-Body Dynamics: Precession Maneuvers

Precession maneuvers with the 50-meter-long wire booms extended are required periodically to track the Sun, with off-Sun angles nominally ranging between 5 and 15 deg. The attitude is kept ahead of the Sun, providing a large margin in case of operational problems. It is desirable that the maneuvers be as large as possible with a single thruster pulse, for a number of reasons: 1) it provides data useful for calibrating the science magnetometer data; 2) due to the non-linear damping of the wire booms, it provides more time at a relatively damped state for other science observations; and 3) it requires fewer maneuvers and, hence, less operational burden and less risk.

Predicted Performance. The simulated dynamics resulting from a 3-deg precession maneuver with the wire booms deployed and resulting 9-deg nutation is depicted in Fig. 8. The plot labeled "NUT" is the main body inertial nutation amplitude versus time, that is, the instantaneous angle between the spacecraft figure axis and the momentum vector. This indicates a half-amplitude damping time constant of the aggregate flexible body of about 1.1 hr (although this damping rate decreases as the boom amplitude and self-damping decreases). The plot labeled "GTO" is the fluid position in the nutation damper, and shows the pendulum-like oscillations of the slug in the least effective laminar-flow "spin-synchronous" mode referred to earlier. The plots labeled "S1S" and "C1D" are side-to-side and up-down motions of sample wire booms, respectively, as in Fig. 6. These show that the initial up-down boom motion relative to the spacecraft is equal to the initial nutation of the main body, which is intuitively satisfying. The effects of beating between the slightly different natural frequencies of the two wire boom pairs, and between the booms and the main body nutation, are also visible in these plots.

It is desirable to perform precessions larger than that shown in Fig. 8 on a regular basis. Other size maneuvers were also simulated, to verify that the system response is linear with impulse, within the range of interest. The limiting quantity is the wire boom motion, which must be kept less than 20 deg. Allowing a 5-deg margin suggests a 15-deg maximum, or extrapolating from Fig. 8, a 5-deg precession as the upper bound. The plan on-orbit was to first perform a series of 2-, 3-, and 4-deg maneuvers to calibrate the dynamics model.

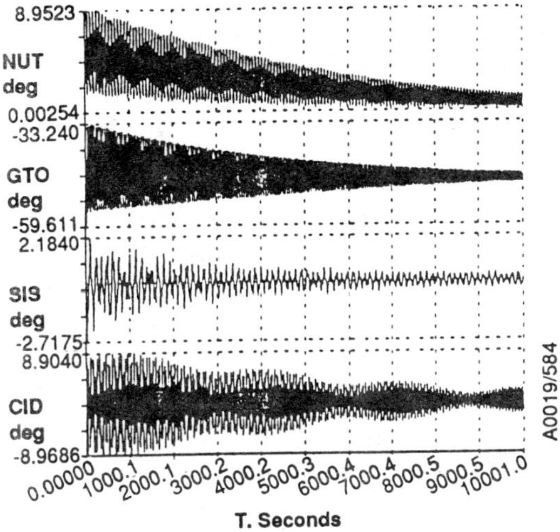

Fig. 8 Simulated nutation damping effects

On-Orbit Performance: Planning Software. Again based on experience with precession maneuvers as a rigid body, the thruster efficiency factors used by the MUS for planning maneuvers were left at the nominal 1.0. The fact that the precession thrusters required no change in calibration, whereas the spin up/down thrusters required a large adjustment (-11%), is due to two reasons. First, the attitude determination accuracies before and after small precession maneuvers do not provide nearly the fidelity in maneuver calibration that the evaluation of spin rate does. Secondly, the precession thruster math models were developed specifically for series of short pulses, whereas the spin rate change thruster models were designed for longer, continuous burns as opposed to the short pulses used for the spin-down doublets. All attitude control maneuvers performed have provided attitudes within 0.5 deg of the expected position, and this residual is due largely to attitude determination errors.

On-Orbit Performance: Dynamics Verification. The 2-deg test precession maneuver was performed to gain confidence in the dynamics modeling. From Fig. 8, it was expected that the main body nutation would be about three times the precession magnitude, or 6 deg. The primary method of verifying this was to be the Sun sensor telemetry. Despite the difficulties in observing nutation with the Sun sensor, it was assumed that the flexible-body coupling means that the maximum nutation angle is nearly observable by the Sun sensor, given a large number of samples, and an additional 0.5 deg was allowed to account for the worst-case LSB effect. Returning to the previous notation, it therefore was assumed that the initial nutation of the main body is approximately 0.25 deg greater than half of the peak-to-peak variation in Sun angle measurement, $\Delta\beta_m$.

After the 2-deg precession pulse was applied, and some telemetry editing discussed below, the maximum Sun angle seen was 8.75 deg. However, the momentum vector was only 5 deg off the Sun, so the (predicted) 6-deg nutation caused the figure axis to occasionally pass back and forth across the Sun. This causes ambiguities in the minimum Sun angle, since it is not signed in this context. That is, the only indication of the figure axis passing across the Sun is an unusually short spin period, which in itself is rather ambiguous because the nutation also causes the Sun crossing times from the Sun sensor to fluctuate considerably. So the observed nutation after this maneuver was estimated to be between 4.25 deg and 7.5 deg. Hence, the wire boom motion was inferred to be between these limits, also.

As another symptom of the figure axis passing through the Sun after this maneuver, there was also one Sun angle telemetered that was "impossibly" large. Additonally, there were two samples of 19.75 deg, which were taken quite seriously at first. These were initially believed because of the repeated measurement, yet they implied a nutation three to four times larger than expected.

In response to the unexpected Sun angles seen after the first maneuver, the second precession size was limited to 2.5 deg. The ambiguous results from the previous maneuver were largely duplicated here, except that no marginally believable Sun angles similar to the 19.75-deg samples were seen, and only one "impossible" value was recorded.

A superior method of measuring nutation was used for the third precession maneuver, which was sized at 3 deg. The maneuver time was chosen to be near perigee, where the stronger ambient magnetic field makes the magnetometer **B** analog telemetry useful. This is sampled every major frame of telemetry, about every 4 sec, or roughly seven times per spin period. Since the magnetometer z-axis is aligned with the spacecraft figure axis, the nutation at the time of each of these vector samples can be calculated from:

$$\theta = \cos^{-1}(B_z / |\mathbf{B}|)$$

This is plotted in Fig. 9 for a 10-minute sample of telemetry taken about 35 minutes after the maneuver. Allowing for the ±1-deg error due to the **B** telemetry LSB at this altitude, the nutation at this time is estimated to be 6 deg or less. This is consistent with the predicted 9-deg initial value for this maneuver size, and the estimated one-hour half-amplitude damping time. The momentum vector after this maneuver was 8 deg off the Sun, so the expected 9-deg nutation might or might not cause the same problems with the Sun sensor interpretation seen before. In practice, the telemetry did not show such ambiguities, with the maximum and minimum Sun angles observed 13.75 and 0.25 deg, respectively. No "impossible" Sun angles were telemetered. Thus, $\Delta\beta_m$ = 13.5 deg and the initial nutation estimated from Sun sensor data was around 7 deg. Consequently, there was sufficient confirmation of the dynamics model to proceed with the maximum 5-deg precession size and to try a third method of nutation measurement.

The 5-deg precession was performed with the final momentum vector 11 deg ahead of the Sun. Thus, the predicted 15-deg nutation would take the figure axis back through the Sun, and was expected to show the same Sun sensor telemetry ambiguities as seen previously. This indeed proved to be the case, so the estimate of $\Delta\beta_m$ was again problematic, and one "impossible" Sun angle was seen. However, the maximum Sun angle seen of 21.25 deg was consistent with the maximum expected possible of about 26 deg. Plots of Sun angle and spin period data following a second 5-deg maneuver similar to this qualification one are shown in Figures 10 and 11, respectively. The Sun-angle plot shows the wide asynchronous scatter of Sun angles for the 150 minutes following the pulse. The aforementioned spin rate variation alleged by the Sun sensor, caused by the large nutation, can also be seen.

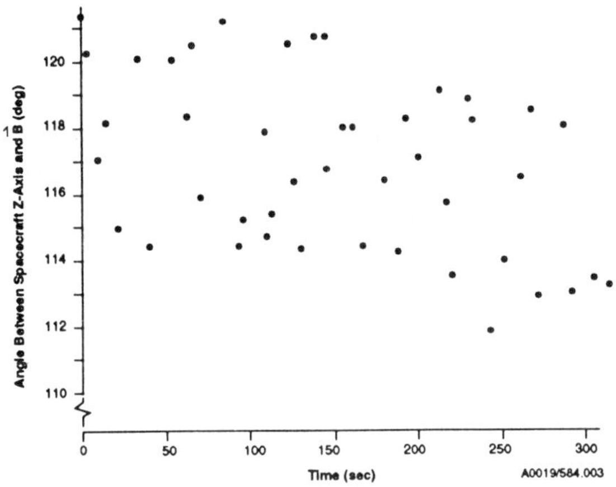

Fig. 9 Observed CRRES nutation from magnetometer approximately 35 minutes after 3-deg precession

Another method of estimating nutation that was implemented for this calibration is the examination of solar array current telemetry. The CRRES solar arrays are oriented normal to the spin axis, so the current coming out of them goes with the cosine of the off-Sun angle. The resolution and sample rate of this telemetry is also quite good, so it may be processed by hand in manner much like the **B** telemetry described above. This analysis showed a maximum off-Sun angle of 24.6 deg, which is again consistent with the predicted maximum of 26 deg, although it has much higher fidelity than the Sun sensor measurements, due to the higher sample rate. This method of using solar array current to measure off-Sun angle works well only when the angle is large enough for the cosine to show some sensitivity to its changes, i.e., the derivative of the cosine is very small for small angles. In contrast, the magnetometer-based measurement of the nutation does not depend on the Sun angle, and may often be the preferred method. In the case of CRRES, the two methods are complementary, since there typically is only one or the other available, due to altitude or Sun-angle limitations.

As a final step in characterizing the unexpected Sun sensor telemetry seen earlier, a 2-deg precession was performed while at a 12.8 deg off-Sun angle. The resulting 6-deg nutation would then not be expected to cause unusual Sun sensor measurements, and indeed did not. We therefore come to the preliminary conclusion that the "impossible" Sun angle measurements are linked with the Sun crossing time ambiguities, when spacecraft nutation causes the figure axis

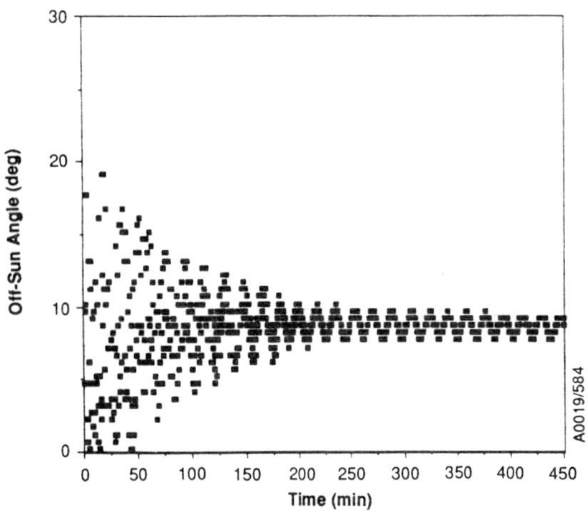

Fig. 10 Nutation observed by Sun sensor following 5-deg precession pulse

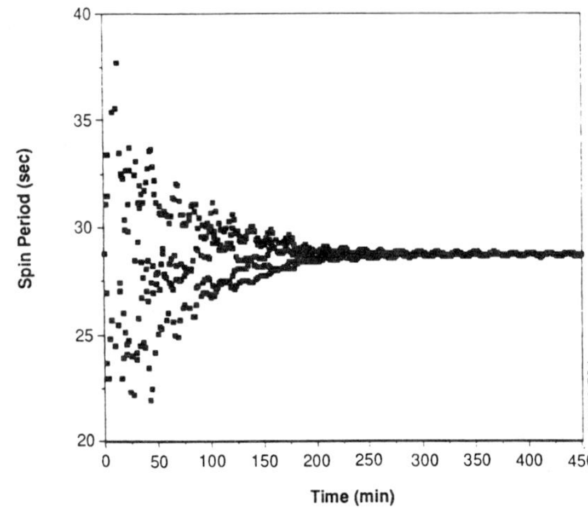

Fig. 11 Spin period oscillations resulting from wire boom coupling and 5-deg precession

to pass back and forth across the Sun. This is surprising, because an optical fence had been deliberately placed directly above the Sun sensor to avoid the threat of glint off of the top edge of the vehicle. There appears to be some optical effect happening that is not yet understood at small Sun angles. However, this is not a normal operational state for CRRES, and so poses only an academic difficulty.

On-Orbit Performance: Nutation Damping. Nutation damping performance of the main body, as driven by the wire booms, has a strong effect on the quality of the science data following attitude control maneuvers. Returning to Fig. 10, the damping time to the first half-amplitude can be seen to approximately agree with the 1.1 hr predicted by Fig. 8. However, a sudden change in damping rate can be seen to occur between 150 and 200 minutes. This is because the coulomb (friction) dampers at the wire boom roots have a deadzone of approximately 1 deg. Since the boom motion and main body nutation amplitudes are roughly the same, it may be expected to see a discontinuity in damping rate when these amplitudes drop below 1 deg, as is indicated by Fig. 10. The damping after this time is primarily due to the solid friction internal to the wire boom cables themselves, and is expected to be nonlinear. This effect is evident in Fig. 10 beyond 200 minutes, where the nutation (as quantized by the Sun sensor) decreases by only 0.25 deg (one-half LSB) during the remaining half of the plot. The long damping time has also been inferred from the science magnetometer data, and shows damping for the first few hours that agrees with Fig. 10. After that time, the magnetometer data indicates damping to roughly 0.5-deg nutation in 14 hr, and 0.3-deg nutation still remaining after approximately 24 hr. This small residual nutation is acceptable to the science experiments on board CRRES.

Fig. 12 and 13 are expanded plots of the last portions of Fig. 10 and 11, respectively. The Sun sensor digitization of the residual nutation mentioned above is evident in Fig. 12. Fig. 12 and 13 also show a coupling that was not expected (nor sought by the simulations), although it is presumably due to the beating between the two different pairs of wire booms. The 1,150-second period seen in these figures is also evident in the science magnetometer data alluded to above. This behavior is within approximately ten percent of that predicted by a special 25,000-second simulation run (not shown). However, it would have been preferable to include a Sun sensor model in the plots shown in Fig. 8, as was done in the spin-down analysis addressed by Fig. 6. This also may have foreshadowed some of the ambiguities discussed above, when using Sun sensor data to verify the flexible-body dynamics.

Flexible-Body Performance: Summary. In review, the flexible-body dynamics predicted to result from precession maneuvers have been verified to within approximately ±10 percent. Ambiguities in using the Sun sensor to perform this calibration have been identified and solved by analysis of other telemetered quantities. Although it is not possible to observe the wire boom motion explicitly, it is inferred from calibration of vehicle nutation that the 5-deg precession maneuvers to be performed for attitude control for the remainder of the CRRES mission will generate less than 20 deg of wire boom motion. Damping of this motion occurs in an acceptable length of time.

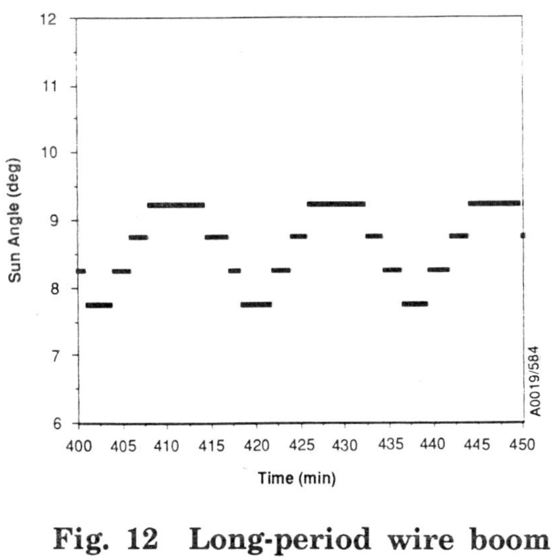

Fig. 12 Long-period wire boom coupling with main body nutation is lightly damped

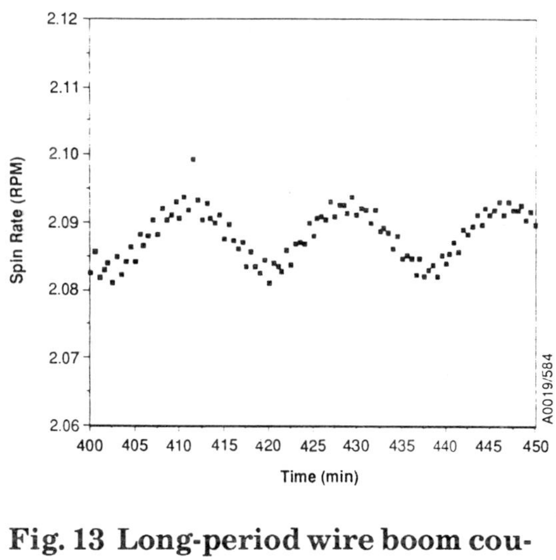

Fig. 13 Long-period wire boom coupling with main body spin rate is lightly damped

ACKNOWLEDGEMENTS

The dynamics analyses and predictions referred to herein were performed by Glen King and Jim Mohl of Ball Aerospace, and are exquisitely documented in a series of 35 internal System Engineering Reports. Their work has been independently verified by Doug Hendrix and Mike Gallaher of MSFC, and Dr. Andre Doran and Gary Wyetzner of Aerospace Corporation. Unfortunately, since none of these analyses have been published in the open literature (with the exception of Ref. 5), they cannot be referenced in a useful manner.

REFERENCES

1. B. Frazier, R. Stone, and P. Thompson, "Selection of Orbits for the CRRES Dual Mission Satellite," *Journal of the Astronautical Sciences*, Vol. 34, No. 3, July-Sept, 1986
2. W. Frazier, R. D. Culp, and G. R. Rosborough, "Semianalytic Study of High-Eccentricity Orbit Evolution," AAS 89-452, AAS/AIAA Astrodynamics Specialist Conference, Aug. 7-10, 1989
3. R. Hopkins, "Launch Window Analysis for the CRRES Mission," AIAA 90-2871, AIAA Astrodynamics Conference, Aug. 20-22, 1990
4. W. Frazier, K. Stewart, and J. Gaiser, "Attitude Determination and Control for the Combined Release and Radiation Effects Satellite," AAS 85-042, Annual Rocky Mountain Guidance and Control Conference, Feb. 2, 1985
5. B. G. King and R. P. Woolley, "Modeling, Tuning, and Effectiveness of Partially-Filled Ring Nutation Dampers," AAS 85-054, Annual Rocky Mountain Guidance and Control Conference, Feb. 2, 1985
6. J. R. Wertz, "Spacecraft Attitude Determination and Control," D. Reidel, 1984

AAS 91-071

INITIAL PERFORMANCE IMPROVEMENTS DUE TO DESIGN MODIFICATIONS FOR THE POINTING CONTROL SYSTEM ON THE HUBBLE SPACE TELESCOPE

G. S. Nurre, J. P. Sharkey and H. B. Waites*

Shortly after the Hubble Space Telescope was deployed on orbit it became apparent that the Pointing Control System was experiencing unexpectedly large disturbances during certain portions of the orbit. While these disturbances were most pronounced during transitions of the Earth's shadow, significant disruptions to the pointing occurred at other times in the orbit. Careful analysis of the flight data has lead to the conclusion that the disturbances are caused by a combination of rapid thermal deformations of the Solar Arrays together with stiction-friction mechanisms within the array that randomly release stored thermal/mechanical energy. This paper describes the baseline pointing control system design and the flight data that characterized the effects of the extraneous disturbances. The paper goes on to describe the procedure for modifying the onboard controller to attenuate the disturbances to levels consistent with the science operations. Flight data demonstrating the improved performance are shown and are tabulated to compare performance prior to and after phase I of the redesign.

INTRODUCTION

The Hubble Space Telescope (HST)† was launched aboard the Orbiter Discovery in April 1990 and one day later (GMT 115:19:38:09) was deployed with the Remote Manipulator System (RMS) into a 332 nautical mile circular orbit.[1] The two Solar Arrays (SA) and two High Gain Antennas (HGA)

* NASA Marshall Space Flight Center, Huntsville, Alabama 35812.

† The Hubble Space Telescope was designed and built under the direction of Marshall Space Flight Center by Lockheed Missiles and Space Company, supplier of the System Support Module, Hughes Danbury Optical Systems (formerly Perkin Elmer), supplier of the Optical Telescope Assembly, and the European Space Agency, supplier of the Solar Array.

were fully deployed prior to release, and the Aperture Door (AD) was opened 42 hours later with the Pointing Control System (PCS) in the sun-pointing mode with the -V1 axis oriented toward the Sun (see figure 1). During the process of configuring the HST to its operational mode, it became apparent that an unexpected disturbance torque was acting on the spacecraft and was producing large pointing disturbances and considerable activity in the torque commands to the Reaction Wheel Assemblies (RWA). The hypothesized nature of these disturbances, their effect on control system pointing performance, and the efforts undertaken to restore the specified performance are the principal topics of this paper. The subsequent section describes the PCS, the analytical model for the HST structural dynamics, and expected performance. Following that same typical flight data are shown, and an explanation is set forth as to why it is felt that the source of the disturbance is very likely associated with thermally induced mechanical excitations of the SA's. The paper goes on to describe the phased approach that was taken to redesign the PCS to accommodate the disturbance and improve pointing stability. The results of initial phase (SAGA I), both design and flight testing, are shown, and the plans for the final phase (SAGA II) are discussed.

PCS BASELINE DESIGN AND POINTING REQUIREMENTS

The pointing requirement for the HST is to maintain an image on the focal plane stable to 0.007 arcseconds (RMS) for observations lasting from 10 seconds to 24 hours. The total error is comprised of a number of contributors ranging from noise-driven sources to thermal/mechanical effects in the optics and focal plane. The error allocated for the total of gyro noise, photon noise, and interface torques from SA's is approximately 0.003 arcseconds (RMS). This number will be compared later with flight data. Since the disturbance torque that is of concern here is thought to originate with the SA's, it is interesting to note that the maximum interface torque from the SA was specified as an interface requirement to be less than 0.002 Nm beginning 5 minutes after the completion of a SA slew. This level of torque is consistent with approximately 0.001 arcseconds of pointing error.

A block diagram of the HST PCS is shown in figure 2. The fast loop which processes information 40 times per second samples gyro data and forms a torque command composed of a position, integral, and derivative (PID) term which is transformed to form commands to the four RWA's. The derivative path includes a Finite Impulse Response filter to "roll off" the response above 3.0 Hz, limiting the effects gyro noise on the torque command and suppressing higher frequency modal response. There is a 7.0 millisecond delay between a gyro measurement and the issuance of the associated torque command. An outer loop that uses a star position measurement from the Fine Guidance Sensors (FGS) updates the attitude reference established by the gyros and computes a short term estimate of gyro drift. This process is performed at a lower bandwidth of 0.02 Hz. The FGS's are optical/electrical/mechanical star tracking devices that use the main optics of the HST and are controlled by their own individual computers. More detailed information on the FGS's is available in reference [2]. References [3, 4] contains additional PCS data on hardware details, mounting arrangements, momentum management, and operational modes.

Selection of the form and constants for the controller is, to a large degree, dependent on the frequencies and gains of the structural modes of the HST. Figure 3 is a tabulation of the calculated modal frequencies and gains for each axis and includes the most significant modal gains for all operational orientations of the SA's. All modes below about 10 Hz are associated with one or more of the five appendages. The Nichols charts for the three spacecraft axes for the baseline gains, tabulated in figure 2, are shown in figure 4 where it is evident that all stability margins are well within the limits of accepted practice.

DISTURBANCE DESCRIPTION

The nature of the disturbance phenomenon can be best understood by observing time histories of the HST attitude error as measured by either the integrated rate gyro signals or the FGS error signals. Figure 5 is such a time history for the three spacecraft axes derived from integrated measurements of the gyro rates and shows a composite orbit made up of two non-contiguous time pieces. Gaps in the data are the result of telemetry drop-outs. The orbital day and night intervals are designated in the figure. It is apparent from the figure that during the transitions from Day to Night (D/N) and Night to Day (N/D) (terminator transitions) a major disruption of the pointing occurs, characterized by peak-to-peak amplitudes of nearly an arcsecond in the V1 axis (the optical axis) and several tenths of an arcsecond in the other two axes. These particular time histories show a difference in the nature of the D/N and N/D transitions; the disturbance associated with the D/N transition has somewhat higher amplitudes and persists for a longer period of time when compared to the N/D transition. In general, this distinction prevails, although each transition is different. During the first half of the orbital day it appears that every minute or two a disturbance occurs, the effects of which lasts for approximately a minute. These midday disturbances are usually less than those of the transitions by about a factor of two. Then toward the end of orbital day the disturbances come less frequently, and finally 10 to 12 minutes before transition to night there are usually none. During orbital night, subsequent to the D/N transition, disturbances occur less frequently and with less amplitude than those during the day. A tabulation of pointing performance for various periods during the orbit are shown in figure 6. The table indicates that for most of the time the pointing errors exceed the specification by considerable amount; however, there are intervals of several minutes that occur near the end of orbital day when the pointing is quite good. Figure 7 shows data similar to figure 5 except that the time and amplitude scales have been expanded to show the details of one particular event that occurred toward the end of orbital day. The excitation of the 0.12 Hz mode is evident particularly in V2 and then decays to what appears to be a time span free of disturbances where the pointing is very good.

In order to determine the source of the disturbance it is essential to ascertain the spectral characteristics of the responses. These are shown in figure 8 in terms of Power Spectral Densities for the spacecraft rate (gyro measurements) for two different orientations of the SA. The two SA positions are 90 degrees apart; in -V1 sun pointing the -V1 axis of the HST is pointed to the Sun so that the plane of the SA is in the V2-V3 plane and in V3 sun pointing the plane of the SA is in the V1-V2 plane. It is apparent from the figure that the dominant frequency is at approximately 0.1 Hz and that, while this frequency appears in V2 in both SA orientations, its occurrence in V1 and V3 is dependent on the SA orientation. There is also a frequency in the 0.6 to 0.7 Hz range that shows up in V1 or V3 as a function of SA position.

Regarding the source of the disturbance, there are two aspects to be considered, the source of the dominant frequencies that are seen in the responses and the mechanism by which the modes associated with those frequencies become excited. As to the former, the dominant frequency in the response of 0.1 Hz is close to the fundamental mode of the SA, an out-of-plane mode predicted to be 0.086 Hz. Actually there are four such modes, a symmetric and an asymmetric mode for each SA, that are very close in frequency and that produce a beating characteristic that is seen in the flight data. All other appendages have much high frequencies. Furthermore, the fact that the 0.1 Hz frequency appears in V1 or V3 depending on SA orientation is a strong indicator that the mode is associated with the SA. For instance, in -V1 sun pointing attitude the 0.1 Hz mode appears in the V3 axis while in V3 sun pointing it appears in V1. Likewise, the mode in the vicinity of 0.6 Hz is an inplane SA mode and dominates the V3 response in V3 sun pointing and the V1 response in -V1 sun pointing. Although other modes can be discerned in the PSD's, the most significant responses seem to be associated the SA's. The other part of the investigation concerns the mechanism by which the modes become excited. The time history of the responses point to a thermally driven source, particularly the very pronounced responses at the terminator transitions. Since the SA's are most susceptible to large thermal deformations, they were a prime target for detailed scrutiny. First a recalculated rate of change of the thermal gradient around the circumference of the STEM booms that support each of the four SA blankets uncovered an error in the previous estimate; it was found that the gradient was established five times faster than was originally thought, giving rise to an equally faster tip deflection. Mathematical modeling of such an inplane deflection showed the beat phenomena seen in the data and produced comparable response amplitudes. However, only part of the characteristics of the transition disturbance can be explained by the initial SA deformation that occurs at terminator transitions. Much of the complex nature of the response at the terminator and during other parts of the orbit are thought to be associated with the storage and release of thermal/mechanical energy in several mechanisms that are part of the deployment system. Each of the eight STEM booms is actually two booms, one inside the other, so that thermally induced deflections will cause sliding between the inner and outer member. Each blanket is attached to the boom tips through a cable/pulley mechanism that allows for a constant tension in the blankets as the booms change length or deform. And lastly the common drum, on which two blankets are rolled up on, has a negator spring that produces the constant tension by allowing the drum to take up or pay out more of blanket. Since the SA's begin to reach thermal equilibrium only toward the end of orbital day, where the pointing becomes quiet, it is not unreasonable to associate the periodic disturbances that occur throughout most of the orbit with a thermally driven stick-slip phenomenon in one or more of these mechanisms. The setting is further complicated when one takes into consideration the inherent distortion in the four blankets as the result of the deployment operation. For more detail on the analysis and isolation of the disturbance phenomena, one is directed to the material in reference [5].

Because of the large disparity between the measured pointing performance and the specification, a fix was required; and since the source of the problem, the SA, is not readily accessible, the choice was made to modify the PCS control laws that reside as software in the flight computer in order to provide better disturbance suppression at the frequencies of concern. The next section describes the redesign procedure.

PCS REDESIGN (SAGA I)

It was decided early in the planning for the control system redesign that a phased approach would be taken. Although it is apparent from the data that the error response contains frequencies up to perhaps 2.0 Hz, the first phase of the redesign, entitled Solar Array Gain Augmentation I (SAGA I), would have the goal of suppressing only the effects of the 0.1 Hz mode. There were two reasons for this approach to the redesign. First, the response of 0.1 Hz was clearly the dominant component, and it was felt that the elimination of that part of the error would greatly expand the breadth of science activities that could be carried out. Secondly, having found significant discrepancies in the analytical SA dynamic model, there was concern that attempting to extend the bandwidth of the controller beyond the baseline with large uncertainties in the higher frequency modal characteristics would entail needless risk. Hence, SAGA I was to suppress the errors at 0.1 Hz by a factor of at least 30 over the baseline, maintain the bandwidth at approximately 1.0 Hz, and maintain acceptable stability margins.

It was determined early that the desired results could not be accomplished by simply changing the gains in the existing system and that more degrees of freedom were needed in the controller. This would require the development of additional code and a subsequent reprogramming of the flight computer. Because of the complexity involved in the reprogramming process, enough flexibility was put into the new code to accommodate future changes as well as SAGA I. Hence, two sixth order digital filters, G_A and G_F, were coded into the controller as shown in figure 9.

The inner loop feedback element, G_F, is designed to provide acceleration feedback with high gain at 0.1 Hz and rolloff characteristics to attenuate high frequency gyro noise and suppress the higher frequency modes. The forward compensator, G_A, is chosen to maintain the basic character of the PID system. Hence, G_A is chosen such that

$$\frac{G_A\, G_P}{1 + G_P\, G_F} = G_P$$

or

$$G_A = 1 + G_P\, G_F.$$

The forms for G_A and G_F in the S-domain for SAGA I are shown in figure 9. The transfer function for θ_ε/T_D for both the baseline PID and the SAGA design is shown in figure 10. With the SAGA I controller there is an additional 35 db of disturbance suppression over the PID controller. The added attenuation decreases rapidly above about 0.12 Hz, and the two controllers have comparable suppression characteristics above 0.3 Hz. The Nichols charts in figure 11 show the controller stability margins. The margins are less than what is normally desired, but were deemed acceptable given the flight experience accumulated over the last eight months and the protection afforded by the SAFEMODE [6] system that had been tested numerous times on orbit. The SAFEMODE system constantly monitors PCS performance; and if a discrepancy is found, it automatically places the PCS in a simpler mode, ultimately going to a sun-pointing attitude. Because the flight computer is a fixed point machine there are selectable limiters on the input and output of both G_A and G_F and in the position path of the PID. Finding the proper set of limits to make the system work in the presence of the varied disturbance environment was not an easy task as will be discussed in the next section on SAGA I flight results.

FLIGHT RESULTS FOR SAGA I

In mid-October 1990, the SAGA I software was uploaded to the flight computer, and series of three tests were run. In each test segment the HST was placed in gyro hold with the acquisition gains installed in the controller. Each test was to run approximately 12 hours so that it would include several terminator transitions, and the SA was oriented in one of three different orientations for each test. The three SA positions were 1) SA plane in the V1-V3 plane (SA angle = 90 degrees), 2) in the V2-V3 plane (SA angle = 0 degree), and 3) midway between at 45 degrees. The 90 degree test went smoothly for six hours at which time two gyro modes switches occurred, and at that point the test segment was terminated. The gyro mode switch is the result of a Safemode test that monitors the rate errors. When the total rate error exceeds 2.0 arcseconds/second for a contiguous period of 5 seconds, the gyros are switched from low to high mode and the controller gains are changed to the maneuver set. During the next test segment with the SA at 0.0 degrees there were seven mode switches that occurred over about a ten hour period. In the last test with the SA at 45 degrees there were no switches until seven hours into the test when three switches occurred during the last three hours.

Although SAGA I did an excellent job of suppressing the 0.1 Hz disturbance, which was the main goal, it was not useful operationally as long as the gyro mode switches persisted. In addition to showing the expected elimination of the 0.1 Hz modal excitation, analysis of the data showed that many of the mode switches occurred at what were previously considered to be quiet times of the orbit, late day and night, and were the result of higher frequency (0.4 - 0.6 Hz) disturbances. These perturbations were large enough to drive certain variables within the controller into their limits and into a limit cycle behavior that produced the high rate errors. Based on these data and some additional simulation work a new set of limits was uplinked, and another test was run in early December for SA angles of 0 degree and 90 degrees. No mode switches occurred for the controller with acquisition gains. The last part of the test was to run the system in the FGS/Observer mode with the science gains installed. When the observer loop was closed, the system went into a limit cycle which persisted for the duration that the FGS's tracked their guide stars, a period of about 20 minutes.

The performance of the SAGA I with acquisition gains is shown in figures 12 and 13. These figures are time histories of attitude error for D/N and N/D transitions. It is apparent from these traces that there is no perceptible response at 0.1 Hz. This is corroborated by the plots in figure 14, which are PSD's of attitude rates in the three spacecraft axes for SA angles at 0 degree and 90 degrees. The suppression of the response at 0.1 Hz, the goal of SAGA I design, is clearly evident. The attitude time histories show that in the case of the N/D transitions there is a short period of time (~ 15 seconds) during which the amplitudes in the individual axes are larger than for the baseline PID controller but for the rest of the transition period are considerably less. For the D/N transitions the amplitudes are appreciably less throughout the period. Although, in general, there is an improvement in performance during most portions of the orbit, the remaining response at the higher frequencies is large compared to the pointing specification and must be dealt with eventually (SAGA II). In addition, further tuning of SAGA I is required to eliminate the limit cycle in the science mode.

CONTINUING WORK

The emphasis at present is to produce a SAGA I design that will work in the science mode so that those science proposals that have been postponed because of the degraded performance can be scheduled. Once a new set of SAGA I parameters has been derived, it will be uplinked, and another series of tests will be carried to verify performance. The next phase is SAGA II which is intended to have disturbance suppression over a wider range of frequencies. It appears from data taken thus far that additional attenuation out to 1.0 Hz is required. However, extending the bandwidth of the controller necessitates a greater sensitivity to the characteristics of higher frequency structural modes. Because of the errors in the initial estimates of the frequencies of even the fundamental modes of the SA's, an on orbit transfer function test has been proposed before finalizing the SAGA II controller. The transfer function is that between torque inputs at the RWA's and rate outputs at the gyros. This test would reduce the uncertainties in the modal parameters that would be applied in the sensitivity analysis for stability margins. The goal of the SAGA II design is to achieve the pointing performance specification throughout the orbit.

SUMMARY

The paper has given a status report on the efforts to modify the HST PCS to meet the pointing specification. Flight data showing the performance with the baseline PID controller when excited by the thermally driven SA disturbance were presented, and a brief hypothesis for the source of the SA perturbations was discussed. Test results for an interim modification, SAGA I, show that the basic goal of eliminating the disturbance associated with the fundamental SA modes at 0.1 Hz is achieved but is not yet compatible with the science gains. This incompatibility is now being worked, and a modified SAGA I will be tested shortly. Finally, the paper discussed the second phase of the redesign, SAGA II, and its purpose.

ACKNOWLEDGEMENT

The form and design analysis of the SAGA controller were the work of Gary A. Beals of the Lockheed Missiles and Space Company, Sunnyvale, CA.

REFERENCES

1. Fienberg, Richard T., "HST: Astronomy's Discovery Machine," Sky and Telescope, Vol. 79, No. 4, April 1990.

2. Nurre, G. S., Anhouse, S. J., and Gullapelli, S. N., "Hubble Space Telescope Fine Guidance Sensor Control System," SPIE Conf., March 1989.

3. Nurre, G. S., and Dougherty, H. J., "The Pointing System for Space Telescope," Optical Platforms National Symposium and Workshop, June 1984.

4. Dougherty, H. J., et al., "Space Telescope Pointing System," J. of Guidance, Control, and Dynamics, Vol. 5, No. 4, pp. 403-409, July-August 1982.

5. Foster, C. L., et al., "The Solar Array-Induced Disturbance of the Hubble Space Telescope Pointing System," 61st Shock and Vibration Symposium, October 1990.

6. Hubble Space Telescope Pointing Control Subsystem, Safing System Description Manual, LMSC Report No. F256969, March 1989.

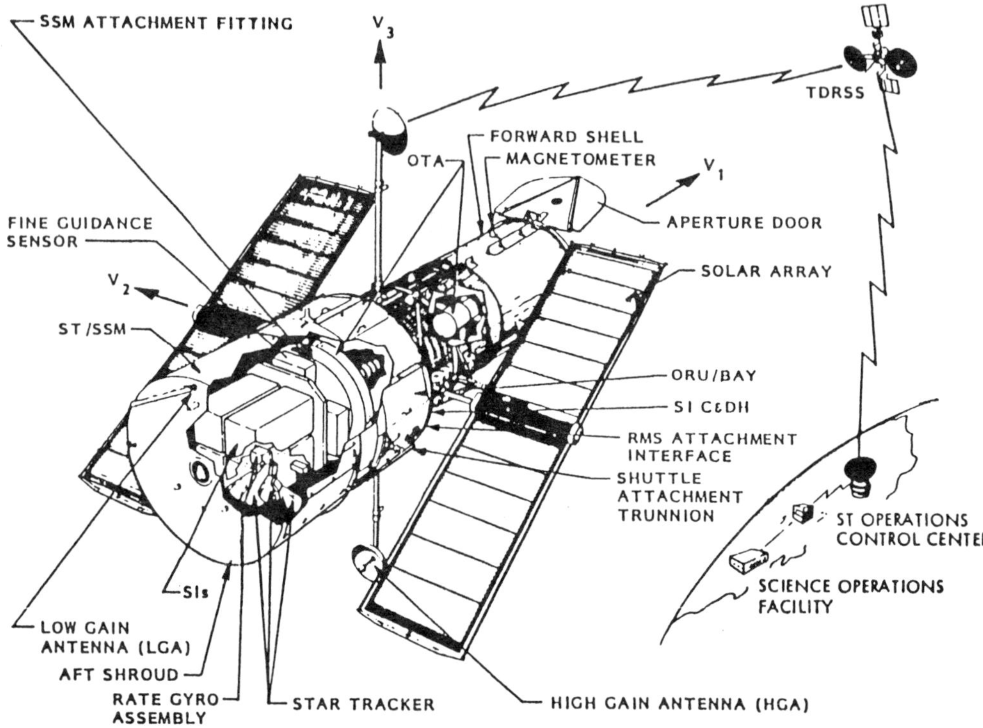

Figure 1 Hubble Space Telescope

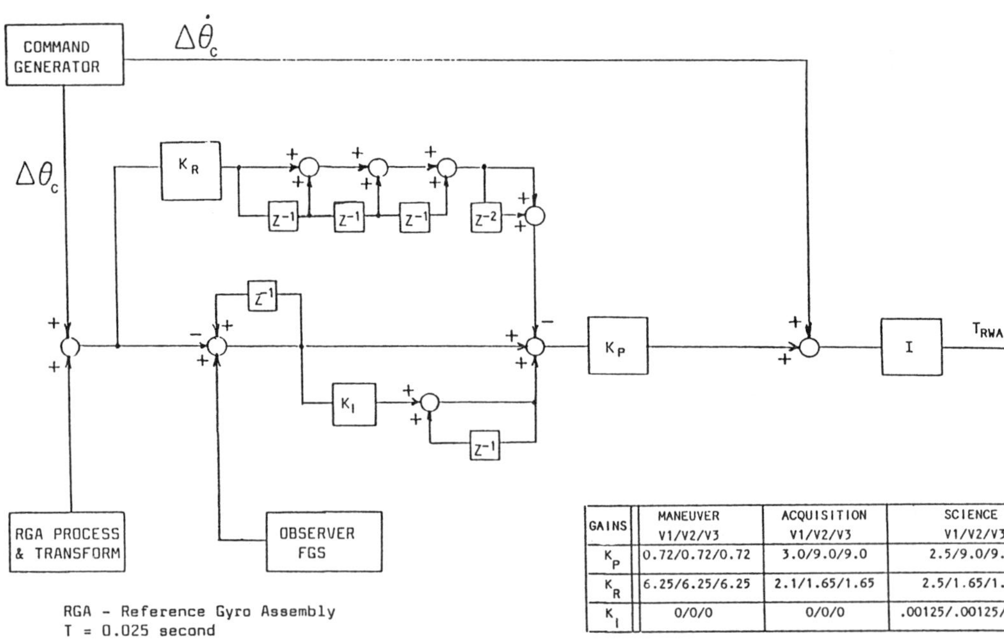

GAINS	MANEUVER V1/V2/V3	ACQUISITION V1/V2/V3	SCIENCE V1/V2/V3
K_P	0.72/0.72/0.72	3.0/9.0/9.0	2.5/9.0/9.0
K_R	6.25/6.25/6.25	2.1/1.65/1.65	2.5/1.65/1.65
K_I	0/0/0	0/0/0	.00125/.00125/.00125

Figure 2 HST PCS Controller - Baseline

V1 Bending Modes

Frequency (Hz)	Modal Gain K_B
0.110 S	0.063
0.600 S	0.211
0.713 H	0.016
1.041 S	0.087
1.155 H	0.094
2.729 S	0.040
2.816 H	0.015
2.863 H	0.059
4.271 S	0.020
10.834	0.048
12.133	0.030
13.201	-0.018
14.285	0.030
15.264	0.076
18.915	-2.161
19.315	-16.792

V2 Bending Modes

Frequency (Hz)	Modal Gain K_B
0.110 S	0.018
0.432 H	0.012
0.912 A	0.057
10.834	0.024
12.133	0.155
13.201	-1.341
14.068	-1.387
14.285	-0.806
15.264	-0.134

V3 Bending Modes

Frequency (Hz)	Modal Gain K_B
0.110 S	0.025
0.600 S	0.079
1.079 S	0.029
2.508 S	0.013
10.834	0.104
12.133	-0.320
13.201	-0.110
14.068	-0.217
14.285	-1.516
15.264	0.170

A - Aperture Door
H - High Gain Antenna
S - Solar Array

$$I = \begin{bmatrix} 31046 & 0 & 0 \\ 0 & 77217 & 0 \\ 0 & 0 & 78754 \end{bmatrix} \quad (\text{kg m}^2)$$

Moment of Inertia Matrix

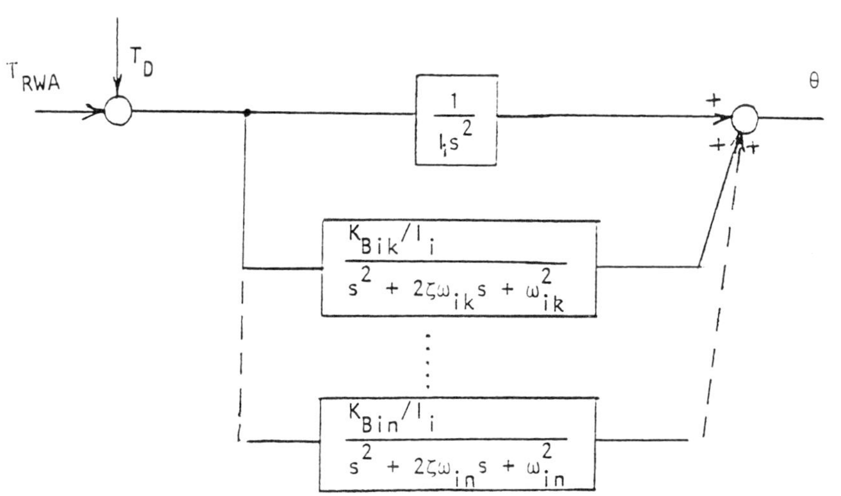

i = Spacecraft Axes (1,2,3)

k = Mode (1,2,....,n) $\zeta = 0.005$

Figure 3 HST Dynamics Model

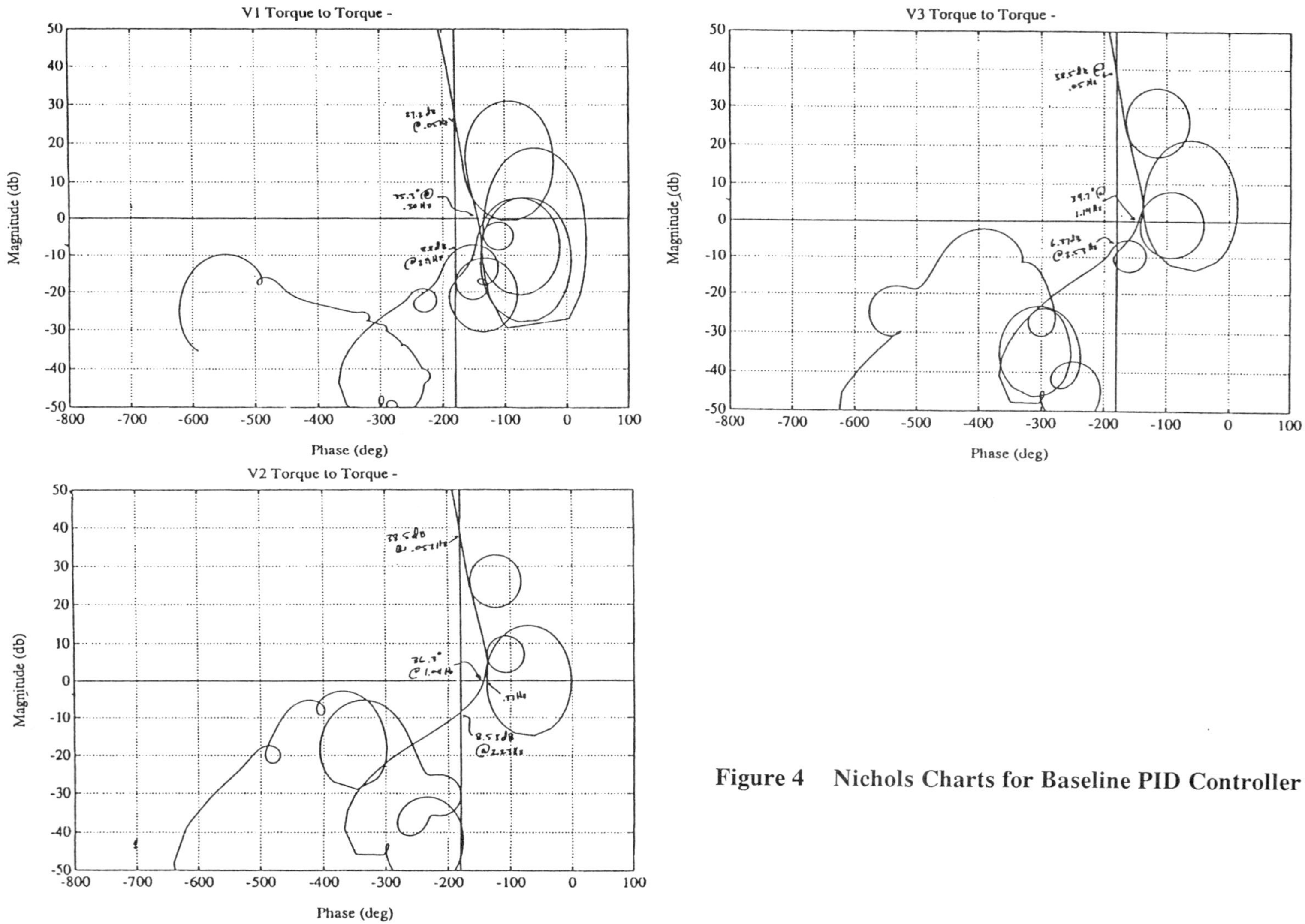

Figure 4 Nichols Charts for Baseline PID Controller

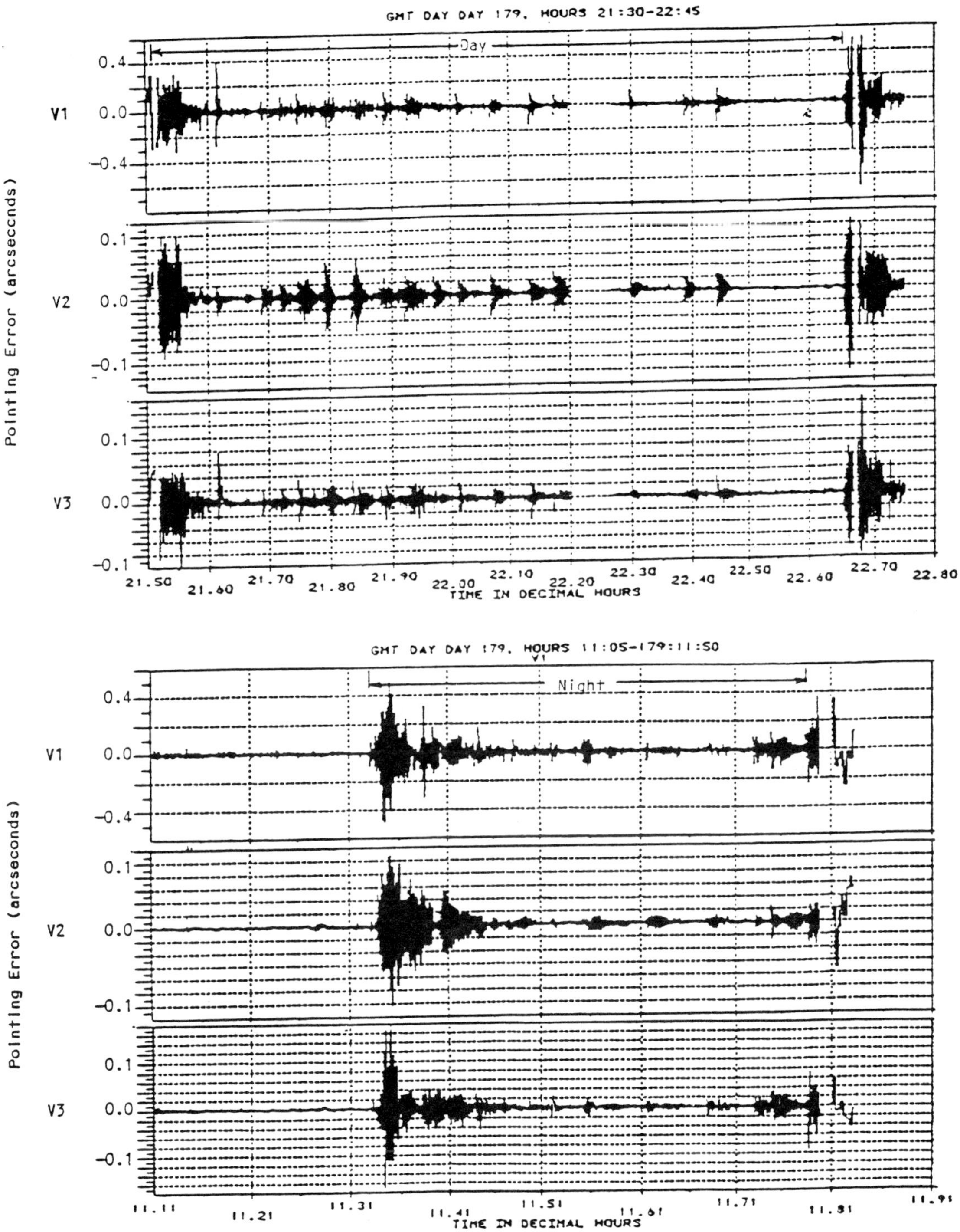

Figure 5 Attitude Time Histories for One Composite Orbit

	DAY/NIGHT TRANSITION (PEAK-PEAK, $\hat{s}$ WORST/BEST)	NIGHT PERIOD (RMS, $\widehat{ms}$)	EARLY DAY PERIOD (RMS, $\widehat{ms}$)	LATE DAY PERIOD (RMS, $\widehat{ms}$)	DAY-QUIET PERIOD (RMS, $\widehat{ms}$)	NIGHT-QUIET PERIOD (RMS, $\widehat{ms}$)
POINTING PERFORMANCE SA ANGLE = 90	v1-3.63/2.54 v2-.254/.180 v3-.325/.200	10.2	16.6	13.3	2.4	5.0
POINTING PERFORMANCE SA ANGLE = 0	v1-1.24/1.01 v2-.248/.165 v3-.361/.254	7.56	26.3	6.56	3.1	7.2

Figure 6 Typical Pointing Performance for Various Periods During Orbit

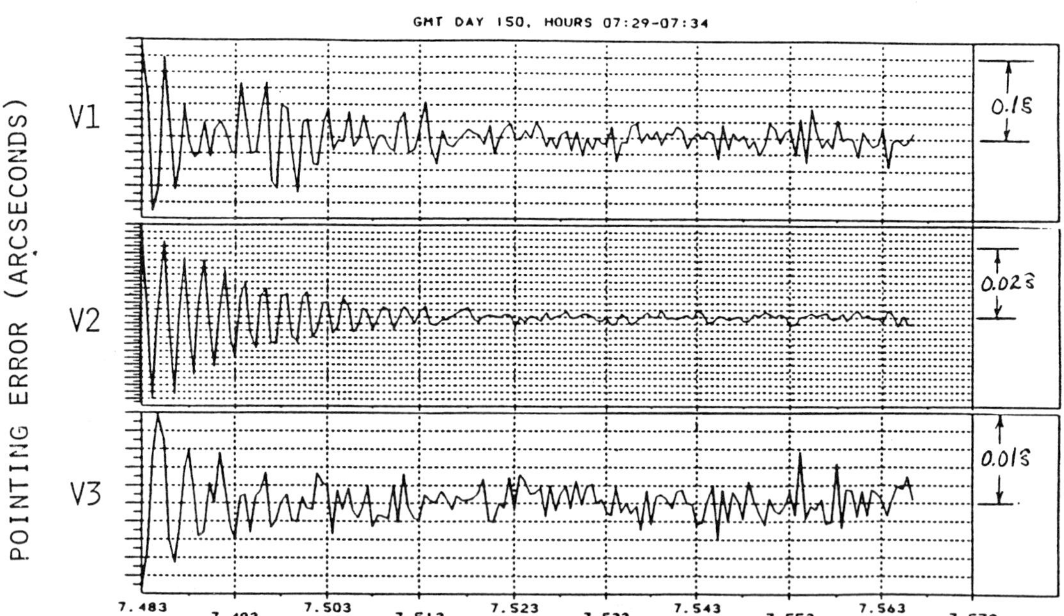

Figure 7 Expanded Time History of Pointing Error for Mid-Day Disturbance

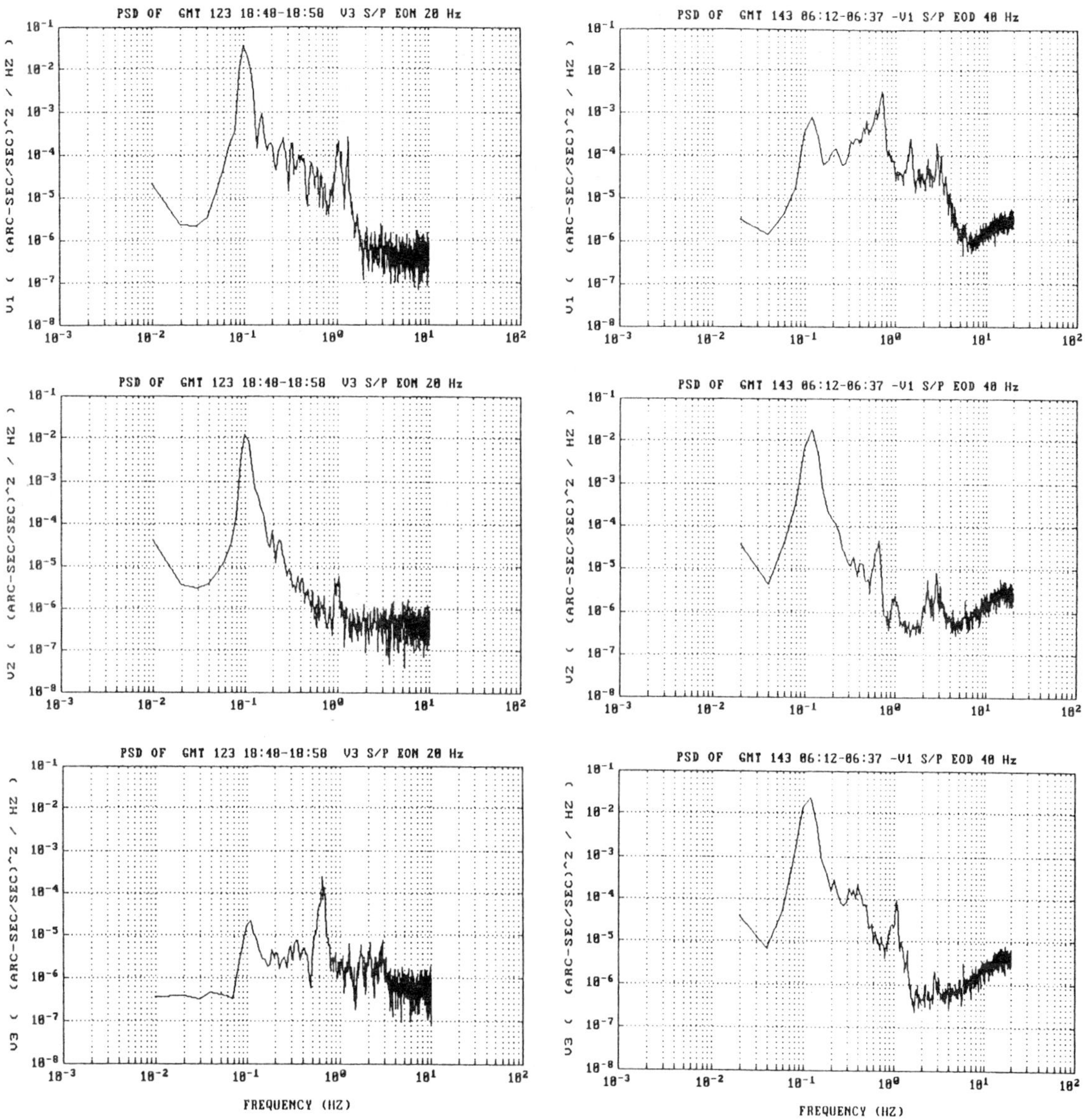

Figure 8 PSD's of Attitude Rate During Transitions - PID

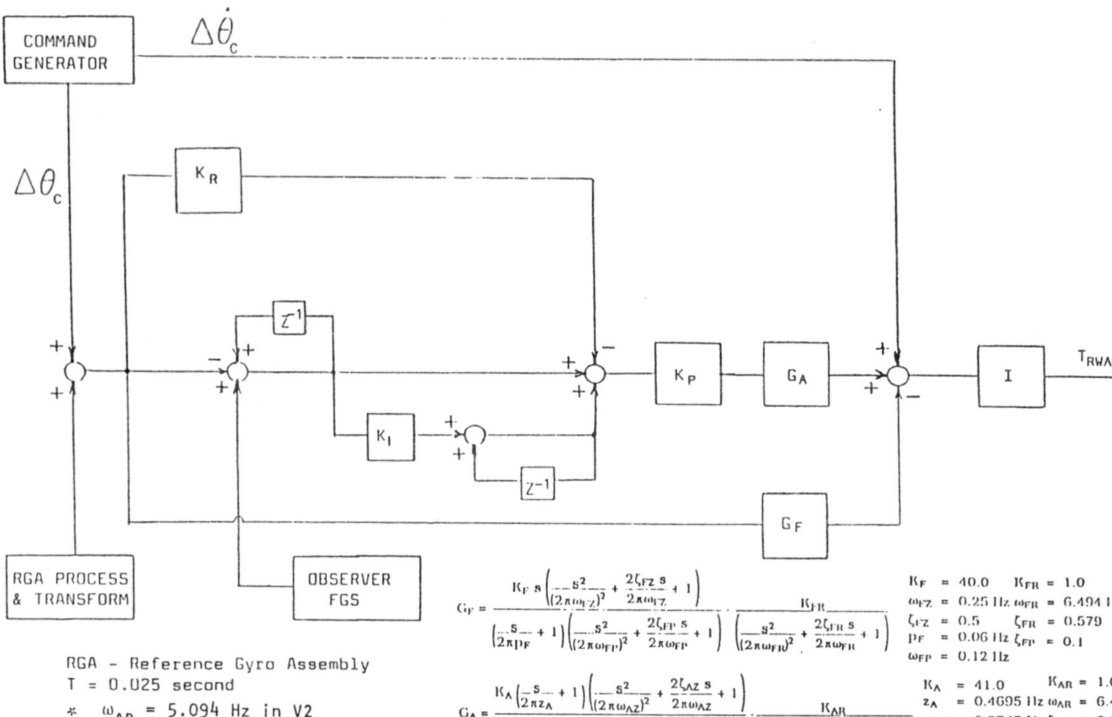

$$G_F = \frac{K_F\, s\left(\frac{s^2}{(2\pi\omega_{FZ})^2} + \frac{2\zeta_{FZ}\, s}{2\pi\omega_{FZ}} + 1\right)}{\left(\frac{s}{2\pi p_F} + 1\right)\left(\frac{s^2}{(2\pi\omega_{FP})^2} + \frac{2\zeta_{FP}\, s}{2\pi\omega_{FP}} + 1\right)} \cdot \frac{K_{FR}}{\left(\frac{s^2}{(2\pi\omega_{FR})^2} + \frac{2\zeta_{FR}\, s}{2\pi\omega_{FR}} + 1\right)}$$

K_F = 40.0		K_{FR} = 1.0	
ω_{FZ} = 0.25 Hz		ω_{FR} = 6.494 Hz	
ζ_{FZ} = 0.5		ζ_{FR} = 0.579	
p_F = 0.06 Hz		ζ_{FP} = 0.1	
ω_{FP} = 0.12 Hz			

$$G_A = \frac{K_A\left(\frac{s}{2\pi z_A} + 1\right)\left(\frac{s^2}{(2\pi\omega_{AZ})^2} + \frac{2\zeta_{AZ}\, s}{2\pi\omega_{AZ}} + 1\right)}{\left(\frac{s}{2\pi p_A} + 1\right)\left(\frac{s^2}{(2\pi\omega_{AP})^2} + \frac{2\zeta_{AP}\, s}{2\pi\omega_{AP}} + 1\right)} \cdot \frac{K_{AR}}{\left(\frac{s^2}{(2\pi\omega_{AR})^2} + \frac{2\zeta_{AR}\, s}{2\pi\omega_{AR}} + 1\right)}$$

K_A = 41.0	K_{AR} = 1.0 *
z_A = 0.4695 Hz	ω_{AR} = 6.494 Hz
ω_{AZ} = 0.2747 Hz	ζ_{AR} = 0.579
ζ_{AZ} = 0.3048	ω_{AP} = 0.12 Hz
p_A = 0.06 Hz	ζ_{AP} = 0.1

Figure 9 Modified HST PCS Controller - SAGA I

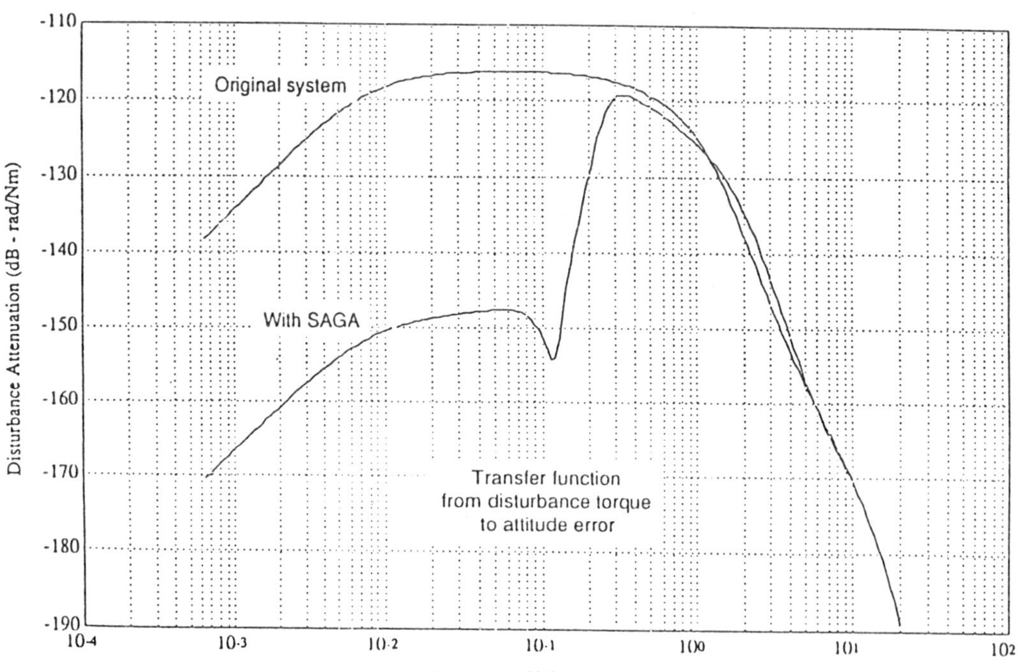

Figure 10 Disturbance Rejection Capability with SAGA

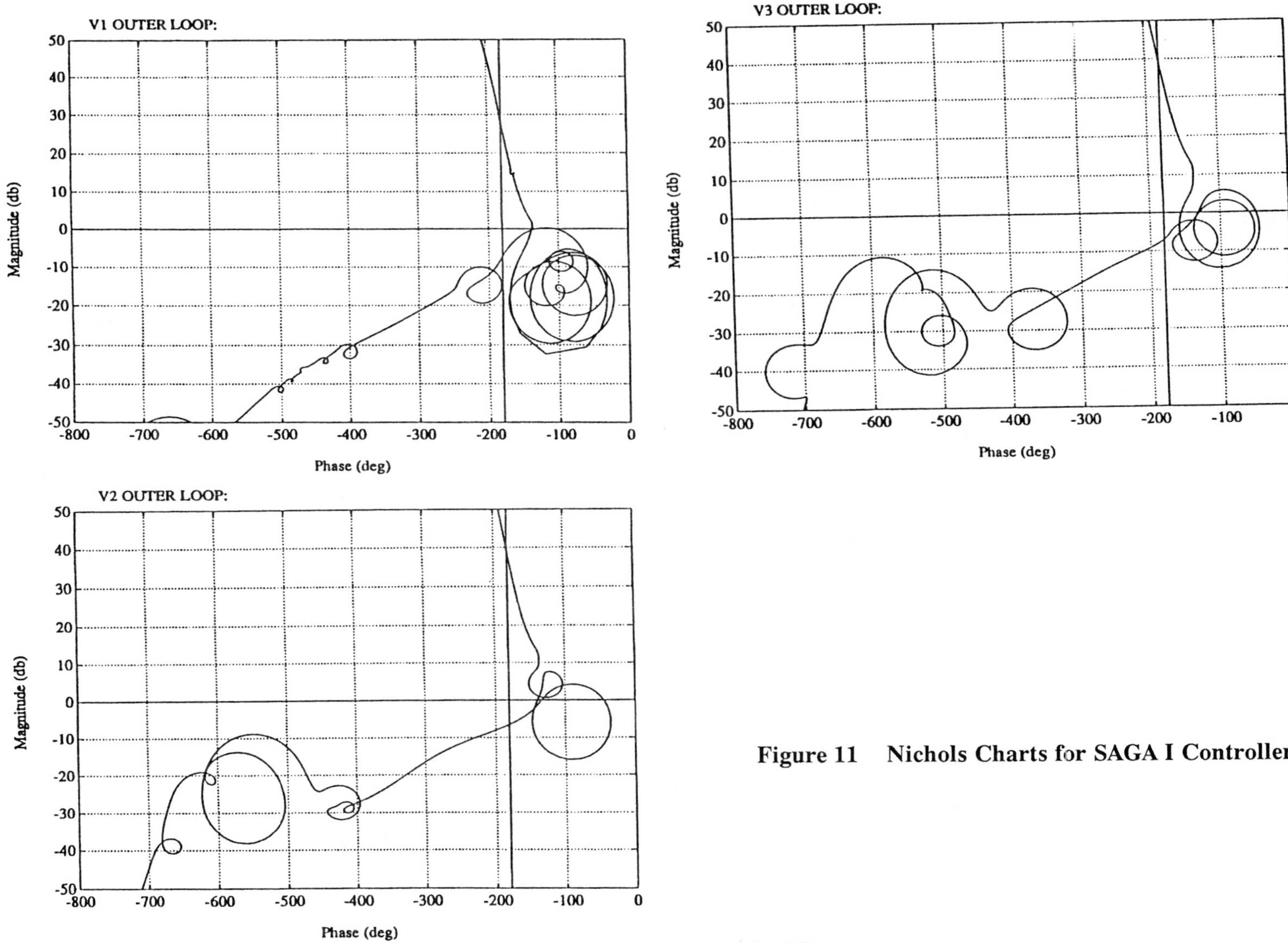

Figure 11 Nichols Charts for SAGA I Controller

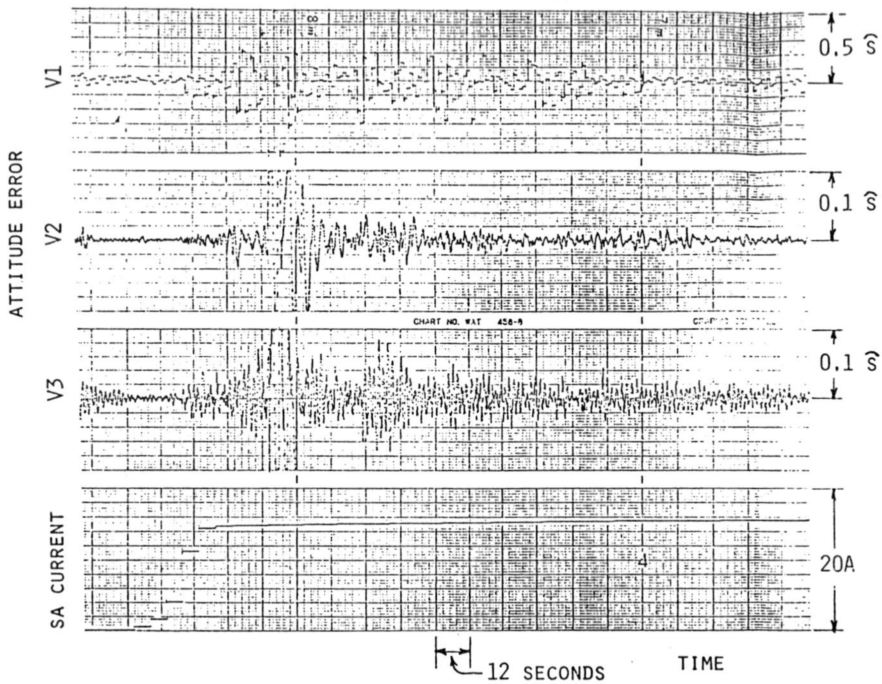

Figure 12 SAGA I Performance, Night to Day Transition, SA = 90°

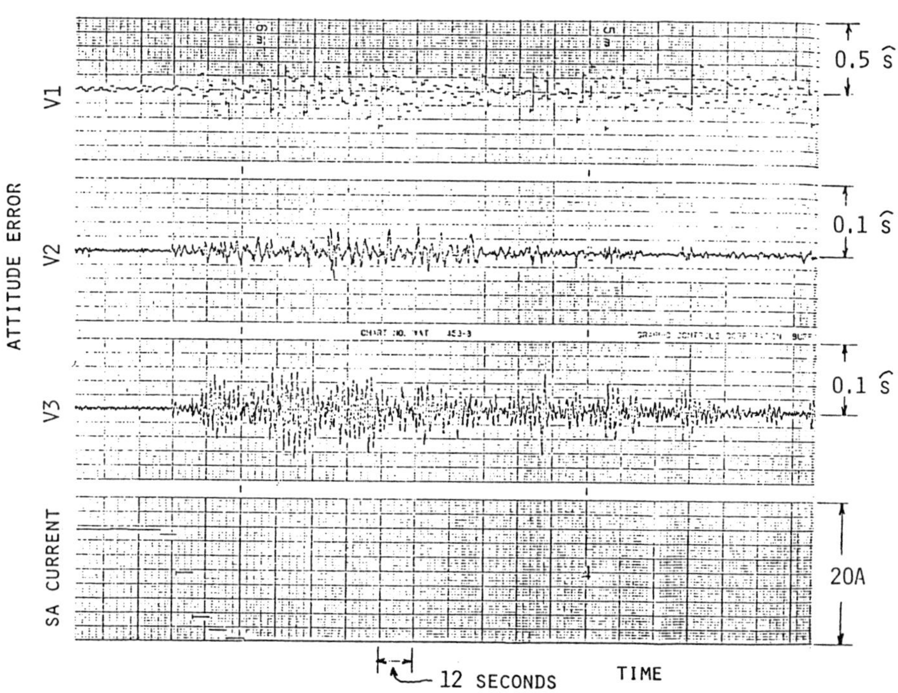

Figure 13 SAGA I Performance, Day to Night Transition, SA = 90°

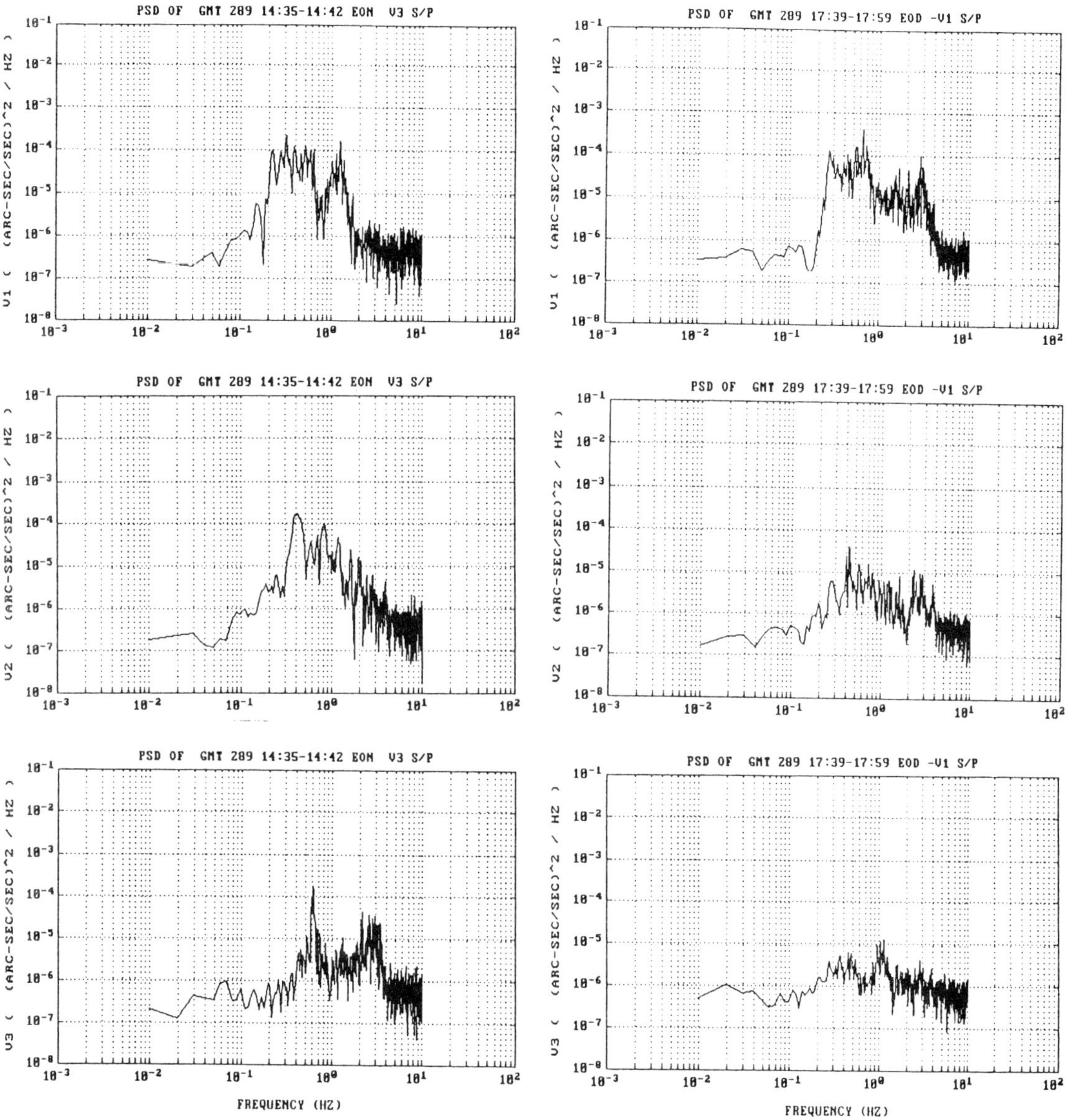

Figure 14 PSD's of Attitude Rate During Transitions - SAGA I

AAS 91-072

MAGELLAN STAR SCANNER EXPERIENCES: WHAT A LONG, STRANGE TRIP IT'S BEEN

Eric H. Seale*

Since its launch to Venus in 1989, the Magellan spacecraft has encountered a variety of interesting phenomena -- several related to its star scanner. After much consternation, it was determined that the scanner events were due to environmental effects on the instrument -- in particular, its response to solar protons and an unanticipated amount of thermal blanket dust (this dust is, in turn, generated and moved by other environmental forces). In short, while those phenomena were originally an operational nuisance, our scanner has unexpectedly shown new use as a particles and fields detector.

Since its simple design makes Magellan's star scanner a good proton detector, a brief description of the charged-particle environment (particularly proton propagation) is also included. Short- and long-term trends in sensor behavior are presented, as are their correlations to the local environment. A summary of results to date is provided in the hope that these may be of help to future operations teams diagnosing similar phenomena.

A summary is also given of methods found to reduce the operational impact of these phenomena.

INTRODUCTION

The Magellan spacecraft was launched on May 4, 1989 aboard the Space Shuttle Atlantis. Its primary mission is to make a global, high-resolution map of Venus using synthetic aperture radar (SAR) imaging. That radar is the only science instrument on-board, and is used to gather radar imaging, altimetry, radiometry, and gravity data of the planet's surface. By gathering this data as Venus rotates beneath Magellan's orbit, the spacecraft will complete one mapping cycle in 243 days. Data from the first cycle alone should result in a map of 70-90% of the planet's surface at a resolution at least an order of magnitude higher than that obtained previously.

Magellan was designed as a low-cost replacement for the VOIR program which was cancelled in 1982. As part of that cost-reduction philosophy, the spacecraft was built largely from government-furnished hardware. This hardware consisted of flight-qualified spare parts from the Voyager, Galileo, Viking, and Ulysses spacecraft (High Gain Antenna, Medium Gain Antenna, thrusters/engines, flight computers, etc.). Additional items were new builds from old designs and tooling (for instance, the Magellan star scanner).

---

* Magellan Flight Team, Attitude and Articulation Control Subsystem, Martin Marietta Astronautics Group, P.O. Box 179, Denver, Colorado 80201.

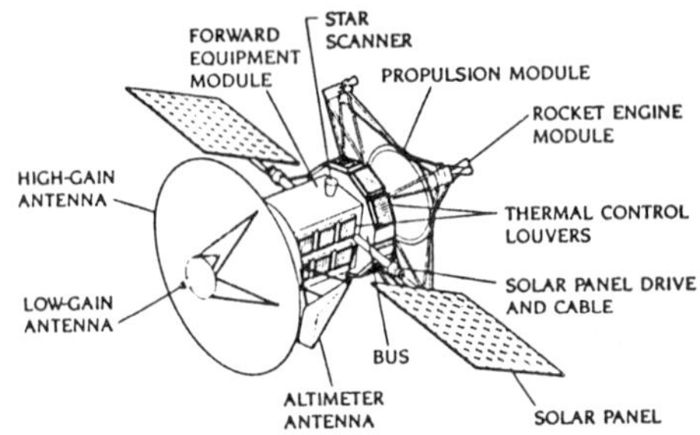

Fig. 1 Magellan On-Orbit Configuration

The spacecraft itself is of a relatively straightforward, 3-axis stabilized design as shown in Fig. 1. Attitude control is accomplished with the use of three reaction wheels (nominally), and four sets of thrusters (for reaction wheel desaturation, and as back-up control). Mechanical rate integrating gyros are used to propagate attitude knowledge; these are updated via a solid-state star scanner (STU). This means, of course, that the spacecraft must maneuver away from its normal playback attitude to scan any stars. The solar panels are articulable about a single axis.

Thermal control is accomplished with the use of a variety of surface coverings -- these are applied according to local requirements of the different parts of the vehicle. Parts of the vehicle that generate a fair amount of heat (the radar electronics, for instance) have louvers; other parts of the vehicle are covered with Optical Surface Reflectors (OSR's), quartz back-surface mirrors. The rest of the vehicle is covered with aluminized Kapton Multiple Layer Insulation (MLI) blankets with Astroquartz outer layers. Note that the outermost layer of aluminum immediately under the astroquartz is not tied to the spacecraft ground as are the other layers.

The Challenger disaster in 1986 forced a 1-year delay of Magellan's launch, and the cancellation of the Centaur G-Prime upper stage (this was to be our "ride" out of LEO). The IUS's lower performance and scheduling constraints (contention with Galileo for launch windows to Venus) forced a change in transfer orbit to Venus. This forced us to fly a Type-IV trajectory, travelling 1-1/2 times around the Sun in 15 months instead of the previously-planned 1/2 orbit in 4 months. The additional time in the cruise phase meant more wear-and-tear on the flight hardware before we even arrived at Venus, but fortuitously it also gave the spacecraft team more "shake-down" time as well.

Magellan relies on a Ball Model CT-203 star scanner to keep its gyros calibrated in flight. This is essentially the same model that was used on several IUS flights in the early 1980's (however the grounding scheme was somewhat modified for Magellan's use). The Magellan photodiode star scanner is, admittedly, a product of 1970's design and the technology of the times[1]. As shown in Fig. 2, the scanner optics image the scanner's field of view on a masked silicon photodiode (p-n junction) detector. This detector is electrically divided into two sensitive portions; each is aligned with a pair of slits (one vertical and one slanted), and is connected to its own channel of electronics (amplifiers, filters, etc.). Note that operationally only one channel is powered at a time. The scanner has commandable reference voltages that function as video detection thresholds; selecting a given threshold value screens out any stimuli that produce voltages lower than the corresponding reference voltage.

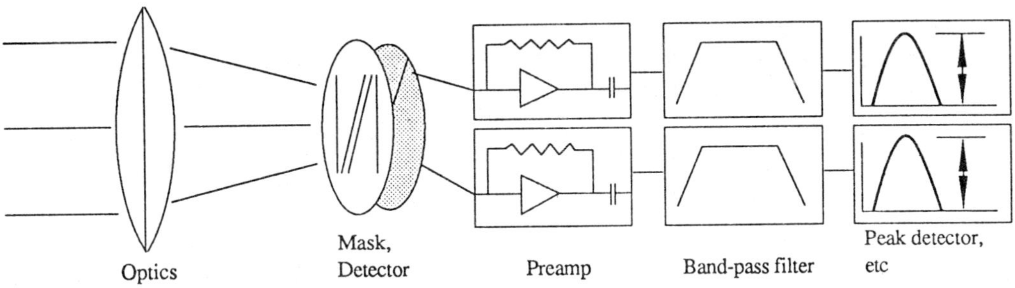

Fig. 2 Star Scanner Simplified Block Diagram

To reduce the area exposed to radiation effects, the scanner's detector was made considerably smaller than in earlier models. Otherwise, no attempts were made in the scanner's design to filter out radiation effects, and the scanner contains no image recognition or pattern recognition capabilities. The scanner is essentially an analog device with simple digital timer circuitry -- modern microprocessor technology was neither available nor feasible when the scanner was designed.

As the spacecraft spins in a star scan, images from the scanner's field of view cross unmasked areas of the powered detector. When the detector is illuminated by a source that exceeds the set threshold, the peak detection circuitry is enabled. As the image drops to a predetermined fraction of the maximum voltage, a timing pulse (interrupt) is output. When the interrupt is output, the peak voltage is also "latched up" for up to 250 milliseconds to allow it to be read by the flight computer. The scanner's digital state machine (used for timing) then zeroes the analog output. Note that any source brighter than the threshold voltage will generate an interrupt, and the instrument saturates at approximately 4.35 volts.

The scanner also has a built-in self-test capability. Here, the flight computer can command an LED inside the scanner to flash on -- this floods the detector which then (nominally) generates an interrupt and voltage as for a star. A self-test is normally commanded at the start of every star scan.

MAGELLAN MAPPING ORBIT PROFILE

To reduce cost, Magellan flies with a minimal set of hardware -- this decision brought with it a complex profile of maneuvers on-orbit. Since the High Gain Antenna (HGA) is used both for communications with Earth and for mapping, the spacecraft must maneuver to move the antenna between these uses. The use of a scanner on a three-axis stabilized spacecraft introduces the need for additional maneuvering.

In fact, a Magellan "star scan" really consists of three maneuvers: a pre-scan maneuver to a suitable attitude, the scan roll itself, and the post-scan maneuver (to point the HGA back at the Earth). Note that Magellan scans stars in pairs -- with the stars in each pair separated by approximately 90°. Also, star scans are performed in alternate directions (positive roll in one scan, negative roll in the next) to reduce the build-up of gyro scale factor error in the spacecraft's attitude knowledge. Fig. 3 schematically shows the 5 maneuvers performed in a normal (3.26 hour) orbit.

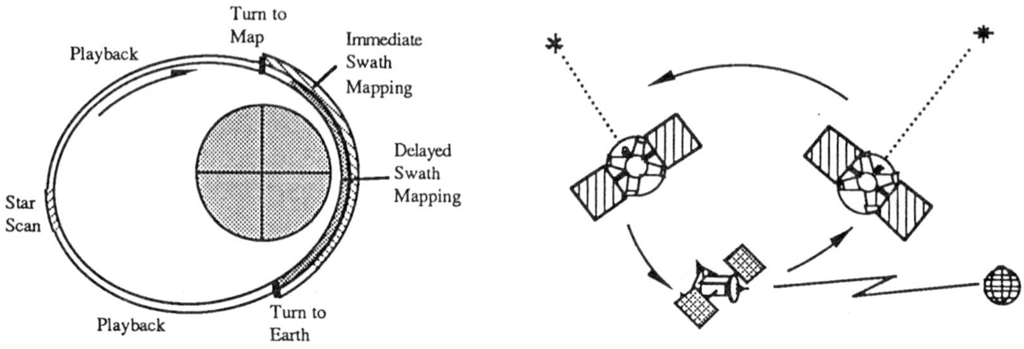

Fig. 3 -- Turns in a Magellan Orbit: Orbit Profile and Star Scan Maneuvers

A QUICK DIVERSION INTO THE WORLD OF SOLAR PHYSICS

"Daddy, where do protons come from?"

Almost all solar proton events are associated with solar flares, so let's start there. As interior heat works its way out to the Sun's surface, convection zones are created -- the Sun is very "lumpy." There are also active regions -- magnetized regions in and around sunspot groups -- here there are very violent processes at work. Meanwhile, the motion of the convection zones twists and stretches the local magnetic field lines. This process stores energy the way a compressed spring would. When this spring "snaps," energy is released explosively in a flare. By the way, flares are most likely in magnetically complex regions, or ones with unusual magnetic configuration.

Active regions are often comparable in size to the distance from the Earth to the Moon; a large flare from one can release the energy equivalent of 2 or 3 billion 1-megaton H-bombs. The explosion creates a plasma at temperatures to 10,000,000°K or more -- a flood of X-rays and extreme ultraviolet light is emitted.

It is coronal material that gets accelerated during a solar flare, so it makes sense that the emitted particles should have a make-up similar to that in the Sun itself. The charged particle flux is dominated by protons (H nuclei) -- this accounts for about 98.5% of particles at 1-20 Mev. These particles can attain velocities greater than 99.5% of the speed of light, and energies up to 10 billion electron volts in a few cases. Aside from electrons, few negatively-charged particles are emitted.

Solar particle events can occur anywhere in the solar cycle, but most events (and the most energetic ones) tend to occur on the rising and falling portions of the solar cycle (in the years surrounding solar maximum). About 96% of all flares do not produce a significant proton enhancement at the Earth. Some of those flares are just too dim, but many (in spite of being visibly bright) are located such that proton effects will be minimal for the observer. It seems that proton activity (as observed at a given point) is very dependent on geometry. And so, we enter the Byzantine world of proton propagation...

Propagation: a day (or two) in the life of a proton

The generation of protons is only the first part of the story. Between the flare and the observer, there are two phases of propagation for our protons: coronal and interplanetary. In Fig. 4, coronal propagation is indicated by the highlighted arc of length theta across the Sun; interplanetary propagation is indicated by the Archimedean spiral between the Sun and spacecraft.

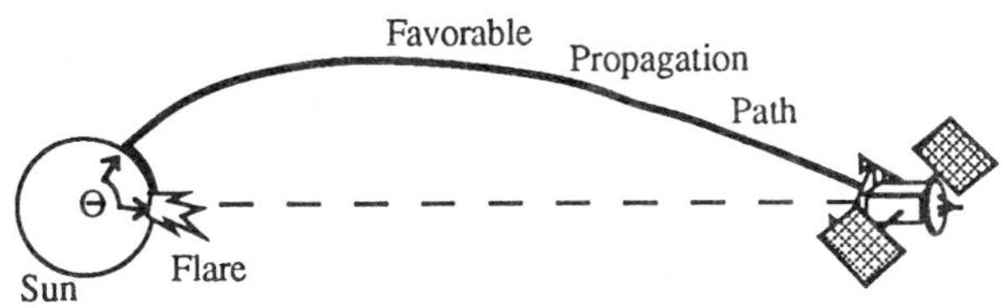

Fig. 4 -- Summary of Charged Particle Propagation.

In the coronal phase of proton propagation, the solar protons diffuse from the flare site to neighboring interplanetary magnetic field lines at their termini in the corona. Diffusion in the solar corona then leads to a solar particle gradient (this is different for each flare, but generally strength drops by an order of magnitude for each radian away from the flare site[2]). Note that coronal propagation is a stochastic process, so it's **slow** -- this phase is the dominant contribution to propagation time.

In the interplanetary phase of proton propagation, the protons (since they're charged) migrate along interplanetary magnetic field lines outward from the sun. As the interplanetary magnetic field lines form Archimedean spirals (they are forced into this configuration by the Sun's rotation), the protons corkscrew out along these lines. Particles travel with a velocity which is a function of the particle energy -- protons can attain velocities exceeding half the speed of light.

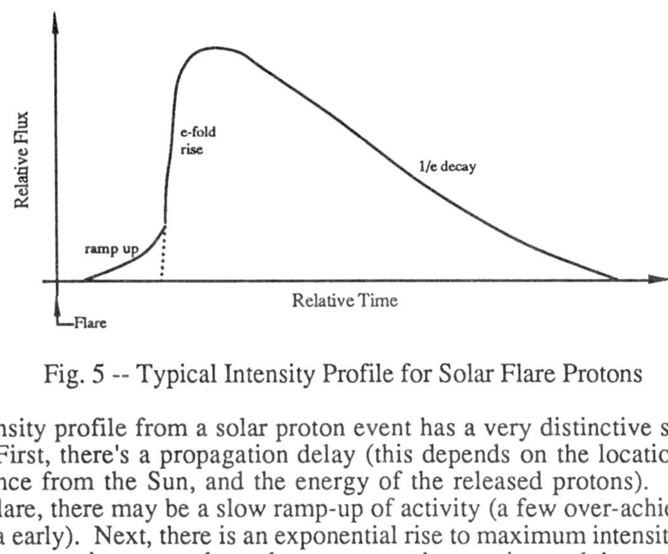

Fig. 5 -- Typical Intensity Profile for Solar Flare Protons

The intensity profile from a solar proton event has a very distinctive shape, as shown in Fig. 5, above. First, there's a propagation delay (this depends on the location of the flare, the observer's distance from the Sun, and the energy of the released protons). Depending on the location of the flare, there may be a slow ramp-up of activity (a few over-achiever protons sneak out of the corona early). Next, there is an exponential rise to maximum intensity -- the rate of this rise and intensity at maximum are dependent on coronal scattering and the geometry of the flare. At Earth, this rise time can be as short as two hours or as long as four days depending on the location of the flare. The proton event winds up with a slow decay (lasting days, sometimes weeks) to background levels.

Thanks largely to coronal scattering, proton emissions are seen sooner and brighter when the observer is situated on a magnetic field line whose "footprint" lies at or near the source flare (these are called "well connected" proton events). Coronal scattering slows and dissipates the proton flux seen along other field lines. As seen in Fig. 6, then, the protons will follow preferred paths after a flare.

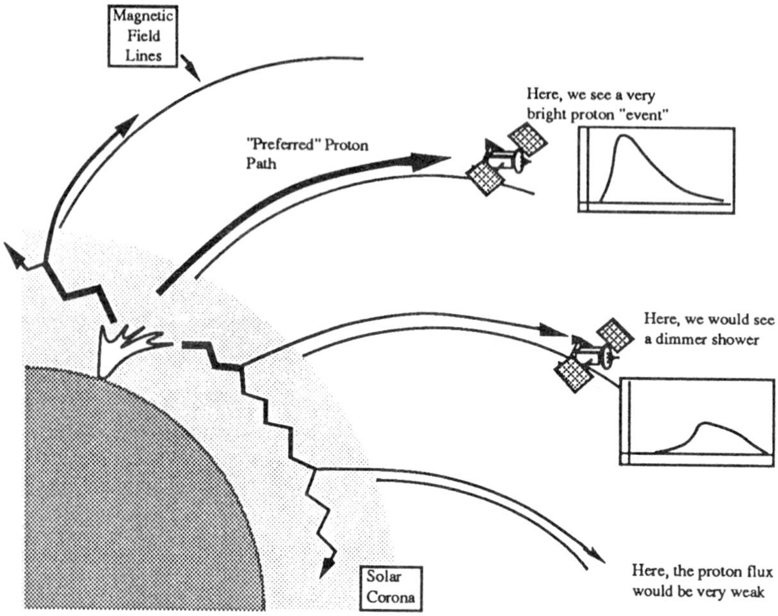

Fig. 6 -- Preferred Propagation Paths for Solar Charged Particle Emissions

The significant points of this phenomenon for spacecraft operations are that proton events last for days, and have a characteristic intensity profile over time. Coronal scattering, and the interplanetary propagation of protons along magnetic field lines leads to a channelling effect on the emitted protons. Also note that all protons are not equal when they deposit energy in a spacecraft -- low-energy protons deposit more energy than high-energy ones. Structures and electronics can even stop some of the least energetic protons -- note that protons deposit much of their energy when they stop. As a result of this, scanners and similar instruments will tend to miss low-energy (due to shielding from the spacecraft structure) and high-energy (little energy will be deposited) flares, and see ones with more moderate energies.

MAGELLAN AND PROTONS -- SO WHAT REALLY HAPPENED?

After Magellan was successfully launched, and kicked off it's IUS booster, the on-board sequence of commands attempted the spacecraft's first star scan on May 5, 1989. The star pair in use at the time consisted of Alpha Bootis (Arcturus) and Gamma Crucis. Unfortunately, this beginning was particularly inauspicious -- two slit crossings of Alpha Bootis were seen (as expected), but the output magnitudes were far from the expected values. In fact, the ratio of magnitudes for that star didn't even agree with what we knew about the hardware. We knew that for a given star, the scanner **should** output a lower voltage from the slanted slit than from the vertical slit -- telemetry indicated just the opposite. Meanwhile, the instrument didn't even detect Gamma Crucis.

It was noticed, though, that "on average" the Arcturus crossing voltages were 31% below expected values -- if the Gamma Crucis crossings were similarly low, they would be far below the threshold voltage that had been set for them. It was decided to drop the onboard "expected magnitudes" for both stars by 31%, and adjust the corresponding detection thresholds accordingly. Basically, we assumed that our understanding of the slit geometry was correct (a reversal of the image from the scanner optics would have explained some of the voltage anomaly), and commanded another scan from the ground. Fortunately this scan worked.

In the next few days, we had sporadic problems (with no particular pattern emerging) with our one-per-day star scans -- primarily on Gamma Crucis (the dimmer of our two stars). In light of our unexpected scan problems, we prepared to do calibration scans of stars later on in cruise (within a week, we had planned scanner calibrations through the first seven months of cruise). Meanwhile, we had seen one crossing value that saturated our scanner. We scrambled to see if the IUS or any celestial body had been in the scanner's field of view during that scan -- but to no avail. To compound the mystery, within days of launch, our scans became suddenly quiet -- with no spurious interrupts seen for weeks. In the interim, our attitude updates had become unstable -- one scan would undo the correction that the previous scan had made. This was not going to be a boring cruise.

By mid June, we had found the source of our attitude update problem -- the ground software that generates our Kalman gains (for attitude updates) and the flight software that uses them indexed the gains differently. This was fixed quickly -- but by now spurious interrupts were causing our scans to fail again. And the bad interrupts had an odd distribution -- as seen in Fig. 7, they were distributed essentially exponentially -- not what you would expect for some random instrument noise. Note in this figure that the distribution is modified by the scanner threshold (set at a range of values, but between 0.5 and 0.7 volts during the time of this data), and the instrument saturation (at approximately 4.35 volts). So points below the threshold will be under-represented; points above saturation will all be shown at the saturation voltage.

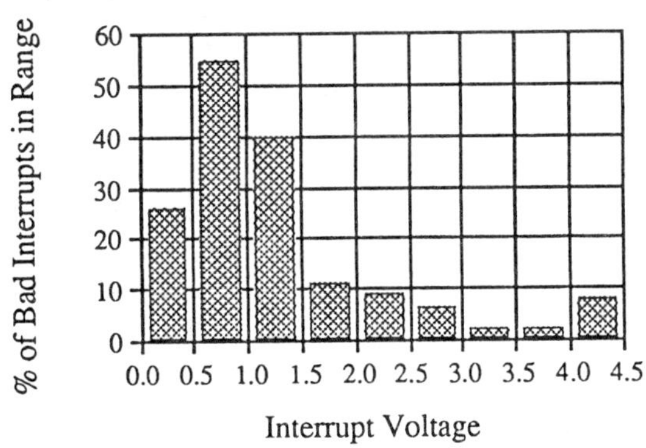

Fig. 7 Distribution of Invalid Interrupts in First 7 Months of Flight

Soon, a correlation between failed scans and solar activity was noticed[3] -- we had launched during a big flare, then the Sun went quiet for more than a month, and now the Sun was picking up steam again. We also noticed that noise in valid crossing voltages seemed to be correlated to solar activity. Most voltages were biased upwards by the noise, but some were skewed downward (some voltages were dropped enough to be below the star's threshold voltage -- these manifested themselves as "missed" crossings).

At this point, a whole range of possible causes for the spurious interrupts was proposed. Was the problem related to electrons (via ElectroStatic Discharges -- ESD) or protons (via Single Event Upsets -- SEU's)? Someone suggested that reaction wheel desaturations were kicking debris out of the thrusters. In the end, mid-energy protons (energies in the range 10 - 50 MeV) were correlated the best with both spurious interrupts and voltage noise.

By early August of 1989, a patch had been written for the flight software -- this would give us essentially real-time telemetry of star scanner voltage. The patch was written to record ten samples per second in high-rate (1200 bits per second) engineering telemetry. Since any voltage that exceeded the scanner threshold would be latched up for 250 ms, this would work for most of what we were interested in (if the voltage spike is below threshold, it will disappear in a few ms[4] -- we would have at best a 1-in-10 chance of seeing it).

The star scanner telemetry patch was finally uplinked to the spacecraft -- even then, it had to be disabled for a while when it didn't work quite right the first time. The timing of the patch's implementation was good, though -- it was finally enabled in August, 1989, just after the start of the largest solar flare in 30 years. In spite of contention for Deep Space Network (DSN) resources, and other occasional data outages, the telemetry patch gave us more than enough data to implicate proton activity as the culprit in most of the failed scans to that point in the mission.

The new telemetry also showed the same exponential distribution of voltages as had been seen previously for spurious interrupts (see Fig. 8, below).

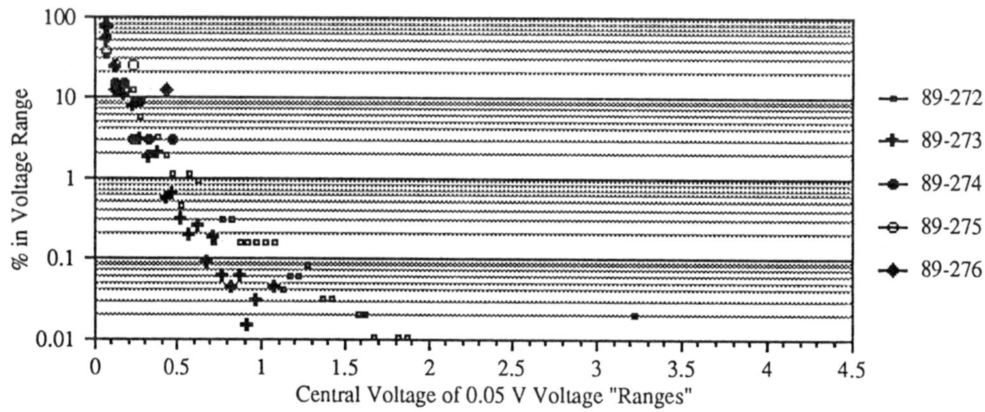

Fig. 8 Flare Voltage Distribution for 89-272 Solar Flare

Most of the proton-induced spikes were below 2 volts (although a few saturation values have been seen). Armed with our new visibility into the scanner's response, we felt we were ready to handle whatever the instrument threw at us. Unfortunately, starting with the August flare, the spacecraft experienced three exceptionally large solar flares in four months. We got concerned.

Short-term effects of solar proton events

The most obvious short-term effect of proton hits is the generation of spurious signals from the star scanner. It is a fact that the Magellan scanner's silicon diode detector is sensitive to high energy particles -- as are most semiconductors[5]. Fig. 9 shows the Magellan scanner's response to the October 1989 flare. During this period of time (before any countermeasures were taken in flight software), approximately 50% of star scans failed due to the large amount of noise present from proton "hits." Note that this plot is of data taken from low-rate (40 bps) telemetry -- here, the sample rate is only 1/30th of what we would normally see. Also note that the brightest star in our scanner's catalog, Sirius, currently generates a 2.1 volt signal in a nominal scan.

We also saw "noise" in valid crossing voltages during these flares. This is basically due to one or more proton hits on the detector while a star is crossing one of the detector slits (the voltages add). Proton hits can also drop voltages if the protons hit while the voltage for the valid star crossing is still "ramping up" -- here, when the scanner's state machine triggers an interrupt and zeroes the plate voltage, it "clears out" some of the energy from the star's crossing.

Long-term effects of solar proton events

As you would expect, there are also longer-term effects of proton impacts on silicon detectors. In particular, they will create Electron / Hole Defects (EHD's) in the detector's crystal matrix -- this can be remedied by forward-biasing the instrument's detector (if the hardware has this capability), or to a lesser degree by cycling power. When Magellan arrived at Venus, it's orbital period was slightly off from that assumed in the on-board sequence of commands. As a result of that mis-synchronization, two days after orbital insertion, the sequence commanded a scanner self-test while the scanner was pointed at the bright limb of Venus. When it noticed the resulting failed self-test, the flight software cycled power on the active scanner channel. After the original hardware configuration was restored (by ground command), the original scanner channel had regained somewhat over 1% of its original performance.

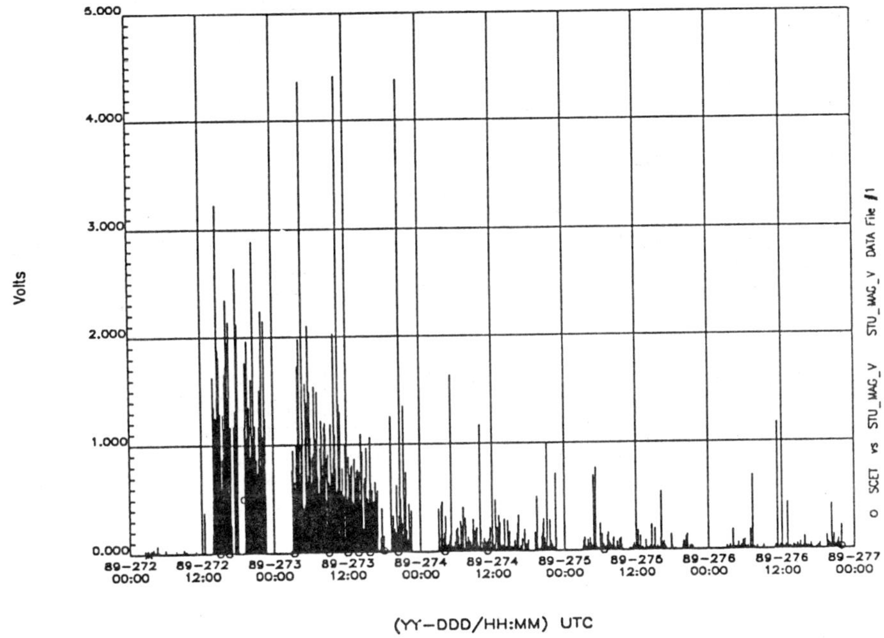

Fig. 9 Scanner Voltage During 89-272 Flare

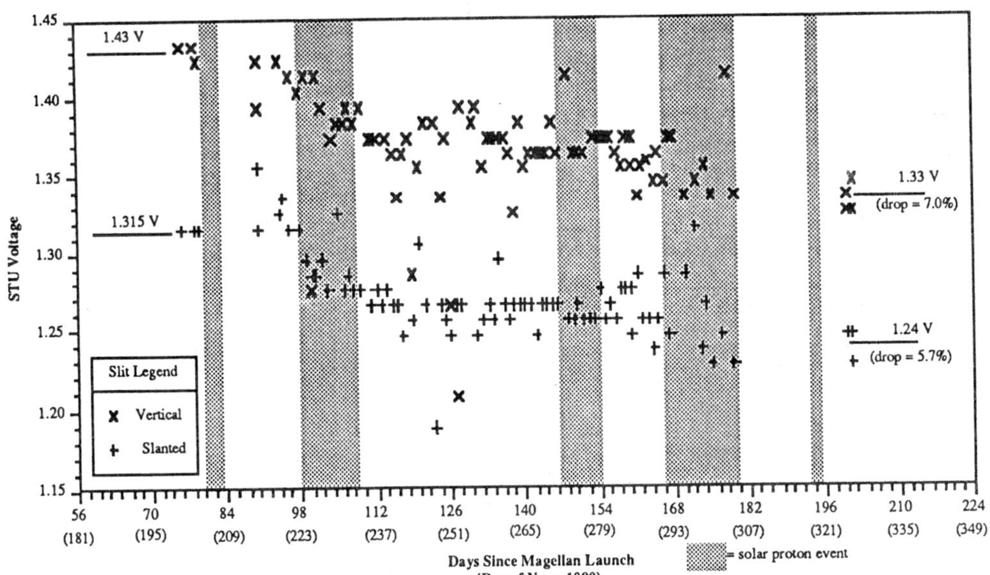

Fig. 10 Proton-Induced Degradation in Star Scanner Response to Canopus

Some proton effects on silicon, though are absolutely permanent -- in particular, crystal defects introduced by the proton impacts. The August 1989 proton event dropped solar panel output on Magellan by approximately 2% over just one day. The star scanner and sun sensors experienced a similar drop in sensitivity. Fig. 10 shows in more detail the long-term effects of proton events on star scanner sensitivity. Here, Canopus was (fortuitously) used as a guide star during much of the early part of Magellan's cruise to Venus. In this period of time, the spacecraft experienced three major solar flares in four months -- the steady use of Canopus unwittingly allowed for much better long-term trending of instrument performance. Since launch, our scanner has experienced approximately a 10% permanent loss of sensitivity -- similar to that seen for the solar panels and sun sensors (the other silicon-based detectors onboard).

BACK TO THE TIMELINE

A few failed star scans, though, weren't explainable by elevated proton levels. After all the commotion over protons had settled, and a little time was available for rational analysis, we also noticed another interesting phenomena. Even during proton-quiet periods, the scanner would often get see large noise spikes after a maneuver -- many of these were even bright enough to saturate the instrument.

Engineering telemetry had been tape-recorded during all maneuvers in the mission. To reduce wear and tear on the recorders, this data was not played back until the recorders had filled with data. When we finally received the first playback of recorded scanner telemetry, it revealed what we had come to expect -- sometimes there was extreme scanner activity during star scans on otherwise calm days. Even more suspiciously, the magnitude of the noise seemed to be dependent on the direction of the scan. Fig. 11, below, shows a fairly mild example of this maneuver-dependent noise during a star scan.

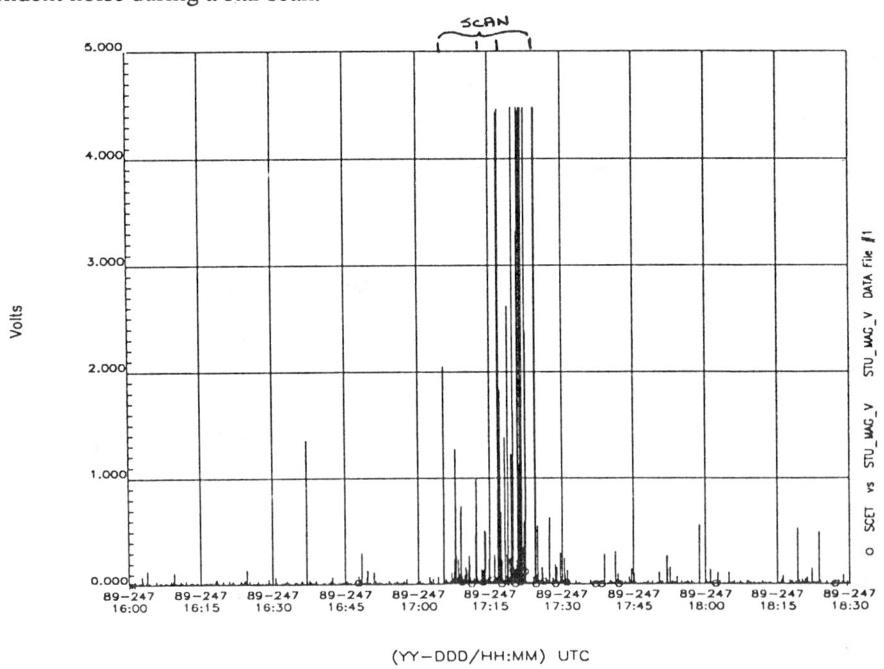

Fig. 11 Star Scanner Activity Surrounding Day 89-247 Star Scan

A whole variety of plots was created to help us see what was so special about the geometry that generated the noise. When our analysis was complete it revealed that the Sun's position was critical. In particular, when a maneuver shifted the +Y (star scanner) face of the spacecraft from sunlight to dark, or from dark to dark, the star scanner showed little or no activity. When a maneuver took the +Y face from dark to light, though, the scanner saw a flurry of activity -- often lasting for minutes after the end of the maneuver. A thermal contingency maneuver was commanded about this time -- to cool spacecraft electronics, Magellan was maneuvered to point the High Gain Antenna at the Sun. The effect of this was that almost all of the spacecraft was put in the HGA shadow. When Magellan maneuvered back to point to the Earth some hours later (putting sunlight on the +Y face in the process), the star scanner experienced a string of saturation pulses for twenty minutes after the maneuver.

By this point, it was obvious that we were dealing with two entirely different phenomena. The first, eventually tagged "Type-A," was our old familiar proton sensitivity. Type-A behavior causes a basically exponential distribution of voltages, with only a few spikes above 2 volts (even for extremely energetic solar proton events). This phenomena is unaffected by either spacecraft attitude or motion.

The second phenomena, "Type-B", is a motion-dependent beast. To create this phenomena, it is necessary to put sunlight on the +Y face of the spacecraft after letting it sit in the dark for a while. The severity of the Type-B behavior is also dependent on the length of time for which that face of the spacecraft experiences "dark-soak." This phenomena, while also having a basically exponential voltage distribution, consists of much higher amplitude pulses -- many are at or above the scanner's saturation level.

A whole parade of mechanisms was proposed to explain the new phenomena. ESD was resurrected for another run; ice in the scanner's sun shade was also proposed, as were dents in the sun shade, complicated reflections from spacecraft structural elements to dust in the shade, and electromagnetic interference from shunt radiators or other electronic components. In November, a new theory was proposed: the scanner could be seeing dust from Magellan's thermal blankets[6]. In the end, this is the only explanation that survived scrutiny.

A SECOND DIVERSION -- INTO THE PROPERTIES OF ASTROQUARTZ

Astroquartz is a woven material where each strand is made up of thousands of 35 micron diameter glass fibers. When bombarded with energetic electrons, Astroquartz conducts these electrons through itself by a process of secondary electron emission. Spacecraft surfaces exposed to the sun will emit photoelectrons that can support a net flux to counter any potential built-up on these surfaces. As a result, illuminated surfaces tend to electric potentials close to that of the environment (plasma potential). Conversely, spacecraft surfaces that are not illuminated do not generate electrons and thus cannot neutralize incoming electron fluxes. As a result, these surfaces tend to float to very high negative potentials[7]. The resulting differential charging on the thermal blankets leads to some interesting results.

There is some suggestion that glass fibers like these can break up in the presence of charged particle bombardment. Additionally, thermal shock and handling damage from before launch should generate dust from the Astroquartz blankets. When the spacecraft is quiescent, this dust sticks to the spacecraft through a combination of electrostatic attraction and, to a lesser degree, gravity. As the spacecraft rotates, illumination on the blankets changes -- thermal shock and differential surface charging on the blankets push dust off the newly-sunlit sides of the spacecraft. Some of that dust "orbits" the spacecraft to redeposit in the dark.

By April of 1990, Cal Tech was doing vacuum chamber tests of Magellan blanket material. Under an imposed electric field, it was found that visible light would prompt no particle release from crumpled Astroquartz, but in the presence of ultraviolet light, a substantial amount of activity was seen. The key factors seen in these experiments to levitate dust were (1) damaged Astroquartz, (2) an ungrounded aluminum backing, and (3) solar levels of ultraviolet light. All three factors are present for Magellan.

Additional confirmation of this hypothesis came from detailed analysis of Type-B events. At high enough "magnification," many of the pulses in those events take the shape shown in Fig. 12, below. The only consistent explanation found for these phenomena (which came to be called "worms" because of their resemblance to inch-worms) is that a bright, defocused object passed in front of the scanner's optics. Note the frequent resets in the signal caused by the scanner state machine.

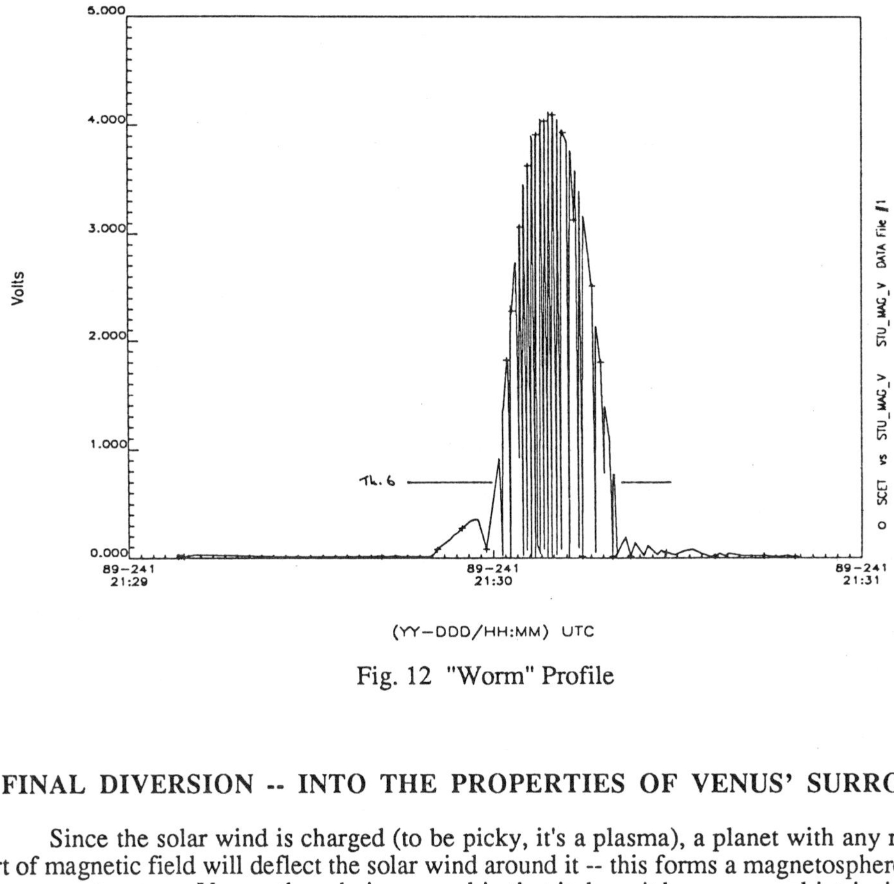

Fig. 12 "Worm" Profile

A FINAL DIVERSION -- INTO THE PROPERTIES OF VENUS' SURROUNDS

Since the solar wind is charged (to be picky, it's a plasma), a planet with any respectable sort of magnetic field will deflect the solar wind around it -- this forms a magnetosphere, bounded by a magnetopause. Venus, though, is unusual in that it doesn't have any real intrinsic magnetic field. So the only obstacle to the movement of the solar wind is the planet itself. What this leads to is that instead of a magnetopause, Venus has an ionopause. Also, Venus' bow shock is much closer to the planet's surface than is the case for the Earth and other planets (the solar wind is deflected by the planet itself instead of by a magnetosphere). Fig. 13 shows the ionopause and bow shock schematically -- but mind you, both of these beasts are very flexible creatures. The locations of the ionopause and bow shock can change fairly broadly depending on solar flux, solar activity, strength of the solar wind, etc.

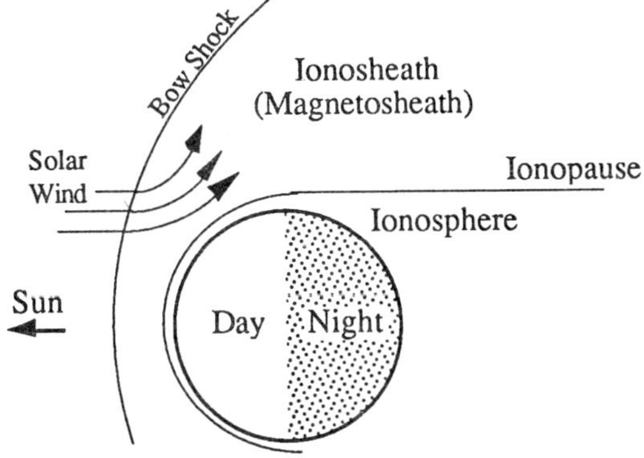

Fig. 13 Venus Electrical Environment

The bow shock is essentially a shock wave formed around the planet. This is largely a creature of the uncharged (mostly Hydrogen) part of the solar wind. The bow shock is formed because the solar wind is moving past Venus faster than pressure waves can move in the interplanetary medium (the same way a supersonic jet forms a shock wave in our atmosphere). Physical conditions will change fairly strongly across the bow shock -- mostly in terms of local particle density -- but electrical properties will change somewhat too (there is a fair amount of plasma bunched up at the shock).

The ionopause is a little trickier. This is the boundary between solar-generated plasma and that from Venus' atmosphere. At the ionopause, the particle density doesn't change as drastically as at the bow shock -- but the electrical state of the local plasma is very turbulent. The weak magnetic fields earlier discovered at Venus are generated by interaction with the solar wind. As this is an induction effect, there should be significant currents in the Venus ionosphere[8].

"Fine," you may say, "but why do we care about all this?" Remember, all that dust on the thermal blankets that causes "Type B" star scanner activity is only very loosely bound to the spacecraft -- static cling just isn't that potent in space. So it doesn't take much of a perturbation to kick this stuff off Magellan. After all the "Type-B" problems we had during cruise, there were suspicions that either the ionopause or the bow shock or both might cause problems -- we were looking for effects ahead of time.

Very preliminary results suggest that bow shock crossings possibly do loosen up a little dust from the blankets. A quick look at some early mapping orbits reveal occasional short-pulse jumps in star scanner voltage (generally greater than 2 volts) for a few seconds at repeatable positions in the orbits. This has not been an operational concern to date since the effect is fairly weak, and occurs in a limited range of orbital geometry. The bow shock is much further from the planet than the ionopause -- Magellan's orbit only crosses it during periods of periapsis occultation (for instance, in and around VOI). Basically, apoapsis has to be nearly along the line from Venus to the Sun.

Note that the positions at which this activity is noted are not correlated with either spacecraft activity (the activity occurs during playback, when the spacecraft is quiescent) or solar activity. It also occurs at similar locations whether the spacecraft is in-bound to or out-bound from periapsis. Fig. 14 shows the locations that have been plotted (for just 10 orbits from early mapping) -- these are shown in Solar Cylindrical coordinates. The location of the bow shock as measured by Pioneer Venus[9] is also included. Note that the Pioneer data was taken during a period near solar minimum (in this case, 1979 - 1981). If the Magellan data are not artifacts, they may show inflation of the ionosheath by increased solar activity.

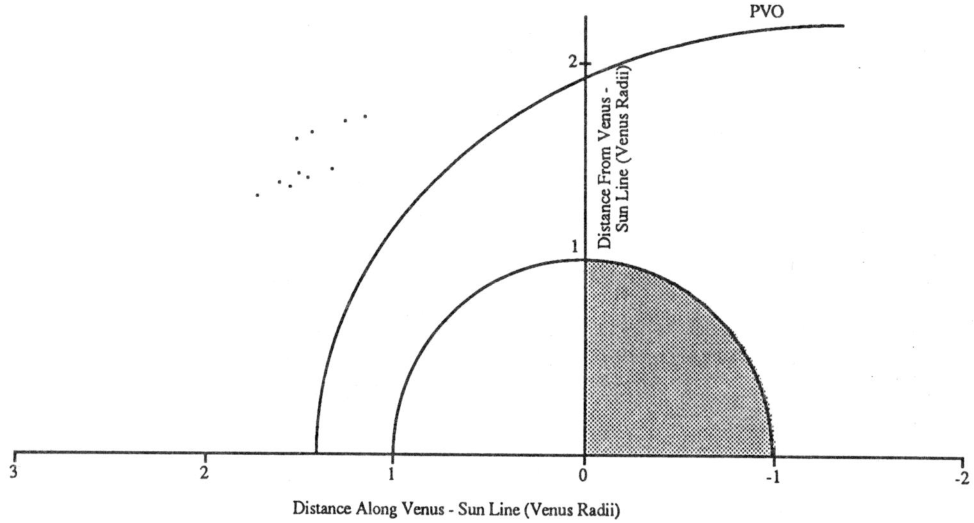

Fig. 14 Uncorrelated Star Scanner Activity During Early Mapping

Similar behavior indicates that crossings of the ionopause are also generating slight dust-related scanner noise. As with bow shock, this data is somewhat preliminary -- only a scattering of days have been analyzed. Similarly, the positions at which this activity is seen are not correlated with either spacecraft activity or solar proton events. Although the effect of ionopause crossings is much stronger than for bow shock crossings, the operational impact is still fairly limited. The ionopause is much closer to the planet than the bow shock -- so close, in fact, that most of the spacecraft's orbit stays outside of it (except at periapsis). For approximately a week on either side of apoapsis occultation, though, the spacecraft is doing star scans at or near the ionopause (this generates the particles) and out of Venus' shadow (this makes sure that they are well-lit). So ionopause crossing only affects scans for two one-week periods in a given Venus year (and then only mildly).

WHAT WE DID, HOW IT WORKS

Our real star scan problems first surfaced two months after launch -- obviously, it was then too late for a remedy in hardware. Our only options were fixes in software and in operations. Our first software patch to help mitigate scan problems was a filter on accepted scanner voltages. This "foreground filter" is a magnitude check. It brings with it an obvious trade-off: if the magnitude bounds are set too broadly the process will accept many spurious interrupts; if they are set too closely, the process will block out real interrupts with voltage noise from proton impacts. We also tightened up period of time that scanner interrupt is enabled -- only allow a fraction of a degree of attitude knowledge error to be seen. Finally, we added another piece of software to the scan analysis processes onboard. This "background filter" screens out crossings that would result in an unusually large attitude update. So instead of accepting the first 2 interrupts for a star and then doing calculations (this was the logic at launch), the processing now accepts up to 8 interrupts (with "reasonable" voltages) for that star and picks the 2 "best" ones in terms of the smallest resulting update.

The motivation behind these filters was to reduce the number of scans that are "thrown out." In actual operation, though, the filters can be somewhat of a two-edged sword: some scans are salvaged that should have been allowed to fail (resulting in small bad updates). In general, though, we now have a 95+% success record for star scans. We saw a moderate solar flare 2 days after Magellan got into Venus' orbit -- the software filters handled it's ill effects without incident. Magellan has experienced another three moderate flares since then -- with only occasional problems. The software filters have so far handled ionosphere and bow shock crossings fairly well also. Bow shock crossings have yet to cause problems with star scans. Ionopause crossing effects are seen in increased filter activity and approximately a 6% failure rate of star scans during times of the critical geometry. In the week before apoapsis occultation, two spurious interrupts managed to make their way past the software filters and resulted in incorrect attitude updates -- this situation had to be corrected by ground command.

Type-B activity tends to create too many interrupts to filter. However, since Magellan only does one star scan per orbit, an operational work-around is fairly straightforward. You just don't maneuver the scanner into light in pre-scan or scan maneuvers. This has presented a significant but "doable" operational impact -- so far, we have been able to work around the Type-B problem via proper selection of stars.

SUMMARY

Recommendations to Future Projects

If at all possible, use a scanner that contains circuitry to screen interrupts based on rise time. Note that this won't protect you against proton noise in the voltage measurement for a valid crossing.

Minimize the size of the active area of any semiconductor-based detector.

Ground all external surfaces to the spacecraft structure whether or not you use Astroquartz, whether or not you think surface charging will be a problem (we only get about 11 volts differential, and it's enough to drive dust around).

ACKNOWLEDGEMENTS

The work described in this paper was performed at the Martin Marietta Astronautics Group, Denver, Colorado under contract with the Jet Propulsion Laboratory.

The author would also like to thank Joan Feynman and Steve Gabriel for their patient efforts at explaining the mysteries of proton propagation.

REFERENCES

1. Magellan Star Scanner Performance Review Report, Ball Aerospace Systems Group, June 1, 1990. pp 13 ff.

2. D.F. Smart, "Predicting the Arrival Times of Solar Particles," in Interplanetary Particle Environment: Proceedings of a Conference, Joan Feynman and Stephen Gabriel, eds. Jet Propulsion Laboratory, Pasadena, 1988.

3. E. Seale, "More STU Trending, or... How I Learned to Stop Worrying and Love Solar Proton Showers." Martin Marietta Interoffice Memorandum, June 26, 1989.

4. Mark McKelvey, "Preliminary Report on UC Davis Proton Tests of Si:Sb 58 x 62 Arrays," Ames Research Center, 1989. IRAC-320

5. Magellan Star Scanner Performance Review Report, Ball Aerospace Systems Group, June 1, 1990 p. 19.

6. Paul A. Robinson Jr., Jet Propulsion Laboratory Interoffice Memorandum # 5217-89-182, "Possible Explanations of the MGN Star Scanner Anomalies," November 22, 1989

7. L.K. Rudolph, "Magellan Spacecraft Charging Assessment," Martin Marietta Interoffice Memorandum, October 11, 1990.

8. *Ibid.*, p. 11

9. C.T. Russell and O. Vaisberg, "The Interaction of the Solar Wind with Venus," in Venus, University of Arizona Press, Tucson, 1983.

AAS 91-073

GN&C FOR THE PEGASUS® AIR-LAUNCHED SPACE BOOSTER: DESIGN AND FIRST FLIGHT RESULTS

Daniel Rovner*

Pegasus is a newly-developed small payload launch vehicle. While the Pegasus design relies heavily on proven technology, the overall concept breaks new ground in several respects. Most notably, Pegasus is launched from a large transport-class carrier aircraft at high altitude. This launch mode confers a number of significant advantages, both for operations and performance. In particular, air-launched performance benefits from the use of aerodynamic lift to reduce gravity losses. Pegasus is unique among launch vehicles in its employment of a wing to fly a lifting ascent trajectory. As a consequence, the Pegasus control system includes features common to both aircraft and conventional launch vehicles.

The stage 1 flight envelope encompasses conditions ranging from a subsonic glide immediately after launch to hypersonic flight under 8 g's of acceleration near the end of the burn. All aerodynamic data used for trajectory and control system design was obtained from analysis. No wind tunnel testing was performed, despite the fact that in contrast to conventional boosters, aerodynamic forces have a major impact on both the performance and stability of Pegasus.

Vacuum attitude control was performed using cold-gas thrusters during coasting periods and a gimbaled nozzle during powered flight.

The vacuum portion of the Pegasus trajectory was determined by a closed-loop guidance system. The Powered Explicit Guidance algorithm originally designed for the Space Shuttle was adapted for use on Pegasus. Significant modifications were made to handle the limitations imposed by an all solid-propellant propulsion system.

The first flight of the Pegasus vehicle took place on April 5, 1990. Two satellites were successfully placed in orbit. The combination of accurate aerodynamic analysis and robust controller design resulted in a stable autopilot, despite the lack of wind tunnel data.

The guidance algorithm also performed well. Vehicle velocity after the first two solid rocket burns was more than 300 ft/sec below pre-flight predictions. This deficit was 50% larger than the performance reserve that had been allocated prior to flight. To compensate, the guidance algorithm re-targeted in real time for a lower orbital alitude. A combination of a slightly overperforming stage 3 motor, and more significantly an algorithmic error in the vendor-supplied inertial navigation system, caused the achieved orbit to be slightly elliptical. Nevertheless all payload requirements were satisfied.

VEHICLE DESCRIPTION

Pegasus is a three-stage, all solid-propellant small payload launch vehicle (Figure 1). Three new graphite composite case motors were developed for Pegasus by Hercules Aerospace Company. Because it is intended to be launched from beneath a large transport-class carrier aircraft, Pegasus also features a wing, with a clipped delta planform. Movable aerodynamic control fins, made like the wing of graphite composite, stabilize the vehicle during atmospheric flight. The first stage has a fixed nozzle, while the nozzles of the upper two stages are equipped with flex-seal gimbals to provide stabilization during vacuum flight. A cold gas reaction control system provides attitude control during coasting periods. The Pegasus avionics system features a distributed-processing telemetry architecture and a powerful flight-control computer.

* Guidance and Control Engineer, Orbital Sciences Corporation, 12500 Fair Lakes Circle, Fairfax, Virginia 22033.

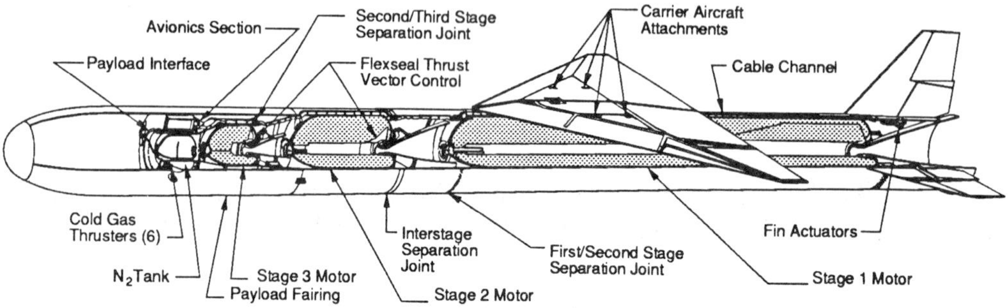

Fig. 1   Pegasus Air-Launched Space Booster

TRAJECTORY DESCRIPTION

The Pegasus flies a low flight path angle trajectory that is unique among launch vehicles (Figure 2). The initial flight regime is a high-subsonic glide at low angle of attack. For payload performance reasons, a pitch-up to a high angle of attack is initiated immediately. By the time of motor ignition 5 seconds after drop, the angle of attack is 12°. The vehicle accelerates through Mach 1 about 4 seconds after ignition. A maximum angle of attack of 20° is sustained for a number of seconds, until the upwards acceleration of the vehicle reaches a desired limiting value. Angle of attack is adjusted to hold the limiting value of upwards acceleration until the flight path angle has rotated up to a target value. The vehicle is then pitched down to an intermediate angle of attack (7° on flight 1).

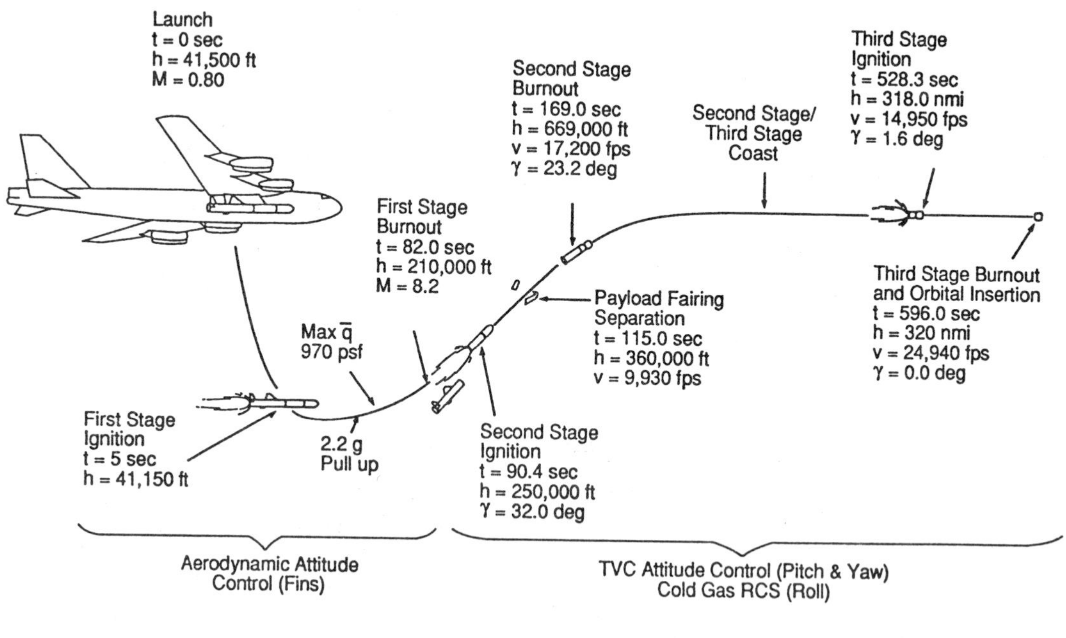

Fig. 2   Pegasus Ascent Trajectory

A gravity turn is initiated about 50 seconds after ignition, and maintained until stage 2 ignition. Some 80 seconds after drop at stage 1 burnout approximate values of altitude, velocity, and Mach are 200,000 ft, 8,000 fps and 8.0 respectively.

The stage 2 burn begins after a short coast to provide clearance between the stage 2 nozzle and the separated lower stage. Guidance calculations are started during the coast so that a converged solution is available in time for stage 2 ignition (No closed-loop guidance is used during atmospheric flight). The guidance system generates vehicle attitude commands throughout the stage 2 burn. Burnout occurs at an altitude of about 700,000 ft and a speed of about 17,000 fps.

A long coast up to the orbital injection altitude follows the stage 2 burn. To avoid any possibility of recontact and damage, the burnt out stage is not separated until just before the ignition of the stage 3 motor.

The stage 3 burn provides the velocity required to achieve the desired orbit. Often the altitude at which the burn occurs will be the perigee of the final orbit.

ATTITUDE CONTROL SYSTEM

Closed-loop control of vehicle attitude about all three body axes is maintained from the time that separation from the carrier aircraft is sensed until post-insertion payload maneuvers are completed. The sequence of different control modes used during flight is summarized in Table 1. Attitude control during atmospheric flight is effected using movable aerodynamic fins. Small fin-mounted auxiliary rockets that augment control authority are ignited as the dynamic pressure drops. During powered flight subsequent to stage 1 separation, vehicle pitch and yaw are controlled by pivoting the main engine nozzle. Vehicle roll is controlled by small cold-gas thrusters mounted on the final stage. During all exoatmospheric coasting flight, the cold-gas thrusters stabilize pitch and yaw as well as roll.

Trajectory Segment	Pitch & Yaw	Roll
Stage 1 Burn	Aerodynamic Fins	Aerodynamic Fins
Stage 1/2 Coast	Cold Gas RCS	Cold Gas RCS
Stage 2 Burn	Thrust Vector Control	Cold Gas RCS
Stage 2/3 Coast	Cold Gas RCS	Cold Gas RCS
Stage 3 Burn	Thrust Vector Control	Cold Gas RCS
Post Burnout	Cold Gas RCS	Cold Gas RCS

Table 1 Autopilot Control Modes

The strap-down inertial measurement unit (IMU) is the only sensor used for attitude control. The pair of two-degree-of-freedom tuned-rotor gyros in the IMU output incremental changes in attitude that are used as approximate measures of angular velocity. Attitude measurements are derived from numerical integration of the angular rates.

Accelerometers in the IMU provide incremental velocity outputs that are used as approximate measures of body accelerations.

The fin and nozzle actuators are both electromechanical, and are of similar design. A small D.C. motor drives through a large ratio gear set to provide the control force required. In the case of the fins, the high dynamic pressure peak combined with even a small mis-estimation of the center of pressure location can induce a substantial hinge moment. A gear ratio of over 600 to 1 was used to obtain the desired torque multiplication. The backlash and friction of the gear sets cause a significant amount of hysteresis. This non-linearity significantly impacted the design of the stage 1 autopilot.

The actuators are powered by thermal batteries. The maximum slew rate achievable by the actuators is reached when the back EMF from the electric motors is essentially equal to the battery voltage. This rate limit is a significant consideration in control law design.

AERODYNAMIC CONTROL SYSTEM

The aerodynamic autopilot provides attitude control from the time that separation from the carrier aircraft is sensed until stage 1 separation. Because the Pegasus launch vehicle features a typical aircraft configuration, analysis and design of the stage 1 autopilot followed aircraft practices. Linearized time-invariant models of the vehicle dynamics were used to evaluate candidate feedback architectures and choose control system gains. Non-linear time-varying simulations were then run to test the performance of the control system under more realistic conditions. During the synthesis phase standard aircraft equations of motion[1] were used, with minor modifications to accommodate non-zero steady-state pitch rates and the effects of jet damping. Pitch plane and lateral dynamics were treated as separate and uncoupled systems, and separate longitudinal and lateral autopilots were designed.

The architecture of the longitudinal autopilot is illustrated in Figure 3. It is a simple system using feedback of pitch attitude, pitch rate, and during some phases of flight, vertical acceleration. A trim integrator is included, mainly to deal with static torques applied by the main engine. Feedforward elevator commands are used to improve trajectory following. Prefilters applied to the pitch and pitch rate commands are intended to reduce the size of the transient at controller turn-on.

The architecture of the lateral autopilot is illustrated in Figure 4. The system features full-state feedback. The gains were extracted from the solution of a steady-state Linear Quadratic Regulator (LQR) problem using a commercial control-design software package. Comparison of LQR and classical controller designs were made during the synthesis process. The LQR was chosen for several reasons. Most importantly, during non-linear simulations the LQR controller displayed better robustness with respect to actuator hysteresis. In addition, the LQR design process was found to be much less labor-intensive than the classical design process. Classical design was greatly impeded by the high degree of coupling between vehicle roll and yaw axes induced by the disparity in respective moments of inertia and the lack of a ventral rudder.

Gain scheduling was used to compensate for the large changes in Mach number, dynamic pressure, angle of attack, vehicle mass, and vehicle moments of inertia encountered during the first stage burn. It was assumed that the shortest time scales associated with any of these changes was still sufficiently large that it could be considered infinite in comparison to the time scales associated with the vehicle dynamics. This approximation, while fairly crude, has been found to be extremely useful in common

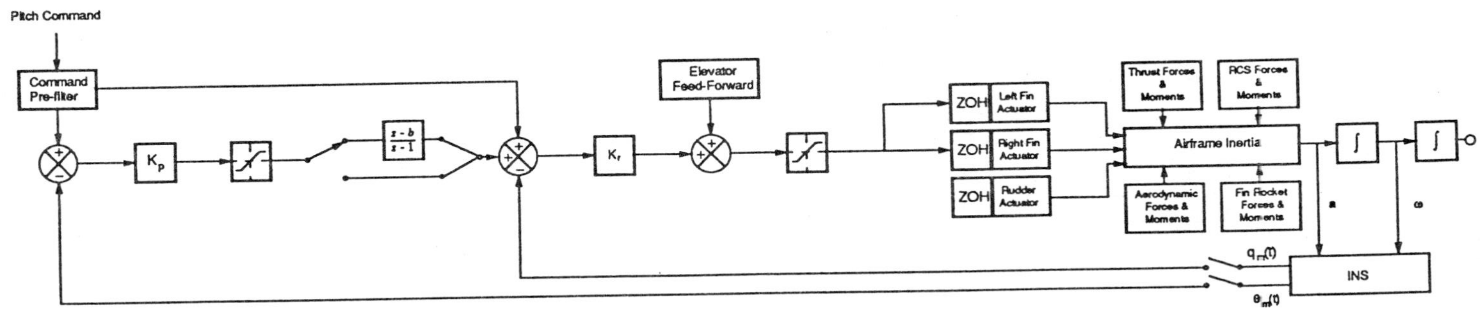

Fig. 3 Longitudinal Aerodynamic Autopilot

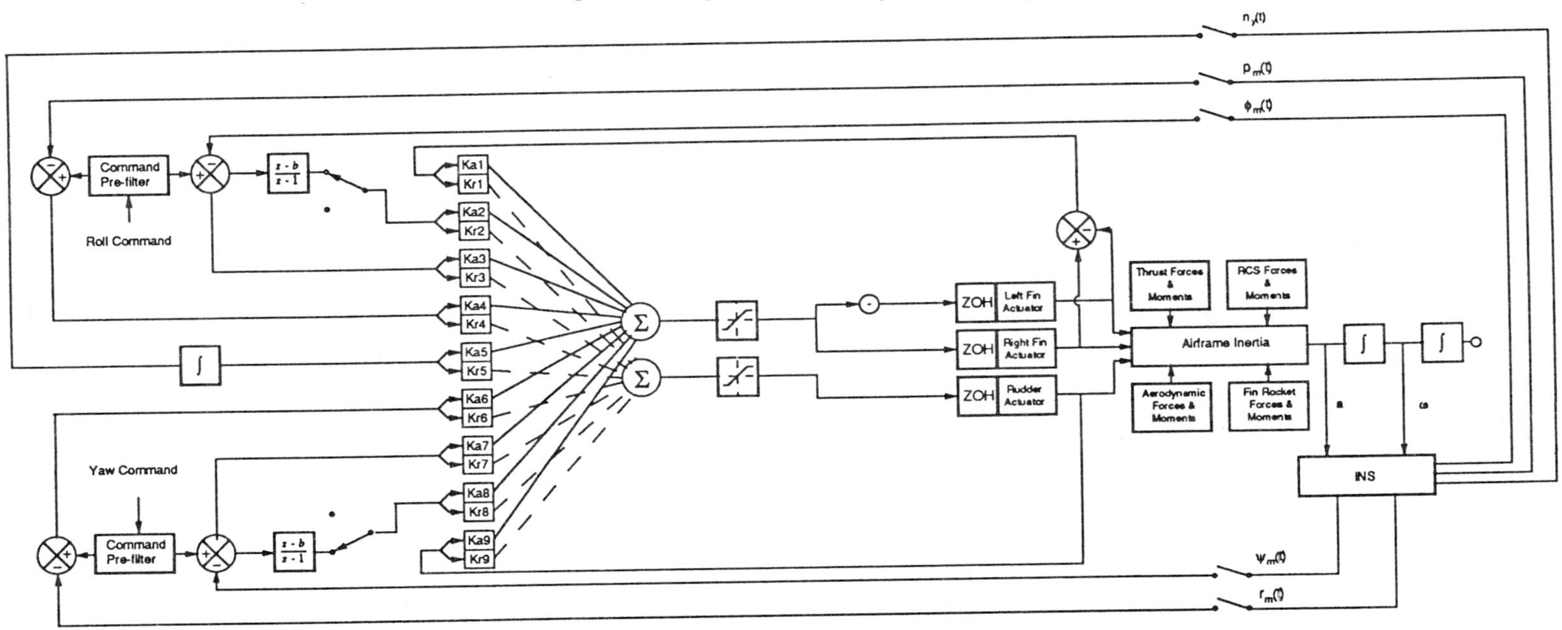

Fig. 4 Lateral Aerodynamic Autopilot

practice. Trajectory points were chosen from the results of non-linear simulations to provide data for the construction of a set of time-invariant dynamic models. A total of 12 points were used along the stage 1 trajectory.

Aerodynamic stability derivatives needed for calculation of longitudinal and lateral dynamic models were obtained from interpolation in tables of data provided by OSC's aerodynamics consultant, N.E.A.R Inc. This was the *sole* source of aerodynamic data used for stability and control analyses. N.E.A.R. estimated total force and moment coefficients as functions of Mach number, angle of attack, angle of sideslip, and control surface deflections. N.E.A.R. relied on aerodynamic prediction codes that essentially automate experimental correlations and simple analytical methods[2]. One of the most remarkable aspects of the Pegasus program is that this data was accurate enough to permit the design of a stable attitude control system.

Figure 5 illustrates the time variation of the resulting longitudinal system matrices. The s-plane locations (frequency and damping) of the typical aircraft short period mode are plotted. As dynamic pressure drops, the short period mode collapses towards the origin, and aerodynamic forces become less important. However, a significant source of disturbance torque, the fixed-nozzle main engine, is unaffected by dynamic pressure. To forestall attitude divergence, small solid-propellant control augmentation rockets were mounted in each of the movable fins. From a linearized standpoint, these rockets have the same effect on control authority as an increase in dynamic pressure. The rockets have sufficient thrust to counteract the torque generated by a main propulsion thrust vector pointed 0.2° away from the center of mass.

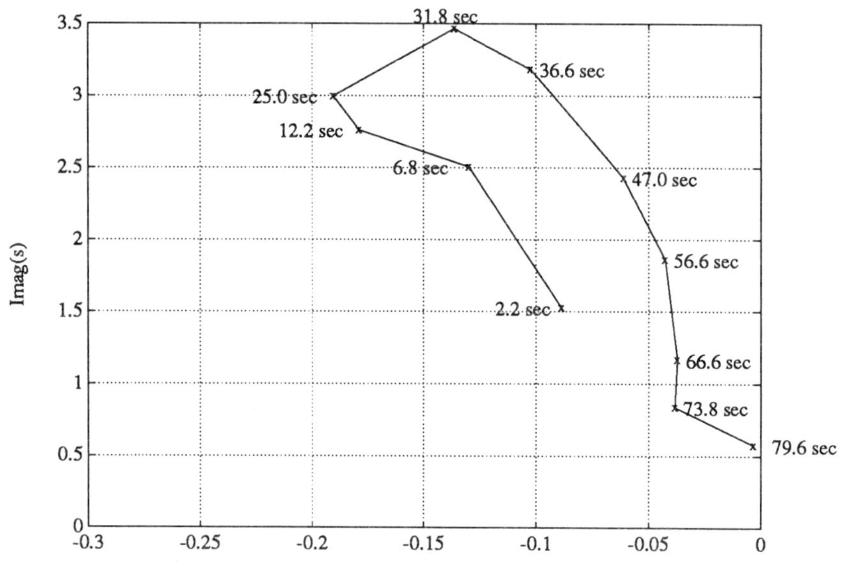

Fig. 5 Time Variation of Longitudinal Short Period Mode

GUIDANCE SYSTEM

Pegasus is equipped with a closed-loop guidance system whose goal is the minimization of errors in vehicle translational state at orbital insertion. Traditionally two basic approaches have been taken towards launch vehicle guidance. Implicit guidance methods utilize a nominal trajectory, and attempt to correct any deviations from that trajectory. Explicit guidance methods determine a trajectory that will take the vehicle from the currently measured state to the desired end-of-burn state. Explicit guidance tends to be more robust, in the sense that larger perturbations from the expected trajectory can be tolerated without appreciable increase in end-of-burn errors. However, it also tends to be computationally intensive.

Because Pegasus uses a reasonably modern flight computer an explicit guidance algorithm based on Space Shuttle Powered Explicit Guidance[3], was developed.

The trajectory constructed by PEG is completely determined by the values of a small set of parameters. An iterative process is used to solve for the values of these parameters that result in a trajectory with a desired end-of-burn state. The first step in the iteration is the prediction of the end-of-burn state for the current guess of the parameter set. This is done by turning the differential equation of motion, $\mathbf{r}''(t) = \mathbf{g}(\mathbf{r}(t)) + \mathbf{f}(t)$, into an integral equation, $\mathbf{r}(t) = \int_0^t du \int_0^u ds\, \mathbf{g}(\mathbf{r}(s)) + \int_0^t du \int_0^u ds\, \mathbf{f}(s)$. Here $\mathbf{r}(t)$ is the vehicle position vector, $\mathbf{g}(\mathbf{r}(t))$ is the gravitational acceleration vector, and $\mathbf{f}(t)$ is the thrust vector. The time integral of the acceleration due to thrust can be approximated very closely using an analytical approach. The time integral of $\mathbf{g}(\mathbf{r}(t))$, the acceleration due to gravity, should of course be taken along the actual trajectory. However, an accurate approximation to the integral can be made without knowing the trajectory solution. The next stage of the iteration compares the predicted end state to the desired end state. In the final, corrector stage, the error between the predicted and desired state is used to update the parameter set. Typically, convergence from an initial crude guess is achieved in less than 5 iterations. Thereafter, iterations are repeated at a low rate during the actual burn to compensate for off-nominal burn conditions.

A core assumption in the PEG algorithm is that the direction of thrust acceleration at any time may be expressed as $\mathbf{u}_f = \text{unit}(\hat{\lambda} + \dot{\lambda}\,(t - t_\lambda))$. Here $\hat{\lambda}$ is a unit vector in the direction of velocity to be gained, $\dot{\lambda}$ is a vector normal to $\hat{\lambda}$ that determines the rate of turning, t is a relative time, and t_λ is a parameter. This equation parameterizes a "linear tangent" turn, so-called because the tangent of the angle between $\mathbf{u}_f$ and $\hat{\lambda}$ is a linear function of time. This steering strategy may be analytically obtained as the solution to a calculus of variations problem involving optimization of final velocity during a change in altitude with assumptions of constant gravity and thrust acceleration.

The iteration proceeds to find the direction of $\hat{\lambda}$. The values of $\dot{\lambda}$ and t_λ are calculated during each pass so that position errors are nulled. Velocity errors are used to update the guess of $\hat{\lambda}$.

Significant changes to the algorithm were made to handle fixed-impulse solid rocket motors, as opposed to liquid motors with thrust cut-off capability. The time-to-go until end of burn is obviously no longer a controllable variable. Because of the high accelerations achieved during a typical Pegasus mission, the predicted end state is very sensitive to the estimated value of the time-to-go. Since propellant temperature variations affect total impulse less than burn time, changes in burn time were inferred by comparing impulse as calculated by accelerometer measurements to impulse as predicted pre-flight. Differences were used to scale the pre-flight predicted burn time. For example, if the total measured impulse n seconds after ignition was x percent greater than expected, the time-to-go was reduced by a suitable factor.

An orbital trajectory flown by a three-stage solid propellant vehicle almost necessarily follows the pattern of the first Pegasus flight, the first two stages are burned in rapid succession, a long coast is used to attain the desired orbital injection altitude, and the final stage provides the required velocity vector at the injection altitude. The primary determinants of injection accuracy are the maximum altitude of the coasting trajectory, which essentially becomes the injection altitude, and the velocity vector added by the final burn, which sets eccentricity and inclination.

In contrast to a liquid propelled vehicle, the high accelerations and short burn times of a solid greatly reduce the changes in position (downrange, crossrange, and altitude) during powered flight. This fact severely restricts the ability of the guidance system to null position errors without drastically affecting the velocity state. Given the significant performance uncertainties attendant to the first launch of a new vehicle, it seemed prudent to provide the capability of using the $\dot{\lambda}$ degrees of freedom to null velocity magnitude errors instead of position errors. This energy-management strategy makes sense if the (small) changes in altitude and longitude of ascending node that result are of less concern than the significant apsidal errors that are induced by an incorrect velocity magnitude. An additional degree of freedom, the time duration of the long coast up to orbital altitude, similarly has a small effect on injection altitude and a significantly larger effect on final velocity magnitude. (A sufficiently "hot" vehicle can be allowed to coast until it has actually attained a negative flight path angle, so that the excess velocity is used up in turning the velocity vector back to a horizontal orientation.) For the first Pegasus flight, this degree of freedom was used to give a "coarse" control of velocity magnitude at end of stage 3 burn, while the $\dot{\lambda}$ degrees of freedom were used to give "fine" control. In fact, an additional feedback was added to the guidance algorithm making the magnitude of $\dot{\lambda}$ a function of the predicted velocity magnitude error.

For the type of trajectory described, extensive simulation experience showed that using energy-management during the preceding stage 2 burn was not desirable. Instead, a fixed turning rate was chosen in advance, and velocity magnitude errors were simply accepted. This strategy had a very significant side effect. Velocity magnitude errors at stage 2 burnout effectively adjust the injection altitude in a direction that compensates for the off-nominal energy state that caused the error. For example, a low-energy vehicle achieves less than the desired velocity magnitude at burnout. The "apogee" of the coasting trajectory is consequently lower than nominal. The smaller change in potential energy required to reach the apogee of the coast almost exactly (it turns out) compensates for the energy deficit that caused the velocity magnitude error in the first place. Accuracy of injection altiude can be sacrificed to a certain extent in order to insure that an accurate injection velocity can be attained.

FIRST FLIGHT RESULTS

The first flight of the Pegasus vehicle took place on April 5, 1990. The carrier aircraft, a NASA-operated B-52, released Pegasus at a point approximately 50 miles off the coast of Monterey, California. Twelve minutes later two satellites were successfully placed into a polar orbit.

AUTOPILOT PERFORMANCE

The combination of accurate aerodynamic analysis and robust controller design resulted in a stable autopilot, despite the lack of wind tunnel data. While there was not sufficient lateral motion of the vehicle to permit identification of stability derivatives, static longitudinal coefficients have been fitted to the flight data. Figures 6 and 7 illustrate the match between predicted and measured values of lift and drag coefficients. Clearly the agreement is excellent. Figure 8 illustrates what appears to be the only consequence of transition to supersonic speeds at a 20° angle of attack, a small roll disturbance.

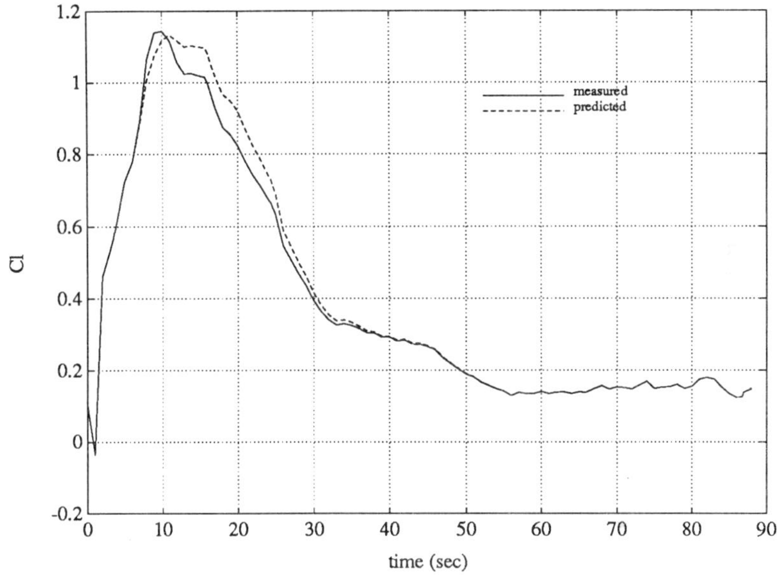

Fig. 6 Comparison of Measured and Predicted Lift

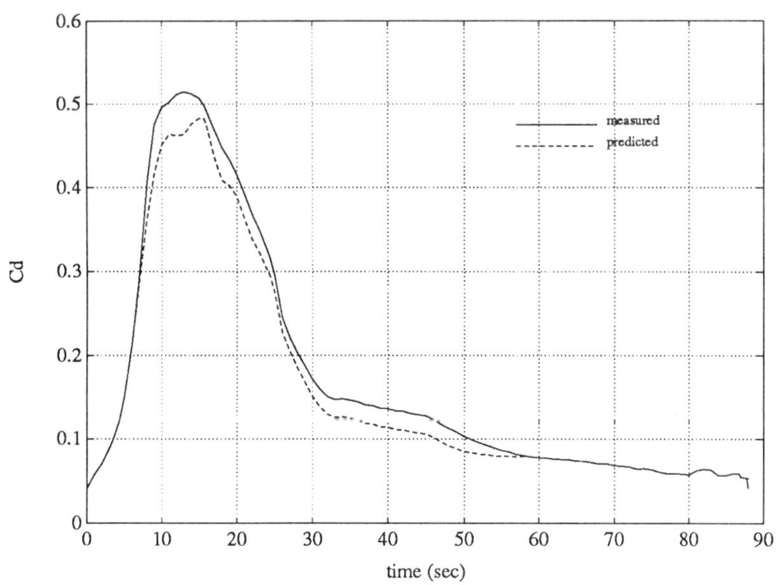

Fig. 7 Comparison of Measured and Predicted Drag

The only significant anomalies seen in the operation of the attitude control system were limit cycles that have been attributed to interactions between the control system and body bending modes. While the frequencies of the modes involved were well-predicted prior to flight, the modal amplitudes appear to have been much larger than expected. The low-pass filters that had been expected to gain-stabilize the bending modes provided inadequate attenuation. For future flights notch filters with greater attenuation will be employed.

No significant anomalies were observed during the operation of the TVC control systems on stages 2 and 3.

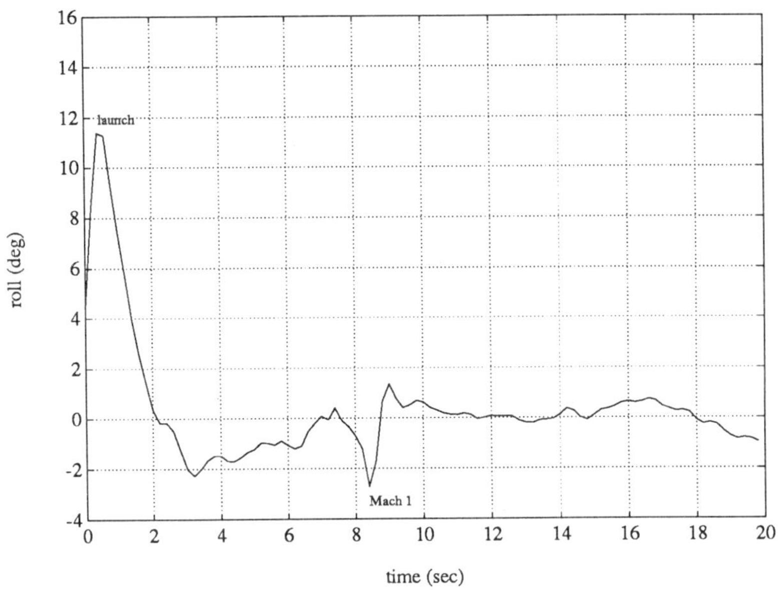

Fig. 8 Roll Disturbance at Mach 1

GUIDANCE PERFORMANCE

The guidance algorithm performed well in the face of significant perturbations. Vehicle velocity after the first two solid rocket burns was more than 300 ft/sec below pre-flight predictions. About two-thirds of this deficit was probably due to higher-than-expected aerodynamic drag. Lower than expected motor impulse accounts for the balance.

The velocity deficit was more than 50% larger than the 180 ft/sec performance reserve that had been allocated prior to flight. As had been expected, however, the apogee of the stage 2/3 coast dropped enough to almost perfectly compensate for the smaller velocity available. As a result, a nominally-performing stage 3 could have circularized at the decreased altitude at which ignition actually occurred. The energy-management planner, faced with the velocity shortfall, cancelled all stage 3 turning maneuvers and shortened the duration of the coast to optimize the stage 3 Δv.

An indication of guidance accuracy performance is obtained by evaluating the constraint errors as measured by the IMU. The constrained quantities for the stage 3 burn were flight path angle and inclination. (Injection speed would have been constrained had it not been for the velocity deficit.) According to the IMU, the inclination error was .002°, and the flight path angle error was -.077°. (In reality the IMU was in error, so that the true constraint errors were 0.15° and .6° respectively.) The main contributor to both constraint errors (as measured by the IMU) was a larger than expected Δv imparted by the stage 3 burn. The velocity vector added by the third stage burn is not collinear with the final velocity, so that a change in magnitude of the stage 3 velocity vector affects both the direction and magnitude of the final velocity vector. However, the angle between the stage 3 velocity vector and the final velocity vector lies mainly in the orbital plane (because of the effects of gravity) so that the inclination was more accurate than the flight path angle.

According to the IMU-measured injection state, the orbital perigee was less than 1 nm below the injection altitude, while the apogee was about 42 nm higher than the injection altitude. This is probably close to the eccentricity the guidance system would have achieved if the IMU had been perfect. The main contributor to the eccentricity of this ideal performance would have been the higher-than-predicted stage 3 Δv.

Comparison of radar and IMU-measured positions and velocities indicate that the IMU accuracy was significantly worse than expected. Post-flight analysis by the IMU vendor uncovered a low-altitude approximation in the kinematics of the navigation equations. This appears to explain most of the observed error. The approximation will be eliminated prior to the next flight to greatly improve guidance accuracy for all future flights.

CONCLUSION

The design and development of the Pegasus launch vehicle was successfully completed in an unusually short period of time and at an unusually low cost. The combination of innovative concepts and proven, off-the-shelf technology made this possible. The Pegasus guidance and control system was likewise developed quickly and efficiently. Innovative use was made of modern design techniques, analysis tools, and simulation resources. The resulting system is extremely simple, but robust. Despite the decision not to perform wind tunnel testing and the limited amount of other test data available, the attitude control system pointed the vehicle with the accuracy required to avoid excessive structural loads and achieve guidance accuracy targets. The guidance system successfully compensated for significant vehicle performance perturbations. As a result, the two payloads were delivered to orbits that supported full achievement of their objectives.

REFERENCES

1. Duane McRuer, Irving Ashkenas, and Dunstan Graham, *Aircraft Dynamics and Automatic Control*, Princeton University Press, Princeton, New Jersey, 1973.

2. Michael Mendenhall *et al*, "Aerodynamic Design of Pegasus," Nielsen Engineering & Research, Inc., Mountain View, CA, 1990.

3. R. L. McHenry *et. al.*, "Space Shuttle Ascent Guidance, Navigation, and Control," *J. of the Astronautical Sciences*, Vol. 27, No. 1, Jan.-March 1979 pp. 1-38

AAS 91-074

EXECUTION OF A SUCCESSFUL ROCKET PLUME OBSERVATION FROM THE LACE SATELLITE

Richard Campion*, Triet Tran*,
D. M. Horan†, H. W. Smathers† and R. E. Palma†

The Ultraviolet Plume Instrument (UVPI) was launched into space aboard the Low-power Atmospheric Compensation Experiment (LACE) satellite in February 1990. The primary purpose of the UVPI is to image rocket plumes in the ultraviolet spectrum. Recently, the UVPI successfully acquired and tracked an upper stage of a sounding rocket. The LACE satellite is gravity gradient stabilized with no orbit or attitude adjustment capability. Operating a precision pointing instrument on such a platform presents many challenging problems including attitude determination, prediction and correction; orbit predictions; and coordination with rocket launch operations. This paper describes the planning and execution of this successful observation and presents preliminary pointing, acquisition and tracking results.

INTRODUCTION

In August of 1990 an imaging instrument aboard an orbiting satellite successfully acquired and tracked the third stage plume of a sounding rocket launched from the NASA Wallops Island Launch Facility. This was the first successful launch observation by the Ultraviolet Plume Instrument (UVPI), which was designed to image rocket plumes in the ultraviolet waveband. The UVPI was designed, built and tested after the host spacecraft was designed and its fabrication begun. Therefore, the UVPI was constrained in size, weight, power consumption, and telemetry to whatever margin remained on the host spacecraft. Also, the requirement that the UVPI be capable of high spatial resolution created the unusual combination in the need for highly accurate instrument pointing on a gravity gradient spacecraft where the instrument did not have direct access to spacecraft attitude measurements. This led to solving many difficult operational problems by building in flexible software control of the instrument. This report will give a brief description of the spacecraft, a description of the UVPI and its capabilities and limitations, and discuss the planning which led to the successful plume observation.

* Bendix Field Engineering Corporation, 711 N. Fayette Street, Alexandria, Virginia 22314.

† Naval Center for Space Technology, Naval Research Laboratory, Washington, D.C., 20375-5000.

Fig. 1 The Lace Satellite

LACE SATELLITE DESCRIPTION

The Low-power Atmospheric Compensation Experiment (LACE) satellite was designed and built by the Naval Research Laboratory (NRL) in Washington, DC. The satellite was launched on February 14, 1990 into a circular orbit with an altitude of 292 nmi. (541 km) and a 43° inclination. The spacecraft weighs 3175 pounds (1440 kg). The body of the spacecraft is basically box shaped, 4 feet (1.2m) by 4 feet (1.2m) and 8 feet high (2.4m). Gravity gradient stabilization is provided by a 150 foot retractable boom with a 200 pound (91kg) tip mass emerging from the top of the spacecraft. The LACE satellite has no orbit adjustment capability. Figure 1 is a drawing of the LACE satellite. The satellite's primary purpose is to provide an orbiting istrumented target board capable of measuring the effects of active compensation of a ground based laser beam propagated through the atmosphere. The LACE spacecraft was designed to support the experiment for 30 months. NRL operates three ground stations to communicate with and control the LACE satellite. Two of the three ground stations are transportable. Built by NRL, the transportable ground stations are designed around two eighteen foot truck trailers. One trailer houses the telemetry, command and radio equipment and the other provides an uninterruptable power supply and work area. The third ground station is permanently located in Maryland. These stations provide all the command and communication links for the LACE spacecraft.

UVPI DESCRIPTION

The Ultraviolet Plume Instrument (UVPI) is carried aboard the LACE spacecraft. Its mission is to collect images of rocket plumes in the ultraviolet waveband. It is also used to collect background image data on the Earth, Earth's limb, and celestial objects. Background object imagery already collected with the UVPI includes the night Earth limb air glow, aurora, sunlit and moonlit clouds, solid Earth scenes with varying solar illumination, cities and stars.

UVPI CAMERAS AND OPTICS

The UVPI, designed and built by Loral Electro Optical Corp., comprises six subassemblies ranging in size from 4.4ft^3(125,900cm^3) to 0.3ft^3 (9,700cm^3). The total weight is about 170 pounds (77kg) and the operating power consumption averages about 170 watts. There are two intensified charged coupled device (CCD) cameras which are boresighted and share a common optical telescope. The tracker camera is used to locate, acquire, and track a target and the plume camera collects target images in the near and mid ultraviolet wavelengths. ***Figure 2*** is a schematic of the UVPI sensor head and ***Figure 3*** is a diagram of the telescope and related optics. The camera images are digitized and transmitted to the ground or are recorded onboard for later transmission. Table 1 lists characteristics of the optics and each camera. Because of the limited telemetry data rate, 3.125 Mbs, the instrument was designed to transmit or store frames in a selectable normal or a high frame rate mode. In the high frame rate, or zoom, mode only the central portion of the CCD image is retained allowing a 30Hz image rate using the same bandwidth. The normal image transmission rate is 5Hz. The photometric range and sensitivity of the UVPI cameras were selected for nighttime operations. That is, it was designed to view and track relatively bright targets against a dark background. The UVPI tracker camera can image stars as dim as visual magnitude 7.

Fig. 2 UVPI Sensor Head

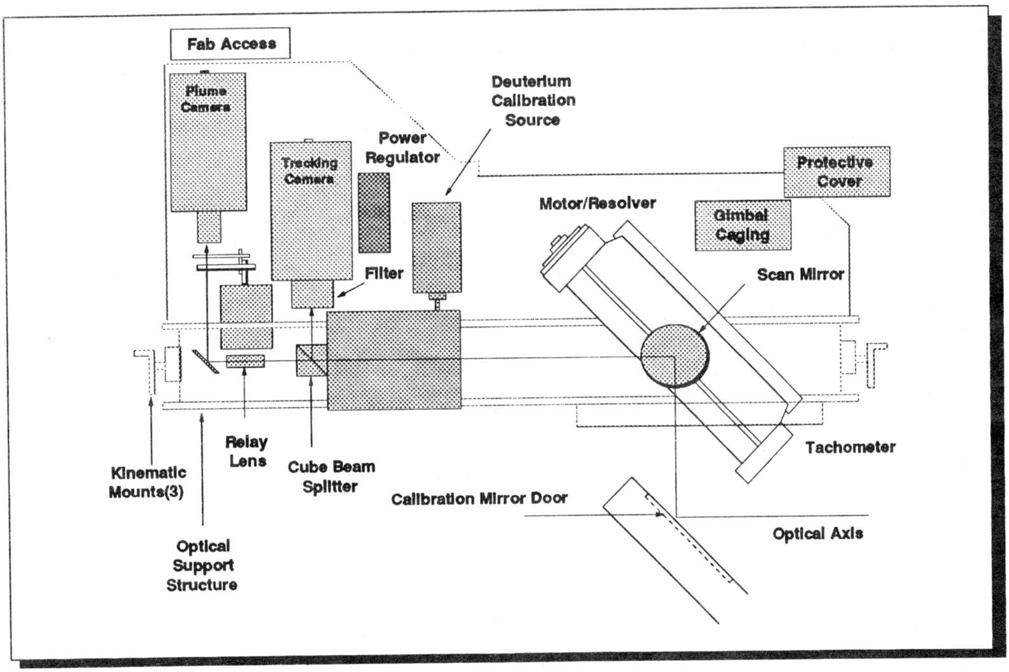

Fig. 3 UVPI Sensor Head Optics Subassembly

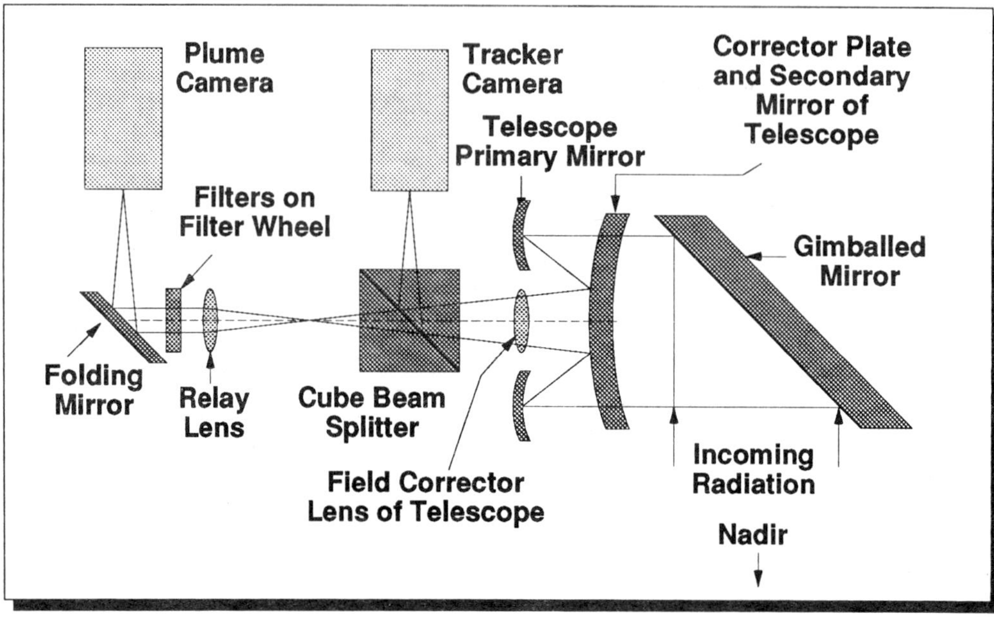

TABLE 1
CAMERA CHARACTERISTICS

PARAMETER	PLUME CAMERA	TRACKER CAMERA
Spectral region	220 - 360 nm	250 - 450 nm
Number of filters	4	1
Time for filter change	1.7 sec	n/a
Frame rate, normal	5 per sec	5 per sec
-- Reduced FOV	30 per sec	30 per sec
Super pixels, normal	251 x 244	251 x 244
-- Reduced FOV	91 x 112	91 x 112
Digitization	8 bits/pixel	8 bits/pixel
Image data rate	2.5 MB/sec	2.5 MB/sec
Field of view	.187 X .137 deg	2.5 x 1.9 deg
Field of regard	100 x 97 deg	100 x 97 deg
Pixel footprint	12.5 x 9.8 μR	182 x 143 μR
System resolution	30 μR	250 x 220 μR
Pixel exposure time	17 ms	17 ms
Frames integrated	1 - 6	1
Exposure range	$>10^{6}$	$>10^{6}$
NEI	$\sim 10^{-16}$ W/cm^2	$\sim 10^{-16}$ W/cm^2
NER	$\sim 10^{-6}$ W/cm^2/sr	$\sim 10^{-8}$ W/cm^2/sr

UVPI Field of Regard

The UVPI is mounted within the satellite and looks through an aperture in the Earth oriented panel. By use of a gimballed mirror, the UVPI has a field of regard of a 50° half-angle cone about the satellite's nadir. When the UVPI is not in use, a door covers the aperture. Attached to the inside of this door is a flat mirror which, when the door is opened part way, allows the UVPI cameras to view celestial objects or the Earth's limb. On-orbit calibration is accomplished by observing stars with a known spectral signature.

Pointing and Tracking

Pointing the UVPI gimballed mirror is possible in either open loop or closed loop modes. In the open loop mode an onboard computer uses a pointing function to calculate a sequence of desired azimuth and elevation angles for the gimbals. The gimbal servo then adjusts the mirror until the mirror angle readouts match the desired angles. The pointing function which calculates the desired angles consists of two segmented polynomials, each of the same order ranging from three to six. Up to 255 pointing functions can be transmitted and stored in the UVPI and the desired function selected by command. In the closed loop mode the UVPI imager determines tracker angles by measureing the angular displacement between a target and the line of sight centor of the tracker field of view. When the UVPI tracker electronics detect a unique target in the tracker camera field of view, the centroid of the target image is computed and the gimballed mirror moved to bring the target image to the center of the field of view. Offset tracking is also and option. The tracker electronics considers a target to be all pixels in the tracker camera which are above a selectable threshold. There are three centroiding algorithms known as intensity, mass, and edge. In the intensity algorithm the centroid is the two dimensional weighted average of the target image pixels where, the weighting function is the pixel intensity. In the mass centroid algorithm all target image pixels above a selectable threshold are averaged with equal weight, and in the edge algorithm the tracking centroid is the leading edge of the target image. Although never used, the tracker also has the capability of tracking a dark target against a bright background. The UVPI has a maximum practical slew rate of about 2° per second and is capable of tracking a target with a line of sight jitter of about ±30 microradian. If the target image is lost while it is being tracked, the UVPI computer will point the gimballed mirror using a second order polynomial extrapolation function based on the recent gimbal angle history. If a target reappears during the extrapolate pointing, the UVPI will reacquire and track the target.

Joystick

As a supplement to the open loop pointing already described, there now exists the capability for a ground based operator to point the UVPI gimballed mirror using a joystick control. This joystick capability was part of the UVPI initial design, but was not fully implemented until after the spacecraft was in orbit. The purpose of the joystick control is to aid the UVPI electronic tracker in bringing a target into the smaller field of view of the plume camera. Although the UVPI was designed to track a relatively bright object against a dark background, it was recognized that many potential rocket targets would be launched during the daylight. It was believed that a person viewing the tracker camera images in real time would be able to recognize the desired target against the cluttered sunlit Earth background and use the joystick to obtain plume camera images at the target. The joystick works by correcting the open loop pointing of the UVPI gimballed mirror. The open loop pointing function serves as a moving reference point from which joystick displacements are subsequently applied. When the target is within the UVPI tracker camera field of view, the joystick operator centers the desired target by adding a small angle or angle rate offset to the open loop pointing function. If the UVPI tracker electronics then signal that the target is being properly centroided the operator can turn the tracking control over to the UVPI.

Remote UVPI Operations

All the necessary commands and pointing functions can be stored in the spacecraft to perform one or several complete UVPI observations over remote parts of the Earth without the need for real time commands. The resulting image data is recorded on a tape recorder consisting of two 3 1/2 minute reels.

Planning, Testing, and Data Evaluation

The LACE project has developed two facilities which help plan, test, troubleshoot, and evaluate UVPI observations. The LACE Operational Test Bed (LOTB) is a combination of hardware and software which simulates the LACE spacecraft, the UVPI and the LACE ground stations. All aspects of the LACE operations can be faithfully simulated including command transmission, execution, and resulting changes in telemetry values and data. In particular the UVPI is simulated using much of the breadboard built during the fabrication and testing of the space instrument. All capabilities of the UVPI can be exercised and tested using the LOTB prior to transmitting commands to the space-born instrument. This includes simulating targets and background scenes and joystick operations. The LOTB is also used to train operators, rehearse for real time commanded UVPI observations, and analyze anomalies discovered in the execution of an observation. The second facility, the UVPI Mission Planning and Assessment Center (UMPAC), provides computers, software, and personnel to help plan and evaluate UVPI observations. In addition, the UMPAC maintains a complete set of all UVPI data collected since launch and supplies all instrument calibration parameters.

Rocket Observation and Results

Target

A three stage sounding rocket to be launched from the NASA Wallops Flight Facility was identified as a cooperative rocket target for a UVPI observation. A cooperative launch is one where the launch time and day are selected to coincide with an overpass of the LACE spacecraft suitable for a UVPI observation, but with the rocket mission otherwise is not related to UVPI. The UVPI observation requirements, which were driven by plume phenomenology requirements, identified the third stage as the primary observation target. The third stage was a solid fueled rocket which could produce an eleven thousand pound thrust. The time span from ignition of the first stage to the burnout of the third stage was 193 seconds. Since a target can remain in the UVPI field of regard for only about three minutes only the second and third stages were viewable. However, optimizing the observation for the third stage, which minimized the slant range while keeping the viewing aspect angle as normal to the rocket plume as possible, precluded viewing the second stage. Optimization of the rocket encounter is controlled by selection of the rocket's launch time.

Target Trajectory and Attitude

The target's trajectory was required for both planning and operations. ***Figure 4*** shows the rocket's height and velocity as functions of time from first stage ignition to third stage burnout. This rocket was unguided and estimates of the deviation of the rocket's position from

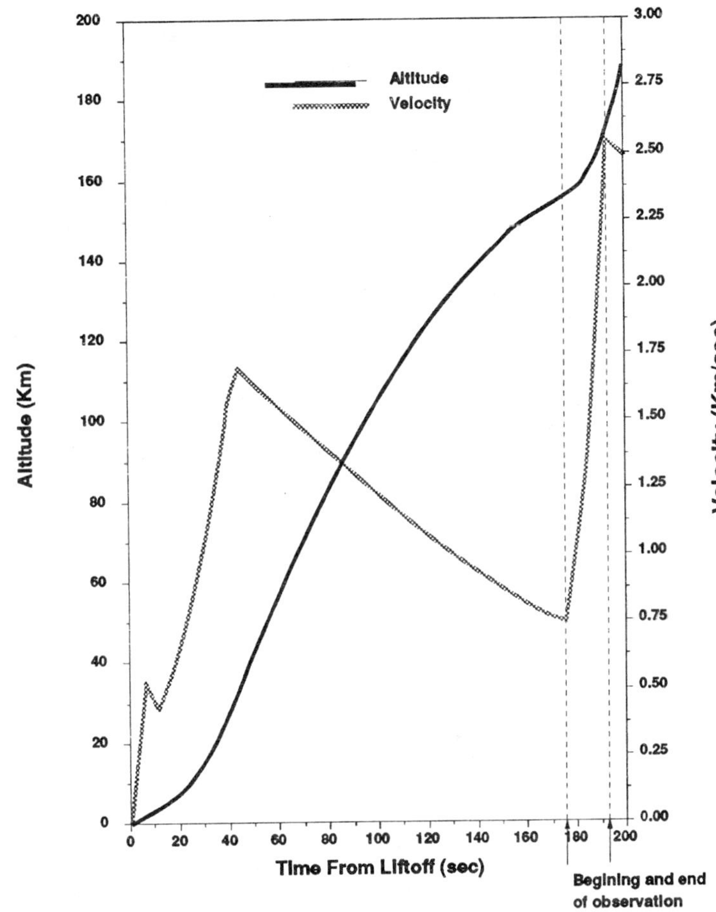

the nominal trajectory ranged as high as 41 km at third stage ignition. These estimates were based on previous flights of similar rockets. The target rocket is visible to the UVPI only by its plume and, therefore, is not visible during the coasting portion of the trajectory.

Fig. 4 Altitude and Velocity Profile of Rocket

LACE Satellite Orbit

Figure 5 shows the ground trace of the LACE orbit on the satellite pass during which the rocket was launched. Also shown is the communication horizon, 5° elevation, of the LACE ground station controlling the spacecraft during this observation. During normal operations, routine LACE orbital elements produced by the Naval Center for Space Surveilance (NAVSPASUR) is used by the LACE project. For critical UVPI observations, an arrangement has been made with NAVSPASUR to provide a precision LACE ephemeris. This

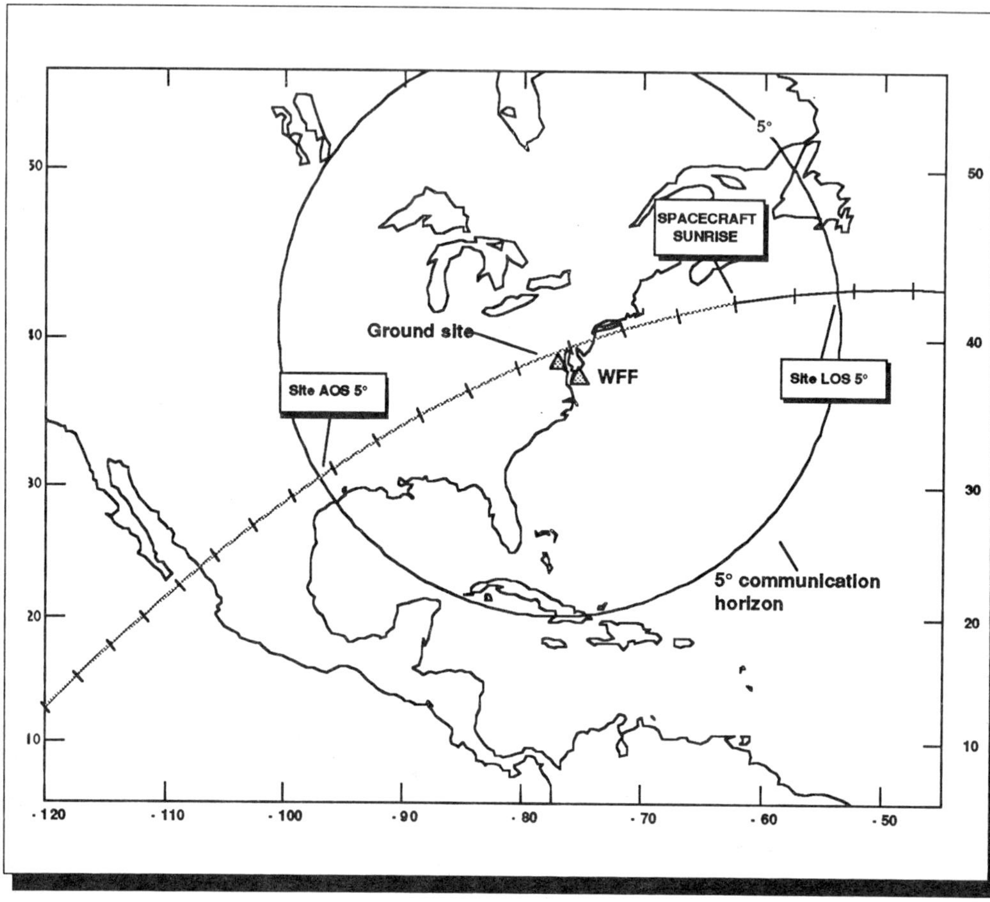

precision ephemeris is provided about three hours prior to a UVPI observation and gives the spacecraft position with an accuracy of about ±1.5 km.

Fig. 5 Ground Trace of the Lace Spacecraft on Launch Observation Pass

LACE SATELLITE ATTITUDE

The LACE spacecraft has three different attitude sensors. A five eye Sun sensor, a conical Earth scanner, and a magnetometer. With these sensors the spacecraft's attitude can be measured to about 0.2° in roll and pitch, and to about 0.5° in yaw when the spacecraft is sunlit. However, since this rocket observation and most potential rocket launches viewable with the UVPI occur when the spacecraft is in darkness, only the conical Earth scanner and magnetometer can be used to determine the attitude. With only these two sensors, the LACE yaw determination can be degraded to an unacceptable level, with yaw uncertainty sometimes exceeding several degrees. The roll and pitch values are determined when the spacecraft is in the dark primarily by the conical Earth scanner and are, therefore not significantly degraded by the absence of sunlight.

To improve the spacecraft yaw measurement, a method was developed to use the UVPI capability to view stars to estimate the LACE yaw value. A bright star in a sparse field of stars is selected for viewing by the UVPI on the same pass as, but prior to, the rocket target observation. The UVPI is commanded to look for this star using a best estimate of the spacecraft attitude. The star is selected to be near the orbit normal where errors in spacecraft pitch and roll values have minimum effect. The UVPI then scans the stellar scene in the yaw direction. When the target star is at the center of the UVPI field of view, the satellite yaw value can be computed from the UVPI gimbal angles.

Measurement of the spacecraft yaw in real time prior to the rocket observation is used as an adjunct to the method of predicting spcecraft attitude. A few hours prior to the rocket observation, the satellite attitude history is collected through one of the LACE ground stations. This data is used to predict what the LACE attitude will be at the time of the rocket observation. When the LACE spacecraft, which is gravity gradient stabilized, is in steady state the roll, pitch, and yaw values oscillate in a nearly sinusoidal pattern about a mean value with a period equal to the orbital period. A technique has been developed to provide a predicted LACE attitude five or six hours into the future with an accuracy of about 0.3° in roll and pitch and about 0.5° in yaw. The error in predicted yaw was further reduced by using a star scan prior to the rocket observation. Predicted roll and pitch values were used.

UVPI POINTING TO THE ROCKET TARGET

Several hours prior to the launch of the rocket, all the LACE and UVPI commands necessary to turn on the instrument, initialize it's computers, open the UVPI door, and conduct the observation were transmitted to and stored in the spacecraft. At acquisition of the spacecraft signal over the ground station located in Maryland minutes prior to the rocket launch, the UVPI was already pointing to a preselected star. Newer UVPI pointing functions to scan the star and to view the rocket target were then transmitted based on the predicted LACE attitude and the precision LACE ephemeris. As the UVPI scanned the star, the spacecraft yaw was computed and another rocket target pointing function was calculated. This pointing function was then transmitted to the UVPI. Immediatly after launch, the exact launch time was relayed to the LACE ground station by telephone from the launch site. A final UVPI rocket target pointing function was then computed based on the best calculation and prediction of the spacecraft attitude, the precision ephemeris, and the actual rocket launch time. Also any offset between the spacecraft clock and the ground satation clock was accounted for in the pointing function calculation. This final pointing function was transmitted to the UVPI seconds prior to the rocket's third stage ignition. The pointing function generated included a circular scan pattern with a 1.1° diameter and a 16 second period superimposed on the nominal pointing function. This scan was used to increase the probability that the third stage of the rocket would come into the UVPI tracker field of view sometime during its seventeen second burn. The inclusion of this scan pattern was motivated by the possibility of a large uncertainty in the target position. The third stage ignited outside the tracker camera field of view and was brought into the field of view by the scan. See ***Figure 6*** for a graphical representation of the path that the rocket image followed in the UVPI tracker camera.

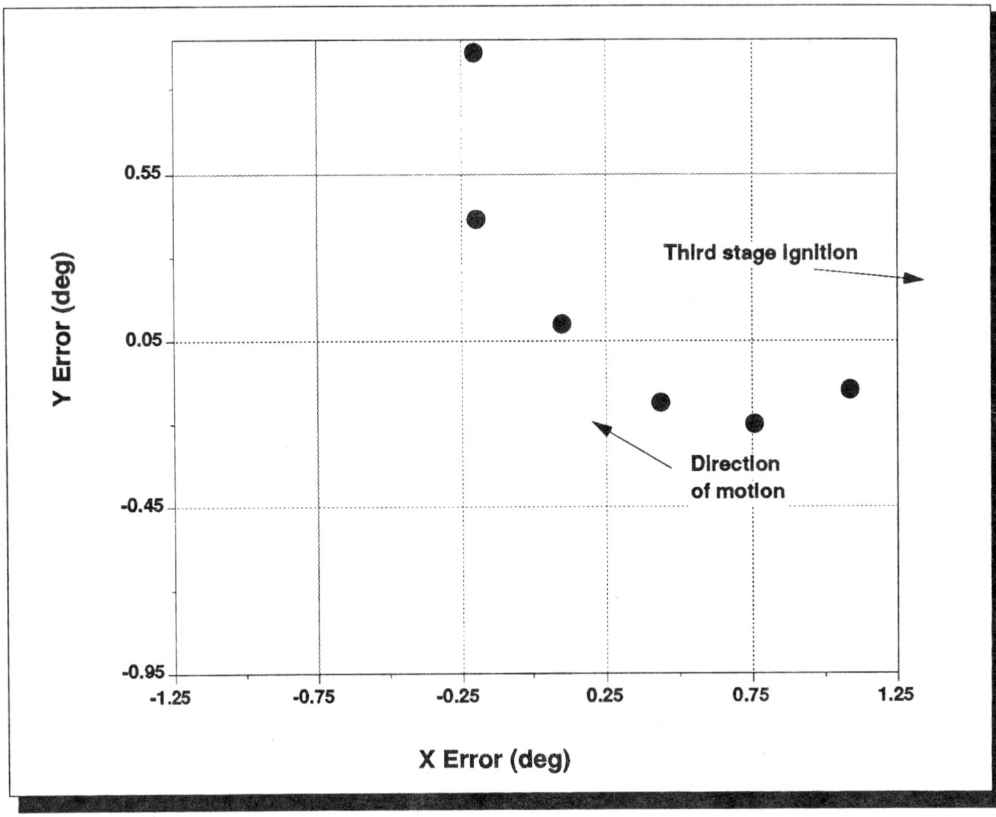

Figure 6: Third Stage Trajectory in the UVPI Tracker Field of View

UVPI TRACKING THE ROCKET TARGET

When the operator at the ground station observed the rocket target in the UVPI tracker camera's field of view, he commanded the UVPI to acquire the target. Each real time command sent to the UVPI takes about 3 seconds from the time it is sent until execution is begun by the UVPI. After execution of the the acquire command is begun, it takes another three-quarters of a second for the UVPI tracker system to bring the target to a steady track in the center of the plume camera field of view. After steady track was achieved, a change in the ratio of transmitted plume-to-tracker camera images from 2-to-8 to 8-to-2 was commanded, and then the command to transmit images at the 30 Hz frame rate was sent. The UVPI observed about 15 seconds of the 17 second burn of the third stage, with about 7 seconds in closed loop tracking. ***Figures 7*** and ***8*** show the slant range and aspect angle between the UVPI and the rocket target during the third stage observation. The aspect angle is the angle between the rocket's longitudinal axis and the UVPI line of sight to the rocket. An aspect angle of 0° means the UVPI is looking at the nose of the rocket.

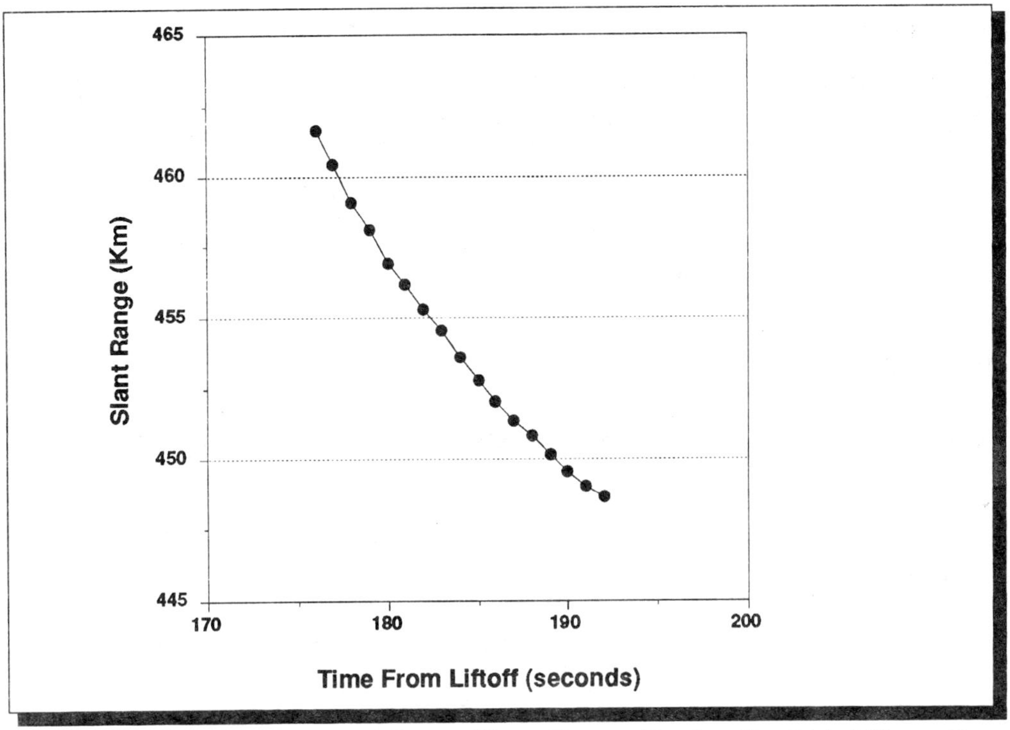

Fig. 7 UVPI to Third Stage Slant Range During the Third Stage Burn

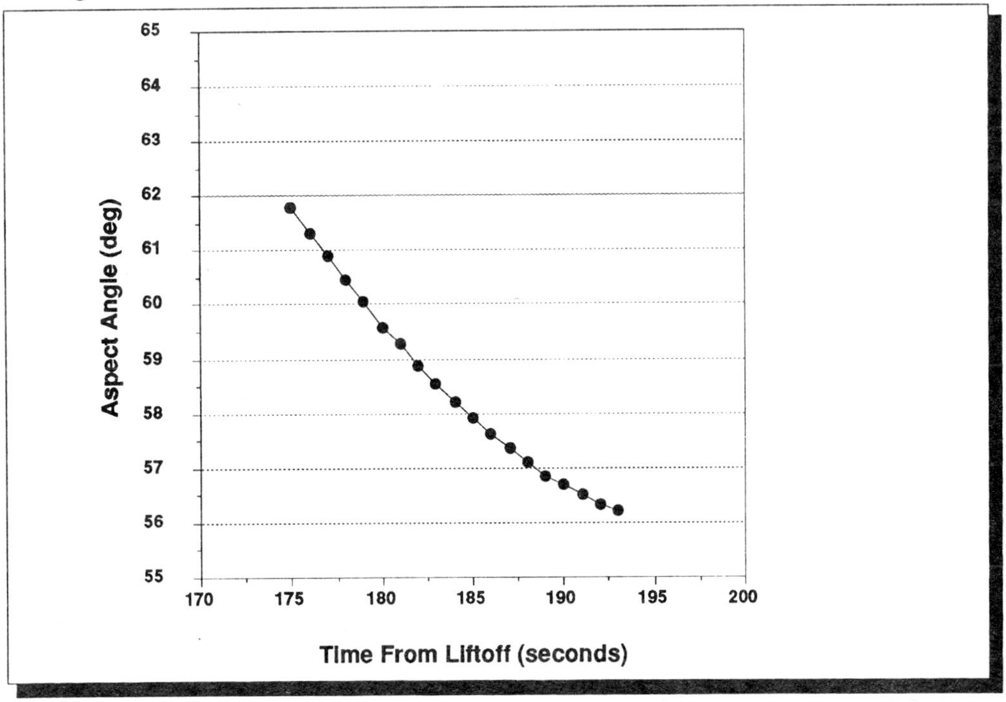

Fig. 8 Aspect Angle Between UVPI LOS and the Longitudinal Axis of the Third Stage

CONCLUSION

The Ultraviolet Plume Instrument successfully pointed to, acquired and tracked the third stage of a sounding rocket commencing 175 seconds into the rocket flight. Fifteen of the 17 second rocket burn was viewed. Ten seconds were tracked during which plume images in the ultraviolet spectrum were collected. This observation was the culmination of planning and operational procedures which were required to overcome pointing challenges which are inherent to the UVPI and its host satellite. The UVPI was used to help determine the spacecraft attitude. The ability to make real time corrections to the UVPI pointing function, making use of real time attitude determination and actual launch time of the rocket, resulted in a successful plume observation.

ACKNOWLEDGMENT

The authors wish to acknowledge the LACE Operations Control Facility, Mr. Michael Horan of Bendix for his spacecraft engineering support, Mr. Robert Prescott of Protasis and Ms. Meriane Werz of Mnemonics for their support planning, testing and real time support and Mr. Jim Gray for publication support.

AAS 91-075

MAGNETIC ATTITUDE DETERMINATION SUBSYSTEM PERFORMANCE ON THE LACE SPACECRAFT

S. M. Fox*, P. K. Pal*, M. L. Psiak† and M. E. Mook‡

The performance of the Magnetic Attitude Determination Subsystem (MADS) has been evaluated with respect to data obtained from the Low-power Atmospheric Compensation Experiment (LACE) spacecraft. MADS was previously developed, analyzed, and tested via computer simulations, with the intent of providing attitude knowledge for the LACE experiments. In the simplest implementation, MADS requires a magnetometer and an on-board computer, but all data processing can be done on the ground. Three-axis magnetometer measurements, in conjunction with the spacecraft ephemeris data and a geomagnetic field model, provide sufficient information to determine / estimate both the spacecraft three-axis attitude and the corresponding attitude rates.

The MADS design incorporates an extended Kalman filter to combine the series of single-vector attitude measurements into a three-axis attitude estimate. The filter estimates an attitude quaternion, angular rates, and constant (slow random walk) disturbance torques. The filter explicitly models the spacecraft attitude dynamics by integrating Euler's equations and the quaternion kinematics. The covariance is propagated via a discrete-time equation, and the state transition matrix of the system is derived via numerical differentiation with respect to the initial conditions of the system dynamics.

The filter's attitude estimate has been compared with conical Earth scanner roll and pitch measurements, and the filter has been re-tuned to reduce its errors. With the data provided for this analysis, the LACE MADS is capable of 3σ accuracies of 1.8° in roll, 2.4° in pitch, and 2.8° in yaw. The limiting factors to accuracy with respect to the disturbances seem to be the ability to correctly model inertia parameters and the magnetically anchored damper. On the measurements side, the accuracy is limited by the ability to correctly model and/or measure the Earth's magnetic field. Filter convergence never seemed to be a problem. Good attitude tracking performance was achieved even for substantial librational motion. This indicates the suitability of such a system for generating feedback signals to an active libration damper or stiffener.

* Ithaco, Inc., 735 W. Clinton St., Box 6437, Ithaca, New York 14851-6437.

† Cornell University, Ithaca, New York 14853.

‡ Naval Research Laboratory, 4555 Overlook Ave., S.W., Washington, D.C. 20375-5000.

INTRODUCTION

The Magnetic Attitude Determination Subsystem (MADS) was developed, analyzed, and simulation tested, as described in Ref. 1. The current work evaluates the post launch performance of MADS with respect to the Low-power Atmospheric Compensation Experiment (LACE) spacecraft. In the simplest implementation, MADS requires a magnetometer and an on-board computer, but all data processing can be done on the ground, as was done for LACE. The three-axis magnetometer is used to measure the local geomagnetic field, which provides the local magnetic field vector in a spacecraft fixed coordinate frame. The magnetic field in the Earth-fixed coordinate system is generated through the use of the available geomagnetic field model and the spacecraft ephemeris data. A comparison of these two magnetic field vectors provides sufficient information to determine / estimate both the spacecraft three-axis attitude and the corresponding attitude rates.

The LACE spacecraft, built by the Naval Research Laboratory (NRL), consists of a collection of detectors that will characterize low-power ground-based laser beams which have propagated through the atmosphere. On February 14, 1990, the satellite was launched by a McDonnell Douglas Delta-2 vehicle into an orbit with a nominal altitude of approximately 535 km (289 nmi), an eccentricity of 0.0016, and inclined 43.1° to the equator. The 1400 kg (3175 lb) satellite has no on-board propulsion system and is stabilized by flying in a gravity gradient attitude. As shown in Fig. 1, the spacecraft main body is a box, 1.4 x 1.4 x 2.4 m (4.6 x 4.6 x 7.8 ft); booms and panels extending from the body were deployed in orbit.

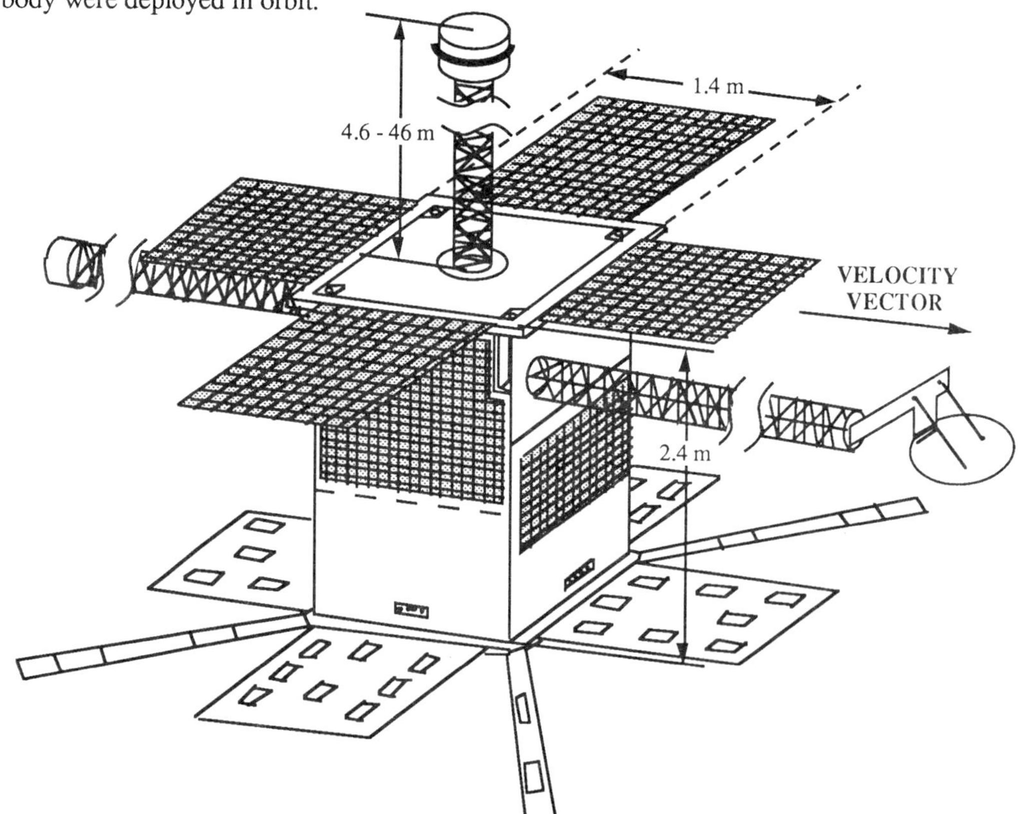

Fig. 1 LACE Spacecraft Flight Configuration

OBJECTIVE

The principal objective of this current analysis was to test and refine the previously designed MADS. This test involved real flight data from the LACE spacecraft magnetometer and Conical Earth Sensor (CES). The baseline MADS performance was to be evaluated and "fine-tuned" as required. The refined MADS filter accuracy was to be assessed with respect to the benchmark attitude data provided by the spacecraft CES. In addition, the ability of MADS to make attitude predictions was to be evaluated.

PRIOR WORK

The original purpose of MADS was to provide attitude knowledge to the LACE experimenters. The accuracy requirement for this knowledge was 1.0° (3σ) in all three axis. MADS was constrained to use only three-axis magnetometer measurements in conjunction with knowledge of the spacecraft ephemeris and a spherical harmonic model of the Earth's magnetic field. An additional requirement was for MADS to be able to make attitude predictions at times beyond the end of the available magnetometer data.

MADS Design

The MADS design incorporates an extended Kalman filter to combine the series of single-vector attitude measurements into a three-axis attitude estimate. Ref. 1 gives the details of this design. The filter estimates an attitude quaternion, angular rates, and constant (slow random walk) disturbance torques. The corresponding state vector, $\tilde{\mathbf{x}}$, is 10-dimensional. Its first three elements consist of the estimates of the inertial angular velocity vector components expressed along body-fixed axes, $\underline{\omega}$. The next four elements consist of the estimates of the quaternion for the rotation from Earth-fixed coordinates to spacecraft-fixed coordinates, $\boldsymbol{q}$. The last three elements of the filter state vector consist of estimates of constant disturbance torques acting about the three spacecraft-fixed axes, $\mathbf{n}_d$.

The spacecraft-fixed coordinate system used in the filter has +X nominally along the velocity vector, +Y nominally opposite to the orbital angular velocity, and +Z nominally along the nadir direction. The Earth-fixed coordinates are defined to have +X through the equator at the Greenwich meridian, +Y through the equator at 90° east longitude, and +Z through the North pole.

The filter explicitly models the spacecraft attitude dynamics by integrating Euler's equations and the quaternion kinematics. The gravity gradient torque and the passive magnetically anchored damper torque are both modelled explicitly. The covariance is propagated via a discrete-time equation, and the state transition matrix of the system is derived via numerical differentiation with respect to the initial conditions of the system dynamics.

Tuning of the filter's steady state performance is achieved by varying the elements of the disturbance torque and measurement-noise covariance matrices. Tuning for fast initial convergence is achieved by varying the elements of the initial state covariance matrix.

Predicted Performance

The pre-launch design and evaluation of this filter indicated good convergence properties and moderate accuracy. Convergence from initial errors of as large as 60° was achievable. The Root Mean Squared (RMS) or 1σ accuracy was predicted to be 0.42° in roll, 0.88° in pitch, and 1.01° in yaw. These accuracy values were based on random and systematic errors. They did not achieve the 1.0°, 3σ design requirement on all three axis, but the overall performance was considered to be worthy of a trial on the LACE spacecraft.

PRESENT WORK

The present work is an off-line evaluation and improvement of the MADS performance with real LACE magnetometer data. MADS can be evaluated in roll and pitch by a direct comparison to the LACE CES attitude measurements.

The work involved reviewing the performance and updating the initial MADS Kalman filter. Also, the LACE magnetometer data was reviewed, validated, and preprocessed. After this initial phase, runs of the filter were made for two sets of magnetometer data. Filter noise and disturbance covariances were modified to improve performance, and the final performance was evaluated with respect to convergence and accuracy.

Preliminary Data Preparation

The initial processing of the LACE attitude sensor data occurs at the LACE Operations Control Facility (LOCF). Here the magnetometer, sun sensor, and CES data, which were received at the ground station, are prepared for the MADS Kalman filter, as well as for other attitude determination programs. NRL personnel format and package the processed magnetometer and CES data along with attitude data calculated by the QUEST attitude determination program. This data is then transferred to ITHACO, where some additional minor preprocessing is performed.

Fig. 2 depicts the flow of the LACE attitude sensor data at the LOCF. The data is down loaded from the spacecraft's Bulk Storage Memory (BSM), stripped from the BSM file, time tagged, and submitted to the attitude determination software. This software consists of several modules, the most important of which are CONVERT and QUEST. The CONVERT module uses numerical models of the attitude sensors to "convert" the sensor data from a binary format to the format required by QUEST. The QUEST module computes the best estimate of the spacecraft attitude based on an arbitrary number of reference and observation vectors, which are the magnetic field, Earth, and sun vectors for this analysis.

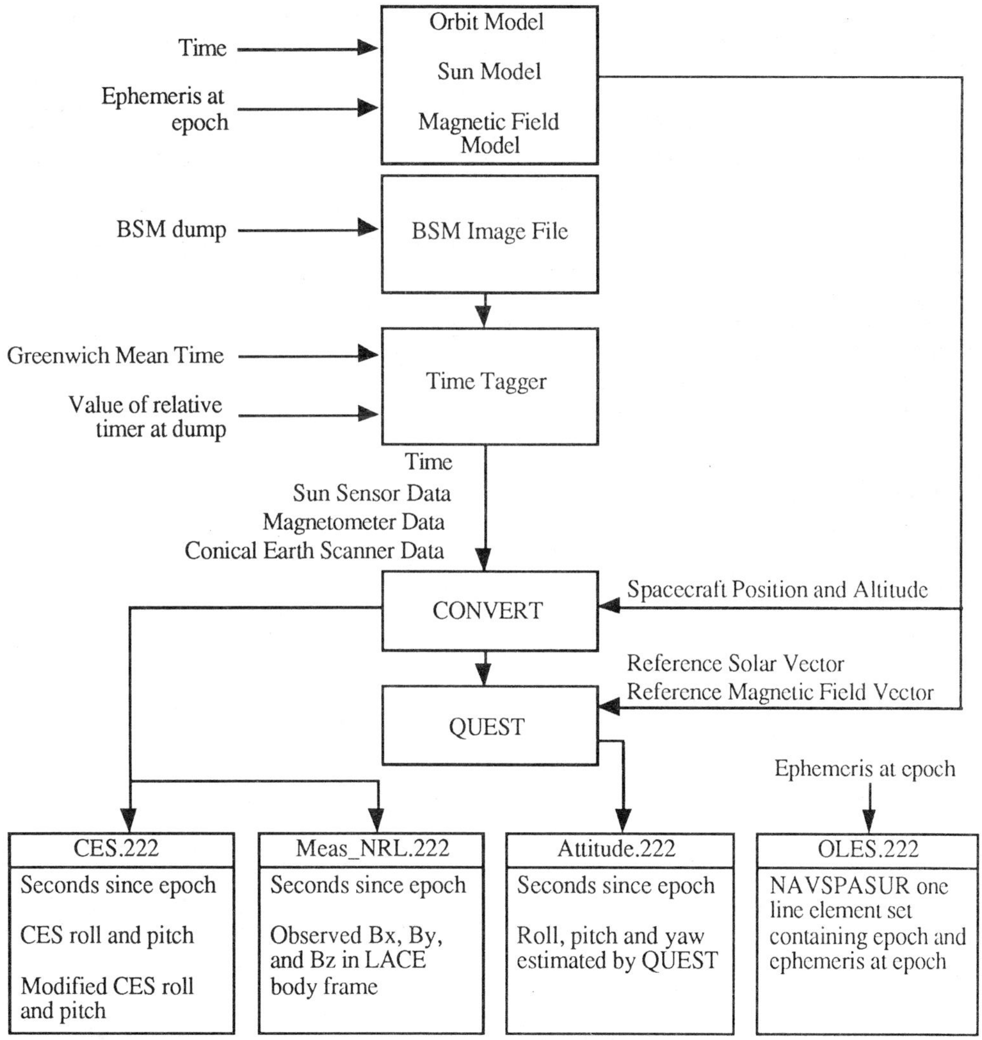

Fig. 2 LOCF Data Processing Flow for the Attitude Sensor Data

In the CONVERT module, the magnetometer model processes the binary magnetometer data to units of nanoTesla. This operation assumes a perfect scale factor linearity over the entire magnetometer measurement range, as well as zero biases. The final magnetometer values are given in the LACE body frame, which is defined with the positive X-axis along the retro-reflector boom (forward), the positive Z-axis along the gravity gradient boom (upwards), and the Y-axis complementing the orthogonal set.

The CES model adjusts the CES output data for variations in the spacecraft altitude and Earth limb, based upon an oblate Earth model and the spacecraft ephemeris. Roll and pitch are referenced to the LACE body frame. Perfect alignment is assumed between the sensor frame and the LACE body frame, but sensor misalignment angles are incorporated when the nadir vector is calculated from the roll and pitch data.

The sun sensor model converts the digital sun sensor output into a unit vector which indicates the sun orientation in the LACE body frame. Alignment angles measured during ground tests are incorporated in the transformation from the sensor frame to the body frame.

The magnetometer data which is provided to the MADS filter, the attitude data from the CES and the output data from QUEST, both of which are needed to evaluate the MADS performance, are all extracted from various points in the data flow and stored in data files, as shown in Fig. 2. These data files are indexed to indicate the orbit revolution on which the BSM dump occurred. At this point, the magnetometer data file is in the proper format for the MADS filter, with the exception of any magnetometer biases which still need to be determined and compensated. The CES data file contains the modified CES data which still needs to be adjusted for sensor misalignment.

Magnetometer Bias / Timing Error Determination

During the course of this study, it became apparent that there might be a timing bias in the magnetometer data. Such a bias would cause the MADS pitch error to have a non-zero mean. Magnetometer biases can be estimated by a batch least squares algorithm that uses the magnetometer measurements, the spacecraft ephemeris, and the spherical harmonic model of the Earth's magnetic field. The algorithm does not depend on spacecraft attitude; it only tries to match the magnitude time history (see Ref. 2, pp. 329-330). Therefore, the magnetometer bias determination algorithm was augmented to make it capable of estimating timing biases as well.

This augmented algorithm calculates magnetometer biases for an assumed timing bias error. For each assumed timing bias and calculated magnetometer bias vector, there is still some residual difference between the modelled field magnitude and the magnitude of the measured / unbiased (or de-biased) field. For each timing bias, the RMS value of this residual magnitude error is calculated. Finally, the timing bias is adjusted to minimize this residual RMS error.

The algorithm was run for three data sets that were generated by NRL: **Data.0222**, **Data.1663**, and **Data.1696**. The first file contains about 5 orbits worth of data, with magnetometer samples spaced at 60 second intervals. The first orbit in this file is LACE revolution (orbit) 222. The last two files contain 5 orbits and 11 orbits worth of data respectively, both with 100 second sampling intervals. The second file starts on revolution 1663; the third file starts on revolution 1696.

The resulting estimates of the magnetometer bias and timing errors for these data sets are reported in Table 1. The magnetometer biases are in the LACE system coordinates. The table also gives standard deviations of the magnetometer bias estimates and a weighted average of the magnetometer biases. To calculate these standard deviations, uncorrelated random errors were assumed: approximately 200 nanoTesla (nT) RMS for the field model magnitude and another 200 nT RMS for the measurement error on each magnetometer channel. While these statistics may not be completely accurate, the standard deviations in the table indicate that enough measurements have been made at enough different conditions to give a good estimate of the biases.

Table 1
BIASES / TIMING ERRORS FOR THREE SETS OF LACE MAGNETOMETER DATA

Data Set	Δt (sec)	$[\mathbf{b}_{bias_{opt}}]_x$ (nT)	$[\mathbf{b}_{bias_{opt}}]_y$ (nT)	$[\mathbf{b}_{bias_{opt}}]_z$ (nT)
0222	-5.0	237.9 ± 32.1	360.4 ± 20.1	770.9 ± 18.6
1663	-7.95	153.8 ± 41.0	466.2 ± 26.8	737.2 ± 25.1
1696	-6.5	129.6 ± 28.0	513.8 ± 19.4	782.1 ± 16.9
Avg	N/A	171.7 ± 18.8	445.4 ± 12.4	769.1 ± 11.2

Corrections for magnetometer and timing biases are important to insure proper MADS operation. The magnetometer biases alone could result in a 1.5° field direction error. After the biases are removed, the likely residual uncertainty in the biases would cause no more than a 0.2° field direction error. The timing biases could cause up to a 0.5° error in the pitch attitude estimate.

Data / Coordinate System Validation

This test compared the unbiased magnetometer output components with the IGRF magnetic field components. For comparison purposes, the IGRF field vector was transformed to orbit following coordinates: +X exactly aligned with the inertial velocity, +Y exactly anti-parallel to the orbital angular velocity, and +Z approximately toward nadir (exactly toward nadir if the orbit is circular). Nominally, the MADS spacecraft axes would align exactly with this frame; therefore, the IGRF component time histories in this frame should behave roughly like the magnetometer time histories.

Data from file **Data.1696** was used for this test. The X-axis components showed good agreement between the measured and modelled values. The components along the Y and Z axes appeared to be almost exactly inverted. This suggested either that the spacecraft was upside down or that the reference frame used to generate the **Data.1696** file was different from that used in the MADS filter. Further discussion verified the latter conjecture.

Review, Update, and Correction of the Spacecraft Dynamics Model

The spacecraft dynamics model had several inaccuracies at the beginning of the post-launch study. For one, the J2 effects on the gravity gradient terms were ignored.

Some of the attitude errors that occurred on many of the performance runs suggested that the filter's model of the passive magnetically anchored damper may be inaccurate. This issue was studied, and it was concluded that the neglect of damper dynamics contributes as much as 0.001 N-m of torque modelling inaccuracy. This damper torque modelling error is mostly in roll and pitch. The pitch torque error is primarily a 2-per-rev error; while the roll torque error is primarily a 1-per-rev error, with some higher harmonics. These error magnitudes probably effect the accuracy of the filter. A third order approximation of the damper dynamics would greatly reduce this contribution to the error, but it would make the filter more complicated and slower. Such a modification was not implemented.

Filter Evaluation and Tuning

Tuning and evaluating the filter are interactive tasks. The evaluation runs suggested possible improvements in the filter weightings. The weighting changes were made and more evaluation runs were made, etc, etc, etc.

Initial Trial Filtering Runs. Once the above mentioned problems had been corrected, the filter was tried on file **Data.1696** with the filter covariances from the pre-launch study. These produced fairly bad results, especially in pitch -- about a 1.4 degree RMS error. This was due partly to filter tuning problems, partly to timing error problems, and partly to uncorrected CES misalignments. The time-varying part of the error was larger than the time-varying part of the spacecraft pitch libration. The measurement noise covariance was increased by a factor of 16, but the pitch errors still remained large, as shown in Fig. 3.

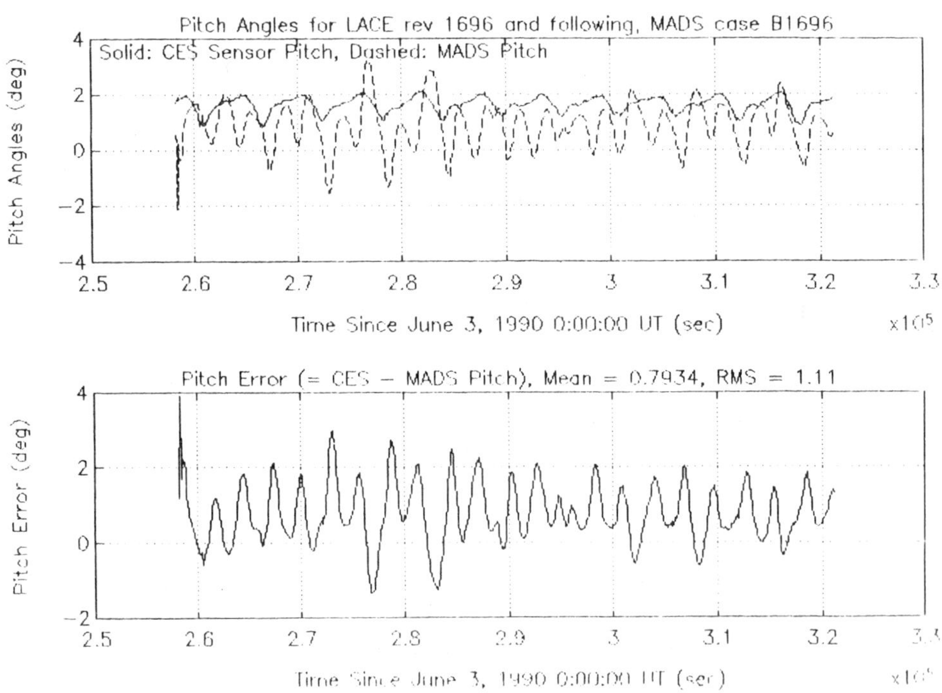

Fig. 3 CES and MADS Pitch Angles and Pitch Error for DATA.1696 With Big Q and Small R (Timing Error Uncorrected)

Linear Analysis of Systematic Errors. A linear, periodic analysis of the filter was performed to see if the errors could be explained by an aliasing-type effect of the diurnal pitch librations. This linear analysis of the steady state periodic filter is similar to that documented in Ref. 1, Section IV. The analysis showed that the diurnal torque variations could not explain the errors. Further analysis suggested the possibility that a timing error could induce a constant part of the pitch error (a non-zero average of the pitch error).

The pitch error also had a strong 2-per-rev component. The filter innovations indicated a much smaller pitch innovation than would be expected purely from a comparison with the CES data, i.e., the filter thought that it was doing much better than it actually was doing. This indicated the possibility of some sort of measurement error, probably at twice the orbital frequency. It should be noted that the residual field magnitude error also has a strong 2-per-rev component.

Re-Navigation via MAGNAV and Re-Filtering for Attitude

A batch least squares algorithm was developed which used the data from file **Data.1696** to estimate the seven orbit parameters (six Keplerian elements plus a decay coefficient) and the three magnetometer biases. (This magnetic navigation or MAGNAV technique is a variation of that described in Refs. 3 and 4.) The strong 2-per-rev error in the field magnitude suggested the possibility that the orbital plane was different than that given by the element set generated via ground tracking. In fact, the batch navigation algorithm yielded an ephemeris for rev 1696 that had an inclination that was about 1 degree larger than that reported by the tracking data.

The new rev 1696 elements and biases were then used in the MADS filter. The resulting performance was different than that achieved with the original data, but it was not much more accurate. The strong 2-per-rev component to the pitch error disappeared, but a strange sort of drift was introduced into the pitch error component, as can be observed in Fig. 4. The results of this portion of the analysis did not yield a significant improvement in the MADS performance, and consequently no further work was done in this area.

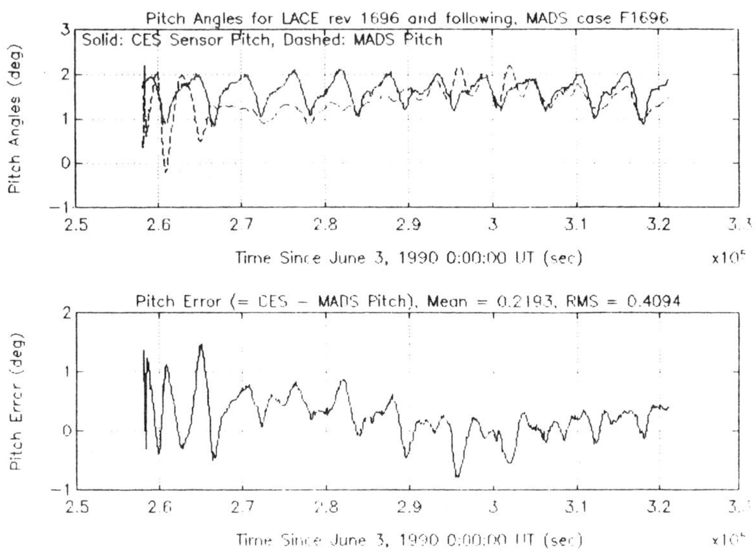

Fig. 4 MADS Pitch Attitude After Re-Navigation Via Batch MAGNAV (Comparison With CES Pitch, DATA.1696)

Filter Tuning. A comparison of the small values of the pitch innovations to the actual pitch errors suggested that the filter needed serious re-tuning. In particular, the filter needed modifications to the covariance matrices, **R** and **Q**. As detailed in Ref. 1, these matrices determine the tradeoff between the rapid tracking of state variations induced via disturbance noise, and the filtering of measurement noise. The **R** matrix needed to be increased to reflect more realistic values of the field model / measurement errors. The **Q** matrix needed to be decreased to reflect more realistic values for the unknown torques and their associated rates of change. The **R** matrix was eventually increased by a factor of 100 from the original values. The (2,2), (3,3), and (8,8) elements of the **Q** matrix were decreased by a factor of 10, as compared to the original values. Both of these changes tend to slow the filter and make it less affected by measurement-type errors, such as the suspected 2-per-rev magnetometer / field model error.

These changes led to a decrease in the pitch error 2-per-rev component, while the roll error maintained good performance. Also, the non-zero mean of the pitch error was reduced by eliminating the timing error and by applying the ground-measured CES pitch alignment to the CES data. Unfortunately, the CES roll alignment term led to an increase in the mean roll error. Fig. 5 depicts the slightly degraded roll performance, and Fig. 6 depicts the much-improved pitch performance.

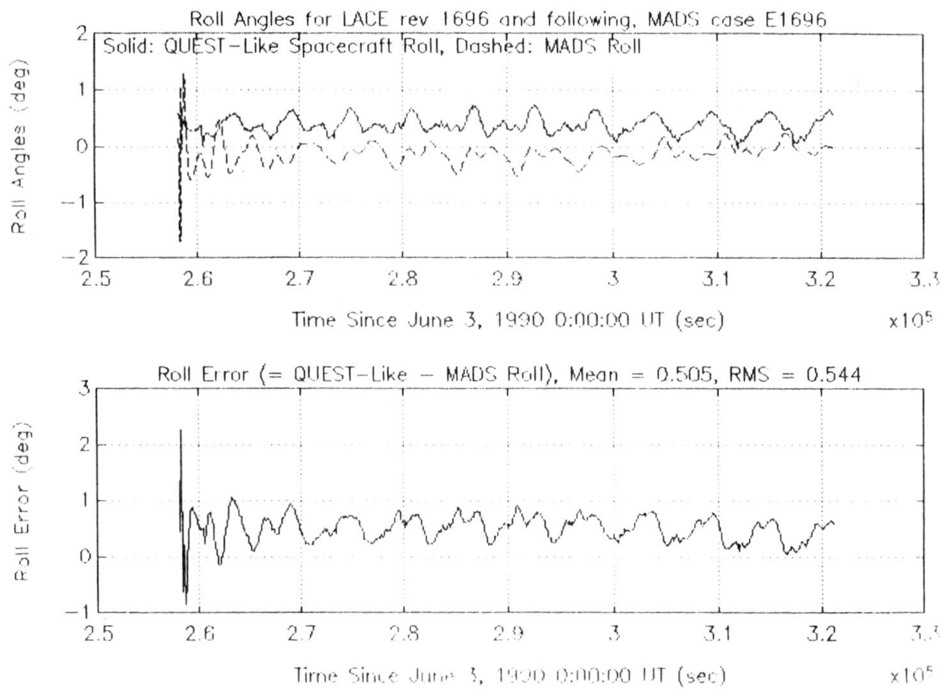

Fig. 5 MADS Roll Angle and Error With Lower Q and Larger R, Compared to CES With Alignment Corrections, DATA.1696

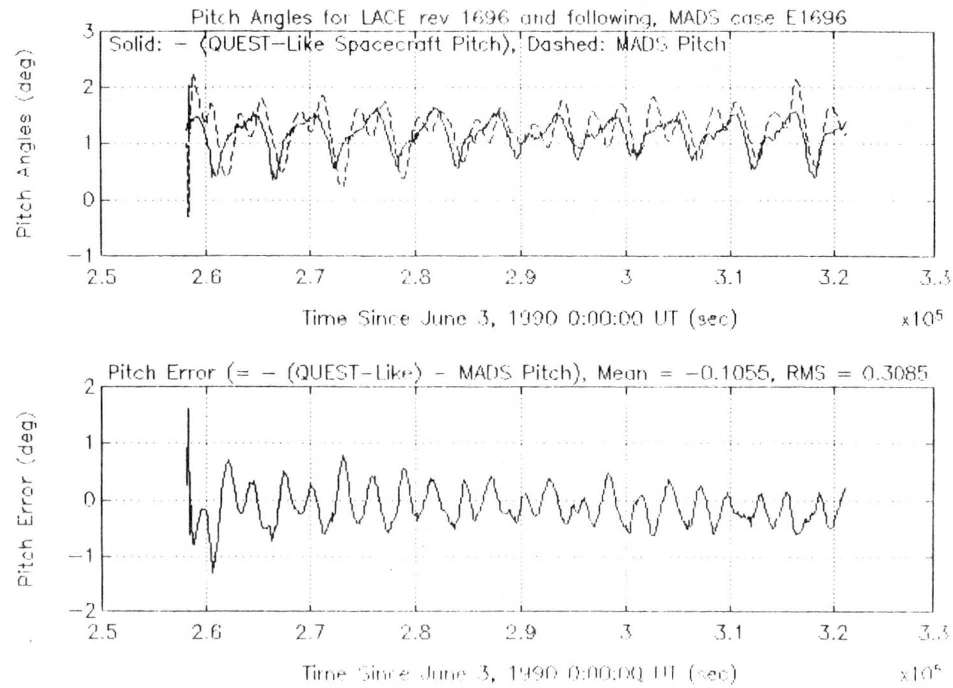

Fig. 6 MADS Pitch Angle and Error With Lower Q and Larger R, Compared to CES With Alignment Corrections, DATA.1696

The data from file **Data.0222** further affected the choice of covariance matrices. An attempt was made at running this case with effectively the same **Q** and **R** matrices as were used to generate Figs. 4 and 5 (**Data.1696**). The data sampling intervals for **Data.0222** and **Data.1696** were 60 and 100 seconds, respectively. Because the sampling interval is smaller for **Data.0222**, the actual **Q** matrix used in this trial was scaled down by a factor of 0.36 (or $[60/100]^2$). The resulting roll accuracy was degraded by a factor of 2 in comparison to that for the **Data.1696** case. The pitch accuracy was degraded by a factor of 4.

The degraded accuracy of the **Data.0222** case was due to the increased libration amplitude, coupled with inaccuracies in such spacecraft parameters as the moments of inertia. The mis-modeling / large libration effect translates into an increased uncertainty in the torques. Consequently, this suggested an increase in the **Q** matrix to inform the filter about the increased uncertainty in the torques.

After these changes in the **Q** matrix values, the MADS filter worked much better on the data from file **Data.0222**. The (1,1), (2,2), and (3,3) elements were increased by factors of almost 1000, but the (7,7), (8,8), and (9,9) elements were actually decreased by factors of about 10. The decreases made the estimated constant torque more "constant", while the increases in the first three diagonal elements left the filter better able to estimate attitude in the face of torque mis-modeling. The resulting roll and pitch accuracies were both improved by a factor of almost 2.

MADS PERFORMANCE

Reference Attitude Determination

In order to evaluate the MADS filter performance, the predicted roll, pitch, and yaw data must be compared to the "true" LACE attitude. In addition to the three-axis magnetometer, the spacecraft has the CES and a sun sensor to provide data for independently estimating the attitude of the spacecraft. The computer code called QUEST is considered to be the most accurate attitude determination routine used by the LACE ground facilities. As previously stated, QUEST computes the best estimate of the LACE attitude based upon the magnetic field, Earth, and sun measurement vectors. After adjustments for the Earth oblateness, spacecraft altitude, and alignment, the roll and pitch data provided by the CES can also be used for comparison.

The basic QUEST algorithm has been presented in several papers (Refs. 5 and 6), and is also described in Ref. 2 (pp. 426-428) as the *q* method of attitude determination. This algorithm can be applied to any number of sensors which provide simultaneous vector observations. QUEST makes use of these N vector observations to determine an optimal or "best" attitude matrix, based upon the following loss function:

$$J.(A) = \sum_{k=1}^{N} w_k \mid \mathbf{u}_{B_k} - A\, \mathbf{u}_{R_k} \mid^2$$

where
w_k = weight of the ith vector measurement
$\mathbf{u}_{B_k}$ = set of $N \geq 2$ vector measurements in the body frame
$\mathbf{u}_{R_k}$ = corresponding set of reference vectors
A = attitude matrix

The weights used here were the inverse of the sensor variances which were assumed to be equal for each output axis of a given sensor.

The weighting of the sensor measurements were initially determined by simulation. Flight data was then used to modify the measurement weights so that the roll and pitch tracked the modified CES data. The sensor variances which, on the average, provided the best in-flight correlation between QUEST and the CES roll and pitch are

CES	0.100°
Sun Sensor	0.144°
Magnetometer	0.400°

Compared to data generated via computer simulations for LACE, QUEST roll and pitch data are expected to be accurate to within ±0.5° (3σ) and the yaw prediction accuracy is better than ±1.0° (3σ).

When QUEST data is not available, the corrected CES data can be used in comparing the MADS determined roll and pitch. The accuracy of the corrected CES data is typically better than ±0.1° (3σ).

Attitude Determination Mode

In this mode of operation, MADS determines attitude based upon magnetometer measurements that continue at regular sampling intervals right up to the time of attitude determination. MADS performs well in this case. Filter convergence to apparent steady-state behavior is achieved in about one orbit. The steady state accuracy of MADS relative to the QUEST output or to the CES, as summarized in Table 2, is better than 1° RMS. It is close to the accuracy predicted in the pre-launch analysis (see Ref. 1). Figs. 7 thru 9 show the MADS performance using the data from file **Data.0222**, and the performance for the **Data.1696** data is depicted in Figs. 10 thru 12. Note that these two cases employ the appropriate disturbance covariances for 60 and 100 second sampling rates, respectively. The roll, pitch, and yaw are given in the MADS coordinate system, i.e., + roll gives right "wing" down, + pitch gives "nose" up, + yaw gives "nose" right.

Table 2
RELATIVE STEADY STATE ACCURACY OF MADS

Quantity	**Data.0222** (QUEST)	**Data.1696**† (CES)
RMS Roll Error	0.59°	0.54°
Mean Roll Error	0.36°	0.46°
RMS Pitch Error	0.78°	0.55°
Mean Pitch Error	-0.12°	0.13°
RMS Yaw Error	0.94°	-
Mean Yaw Error	-0.10°	-

† QUEST values were not available for the data from file **Data.1696**. For a comparison, CES values for roll and pitch, with alignment corrections provided by NRL, were used.

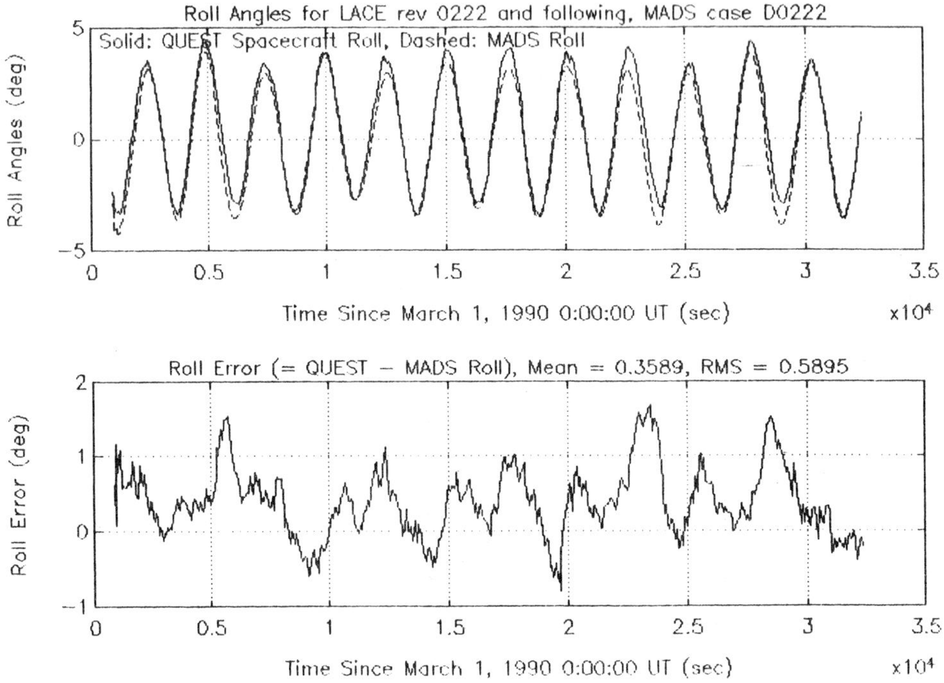

Fig. 7 MADS Roll Angle and Error With Larger R and Moderate Q For Larger Librations, In Comparison With QUEST, DATA.0222

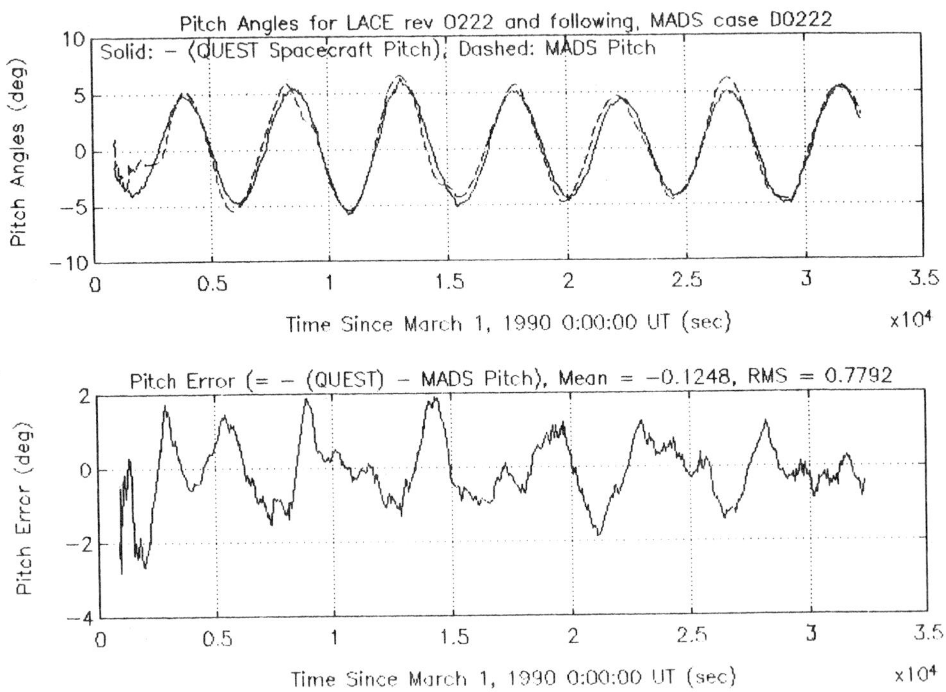

Fig. 8 MADS Pitch Angle and Error With Larger R and Moderate Q For Larger Librations, In Comparison With QUEST, DATA.0222

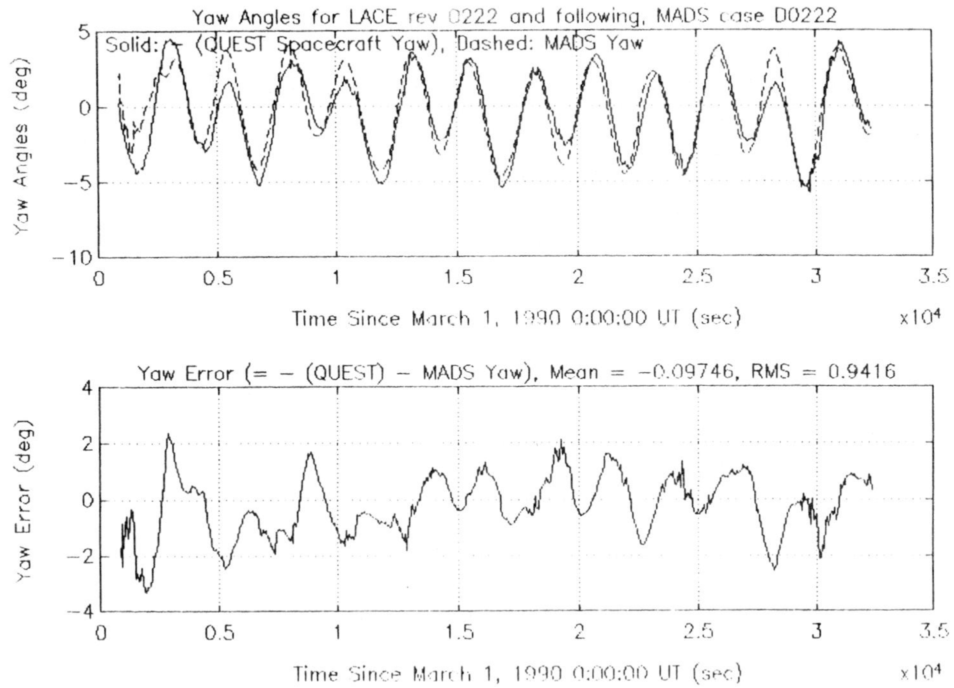

Fig. 9 MADS Yaw Angle and Error With Larger R and Moderate Q For Larger Librations, In Comparison With QUEST, DATA.0222

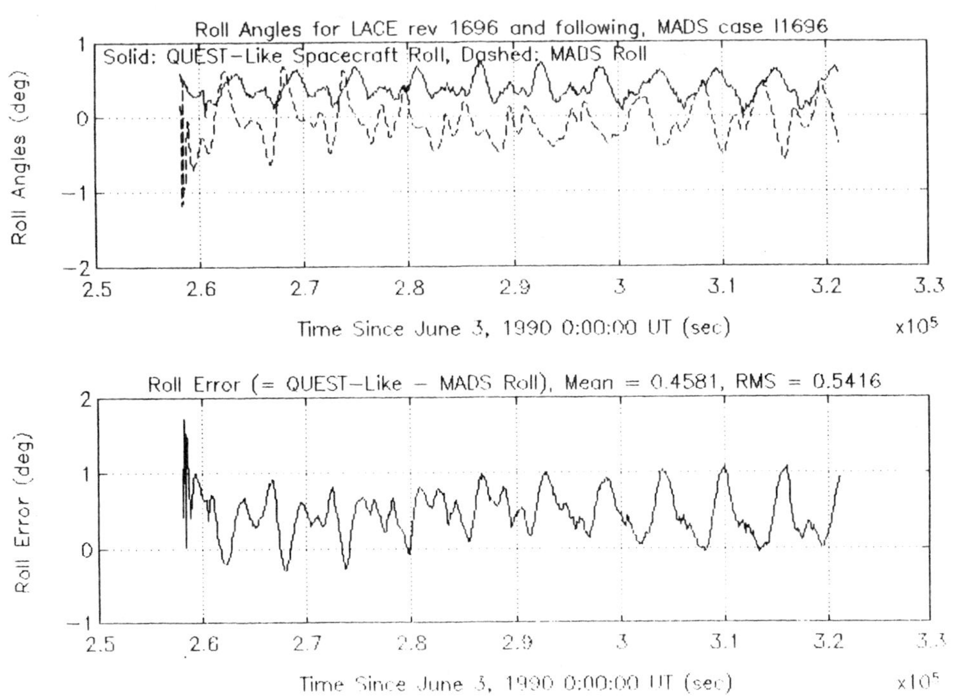

Fig. 10 MADS Roll Angle and Error With Larger R and Moderate Q For Small Librations, In Comparison With QUEST, DATA.1696

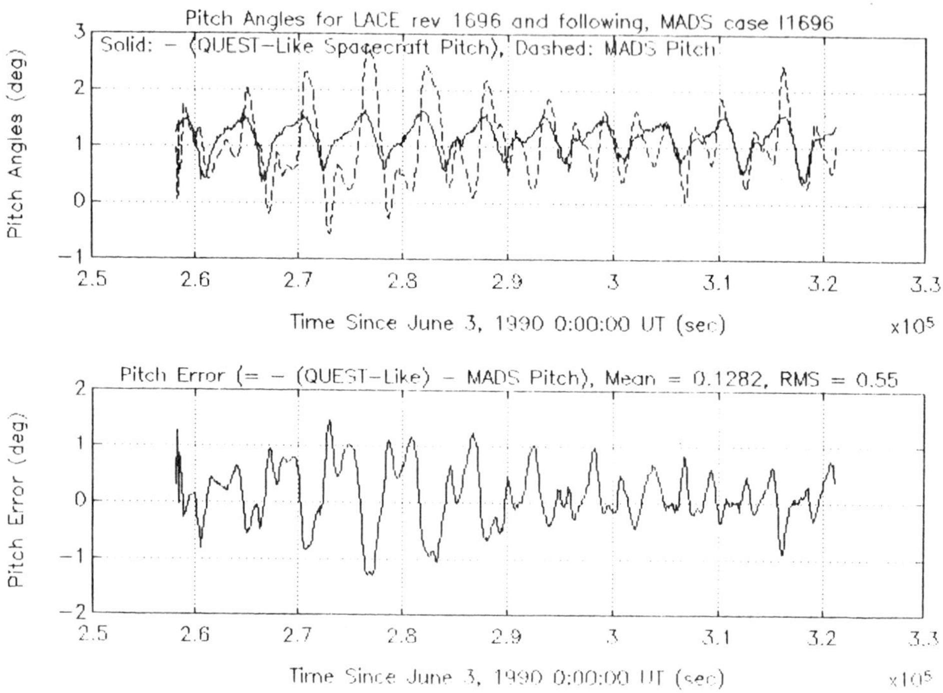

Fig. 11 MADS Pitch Angle and Error With Larger R and Moderate Q For Small Librations, In Comparison With QUEST, DATA.1696

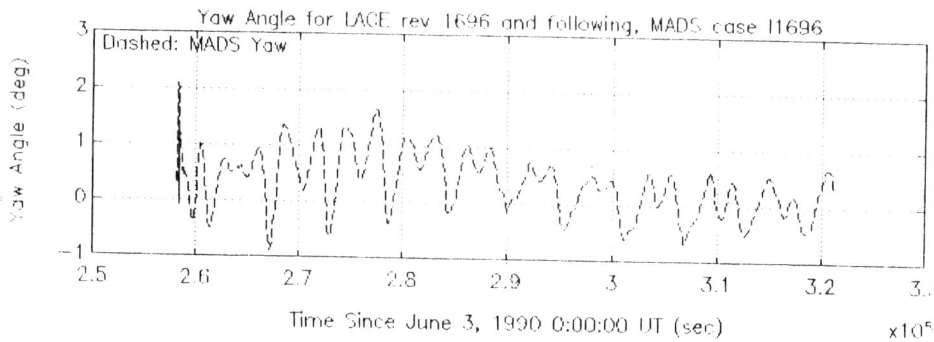

Fig. 12 MADS Yaw Angle With Larger R and Moderate Q For Small Librations, DATA.1696

Figs. 13 and 14 show the disturbance torque estimates that go with these cases. As expected, the dominant torque is on the pitch axis. It is in a nose-up direction. The magnitude is similar in the two cases, between 0.004 and 0.006 N-m, which seems reasonable in light of earlier studies.

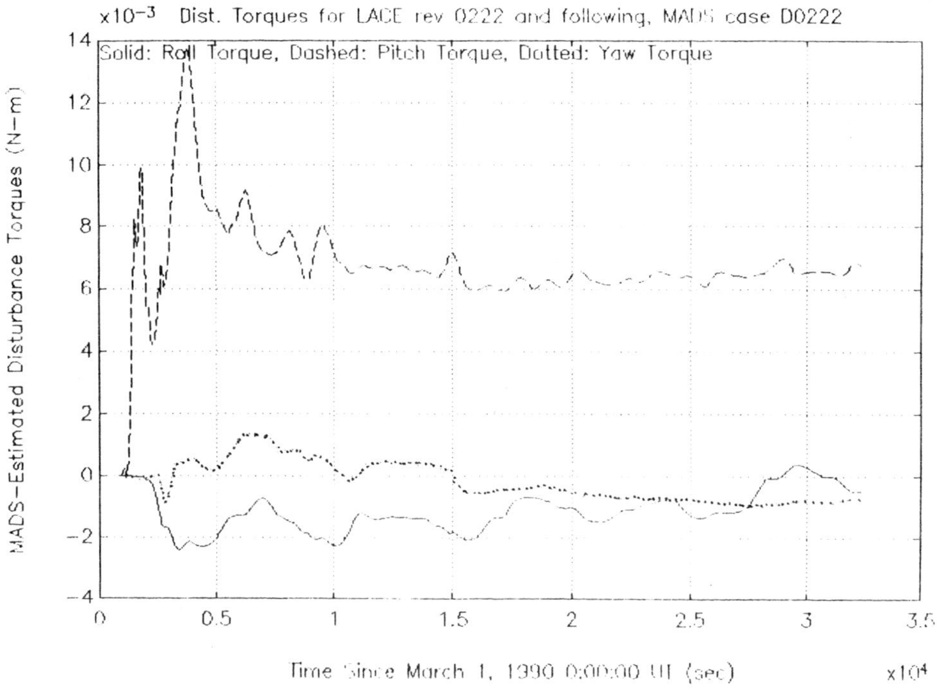

Fig. 13 Roll, Pitch, and Yaw Disturbance Torque Estimates During Larger Librations, MADS With a Larger R and a Moderate Q, DATA.0222

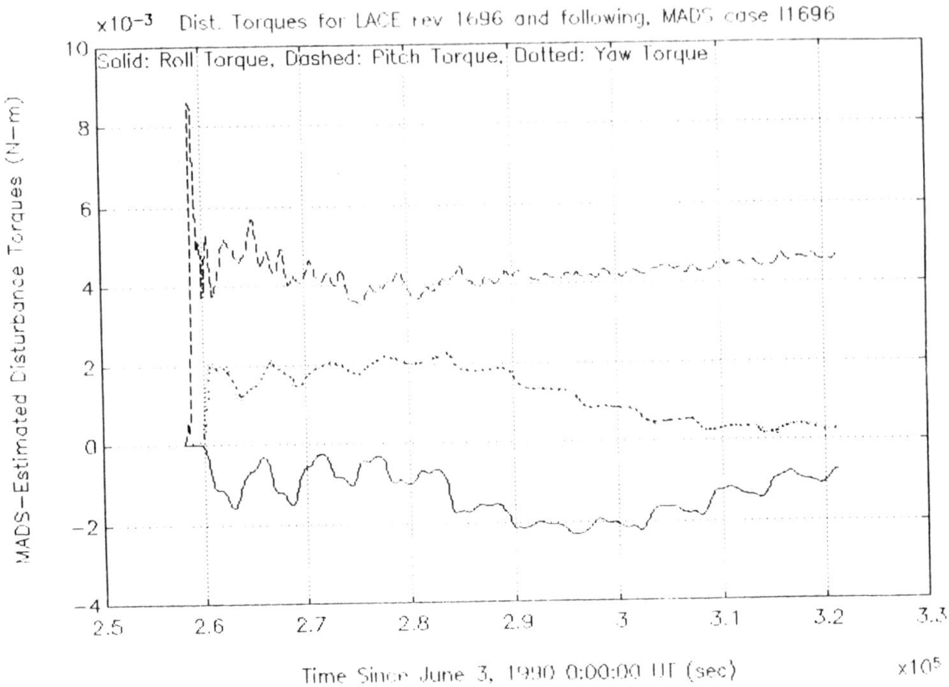

Fig. 14 Roll, Pitch, and Yaw Disturbance Torque Estimates During Small Librations, MADS With a Larger R and a Moderate Q, DATA.1696

It is believed that the attitude errors in **Data.0222** are due largely to torque mis-modelling errors. By examining the capability of MADS to predict attitude (see the following section), it is clearly demonstrated that there is an inertia modelling error in the **Data.0222** case. This error could explain the tendency of the filter's estimate to lead the actual librations, which is probably the dominant error source for **Data.0222**.

The attitude errors in case **Data.1696** are quite different. They have a strong 2-per-rev component. The residual modelling error in the field magnitude has this same 2-per-rev component. Both of these 2-per-rev errors are believed to stem from some common error associated with the field modelling or the magnetometer measurements.

Attitude Prediction Mode

MADS can attempt to make attitude predictions that apply to times well beyond the last processed magnetometer measurement. This is done by taking the post-update state estimate at the last measurement and propagating it forward via numerical integration of the rotational equations of motion.

This prediction capability has been tried for both sets of data. In each case MADS filtered 2 orbits worth of data (11,400 seconds worth of data) and predicted attitude for the remaining orbits. Figs. 15 thru 17 show the prediction performance for case **Data.0222**, and that for case **Data.1696** is shown in Figs. 18 and 19. These plots show that MADS as a predictor works quite differently in the two cases.

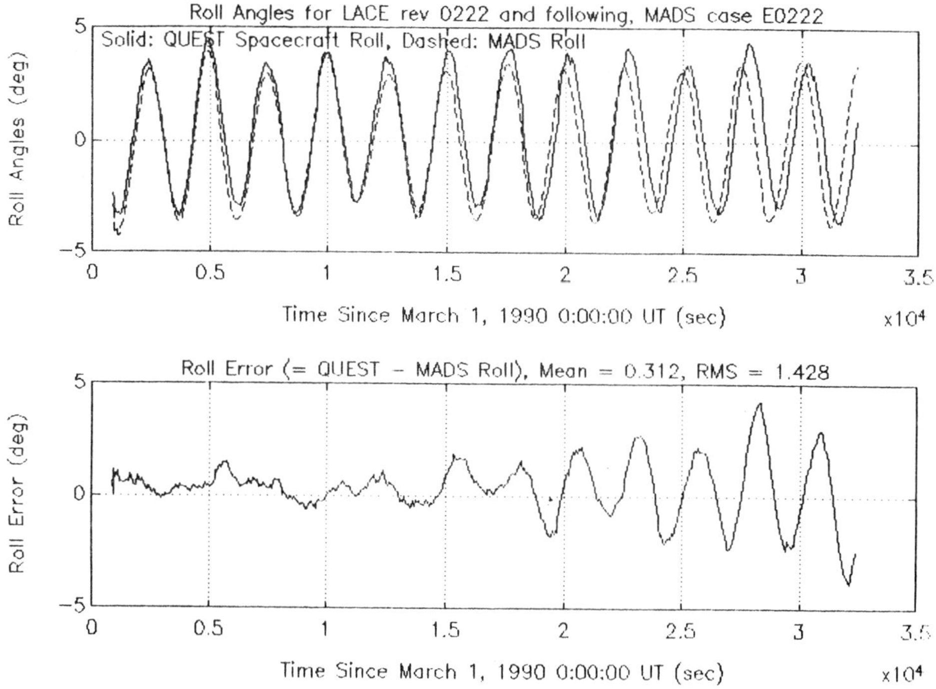

Fig. 15 MADS Roll Angle Predictions For Larger Librations, In Comparison With QUEST Estimates, DATA.0222

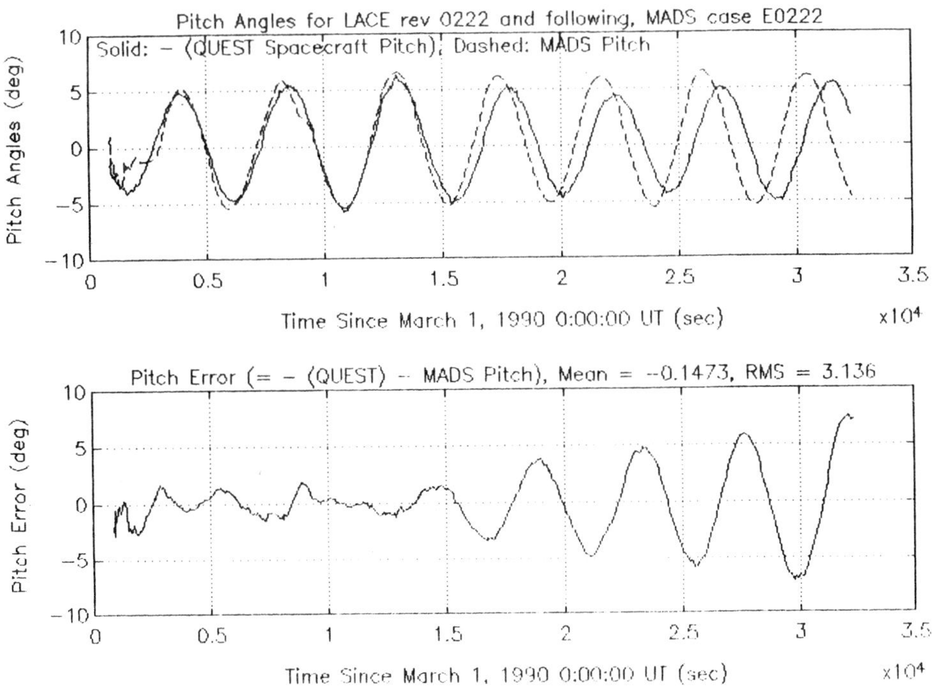

Fig. 16 MADS Pitch Angle Predictions For Larger Librations, In Comparison With QUEST Estimates, DATA.0222

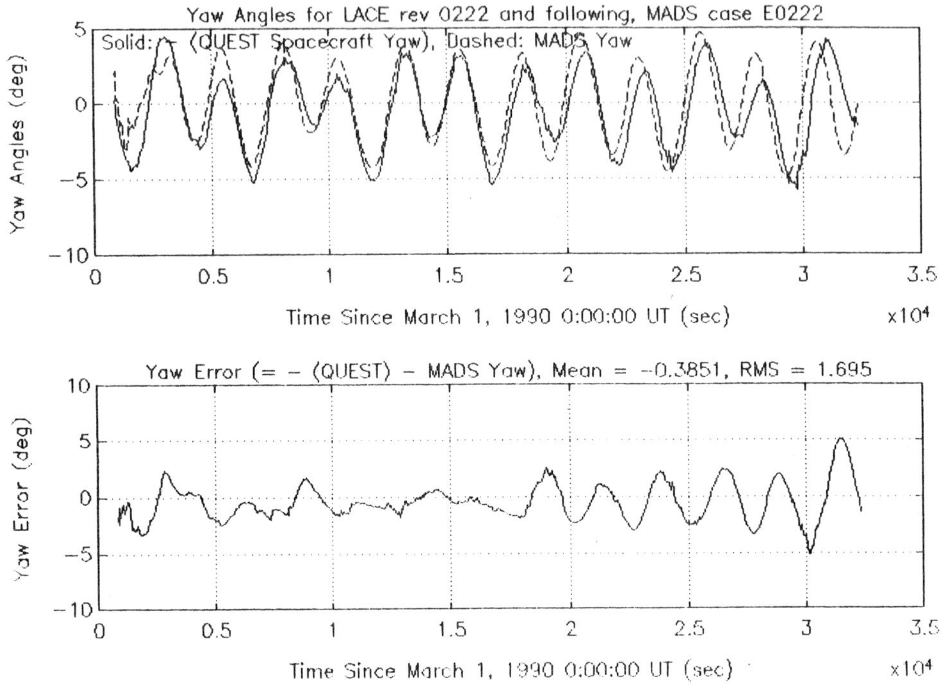

Fig. 17 MADS Yaw Angle Predictions For Larger Librations, In Comparison With QUEST Estimates, DATA.0222

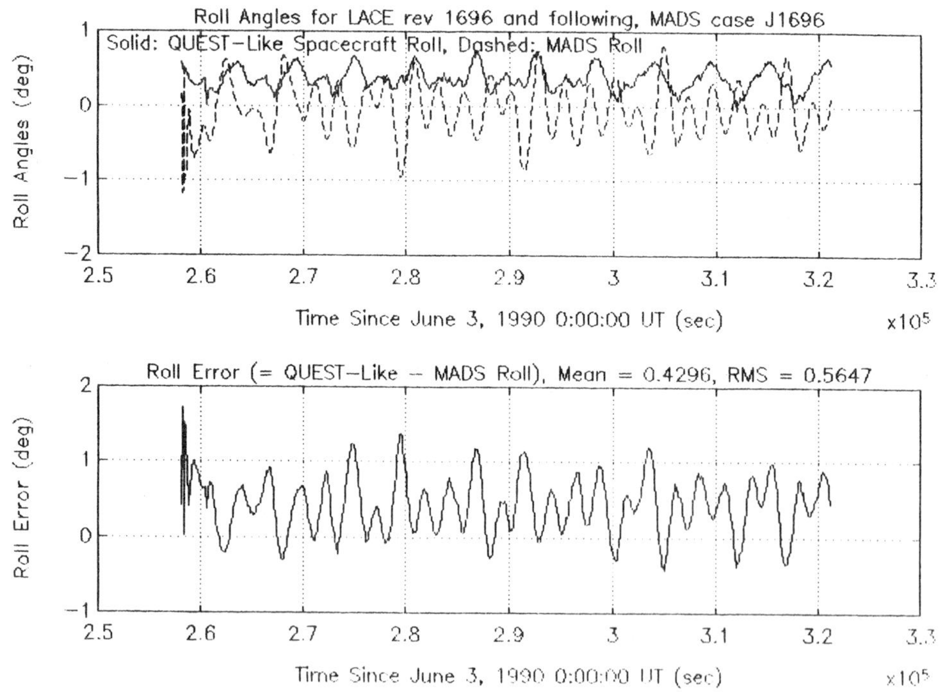

Fig. 18 MADS Roll Angle Predictions For Small Librations, In Comparison With QUEST Estimates, DATA.1696

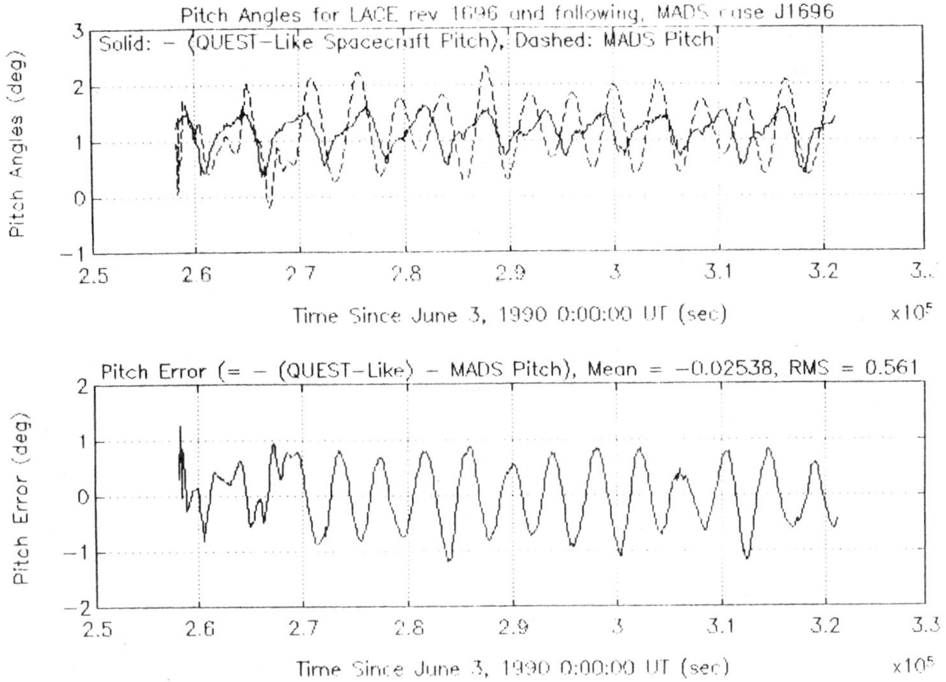

Fig. 19 MADS Pitch Angle Predictions For Small Librations, In Comparison With QUEST Estimates, DATA.1696

In the **Data.0222** case, prediction works for about one orbit following the end of the filtered data. Afterwards the predicted values (dashed curves) get too far ahead of the actual values in libration phase. Errors on the order of the libration magnitude rapidly develop. The error curves grow like a beat pattern produced by sine waves of nearly equal frequencies. By examining Figs. 15 thru 17 carefully, one can see that the predicted libration periods are all shorter than the actual libration periods. This implies that there is an inertia matrix modelling error.

The **Data.1696** predictions are more accurate than the **Data.0222** predictions because the librations for the **Data.1696** case are much smaller. The predicted librations do not resemble the actual librations. This is because the actual librations are due to diurnal forcing effects, while the predicted librations are due only to the initial conditions and the natural librational modes of the spacecraft. Even so, the prediction accuracy is fair-to-good because neither libration is very large.

CONCLUSIONS AND RECOMMENDATIONS

An extended Kalman filter for determining 3-axis spacecraft attitude solely from magnetometer measurements has been tested. This system is used as a backup attitude knowledge system for experimenters on the LACE project. The filter's attitude estimate has been compared with conical Earth scanner roll and pitch measurements, and the filter has been re-tuned to reduce its errors.

With the data provided for this analysis, the LACE MADS is capable of 3σ accuracies of 1.8° in roll, 2.4° in pitch, and 2.8° in yaw. The limiting factors to accuracy with respect to the disturbances seem to be the ability to correctly model inertia parameters and the magnetically anchored damper. On the measurements side, the accuracy is limited by the ability to correctly model and/or measure the Earth's magnetic field.

The errors due to field modelling / measurement noise seemed very systematic. They also seemed to be too large for the predicted uncertainties in the field model coefficients. Perhaps the IGRF model is not the best model to use for spaceflight applications. Additional investigations with respect to these areas of concern are recommended.

During this study, a considerable amount of time was spent in tracking down the timing error in the magnetometer data. If this system is to be used in the future, it is necessary to make very specific demands of the ground station or on-board systems concerning the absolute accuracy of the measurement time tags.

A pre-launch measurement of the magnetometer alignment would have helped this study. The CES had a mounting skew angle of 0.5°. Certainly, the magnetometer could have a similar offset angle. If these misalignments had been measured, perhaps the mean MADS errors would have been reduced. This would also have reduced the RMS errors, because the reported RMS values included the contributions of the respective mean values.

Filter convergence never seemed to be a problem. Good attitude tracking performance was achieved even for substantial librational motion. This indicates the suitability of such a system for generating feedback signals to an active libration damper or stiffener.

ACKNOWLEDGEMENTS

This work was supported in part by the Naval Center For Space Technology and Bendix Field Engineering, under Contract Number N00014-89-C-2422.

REFERENCES

1. M. Psiaki, F. Martel, P. Pal, "Three-Axis Attitude Determination via Kalman Filtering of Magnetometer Data", **Journal of Guidance, Control, and Dynamics**, Volume 13, Number 3, May-June 1990.

2. Wertz, J.R. ed., **Spacecraft Attitude Determination and Control**, D. Reidel Pub. Co., (Boston, 1978).

3. M. Psiaki and F. Martel, "Autonomous Magnetic Navigation for Earth Orbiting Spacecraft", Third Annual AIAA/Utah State University Conference On Small Satellites, September 27, 1989.

4. S. M. Fox, P. K. Pal, and M. Psiaki, "Magnetometer-Based Autonomous Satellite Navigation (MAGNAV)", 13th Annual AAS Guidance and Control Conference, February 3-7, 1990.

5. M. D. Shuster and S. D. Oh, "Three-Axis Attitude Determination From Vector Observations", **Journal of Guidance, Control, and Dynamics**, Volume 4, Number 1, January-February 1991.

6. M. D. Shuster, "Maximum Likelihood Estimation of Spacecraft Attitude", **Journal of the Astronautical Sciences,** Volume 37, Number 1, January-March 1989.

AAS 91-076

DISCOVERY OF BE-7 ACCRETION IN LOW EARTH ORBIT

G. W. Phillips[*], S. E. King[*], R. A. August[*], J. C. Ritter[*]
J. H. Cutchin[†], P. S. Haskins[‡], J. E. McKisson,[**] D. W. Ely,[**]
A. G Weisenberger[**], R. B. Piercey[††] and T. Dybler[††]

We observed strong ^{7}Be activity on the surface of the Long Duration Exposure Facility (LDEF) during the first complete gamma-ray survey of a large spacecraft after its return to earth. This is the first known evidence for accretion of a radioactive isotope onto an orbiting spacecraft. ^{7}Be is produced by spallation reactions of cosmic rays on nitrogen and oxygen in the upper atmosphere. However, the observed density is much greater than expected due to cosmic ray production *in situ*; this is significant for models of atmospheric mixing. ^{7}Be may be a valuable tracer in future studies of the upper atmosphere. Other isotopes seen during the survey, the strongest being ^{22}Na, are due to activation of spacecraft components while in orbit. We discuss the likely accretion of other cosmic-ray produced isotopes and their possible effects on spacecraft in low-earth orbit.

THE LDEF SPACECRAFT

The Long Duration Exposure Facility (LDEF) was launched by the Space Shuttle Challenger on 7 April 1984. It was retrieved in orbit by the Shuttle Columbia on 12 January 1990 and brought back to Earth on 20 January 1990. The LDEF carried a broad range of passive or low-powered experiments designed to study the space environment in low-earth orbit and to determine the effects of the environment on various materials, coatings, spacecraft components, and lastly, 12 million tomato seeds. The spacecraft is a 12-sided cylindrical aluminum structure, 9.1 m long by 4.2 m diameter, with a total weight of about 9700 kg. Along the sides and on both ends were 86

* Naval Research Laboratory, 4555 Overlook Ave., S.W., Washington, D.C. 20375-5000.

† Sachs/Freeman Associates, Washington, D.C.

‡ University of Florida, Gainesville, Florida 32611.

** Institute for Space Science and Technology, Gainesville, Florida.

†† Mississippi State University, Mississippi State, Mississippi 39762.

experimental trays, designed to be brought back to earth and analyzed in the laboratory by investigators from across the US, Canada, and Europe. The tomato seeds were distributed to school children to be grown and studied as science projects.

The LDEF was to have remained in orbit for a year, but recovery was delayed due to the Challenger accident, and exposure lasted over five and one half years. It was launched into a nearly circular orbit at an altitude of 480 km and an inclination of 28.5 degrees, where it was exposed continuously to cosmic rays, interplanetary dust and the residual atmosphere. In addition, the orbit took it through the South Atlantic Anomaly (SAA) exposing the spacecraft to energetic trapped protons and electrons. In the months prior to retrieval, the orbit was decaying rapidly and LDEF was down to an altitude of 310 km when recovered by the shuttle.

The LDEF spacecraft was gravity-gradient stabilized while in orbit so that its axis was aligned to the Earth's radius vector, with one end always pointed toward space and the other end toward Earth. Also, rotation about this axis was stabilized with respect to the orbital velocity so that the leading edge was always side number 9 (plus about 8 degrees). There were a number of duplicate experimental trays positioned around LDEF in order to get information about the differential flux of particles and micrometeroids. The LDEF orbital velocity (7.8 km/s at retrieval) exceeded the average thermal velocity of the rarified atmosphere so that exposure to the atmosphere was primarily on the leading edge of the spacecraft.

GAMMA-RAY SURVEY

After landing, LDEF was returned to Kennedy Space Center (KSC) for post-recovery examination. There, the spacecraft was mounted on a stand so that it could be rotated about its axis for inspection. During this period, an array of high-purity germanium detectors from the Naval Research Laboratory (NRL) and single detectors from the Institute for Space Science and Technology (ISST), were used to conduct the first detailed gamma-ray survey of a large spacecraft after exposure in low-earth orbit. The residual gamma-ray emission depends both on the flux of high-energy particles to which LDEF was exposed and on the particular materials in each experimental tray. To observe the distribution of gamma-ray activity about the spacecraft, we set up the array with detectors facing each tray position along one side of LDEF. The single detectors were positioned at each end facing one of the experimental trays. The distance from the detectors to each tray was about 0.6 m. Background spectra were taken prior to the arrival of LDEF, and the detectors were calibrated in place using known gamma-ray sources.

After the arrival and setup of LDEF, spectra were taken along each side for a minimum of 12 hours. The LDEF was then rotated so that a new side faced the array and new trays faced the detectors at each end. In this manner the entire spacecraft was surveyed over a period of about two weeks. During the disassembly period which followed, spectra were taken of selected experimental trays after they were removed from LDEF.

^{22}NA AND ^{7}BE ACTIVITIES

We expected to see gamma rays from the decay of isotopes produced by the long bombardment of energetic protons, neutrons, and heavier cosmic rays. The highest activity was expected from ^{22}Na, which has a halflife of 2.6 years and is produced by spallation from high-energy protons on the aluminum of the spacecraft body and experimental trays. At equilibrium, the activity from ^{7}Be, another spallation product with a 53 day halflife, was expected to be lower than ^{22}Na in intensity by two orders of magnitude. Data analysis capabilities during the survey were limited and consisted mainly of plotting spectra and identifying peaks. As expected, the strongest gamma-ray line at 1274 keV was from ^{22}Na. The 478 keV line from ^{7}Be was unexpectedly strong at some positions around the LDEF and much weaker at other positions. After the survey was complete, a preliminary plot of count rate versus position around the spacecraft showed a strong enhancement of ^{7}Be at the leading edge compared to the trailing edge. During post-collection data analysis, spectra were analyzed for each detector and peak intensities were extracted using the computer program HYPERMET[1]. Fig. 1 shows the ^{22}Na and ^{7}Be activities for each side of LDEF, corrected for decay from the date of retrieval of the spacecraft.

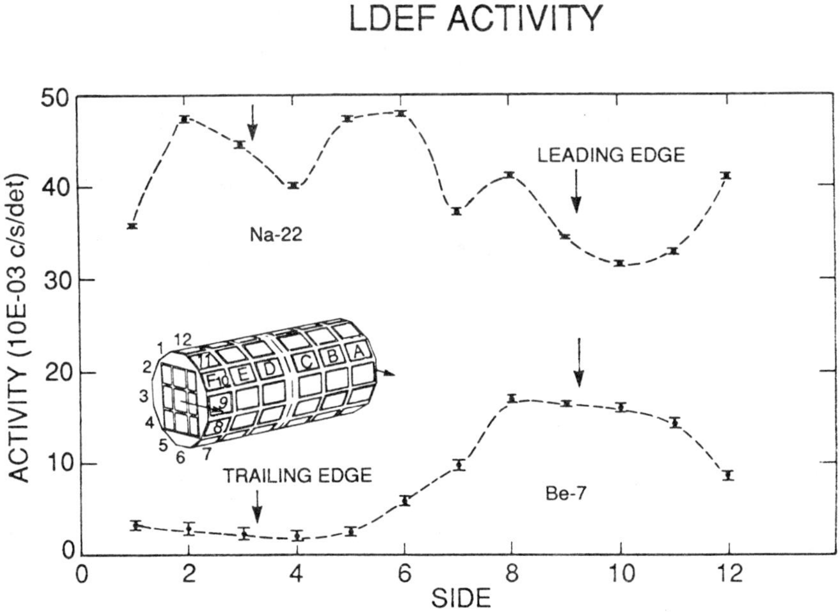

Fig. 1. Comparison of Activities of ^{7}Be (Lower) and ^{22}Na (Upper) Seen During the Gamma-Ray Survey of the LDEF Spacecraft. The average counts per second per detector are shown for each side of LDEF for an average detector efficiency of 38.8% at 1332 MeV relative to a 7.6 x 7.6 cm diameter NaI(Tl) detector. The error bars include statistical and peak-fitting uncertainties. As a visual aid, dashed curves are drawn connecting the data points. The overlay is a diagram of LDEF.

The distribution of ^{22}Na activity in Fig. 1 shows a small trailing edge enhancement, although there is some variance due to the distribution of aluminum and other activation material around the spacecraft. The trailing edge enhancement might be explained by the asymmetry in the trapped proton flux in the SAA. This flux is strongly peaked from the westward direction, the trailing direction in orbit[2]. Although many of the trapped protons are energetic enough to penetrate LDEF, they could produce the asymmetry seen in the activation data. Several other less intense activation lines were also observed and will be reported elsewhere[3].

In contrast to the ^{22}Na distribution, a strong leading edge enhancement for the ^{7}Be activity is evident in Fig. 1. The weak activity seen from the trailing edge can be wholly accounted for by penetration of gamma rays from the opposite side of the hollow spacecraft. This distribution of the ^{7}Be activity is not consistent with any known mechanism for activation of the spacecraft materials. It can only be explained by accretion of the isotope onto the leading surfaces of LDEF as it moved through the thin upper atmosphere in orbit.

The overlay in Fig. 1 gives a diagram of LDEF. Each experimental tray position around the cylinder is identified by the side, numbered 1 through 12, and the row lettered A through F. The view is toward the space end and the leading edge; the arrow vectors indicate the direction of the orbital velocity. Fig. 2 shows a two dimensional mapping of the ^{7}Be and ^{22}Na activity. The mapping shows the data as it would appear after cutting the cylinder between rows 1 and 12 and unrolling it flat. The data for each tray position is plotted by side (running from 1 to 12) along the right axis and by row (running from A to F) along the left axis. The leading and trailing edges in orbit are identified by the dashed lines in the figure. The ^{7}Be activity is shown to be distributed along the entire leading edge and not confined to a single tray. The data tend to be somewhat higher in the middle compared to the edges of the spacecraft. This can be explained by gamma rays from adjacent trays penetrating the 3 mm lead collimators which surrounded the detectors. Similarly, the weak trailing edge activity can be explained by penetration of gamma rays from the opposite leading edge.

The absence of ^{7}Be activity on the trailing edge was confirmed by measurements of gamma-ray spectra from individual experimental trays after they were removed from LDEF. Fig. 3 shows a comparison of spectra from nearly identical trays from the leading and trailing edges, containing germanium plates covered with a thin foil designed to capture interplanetary dust particles. The ^{7}Be peak is indicated in the figure and appears only on the tray from the leading edge.

Further confirmation of the lack of ^{7}Be on the trailing edge of LDEF came from low-level activity measurements of aluminum plates and tray clamps by NASA/Marshall Space Flight Center[4] (MSFC), which showed ^{7}Be activity only on the parts from the leading edge. In addition, they showed that an acid etch of an aluminum plate from the leading edge removed most of the activity, demonstrating that most of the ^{7}Be is found on the surface.

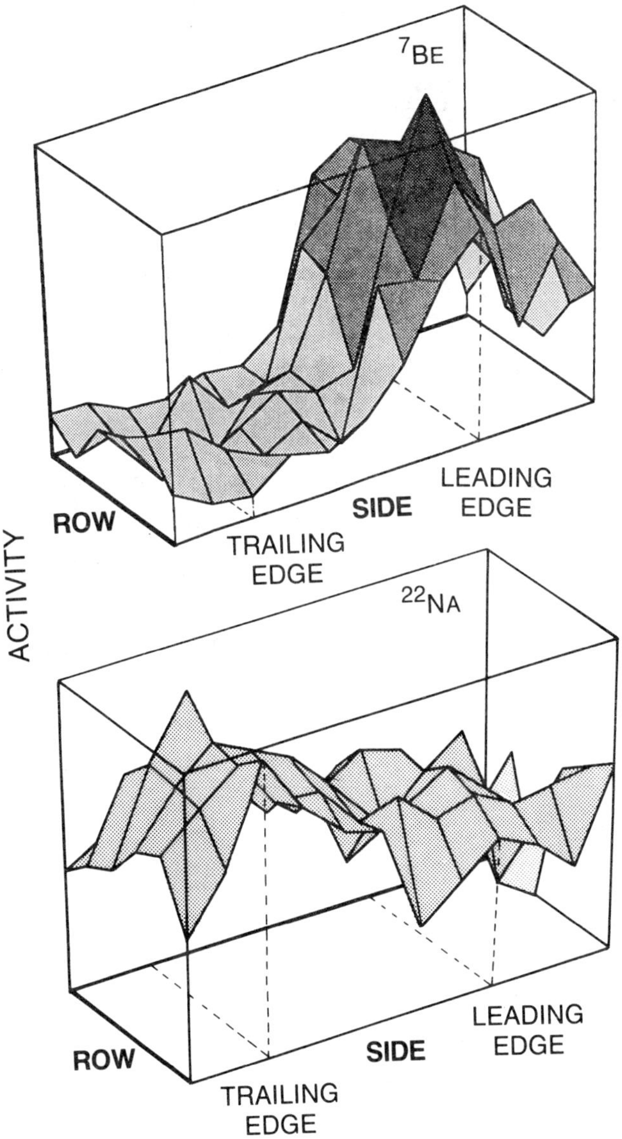

Fig. 2. Map of the Distribution of ^{7}Be and ^{22}Na activities around the LDEF spacecraft. There are 12 sides along the right axis and six rows along the left axis, with data from one experimental tray plotted for each row and side. The dashed lines indicate the positions of the leading and trailing edges. The ^{7}Be activity is strongly peaked along the leading edge, while the ^{22}Na activity is higher along the trailing edge.

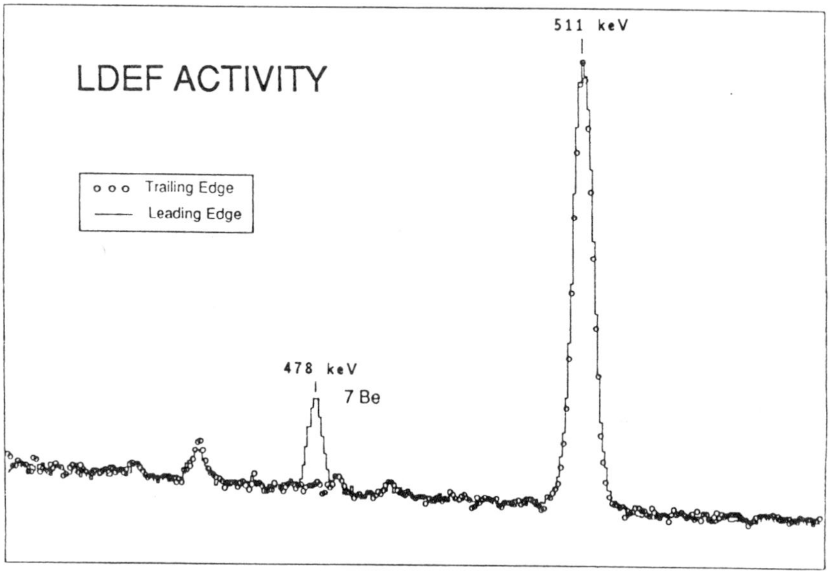

Fig. 3. Comparison of Gamma-Ray Spectra of Germanium Plates from Trays E3 and E8 After Their Removal from LDEF. Shown is the region including the 478 keV gamma ray from ^{7}Be which is seen on tray E8 near the leading edge and not on tray E3 at the trailing edge. The 511 keV peak due to positron annihilation is seen both in the background and from ^{22}Na. The weaker unlabeled peaks are all in the background.

DENSITY OF ^{7}BE IN THE UPPER ATMOSPHERE

Assuming that the ^{7}Be was accreted onto the surface of LDEF in low-earth orbit, the question arises: how did it get there with such intensity? ^{7}Be is produced by spallation interactions of high-energy cosmic rays primarily on nitrogen and oxygen in the upper atmosphere. (These complex nuclear disintegrations are called "stars" from their appearance in photographic emulsions.) From the literature, we can estimate the ^{7}Be density at 310 km due to cosmic ray production for comparison to our measurements. Using curves of cosmic ray interaction rates derived from measurements during a period of high solar activity[5] and trapped proton fluences, we obtain an estimated ^{7}Be density of 5.4 x 10^{-5} atoms/m^3, due to production in situ (see the Appendix for details.) From our measurements we can derive a minimum ^{7}Be density in orbit of 0.10 ± 0.03 atoms/m^3. This exceeds the estimated in-situ production by a factor of 1800.

It is difficult to explain such a large enhancement in the ^{7}Be density. One possibility is the mixing of air from the poles where the star production rate is higher than at lower latitudes, which are partially shielded from cosmic rays by the Earth's magnetic field. Measurements in the stratosphere[6] imply significant mixing between polar and low-latitude air, showing increases by a factor of 2 to 5 over the equilibrium value at 31° N. In the upper atmosphere, above 120 km, the polar production rate is about a factor of 10 higher than the average rate from 0° to 30° latitude[5]. Thus, complete displacement by polar air would still leave a factor of 180 unexplained.

A second possible source of increased activity is diffusion or convection of ^{7}Be from air at lower altitudes where production rates are higher due to increased atmospheric density. During periods of high solar activity[7], the estimated in-situ production rate increases by a factor of 300 from an altitude of 310 km down to 120 km, which is approximately the altitude where diffusive equilibrium begins[8]. Because ^{7}Be is considerably lighter than the mean atmospheric molecular weight, it will tend to diffuse upward; because the atmosphere is so thin, diffusion will be rapid. The equilibrium distribution of atmospheric molecules due to diffusion is a decreasing exponential function of the altitude with a scale height[9] which is proportional directly to the temperature and inversely to the atomic weight. The mean global temperature[10] rises rapidly from about 380K at 120 km to 1040K above 200 km. During periods of high solar activity, temperatures as high as 1700K have been measured[11,12]. For ^{7}Be, this corresponds to a scale height of 206 km and an average thermal velocity of 2.5 km/s. Several large solar flares occurred in 1989, including the late September-early October flare which was the largest in 33 years and had a very hard spectrum[13]. Such events cause heating and expansion of the upper atmosphere, where winds have been measured at several hundred meters per second[11,12], driven by solar activity, diurnal solar heating and geomagnetic storms. These act both to mix polar and lower latitude air and to transport air upward from lower altitudes[9,11].

CONCLUSIONS AND IMPLICATIONS FOR FUTURE SPACECRAFT

Our observations of ^{7}Be activity on the leading surfaces of the LDEF spacecraft imply a minimum density for ^{7}Be in low-earth orbit which greatly exceeds the local equilibrium due to cosmic-ray production in situ. One possible explanation would require the transport by diffusion or atmospheric mixing of ^{7}Be from much lower altitudes and higher latitudes into the LDEF orbit. Thus, the current results should be important for validating and refining models of the upper atmosphere. With more extensive measurements, ^{7}Be should prove valuable as a natural tracer for studies of upper atmospheric mixing. Our next step is to use existing atmospheric circulation models and a calculation of ^{7}Be production rates at lower altitudes to predict the upward transport of ^{7}Be. Future observations should focus on sampling at both lower and higher altitudes and should extend to polar latitudes. These should be closely correlated with data on wind, temperature, pressure and solar activity.

In addition, the observation of the accretion of significant quantities of ^{7}Be is an indication of possibly similar behavior for other light cosmic-ray produced isotopes. Table 1 gives the spallation yields[5] for all light isotopes with yields greater than or of the order of ^{7}Be. Also given are their halflives and decay modes. ^{3}He is stable and non-reactive. The remaining isotopes, other than ^7Be, are all pure beta emitters and thus would not be seen in the present survey. They could, however, be significant sources of noise for low-level sensors on spacecraft in low-earth orbit and could slowly degrade other components by coating or by surface reactions. Lithium, the decay product of ^7Be, could affect exposed semiconductor sensors even in very low concentrations. As a result of our observations, other groups are currently looking for trace amounts of Li and ^{14}C on LDEF components[14,15].

Table 1
COSMIC RAY PRODUCED LIGHT ISOTOPES

Isotope	Yield/Star	Half-life	Decay Modes
^{14}C	1.5*	5.7 x 10^4 y	beta
^3H	0.14	12.3 y	beta
3He	0.12	stable	none
^7Be	0.045	53 d	ec+, gamma
^{10}Be	0.025	1.6 x 10^6 y	beta

* Relative yield, produced mainly by thermal neutrons
\+ Electron capture

APPENDIX

<u>Equilibrium ^7Be Density from Cosmic-Ray Production.</u> Lal and Peters[5] provide curves of cosmic-ray produced interaction rates (stars per gram of air per second) versus latitude and altitude, using a model derived from measurements during a period of high solar activity. From these curves and the known spallation yields[5], we obtain a ^7Be production rate per gram of air between 0° and 30° latitude of 9.0 x 10^{-5} atoms/g-s at the "top" of the atmosphere (above 120 km). The mean atmospheric density during periods of high solar activity[7] at an altitude of 310 km is about 6.1 x 10^{-8} g/m^3. This gives an in-situ production rate for ^7Be at 310 km of 5.5 x 10^{-12} atoms/m^3-s. Multiplying by the equivalent in seconds of the 77 day mean ^7Be lifetime gives an equilibrium density for ^7Be of 3.6 x 10^{-5} atoms/m^3, due to production in situ.

The trapped proton flux provides an added production source of ^7Be at 310 km. Using the trapped proton fluence for solar maximum given by Stassinopoulos[16], production cross sections[5], and an atmospheric density for November/December 1989 calculated by Hickey[17], the average equilibrium density of ^7Be is 1.8 x 10^{-5} atoms/m^3. The total density is then 5.4 x 10^{-5} atoms/m^3. However, the density calculated for the trapped protons is somewhat misleading since virtually all the production occurs in the South Atlantic Anomaly where the density would be considerably higher.

Minimum ^{7}Be Density from Our Measurements. Our measurements of the ^{7}Be activity on the LDEF leading edge give an average surface density for ^{7}Be of (5.4 ± 1.4) x 10^9 atoms/m^2, corrected to the date of retrieval of the spacecraft. With an orbital velocity of 7.8 km/s, LDEF traveled a distance of 5.2 x 10^{10} m during one mean lifetime of ^{7}Be. Assuming 100% adherence of ^{7}Be to the surface of LDEF, this implies a minimum density in orbit for ^{7}Be of 0.10 ± 0.03 atoms/m^3. (Less than 100% adherence would imply an even greater ^{7}Be density in orbit.)

REFERENCES

1. G.W. Phillips and K.W. Marlow, Nucl. Instr. and Methods 137 (1976) 525-536.
2. J.W. Watts, T.A. Parnell, and H.H. Heckman, AIP Conference Proceedings 186, High Radiation Background in Space, ed. A.C. Rester, Jr., and J.I. Trombka, AIP, New York (1987) 75-85.
3. S.E. King et al., submitted to IEEE Transactions in Nuclear Science.
4. G.J. Fishman, et al., submitted to Nature.
5. D. Lal and B. Peters, Encyclopedia of Physics, 46/2, ed. K. Sitte, Springer, New York (1967) 551-612.
6. N. Bhandari, J. Geophys. Res. 75 (1970) 2927-2930.
7. Appendix C, COSPAR International Reference Atmosphere: 1986, Part 1: Thermospheric Models, ed. D. Rees, Pergamon Press, New York (1989), 315-470 (see Table 5.12.)
8. I. Harris and N.W. Spencer, Introduction to Space Science, ed. W.N. Hess and G.D. Mead, Gordon and Breach, New York (1968) 93-131.
9. J.K. Hargreaves, The Upper Atmosphere and Solar-Terrestrial Relations, Van Nostrand Reinhold, New York (1979).
10. A.E. Hedin, COSPAR International Reference Atmosphere: 1986, Part 1: Thermosphere Models, ed. D. Rees, Pergamon Press, New York (1989) 9-25.
11. N.W. Spencer and G.R. Carignan, ibid., 107-117.
12. G. Hernandez and T.L. Killeen, ibid., 149-213.
13. J.H. Allen, "1989 Solar Activity and its Consequences at Earth and in Space", presented at the DOD-DOE-NASA Seventh Symposium on Single Event Effects, May 1990.
14. N. Tolk, Vanderbilt University, Brentwood, TN, private communication.
15. A.J.T. Jull, University of Arizona, Tucson, AZ, private communication.
16. E.G. Stassinopoulos and J.M. Barth, Goddard Space Flight Center Report No. X-600-87-7, Greenbelt, MD (1987).
17. M. Hickey, Marshall Space Flight Center, private communication, June, 1990.

Section VI
LOW-EARTH ORBIT SPACE ENVIRONMENT

TUTORIAL SESSION

Local Chairpersons: Eric Lander
Martin Marietta Astronautics

Tom Kroncke
Ball Aerospace Systems Group

The following paper was not available for publication:

AAS 91-014 "Magnetics in the Low-Earth Orbit Space Environment", Gary Heckman, NOAA

The following paper numbers were not assigned:

AAS 91-001 to -009
AAS 91-016 to -019
AAS 91-083 to -099

AAS 91-010

GRAVITATIONAL EFFECTS ON LOW-EARTH ORBITERS

George W. Rosborough*

A brief review of the orbital perturbations produced by the Earth's gravity field is given. The spherical harmonic representation of the gravitational potential is presented along with the relations for using normalized coefficients. Recurrence relations for generating the Associated Legendre Functions are also included. Simple expressions for evaluating the dominant effects on orbital motion are derived from the results of Kaula. Example results are discussed for the particular orbits of the Lageos and Landsat satellites and using the recent GEM-T1 gravity model.

1 Introduction

Satellites in low-Earth orbit are affected by a broad spectrum of perturbations due to the Earth's gravity field. The largest of these perturbations are produced by the Earth's oblateness. Beyond the oblateness, there exist much smaller undulations in the gravity field. These variations produce much smaller, but certainly observable, effects on low-Earth orbiters.

The modeling of the Earth's gravity field using spherical harmonics is convenient for both numerical integration of satellite trajectories as well as analytic developments for the orbital perturbations. Both of these tasks are briefly reviewed. The equations of motion of a satellite moving in the gravitational field are derived. Computer implementation of these equations is facilitated by recurrence relations for the Associated Legendre Functions which are also given.

The principal perturbations produced on a satellite orbit are then defined and approximate expressions quantifying the perturbations are found using Kaula's approach. A current Earth gravity model (GEM-T1) is discussed and then used to evaluate the effects on two example satellite orbits (Lageos and Landsat).

* Colorado Center for Astrodynamics Research, Department of Aerospace Engineering Sciences, Campus Box 429, University of Colorado, Boulder, Colorado 80309.

2 Mathematical Model

2.1 Gravitational Potential

The common approach for modeling the gravitational field of a planetary body is through the spherical harmonic representation,

$$U = \frac{\mu}{r} \sum_{\ell=0}^{\infty} \left(\frac{a_e}{r}\right)^{\ell} \sum_{m=0}^{\ell} P_{\ell,m}(\sin\phi)\left[C_{\ell,m}\cos m\lambda + S_{\ell,m}\sin m\lambda\right] \tag{1}$$

where

μ	$\equiv$	product of the universal constant of gravitation, G, and the mass of the Earth, M
a_e	$\equiv$	semi-major axis of the Earth's reference ellipsoid
r, ϕ, λ	$\equiv$	satellite distance, latitude and longitude, respectively, in a body-fixed coordinate system
$C_{\ell,m}$, $S_{\ell,m}$	$\equiv$	spherical harmonic coefficients of degree ℓ and order m
$P_{\ell,m}$	$\equiv$	Associated Legendre Functions of degree ℓ and order m

A gravitational model consists of a set of constants that specify μ, a_e and the $C_{\ell,m}$, $S_{\ell,m}$ coefficients. It should also be noted that such a set of constants also implicitly defines a body-fixed coordinate system. The coordinate system defined is precisely that which was used in the solution of the spherical harmonic coefficients.

This representation of the geopotential can be thought of as consisting of three constituent parts,

$$U = U_1 + U_2 + U_3 \tag{2}$$

The first part is simply the leading term of the expansion corresponding to the degree and order zero term. The Associated Legendre Function, P_{00} has a value of one as does the C_{00} coefficient. So the leading term is simply,

$$U_1 = \frac{\mu}{r} \tag{3}$$

This is the familiar potential resulting from treating the body as point mass and that used for deriving the fundamental results of two body motion.

The second part of the spherical harmonic representation are those terms (besides the above two body term) which do not have a longitude dependence. These are the terms corresponding to $m = 0$ and are denoted as the zonal contribution to the potential,

$$U_2 = \frac{\mu}{r} \sum_{\ell=1}^{\infty} \left(\frac{a_e}{r}\right)^{\ell} P_{\ell,0}(\sin\phi)C_{\ell,0} \tag{4}$$

The degree 2 zonal term models the contribution due to the planetary oblateness. As such, it is the second largest contributor to the overall potential following the central body contribution. (The degree 1 term is zero assuming that the center of the Earth-fixed coordinate system coincides with the center of mass of the Earth.) The notation J_ℓ is often used for the zonal coefficients instead of the above $C_{\ell,0}$. The two notations simply differ in sign,

$$J_\ell = -C_{\ell,0} \tag{5}$$

so that the zonal part of the potential could also be written in the form,

$$U_2 = -\frac{\mu}{r}\sum_{\ell=1}^{\infty}\left(\frac{a_e}{r}\right)^{\ell} P_{\ell,0}(\sin\phi)J_\ell \tag{6}$$

The C notation will be used throughout this paper.

The remaining part of the spherical harmonic representation is that part depending on longitude,

$$U_3 = \frac{\mu}{r}\sum_{\ell=1}^{\infty}\left(\frac{a_e}{r}\right)^{\ell}\sum_{m=1}^{\ell} P_{\ell,m}(\sin\phi)\left[C_{\ell,m}\cos m\lambda + S_{\ell,m}\sin m\lambda\right] \tag{7}$$

The largest longitudinal contributor to the potential is usually the degree 2 and order 2 terms. These terms represent the amount that the planet is "out of round" about the equator. (As with the degree 1 zonal coefficient, the degree 1 and order 1 coefficients will be zero under the assumption that the center of the coordinate system coincides with the center of mass.)

The spherical harmonic representation of the potential (Eq. 1) can then be written as,

$$\begin{aligned} U &= \frac{\mu}{r} \\ &+ \frac{\mu}{r}\sum_{\ell=1}^{\infty}\left(\frac{a_e}{r}\right)^{\ell} P_{\ell,0}(\sin\phi)C_{\ell,0} \\ &+ \frac{\mu}{r}\sum_{\ell=1}^{\infty}\left(\frac{a_e}{r}\right)^{\ell}\sum_{m=1}^{\ell} P_{\ell,m}(\sin\phi)\left[C_{\ell,m}\cos m\lambda + S_{\ell,m}\sin m\lambda\right] \end{aligned} \tag{8}$$

In the general case, where temporal variations in the potential exist (e.g., tides), the spherical harmonic representation is still valid though the geopotential coefficients $(C_{\ell,m}, S_{\ell,m})$ then become time dependent.

2.1.1 Spherical Harmonics

To better understand the utility of using a spherical harmonic representation of the geopotential a closer look at the spherical harmonic functions is required. The spherical harmonic functions are those functions formed by the product of the Associated Legendre Functions with $\cos m\lambda$ and $\sin m\lambda$ which appear in Eq. 1,

$$A_{\ell,m}(\phi,\lambda) = P_{\ell,m}(\sin\phi)\cos m\lambda \quad \text{and} \quad B_{\ell,m}(\phi,\lambda) = P_{\ell,m}(\sin\phi)\sin m\lambda \tag{9}$$

These functions are orthogonal. Thus, each function (for a given degree and order) can be thought of as contributing independent information with an amplitude given by their respective $C_{\ell,m}$ and $S_{\ell,m}$ coefficients.

In addition to being orthogonal, the qualitative shapes of the spherical harmonics are easily visualized. The zonal harmonics (corresponding to $m = 0$) have no longitude dependence and have ℓ zeroes between ± 90 degrees in latitude. So the even degree zonals are symmetric about the equator and the odd zonal are asymmetric. Note also that as the degree increases the number of zeroes in latitude increases and the harmonics represent finer and finer latitudinal variations in the potential. If only large scale (such as the oblateness) variations need to be modeled then only the lowest degree zonal terms need to be used.

The non-zonal harmonics all have longitudinal variations. The presence of the $\cos m\lambda$ and $\sin m\lambda$ give the functions $2m$ zeroes in longitude. And the Associated Legendre Functions have $\ell - m$ zeroes in latitude. So, similar to the zonals, the higher degree and order harmonics represent finer and finer spatial detail of the gravitational potential. The non-zonal coefficients are called tesserals and for the specific case of $\ell = m$ they are referred to as sectorials.

Generally, the spherical harmonics can be thought of as representing variations in the gravitational potential that have wavelengths of the circumference of the Earth divided by m in longitude, and divided by $\ell - m$ in latitude.

2.1.2 Normalization

The spherical harmonic coefficients appearing in Eq. 1 are unnormalized. These coefficients tend to very small values as the degree increases. This is partly a consequence of the nature of the Earth's gravity field but is for the most part due to the fact that the Associated Legendre Functions tend to large values as degree increases. Thus it is numerically advantageous to normalize the Associated Legendre Functions and the coefficients. The normalization is achieved by multiplying the Legendre functions by a scale factor depending on the degree and order of the function. Denoting normalized values by an overbar, the normalized Associated Legendre Functions are,

$$\bar{P}_{\ell,m} = \left[(2 - \delta_{m0})(2\ell + 1)\frac{(\ell - m)!}{(\ell + m)!}\right]^{1/2} P_{\ell,m} \tag{10}$$

where the Kronecker delta, δ_{m0}, is equal to 1 if m is zero and equal to 0 if m is greater than zero. The geopotential coefficients, $C_{\ell,m}$ and $S_{\ell,m}$, are normalized by the inverse of this scale factor,

$$\left\{ \begin{array}{c} \bar{C}_{\ell,m} \\ \bar{S}_{\ell,m} \end{array} \right\} = \left[\frac{1}{(2 - \delta_{m0})(2\ell + 1)}\frac{(\ell + m)!}{(\ell - m)!}\right]^{1/2} \left\{ \begin{array}{c} C_{\ell,m} \\ S_{\ell,m} \end{array} \right\} \tag{11}$$

The spherical harmonic expansion of the geopotential (Eq. 1) can now be written in terms of normalized quantities,

$$U = \frac{\mu}{r}\sum_{\ell=0}^{\infty}\left(\frac{a_e}{r}\right)^{\ell}\sum_{m=0}^{\ell}\bar{P}_{\ell,m}(\sin\phi)\left[\bar{C}_{\ell,m}\cos m\lambda + \bar{S}_{\ell,m}\sin m\lambda\right] \tag{12}$$

This is usually the preferred formulation for numerical implementations of the spherical harmonic representation. For many analytical developments (such as the effect on orbital motion) it is easier to work with the unnormalized form (Eq. 1).

2.1.3 Associated Legendre Functions

Evaluation of the spherical harmonic expansion requires evaluating the Associated Legendre Functions. This evaluation is most conveniently performed using recurrence relations. If only a few terms are needed (low degree and order) then explicit coding of the functions may be more desirable. In general the Associated Legendre Function of degree ℓ and order m is,

$$P_{\ell,m}(x) = \frac{(1-x^2)^{m/2}}{2^\ell} \sum_{k=0}^{\frac{\ell-m}{2}} (-1)^k \frac{(2\ell-2k)!}{k!(\ell-k)!(\ell-m-2k)!} x^{\ell-m-2k} \tag{13}$$

The first few of these functions are given in Table 1.

Table 1

ASSOCIATED LEGENDRE FUNCTIONS $P_{\ell,m}(x)$

	Order m			
Degree ℓ	0	1	2	3
0	1			
1	x	$(1-x^2)^{1/2}$		
2	$\frac{1}{2}(3x^2-1)$	$3x(1-x^2)^{1/2}$	$3(1-x^2)$	
3	$\frac{1}{2}(5x^3-3x)$	$\frac{3}{2}(5x^2-1)(1-x^2)^{1/2}$	$15x(1-x^2)$	$15(1-x^2)^{3/2}$

Recurrence relations for evaluating these functions are generally in one of two forms. The difference is due to whether the recurrence is done holding the degree fixed or holding the order fixed. Either approach allows the computation of all the needed Legendre functions. In both cases, the starting values for the recurrences are the $\ell = m$ and $\ell = m+1$ functions which are easily computed from,

$$P_{\ell,\ell}(x) = \frac{(2\ell-1)!}{2^{\ell-1}(\ell-1)!}(1-x^2)^{\ell/2} \tag{14}$$

$$P_{\ell,\ell-1}(x) = \frac{x}{(1-x^2)^{1/2}} P_{\ell,\ell}(x) \tag{15}$$

One recursion is to compute $P_{\ell,\ell}$ and $P_{\ell,\ell-1}$ and then compute the functions for all lower orders of degree ℓ using,

$$P_{\ell,m}(x) = \frac{1}{(\ell-m)(\ell+m+1)} \left[2(m+1)\frac{x}{(1-x^2)^{1/2}} P_{\ell,m+1}(x) - P_{\ell,m+2}(x) \right] \tag{16}$$

The alternate recursion is to compute $P_{\ell,\ell}$ and $P_{\ell+1,\ell}$ and then compute the functions for all higher degrees using,

$$P_{\ell,m}(x) = \frac{1}{\ell-m} \left[(2\ell-1)xP_{\ell-1,m}(x) - (\ell+m-1)P_{\ell-2,m}(x) \right] \tag{17}$$

The recurrence relations can also be rewritten to directly work with the normalized Associated Legendre Functions. The recurrence equivalent to Eq. 16 in normalized form is,

$$\bar{P}^*_{\ell,m}(x) = 2(m+1)\left[\frac{1}{(\ell+m+1)(\ell-m)}\right]^{1/2}\frac{x}{(1-x^2)^{1/2}}\bar{P}^*_{\ell,m+1}(x) - \left[\frac{(\ell+m+2)(\ell-m-1)}{(\ell+m+1)(\ell-m)}\right]^{1/2}\bar{P}^*_{\ell,m+2}(x) \tag{18}$$

where $\bar{P}_{\ell,m} = \bar{P}^*_{\ell,m}$ if $m > 0$ and $\bar{P}_{\ell,0} = \frac{1}{\sqrt{2}}\bar{P}^*_{\ell,0}$ when $m = 0$.

The recurrence equivalent to Eq. 17 in normalized form is,

$$\bar{P}_{\ell,m}(x) = \left[\frac{(2\ell-1)(2\ell+1)}{(\ell-m)(\ell+m)}\right]^{1/2} x\bar{P}_{\ell-1,m}(x) - \left[\frac{(2\ell+1)(\ell+m-1)(\ell-m-1)}{(2\ell-3)(\ell+m)(\ell-m)}\right]^{1/2}\bar{P}_{\ell-2,m}(x) \tag{19}$$

2.2 Gravitational Acceleration

The gravitational acceleration at any given location is obtained by computing the gradient of the potential. Since the potential is given as a function of Earth-fixed spherical coordinates, it is most convenient to compute the gradient in the same system. In Earth-fixed spherical coordinates, this gradient is,

$$\vec{a} = \nabla U = \frac{\partial U}{\partial r}\vec{u}_r + \frac{1}{r}\frac{\partial U}{\partial \phi}\vec{u}_\phi + \frac{1}{r\cos\phi}\frac{\partial U}{\partial \lambda}\vec{u}_\lambda \tag{20}$$

where $\vec{u}_r$, $\vec{u}_\phi$ and $\vec{u}_\lambda$ are unit vectors in the r, ϕ, λ basis. This basis has $\vec{u}_r$ pointing along the radius vector to the satellite, $\vec{u}_\phi$ is in the direction of increasing north latitude and $\vec{u}_\lambda$ is in the direction of increasing East longitude. The acceleration vector obtained from this expression will be the inertial acceleration for the point of interest. Though, as noted, the components of the acceleration are given in the Earth-fixed coordinate system. For most applications it will be desired to have the components of the acceleration expressed in an inertial (non-rotating) coordinate system. This is accomplished by applying the appropriate coordinate transformation from the spherical coordinates to the desired coordinate system. So as a first step, the components of the inertial acceleration in the Earth-fixed (rotating) coordinate system are obtained. Substituting for the gravitational potential (Eq. 1) and taking the indicated partials in Eq. 20 gives the acceleration vector,

$$\begin{aligned}
\vec{a} &= \left\{-\frac{\mu}{r^2}\sum_{\ell=0}^{\infty}(\ell+1)\left(\frac{a_e}{r}\right)^{\ell}\sum_{m=0}^{\ell}P_{\ell,m}(\sin\phi)\left[C_{\ell,m}\cos m\lambda + S_{\ell,m}\sin m\lambda\right]\right\}\vec{u}_r \\
&+ \left\{\frac{\mu}{r^2}\sum_{\ell=1}^{\infty}\left(\frac{a_e}{r}\right)^{\ell}\sum_{m=0}^{\ell}\frac{\partial P_{\ell,m}(\sin\phi)}{\partial\phi}\left[C_{\ell,m}\cos m\lambda + S_{\ell,m}\sin m\lambda\right]\right\}\vec{u}_\phi \\
&+ \left\{\frac{\mu}{r^2}\sum_{\ell=1}^{\infty}\left(\frac{a_e}{r}\right)^{\ell}\sum_{m=1}^{\ell}m\frac{P_{\ell,m}(\sin\phi)}{\cos\phi}\left[-C_{\ell,m}\sin m\lambda + S_{\ell,m}\cos m\lambda\right]\right\}\vec{u}_\lambda
\end{aligned} \tag{21}$$

Notice that the leading term of the radial component (degree and order equal to zero) is simply the expected two-body gravitational acceleration $-\mu/r^2$. Also, if only zonal terms are used ($m = 0$), then the longitudinal component of the acceleration is zero.

Next, the Earth-fixed cartesian components of the acceleration can be obtained by rotating from the spherical coordinates to the x, y, z basis. Let the components of the acceleration in spherical coordinates be represented by,

$$\vec{a} = a_r \vec{u}_r + a_\phi \vec{u}_\phi + a_\lambda \vec{u}_\lambda \tag{22}$$

where the components a_r, a_ϕ and a_λ are given in Eq. 21. The acceleration vector in cartesian coordinates can be written as,

$$\vec{a}_{xyz} = a_x \vec{u}_x + a_y \vec{u}_y + a_z \vec{u}_z \tag{23}$$

where $\vec{u}_x$, $\vec{u}_y$ and $\vec{u}_z$ are the cartesian unit vectors in the Earth-fixed (rotating) coordinate system. The cartesian components of the acceleration can be obtained from the spherical coordinate components through the standard transformation,

$$\begin{vmatrix} a_x \\ a_y \\ a_z \end{vmatrix} = \begin{vmatrix} \cos\phi\cos\lambda & -\sin\phi\cos\lambda & -\sin\lambda \\ \cos\phi\sin\lambda & -\sin\phi\sin\lambda & \cos\lambda \\ \sin\phi & \cos\phi & 0 \end{vmatrix} \begin{vmatrix} a_r \\ a_\phi \\ a_\lambda \end{vmatrix} \tag{24}$$

Having obtained the Earth-fixed cartesian components of the acceleration one further coordinate transformation is necessary to obtain the acceleration components in the defined inertial coordinate system. If the matrix T represents the coordinate transformation from the Earth-fixed system to the inertial coordinate system, then the acceleration components in the inertial system will be,

$$\vec{a}_{XYZ} = T\vec{a}_{xyz} \tag{25}$$

where $\vec{a}_{XYZ}$ is the inertial acceleration vector in inertial coordinates,

$$\vec{a}_{XYZ} = a_X \vec{u}_X + a_Y \vec{u}_Y + a_Z \vec{u}_Z \tag{26}$$

with $\vec{u}_X$, $\vec{u}_Y$ and $\vec{u}_Z$ being the unit vectors of the cartesian inertial coordinate system. In component form, this final transformation will have the structure,

$$\begin{vmatrix} a_X \\ a_Y \\ a_Z \end{vmatrix} = \begin{vmatrix} T_{11} & T_{12} & T_{13} \\ T_{21} & T_{22} & T_{23} \\ T_{31} & T_{32} & T_{33} \end{vmatrix} \begin{vmatrix} a_x \\ a_y \\ a_z \end{vmatrix} \tag{27}$$

The actual elements of the transformation matrix T depend on the inertial coordinate system being used. In the most general case, this transformation will account for polar motion (the motion of the spin axis with respect to the Earth crust), Earth rotation (the largest effect) and, precession and nutation (the motion of the spin axis with respect to the stars). In the simplest case, all of these effects are neglected except for Earth rotation. This defines a coordinate system with the same z axis as the Earth-fixed system but not rotating with the Earth. For many applications such a system is effectively inertial. The transformation from the Earth-fixed system to this non-rotating system is simply,

$$T = \begin{vmatrix} \cos\theta & -\sin\theta & 0 \\ \sin\theta & \cos\theta & 0 \\ 0 & 0 & 1 \end{vmatrix} \tag{28}$$

where θ is the Greenwich Hour Angle (the angle from a reference direction, usually the Vernal Equinox, to the Greenwich meridian).

2.2.1 Example Equations of Motion

The equations of motion for a satellite moving in a gravitational field modeled in spherical harmonics are given by Eq. 27. In the most general case, the components of the acceleration in spherical coordinates are evaluated (Eq. 21) complete to the highest degree and order of the gravity model being used. These components are then rotated to Earth-fixed cartesian components (Eq. 24) and finally to inertial coordinates as shown in Eq. 27.

As an example, the resulting equations of motion due to the second and third degree zonal will be presented. The second degree zonal is the most important (largest) term of the potential and is necessary for the most rudimentary modeling of orbital motion. The third degree zonal is responsible for long period (on the order of 100 days) variations in the elements of an orbiting spacecraft and should be included for any long term orbit propagation studies. Assume that the inertial coordinate system to be used is one neglecting polar motion, precession and nutation. That is, the matrix defined in Eq. 28 is the transformation from Earth-fixed (rotating) coordinates to the inertial (non-rotating) system. Letting, X, Y, Z represent the location of the satellite in this inertial system, the equations of motion for the satellite are,

$$\begin{aligned} \ddot{X} = & -\frac{\mu}{r^2}\frac{X}{r} - \frac{\mu}{r^2}\left(\frac{a_e}{r}\right)^2 C_{2,0}\left[\frac{15}{2}\left(\frac{Z}{r}\right)^2 - \frac{3}{2}\right]\frac{X}{r} \\ & -\frac{\mu}{r^2}\left(\frac{a_e}{r}\right)^3 C_{3,0}\left[\frac{35}{2}\left(\frac{Z}{r}\right)^3 - \frac{15}{2}\frac{Z}{r}\right]\frac{X}{r} \end{aligned} \tag{29}$$

$$\begin{aligned} \ddot{Y} = & -\frac{\mu}{r^2}\frac{Y}{r} - \frac{\mu}{r^2}\left(\frac{a_e}{r}\right)^2 C_{2,0}\left[\frac{15}{2}\left(\frac{Z}{r}\right)^2 - \frac{3}{2}\right]\frac{Y}{r} \\ & -\frac{\mu}{r^2}\left(\frac{a_e}{r}\right)^3 C_{3,0}\left[\frac{35}{2}\left(\frac{Z}{r}\right)^3 - \frac{15}{2}\frac{Z}{r}\right]\frac{Y}{r} \end{aligned} \tag{30}$$

$$\begin{aligned} \ddot{Z} = & -\frac{\mu}{r^2}\frac{Z}{r} - \frac{\mu}{r^2}\left(\frac{a_e}{r}\right)^2 C_{2,0}\left[\frac{15}{2}\left(\frac{Z}{r}\right)^2 - \frac{9}{2}\right]\frac{Z}{r} \\ & -\frac{\mu}{r^2}\left(\frac{a_e}{r}\right)^3 C_{3,0}\left[\frac{35}{2}\left(\frac{Z}{r}\right)^4 - 15\left(\frac{Z}{r}\right)^2 + \frac{3}{2}\right] \end{aligned} \tag{31}$$

The Greenwich Hour Angle, θ, appearing in the transformation matrix (Eq. 28) does not appear in these equations of motion, because there is not any longitudinal variation in the potential when only zonal terms are used. So the acceleration of the spacecraft is independent of it's longitudinal location with respect to Greenwich.

3 Earth's Gravity Field

Numerous spherical harmonic models of the Earth's gravity field have been developed. These models are primarily based on the Earth-based tracking of low-Earth orbit satellites. Other data types that are valuable in estimating the Earth's gravity field include surface gravity measurements, satellite-to-satellite tracking and more recently, satellite radar-altimeter measurements of the ocean surface. The maximum degree (ℓ) of the spherical harmonic representations of the Earth is more than 300 in some models. Models based solely on satellite tracking data usually have a maximum degree of approximately 50. Indeed, for most satellite applications, high degree models are not needed due to the insensitivity of the satellite motions to the small scale features represented in such models.

A good general purpose model for satellite applications is the GEM-T1 model developed at the NASA Goddard Space Flight Center (GSFC) [*Marsh et al.*, 1988]. This model is based solely on the tracking data of Earth satellites and is complete to degree and order 36. The accuracy of this model is greatly improved with respect to earlier Goddard Earth Models (GEM). In addition to estimating the spherical harmonic coefficients of the gravity field, GSFC also estimates the accuracy of those coefficients. Such accuracy estimates are very valuable when attempting to estimate the orbit error which may be induced when using the model for orbit propagation.

Recently, a newer GEM (GEM-T2) model has been published [*Marsh et al.*, 1990]. This model is based on more satellite tracking data and includes some coefficients beyond degree 36. The overall accuracy is also improved with respect to GEM-T1. For purposes of discussion the GEM-T1 model will be used for the analyses presented in this paper.

Table 2

GEM-T1 GRAVITY MODEL
NORMALIZED ZONALS ($\bar{C}_{\ell,0}$)
(units of 1 x 10^{-9})

ℓ	m	$\bar{C}_{\ell,m}$	σ	ℓ	m	$\bar{C}_{\ell,m}$	σ	ℓ	m	$\bar{C}_{\ell,m}$	σ
2	0	−484164.97	0.4	14	0	−19.73	4.0	26	0	1.84	4.7
3	0	957.24	0.1	15	0	1.87	4.1	27	0	4.12	4.3
4	0	538.73	1.1	16	0	−9.38	4.5	28	0	−5.85	5.0
5	0	68.78	0.3	17	0	20.40	4.0	29	0	−3.91	4.7
6	0	−148.10	1.5	18	0	11.29	4.1	30	0	−0.27	4.6
7	0	90.53	0.8	19	0	−4.61	3.4	31	0	5.12	4.7
8	0	45.90	2.4	20	0	15.31	5.1	32	0	0.08	4.6
9	0	28.38	1.5	21	0	9.78	2.8	33	0	2.23	4.5
10	0	57.22	2.6	22	0	−4.84	4.8	34	0	−2.48	4.0
11	0	−51.26	2.5	23	0	−24.13	2.6	35	0	1.27	4.7
12	0	32.08	3.5	24	0	−0.96	4.9	36	0	0.74	4.1
13	0	42.23	3.5	25	0	6.89	3.4				

The GEM-T1 normalized zonal coefficients, complete to degree 36, are given in Table 2. The non-zonal coefficients, up to degree and order 20, are given in Table 3.

Along with the coefficient values in each table, the estimated uncertainty (σ) of the individual coefficients is also given. The gravitational constant and equatorial radius specified for the GEM-T1 model are,

$$\mu = GM = 3.98600436 \times 10^{14}\ \mathrm{m}^3/\mathrm{s}^2 \qquad \text{and} \qquad a_e = 6378137.\ \mathrm{m} \tag{32}$$

Several points can be made by examining the coefficient values and their uncertainties. Foremost, the value of the second degree zonal coefficient is seen to be more than two orders of magnitude larger than any other coefficient. The next largest coefficients are those of degree 2 and order 2. Analogous to the second degree zonal which represents the oblateness of the Earth, these coefficients correspond to the ellipticity about the equator.

Also evident is that the magnitude of the coefficients decreases significantly as the degree increases (keep in mind that these are normalized coefficients and effectively have equal weight in their total contribution to the gravitational potential). This characteristic has been formalized in the so-called "Kaula's Rule" [*Kaula*, 1966]. This rule gives the expected size of the Earth's normalized harmonic coefficients of degree ℓ to be $\pm 10^{-5}/\ell^2$. This rule of thumb allows one to estimate the expected magnitude of a gravitational coefficient if a value is not otherwise known (this is particularly convenient for higher degree coefficients for which accurate estimates have generally not been obtained). The magnitude of the GEM-T1 coefficients (complete to degree and order 36) are plotted in Figure 1 along with Kaula's Rule.

Table 3

GEM-T1 GRAVITY MODEL
NORMALIZED SECTORIALS AND TESSERALS ($\bar{C}_{\ell,m}$ and $\bar{S}_{\ell,m}$)
(units of 1 x 10^{-9})

ℓ	m	$\bar{C}_{\ell,m}$	σ	$\bar{S}_{\ell,m}$	σ	ℓ	m	$\bar{C}_{\ell,m}$	σ	$\bar{S}_{\ell,m}$	σ
						2	2	2438.93	0.3	−1399.84	0.3
3	1	2029.77	1.0	249.59	1.1	3	2	903.55	1.1	−620.42	1.2
4	1	−533.43	0.9	−475.12	0.8	4	2	347.00	1.5	664.03	1.5
5	1	−58.95	2.5	−95.54	2.8	5	2	655.79	2.5	−323.41	2.9
6	1	−81.38	1.8	23.89	1.7	6	2	51.61	3.2	−375.00	3.2
7	1	277.10	3.9	97.82	4.2	7	2	317.71	4.4	91.61	5.2
8	1	28.86	3.1	54.72	3.0	8	2	70.38	5.2	68.45	5.1
9	1	148.04	5.8	24.53	6.0	9	2	31.14	5.9	−32.39	7.4
10	1	76.97	4.4	−138.11	4.3	10	2	−80.52	6.7	−51.34	6.7
11	1	9.50	7.0	−27.81	6.6	11	2	9.05	7.2	−99.24	9.0
12	1	−49.26	5.6	−49.65	5.6	12	2	7.64	7.0	34.92	6.9
13	1	−54.06	6.9	43.46	6.0	13	2	53.44	8.7	−57.58	10.3
14	1	−18.75	6.3	23.22	6.7	14	2	−34.81	7.1	−6.07	7.2
15	1	8.29	6.5	14.21	6.5	15	2	−21.63	9.5	−36.44	10.5
16	1	31.71	6.2	17.35	6.9	16	2	−15.64	9.0	24.54	9.3
17	1	−30.94	8.0	−26.85	9.0	17	2	−5.78	8.8	17.12	9.3
18	1	−0.23	6.1	−45.61	6.8	18	2	8.41	9.9	16.84	10.5
19	1	−11.59	8.9	5.38	10.2	19	2	8.44	9.0	−10.47	9.4
20	1	14.51	6.7	−21.27	6.9	20	2	19.88	9.6	3.23	9.9
3	3	720.99	1.2	1413.17	1.2						
4	3	990.98	0.9	−200.62	0.9	4	4	−190.03	0.9	308.46	0.9
5	3	−448.20	2.0	−215.14	2.1	5	4	−294.82	1.7	52.41	1.7
6	3	61.97	2.5	4.64	2.6	6	4	−92.80	1.9	−473.31	1.9
7	3	250.74	4.1	−209.16	4.3	7	4	−273.74	3.1	−122.02	3.1
8	3	−19.97	4.6	−86.94	4.7	8	4	−246.06	3.6	67.75	3.7
9	3	−155.37	6.6	−84.02	7.0	9	4	−12.83	5.7	23.26	5.5
10	3	1.31	7.1	−161.48	7.5	10	4	−97.31	5.9	−69.38	6.3

Table 3

GEM-T1 GRAVITY MODEL
NORMALIZED SECTORIALS AND TESSERALS ($\bar{C}_{\ell,m}$ and $\bar{S}_{\ell,m}$)

(units of 1×10^{-9})

ℓ	m	$\bar{C}_{\ell,m}$	σ	$\bar{S}_{\ell,m}$	σ	ℓ	m	$\bar{C}_{\ell,m}$	σ	$\bar{S}_{\ell,m}$	σ
11	3	−28.89	8.3	−132.50	8.6	11	4	−33.21	8.9	−70.00	8.7
12	3	32.42	8.7	17.94	9.5	12	4	−65.30	8.5	−3.01	8.9
13	3	−14.03	8.4	83.66	8.8	13	4	−8.82	10.8	−0.37	10.4
14	3	36.93	8.9	22.42	9.6	14	4	−8.83	10.2	1.88	10.7
15	3	44.63	9.2	26.54	9.8	15	4	−44.38	10.5	12.64	10.1
16	3	−32.08	8.3	−45.03	8.5	16	4	36.51	10.5	43.86	11.0
17	3	10.12	10.9	9.89	11.3	17	4	12.59	10.1	31.22	10.0
18	3	−1.00	9.6	−7.05	9.6	18	4	43.42	9.0	6.09	9.4
19	3	1.44	11.0	14.20	11.3	19	4	2.57	10.7	7.67	10.8
20	3	8.27	10.6	13.72	10.6	20	4	−1.80	9.2	0.83	9.6
5	5	177.76	2.2	−666.03	2.3						
6	5	−265.76	1.3	−537.75	1.3	6	6	9.06	0.8	−236.33	0.8
7	5	3.47	2.7	19.65	2.8	7	6	−357.85	0.5	150.92	0.5
8	5	−24.93	2.8	85.30	2.8	8	6	−66.42	2.0	312.83	1.9
9	5	−14.11	4.3	−60.06	4.4	9	6	70.53	3.2	216.63	3.2
10	5	−50.44	5.3	−43.83	5.5	10	6	−34.74	3.5	−77.72	3.3
11	5	45.91	6.9	55.28	7.3	11	6	8.47	5.0	24.29	5.1
12	5	30.60	8.0	−1.47	8.1	12	6	1.39	6.2	45.83	6.1
13	5	59.62	10.0	57.45	10.5	13	6	−22.39	7.7	−11.84	7.7
14	5	22.80	10.1	−11.61	10.0	14	6	−3.19	9.2	6.51	9.1
15	5	16.07	11.5	10.89	12.1	15	6	27.23	10.5	−51.71	10.4
16	5	−7.72	11.3	−1.68	10.9	16	6	17.99	11.1	−26.78	11.0
17	5	−11.15	10.6	−5.60	11.2	17	6	0.29	12.1	−30.42	12.0
18	5	1.74	11.7	21.13	11.4	18	6	31.20	11.0	−8.56	11.0
19	5	−2.45	10.7	17.38	10.8	19	6	−6.25	11.1	3.92	10.8
20	5	−10.42	10.9	0.30	10.7	20	6	12.76	10.6	0.97	10.5
7	7	1.60	2.1	22.00	2.0						
8	7	70.42	1.9	74.86	1.9	8	8	−118.88	3.9	122.33	3.8
9	7	−118.62	3.5	−100.55	3.3	9	8	184.50	3.2	−1.85	3.2
10	7	9.75	3.4	−4.29	3.5	10	8	43.75	3.0	−92.48	2.9
11	7	9.61	4.4	−91.89	4.3	11	8	−6.35	4.1	22.58	4.0
12	7	−12.70	5.2	34.83	5.3	12	8	−21.22	3.0	16.90	3.1
13	7	3.57	6.5	−6.62	6.3	13	8	−12.30	5.0	−11.09	5.0
14	7	37.48	7.4	−4.36	7.6	14	8	−32.94	5.1	−13.18	5.0
15	7	66.71	8.8	11.45	8.8	15	8	−40.67	6.7	24.73	6.7
16	7	3.05	10.1	−9.07	10.1	16	8	−13.44	8.0	2.28	7.9
17	7	22.95	10.4	−11.98	10.5	17	8	31.16	9.2	8.77	9.2
18	7	−0.80	11.3	6.72	11.3	18	8	45.72	10.8	0.43	10.5
19	7	5.14	11.5	−1.66	11.5	19	8	14.86	10.6	−11.33	10.5
20	7	−7.79	11.4	4.90	11.4	20	8	−2.01	10.8	−1.29	10.7
9	9	−55.55	5.9	97.59	5.7						
10	9	128.18	3.5	−48.19	3.5	10	10	94.56	4.3	−20.10	4.2
11	9	−38.78	4.1	40.28	4.0	11	10	−52.06	4.3	−17.61	4.3
12	9	46.94	4.4	13.22	4.5	12	10	−9.13	2.8	31.68	2.7
13	9	20.38	4.4	45.78	4.4	13	10	43.30	4.7	−38.04	4.7
14	9	37.16	5.1	17.93	5.2	14	10	37.00	2.8	−2.80	2.8
15	9	13.44	5.7	41.02	5.7	15	10	9.59	4.9	16.08	4.9
16	9	−16.58	6.2	−50.98	6.3	16	10	−10.42	4.2	6.61	4.1
17	9	−3.20	6.9	−34.32	7.0	17	10	2.10	5.8	20.12	5.8
18	9	−13.52	8.1	19.24	8.1	18	10	9.01	6.0	−10.86	6.0
19	9	1.76	8.6	8.66	8.6	19	10	−35.35	7.4	−2.66	7.3
20	9	22.81	10.0	7.24	10.2	20	10	−22.42	8.0	−8.09	8.0
11	11	54.33	6.2	−54.73	6.3						
12	11	5.41	3.0	−9.52	2.9	12	12	−3.53	2.5	−11.80	2.5
13	11	−40.19	4.4	5.50	4.4	13	12	−28.01	2.5	86.41	2.5
14	11	8.08	3.6	−41.36	3.6	14	12	8.97	1.6	−32.07	1.5
15	11	1.72	4.3	28.93	4.2	15	12	−28.33	3.4	12.49	3.4
16	11	14.02	3.9	−6.44	3.9	16	12	20.88	1.6	5.74	1.6
17	11	−17.11	4.7	17.50	4.7	17	12	34.27	4.1	17.26	4.1

Table 3

GEM-T1 GRAVITY MODEL
NORMALIZED SECTORIALS AND TESSERALS ($\bar{C}_{\ell,m}$ and $\bar{S}_{\ell,m}$)
(units of 1×10^{-9})

ℓ	m	$\bar{C}_{\ell,m}$	σ	$\bar{S}_{\ell,m}$	σ	ℓ	m	$\bar{C}_{\ell,m}$	σ	$\bar{S}_{\ell,m}$	σ
18	11	−12.80	4.0	−0.60	3.9	18	12	−26.18	2.4	−16.53	2.3
19	11	16.48	4.7	13.47	4.7	19	12	3.20	4.7	4.33	4.7
20	11	11.38	4.4	−23.93	4.3	20	12	−4.06	3.2	17.30	3.1
13	13	−61.55	1.3	68.27	1.3						
14	13	31.53	0.9	44.62	0.9	14	14	−50.57	1.8	−6.37	1.8
15	13	−28.11	1.0	−4.98	1.0	15	14	6.17	1.0	−25.61	0.9
16	13	13.08	1.1	0.61	1.1	16	14	−19.12	0.8	−38.29	0.8
17	13	16.91	1.2	20.11	1.2	17	14	−13.34	0.8	11.76	0.8
18	13	−6.58	1.1	−35.16	1.1	18	14	−9.28	1.2	−10.94	1.2
19	13	−6.09	1.6	−29.17	1.6	19	14	−5.12	0.7	−12.64	0.7
20	13	26.65	1.3	4.89	1.3	20	14	10.32	1.8	−11.76	1.8
15	15	−18.09	3.2	−8.09	3.2						
16	15	−12.53	3.9	−32.30	3.9	16	16	−32.41	7.1	−4.37	7.1
17	15	4.94	1.4	5.75	1.4	17	16	−29.07	5.5	1.88	5.5
18	15	−37.76	3.8	−19.82	3.8	18	16	9.79	4.0	5.00	4.0
19	15	−18.32	2.7	−12.77	2.7	19	16	−19.90	5.0	−11.93	5.0
20	15	−22.73	3.5	−0.41	3.5	20	16	−10.67	4.3	1.69	4.3
17	17	−38.31	7.5	−20.62	7.6						
18	17	6.11	7.4	8.77	7.4	18	18	−4.45	11.4	−5.06	11.4
19	17	27.95	4.0	−10.88	4.0	19	18	21.65	9.2	−3.11	9.1
20	17	4.29	7.0	−8.98	7.1	20	18	10.58	6.5	1.30	6.5
19	19	6.46	10.5	10.42	10.5						
20	19	−7.10	9.4	8.46	9.5	20	20	1.71	13.1	−13.51	13.1

The estimated accuracy of the various coefficients shown in Tables 2 and 3 show that the lower degree coefficients are the best determined and the accuracy degrades as the degree increases. This variation in accuracy is a reflection of the fact that satellite tracking data was used to solve for the coefficients. As discussed in later sections, the sensitivity of the satellites to the harmonic coefficients decreases as the degree increases. That is, the low degree coefficients produce large perturbations to the orbital motion and the high degree coefficients produce much smaller perturbations. The ability to recover high degree coefficients is a direct function of the accuracy of the tracking data being utilized and also the geographic distribution of that data. Since the high degree coefficients represent fine scale features in the gravity field it is necessary to have wide geographic coverage, and accurate tracking data, to completely capture such details. The low degree coefficients on the other hand represent the large scale features (continental in size) and it is possible to accurately model such details given sparser geographic coverage.

Overall, the accuracy of the geopotential acceleration is dependent on the constants μ, $\bar{C}_{\ell,m}$ and $\bar{S}_{\ell,m}$ and the accuracy of the coordinate transformation from body-fixed to inertial coordinates. The accuracy is also clearly dependent on the degree of truncation of the infinite series describing the potential. It should be noted that the quantity a_e enters the potential strictly as a scaling factor and thus does not affect the accuracy of the geopotential computation. Of these possible error sources, the accuracy of the $\bar{C}_{\ell,m}$ and $\bar{S}_{\ell,m}$ coefficients is currently the limiting factor in precise low-Earth orbit determination.

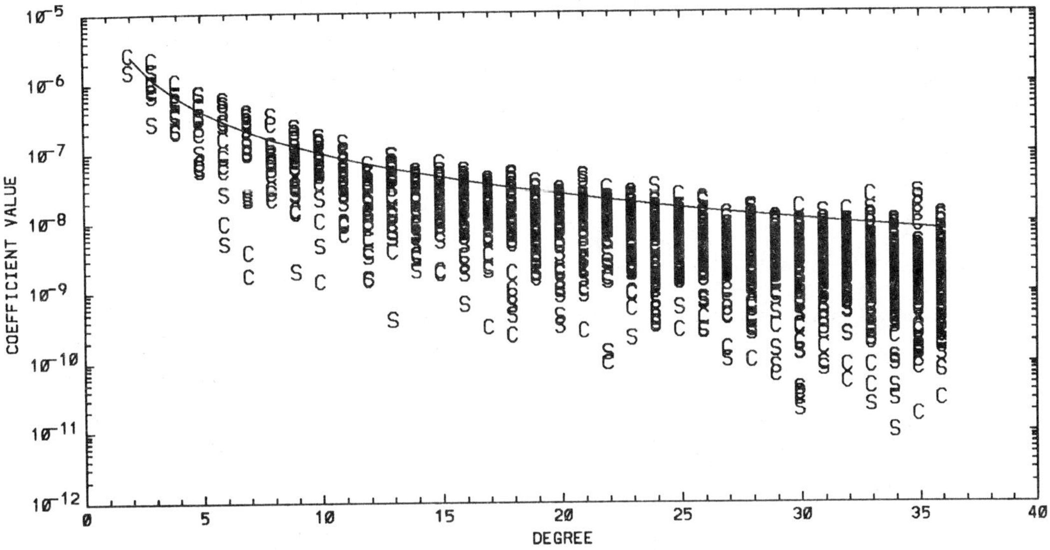

Figure 1: Absolute value of the normalized GEM-T1 coefficients and Kaula's Rule.

4 Orbit Perturbations

A satellite moving in the low-Earth orbit environment does not follow the idealized elliptical path resulting from the consideration of the motion of a particle about a point mass. Instead, the satellite will deviate from the idealized elliptical path as the various acting forces perturb the satellite. In this discussion, only the perturbing forces resulting from the Earth's gravity field are considered. Though it should be emphasized that depending on the actual orbital geometry and spacecraft design, other forces (principally, atmospheric drag, solar radiation pressure and, lunar and solar gravity) may be more important.

The general problem of determining the motion of a satellite orbiting in a gravitational field (described by spherical harmonics) can only be solved through numerical integration. The equations presented in section 3 can be used for this purpose. However, it is possible through various degrees of approximation to analytically determine the important characteristics of the motion.

The largest effect on a satellite orbiting the Earth is that due to the oblateness (second degree zonal). This is not surprising given the size of the second degree zonal with respect to all other coefficients of the gravity field (Tables 2 and 3). The most noticeable effect of the oblateness is to cause the right ascension of the ascending node and argument of periapse of the orbit to vary linearly in time. In the two-body problem these quantities remain constant. The oblateness also causes a small change in the orbital period. Besides these secular effects (linear changes in time) the oblateness causes (relatively) large periodic changes in all of the orbital elements.

A satellite orbiting the Earth can be reasonably approximated by a secularly precessing ellipse. This is an ellipse that has constant values for the semi-major axis, eccentricity and inclination, and values that vary linearly in time for the right ascension of the ascending node, argument of periapse and mean anomaly. The actual motion of the satellite then consists of small deviations away from this secularly precessing ellipse. To characterize these deviations away from this precessing ellipse the solution presented by Kaula [1966] is very useful.

4.1 Kaula's Solution

Kaula's approach was to define a reference orbital path corresponding to the secularly precessing ellipse. The orbital elements that define this reference orbital path are referred to as mean (or averaged) orbital elements. Kaula determined (to first order) the deviation from these mean orbital elements caused by the spherical harmonic coefficients of the gravity field. That is, it is assumed that the orbital elements can be written in the form,

$$\text{semi-major axis} \quad a = \bar{a} + \Delta a \tag{33}$$

$$\text{eccentricity} \quad e = \bar{e} + \Delta e \tag{34}$$

$$\text{inclination} \quad i = \bar{\imath} + \Delta i \tag{35}$$

$$\text{right ascension of the ascending node} \quad \Omega = \dot{\bar{\Omega}} t + \bar{\Omega}_o + \Delta\Omega \tag{36}$$

$$\text{argument of periapse} \quad \omega = \dot{\bar{\omega}} t + \bar{\omega}_o + \Delta\omega \tag{37}$$

$$\text{mean anomaly} \quad M = \dot{\bar{M}} t + \bar{M}_o + \Delta M \tag{38}$$

where the bar quantities represent the orbital elements of the reference ellipse, t is time from a reference epoch and $\bar{\Omega}_o$, $\bar{\omega}_o$ and $\bar{M}_o$ are the values at the reference epoch. The deviations from the reference ellipse will be purely periodic (assuming the mean orbital elements and rates have been accurately defined).

Kaula found that the periodicities of the orbital element variations caused by the gravity field can be expressed in the form,

$$\dot{\Psi} = (\ell - 2p)\dot{\bar{\omega}} + (\ell - 2p + q)\dot{\bar{M}} + m(\dot{\bar{\Omega}} - \dot{\theta}) \tag{39}$$

where $\dot{\theta}$ is the Earth rotation rate, ℓ and m are the spherical harmonic degree and order, p is an integer that can take on the values 0 to ℓ, and q is an integer that can take on the values $\pm\infty$. Kaula's solution is obtained through an expansion in eccentricity which is the reason the index q has an infinite extent. For orbits with small eccentricities only a small range of q values needs to be considered. High eccentricity orbits require the consideration of much greater range in q and Kaula's solution may not be the best approach in such cases.

Analysing the form of the general frequency (Eq. 39) shows that there are 5 categories that the orbital element variations can be placed in. Each of these categories are discussed in the following subsections.

4.1.1 Secular Effects

As mentioned, the Earth's oblateness produces linear changes in time of the node and periapse. For a low altitude orbiter these changes are on the order of a few degrees per day. In addition, the mean motion (time rate of change of mean anomaly) is also altered as compared to it's two-body value. These effects correspond to the particular case of the integer multipliers in the general frequency (Eq. 39) being zero,

$$\ell = 2p \tag{40}$$

$$m = 0 \tag{41}$$

$$q = 0 \tag{42}$$

Thus the terms in the gravity field that can produce such effects are the even degree zonals ($\ell = 2p$, $m = 0$). Since the second degree zonal is several orders of magnitude larger than all other even zonal coefficients, the secular effects produced by this zonal are dominant. The linear rates due to the second degree zonal are (to first order),

$$\frac{d\omega}{dt} = -\frac{3}{4}n\left(\frac{a_e}{a}\right)^2 \frac{(5\cos^2 i - 1)}{(1-e^2)^2} C_{2,0} \tag{43}$$

$$\frac{d\Omega}{dt} = \frac{3}{2}n\left(\frac{a_e}{a}\right)^2 \frac{\cos i}{(1-e^2)^2} C_{2,0} \tag{44}$$

$$\frac{dM}{dt} = n - \frac{3}{4}n\left(\frac{a_e}{a}\right)^2 \frac{(3\cos^2 i - 1)}{(1-e^2)^{3/2}} C_{2,0} \tag{45}$$

where n is the two-body mean motion ($\sqrt{\mu/a^3}$) and $C_{2,0}$ is the unnormalized second degree zonal coefficient. While not explicitly noted (for notational convenience) the right hand side of the equations are evaluated using the mean orbital elements of the reference ellipse.

4.1.2 Long Period Perturbations

If the integer multipliers $\ell - 2p + q$ and m are zero in the general frequency expression (Eq. 39) then the frequency of the orbital elements perturbation will be a multiple of the argument of periapse rate,

$$\dot{\Psi} = -q\dot{\omega} \tag{46}$$

where as noted, q is an integer in the range $\pm\infty$. The longest period resulting in this case will be for $q = \pm 1$. In this case the value of ℓ is constrained to be,

$$\ell = 2p - 1 \quad \text{or} \quad \ell = 2p + 1 \tag{47}$$

which states that the degree must be odd. Since $m = 0$ this states the odd degree zonal coefficients will produce long period perturbations.

As an example, the perturbations produced by the degree 3 zonal coefficient are,

$$\Delta a = 0 \tag{48}$$

$$\Delta e = -\frac{1}{2}\frac{a_e}{a}\sin i\frac{C_{3,0}}{C_{2,0}}\sin\omega \tag{49}$$

$$\Delta i = \frac{1}{2}\frac{a_e}{a}\frac{e}{1-e^2}\cos i\frac{C_{3,0}}{C_{2,0}}\sin\omega \tag{50}$$

$$\Delta\Omega = -\frac{1}{2}\frac{a_e}{a}\frac{e}{(1-e^2)}\frac{\cos i\,(15\sin^2 i-4)}{\sin i\;(5\sin^2 i-4)}\frac{C_{3,0}}{C_{2,0}}\cos\omega \tag{51}$$

$$\Delta\omega = \frac{1}{2}\frac{a_e}{a}\frac{1}{(1-e^2)}\left[\frac{\cos^2 i\,(15\sin^2 i-4)}{\sin i\;(5\sin^2 i-4)}e-\sin i\frac{(1+4e^2)}{e}\right]\frac{C_{3,0}}{C_{2,0}}\cos\omega \tag{52}$$

$$\Delta M = \frac{1}{2}\frac{a_e}{a}\frac{(1-4e^2)}{e(1-e^2)^{1/2}}\sin i\frac{C_{3,0}}{C_{2,0}}\cos\omega \tag{53}$$

All of the perturbations are sinusoidal functions of the argument of periapse and thus will have periods given by $2\pi/\dot{\bar{\omega}}$.

It should be emphasized that these are *linear* perturbations with respect to the defined reference secularly precessing ellipse. The true orbit evolution will differ slightly from these results due to the nonlinear coupling between orbital elements.

4.1.3 *m*-Daily Perturbations

If the integer multiplier of $\dot{\bar{M}}$ in the fundamental frequency (Eq. 39) is zero ($\ell-2p+q=0$) then the frequency is of the form,

$$\dot{\Psi} = -q\dot{\bar{\omega}} + m(\dot{\bar{\Omega}} - \dot{\theta}) \tag{54}$$

Since $\dot{\bar{\omega}}$ and $\dot{\bar{\Omega}}$ are on the order of a few degrees per day and $\dot{\theta}$ (the Earth rotation rate) is a few hundred degrees per day, the magnitude of this frequency is approximately given by $m\dot{\theta}$. So these frequencies are approximately m cycles per day and thus are referred to as the m-daily effects. Both even and odd degree non-zonal coefficients produce such effects. Order 1 produces perturbations with frequencies of approximately once per day, order 2 produces twice per day perturbations, and so on. Since the gravitational coefficients get smaller with increasing order the low order terms produce the most significant effects.

4.1.4 Short Period Perturbations

The effects previously discussed correspond to particular cases where the multiplier of the mean anomaly rate ($\dot{\bar{M}}$) is zero. If the multiplier ($\ell-2p+q$) is non-zero then the resulting frequency is likely to be short period, that is, a multiple of the orbital frequency. (The exception to this is cases of resonance discussed in the following section.) In fact, the vast majority of the perturbation spectrum consists of such short period effects.

The short period variations are for the most part much smaller than the other effects. The exception to this are those short period effects produced by the second degree zonal. Due to the size of the second degree zonal coefficient there are large variations in the orbital elements. For orbits of small (non-zero) eccentricity the largest of these variations are given by,

$$\Delta a = -\frac{3}{2}na\left(\frac{a_e}{a}\right)^2 \sin^2 i \frac{\cos 2(\omega+M)}{\dot{\omega}+\dot{M}} C_{2,0} \tag{55}$$

$$\begin{aligned}\Delta e = & -\frac{3}{4}n\left(\frac{a_e}{a}\right)^2 (2-3\sin^2 i)\frac{\cos M}{\dot{M}} C_{2,0} \\ & -\frac{3}{8}n\left(\frac{a_e}{a}\right)^2 \sin^2 i \frac{\cos(2\omega+M)}{2\dot{\omega}+\dot{M}} C_{2,0} \\ & -\frac{21}{8}n\left(\frac{a_e}{a}\right)^2 \sin^2 i \frac{\cos(2\omega+3M)}{2\dot{\omega}+3\dot{M}} C_{2,0}\end{aligned} \tag{56}$$

$$\Delta i = -\frac{3}{4}n\left(\frac{a_e}{a}\right)^2 \sin i \cos i \frac{\cos 2(\omega+M)}{\dot{\omega}+\dot{M}} C_{2,0} \tag{57}$$

$$\Delta \Omega = -\frac{3}{4}n\left(\frac{a_e}{a}\right)^2 \cos i \frac{\sin 2(\omega+M)}{\dot{\omega}+\dot{M}} C_{2,0} \tag{58}$$

$$\begin{aligned}\Delta \omega = & -\frac{3}{4}n\left(\frac{a_e}{a}\right)^2 \frac{(2-3\sin^2 i)}{e}\frac{\sin M}{\dot{M}} C_{2,0} \\ & +\frac{3}{8}n\left(\frac{a_e}{a}\right)^2 \frac{\sin^2 i}{e}\frac{\sin(2\omega+M)}{2\dot{\omega}+\dot{M}} C_{2,0} \\ & -\frac{21}{8}n\left(\frac{a_e}{a}\right)^2 \frac{\sin^2 i}{e}\frac{\sin(2\omega+3M)}{2\dot{\omega}+3\dot{M}} C_{2,0}\end{aligned} \tag{59}$$

$$\begin{aligned}\Delta M = & \frac{3}{4}n\left(\frac{a_e}{a}\right)^2 \frac{(2-3\sin^2 i)}{e}\frac{\sin M}{\dot{M}} C_{2,0} \\ & -\frac{3}{8}n\left(\frac{a_e}{a}\right)^2 \frac{\sin^2 i}{e}\frac{\sin(2\omega+M)}{2\dot{\omega}+\dot{M}} C_{2,0} \\ & +\frac{21}{8}n\left(\frac{a_e}{a}\right)^2 \frac{\sin^2 i}{e}\frac{\sin(2\omega+3M)}{2\dot{\omega}+3\dot{M}} C_{2,0}\end{aligned} \tag{60}$$

As the sinusoidal dependence on mean anomaly shows, these short period variations occur at frequencies of (approximately) once, twice and three times per revolution. While the results are valid for orbit of all eccentricities, orbits of high eccentricities will have additional variations of similar magnitude. For orbits of small eccentricity these are the largest effects.

4.1.5 Resonant Perturbations

Situations of resonance can arise when a satellites path on the surface of the Earth exactly or nearly exactly repeats after some number of revolutions. The fewer revolutions it takes for the satellite to nearly repeat the larger the resonant effect. Such repeats cause the satellite perturbations to be reinforced and can produce long period variations in the orbital element evolution. In terms of the fundamental frequency, this is characterized by the frequency becoming very small,

$$\dot{\Psi} = (\ell - 2p)\dot{\omega} + (\ell - 2p + q)\dot{M} + m(\dot{\Omega} - \dot{\theta}) \approx 0 \tag{61}$$

This situation is to be distinguished from the previously discussed long period variations where the fundamental frequency was,

$$\dot{\Psi} = -q\dot{\omega} \tag{62}$$

and corresponds to a small frequency since $\dot{\omega}$ is only a few degrees per day. In the present situation the multipliers $\ell - 2p + q$ and m are non-zero and it is by virtual of the values of the orbital rates that the frequency becomes small. Though it should be noted that resonant perturbations can have periods as long or longer than the so-called long period effects. A distinction is made because the two effects arise from two completely different sources.

To see what the condition for resonance is, first rewrite the frequency as,

$$\dot{\Psi} = -q\dot{\omega} + (\ell - 2p + q)(\dot{\omega} + \dot{M}) + m(\dot{\Omega} - \dot{\theta}) \approx 0 \tag{63}$$

since $-q\dot{\omega}$ is small compared to the other two terms it will be neglected. So that the condition for resonance can be written as,

$$m(\dot{\theta} - \dot{\Omega}) \approx (\ell - 2p + q)(\dot{\omega} + \dot{M}) \tag{64}$$

The smallest order m for which this condition will be approximately satisfied will be when $\ell - 2p + q = 1$ so that,

$$m \approx \frac{\dot{\omega} + \dot{M}}{\dot{\theta} - \dot{\Omega}} \tag{65}$$

In general the right hand side will not be an integer (it will be for the special cases of orbits that exactly repeat their ground traces) so the integer closes to the obtained real number will be the coefficient order that produces resonance effects on the satellite. Higher order resonances will also exist and can be found by setting $\ell - 2p + q$ to progressively higher values.

4.2 Examples

The preceding only provides a very limited look at the orbit perturbations produced by the gravity field. Kaula's solution can be used to determine the complete spectrum of orbital variations produced by the spherical harmonic coefficients of the gravity field. However, a presentation of the full solution is beyond the scope of this paper. Example results computed using the full solution will be presented to better illustrate the spectrum of the orbit perturbations.

The orbital perturbations of the Landsat and Lageos satellites orbits will be considered. Landsat is at an altitude of 700 km and is in a sun-synchronous orbit ($\dot{\Omega} \approx 1$ degree per day). Lageos is at approximately 1 Earth radii in altitude and an inclination of 110°.

4.2.1 Lageos

Due to the altitude of the Lageos orbit it is only appreciatively perturb by the lowest degree coefficients of the gravity field. This is favorable from an introductory analysis standpoint since it results in a relatively uncrowded perturbation spectrum. Figure 2 presents the perturbation spectrum of the Lageos eccentricity (Δe). This figure plots the amplitude of the eccentricity perturbation versus the frequency of the perturbation in cycles per day (the Lageos mean motion is approximately 6.4 cycles per day). Recall that the frequencies are all of the form,

$$\dot{\Psi} = (\ell - 2p)\dot{\omega} + (\ell - 2p + q)\dot{M} + m(\dot{\Omega} - \dot{\theta}) \tag{66}$$

and for Lageos,

$$\begin{aligned} \dot{\omega} &= -.21 \text{ degrees per day} \\ \dot{\Omega} &= +.34 \text{ degrees per day} \\ \dot{M} &= 2299.20 \text{ degrees per day} \end{aligned}$$

The spectrum shown in Figure 2 illustrates the 4 types of periodic variations previously discussed. The spectral line furthest to the left (lowest frequency) is the long period variation due to the circulation of the periapse and produced by the odd degree zonals. The frequency of the perturbation is precisely $\dot{\omega}$ (.21 degrees per $day or 5.8 \times 10^{-4}$ cycles per day.

The two spectral lines between the long period perturbation and the line at 1 cycle per day are the resonant perturbations. Evaluating Eq. 65 to find the harmonic order with which Lageos is in resonance gives

$$m \approx \frac{\dot{\omega} + \dot{M}}{\dot{\theta} - \dot{\Omega}} = 6.4$$

This indicates Lageos will be in near resonance with orders 6 and 7 with the stronger perturbations occurring for order 6. Indeed, the larger of the two spectral lines is that due to order 6.

Immediately to the right of the two resonance lines is a line at 1 cycle per day. This is the dominant m-daily perturbation produced by the order 1 coefficients. There are

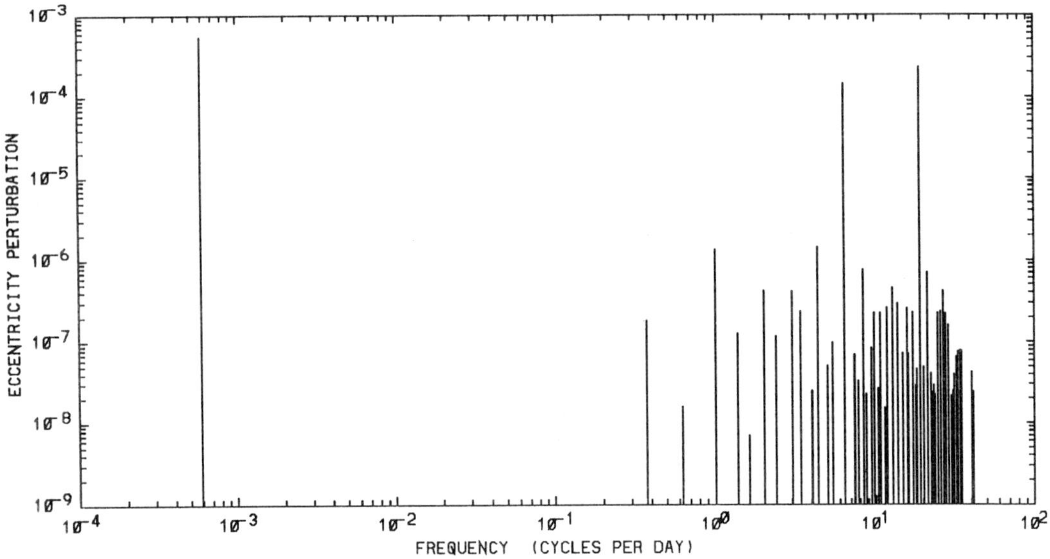

Figure 2: Lageos eccentricity perturbation based on the GEM-T1 gravity model.

also lines at 2, 3, 4 and 5 times per day. Interspersed with these lines are other lines which are some of the numerous short period perturbations.

The two largest short period perturbations are at once and three times per revolution (see Eq. 56). These in fact are the largest perturbations next to the long period perturbation.

Spectrums of the other five orbital element are qualitatively similar in nature (except for the notable exception that semi-major axis does not have a long period variation (see Eq. 48)).

An alternative way of examining the perturbations is to convert the orbital element variations into the equivalent position and velocity perturbations. Depending on the application, knowing the actual position and velocity perturbation spectrum may be more useful than the orbital element spectrums. As an example, Figures 3 and 4 show the magnitude of the radial perturbations affecting the Lageos orbit. Instead of plotting the spectrums, the RMS (root mean square) of the perturbations produced by each degree (Figure 3) or order (Figure 4) of the gravity field are given. This type of analysis gives a clear indication of the size of gravity field that would be required to model satellite motion to a given level. If it is desired to model the motion of Lageos down to the meter level than a gravity field complete to degree and order 6 would be required.

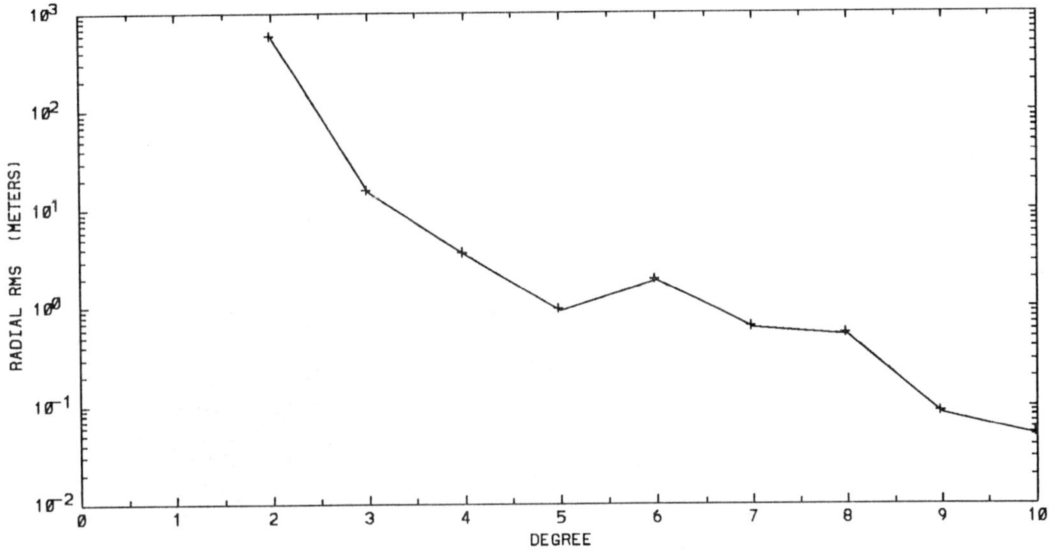

Figure 3: RMS of the Lageos radial orbit perturbations for each coefficient degree. Based on the GEM-T1 gravity model.

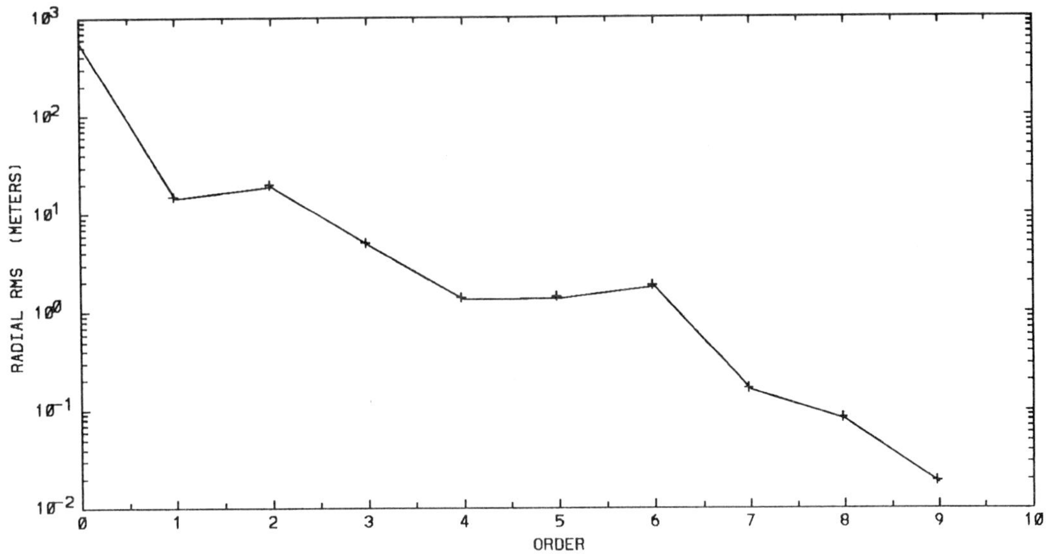

Figure 4: RMS of the Lageos radial orbit perturbations for each coefficient order. Based on the GEM-T1 gravity model.

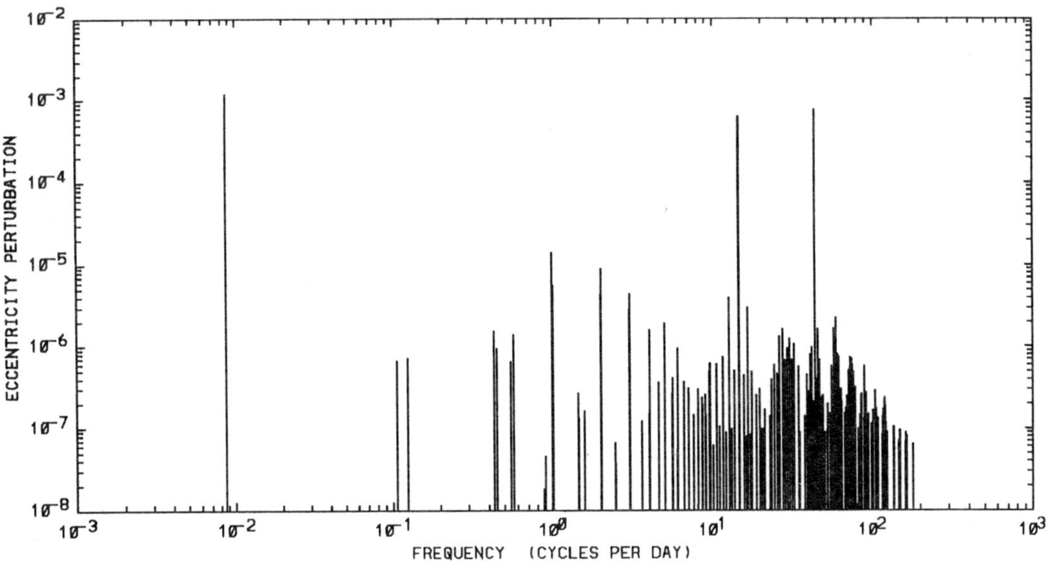

Figure 5: Landsat eccentricity perturbation based on the GEM-T1 gravity model.

4.2.2 Landsat

The Landsat orbit is much lower than Lageos (700 km versus 6000 km in altitude) and thus is much more sensitive to the gravity field. In addition to larger perturbations there are also many more perturbation frequencies (mostly short period). This is evident in the eccentricity perturbation spectrum for Landsat shown in Figure 5. The principal features discussed in the Lageos result are also evident here.

The radial perturbations of the Landsat orbit are given in Figures 6 and 7. These figures clearly indicate the higher sensitivity of the satellite to the gravity field as compared to Lageos. Whereas Lageos could be modeled to the 1 meter level with a gravity model field complete to degree and order 6, Landsat would need a model extending to degree and order 30.

Figure 7 also clearly shows the effect of resonances. For the most part the perturbations get smaller with increasing order of the gravity field. The exception to this is at the resonant orders. For Landsat there is a near resonance effect with orders 14 and 15 and then another resonance at order 29.

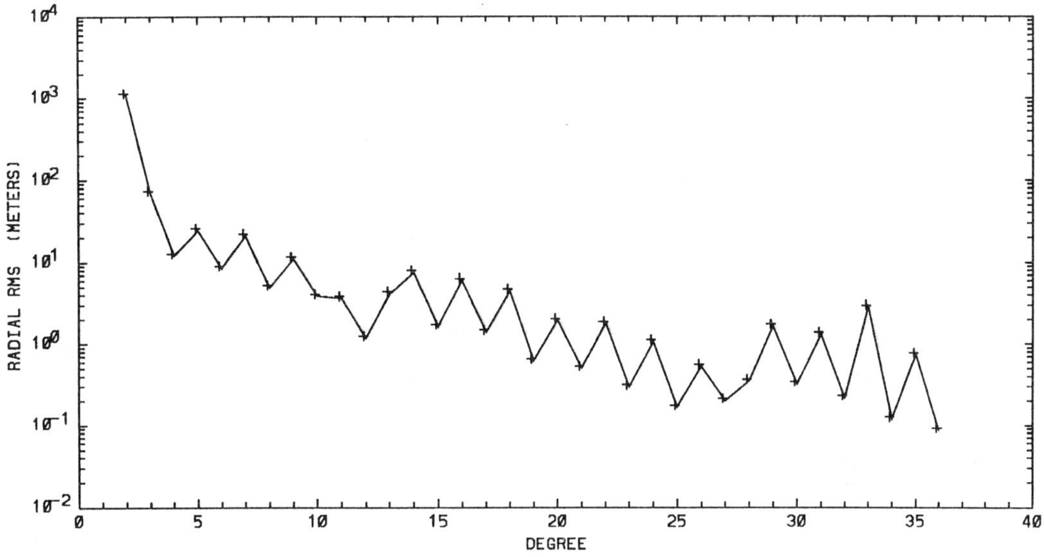

Figure 6: RMS of the Landsat radial orbit perturbations for each coefficient degree. Based on the GEM-T1 gravity model.

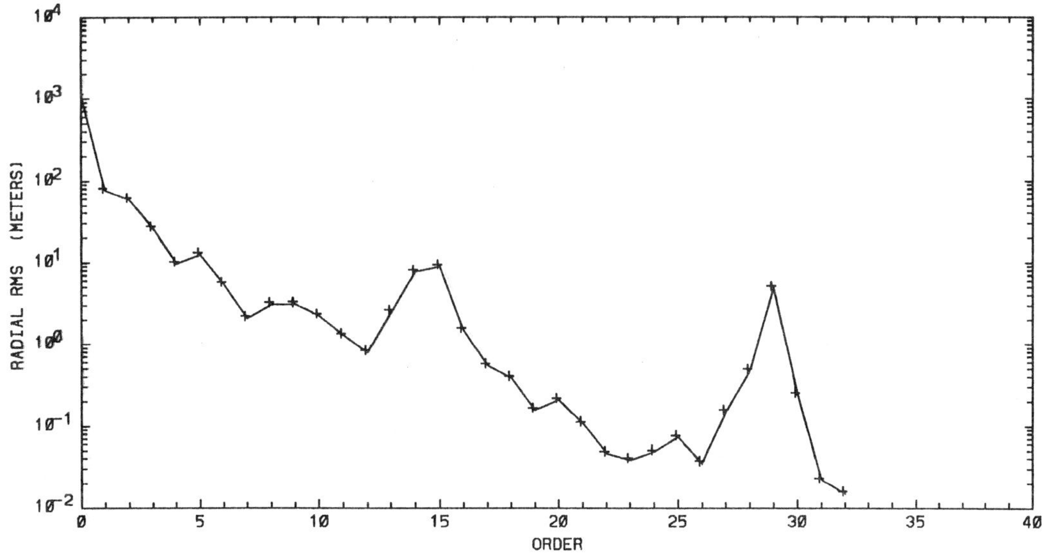

Figure 7: RMS of the Landsat radial orbit perturbations for each coefficient order. Based on the GEM-T1 gravity model.

References

[*Kaula,* 1966] Kaula, W. M., *Theory of Satellite Geodesy,* Blaisdell Publishing Company, Waltham, Massachusetts, 1966.

[*Marsh et al.*, 1988] Marsh, J. G., F. J. Lerch, B. H. Putney, D. C. Christodoulidis, D. E. Smith, T. L. Felsentreger, B. V. Sanchez, S. M. Klosko, E. C. Pavlis, T. V. Martin, J. W. Robbins, R. G. Williamson, O. L. Colombo, D. D. Rowlands, W. F. Eddy, N. L. Chandler, K. E. Rachlin, G. B. Patel, S. Bhati and D. S. Chinn, "A New Gravitational Model for the Earth From Satellite Tracking Data: GEM-T1," *Journal of Geophysical Research*, Vol. 93, pp. 6169–6215, June 1988.

[*Marsh et al.*, 1990] Marsh, J. G., F. J. Lerch, B. H. Putney, T. L. Felsentreger, B. V. Sanchez, S. M. Klosko, G. B. Patel, J. W. Robbins, R. G. Williamson, T. L. Engelis, W. F. Eddy, N. L. Chandler, D. S. Chinn, S. Kapoor, K. E. Rachlin, L. E. Braatz, and E. C. Pavlis, "The GEM-T2 Gravitational Model," *Journal of Geophysical Research*, Vol. 95, pp. 22,043–22,071, December 1990.

AAS 91-011

ATMOSPHERIC DRAG

R. G. Roble*

The Earth's thermosphere is driven primarily by the absorption of solar ultraviolet radiation at wavelengths less than 200 nm. In the upper thermosphere, solar heating drives a diurnal circulation from the dayside to the nightside of the Earth. In the lower thermosphere, a semi-diurnal circulation dominates. It is driven by semi-diurnal in-situ heating components and upward propagating tides from the lower atmosphere. Large amounts of solar wind energy are also deposited in the high latitude thermosphere through auroral processes. This energy sometimes exceeds the locally absorbed solar electromagnetic energy. As a result, the thermosphere is a dynamically active region with large variations about its global mean state caused by variations in the solar ultraviolet radiation and solar wind energy transferred into the Earth's atmosphere through auroral processes.

A thermosphere/ionosphere general circulation model (TIGCM) has been developed to describe the basic structure and circulation of the coupled thermosphere/ionosphere system and its response to solar UV and auroral variability. Time-dependent simulations of the thermosphere and ionosphere response to variable auroral inputs display a wide variety of dynamic phenomena that depend upon the nature and time history of the forcings. Results from a TIGCM simulation of the geomagnetically disturbed period, March 20 through April 10, 1979, a period of moderate geomagnetic activity, are described. The results show considerable density variability associated with changes in the mean motion and large amplitude waves that can alter the quiet-time atmospheric drag experienced by orbiting spacecrafts.

INTRODUCTION

The thermosphere is that rarefied portion of our atmosphere, lying between 80 and 500 km above the Earth's surface, where the temperature increases dramatically with altitude. Many satellites orbit in the upper thermosphere, experiencing a drag that eventually brings them down to burn up in the lower atmosphere. The variable properties of the thermosphere

* Senior scientist at the High Altitude Observatory, National Center for Atmospheric Research, Box 3000, Boulder, Colorado 80307. The National Center for Atmospheric Research is sponsored by the National Science Foundation.

make accurate predictions of satellite drag difficult; witness the case of Skylab, which plunged to a well-publicized death in Australia several years earlier than predicted.

Embedded within the thermosphere is the ionosphere, a weakly ionized plasma. The charged particles that comprise the ionosphere are strongly influenced by electric and magnetic fields. These fields, in turn, are affected by the distributions and motions of the charged particles. Mutual neutral gas/plasma interactions affect the overall structure of both the ionosphere and thermosphere. Of practical importance are the ionosphere's influence on radio waves and the capability of ionospheric electrical currents to produce magnetic disturbances on the ground.

In this tutorial the progress made in modelling these upper atmosphere regions is reviewed. Particular emphasis will be placed on results obtained from the National Center for Atmospheric Research (NCAR) thermosphere/ionosphere general circulation model (TIGCM) in describing thermospheric density variation that affect atmospheric drag. The possibility of using such a model for real-time forecasting of atmospheric density variations for satellite orbital determinations will also be discussed.

GLOBAL MEAN THERMOSPHERIC STRUCTURE

The Earth's atmosphere below 80 km is almost entirely molecular: about 78% molecular nitrogen, 21% molecular oxygen, and 1% other minor gas constituents, such as ozone, carbon dioxide, and argon. Above 80 km, in the thermosphere, these molecular species are subjected to intense solar UV radiation. The molecules absorb both solar EUV radiation, (with wavelength's less than 100 nm) and solar UV radiations (with wavelength's between 100 nm and 200 nm) because of the large absorption and ionization cross sections of thermospheric atoms and molecules at these wavelengths. Within the thermosphere, a solar photon with a wavelength shorter than about 200 nm, which is absorbed by either N_2, O_2, or O eventually results in a molecular oxygen molecule being photodissociated into two oxygen atoms, even though many complex chemical reactions may be involved in the chain of intermediate processes. For two oxygen atoms to recombine chemically back to molecular oxygen, they must collide with a third body, M, $(O + O + M \rightarrow O_2 + M)$. Above 100 km, such a reaction proceeds slowly because the air density is low. Thus, atomic oxygen produced in the high thermosphere must move downwards into the denser regions of the atmosphere, where collisions occur rapidly enough for the recombination rate to exceed the photodissociation rate. Above about 150 km, atomic oxygen is the major atmospheric constituent of the thermosphere.

In the tenuous upper thermosphere, above about 110 km, atmospheric turbulence dies away and molecular diffusion processes become important. As a first approximation, the atmospheric constituents adjust themselves in the Earth's gravitational field according to their molecular weights as shown in Fig. 1. Heavier gases, such as N_2 and O_2 occupy the lower thermosphere and the lighter gases, O, He, and H, dominate in the higher altitudes. One of the fundamental study areas in thermospheric research concerns the cycle of atomic oxygen as illustrated in Fig. 2 – its formation, transport by molecular and turbulent eddy mixing processes and by the global circulation of the thermosphere and recombination back to molecular oxygen – for these processes largely determine the density and compositional structure of the atmosphere above 80 km.

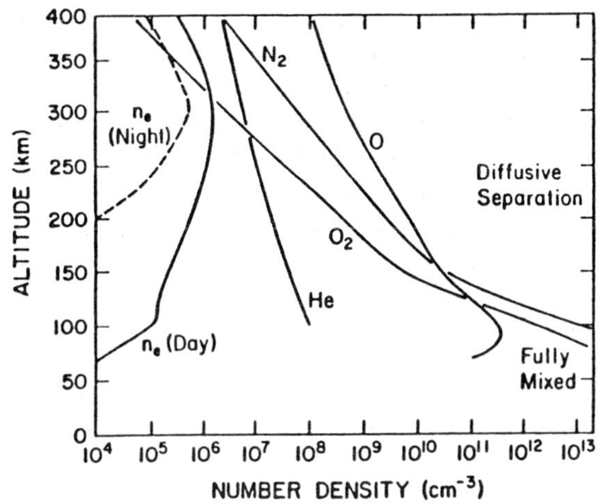

Fig. 1. Distribution of some important constituents in the thermosphere: atomic oxygen, O; molecular nitrogen, N_2; molecular oxygen, O_2; helium, He and ionospheric electrons n_e for both daytime and nighttime conditions.

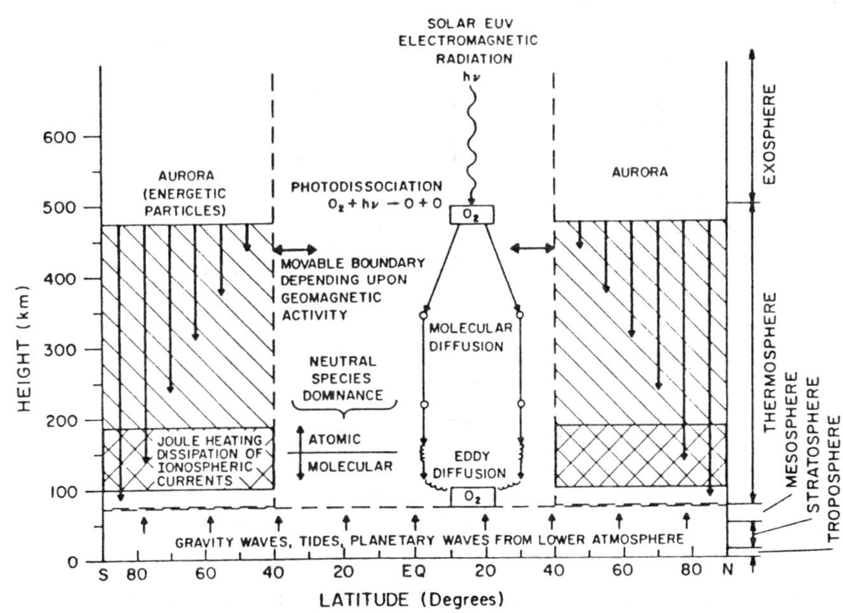

Fig. 2. Schematic illustrating the main processes and energy sources that govern thermospheric structure and dynamics.

About 60% of the total solar energy absorbed in the thermosphere and ionosphere heats the ambient neutral particles; 20% is radiated out of the thermosphere as UV airglow from excited species; and 20% is stored in the chemical energy of the dissociated oxygen atoms, which is released later when recombination occurs in the lower thermosphere. Most of the neutral-gas heating is derived from excess energy released by the products of ion-neutral and neutral-neutral chemical reactions. Collision between these reaction products and ambient neutral particles result in heating of the ambient ion and neutral gases.

Atomic and molecular gases in the thermosphere do not effectively radiate heat to space by infrared processes. The average vertical temperature structure is determined mainly by a balance of local heating with downward thermal conduction to the region of minimum temperature near an altitude of 80 km. There, energy is radiated to space by optically active constituents such as carbon dioxide and ozone, which cannot exist at higher altitudes because of the intense photodissociation of these species in the thermosphere. Infrared nitric oxide radiation at 5.3 μm and atomic oxygen at 63 μm does participate in radiational cooling of the thermosphere above 120 km but they are relatively small compared to downward molecular thermal conduction.

For heat to be conducted downward in the thermosphere, its temperature must increase with altitude. In fact, the global mean thermospheric temperature increases from about 200 K near 80 km to between 600 K and 1200 K above 300 km, depending upon the level of solar EUV and UV output as shown in Fig. 3. At that altitude, molecular thermal conduction is so fast that vertical temperature differences are largely eliminated. The isothermal temperature in the upper thermosphere is called the "exospheric temperature." Above about 500 km, in the exosphere, collisions between atmospheric gases are so rare that energetic particles can either escape from the Earth's gravitational pull or can obtain ballistic trajectories in their transit.

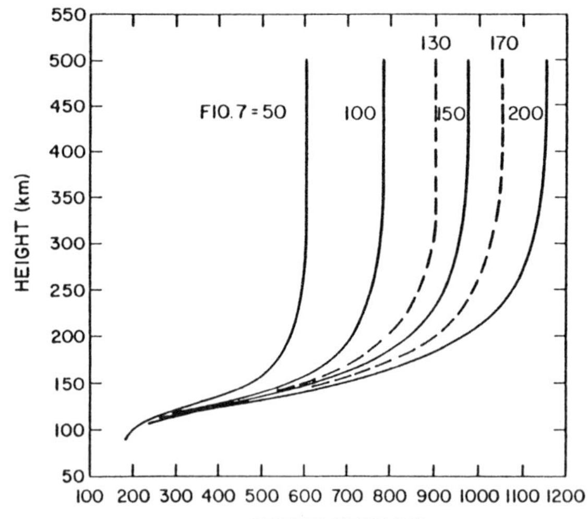

Fig. 3. The global mean temperature profile in the thermosphere for various levels of solar activity which is represented as the level of the 10.7 cm radio emission from the sun (F10.7) in units of $10^{-22}Wm^{-2}Hz^{-1}$. An F10.7 of about 70 represents solar cycle minimum whereas an F10.7 of 200 is representative of solar cycle maximum activity.

The temperature profile is strongly dependent on the intensity of solar EUV radiation, which varies with time and correlates with the sunspot cycle (approximately 11 years). During sunspot minimum, solar EUV output produces an exospheric temperature between 500 and 700 K. At sunspot maximum, the solar EUV output can be as much as 10 times stronger at short wavelengths and 2–3 times stronger at longer EUV wavelengths, producing a global mean exospheric temperature of about 1200 K as shown in Fig. 3. The solar $F_{10.7}$ index on the curves is a measure of the solar 10.7 cm radio flux that can be measured from the ground and has been shown to be a measure of the solar EUV output. Typical values during solar minimum are 70 ($\times 10^{-22} Wm^{-2} Hz^{-1}$) and 250 during solar maximum.

The chemical composition and densities of the thermosphere and ionosphere also vary with the solar EUV output and temperature profile. The intense radiation at sunspot maximum dissociates more molecular oxygen, thereby increasing the magnitude of the atomic oxygen cycle. It also increases the magnitude of various other chemical cycles and causes larger electron and ion densities. In addition, the increased heating causes the atmosphere to expand outward from the Earth. Satellites at a given altitude experience variations in atmospheric drag as the atmosphere expands and contracts in response to changing solar EUV output.

THERMOSPHERIC CIRCULATION

As the Earth rotates about its axis, absorption of solar energy in the thermosphere undergoes a daily variation. Dayside heating causes the atmosphere to expand; the loss of heat at night contracts it. This heating pattern creates pressure differences that drive a global thermospheric circulation, transporting heat from the warm dayside to the cool nightside.

The meteorology of the thermosphere differs from the familiar weather we experience near the Earth's surface. The fluid motions are governed by the same equations as those used by meteorologists studying weather systems in the lower atmosphere; however there are notable differences between some of the physical and chemical processes in the upper and lower atmosphere. In the troposphere, the temperature decreases with altitude, while in the thermosphere the temperature increases with altitude. This difference makes the thermosphere more stable than the troposphere. Two forces that are important at thermospheric heights are molecular viscosity and ion drag. Viscosity is small in the lower atmosphere, but within the thermosphere it increases exponentially with altitude by several orders of magnitude as thermospheric density decreases. Viscosity primarily transfers momentum between various altitudes, smoothing out vertical gradients in wind velocity. Above 300 km, the atmospheric layers are coupled strongly by viscosity and wind velocity becomes nearly constant with altitude.

The ion drag force is caused by collisions between the charged particles of the ionosphere and the neutral particles of the thermosphere. Above 125 km, where the ion gyrofrequency (the frequency at which an ion spirals around a magnetic field line) is greater than the ion-neutral collision frequency; ions are essentially locked to the Earth's geomagnetic field lines and can move across them only in response to an electric field. A neutral-gas wind flowing through the relatively immobile ions experience a collisional drag that is largest at the peak of the ionospheric layer, providing the main resistance that balances the driving pressure force within the upper thermosphere. In the lower atmosphere, the pressure force is primarily balanced by the Coriolis force, and winds flow perpendicular to pressure gradients. In the upper thermosphere, on the other hand, the air motion is nearly parallel to the horizontal pressure gradient, flowing from the dayside high to the nightside low.

Another difference between tropospheric meteorology and thermospheric meteorology is the presence in the thermosphere of large-amplitude waves that propagate vertically with periods ranging from several minutes to one day. The amplitude of these waves is large because they tend to grow exponentially as they propagate upward into increasingly rarefied regions. Waves with periods of 12 and 24 hours are produced by the daily variation of solar UV absorption in the Earth's ozone layer but propagate upward into the thermosphere. These waves, called thermal tides, can produce winds that move at about 100 m/s in the thermosphere. The thermal tide is produced by the heating of atmospheric gases, rather than, as with ocean tides, by the gravitation attraction of the moon and sun.

Many important features of thermospheric circulation can be illustrated by describing the longitudinally averaged vertical and north-south components of motion as shown in Fig. 4. During spring and fall equinoxes, solar radiant heating reaches a maximum near the equator. The heated air expands and rises, flowing northward and southward towards the poles which are cooler regions of less pressure. The mean solar driven circulation in the upper thermosphere during equinoxes is from the equator toward both poles, as shown in the upper left panel of Fig. 4.

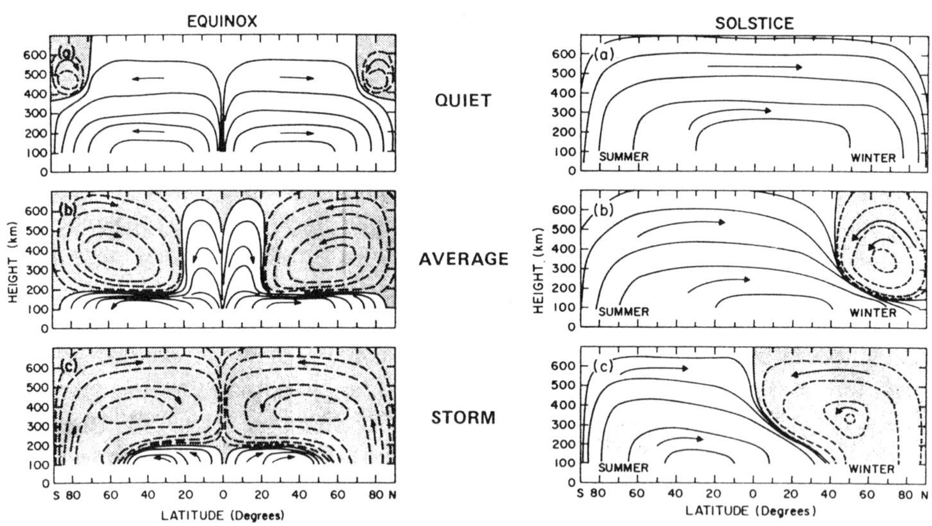

Fig. 4. Schematic diagram of the zonal mean meridional circulation in the Earth's thermosphere during equinox and solstice for various levels of auroral activity: (a) quiet auroral activity where circulation is primarily driven by solar UV heating, (b) average auroral activity where heating is 10^{11} watts, and (c) during geomagnetic storms where the heating is about 10^{12} watts. The contours schematically illustrate the mass flow and the arrows indicate the direction of the motion.

During solstices, the longitudinally averaged solar heating is greatest in the summer hemisphere and least in the winter hemisphere. This hemispherical asymmetry in solar heating drives a summer-to-winter circulation, an upward motion over the summer hemisphere and a downward motion over the winter hemisphere, accompanied by summer-to-winter hemisphere wind as shown in the upper right panel of Fig. 4.

In addition to solar radiative heating driving the basic thermospheric circulation, there are important heat and momentum sources that are deposited in the high latitude thermosphere by auroral processes. The aurora is driven by plasma processes associated with the interaction of the solar wind with the Earth's geomagnetic field. Complex interactions between the solar wind and magnetosphere energize the plasma causing some of it to be transferred into the ionosphere and thermosphere. Energetic electrons bombard the atmosphere in the auroral ovals, in magnetic conjugate hemispheres, of the polar cap. The electrons are sufficiently energetic to ionize atoms and dissociate molecules and produce copius amounts of atmospheric emissions that can be seen from the ground and by satellites from space. The electrons flowing into the ionosphere and thermosphere carry an electrical current of about a million amperes. During intense geomagnetic or auroral storms, the electric currents and energetic electrons may deposit as much energy at high latitudes as that from the absorption of solar UV radiation. Typically, the global energy input during very quiet auroral activity is about 10^{10} W. This increases to 10^{11} W during moderate auroral activity and to 10^{12} W for short periods during intense auroral storms.

The thermosphere responds to this energy input in a number of ways. The strong and rapid changes in energy input generate a variety of atmospheric waves, ranging from subaudible sound waves, which can be detected on the ground, up to waves of with periods of hours and horizontal wavelengths of thousands of kilometers which can propagate throughout the entire thermosphere. These larger scale waves have wind speeds of up to several hundred meters per second and temperature enhancements of hundreds of degrees K associated with them and they can produce major density perturbations. A longer-lasting response of the thermosphere, however, is produced by Joule heating, due to the dissipation of ionospheric current systems and the Ampére acceleration of the winds due to momentum transfer to the neutral gas by the strong currents in the auroral zone. Joule heating can raise the air temperature by hundreds of degrees K in the auroral zones at high latitudes and drive a circulation that tends to oppose the circulation driven by solar radiant heating during equinox as shown in the middle panel of Fig. 4. During average auroral activity, the circulation below 150 km is dominated by the solar UV heating, giving an equator-to-pole flow; however, above that altitude, the high-latitude heat source becomes important and the flow is equatorward to about 20 degrees latitude. During large auroral storms, the equatorward circulation increases greatly, extending to the equator and to lower latitudes as shown in the lower left panel in Fig. 4.

During solstice, the high latitude heat source associated with average auroral activity reinforces the solar driven circulation in the summer hemisphere but forces a reverse circulation in the winter hemisphere as shown in the right panels of Fig. 4. Below about 150 km, however, the summer-to-winter circulation is maintained at all latitudes. During auroral storms at solstice, the circulation is equatorward in both hemispheres above 300 km with an asymmetry at lower altitudes. Thus it is evident that the mean motions in the thermosphere are in a constant state of agitation, depending upon the amount and duration of high latitude heating from auroral processes.

The alteration of thermospheric circulation produced by magnetic storms produces other changes in the state of the upper atmosphere. One is a world-wide increase in the thermospheric temperature. This temperature increase – hundreds of degrees – causes the atmosphere above 100 km to expand. Consequently, the atmospheric drag on satellites is increased during auroral storm periods, shortening their orbital lifetimes.

Stormtime circulation also stirs up the chemically heterogeneous thermosphere. Upward motion in the auroral zones lifts the heavier atmospheric species N_2 and O_2 into the upper thermosphere, increasing their concentration many times over the quiet-time values. At lower latitude, two processes affect the composition. First, the sinking motion tends to deplete the heavier constituents by replacing them with the lighter atomic oxygen, helium and hydrogen from higher altitudes. Second, if the disturbed circulation persists for many hours or days, the high-latitude air rich in N_2 and O_2 will be transported equatorward by the winds at 125 km. All the while, the stirred-up air tends to diffuse back toward its quiet-state, with the heavier constituents sinking toward lower altitudes. The diffusion time is roughly one day at 125 km, but becomes faster with increasing altitude.

TIGCM

Over the past decade considerable progress has been made in the development of a realistic global time-dependent numerical model of the coupled thermosphere-ionosphere system. One of these models has been developed at the National Center for Atmospheric Research (NCAR) and it is known as a thermosphere/ionosphere general circulation model (TIGCM).

The TIGCM is based on the pressure coordinate primitive equations of lower atmosphere dynamic meteorology. The physical processes incorporated into the model, however, are those appropriate to thermospheric heights. The model solves prognostic thermodynamic, eastward and northward momentum equations, and diagnostic equations of state and continuity. To solve the model equations, we use the basic framework of the NCAR fourth-order general circulation model developed to study meteorology of the lower atmosphere (Washington and Williamson, 1977) but appropriately modified to account for the fast vertical diffusion of heat and momentum at thermospheric heights. The TIGCM has an effective 5° latitude-longitude grid with 24 constant pressure surfaces in the vertical extending from approximately 95 km to 500 km altitude. The model time step is typically 5 minutes and the current version of the TIGCM that includes coupled thermospheric and ionospheric physics takes 20 minutes of CRAY-Y-MP 8/64 computer time per simulated model day. The TIGCM has evolved considerably over the past decade. The basic model and the thermospheric response to solar EUV and UV radiational forcings was described by Dickinson *et al.* (1981). Roble *et al.* (1982) investigated the effect of ionospheric convection at high latitudes on the thermospheric structure and circulation. The model was then extended to include the couplings between dynamics and the major neutral composition of the thermosphere (Dickinson *et al.*, 1984), the effects of upward propagating tides from the middle atmosphere (Fesen *et al.*, 1986) and auroral particle precipitation (Roble and Ridley, 1987). Roble *et al.* (1987a) designed a self-consistent aeronomic scheme and tested it in a global mean model of the coupled thermosphere and ionosphere system and then later incorporated that scheme into the TIGCM (Roble *et al.*, 1988). This new model includes a self-consistent Eulerian model of the ionosphere. Mutual coupling between the thermospheric neutral gas and ionospheric plasma occur at each model time step and at each point of the geographic grid. Recently, a global dynamo model has been included in the TIGCM to allow for electrodynamic couplings between the thermosphere/ionosphere and magnetosphere. This new version of the model , a thermosphere/ionosphere/electrodynamic general circulation model (TIE-GCM) now calculates the coupled thermosphere/ionosphere/electrodynamic structure in a self-consistent manner considering only external sources.

In the next sections I describe results obtained from the TIGCM for both steady state and time-dependent simulations. A description of results from the new TIE-GCM is deferred to a later paper.

BASIC 3-D TEMPERATURE STRUCTURE AND CIRCULATION

The TIGCM has been run for equinox conditions during solar cycle minimum for steady moderately disturbed geomagnetic conditions and the calculated steady-state, diurnally reproducible temperature structure and circulation are shown in Fig. 5. The upper panel shows temperature contours and arrows, representing the instantaneous wind pattern at 300 km, at 1200 UT for March solar minimum conditions. The lower panel show the same fields but at a constant pressure surface near 120 km. In the upper thermosphere the maximum daytime temperature occurs near 1600 local time in the vicinity of the equator with a minimum temperature occurring near 0400 local time also near the equator. The circulation at these altitudes is basically from the warm dayside to the cool nightside of the Earth. At high latitudes, however, the circulation is greatly altered by auroral processes. The momentum source of ionospheric convection at high latitudes greatly accelerates thermospheric winds to about 400 m/s at these altitudes, exceeding the typical solar driven circulation at low and mid-latitudes of about 100 m/s. This pattern remains fixed to the dayside of the Earth, and therefore it revolves over the surface of the Earth in a 24-hr period. There is a slight wobble to the pattern over the day that is caused by the Earth's geomagnetic pole not being coincident with the geographic pole. If the solar and geomagnetic forcings are steady with time, this pattern remains nearly the same but rotates with the Earth. Unlike the lower troposphere of the Earth where weather system evolve due to natural instabilities in the atmosphere, the thermospheric circulation is relatively stable and diurnally reproducible. Variability to the basic circulation is introduced into the system primarily by variable solar EUV and UV and auroral inputs.

In the lower thermosphere near 120 km, the basic circulation and temperature structure is considerably different than at 300 km as seen in the lower panel of the figure. At this altitude the basic circulation and temperature structure are controlled by the semi-diurnal tides that are excited by solar UV absorption in the ozone layer of the upper stratosphere and lower mesosphere that propagate up into the thermosphere from below. Semi-diurnal temperature oscillations of about 75 K and wind speeds of 75 m/s are prevalent in the region. Density variations are also semi-diurnal at any geographic location.

In addition to the basic structure described here for equinox conditions there are seasonal and solar cycle variations that provide considerable variability to the basic fields.

HIGH LATITUDE THERMOSPHERIC DYNAMICS

At high latitudes, the solar-driven circulation is modified greatly by ionospheric convection and auroral particle precipitation. The ion-drag momentum source of

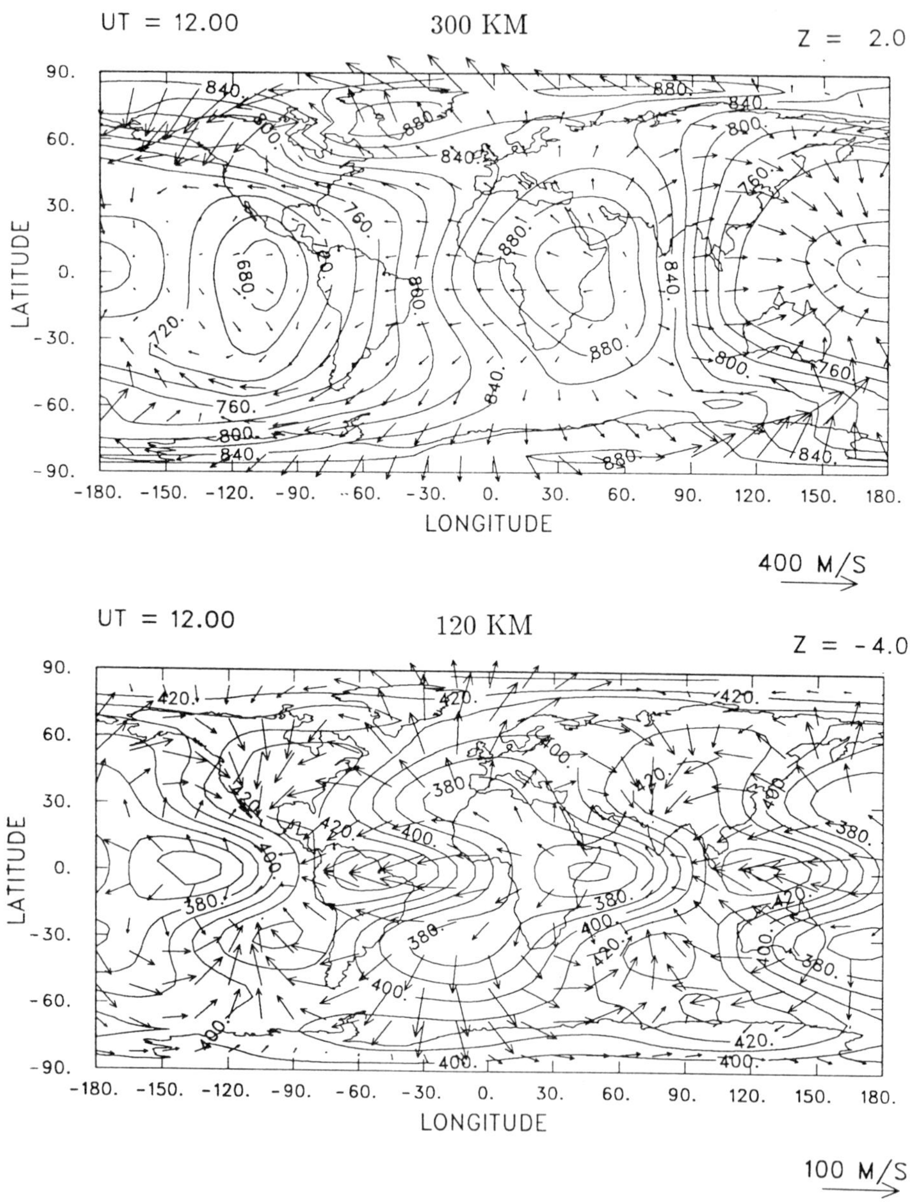

Fig. 5. TIGCM calculated contours of neutral gas temperature (K) and winds, as represented by arrows indicating direction and speed for equinox conditions during solar cycle minimum at 1200 UT. The upper panel shows the temperature pattern and winds along a constant pressure surface near 300 km and the lower panel gives the same fields along a constant pressure surface near 120 km.

ionospheric convection drives a largely rotational, nondivergent, double vortex wind system in the upper thermosphere that essentially follows the two-cell pattern of ionospheric convection that is driven by the interaction of the solar wind with the Earth's geomagnetic field. Wind velocities vary with the magnitude of the cross-polar cap potential drop that drives the ion convection pattern and collisions between ions and neutrals produce wind magnitudes greater than 500 m/s during moderate levels of geomagnetic activity. The circulation and perturbation temperature (difference between disturbed and quiet geomagnetic conditions) structure over the summer hemisphere (southern for December solstice) polar cap for three different cases are shown in Figs. 6 and 7. At 300 km , the solar only driven circulation, shown in Fig. 6a, has a wind speed in the polar cap of about 100 m/s and it flows from the dayside to the nightside over the polar cap. With ionospheric convection forcing, with a cross-polar cap potential drop of 20 kV, the wind speeds in the polar cap increase to 200 m/s with the two cell pattern being superimposed of the solar-driven flow. At 60 kV cross-polar cap potential drop, the two cell pattern clearly stands out in the neutral wind pattern with speeds approaching 400 m/s. Joule heating raises the perturbation temperature from 200 K for the solar only case to 300 K to 450 K for the two other cases respectively. Corresponding density increases occur than can affect satellite drag in the polar regions.

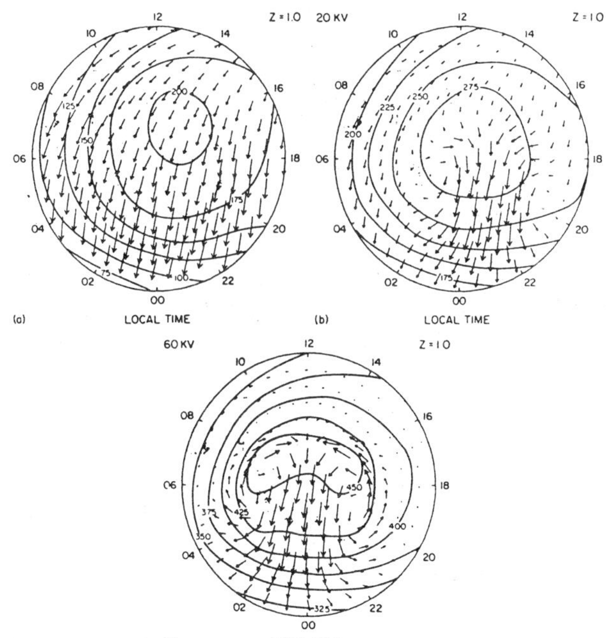

Fig. 6. Polar plots giving the direction and magnitude of the calculated southern hemisphere (summer) high-latitude circulation and contours of perturbations temperature (K) along the Z=+1 (~ 300 km) constant pressure surface at December solstice for (a) solar heating only, (b) solar heating plus magnetospheric convection with a cross-polar cap potential drop of 20 kV, and (c) same as (b) except for 60 kV. The wind speed associated with the maximum arrow is 110 ms^{-1} in (a), 200 ms^{-1} in (b), and 380 ms^{-1} in (c), respectively (Roble *et al.*, 1983).

The vorticies generated by ionospheric convection extend downward into the lower thermosphere, as shown in Fig. 7, with a cold low pressure cyclonic circulation near the dawn terminator and a warm high pressure anti-cyclonic circulation in the evening sector. The strength of the vorticies increases with the cross-polar cap potential drop as shown in the figure.

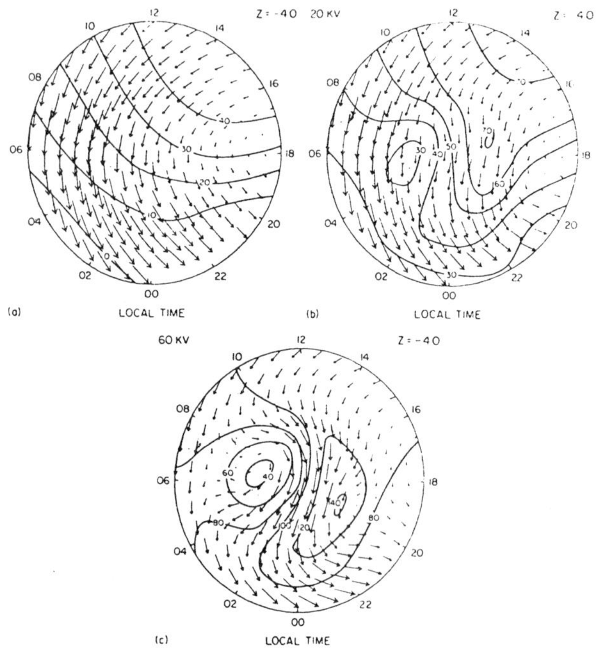

Fig. 7. Same as in Fig. 6 expect along the Z=-4 ($\sim$ 130 km) constant pressure surface. The wind speed associated with the maximum arrow is 75 ms^{-1} in (a), 77 ms^{-1} in (b), and 89 ms^{-1} in (c), respectively (Roble *et al.*, 1983).

TIME-DEPENDENT TIGCM SIMULATIONS

The NCAR TIGCM has been modified to accommodate time-dependent auroral forcings and used to simulate the time-dependent thermospheric and ionospheric circulation, temperature and compositional structure for specific geomagnetic storm events or days. Auroral oval particle precipitation patterns are based on the empirical model of Fuller-Rowell and Evans (1987) and the ionospheric convection pattern is based on the empirical model of Heelis *et al.* (1982). These auroral forcings vary with time depending upon satellite measurements of auroral particle precipitation and the properties of the solar wind as discussed by Roble and Ridley (1987). Roble *et al.* (1987b) have used the TIGCM to study the thermospheric response to the geomagnetically disturbed period 22 March 1979. The magnetospheric and ionospheric characteristics of this storm have been studied intensively by the CDAW-6 coordinated workshop (McPherron and Manka, 1985). The time-dependent auroral forcings were estimated from these studies and used as inputs for the TIGCM. The calculated thermospheric temperature and wind response on a constant pressure surface near 300 km are shown in Fig. 8. Geomagnetic activity was low for the period preceding the storm as illustrated by the auroral electrojet index AE shown in the bottom panel, and the wind velocities are generally small with a uniform temperature over the polar cap as shown in Fig. 8a. At 1230 UT the storm intensified and both the wind velocities and temperature increased as shown in Fig. 8b. By 1530 UT, near the peak of the storm when the auroral electrojet index AE was near 1600, the wind velocities approached 1 km/s with temperatures exceeding 2000 K near the throat of the ionospheric convection pattern as shown in Fig. 8c. By 2030 UT (Fig. 8d) both the wind velocities and temperature decreased but there was still residual heating and dynamics lingering after the intense storm period.

In the lower thermosphere, near 120 km, dawn and dusk vorticies were spun up during the storm period as shown in Fig. 9. Once the vorticies were spun up, they decayed slowly with time with some residual dynamics remaining at 2400 UT, 9 hours after the peak of the storm.

Because of the impulsive nature of the storm, large-scale thermospheric disturbances were generated at ionospheric F-region heights, near 300 km, that propagated equatorward from both auroral zones. To illustrate the stormtime response, two TGCM simulations were made and difference fields were constructed: TIGCM storm simulation minus TGCM simulation where auroral forcing remained constant at its pre-storm values. The auroral generated wind and temperature disturbances in magnetic conjugate hemispheres propagated equatorward and both reached the equator in about 2 hours, as shown in Fig. 10, where they interacted. Converging waves are associated with sinking motion and adiabatic heating and subsequent diverging motion is associated with upwelling and adiabatic cooling. As a result, the temperature at the equator first increases by about 160 K and then decrease by about 50 K. These interacting waves were shown by Fesen *et al.* (1987) to have an important influence on the equatorial ionosphere especially in the midnight sector where impulsive waves travel effectively toward lower latitudes. There are also density variations that a satellite would experience during the storm that are discussed in detail by Forbes *et al.* (1987).

Although the study by Roble *et al.* (1987b) and Forbes *et al.* (1987) only considered thermospheric dynamics for one day on March 22, 1979, Roble and Marcos (1991) continued the TIGCM study for 20 additional days covering the period of 20 March – 10 April, 1979. This was a period of moderate geomagnetic activity and the TIGCM calculated fields displayed considerable density variability throughout the period as both the thermosphere and ionosphere responded to variable geomagnetic activity. Comparisons of thermospheric density from the TIGCM were made with density measurements made by an orbiting satellite in the vicinity of 200 km altitude and the results showed that the TIGCM was able to reproduce and follow the measured thermospheric density variations to within ± 10-15% on a point-by-point basis along the satellite track. It was also shown that if improvements in the specification of the auroral inputs and its time history could be improved, by using real time data, the TIGCM had the capability of greatly improving the instantaneous prediction of thermospheric density and that could be used for the prediction of satellite drag and orbital perturbations.

In addition to the above storm period the TIGCM has been used by Crowley *et al.* (1989a, b) to investigate the thermospheric response to a geomagnetic storm period in September 1984 and by Roble *et al.* (1987b) to investigate a period of moderate geomagnetic

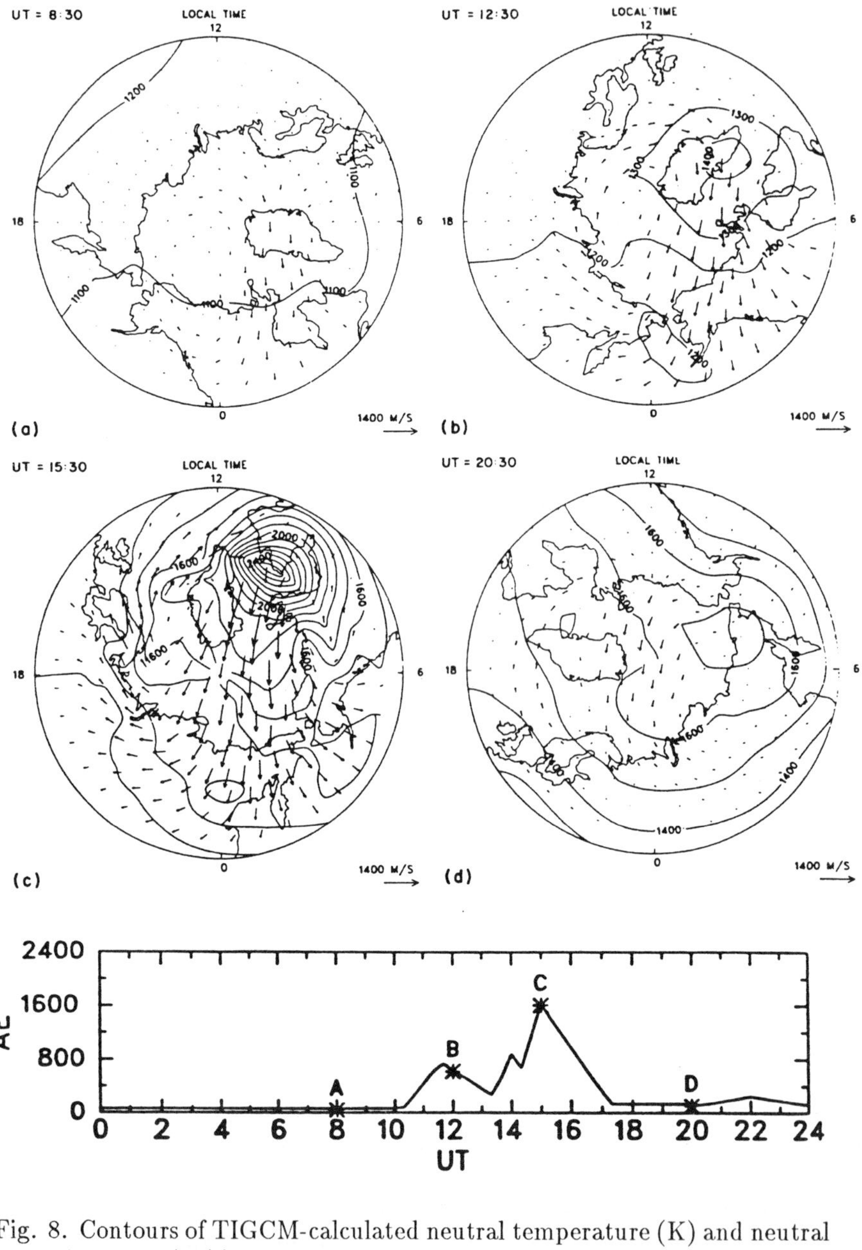

Fig. 8. Contours of TIGCM-calculated neutral temperature (K) and neutral wind vectors (m/s) on a constant pressure surface near 300 km over the northern hemisphere polar cap for four different Universal Times during the geomagnetic storm on March 22, 1979: (a) 0830 UT, (b) 1230 UT, (c) 1530 UT, and (a) 2030 UT. The length of the maximum arrow is 1400 m/s and it is shown for reference. The arrows in each of the figures are scaled in relation to the maximum arrow.

activity in November 1981. Comparisons of model results with model simulations have been presented by Roble and Killeen (1987) and Killeen and Roble (1987).

CONCLUDING REMARKS

Considerable progress has been made in recent years to improve the physical and chemical processes in thermospheric and ionospheric general circulation models. The NCAR TIGCM has developed into a self-consistent coupled thermosphere/ionosphere model that is capable of predicting time-dependent variations that are in reasonable agreement with satellite and ground-based observations. It appears that the TIGCM with further development should be able to produce a sufficiently realistic forecast of thermospheric and ionospheric weather to improve forecasts of thermospheric density variation, satellite drag and point density values at specific locations around the globe with errors less than 10%.

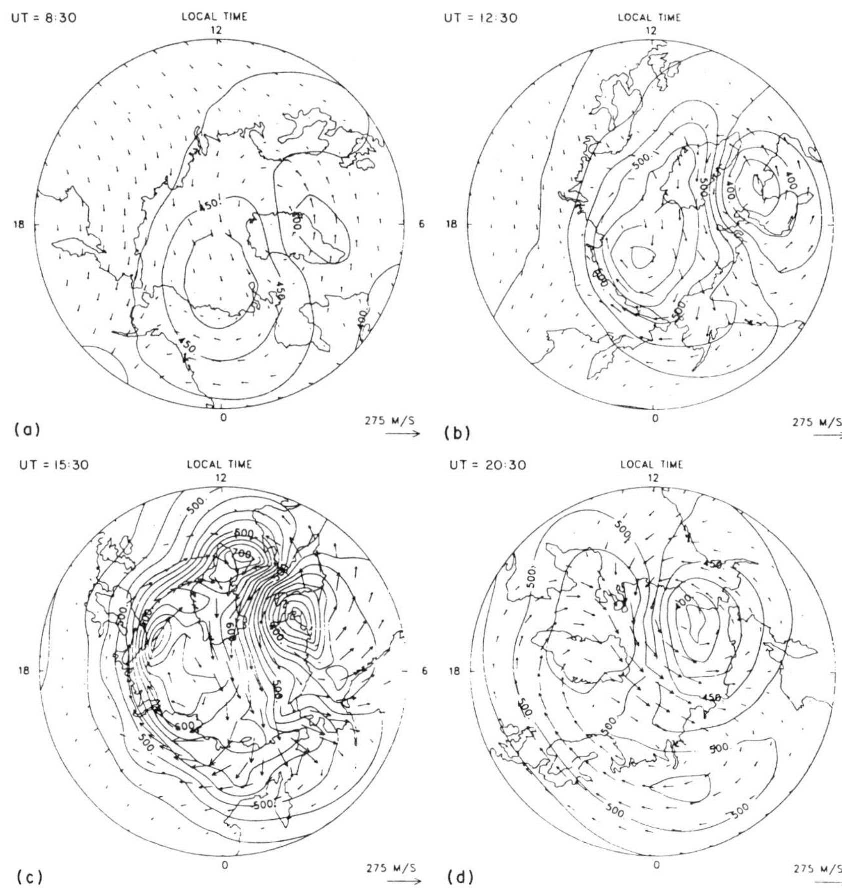

Fig. 9. Same caption as in Fig. 8 except for a constant pressure surface near 120 km. The length of the maximum arrow is 275 ms^{-1}, and all arrows are scaled in relation to this maximum value.

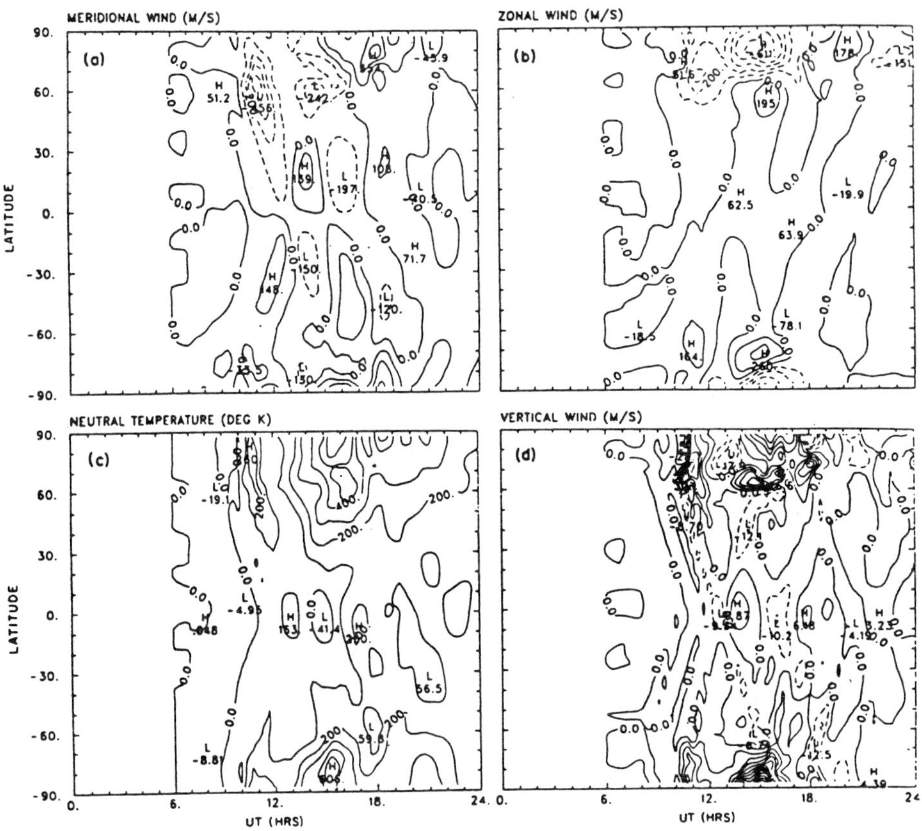

Fig. 10. Contours of the time variation of the (a) meridional wind (m/s), (b) zonal wind (m/s), (c) temperature (K), and (d) vertical wind (m/s) difference fields (TIGCM run with storm forcings minus TIGCM quiet forcings) along the -140° (W) longitude circle and along a constant pressure surface near 300 km for the March 22, 1979 geomagnetic storm. Contour intervals are (a and b) 100 m/s, (c) 100° K, and (d) 5 m/s.

REFERENCES

1. G. Crowley, B.A. Emery, R.G. Roble, H.C. Carlson, Jr., and D.J. Knipp, Thermospheric dynamics during September 18-19, 1984. 1. Model simulations, *J. Geophys. Res., 94*, 16,925-16,944, 1989a

2. G. Crowley, B.A. Emery, R.G. Roble, H.C. Carlson, Jr., J.E. Salah, V.B. Wickwar, K.L. Miller, W.L. Oliver, R.G. Burnside, and F.A. Marcos, Thermospheric dynamics during September 18-19, 1984. 2. Validation of the NCAR thermospheric general circulation model, *J. Geophys. Res., 94*, 16,945-16,959, 1989b

3. R.E. Dickinson, E.C. Ridley, and R.G. Roble, A three-dimensional general circulation model of the thermosphere, *J. Geophys. Res., 86*, 1499-1512, 1981

4. R.E. Dickinson, E.C. Ridley, and R.G. Roble, Thermospheric general circulation with coupled dynamics and composition, *J. Atmos. Sci., 41*, 205-219, 1984

5. C.G. Fesen, R.E. Dickinson, and R.G. Roble, Simulation of thermospheric tides at equinox with the NCAR thermospheric general circulation model, *J. Geophys. Res., 91*, 4471-4489, 1986

6. C.G. Fesen, R.G. Roble, and G. Crowley, Ionospheric effects at low latitudes during the March 22, 1979 geomagnetic storm, *J. Geophys. Res., 94*, 5405-5417, 1989

7. J.M. Forbes, R.G. Roble, and F.A. Marcos, Thermospheric dynamics during the March 22, 1979, magnetic storm (2) Comparison of model predictions with observations, *J. Geophys. Res., 92,* 6069-6081, 1987

8. T.J. Fuller-Rowell and D.S. Evans, Height-integrated Pedersen and Hall conductivity patterns inferred from the TIROS-NOAA satellite data, *J. Geophys. Res., 92,* 7606-7618, 1987

9. R.A. Heelis, J.K. Lowell, and R.W. Spiro, A model of the high-latitude ionospheric convection pattern, *J. Geophys. Res., 87,* 6339-6345, 1982

10. T.L. Killeen and R.G. Roble, Thermosphere dynamics: Recent contributions from Dynamics Explorer-2, *Adv. Astronautical Sci., 65*, 1625-1648, 1987

11. R.L. McPherron and R.H. Manka, Dynamics of the 1054 UT March 22, 1979, substorm event: CDAW 6, *J. Geophys. Res., 90,* 1175-1190, 1985

12. R.G. Roble, R.E. Dickinson, and E.C. Ridley, The global circulation and temperature structure of the thermosphere with high-latitude plasma convection, *J. Geophys. Res., 87,* 1599-1614, 1982

13. R.G. Roble, R.E. Dickinson, E.C. Ridley, B.A. Emery, P.B. Hays, T.L. Killeen, and N.W. Spencer, The high-latitude circulation and temperature structure of the thermosphere near solstice, *Planet. Space Sci., 31*, 1479-1499, 1983

13. R.G. Roble and E.C. Ridley, An auroral model for the NCAR thermospheric general circulation model, *Annales. Geophysicae, 5A,* (6), 369-382, 1987

14. R.G. Roble and T.L. Killeen, Modeling the global structure of the thermosphere: Circulation, temperature and density, *Adv. Astronautical Sci., 65*, 1683-1700, 1987.

15. R.G. Roble, E.C. Ridley, and R.E. Dickinson, On the global mean structure of the thermosphere, *J. Geophys. Res., 92*, 8745-8758, 1987a

16. R.G. Roble, J.M. Forbes, and F.A. Marcos, Thermospheric dynamics during the March 22, 1979 magnetic storm (a) Model simulations, *J. Geophys. Res., 92,* 6045-6068, 1987b

17. R.G. Roble, E.C. Ridley, A.D. Richmond, and R.E. Dickinson, A coupled thermosphere/ionosphere general circulation model, *Geophys. Res. Lett., 15*, 1325-1328, 1988

18. R.G. Roble and F.A. Marcos, Thermospheric and ionospheric response to the geomagnetically disturbed period 20 March – 10 April 1979, *J. Geophys. Res.,* in preparation, 1991

19. W.M. Washington and D.L. Williamson, A description of the NCAR global circulation models, in *Methods in Computational Physics, Vol. 17, General Circulation Models of the Atmosphere*, (J. Chang, ed.), 111-172, Academic Press, New York, 1977

AAS 91-012

ORBITAL DEBRIS*

Robert D. Culp[†]

In this paper, which consists of presentation charts, statistical information is given on orbiting objects, a breakdown of space debris sources, including sources from normal operations, fragmentations of orbiting objects and untrackable debris from these sources. The paper discusses debris cloud evolution and the detection of untrackable debris and detection capability, the LEO orbital debris environment, protection and prediction of untrackable debris.

Information on orbital debris models for general use is provided. Charts are given for orbital debris regimes, a NASA flux model, satellite breakup inclination and altitude, catalog growth by orbital regime, relative velocity at impact, including impact probabilities and hazards. Protection from debris, including methods of cleaning up debris and protective actions, is discussed. Many reference sources for space debris activities are given.

---

* This abstract was prepared by the editors.

† Professor, Department of Aerospace Engineering Sciences, University of Colorado, Campus Box 429, Boulder, Colorado 80309-0429.

ORBITAL DEBRIS

KNOWN ORBITING OBJECTS

as of October 1, 1990

Total Ever Catalogued
20,832

payloads: 4,081
debris: 16,751

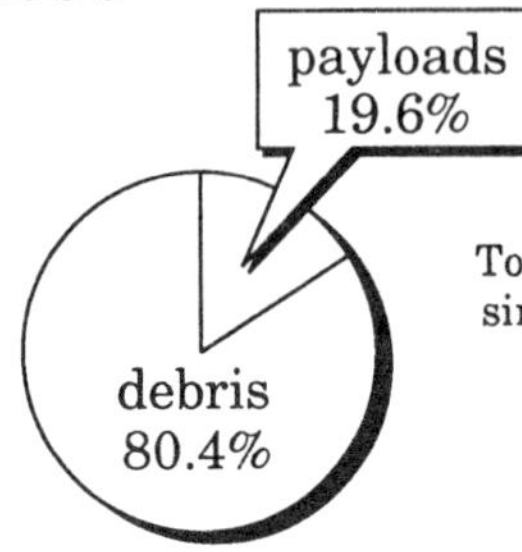

Total catalogued since 1957

Current Orbiting Objects Catalogued
6,645

payloads: 1,946
debris: 4,699

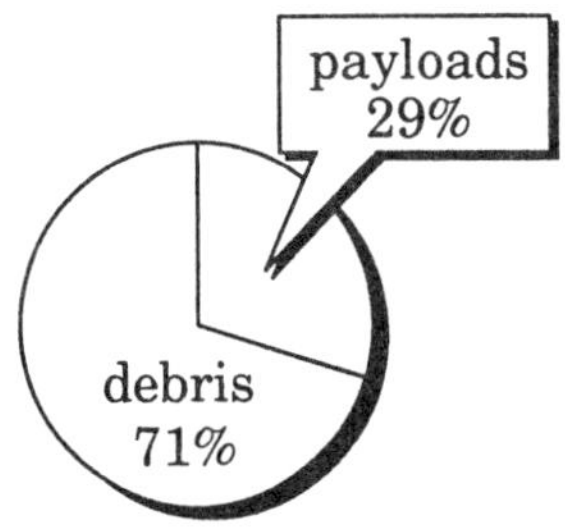

ORBITAL DEBRIS

SOURCES OF SPACE DEBRIS

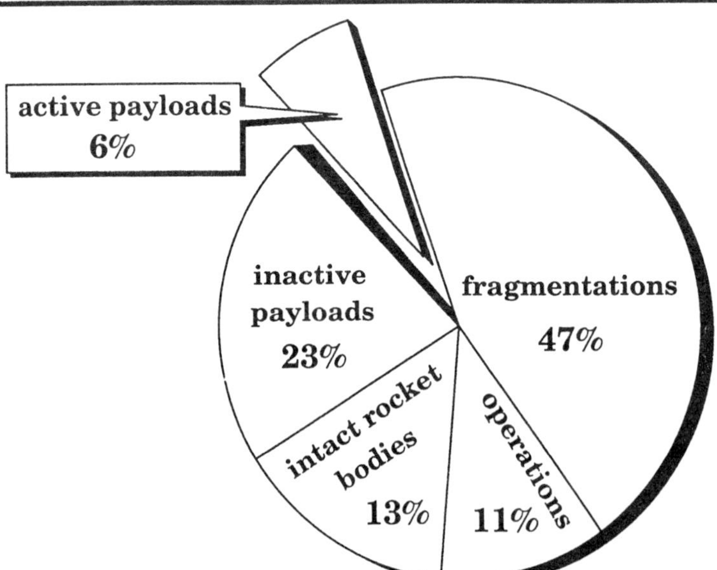

ORBITAL DEBRIS

SOURCE: NORMAL OPERATIONS

- **LAUNCH OPERATIONS**

  Typically generate 1 - 10 pieces of catalogued debris

- **PAYLOAD DEPLOYMENT AND MANEUVERS**

  Separation devices, spinup devices, temporary protection

- **SPACE STATION AND SPACE SHUTTLE ACTIVITY**

  Salyut 6 generated more than 100 pieces of catalogued debris
  Salyut 7 generated at twice the rate of Salyut 6

  Notable:
  - Ed White's glove (1965)
  - Powered hand-tool lost during repair of Solar Max (1984)

- **INTACT ROCKET BODIES, UPPER STAGES, AND DEAD SATELLITES**

ORBITAL DEBRIS

SOURCE: FRAGMENTATIONS OF ORBITING OBJECTS

100 space fragmentation events, most recent on November 30, 1990

CAUSE

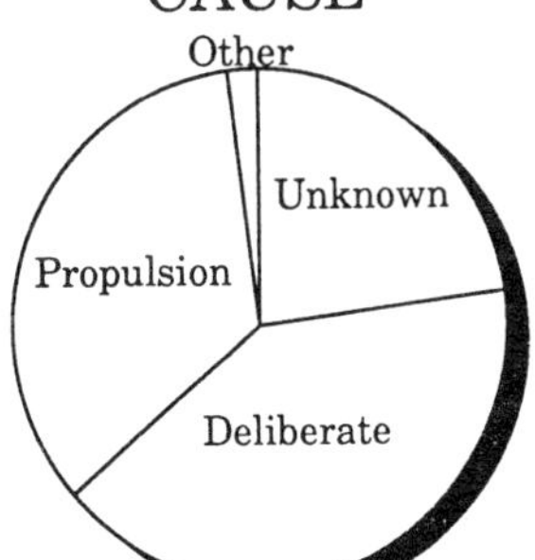

Unknown: 24%
Deliberate: 42%
Propulsion: 32%
Other: 2%

TEN WORST FRAGMENTATIONS

7 Rocket Body Explosions
3 Payloads

WORST: Ariane Spot 1 Rocket Body
November 13, 1986

FIRST: Transit 4A Rocket Body
June 29, 1961 (still 5th worst)

NOTABLE: 7 Delta Rocket Second Stage Fragmentations
1973 - 1981

More than 30 USSR Deliberate Fragmentations

ORBITAL DEBRIS
UNTRACKABLE DEBRIS: THE MAJOR CHALLENGE

- Catalog contains only easily tracked objects, typically no smaller than 10 cm in characteristic length.

- Catalog contains nearly all the mass (99.9%) but only approximately 7000 pieces.

- Untrackable orbiting pieces

1 cm - 10 cm	10,000 - 100,000 pieces	(potentially catastrophic)
1 mm - 1 cm	1,000,000 pieces	(seriously damaging)
0.1 mm - 1 mm	millions	(mission degrading)

- Challenges: Detection
  Protection
  Prediction

ORBITAL DEBRIS
UNTRACKABLE DEBRIS FROM NORMAL OPERATIONS

- Solid Rocket Effluent (primarily Al_2O_3)
- Staging and reentry events that are explosive
- Planned experiments (radar calibration, Westford needles)
- Deterioration of satellites (paint flecks, insulation)
- Waste and refuse from space stations and shuttles

ORBITAL DEBRIS

UNTRACKABLE DEBRIS FROM FRAGMENTATIONS

Fragmentations produce many more untrackable particles than trackable.

Example: The worst fragmentation: Ariane Spot 1 Rocket Body, November 13, 1986

Catalogued debris: Almost 500 pieces

Estimated 1 cm - 10 cm pieces: 1000 - 5000
Estimated 1 mm - 1 cm pieces: 1000 - 5000
Estimated 0.1 mm - 1 mm pieces: more than 10,000

Example: The first fragmentation: Transit 4A Rocket Body, June 29, 1961

Catalogued debris: 286 pieces

Estimated 1 cm - 10 cm pieces: 500 - 3000
Estimated 1 mm - 1 cm pieces: 500 - 3000
Estimated 0.1 mm - 1 mm pieces: 5000

ORBITAL DEBRIS

DEBRIS CLOUD EVOLUTION

- A cloud of fragments emanates from the breakup position.
- A cloud becomes an ellipsoid following the original orbit for a few revolutions.
- The ellipsoid elongates along the orbital path until (days) a torus is formed.
  Cause: Differing periods of particles
- The torus spreads and gradually (months) becomes a belt with the same ground coverage as the original satellite.
  Cause: Differing plane precession of particles
- The debris becomes (years) evenly distributed background particles.
- The debris rains down through lower elevations, beginning at once, continuing for decades.
  Cause: Atmospheric drag

ORBITAL DEBRIS

DEBRIS CLOUD EVOLUTION

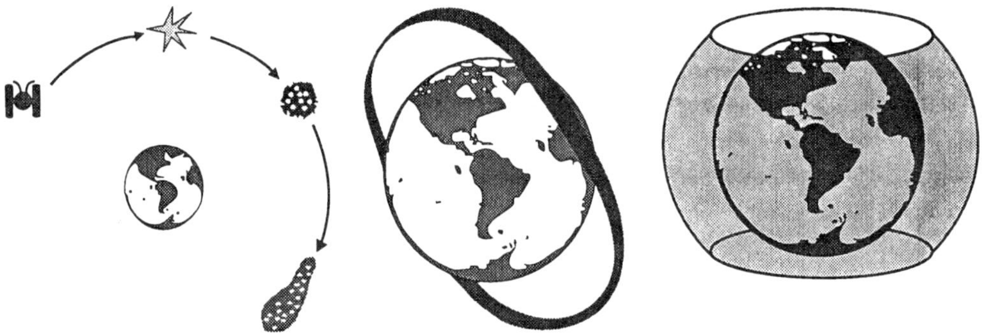

ORBITAL DEBRIS

DETECTION OF UNTRACKABLE DEBRIS

SAMPLING

Returned Surfaces (less than 1 mm size particles)
- LDEF
- Solar Max
- Space Shuttle
- future surface experiments

Optical (no smaller than 1 cm size particles)
- GEODSS (electro-optical)
- Baker-Nunn cameras
- future space-based optical sensors

Radar (perhaps as small as 0.5 cm size particles)
- Existing phased-array radar (PARCS, PAVE PAW, COBRA DANE)
- NAVSPASUR radar fence
- Debris Environment Characterization Radar [DECR] (planned)
- Haystack long-range imaging radar
- future space-based radar

ORBITAL DEBRIS

SPACE OBJECT DETECTION CAPABILITY

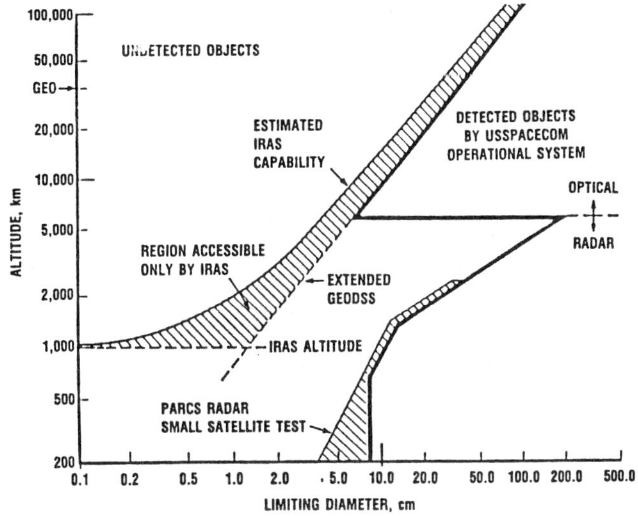

SOURCE: V.A. Chobotov, Aerospace Corporation

LEO ORBITAL DEBRIS ENVIRONMENT

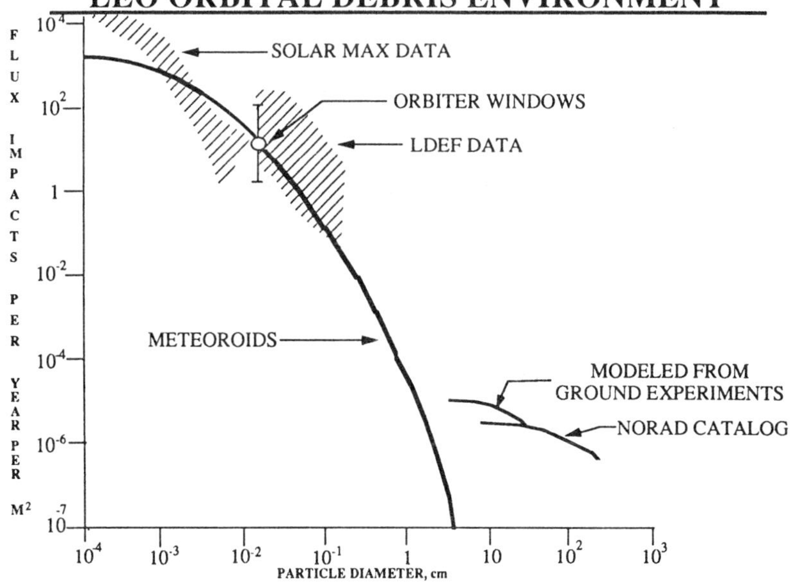

ORBITAL DEBRIS

PROTECTION FROM UNTRACKABLE DEBRIS

- Trackable debris is avoided by mission planning and maneuvers
- Protection from untrackable debris must come from (currently feasible and worthwhile):
    1) shielding
    2) spacecraft configuration
    3) mission planning

  (future development, and of questionable feasibility):
    4) short-term detection and warning systems that allow
        a) temporary shielding
        b) avoidance or minimization maneuvers
        c) destruction or interception of debris

ORBITAL DEBRIS

PREDICTION OF UNTRACKABLE DEBRIS

Models of Current Untrackable Debris Environment

- Based on analysis of sources of debris
- Constructed using available detection of debris
- Validated by continual debris sampling

Models for Future Untrackable Debris Environment

- Breakup models
- Debris cloud evolution models
- Scenario construction

ORBITAL DEBRIS

ORBITAL DEBRIS MODELS FOR GENERAL USE

Space Environment Debris Models

THE NASA MODEL - (Kessler)

NASA TM 100471 "Orbital Debris Environment for Spacecraft Designed to Operate in Low Earth Orbit", April 1989

TELEDYNE BROWN ENGINEERING (Johnson, Nauer)

UNIVERSITY OF COLORADO

Fragmentation-based model (Maclay)
Catalog-based model (Madler)

TECH. UNIV. OF BRAUNSCHWEIG (Rex, Eichler)

Breakup and Debris Cloud Evolution

THE AEROSPACE CORPORATION (Chobotov, Wolfe)

IMPACT and DEBRIS

KAMAN SCIENCES (McKnight)

SCREEN

NASA JOHNSON SPACE CENTER (Kessler)

LOCKHEED ESC

UNIVERSITY OF COLORADO

XON TECH., INC. (Bohannon, Young)

A diskette with the University of Colorado implementation of the NASA Debris Flux Model can be obtained from:

Madler/Maclay/McNamara
% Professor Robert D. Culp
Department of Aerospace Engineering Sciences
C.B. 429
University of Colorado
Boulder, CO 80309-0429

A $20.00 donation is requested. The money will be used for student support and functions.

Checks should be made payable to The University of Colorado.

ORBITAL DEBRIS
SPATIAL DENSITY FOR TRACKABLE AND UNTRACKABLE SIZE REGIMES,1988

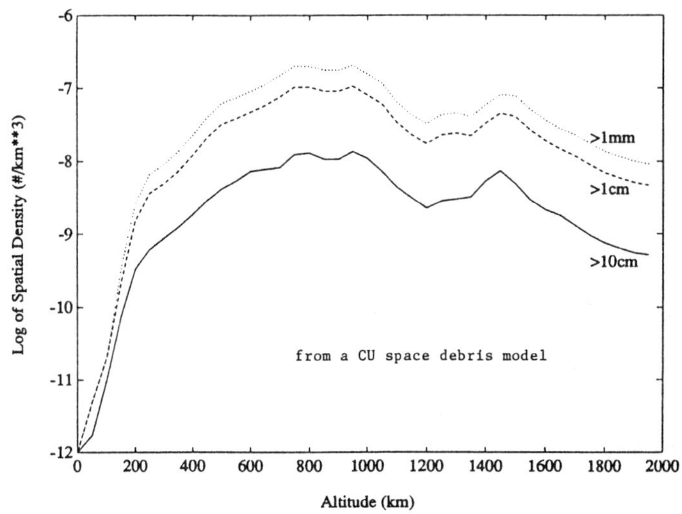

ORBITAL DEBRIS
FLUX FROM NASA FLUX MODEL FOR VARIOUS SIZES FOR 1988

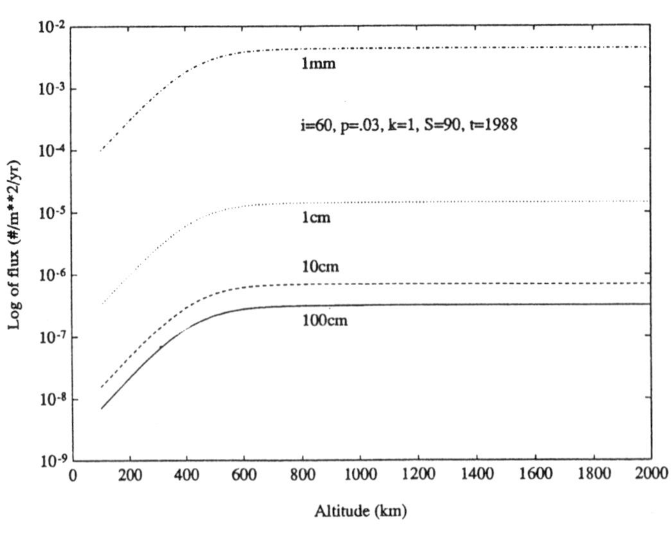

ORBITAL DEBRIS

SATELLITE BREAKUP INCLINATION AND ALTITUDE

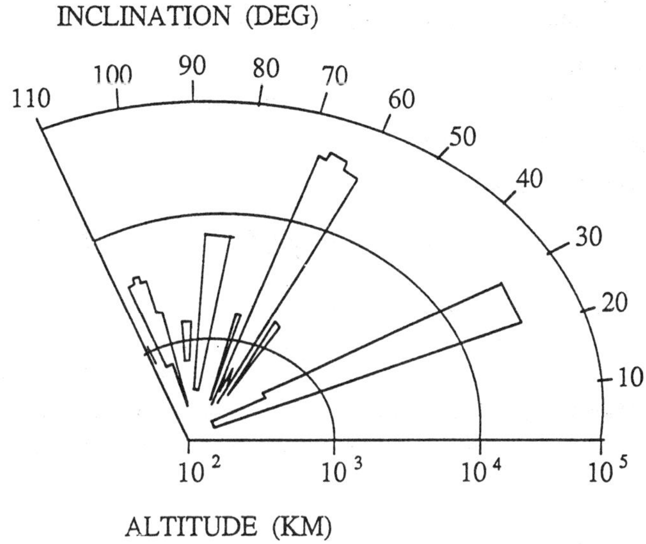

ORBITAL DEBRIS

CATALOG GROWTH BY ORBITAL REGIME

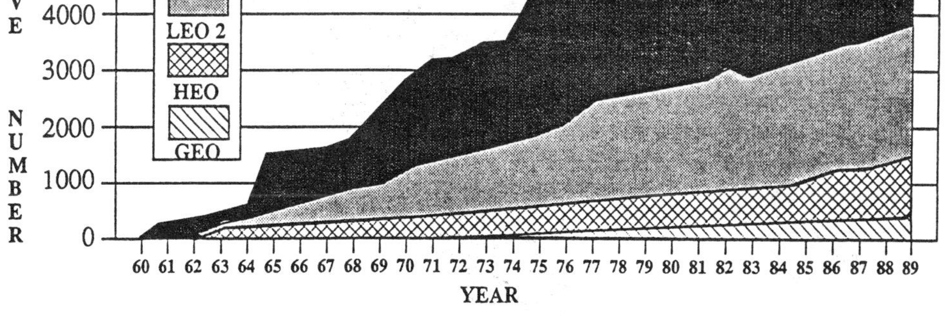

ORBITAL DEBRIS

RELATIVE VELOCITY AT IMPACT

Meteoroids

Hyperbolic velocity with respect to Earth: 15 - 20 km/sec

All possible inclinations, randomly distributed

Space Debris

Between circular and escape velocity: 7 - 10 km/sec

Nearly all in direct orbits

Relative Velocity

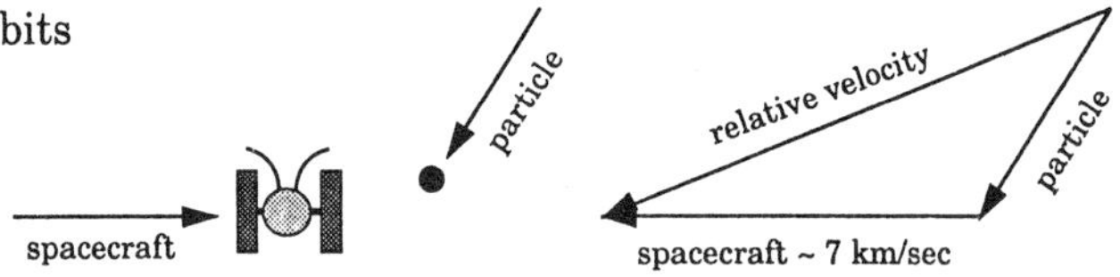

Meteoroid relative velocity:	8 - 27 km/sec	
Space debris relative velocity:	0 - 17 km/sec	in theory
	3 - 12 km/sec	(because of inclination distribution)

SHADOWING: There will be 10 - 20 times more impacts on the leading surfaces than on the trailing surfaces

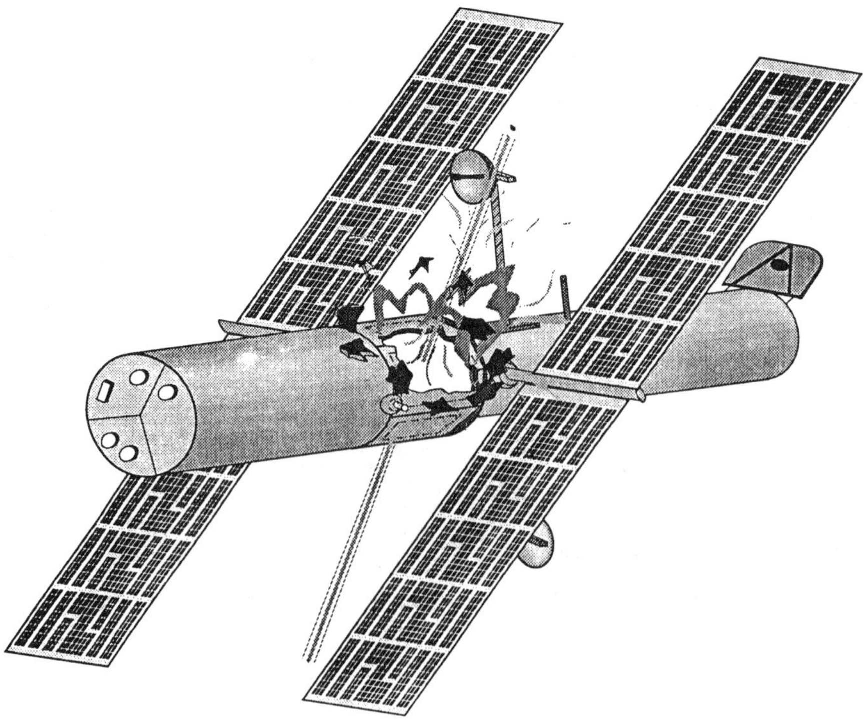

ORBITAL DEBRIS

IMPACT PROBABILITIES AND TIME BETWEEN IMPACTS

Flux **F** (impacts per square meter per year), on cross-sectional area, **A**
In time **T**, average number of impacts **N**, on **A** is:

$$N = \int_0^T FA\,dt$$

Probability of no impacts, P_0, is:

$$P_0(T) = e^{-N}$$

Probability of at least one impact, $P_1(T)$, is:

$$P_1(T) = 1 - P_0(T)$$

For infrequent impacts:

$$P_0(T) \approx 1 - N$$

$$P_1(T) \approx N$$

and time between impacts is:

$$\Delta T \approx \frac{1}{N}$$

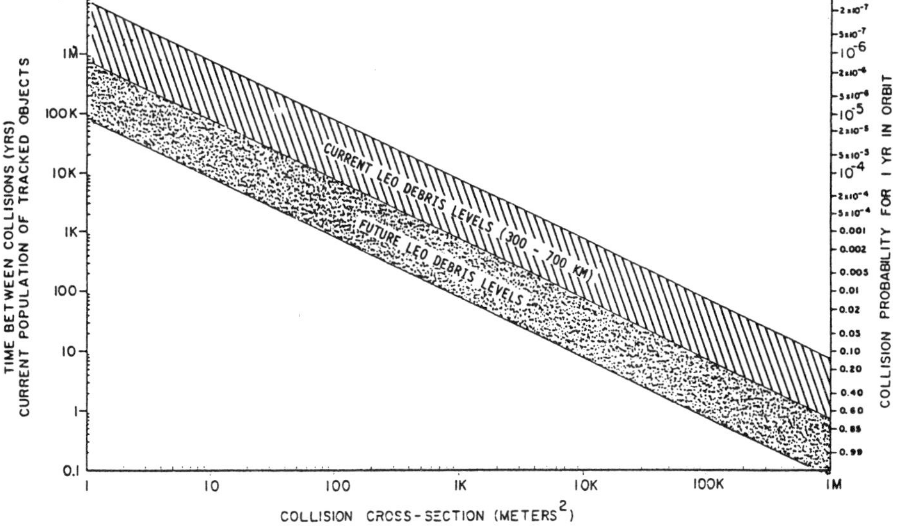

ORBITAL DEBRIS

HAZARD FROM UNTRACKABLE SPACE DEBRIS

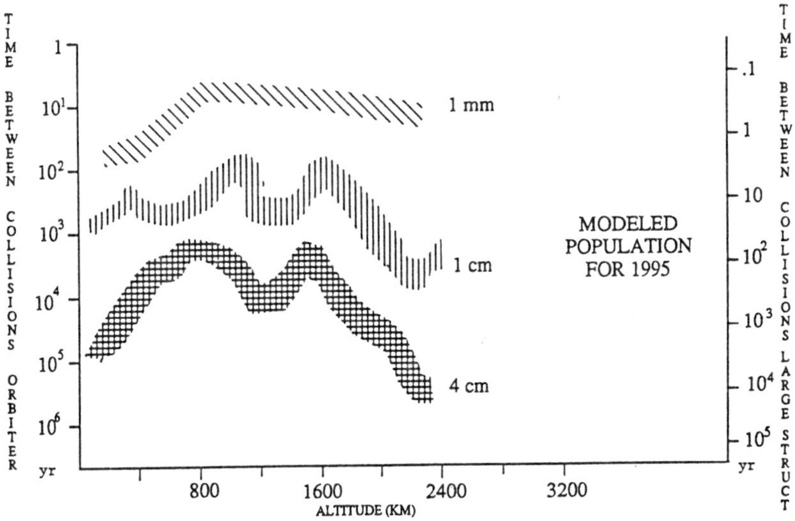

ORBITAL DEBRIS

DEBRIS PROTECTION REQUIREMENTS FOR SPACE STATION

10 year requirements for Space Station (both meteoroid and debris)

- Greater than 0.9955 probability of no penetration of critical element (includes all pressurized volumes)
- Greater than 0.95 probability of maintaining space station integrity
- Greater than ? probability of no space suit penetration

Design and mission planning implications for Space Station

- Pressurized volumes must be shielded against at least 1 cm debris
- Critical elements must be shielded against at least 1 cm debris
- Shielding or redundancy of various noncritical systems
- Replacement of damaged systems
- Restrictions on extra-vehicle activity
- Collision avoidance of tracked population

ORBITAL DEBRIS

1990's SPACE DEBRIS ENVIRONMENT

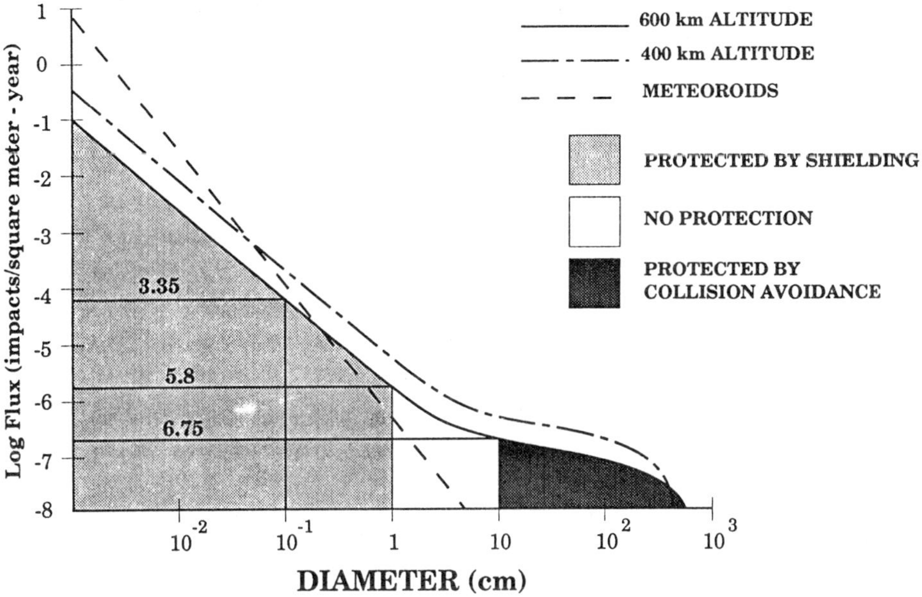

SOURCE: Space Station Design Document JSC-20001 D. Kessler

ORBITAL DEBRIS

FUTURE HAZARD FROM CURRENT SATELLITES

SOURCE:

- 3,000,000 kg of mass in Earth orbit
- >90% of the mass is below 5000 km altitude
- >70% of the mass is between 500 km and 2000 km altitude

SCENARIO:

- Gradual deterioration of orbiting objects yields many smaller objects
- Collisions of small objects with large masses yields large fragmentation events
- Increased debris population escalates the process
- Finally, most of the mass becomes small debris

RESULT:

- The current orbiting population yields an increased untrackable debris flux 10,000 times the current flux

ORBITAL DEBRIS

MODEL FORCASTS FOR LONG-TERM CONSEQUENCES

- Growth of small debris flux from breakup of current satellite population renders LEO extremely hazardous for extended missions
- Planned additions of large masses in LEO will greatly increase this effect
- All models show altitudes of 500 - 1500 km becoming unusable by mid-21st century
- Control of debris sources can delay, not prevent, this condition

ORBITAL DEBRIS

METHODS OF CLEANING UP SPACE DEBRIS

All methods are economically and technologically unfeasible

Large debris

Suggested cleanup methods:

- Remove from orbit by reentry or escape
- Salvage
- Breakup

Untrackable debris

Suggested cleanup methods:

- Nerf ball sweeper
- Directed-energy beam vaporization

ORBITAL DEBRIS

REALISTIC ACTIONS TO MINIMIZE DEBRIS AND PROTECT MISSIONS

CURRENT:

- Reduce operational debris
- Design deorbit mechanisms into spacecraft
- Plan missions to avoid high-concentrations of debris
- Design spacecraft to minimize impact damage

FUTURE:

- On-orbit detection, active shielding, interception, and avoidance techniques
- Use higher orbits for Earth-observing missions
- Seek feasible large-object deorbit methods

ORBITAL DEBRIS

KEY PLAYERS AND ORGANIZATIONS

Donald J. Kessler, Project Scientist for Space Debris
Space Sciences Branch SN3
NASA Johnson Space Center
Houston, Texas 77058
ph: (713) 483-5313

Vladimir A. Chobotov, Manager, Space Hazards Section
Astrodynamics Department
The Aerospace Corporation
El Segundo, CA 90009-2957
ph: (213) 615-4280

Nicholas L. Johnson, Advisory Scientist
Teledyne Brown Engineering
1250 Academy Park Loop, Suite 240
Colorado Springs, CO 80910-3799
ph: (719) 574-7270

University of Colorado, Boulder
Aerospace Engineering Sciences
Basic Research and Graduate Education

ORBITAL DEBRIS
ADDITIONAL ORGANIZATIONS AND CONTACTS

Naval Research Laboratory (Kaufman, Heard)

NASA Marshall Space Flight Center

NASA Langley Research Center (Kinard)

Wright Laboratory (Eglin AFB - Jerome)

Phillips Laboratory (Kirtland AFB - Lillie, Campbell; Hanscom AFB)

Defense Nuclear Agency (Tedeschi, Connell)

Kaman Sciences Corp. (McKnight)

Xon Tech, Inc. (Bohannon, Young)

Lockheed - ESC (Houston, - Anz-Meador, Talent)

General Research Corp. (Isbell, Carlyle)

Technical University of Braunschweig (Rex, Eichler)

ORBITAL DEBRIS
RESOURCES

ORBITAL DEBRIS MONITOR

3004 Robin Ridge Court
Fairfax, Virginia 22031
Quarterly: $50.00 per year
Dr. Darren S. McKnight, Editor

"Space Debris: A Potential Threat to Space Station and Shuttle"

Samuel W. Bowlen, Editor, April, 1990
GAO/MTEC - 90 - 18
U.S. General Accounting Office
P.O. Box 6015
Gaithersburg, Maryland 20877
(202) 275-6241 FREE

"Report on Orbital Debris"

by Interagency Group (Space), February, 1989
National Security Council
Washington, D.C.

Artificial Space Debris

Nicholas L. Johnson and Darren S. McKnight, 1987
Orbit Book Company, Inc.
2005 Township Road
Malabar, Florida 32950

ORBITAL DEBRIS

RESOURCES AND REFERENCES

LDEF Spaceflight Environment Effects Newsletter

William Kinard, Editor-in-Chief
LDEF Systems Special Investigation Group
P.O. Box 10518
Silver Spring, Maryland 20914

"Orbital Debris Environment for Spacecraft Designated to Operate in Low Earth Orbit"

D.J. Kessler, R.C. Reynolds, P.D. Anz-Meador, April, 1989
NASA TM 100471

"History of On-Orbit Satellite Fragmentations"

Fourth Edition, Nicholas L. Johnson and David J. Nauer
TBE Technical Report CS90-TR-JSC-002D, January, 1990
Teledyne Brown Engineering, Colorado Springs, CO 80910

"Hazard Analysis of Commercial Space Transportation"

Volume I: Operations
Volume II: Hazards
Volume III : Risk Analysis
U.S. Department of Transportation
Office of Comercial Space Transportation
Transportation Systems Center
Cambridge, MA 02142

ORBITAL DEBRIS

RESOURCES AND REFERENCES

"Dynamics of Debris Motion and the Collision Hazard to Spacecraft Resulting from Orbital Breakup"

V.A. Chobotov, et al,
Report SD-TR-88-96, January, 1988, The Aerospace Corporation
El Segundo, CA 90245

"Orbiting Debris: A Space Environmental Problem"

Ray A. Williamson, Project Director
Office of Technology Assessment (GPO No. 052-003-01215-4)
available from Superintendent of Documents
Government Printing Office
Washington, D.C. 20402-9325
price: $3.00

EnviroNET, The Space Environment Information Service

User Guide, Third Edition, April 2, 1990
Mail Code 410.1
NASA Goddard Space Flight Center
Greenbelt, MD 20771
ph: (301) 286-5690

AAS 91-013

RADIATION EFFECTS ON SPACECRAFT

James C. Ritter*

Space systems can be made to survive and operate in the earth's radiation belts. But to make them survive and operate one has to understand them as systems, not just as a collection of individual piece parts. Whereas individual parts may upset or degrade in gain or even, in extreme cases, burn out, the system (properly designed) may continue to survive and operate. The same considerations also apply to individual guidance, navigation and control subsystems which may be designed in a modular form and placed in a variety of spacecraft. They must, of course, be designed to survive and operate in the worst case radiation environment in which they will be used. Such subsystems may be for acquisition, tracking and pointing, or they may be star trackers, or autonomous attitude determination and navigation subsystems.

This talk will address the over-all issue of how to make space systems survive and operate in natural radiation environments. In this talk, selected portions of space survivability history will be discussed to see how the major radiation effects were first observed and to contrast them with our present understanding of the phenomena and how to harden against them. The radiation environments in which space systems have to operate will be described and which specific radiation effects those environments cause will be shown, briefly describing each radiation effect. Then, how one goes about hardening a spacecraft, what steps are taken on individual devices, on circuits and on subsystems and what testing is needed to assure the survivability and operability of the subsystem or the entire spacecraft will be shown. Finally, the challenges in hardening future spacecraft with their new technologies and miniature devices will be discussed briefly.

* Naval Research Laboratory, 4555 Overlook Avenue, S.W., Washington, D.C. 20375.

HISTORY OF THE PROBLEM

The history of the observation of radiation effects in space systems began shortly after Sputnik I was launched on October 4, 1957. The first successful U.S. spacecraft was Explorer I, launched on January 31, 1958. A geiger counter experiment put on board by J.A. Van Allen discovered a high count rate region which was caused by charged particles trapped in the earth's radiation belts (now called the Van Allen belts). Early space experiments mapped the radiation belts and measured the population of trapped particles. The unexpected presence of intense radiation belts demonstrated the need for hardening spacecraft to the natural space environment.

Nuclear weapon experiments conducted from 1958 to 1962 demonstrated that electrons from beta decay of fission fragments could also be trapped by the earth's magnetic field to form artificial radiation belts. The Starfish test, performed on July 9, 1962, produced an intense, artificial belt, the effects of which lasted until the early 1970's. The Starfish electrons and redistributed protons produced premature degradation and failure in the solar arrays of several satellites in low-altitude orbits at the time. Three satellites, Transit 4B, Traac and Ariel became inoperable because of degradation in their solar cell power systems.[1] One failed in 4 days, one in 24 and a third in 36 days. The solar cell output as a function of time in orbit is shown for Transit 4B and Traac[2] in Fig. 1. This figure demonstrates the dramatic effects of Starfish. In all, Starfish caused serious problems or failure in up to seven spacecraft and dramatically pointed out the need to study and understand radiation effects in solid state devices used in satellites such as solar cells and microelectronic devices. Solar cells were shown to be exceptionally vulnerable because they are on the outside of the spacecraft where they do not benefit from inherent shielding. The above example of the loss of several spacecraft from solar cell damage is the first example of displacement damage causing the loss of a space system.

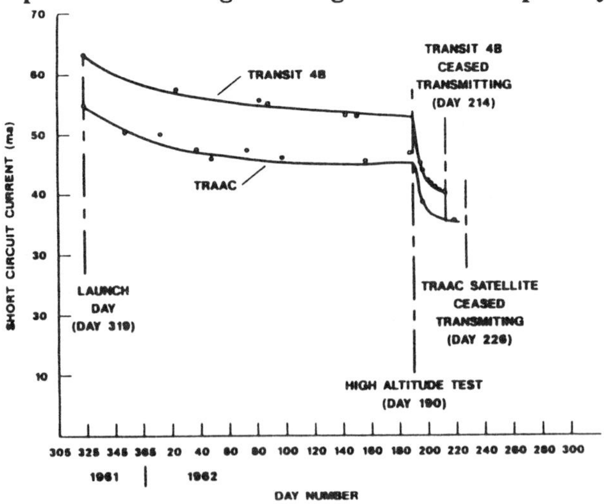

FIG. 1 SOLAR CELL OUTPUT AS A FUNCTION OF TIME FOR TRANSIT 4B AND TRAAC SATELLITES

To see what the state of radiation effects and hardening was in the early 1960's, let us look at a rather innovative satellite which was launched just one day after Starfish, Telstar I. Telstar was launched into a 515 nmi by 3044 nmi orbit with 44.8 degrees of inclination. Telstar is an interesting example.

Solar cell damage caused by electrons and protons in the belts is a serious problem which can limit the lifetime of spacecraft. This damage is caused by reduction of the minority-carrier lifetime in the material which decreases the probability that electron-hole pairs separated in the device will cross the junction, contributing to the useful output current of the cell. Starfish caused some additional damage to Telstar's solar cells, beyond that expected from the natural environment, but a good design did not make that critical for the satellite. Telstar's designers had, fortunately, recently found out about an experiment which showed that cells which were manufactured with an n layer on top of a p substrate (n on p cells) were nearly an order of magnitude harder than the conventional p on n cells used on every other U.S. satellite at the time. Loferski, Rappaport and Scott-Monck[3] at RCA, working under a NASA contract had just performed the experiment using some recently developed n on p cells and compared them to the conventional cells. Accordingly, AT&T employed what is now a standard hardening technique, part substitution, using n on p cells on Telstar I.

AT&T also used two other hardening techniques that are now standard, shielding and use of design margin, for the solar cells. The space environment contains a very large number of low energy electrons and protons which have only a limited range in materials. To reduce the damage produced by these particles, cover slips are used over the solar cells. The covers protect the cells by absorbing these particles. On Telstar, 30 mil sapphire covers were used because of the severe environment for its orbit (6-mil covers are used on low- and geosynchronous-orbit spacecraft which are below or above the most intense part of the belts). The solar power supply was also designed (as is done on every spacecraft) for an end of life (EOL) power rather than a beginning of life (BOL) power. The EOL power takes into account, among other things, the expected degradation in the cells caused by radiation.

An important experiment was also included onboard Telstar I. A group of five specially fabricated n-p-n diffused silicon transistors were used to provide an independent measurement of displacement damage for comparison to the solar cells.[4] The transistors were placed behind different amounts of shielding to produce curves of damage (gain degradation) as a function of accumulated particle fluence (time in orbit). The damage observed[5] correlated well with simultaneous solar cell data. A single damage rate vs. shielding thickness curve was determined from the composite solar cell vs. cover slip thickness data plus the above transistor damage data and compared to the damage expected vs. shielding for electron and proton fluences as measured. It was concluded by Brown et. al.[5] that about one half of the damage for n on p solar cells shielded by 30 mil (0.30 g/cm^2) sapphire covers was due to protons and one half due to electrons. This additional factor of about two in the total equivalent fluence incident on the solar cells from the unexpected high energy Starfish electrons was within the ordinary conservative design margins used on spacecraft and hence Starfish did not destroy Telstar's solar arrays.

The conclusions drawn[5] from the above data were that the n on p solar cells used on Telstar would degrade to 68% of their initial value in two years in the space environment (including Starfish). If 6-mil sapphire shields had been used, the same degradation would have been reached in about 2.5 months; if, in addition, p on n cells had been used, that point would have been reached in about two or three weeks! The effects of part substitution, shielding and designing for degraded performance have paid off for spacecraft survivability since the earliest days of spacecraft.

The Starfish tests greatly increased the dose observed on Telstar I. This additional dose, two orders of magnitude greater than expected, produced what was probably the first instance of the loss of a spacecraft due to total ionizing dose damage in a semiconductor device. The command decoder on Telstar I, consisting of diodes and narrow base transistors is not sensitive to displacement damage, but is sensitive to total ionizing dose. Telstar was launched on July 10, 1962. In mid November, the command decoder became sluggish and on November 24, it failed. A series of creative steps were taken and it eventually recovered its function, but again failed, this time permanently, on February 21, 1963. Ground tests on two transistors with the same date code showed failure after 0.6 and 2.3 megarads and were consistent with the dose observed on Telstar. The failure behavior was also the same and it was concluded by Mayo et. al.[6] that the failure was caused by surface damage to the Command Decoder caused by enhanced radiation from the Van Allen belts.

Another effect in space systems which was first conceived in 1962 is single event upset (SEU). The history of single event upsets is interesting. It started with a paper by Wallmark and Marcus[7], of RCA, published in 1962, in which the authors speculated that cosmic rays would be able to produce upsets in microcircuits as the microminiaturization trend continued and that cosmic ray upsets would set the ultimate limit on how small a device could be made. Their paper, however, was so far ahead of its time that it was largely overlooked. The next significant event took place 13 years later, in 1975, when Binder, Smith and Holman[8], of the Hughes Corp., observed unexpected triggering of digital circuits in an operating satellite and postulated that the bipolar flip-flop circuits were being upset by charge collected from the dense ionization track of an energetic heavy ion cosmic ray passing through an individual transistor. Binder, Smith and Holman calculated an upset rate which was within a factor of two of the observed results. This paper too was ahead of its time and did not awaken the community to the potential seriousness of the problem.

In 1978, May and Woods[9], of Intel Corp., discovered that anomalous upsets which were occurring in dynamic random access memories, were being caused by alpha particles emitted from trace amounts of thorium and uranium in the materials from which the devices were made. In 1978 also, Pickel and Blandford[10], of the Rockwell Corp., studied anomalous upsets which were being observed in NMOS dynamic RAMs used in a satellite system and concluded that these upsets were due to heavy ion cosmic rays. They also developed an approach for calculating upset rates to be expected in a cosmic ray environment. The Pickel and Blandford paper finally alerted the radiation effects community to the problem of heavy ion cosmic ray upsets in microcircuits.

In 1978 Guenzer, Wolicki and Allas, of NRL, realized that if cosmic ray ionization could upset a microcircuit then, possibly, nuclear reactions induced by energetic neutron and protons could also. They[11] then discovered that high energy neutrons and protons could, indeed, produce upsets in dynamic RAMs. Wyatt, McNulty, et. al., of Clarkson Univ., reported on high energy proton induced upsets at the same time[12]. Other results were obtained by Kolasinski, Blake, Anthony, Price and Smith[13], who reported that soft errors and single event latchup (SEL) were produced in static RAMs by BeV heavy ions which were obtained from the Berkeley Bevalac accelerator.

Single event effects have been a serious problem for satellite systems ever since the late 1970's when microelectronic devices first became small enough in feature size to be sensitive to them. There is a large number of spacecraft programs which have had to spend extra time and money to study, design or retrofit as a result of single event effects. In the Galileo Program, for example, about \$17 M and nearly two years was spent to solve a potential SEL problem.

A final natural environment radiation effect in space systems which will be discussed, spacecraft charging, was first recognized about 1975 as a potential cause of a problem which had been plaguing spacecraft since the early 1970's. "Spacecraft anomalies" such as upsets, switching errors and resets had been occurring in some space systems[14]. Other work by DeForest[15] indicated that a synchronous satellite was probably being charged to high potentials by particles. In 1975, it was noticed[14] that an unusually large fraction of the spacecraft anomalies occurred in the midnight to six a.m. quadrant and at high altitudes, near geosynchronous altitude (See Fig. 2). The probability that 31 out of 47 "anomalous events" would occur at random in a single quadrant was calculated as 3 chances in 10^9. The explanation suggested was that events occur preferentially just after midnight because the spacecraft passes through the geomagnetic tail where electrons, accelerated by geomagnetic activity, are injected and drift from midnight toward dawn. This problem was viewed as serious enough that a spacecraft, SCATHA, was designed and launched specifically to study spacecraft charging.

UPSETS SHOWN IN LOCAL TIME

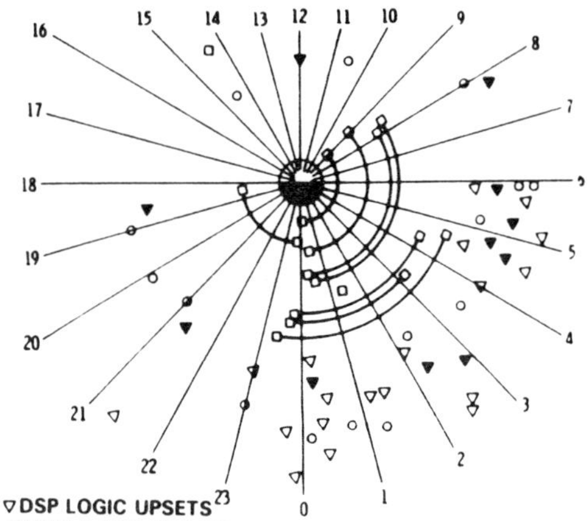

FIG. 2. LOCAL TIME DEPENDENCE OF CIRCUIT UPSET FOR SEVERAL DoD AND COMMERCIAL SATELLITES. (TAKEN FROM REF. 14)

NATURAL SPACE RADIATION ENVIRONMENT

As discussed above, the earth's radiation belts constitute a significant hazard to the operation and survivability of space systems. The space radiation environment will first be discussed briefly then the effects it produces on space systems will be described. The earth's radiation belts are made up primarily of electrons and protons and extend from low altitude orbits of several hundred nautical miles to beyond geosynchronous altitude (19,323 nmi) orbits. Because of the misalignment of the earth's magnetic field by 11 degrees with respect to the north pole and its slight offset from the geometric center of the earth, the radiation belts dip down into the upper atmosphere near South America causing what is known as the "South Atlantic Anomaly". Low altitude spacecraft receive an intense burst of particles when they pass through that region.

The electron integral fluences based on NASA models AE6 and AE4[16,17] are shown in Fig. 3 for electron energies greater than 0 MeV (the actual lower limit is 40 KeV) and 1 MeV as a function of satellite altitude and orbital inclination. The curves marked E >0 MeV are the fluence incident upon the outer skin of the satellite. A typical spacecraft has about a 30 mil skin and electronic circuits are usually placed inside of boxes made of about 50 mils of aluminum making a total of about 80 mils Al shielding between the outside surface and the closest electronics device. One MeV electrons are absorbed by approximately 80 mils of aluminum so the curves marked E >1 MeV may be thought of roughly as the fluence incident on the electronics inside of the spacecraft boxes. It is clear that the skin and the electronic boxes provide substantial protection for the electronics from the earth's radiation belts at all altitudes. The dose per year produced by these electrons behind a 100 mil aluminum slab for the AE8 NASA model is shown as a function of altitude and orbital inclination in Fig 4. The shading shows the uncertainties in the environments[18]. Note that doses as high as 10^5 rads per year can be seen for a worst case orbit but most satellites do not operate in the heart of the belts; they operate either below or above the main parts of the belts.

Fig. 5 shows similar curves for earth's radiation belt protons[19] based on NASA model AP8. 80 mils of Al is about the range of a 20 MeV proton, so the curves marked E >0 MeV represent the fluence incident on the outer surface of the spacecraft and the curves marked E >20 represent, again roughly, the fluence just inside the electronics boxes. The skin and electronic boxes provide a great deal of protection for electronics from protons at high altitudes but less at low altitudes.

As can be seen from the last three figures, the earth has intense radiation belts which have a broad peak between an altitude of 2000-8000 nmi. To avoid the radiation induced problems caused by these belts, spacecraft are generally flown either below or above the main parts of the belts unless there are over-riding considerations which require operating in the heart of the belts. In general, the satellites "footprint", i.e. the area of the earth that it can see directly, gets larger rapidly as the satellite's altitude increases, thus increasing the coverage. The disadvantage is that the higher the spacecraft goes, the higher the power required for communication, surveillance or mapping. For a radar satellite, the tradeoff becomes acute. Since the reflected power at earth decreases inversely as the square of the spacecraft altitude, the received power at the spacecraft depends inversely on the 4th power of altitude. For this reason radar satellites are prime candidates for operating in the heart of the belts. They need to be high enough so that good coverage can be obtained with a minimum number of satellites, yet not so high that their power requirements are overwhelming.

There are two more near earth natural space environments of importance, namely cosmic rays and solar flares. The total dose deposited by cosmic rays is not significant compared to the dose deposited by electrons and protons from the belts but, nevertheless, cosmic rays can produce significant effects in spacecraft electronics through single event phenomena. The solar flare protons can produce SEU's and they can also do serious damage to solar cells particularly during anomalously large events which can occur several times per 11 year solar cycle. There were three anomalously large (AL) events in solar cycle 19, one in cycle 20 and none in cycle 21. Such AL events constitute the major portion of the total fluence for the entire cycle. For example, in solar cycle 20, the one AL event which occurred in August, 1972 produced 84% of the total solar proton fluence for the 20th cycle.

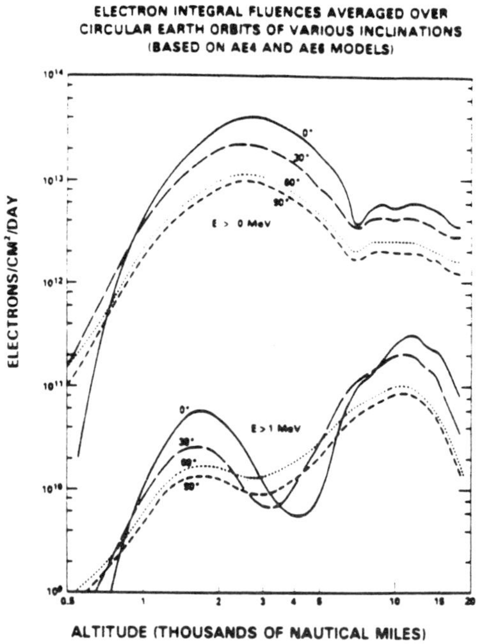

FIG. 3 ELECTRON INTEGRAL FLUENCES IN ELECTRONS/CM²/DAY AVERAGED OVER CIRCULAR EARTH ORBITS OF VARIOUS INCLINATIONS (BASED ON NASA AE4 AND AE6 RADIATION BELT MODELS). THE ELECTRON THRESHOLD WAS 40 KeV FOR THE MEASUREMENTS.

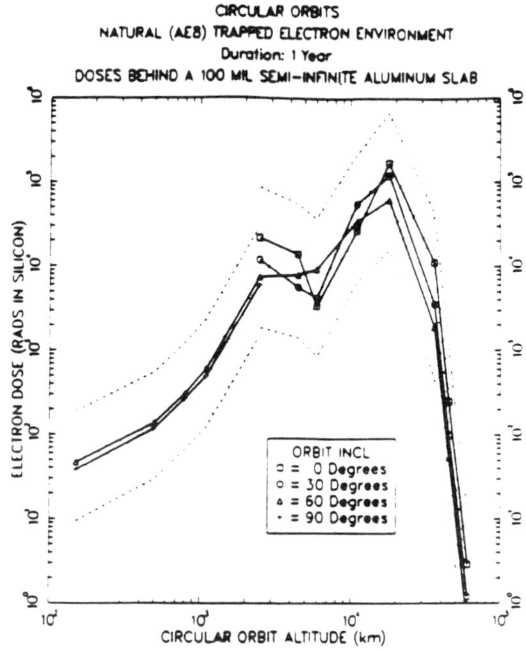

FIG 4. ONE YEAR ACCUMULATED DOSE IN THE NATURAL ENVIRONMENT FOR VARIOUS CIRCULAR ORBITS.

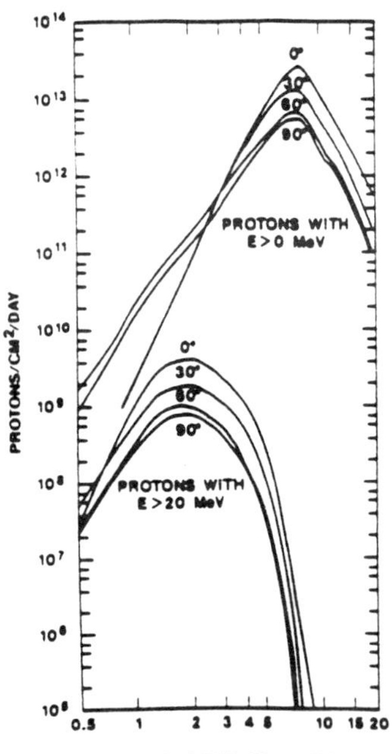

FIG. 5 PROTON INTEGRAL FLUENCES IN PROTONS/CM²/DAY AVERAGED OVER CIRCULAR EARTH ORBITS OF VARIOUS INCLINATIONS (BASED ON NASA AP8 MIN MODEL).

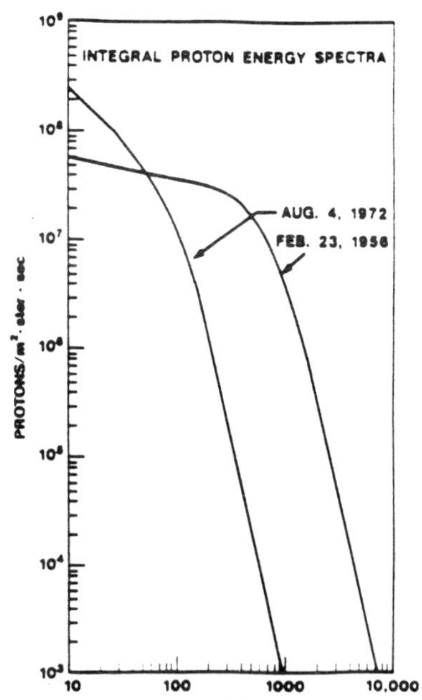

FIG. 6 THE INTEGRAL PROTON SPECTRA FOR THE PEAKS OF THE AUGUST 1972 AND FEBRUARY 1956 SOLAR FLARES. (TAKEN FROM REF. 21)

Solar flares generate an uncertain environment for spacecraft. In general, they are not an especially severe problem, yet they might cause a very high number of SEU's during an AL event and the degradation they produce might require an oversize solar cell array for a long mission. Nevertheless, for a short mission it is quite likely that not even a single AL event will occur. Because the AL events produce such a large portion of the total fluence, and because their probability is so small per year, it is necessary to specify the expected fluence from solar flares during a spacecraft mission very carefully. It is generally specified as a small number of AL events for the mission and the spectrum of the events is also given. King[20] has calculated the probabilities of solar flares per year of spacecraft mission. The spectrum generally selected is that of the event of August, 1972. For comparison, integral proton spectra are shown in Fig. 6 for the August, 1972 event and for one of the highest energy events yet recorded, the February, 1956 event[21]. Hardening against solar flares requires making a tradeoff of risk vs. cost and weight penalties. One consideration for solar array size on any spacecraft is the risk. The System Program Office (SPO) decides what risk he is willing to take, i.e., a 1% or 10% chance that more AL events will occur than he has designed for and that risk determines his design. SPO's sometimes simply ignore the increased SEU risk, arguing that one bad week of operation would not be a disaster and that a small reduction in power could be tolerated near the end of life of the spacecraft.

The mission determines the designer's response. For example, the Combined Release and Radiation Effects Satellite (CRRES) has a mission to measure the SEU's in a large number of devices as a function of time and position in orbit. For this mission an AL event represents, not a problem, but a rare opportunity! Accordingly, the CRRES Microelectronics Experiment was designed quite carefully to count as many SEU's as possible and record their address locations within each device during a flare before being forced by a high SEU rate into a mode where it will record only the total number of SEU's observed by all of the devices.

Cosmic rays are the final near-earth environment of interest which will be discussed. The relative abundance of cosmic rays and their spectrum is shown in Fig. 7. Fig. 8 shows the cosmic rays in comparison to an AL event[22]. Whenever an AL event is present it clearly dominates, however AL events rarely happen. The major effect of cosmic rays on spacecraft is to produce SEU's; their contribution to total dose is negligible. Cosmic rays cannot be shielded. This completes the description of the near-earth radiation environment.

RADIATION EFFECTS IN DEVICES AND MATERIALS

There are a variety of radiation effects or damage mechanisms in semiconductor materials and devices produced by the natural space environments described above. Before describing each effect, it is useful to make a differentiation between the environment itself and the effects or damage it produces. The total spacecraft

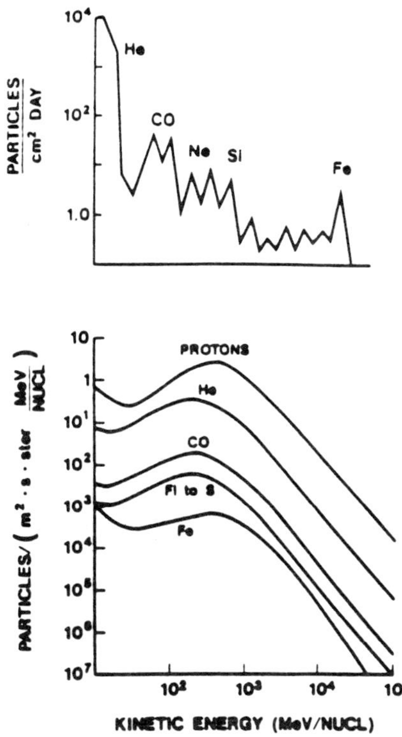

FIG. 7 COSMIC RAY RELATIVE ABUNDANCE (TOP) AND SPECTRA (BOTTOM) FOR SELECTED ELEMENTS.

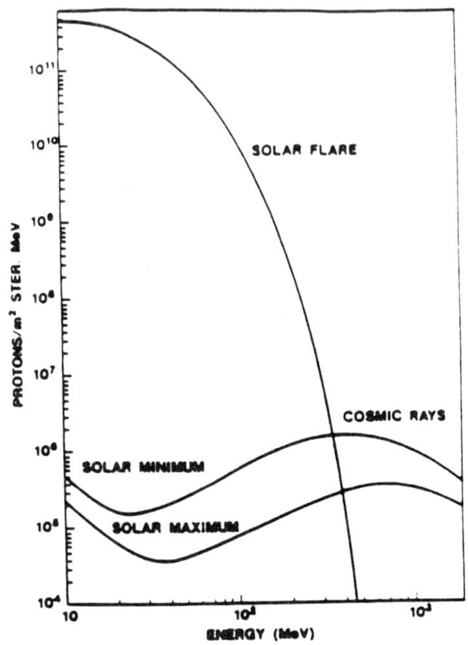

FIG. 8 THE ACCUMULATED SOLAR FLARE PROTON DIFFERENTIAL ENERGY SPECTRUM FOR THE WEEK OF AUGUST 4 - 11, 1972. THIS IS COMPARED TO THE GALACTIC COSMIC RAY PROTON SPECTRA FOR SOLAR MINIMUM AND MAXIMUM ALSO ACCUMULATED FOR ONE WEEK. (TAKEN FROM REF. 22)

radiation environment is determined by specifying the intensity for each of the individual environments. When the vulnerability or hardness of a device or material is discussed, it is not useful to state it in terms of individual radiation environments. Several different environments may be present and they must be combined; they may also affect the device in different ways. Effects or damage in devices or materials are given in terms of the damage mechanisms as shown in Fig. 9. For example trapped protons, solar flares and cosmic rays all contribute to single event phenomena and all of the environments contribute, to some extent, to the total ionizing dose received, but the cosmic ray contribution to total dose is negligible. The device hardness and damage parameters are therefore specified in terms of the total ionizing dose and displacement damage the device can withstand without permanent damage or the SEU rate it would sustain in a given orbit.

NATURAL SPACE ENVIRONMENT DAMAGE MECHANISMS

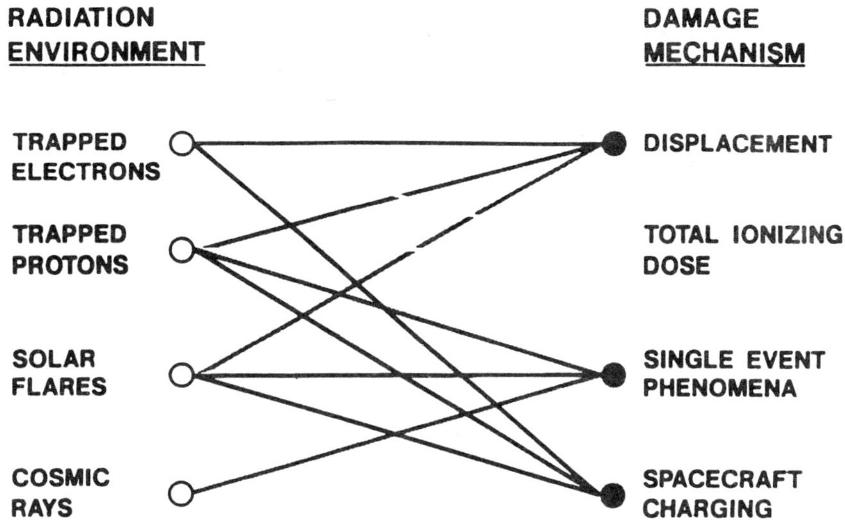

FIG. 9 RADIATION EFFECTS OR DAMAGE MECHANISMS IN SPACECRAFT DEVICES AND MATERIALS PRODUCED BY THE NATURAL SPACE ENVIRONMENT. THE FIGURE PROVIDES A TRANSLATION BETWEEN THE VARIOUS RADIATION ENVIRONMENTS AND THE DAMAGE MECHANISMS WHICH THEY PRODUCE IN DEVICES. THE EFFECTS OF COMBINED ENVIRONMENTS ARE SHOWN.

Each of the damage mechanisms will now be described. When protons or high energy electrons collide with atoms of a semiconductor displacing them from their lattice sites, displacement damage is created. This damage can anneal to some extent at room temperature. Displacement damage generally results in gain degradation of a transistor or solar cell power loss and can cause failure in the circuit if the gain degradation goes beyond the permitted operating range for the device as used in the circuit.

Note that in designing a subsystem for a spacecraft, there is a tradeoff between gain required for correct operation of the circuit and vulnerability to radiation effects. If the circuit is designed to require only a much lower gain to operate correctly than the initial gain of the device, it will be insensitive to radiation gain degradation. Good design practice takes advantage of this circumstance.

For many years it was believed that displacement damage would depend strongly on the type of particle producing the damage and on its rate of energy loss. That is, the damage was expected to be determined by the way in which defects clustered along the particle path, which depends on the particle type and energy loss per unit path length. Clusters of defects near the end of the path were supposed to behave differently than isolated defects along the path. Recent research[23-26] has shown, however, that displacement damage depends linearly on a simple parameter called the nonionizing energy loss of the particle and is independent of cluster effects. This is a surprising result. It is true over as many as six orders of magnitude in nonionizing energy deposition and has been observed for particles as heavy as copper ions and as light as electrons. Copper ions and electrons differ by as much as six orders of magnitude in the total rate of energy loss. This linear dependence has been shown for three semiconductor materials, viz., silicon, germanium, and gallium arsenide and more recently it has also been shown[27,28] for three superconducting materials, namely $YBa_2Cu_3O_7$, $Bi_2Ca_2Sr_2Cu_3O_{10}$ and $Tl_2Ca_2Ba_2Cu_3O_{10}$ and appears to be a quite general property.

Ionizing radiation causes the accumulation of trapped charge in the insulators that all electronic devices contain. In MOS devices, the most serious problems are associated with the gate oxide and the shifts in threshold voltage that the trapped charge causes. If the threshold voltage (the voltage needed to switch a transistor) shifts beyond the range of operation of the device, then device failure occurs. In bipolar devices, the problem is likely to be an increase in leakage current caused by trapped charge in the field oxide used to passivate the surface or in an isolation oxide used to separate one device from an adjacent one. Increased leakage currents or increases in such parameters as propagation delay can also cause the device to fail.

The total ionizing dose effects described above are often permanent, but time dependent effects, such as annealing of trapped holes and generation of interface states, are also common. Time dependent effects are one of the most urgent radiation effects problems now being studied. Because of time dependent effects, device damage can continue to increase after irradiation or recovery can occur. Time dependent effects make dose rate considerations in testing of paramount importance.

Note once again, that spacecraft designers are not without tradeoffs to improve survivability. If circuits are designed which are tolerant to threshold voltage shifts, increased leakage current and to increases in propagation delay, then a higher level of survivability can be achieved.

When a single particle produces ionization directly or a nuclear reaction resulting in ionization along a narrow track or tracks in micro electronic devices, the charge collected from these tracks can cause transient upsets (SEU), multiple upsets or single event latchup (SEL). Single event phenomena can effect memories, logic circuits and power MOSFETs.

SEL is a problem which generally must be solved by part substitution, particularly if it results in such a large current that the device can burn out. Thus part substitution was also the only practical solution for the spacecraft Galileo, where a latchup near Jupiter in the original part could have caused mission failure. Deep space missions present a special problem due to the transit time of several hours required to discover and cure a latchup.

SEU can be ignored if it merely affects a few bits of data in a large collection where errors will be unimportant or will stand out and be detectable. Often it is also possible, at a modest penalty, to insert an error detection and correction circuit in the subsystem and remove the errors (if the error rates are not too high). The solutions for logic circuits and microprocessors are not so simple and, in some cases may be extremely difficult to achieve.

Spacecraft charging is caused by electrons and protons in the magnetosphere stopping in a spacecraft material, depositing charge. This buildup of charge within dielectrics can cause spontaneous destructive discharges within the spacecraft. This effect is particularly important in geosynchronous orbits. Transient currents, voltages and fields can result from dielectric breakdown and these can get coupled into other circuits. It is particularly difficult to deal with when it occurs in a coaxial cable since signals can get coupled directly into sensitive circuits. It can be mitigated by making dielectrics slightly conducting, putting terminal protective devices such as Zener diodes on the inputs of electronic boxes and designing circuits which are less upset sensitive.

METHODS OF ACHIEVING SURVIVABILITY

Achieving survivability for a space system is an iterative process. From the specified environment, simple, "worst case" calculations of the dose in the electronics are made. Worst case means calculations made, often with many simplifying assumptions, that result in answers which are accurate or too high, but not too low. For example, the dose may be calculated just inside the least shielded part of the box and used as though it applied to all parts in the box. This procedure is often useful. If the parts available are harder than this level (including an appropriate safety factor) then no further shielding is needed, and one can move on to other problems. Other worst case calculations are also made such as the worst case current in a circuit or the worst case voltage on a cable. When these worst case calculations produce a result which is still satisfactory, i.e., the device, circuit or subsystem will operate properly for its mission lifetime, then no further calculations of that type are needed. This technique is desirable because it saves making a lot of very complicated, expensive calculations. This worst case technique finds many applications in survivability analysis.

A critical technique for achieving survivability is the selection of hardened parts. Parts are hardened for total ionizing dose or displacement damage effects. They are also hardened to SEU and single event latchup (SEL) effects and sometimes to dose rate effects are also included. If parts can be selected to meet the worst case dose (including safety factor), then that is clearly a desirable solution. If they cannot, then there are a variety of design and shielding tradeoffs and other options available, including making more sophisticated calculations in place of the simpler, worst case calculations to determine if a lower level of hardness will suffice.

To determine the hardness of parts, radiation tests are carried out with a sample size of generally five or more devices from the same lot and wafer. Overtests are used to increase confidence that the part can be used safely at the actual level it will be used in the spacecraft. The larger the sample size is, the smaller the overtest factor has to be to produce the same confidence. A comprehensive piece part testing program is a sine qua non for achieving survivability in a space system. Piece part testing is both a method of achieving survivability and a method of validation testing to see if it has been achieved. Since it is a part of the testing program to achieve confidence that an adequate hardening job has been done, it will be discussed there. Nevertheless, part testing is the source of part hardness data, device derating factors, device functional response data and other data needed in the hardening program.

Another critical method of achieving survivability is by circuit level analysis. In circuit analysis, parts data and a variety of codes such as Spice and Radspice are used to calculate the current, voltage and functional response of all circuits in the radiation environment. Device derating factors are calculated and design margins are determined, then design tradeoffs are explored to achieve the desired level of survivability with the minimum cost, weight and schedule delay penalties possible.

Hardened designs are yet another method of achieving survivability. When circuits are designed, hardening techniques such as current limiting, use of conservative device derating factors, and use of terminal protective devices are a part of the process. If a component or subsystem is critical to the operation of the space system, redundancy may be used as a further protection for the system, especially if the subsystem is believed to be vulnerable or even vulnerable at certain times. The redundant subsystem may be powered or unpowered until needed. If it is left unpowered it will receive significantly less radiation damage. It should be noted that with proper hardened designs, even though a component or even a subsystem may fail, the system may nevertheless survive!

Circuits are designed then checked to see what the effect would be if an upset were to occur in the circuit. They must be designed so that no functional failure can occur if an upset were to occur. SEU's can be eliminated or their probability reduced through use of parts with very low upset crossections and addition of error detection and correction circuits. In extreme cases, when critical data must be preserved or when an upset simply cannot be tolerated, techniques such as triple modular redundancy or other voting schemes are used. Frequently space systems must satisfy requirements such as upsets being permitted but the system must recover within a specified time, without taking any unintended action and without a ground command being required. These requirements are all a part of the design process.

The final method of achieving survivability which will be discussed is that of shielding. External devices such as solar cells, sensors and antennas are outside the skin of the spacecraft and hence do not benefit from any intrinsic shielding supplied by the spacecraft; any shielding they need must be applied directly. The spacecraft skin is typically about the equivalent of 30 mils of aluminum and electronics is usually found inside an aluminum box which is about 50 mils thick. Any additional shielding required must be supplied with an accompanying weight penalty.

When it is determined through analysis that shielding is needed for survivability, the question becomes, "What is the amount needed and where should it be placed to optimize the survivability of the spacecraft and minimize the additional weight?" There are, in general, two major reasonable choices, spot and comprehensive shielding. Brute force techniques such as applying a uniformly thick shield over the whole spacecraft incur weight penalties that are nearly always prohibitive and, hence, are rarely used. If only a few devices are too sensitive, the best solution is to apply spot shields to just those devices. On the other hand, when sensitive devices are distributed throughout the electronics, then shielding which takes advantage of the inherent shielding of the spacecraft to save weight should be used. An optimal solution is to apply it only in those places where it is needed to reduce the dose in the entire area to the required level is the optimal solution. How to determine exactly the "right" places to shield in anything as complex as a modern spacecraft, however, is not obvious. For example, if a battery or a massive part of the spacecraft structure is nearby or if a box is deeply buried, then less shielding will be required to reach a given dose. Determining the optimum shielding is complex and requires use of detailed spacecraft modeling and large computer codes to calculate.

Excellent computer codes are available for optimizing the placement of shielding within spacecraft. They were developed at the Air Force Weapons Laboratory (AFWL) under the direction of Joseph Janni in 1977. The immediate impetus was provided by a spacecraft which had developed a serious problem. It had a design lifetime of only 18 months and made extensive use of a device which had a failure level above 50 krads. Unfortunately, an alert had just been issued notifying potential users that the manufacturer had apparently changed something in the processing of the device and the current hardness was about 5 krads. This news came at a particularly bad time. The devices were already installed, there was a disagreement between NASA and The Aerospace Corporation about whether the environment might be more intense than previously thought, and the spacecraft had an extremely small weight margin. Preliminary calculations showed that if shielding were done in the (then) conventional way, it would weigh over 80 pounds to obtain an acceptable spacecraft lifetime. With this strong incentive, AFWL developed the codes for shielding optimization in record time. Using the newly developed codes they were able to reduce the shielding required from 80 pounds to 17 pounds and thus obtain a reasonable lifetime for the spacecraft. These codes have proved to be very useful in shielding computations for other spacecraft.

VALIDATION TESTING

No space system hardening process can be considered to be complete without a significant amount of validation testing. Testing can be done at several different levels; namely, piecepart, circuit, box and at the whole space system level. A testing program begins at the piecepart level. Individual parts are tested using cobalt-60 sources and with particle accelerators. Parts from the same diffusion lot and wafer as those to be used in the space system are used. The larger the sample size available for testing, the lower the overtest factor required to give a sufficiently high level of confidence will be. The sample size chosen is generally five or more. If the part survives an overtest of ten, then five parts is considered sufficient. If, however, the part only survives an overtest of two or three or if there is a substantial variation in the failure level for various parts from the same group, then a larger sample size should be used. Displacement damage tests result in gain degradation of the device. The failure level for the device cannot be specified without knowing what minimum gain is actually required for correct operation of the circuit. If a device has a pre-irradiation gain of 100, and the expected dose reduces the gain to 50, it can perform quite well in a circuit designed for a minimum gain of 20. If the post-irradiation gain is too low, an acceptable option is to redesign the circuit to tolerate lower gain. A 100 percent screen of all latchup-sensitive devices is generally done because latchup frequently carries serious system penalties if it occurs.

The next level of testing is at the circuit level. Circuits which are critical to the operation of the system are tested when necessary. Overtest margins at the circuit level are lower than at the piece part level, typically about a factor of three. Box level tests are usually done and they usually have overtest factors of one or two as do whole spacecraft tests, but the latter are rarely, if ever, done.

Validation testing is a crucial part of space system survivability. It is necessary to provide the required level of confidence that the system will perform as designed and for the desired lifetime.

FUTURE CHALLENGES

Future space systems will include laser or 60 GHz crosslinks for rapid data transmission between spacecraft. The degree of autonomy in spacecraft will also increase, to reduce dependence on ground stations. More complex onboard data processing will also be done. All of these tasks will require faster, more complex and advanced devices and even larger memories, probably more than 10^9 bits on a spacecraft.

As future microelectronic devices become more complex with smaller feature sizes (and smaller critical charges) and higher speeds permitting more to be done on the same size chip, the capability of spacecraft using these devices will increase greatly. Unfortunately, the vulnerability to radiation damage and SEU will also increase greatly. At 10 microns the charge necessary to upset the device is about 1 pico Coulomb and very few devices with that large a critical charge can be upset even by highly ionizing particles. At 1 micron (the current state of the art) the critical charge has dropped by two orders of magnitude to 0.01 pico Coulomb and even lightly ionizing particles can upset the device. When radiation belt protons can upset devices by direct ionization instead of through a nuclear reaction, then the upset rate will increase by several orders of magnitude. Unfortunately recent experiments[29] have shown that, with the most sensitive GaAs devices (smallest feature size), we have already reached that point! A three order of magnitude increase in upset rate has been observed. This new discovery is ominous because future devices will also have very small feature sizes and correspondingly small critical charges for upset. As these new devices are put into spacecraft their SEU problems will truly be astronomical.

Most solar cells used in the current generation of satellites are made of silicon (Si). Several years ago gallium arsenide (GaAs) cells were introduced. Such cells are more efficient and degrade less than Si cells in a radiation environment. Recently an even more radiation resistant solar cell, indium phosphide (InP), has been developed. InP can absorb 30 to 50 times more fluence for the same damage[30]. In addition InP solar cells have been reported to anneal most of their radiation damage in sunlight. It therefore appears that solar cells may be even harder in the future than they are at present.

As new technologies such as high temperature superconducting devices are introduced into spacecraft, the spacecraft's capability will increase substantially. Superconducting devices are to be flown on the first High Temperature

Superconductivity Space Experiment (HTSSE) currently planned for launch in 1992. This space experiment[31] is designed to demonstrate the utility of superconductivity to improve the operational capability of space systems. These first generation devices will offer much higher Q's in narrow band-pass filters; dispersionless, lossless and far more compact delay lines; the potential for extremely broad-band IR sensors; and masers with superconducting cavities will have an order of magnitude improvement in accuracy. Second generation devices will move from primarily passive microwave devices to active electronic devices as soon as a Josephson junction device can be demonstrated in high T_c materials. Such devices offer the promise of the equivalent of a Cray in a very small volume and using only a small fraction of the power.

There is, however, at present little radiation effects data on superconductors, and low temperature superconductor Josephson junctions have been shown to be susceptible to SEU[32]. Furthermore, as superconducting devices such as IR sensors (very radiation sensitive) start to be used on spacecraft more, they will operate at low temperature and will need to be interfaced to conventional electronics. This means that we will need to understand how to harden conventional devices to operate at low temperature where thermal annealing is negligible.

From the above brief illustrations it is clear that much exciting research remains to be done in the area of radiation effects in space systems. Smaller devices and new types of device will be used on ever more complex systems. Autonomous subsystems for guidance, control and navigation will become more complicated and they may become the heart of some new space systems such as Brilliant Pebbles or Brilliant Eyes. It appears very likely that many survivability challenges will present themselves.

ACKNOWLEDGEMENTS

This paper is a revised, updated and shortened version of a paper "Space Systems Survivability" which was presented at the IEEE Nuclear and Space Radiation Effects Conference as a part of the Short Course on Survivability and Hardening of Space Systems in Marco Island, Florida on July 24, 1989. The paper has not been previously published in any Journal.

REFERENCES

1. E. Wenaas, "Spacecraft Charging Effects on Satellites Following the Starfish Event," CSC Report H/3-J-280, February 1978.

2. R.E. Fischell, "ANNA-1B Solar Cell Damage Experiment," Transcript of the Photovoltaic Specialists Conference, April 10, 1963, Washington D.C. (Available from Defense Documentation Center as PIC-SOL 209/3 July 1963).

3. J.J. Loferski, P. Rappaport and J. Scott-Monck, "Radiation Damage to Silicon Solar Energy Converters," R.C.A. Labs Quarterly Report No. 1, NASA Contract NAS5-47, October 15, 1960.

4. W.L. Brown, T.M. Buck, L.V. Medford, E.W. Thomas, H.K. Gummel, G.L. Miller and F.M. Smits, "The Spacecraft Radiation Experiments," NASA SP-32, Vol. 1, June 1963.

5. W.L. Brown, J.D. Gabbe and W. Rosenzweig, "Results of the Telstar Radiation Experiments," NASA SP-32, Vol. 2, June 1963.

6. J.S. Mayo, H. Mann, F.J. Witt, D.S. Peck, H.K.Gummel, and W.L. Brown, "The Command System Malfunction of the Telstar Satellite," NASA SP-32, Vol. 2, June 1963.

7. J. T. Wallmark and S. M. Marcus, "Minimum Size and Maximum Packaging Density of Nonredundant Semiconductor Devices," Proc. IRE, 50, 286(1962).

8. D. Binder, E. C,. Smith, and A. B. Holman, "Satellite Anomalies From Galactic Cosmic Rays," IEEE Trans. Nuc. Sci., NS-22, 2675(1975).

9. T. C. May and M. H. Woods, "Alpha Particle Induced Soft Errors in Dynamic Memories," IEEE Trans. Elec. Dev. ED-26, 2(1979).

10. J. C. Pickel and J. T. blandford, "Cosmic Ray Induced Errors in MOS Memory Cells," IEEE Trans. Nuc. Sci., NS-25, 1166(1978).

11. C. S. Guenzer, E. A. Wolicki, and R. G. Allas, "Single Event Upset of Dynamic RAMS by Neutrons and Protons," IEEE trans. Nuc. Sci., NS-26, 5048(1979)

12. R. C. Wyatt, P. J. McNulty, P. Toumbas, P. L. Rothwell, and R. C. Filz, "Soft Errors Induced by Energetic Protons," IEEE Trans. Nuc. Sci., NS-26, 4905(1979).

13. W. A. Kolasinski, J. B. Blake, J. K. Anthony, W. E. Price, and E. C. Smith, "Simulation of Cosmic Ray Induced Soft Errors and Latchup in Integrated Circuit Computer Memories," IEEE Trans. Nuc. Sci., NS-26, 5087(1979)

14. D. A. McPherson and D. R. Schober, "Spacecraft Charging at High Altitudes: The SCATHA Satellite Program," presented at the AIAA 13th Aerospace Sciences Meeting, Pasadena, CA, January 20-22, 1975.

15. S.E. DeForest, "Spacecraft charging at synchronous orbits," Journal of Geophysical Research, Vol. 77, 651 (1972)

16. M.J. Teague, K.W. Chan, J.I. Vette, "AE:6, A Model Environment of Trapped Electrons for Solar Maximum," NSSDC/WDC-A-R&S 76-04, GSFC, Greenbelt, MD,(May 1976).

17. G.W. Singley, J.I. Vette, "The AE-4 Model of the Outer Radiation Zone Electron Environment," NSSDC 72-06, GSFC, Greenbelt, MD (Aug 1972).

18. R. Leong, D. Lahti, K. Murphy and S. Rieco, "AFWL Exoatmospheric Nuclear Radiation Environment Prediction and Shielding Codes Applicable to the Strategic Defense Initiative," AFWL Draft Report, November, 1985.

19. D.M. Sawyer and J.I. Vette, "AP-8 Trapped Proton Environment for Solar Maximum and Solar Minimum," NSSDC/WDC-A-R&S 76-06, GSFC, Greenbelt, MD, (Dec 1976).

20. J.H. King, "Solar Proton Fluences as Observed During 1966-1972 and as Predicted for 1977-1983 Space Missions," NASA X-601-73-324, (Oct 1973).

21. J. H. Adams and A. Gelman, "The Effects of Solar Flares on Single Event Upset Rates," IEEE Trans. Nuc. Sci., NS-31, 1212 (1984).

22. J.H. Adams, Jr., R. Silberberg, and C.H. Tsao, "Cosmic Ray Effects on Microelectronics, Part I: The Near-Earth Particle Environment," NRL Memorandum Report 4506, (1981).

23. C.J. Dale, P.W. Marshall, E.A. Burke, G.P. Summers, and E.A. Wolicki, "High Energy Electron Induced Displacement Damage In Silicon," IEEE Trans. Nucl. Sci. NS-35, 1208 (1988).

24. G.P. Summers, E.A. Burke, C.J. Dale, E.A. Wolicki, P.W. Marshall, and M.A. Gehlhausen, "Correlation of Particle-Induced Displacement Damage in Silicon," IEEE Trans. Nucl. Sci. NS-34, 1134 (1987).

25. P.W. Marshall, C.J. Dale, G.P. Summers, E.A. Burke, and E.A. Wolicki, "Proton, Neutron, and Electron Induced Damage in Germanium," to be published, IEEE Trans. Nucl. Sci., (Dec 1989).

26. G.P. Summers, E.A. Burke, M.A. Xapsos, C.J. Dale, P.W. Marshall, and E.L. Petersen, "Displacement Damage In GaAs Structures," IEEE Trans. Nucl. Sci. NS-35, 1221 (1988).

27. G.P. Summers, E.A. Burke, D.B. Chrisey, M. Nastasi and J.R. Tesmer, "Effect of Particle-Induced Displacements on the Critical Temperature of $YBa_2Cu_3O_{7-\delta}$," Appl. Phys. Lett. 55, 14, October (1989).

28. B.D. Weaver, G.P. Summers, E.A. Burke, "Displacement Damage in Cuprate Superconductors," to be submitted to Appl. Phys. Lett.

29. P.G. Coakley, R.A. Poll and H. Fasig, "Proton End of Range Ionization Upset of Submicron GaAs Static RAM," Presented at 7th Annual HEART Conference, Feb. 1989.

30. I. Weinberg, C.K. Swartz, R.E. Hart and R.L. Statler, "Radiation and Temperature Effects in Gallium Arsenide, Indium Phosphide and Silicon Solar Cells," Nineteenth IEEE Photovoltaic Specialists Conference-1987, IEEE Catalog No. 87CH2400-0 pp 548-557.

31. J.C. Ritter, M. Nisenoff, G. Price and S.A. Wolf, "High Temperature Superconductivity Space Experiment (HTSSE)," to be published in the IEEE Transactions on Magnetics MAG-27, April 1991.

32. R. Magno, M. Nisenoff, R. Shelby, A.B. Campbell, and J. Kidd, "Upset Events in Josephson Digital Devices Under Alpha Particle Irradiation," IEEE Trans, Nucl. Sci. NS-28, 3994 (1981).

AAS 91-015

INITIAL RESULTS FROM LONG DURATION EXPOSURE FACILITY (LDEF) POSTRETRIEVAL OBSERVATIONS

William H. Kinard*, Glenna D. Martin† and Robert L. O'Neal‡

Initial postretrieval observations of the Long Duration Exposure Facility and the 57 onboard experiments are reported in this paper. These observations have revealed that the facility and experiments, which had remained in space for 68 months prior to being retrieved in January 1990, can be used to generate "gage block" type data to verify the existing models of the low Earth orbit environments and, if necessary, the data to correct errors in these existing models. The LDEF observations can also be used to update and revise the existing models used by spacecraft designers to predict the effects of long exposures in these environments on spacecraft materials and components.

INTRODUCTION

The LDEF was designed and built by NASA's Langley Research Center (LaRC) for NASA's Office of Aeronautics, Exploration and Technology initially to provide opportunities for technology experiments in space. The performance of science and applications experiments was later added as an LDEF objective. Specifically the LDEF was designed to allow experiments to be transported into space via the Space Shuttle; to free fly in orbit in a stable manner, thus allowing the onboard experiments to be exposed in space for extended periods of time; and to allow the experiments to be retrieved, after the desired exposure in space, via the Shuttle and returned to the investigators for postexposure analyses in their laboratories.

* Dr. Kinard is LDEF Chief Scientist and Head, LDEF Science Office, Mail Stop 404, NASA Langley Research Center, Hampton, Virginia 23665-5225.

† Special Assistant to Chief Scientist, LDEF Science Office, Mail Stop 404, NASA Langley Research Center, Hampton, Virginia 23665-5225.

‡ Mr. O'Neal, a retired NASA past LDEF Project Manager, is affiliated with Lockheed Engineering and Management Services Company, Inc., Hampton, Virginia, and is currently under contract to NASA to manage all LDEF photographic documentation for the LDEF Science Office.

LDEF DESCRIPTION

The LDEF is a 12-sided, 15-foot diameter, 30-foot long aluminum frame. The flight hardware required for LDEF experiments is accommodated in trays that are approximately 4 feet long and 3 feet wide, with depths varying from 3 inches to 12 inches. Eighty-six of these trays can be attached to the outer surfaces of the 12 sides and the two ends of the aluminum frame.

A viscous magnetic rate damper is mounted on the LDEF structure to accelerate the damping of unwanted motions at the time of deployment in orbit. The only active electronic system on the facility is an experiment initiate system which can turn on the active experiments after achieving orbit and which can turn off these experiments after capture of the LDEF during retrieval. The LDEF has no central power or data systems to support experiments. The active experiments do have these systems as a part of their unique in-tray hardware.

The LDEF is shown in Fig. 1 on its LDEF Assembly and Transportation System (LATS) prior to the mounting of the experiments. The insert in the figure shows a typical experiment tray. The LATS is a specially designed combination transportation and work platform system which allows the LDEF to be rotated much like a ferris wheel for the installation and removal of experiments and for detailed inspections and testing of the experiments after they are mounted on the facility.

Fig. 1. LDEF on LDEF Assembly and Transportation System (LATS) Prior to Installation of Experiments. Insert Shows a Typical Experiment Tray.

LDEF MISSION

During the launch and retrieval flight operations, the LDEF occupies one half of the Shuttle's cargo bay. On orbit, the Shuttle's Remote Manipulator System lifts the LDEF from the cargo bay and places it in the proper attitude for release and free flight. Once free flying, the LDEF is passively stabilized by its mass distribution. One end of the spacecraft always points toward the center of the Earth, the other end always points out into space; one of the 12 sides always points forward in the direction of travel, and the opposite side always points in the trailing direction as indicated in Fig. 2. This stable orientation is necessary for the LDEF to be later captured and retrieved by the Shuttle and, in addition, it allows experiments to be exposed to selected elements of the space environment.

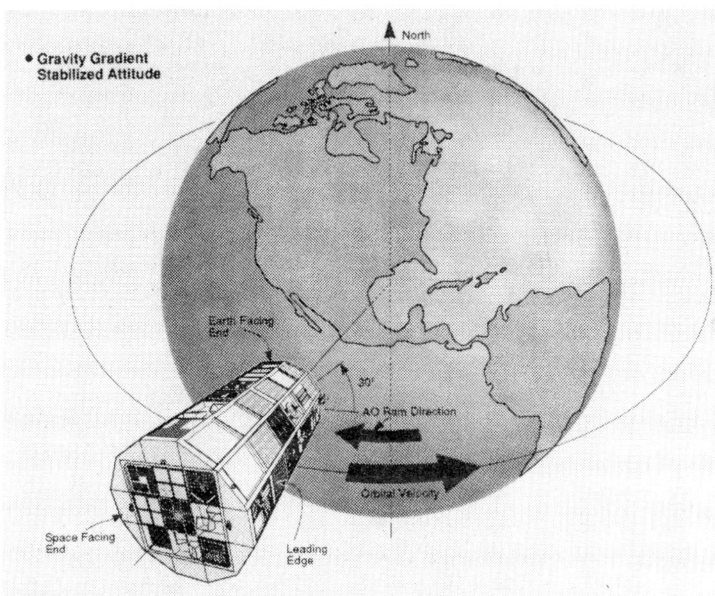

Fig. 2. LDEF Orbital Orientation.

The LDEF was launched into a 257 nautical mile, 32° inclination, near-circular orbit by the Shuttle Challenger on April 7, 1984. Plans at that time were to retrieve it 1 year later; however, because of Shuttle manifesting problems initially and later the Challenger accident, the LDEF remained in orbit for almost 6 years before it was retrieved by the Shuttle Columbia during STS Mission 32 on January 12, 1990. The LDEF orbit had decayed to an altitude of approximately 180 nautical miles before it was recovered.

LDEF EXPERIMENTS

Fifty-seven science, technology, and applications experiments flew on the LDEF during the mission (Ref. 1). They generally measured specific space environments such as meteoroid, man-made debris, and radiation level, or, they measured the effects of the space environments on typical spacecraft hardware. A few of the experiments measured space environmental effects on simple forms of life (seeds, spores, and eggs) and one experiment investigated the growth of crystals in reduced gravity. The experiments were selected to take advantage of the LDEF-provided opportunity for postflight laboratory analysis of retrieved test specimens.

Fifty percent of the LDEF experiments were completely passive with the data being derived entirely from comparisons of preflight and postflight observations of the test specimens. The other experiments on board were active involving either in-flight data measurements and data storage systems (no telemetry) or exposure control systems, or both. The active experiments were also designed to take full advantage of postflight studies of the retrieved test specimens to expand their data return.

The LDEF experiments exposed more than 10,000 test specimens to the environments of space during the 68-month stay on orbit. All of the active experiment systems were powered and programmed to function for only 1 year since the mission was initially planned as a 1-year mission. The power for the active experiments, with one exception, was provided by specially designed lithium batteries. A solar array and NiCa batteries provided the power for one experiment.

Since the LDEF and all of the onboard experiment hardware remained in space for almost 6 years, valuable knowledge of space environmental effects can be gained from detailed examinations of every piece of retrieved hardware--not just the 10,000 test specimens that were originally planned for study by the experiment principal investigators (P.I.'s). To facilitate the investigations of the non-test specimen hardware, four Special Investigation Groups (Ionizing Radiation, Materials, Meteoroid and Debris, and Systems) were established. These special investigation groups were also chartered to generate combined data bases which, when completed, will contain the data each group generates and the data the experiment principal investigators (P.I.'s) in related disciplines generate.

There are several hundred investigators from universities, industries, and government laboratories in the United States and 8 foreign countries associated with the 57 LDEF experiments, and, there are several hundred other investigators associated as members of the LDEF Special Investigation Groups. In addition, approximately 3,500,000 students around the world are involved in investigations of the 13,000,000 tomato seeds which flew in the LDEF SEEDS Experiment.

FIRST OBSERVATIONS ON-ORBIT

The first on-orbit looks at the LDEF by the astronauts flying the retrieval mission revealed that the facility was intact and structurally sound, and that it was stable and properly oriented--essential factors to perform the retrieval. The astronauts' first looks also revealed that the experiments had weathered during the prolonged stay in space. Fig. 3 is an early photograph which shows the weathering of the experiments on the space end of the facility. This photograph was taken by the crew as the Shuttle descended to capture the LDEF for retrieval. Tears in the Mylar thermal blankets covering two trays

Fig. 3. First On-orbit View of LDEF.

of the Naval Research Laboratory's Heavy Ions Experiment are clearly visible. These thermal blankets were originally white. They are now a dingy brown. The astronauts also reported metallic flakes floating near some of the experiment trays. Kapton thermal blankets in these trays had been eroded away freeing flakes of metallized aluminum coatings.

At one point the astronauts saw and photographed a small segment of a solar cell array flying in formation with the LDEF. This array segment was later identified as originally being a part of the Solar Cell Materials Experiment that was provided by the NASA Marshall Space Flight Center (MSFC). The Kapton cover film that had retained the solar cell array on the experiment had been eroded away by the impingement of atomic oxygen.

The LDEF investigators had expected that many of the exposed experiment surfaces would be deteriorated after the prolonged stay in space. They also had expected that the deteriorated surfaces would be extremely fragile and that they might be altered by the retrieval operations. Recognizing that if the surfaces were altered during the retrieval it would be difficult to establish the space exposure effects after retrieval, the STS 32 crew performed a very detailed visual and photographic survey of each experiment tray immediately after the LDEF was captured. Fig. 4 is a photograph of the LDEF as the

Fig. 4. Captured LDEF Positioned for Photography of Leading Surface Experiments.

crew oriented it to photograph the experiments on the leading side. Fig. 5 is a photograph of a single tray of the Aerospace Corporation's Space Environment Effects on Spacecraft Materials Experiment. The deteriorated test specimens in this tray are clearly visible. Photographs of this type were taken of each tray, and these photographs and the retrieval crew notes have provided excellent documentation of the on-orbit condition of the spacecraft and the experiments.

After the astronauts completed the on-orbit inspections and the photographic documentation, they berthed and latched the LDEF in the cargo bay of the orbiter and proceeded to complete the other activities planned for the on-orbit phase of the 11-day STS 32 Mission. The on-orbit phase ended as the Orbiter Columbia landed at Edwards Air Force Base in California. The Columbia, with the LDEF still in its cargo bay, was then ferried back to the

Fig. 5. Close-up Photograph of Aerospace Corporation's Materials Experiment on LDEF Leading Surface.

NASA Kennedy Space Flight Center (KSC) in Florida on the back of the specially equipped 747 aircraft. Only after the Columbia was placed in the Orbiter Processing Facility at KSC could its cargo bay doors be opened and the LDEF removed.

POSTRETRIEVAL OBSERVATIONS AT KSC

After the LDEF was removed from the orbiter and again mounted on its LATS (Fig. 6), a postretrieval photographic survey of all of the exposed LDEF surfaces was performed. Fortunately, these photographs revealed that the retrieval and the reexposure to the Earth's atmosphere had caused very little immediate change to the spacecraft. The surfaces looked essentially as they did in the on-orbit photographic survey pictures.

The experiment P.I.'s and the special investigation groups examined every cm of the retrieved LDEF and experiment surfaces before they began removing the experiment trays from the facility. After the trays were removed, they were again examined close up in more detail and status tests of systems in the active experiments were performed before the experiment hardware was shipped from KSC back to the respective P.I.'s around the world.

The following paragraphs provide brief descriptions of some of the more significant observations and measurements that were made during the processing of the LDEF and the onboard experiments at KSC, and during early examinations of the retrieved experiment hardware after it was shipped back to the P.I.'s.

Fig. 6. Investigators Inspecting Retrieved LDEF on LDEF Assembly and Transportation System (LATS) at KSC.

It should be mentioned that the LDEF investigators received invaluable assistance during the processing activities at KSC from both LaRC and KSC support personnel. The dedicated efforts of these support personnel contributed greatly to the quality and quantity of the data that have been and will be gathered from the LDEF.

<u>Ionizing Radiation:</u> Fifteen of the LDEF experiments were instrumented with radiation dosimetry devices. In addition, the Ionizing Radiation Special Investigation Group surveyed the induced radiation levels in each of the experiment trays and the LDEF structural elements at KSC (Fig. 7). The group also selected and shipped specific material specimens from the experiments and facility to numerous low-background radiation-counting facilities around the country.

The initial induced radiation measurements made at KSC identified radioactive beryllium 7 on the leading surfaces of the spacecraft (Ref. 2). This beryllium 7 was probably generated by protons and cosmic rays colliding with the oxygen and nitrogen atoms which are present at the altitude of the LDEF orbit and it then collected on leading LDEF surfaces.

The group's induced radiation measurements of aluminum facility parts, which were made in low background counting facilities, have confirmed existing predictions of directional effects of trapped radiation. They found that the trailing side of the LDEF received almost three times the leading side's radiation from measurements of the levels of Na^{22} in the aluminum (Ref. 3).

Fig. 7. Measuring Induced Radiation Levels in Retrieved LDEF.

Initial analyses made at the University of San Francisco by the P.I.'s of the LDEF Linear Energy Transfer Spectrum Measurements Experiment have indicated that the current prediction of 1600 rads in 30 years behind 0.5 cm of aluminum is adequate for the design of Space Station Freedom, if the Station is operated at a constant density altitude. Higher radiation levels which have been considered would require radiation hardening of more of the Station electronic components.

Many of the materials test specimens on the LDEF contain evidence of radiation effects such as darkening. The Ionizing Radiation Special Investigation Group will support the associated P.I.'s and members of the Materials and Systems Special Investigation Groups in future analysis of these specimens.

In summary, the orientation and the prolonged exposure in space made the LDEF a unique facility to investigate the near-Earth ionizing radiation environment, an environment consisting primarily of energetic cosmic rays and protons trapped in the South Atlantic Anomoly. When all of the LDEF data pertaining to ionizing radiation are gathered, they will certainly provide "gage block" data to check the accuracy of all current radiation environment and transport computer codes and, when it is necessary, to make revisions.

Meteoroid and Debris: Eight of the experiments flown on the LDEF had as their primary objective the study of meteoroids and/or man-made space debris. A number of the P.I.'s associated with these experiments have already published results from their early analysis (Ref. 4, 5 and 6). It is a fact, however, that the entire exposed LDEF is providing meteoroid and debris information and as such the entire LDEF should be considered a meteoroid and debris experiment.

Members of the Meteoroid and Debris Special Investigation Group examined, while at KSC, every square cm of the exposed surfaces on the LDEF and, when possible, they also examined surfaces behind the thinner penetrated exposed surfaces. During these examinations members of the group documented and took stereo sets of color pictures of more than 5,000 of the larger impact craters (Ref. 7) for future studies. These documented craters ranged in size from approximately 0.5 mm to 5 mm. There are also tens of thousands of smaller craters on these surfaces which hopefully will also be the subject of later investigations. The meteoroid and debris group acquired many of the exposed non-test specimen surfaces during the processing at KSC, and, they hope to acquire many other surfaces in the future after the investigators who own the surfaces are finished with them. All of the surfaces acquired by the group will be archived at NASA Johnson Space Center (JSC) and made available to qualified and interested researchers in the future.

The 0.005-inch thick second surface silvered Teflon thermal blankets on several trays located on the leading side presented the most prominant displays of meteoroid and debris impacts that were observed on the LDEF (Fig. 8). These impacts delaminated the silver from the rear surface of the Teflon over a rather large region surrounding the craters. Atomic oxygen then oxidized the exposed silver producing the visually prominent dark areas around each crater. These Teflon blankets also provided impressive views of the effectiveness of "Meteoroid Bumper Shields" (Fig. 9). Hundreds of complete penetrations of the blanket can be seen when it is backlighted, yet there is almost no evidence of particle fragments impacting the black painted pressurized cylinders which were 7 to 15 cm behind the blankets.

Almost every conceivable type of spacecraft surface was exposed on LDEF and examinations of the meteoroid and debris damage to these different surfaces provide valuable data for spacecraft designers. For example, the impacts shown in Fig. 10 illustrate the effect of skin thickness and surface condition on the resulting damage. The crater on the left in this figure is in a yellow painted, thick aluminum, scuff plate that was around one of the LDEF trunnion pins. The paint immediately around the crater has been spalled away. There is no damage outside of this spall region. The crater on the right is in a thin aluminum electronics package cover plate that was painted white. The ring patterns outside the immediate spall region resulted

Fig. 8. Meteoroid Penetrations of Silvered Teflon Thermal Blanket on LDEF Leading Side

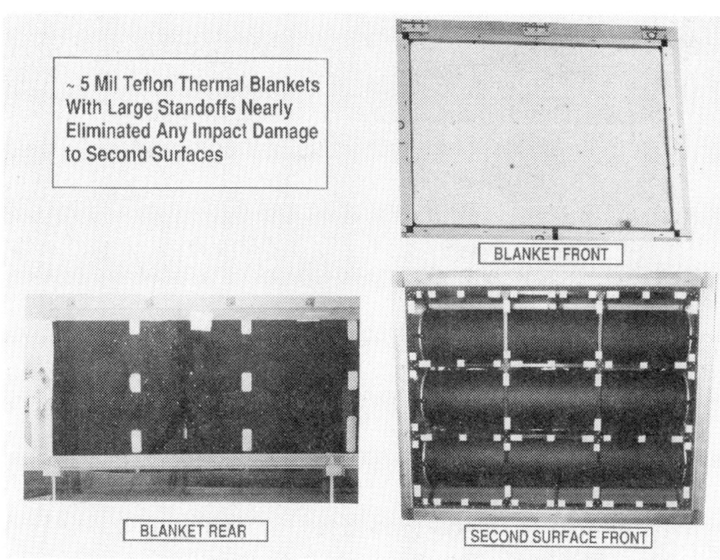

Fig. 9. Thermal Blankets Showed "Meteoroid Bumpers" to be Effective.

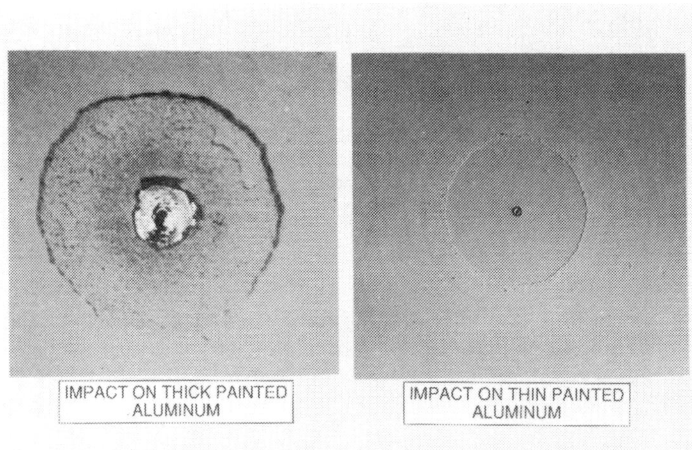

Fig. 10. Impact Damage to Thick and Thin Painted Aluminum Plates.

from the ejection of weathered paint pigments on the surface by the impact shock and reflections of this shock from the rear surface of the thin plate. This impact occurred late in mission. There were craters of the same size observed on this cover plate that did not have the ringed patterns--only the crater and a small paint spall region around the crater. These ringless craters probably occured early in the mission before the paint binder was deteriorated by the space exposure.

Examples of impact damage on other types of LDEF surfaces are illustrated in Fig. 11. Most of the craters in bare aluminum surfaces were hemispherical and generally shaped like hypervelocity craters produced in laboratory impact test facilities. Close examinations of some of these craters, however, reveal detail features that are not seen in craters produced in the laboratory. There were some impact craters observed that appeared to have resulted from either a very small cluster of particles or a particle that consisted of a low-density matrix with high density particles embedded in it.

The LDEF active Interplanetary Dust Experiment (IDE) recorded the direction and the precise time thousands of impact events occured (Fig. 12). IDE detectors mounted on the leading side of the LDEF recorded an order of magnitude more impacts than the detectors mounted on the trailing side. This distribution of impacts is consistent with the distribution of crater counts made on other surfaces around the facility. The IDE data also indicated that impact events are not random in time. The data indicate clustering of the IDE-detected particles. Some of these clusters are very probably the result of manned operations in space and others may prove to be particles in comet streams. It is expected that future analysis of the IDE data will provide excellent clues to the orbital dynamics of small particles near Earth.

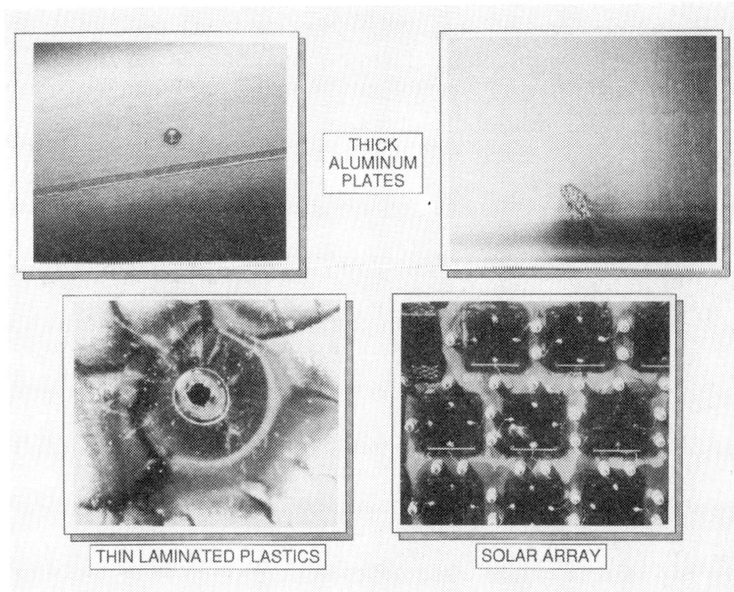

Fig. 11. Examples of Impact Damage Observations.

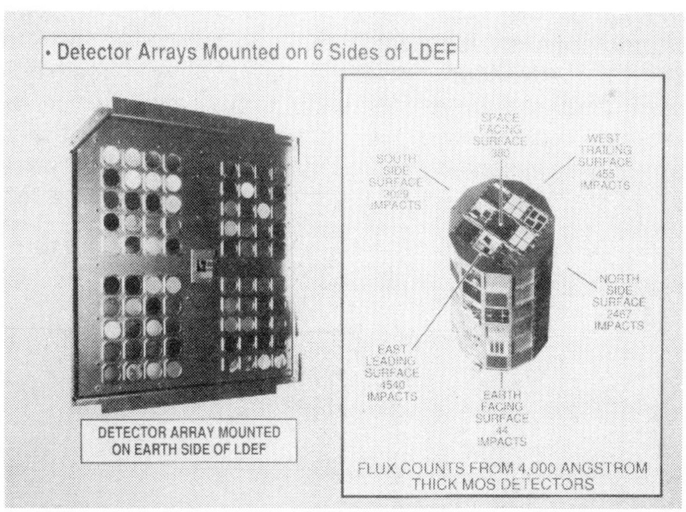

Fig. 12. Interplanetary Dust Experiment Data.

In summary, the initial observations of the craters on the LDEF indicate that the existing meteoroid and debris environment models probably do not have gross (orders of magnitude) errors in the smaller particle size range. The complete LDEF story on meteoroid and debris, however, will not be available for a number of years since many diverse studies and analyses must be done to complete the story. The detail characteristics of the craters must be studied for a better insight into cratering mechanics at meteoroid speeds. Laboratory impact testing and analytical modeling will be required to support these cratering studies. Chemical analysis of the impactor residues in the LDEF craters must be made to separate the flux of natural meteoroids encountered from the flux of the debris particles and to better understand the physical characteristics of each type of particle. Exobiologists will attempt to identify the molecular chemistry of the impactor residues looking for evidence of chemistry in space that may have led to the evolution of life. The distribution of the craters on the LDEF surfaces and the data on the time of impacts from the active IDE must be analyzed to better understand the orbits of the particles.

When these diverse analyses are completed, they will provide the most significant body of flight-derived information available on meteoroid and debris in near-Earth space and on the spacecraft damage to expect from impacts with these particles.

The LDEF data will also be "gage blocks" to confirm meteoroid and debris environment and spacecraft impact damage models.

Atomic Oxygen: Atomic oxygen impingement was concentrated on the leading side surfaces of the LDEF. This fact is vividly illustrated in Fig. 13 which shows photographs of identical thermal blankets on two trays of the Dublin Institute for Advanced Studies' Ultraheavy Cosmic-Ray Nuclei Experiment. These blankets were .005-inch thick Teflon which was silvered on the rear surface. The tray on the right, which was on the trailing side surface of LDEF, looks essentially as it did before launch. The tray on the left, which was on the leading side surface of the LDEF, now looks diffused

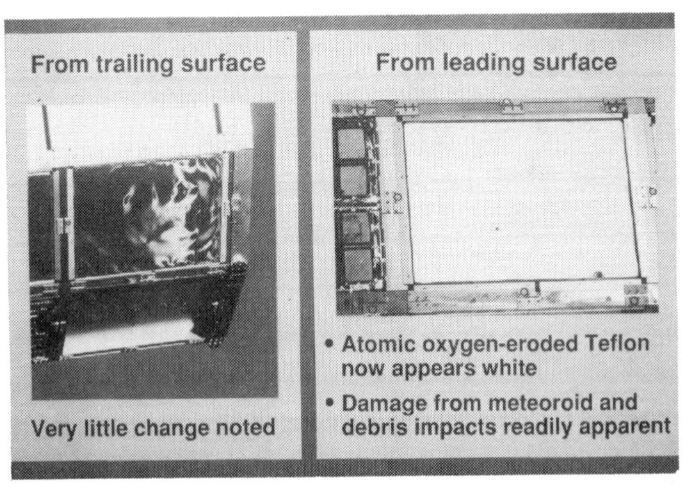

Fig. 13. Atomic Oxygen Erosion of Silvered Teflon on Leading and Trailing Surfaces of the LDEF.

and white as a result of the erosion of the surface by atomic oxygen. The electron microscope photograph in Fig. 14 shows the erosion in detail. A salt crystal, which deposited on this blanket prior to launch, protected a small region of the blanket and allowed a precise measurement of the .001-inch erosion depth. This erosion rate is approximately 8 times that expected. Fortunately, the thermal properties of the blankets were changed very little by the erosion (only a slight increase in the solar absorption) and thus the temperatures of the experiments behind these blankets remained within their design limits for the entire mission.

Fig. 14. Electron Microscope Photograph of Teflon Surface Eroded by Atomic Oxygen.

Kapton films, .005-inch thick, on several experiments were eroded away completely on the leading side surfaces on LDEF. Fig. 15 shows photographs of two trays of the TRW Space and Technology Group's Space Plasma High Voltage Drainage Experiment. The Kapton on the leading side tray of this experiment is gone, and the Kapton on the trailing side tray is undamaged.

In some instances the atomic oxygen effects were observed to be beneficial. For example, a number of white dots that were painted on experiment tray clamp blocks located on the trailing side of LDEF were darkened by exposure to the Sun's ultraviolet radiation. On the leading side the atomic oxygen continually eroded the surface pigments of the paint away before they could be darkened.

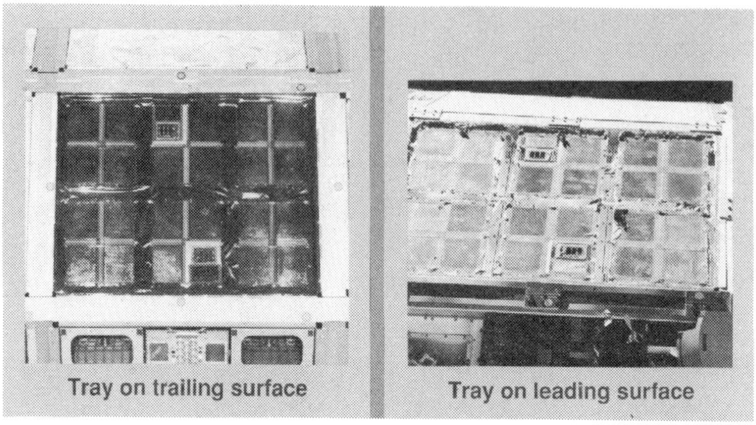

Fig. 15. Kapton Eroded From TRW's High Voltage Drainage Experiment.

<u>**Contamination:**</u> Observations of the retrieved LDEF have revealed a wealth of information on spacecraft contamination and the synergistic effects of the space environments on contaminated surfaces. Silicon deposits were detected on most of the exposed LDEF surfaces. The source may have been outgassing from adhesives used in a number of experiments. The atomic oxygen impingement in some cases partially removed these deposits and in other cases it darkened the deposits as illustrated in the photographs of LDEF structural members shown in Fig. 16. The end structural members

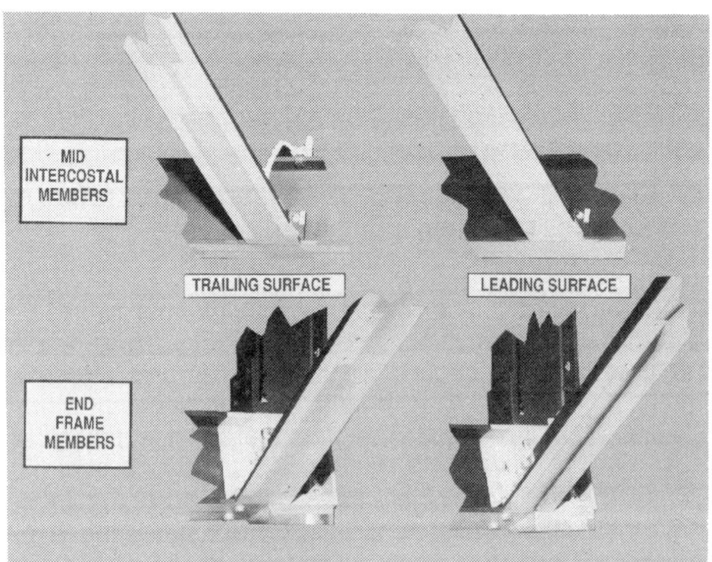

Fig. 16. Contamination of LDEF Structural Members on Leading and Trailing Surfaces.

were exposed to large amounts of contamination venting under thermal covers. These end frame deposits were darkened on the leading surface and the deposits on mid members were lightened on the leading surface by atomic oxygen.

There are a number of places on LDEF where it appears that electrical fields have affected contamination deposits. Fig. 17 is a photograph of the JSC's Interstellar Gas Experiment. This experiment had high voltage grids (several thousand volts) on the front and heavy contamination deposits are visible on surfaces around the leads to these grids.

Fig. 17. Contamination Deposits on Interstellar Gas Experiment.

Cracks and holes between the LDEF structure and experiment trays and covers allowed atomic oxygen to impinge on interior surfaces leaving interesting patterns as illustrated in Fig. 18. The effects of residual adhesive left from masking tape used during the construction of this tray are also clearly visible in this photograph. In a number of other locations it is apparent that the cleaning procedures were not adequate. Fig 19 is a photograph of the MSFC's Thermal Control Surfaces Experiment. A trace of the adhesive used to bond the silvered Teflon thermal control blanket to the aluminum substrate remained on the surface of the Teflon in this experiment. This trace, which was not visible before launch, is now obvious after the exposure to atomic oxygen.

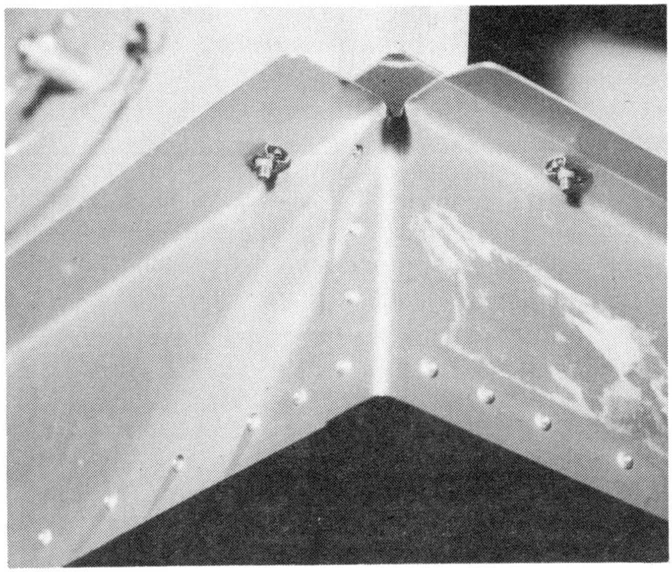

Fig. 18. Effects of Contamination and Atomic Oxygen on Interior Side of Experiment Tray.

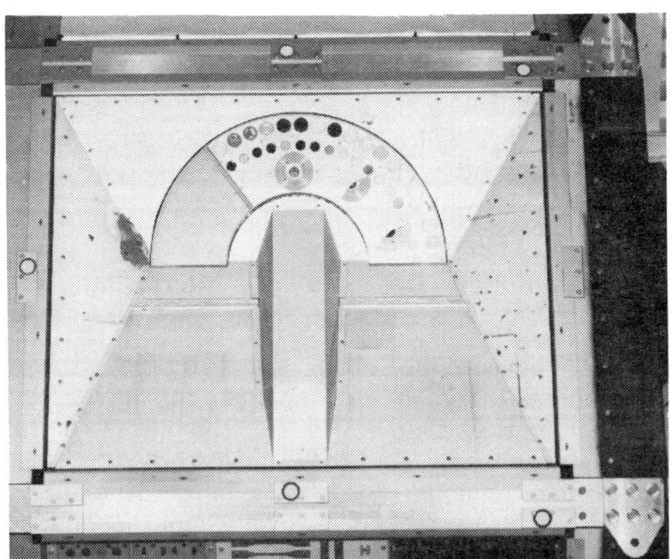

Fig. 19. Contamination Effects on Thermal Control Surface of MSFC's Active Coatings Experiment.

Materials and Coatings Observations: Almost every basic type of material and coating that can be used in the construction of future spacecraft was flown on board the LDEF--in either the 10,000 test specimens, or as a part of other experiment-unique hardware, or as a part of the LDEF itself. The data obtainable on the condition of these materials after almost 6 years in space will form an unprecedented data base for spacecraft designers.

Changes in the optical properties of a number of coatings measured by the MSFC Thermal Control Surfaces Experiment are illustrated in Fig. 20. The

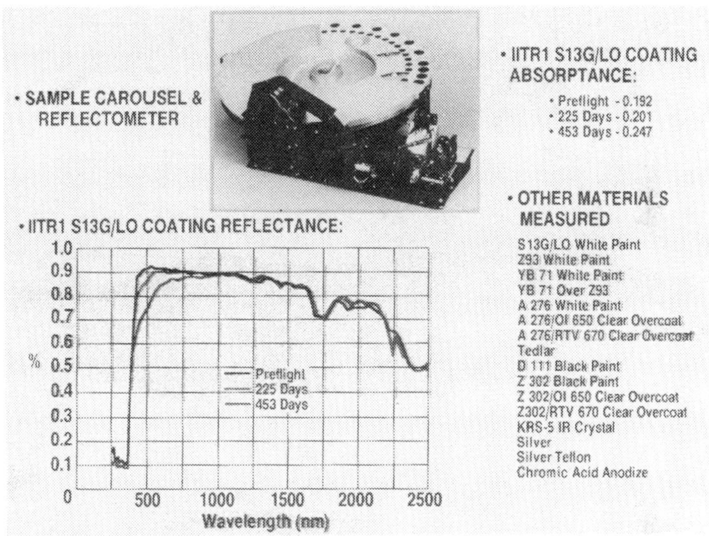

Fig. 20. Data From MSFC's Active Coatings Experiment.

LaRC Space Exposure of Composite Materials for Large Space Structures Experiment and the LaRC Exposure of Spacecraft Coatings Experiment, shown after the mission in Fig. 21, are typical LDEF passive materials experiments. These two LaRC experiments did employ an active exposure control canister which opened on orbit and remained open for approximately 11 months before it closed.

The damage to a number of thin film polymer specimens is apparent in the LaRC experiments. Fig. 22 shows a close-up photograph of one composite specimen in this experiment. Atomic oxygen eroded both the matrix material and the fibers in the composite specimen. The LaRC P.I. for this experiment reported a loss of strength in many of these composite specimens that is consistent with the reduction in thickness.

The University of Toronto's active composite materials experiment, shown in Fig. 23, measured stress and strain of composites while they were exposed in space. The P.I.'s of this experiment have reported that the active data

Fig. 21. LaRC's Materials and Coatings Experiment.

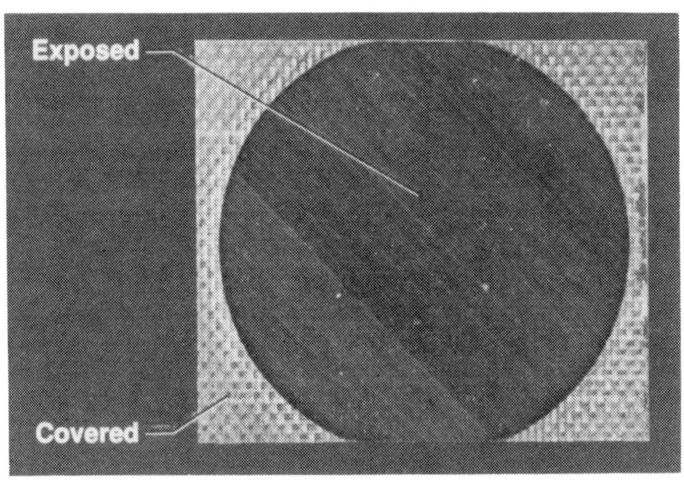

Fig. 22. Atomic Oxygen Erosion of Composite Specimen on LaRC's Materials Experiment.

Fig. 23. University of Toronto's Active Composite Materials Experiment.

measurements show a loss in strength and a substantial shift in the stress strain curves which occurred in their specimens after a relatively short exposure in space (months).

The Texas A & M University's Balloon Materials Degradation Experiment, which is shown in the postretrieval photograph in Fig. 24, was mounted on a side of LDEF where it was not exposed to atomic oxygen. The damaging

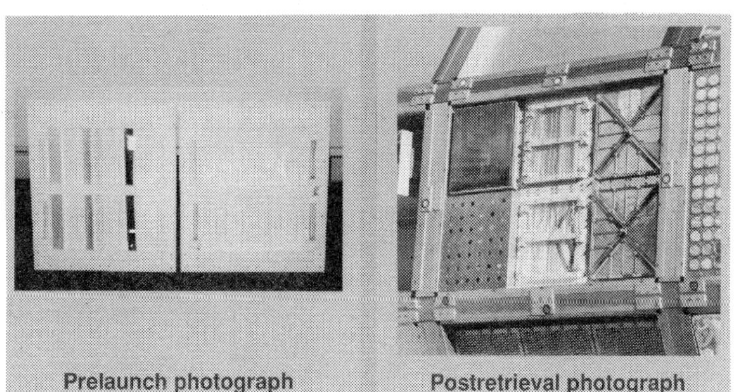

Fig. 24. Texas A & M's Balloon Materials Experiment.

effects of ultraviolet radiation on the polymer specimens in this experiment can be seen. In general, unprotected polymers did not survive very well on LDEF.

LDEF SYSTEMS OBSERVATIONS

A number of the LDEF experiments had specific objectives to test various systems in the environments of space. Fig. 25 is a photograph of one such experiment--the Air Force Weapons Laboratory's Space Environment Effects on Fiber Optic Systems Experiment. This experiment measured the ability of a number of different types of fiber systems to transmit high-speed data streams over the first year's exposure in space. In addition, the postretrieval observations of the experiment hardware are revealing the effects of remaining in space for almost 6 years. There were three other experiments on LDEF which tested fiber optic systems. In general the P.I.'s are reporting that the fiber systems held up and functioned well. Darkening of the outer sheath of the cables in the Weapons Laboratory's experiment can be seen. One of these fibers also suffered a direct hit from either a meteoroid or a man-made debris fragment and it is still operational.

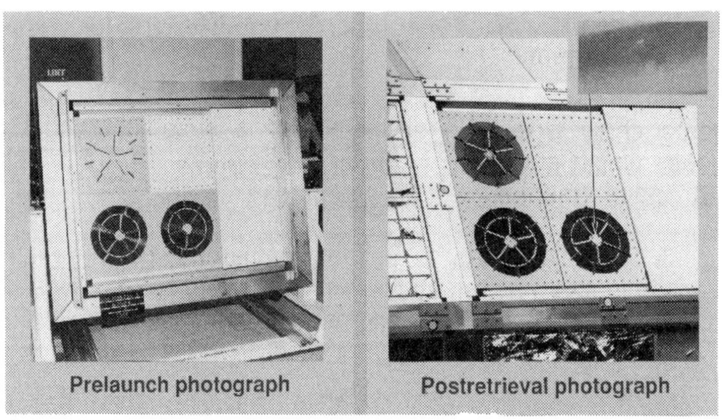

Fig. 25. Air Force Weapons Laboratory's Fiber Optic Experiment.

With the exception of light contamination deposits in some areas, the interior of the LDEF looked like new when it was inspected at KSC (Fig. 26). In postretrieval tests, the LDEF systems were all found to be still capable of operating and functioning within their original performance specifications. Some of the mechanical fasteners on the LDEF and experiments were difficult to remove and a number of them broke during the removal process. Initial thoughts are that these fastener removal problems are not a result of the prolonged space exposure but rather they were a result of the locking techniques employed in the fastener designs.

Fig. 26. Interior of LDEF.

The LDEF and the experiment temperatures were controlled primarily by anodized aluminum panels located on all sides of the LDEF. These thermal panels did double duty in that they also served as meteoroid and debris detector panels. The postflight observations of the anodized aluminum showed that it was essentially uneffected by the stay in space. Even the American flag decal (Fig. 27) on one of the space end thermal panels survived undamaged.

CONCLUSIONS

As has been stated many times throughout this paper, the LDEF can provide "gage block" type data to check existing models of the environments encountered in low Earth orbit and to check the damage these environments will inflict on spacecraft. The LDEF data will also allow new higher fidelity models to be generated.

In summary, the LDEF-derived data will serve as "product assurance" for a great many future spacecraft.

Fig. 27. Anodized Aluminum Thermal Control Panel from Space End of LDEF.

It should also be mentioned that the data return from the first LDEF mission has illustrated the cost effectiveness of low cost LDEF-type missions. One can only hope that LDEF-type opportunities will be a part of U. S. space plans in the future.

REFERENCES

1. Lenwood G. Clark, William H. Kinard, David J. Carter, Jr., and James L. Jones, Jr., "The Long Duration Exposure Facility Mission 1 Experiments," NASA Special Publication 473, 1984

2. G. W. Phillips, S. E. King, R. A. August, J. C. Ritter, J. H. Cutchin, P. S. Haskins, J. E. McKisson, D. W. Daly, A. G. Weisenberger, R. B. Piercey, and T. Dybler, "Observations of Beryllium 7 on the LDEF Spacecraft," ESA Space Environment Analysis Workshop, ESTEC, Noordwijk, The Netherlands, October 9-12, 1990

3. Watts, J. W., Parnell, T. A., Heckman, H. H., Approximate Angular Distribution and Spectra for Trapped Protons in Low Earth Orbit, AIP Conference Proceedings, 1989

4. J. A. M. McDonnell, K. Sullivan, S. F. Green, T. J. Stevenson, and D. H. Niblett, "Dynamic Modelling Transformations for the Low Earth Orbit Satellite Particulate Environment," Proceedings of IAU Colloquium No.126, Kluwer Academic Publishing Company, 1991

5. H. A. Zook, "Flux vs Direction of Impacts on LDEF by Meteoroids and Orbital Debris," Proceedings of 20th Lunar and Planetary Sciences Conference, (in press)

6. S. F. Singer, J. E. Stanley, P. C. Kassel, W. H. Kinard, J. J. Wortman, J. L. Weinberg, J. D. Mulholland, G. Eichhorn, W. J. Cook, and N. Montague, "First Spatio-Temporal Results from the LDEF Interplanetary Dust Experiment," Advanced Space Research (in press)

7. T. See, M. Allbrooks, D. Atkinson, C. Simon, and M. Zolensky, "Meteoroid and Debris Impact Features Documented on the Long Duration Exposure Facility," NASA Johnson Space Center Publication #24608

APPENDICES

PUBLICATIONS OF THE AMERICAN ASTRONAUTICAL SOCIETY

Following are the principal publications of the American Astronautical Society:

JOURNAL OF THE ASTRONAUTICAL SCIENCES (1954 -)
Published quarterly and distributed by AAS Business Office, 6352 Rolling Mill Place, Suite #102, Springfield, Virginia 22152. Back issues available from Univelt, Inc., P.O. Box 28130, San Diego, California 92198.

SPACE TIMES (1986 -)
Published bi-monthly and distributed by AAS Business Office, 6352 Rolling Mill Place, Suite #102, Springfield, Virginia 22152.

AAS NEWSLETTER (1962 - 1985)
Incorporated in *Space Times*. Back issues available from AAS Business Office, 6352 Rolling Mill Place, Suite #102, Springfield, Virginia 22152.

ASTRONAUTICAL SCIENCES REVIEW (1959 - 1962)
Incorporated in *Space Times*. Back issues still available from Univelt, Inc., P.O. Box 28130, San Diego, California 92198.

ADVANCES IN THE ASTRONAUTICAL SCIENCES (1957 -)
Proceedings of major AAS technical meetings. Published and distributed for the American Astronautical Society by Univelt, Inc., P.O. Box 28130, San Diego, California 92198.

SCIENCE AND TECHNOLOGY SERIES (1964 -)
Supplement to *Advances in the Astronautical Sciences*. Proceedings and monographs, most of them based on AAS technical meetings. Published and distributed for the American Astronautical Society by Univelt, Inc., P.O. Box 28130, San Diego, California 92198.

AAS HISTORY SERIES (1977 -)
Supplement to *Advances in the Astronautical Sciences*. Selected works in the field of aerospace history under the editorship of R. Cargill Hall. Published and distributed for the American Astronautical Society by Univelt, Inc., P.O. Box 28130, San Diego, California 92198.

AAS MICROFICHE SERIES (1968 -)
Supplement to *Advances in the Astronautical Sciences*. Consists principally of technical papers not included in the hard-copy volume. Published and distributed for the American Astronautical Society by Univelt, Inc., P.O. Box 28130, San Diego, California 92198.

Subscriptions to the *Journal of the Astronautical Sciences* and the *Space Times* should be ordered from the AAS Business Office. Back issues of the *Journal* and all books and microfiche should be ordered from Univelt, Incorporated.

ADVANCES IN THE ASTRONAUTICAL SCIENCES SERIES (1957-)

ISSN 0065-3438, LIBRARY OF CONGRESS CARD NO. 57-43769

Proceedings of Major AAS Technical Meetings

Vol. 1 Third Annual AAS Meeting, Dec. 6-7, 1956, New York, NY, 1957, 184p., ed. Norman V. Petersen, Microfiche only, $20 *(ISBN 0-87703-002-2)*

Vol. 2 Fourth Annual AAS Meeting, Jan. 29-31, 1958, New York, NY, 1958, 440p., eds. Norman V. Petersen, Horace Jacobs, Microfiche only, $20 *(ISBN 0-87703-003-0)*

Vol. 3 First Western National AAS Meeting, Aug. 18-19, 1958 530p., eds. Norman V. Petersen, Horace Jacobs, Microfiche only, $20 *(ISBN 0-87703-004-9)*

Vol. 4 Fifth Annual AAS Meeting, Dec. 27-31, 1958, Washington, D.C., 1959, 462p., ed. Horace Jacobs, Microfiche only, $20 *(ISBN 0-87703-005-7)*

Vol. 5 Second Western National AAS Meeting, Aug. 4-5, 1959, Los Angeles, CA, 1960, 364p., ed. Horace Jacobs, Microfiche only, $20 *(ISBN 0-87703-006-3)*

Vol. 6 Sixth Annual AAS Meeting, Jan. 18-21, 1960, New York, NY, 1961, 968p., eds. Horace Jacobs and Eric Burgess, Hard Cover $45 *(ISBN 0-87703-007-3)*

Vol. 7 Third Western National AAS Meeting, Aug. 4-5, 1960, Seattle, WA, 1961, 464p., eds. Horace Jacobs and Eric Burgess, Microfiche only, $20 *(ISBN 0-87703-008-1)*

Vol. 8 Seventh Annual AAS Meeting, Jan. 16-18, 1961, Dallas, TX, 1963, 602p., ed. Horace Jacobs, Microfiche only, $20 *(ISBN 0-87703-009-X)*

Vol. 9 Fourth Western Regional AAS Meeting, Aug. 1-3, 1961, San Francisco, CA, 1963, 910p., ed. Eric Burgess, Hard Cover $45 *(ISBN 0-87703-010-3)*

Vol. 10 Manned Lunar Flight (AAS/AAAS Symposium) Dec. 19, 1961, Denver, CO, 1963, 310p., eds. George W. Morgenthaler and Horace Jacobs, Hard Cover $35 *(ISBN 0-87703-011-1)*

Vol. 11 Eighth Annual AAS Meeting, Jan. 16-18, 1962, Washington, D.C., 1963, 808p., ed. Horace Jacobs, Hard Cover $45 *(ISBN 0-87703-012-X)*

Vol. 12 Scientific Satellites - Mission and Design (AAS/AAAS Symposium), Dec. 27, 1962, Philadelphia, PA, 1963, 262p., ed. Irving E. Jeter, Hard Cover $25 *(ISBN 0-87703-013-8)*

Vol. 13 Interplanetary Missions, 9th Annual AAS Meeting, Jan. 15-17, 1963, Los Angeles, CA, 1963, 690p., ed. Eric Burgess, Hard Cover $45 *(ISBN 0-87703-014-6)*

Vol. 14 Second AAS Symposium on Physical and Biological Phenomena under Zero G Conditions, Jan. 18, 1963, Los Angeles, CA, 1963, 382p., eds. Elliot T. Benedikt and Robert W. Halliburton, Hard Cover $30 *(ISBN 0-87703-015-4)*

Vol. 15 Exploration of Mars Symposium, Jun. 6-7, 1963, Denver, CO, 1963, 634p., ed. George W. Morgenthaler, Hard Cover $45 *(ISBN 0-87703-016-2)*

Vol. 16 Space Rendezvous, Rescue, and Recovery Symposium, Sept. 10-12, 1963, Edwards, CA 1963, 1408p., ed. Norman V. Petersen, Hard Cover, **Part 1**, 1028p., $45 *(ISBN 0-87703-017-0)*; **Part 2**, 380p., $30 *(ISBN 0-87703-018-9)*

Vol. 17 Bioastronautics - Fundamental and Practical Problems (AAS/AAAS Symposium), Dec. 30, 1963, Cleveland, OH, 1964, 128p., ed. William C. Kaufman, Microfiche only, $10 *(ISBN 0-87703-019-7)*

Vol. 18 Lunar Flight Programs, 10th Annual AAS Meeting, May 4-7, 1964, New York, NY, 1964, 630p., ed. Ross Fleisig, Hard Cover $45 *(ISBN 0-87703-020-0)*

Vol. 19 Unmanned Exploration of the Solar System Symposium, Feb. 8-10, 1965, Denver, CO, 1965, 1000p., eds. George W. Morgenthaler, Robert G. Morra, Hard Cover $45 *(ISBN 0-87703-021-9)*

Vol. 20 Post Apollo Exploration, 11th Annual AAS Meeting, May 3-6, 1965, Chicago, IL, 1966, 1220p., ed. Francis Narin, Microfiche only, **Part I**, 572p., $30 *(ISBN 0-87703-022-7)*; **Part 2**, 648p., $35 *(ISBN 0-87703-023-5)*

Vol. 21 Practical Space Applications Symposium, Feb. 21-23, 1966, San Diego, CA, 1967, 508p., ed. Lawrence L. Kavanau, Hard Cover $40 *(ISBN 0-87703-024-3)*

Vol. 22 The Search for Extraterrestrial Life, 12th Annual AAS Meeting, May 23-25, 1966, Anaheim, CA, 1967, 388p., ed. James S. Hanrahan, Microfiche only $30 *(ISBN 0-87703-025-1)*; Microfiche Suppl. (Vol. 1 AAS Microfiche Series) $12 *(ISBN 0-87703-132-0)*

Vol. 23 Commercial Utilization of Space, 13th Annual AAS Meeting, May 1-3, 1967, Dallas, TX, 1968, 512p., eds. J. Ray Gilmer, Alfred M. Mayo, Ross C. Peavey, Hard Cover *(ISBN 0-87703-026-X)*; plus Microfiche Suppl. (Vol. 3 AAS Microfiche Series) $60 *(ISBN 0-87703-216-5)*

Vol. 24 Exploitation of Space for Experimental Research, 14th Annual AAS Meeting, May 13-15, 1968, Dedham, MA, 1968, 363p., ed. Harry Zuckerberg, Hard Cover $30 *(ISBN 0-87703-027-8)*

Vol. 25 Advanced Space Experiments, Sept. 16-18, 1968, Ann Arbor, MI, 1969, 530p., eds. O. Lyle Tiffany and Eugene M. Zaitzeff, Hard Cover $40 *(ISBN 0-87703-028-6)*

Vol. 26 Planning Challenges of the 70's in Space, 15th Annual AAS Meeting, Jun. 17-20, 1969, Denver, CO, 1970, 470p., eds. George W. Morgenthaler and Robert G. Morra, Hard Cover $35 *(ISBN 0-87703-053-7)*; Microfiche Suppl. (Vol. 14 AAS Microfiche Series) $20 *(ISBN 0-87703-130-4)*

Vol. 27/28 Space Stations (**v27**) and Space Shuttles and Interplanetary Missions (**v28**), 16th Annual AAS Meeting, Jun. 8-10, 1970, Anaheim, CA, 1970, **Vol. 27**, eds. Lewis Larmore and Robert L. Gervais, 606p., Hard Cover $45 *(ISBN 0-87703-054-5)*; **Vol. 28**, eds. Lewis Larmore and Robert L. Gervais, 488p., Hard Cover $35 *(ISBN 0-87703-055-3)*

Vol. 29 The Outer Solar System, 17th Annual AAS Meeting, Jun. 28-30, 1971, Seattle, WA, 1971, 1358p., $85; ed. Juris Vagners, Hard Cover, **Part 1**, 618p., $40 *(ISBN 0-87703-059-6)*; **Part 2**, 740p., $45 *(ISBN 0-87703-060-X)*

Vol. 30 International Congress of Space Benefits, 19th Annual AAS Meeting, Jun. 19-21, 1973, Dallas, TX, 1974, 528p., ed. Francis S. Johnson, Hard Cover $40 *(ISBN 0-87703-065-0)*

Vol. 31 The Skylab Results, 20th Annual AAS Meeting, Aug. 20-22, 1974, Los Angeles, CA, 1975, 1174p., eds. William C. Schneider and Thomas E. Hanes, Microfiche only (ISBN 0-87703-072-3); Plus Microfiche Suppl. (Vol. 22 AAS Microfiche Series) $60 *(ISBN 0-87703-043-6)*

Vol. 32 Space Shuttle Missions of the 80's, 21st Annual AAS Meeting, Aug. 26-28, 1975, Denver, CO, 1977, 1364p., eds. William J. Bursnall, George W. Morgenthaler, Gerald E. Simonson, Hard Cover, **Part 1**, 598p., $40 *(ISBN 0-87703-078-2)*; Hard Cover, **Part 2**, 766p., $55 *(ISBN 0-87703-087-1)*; Microfiche Suppl. (Vol. 25 AAS Microfiche Series $65 *(ISBN 0-87703-133-9)*

Vol. 33 AAS/AIAA Astrodynamics Conference, July 28-30, 1975, Nassau, Bahamas, 1976, 390p., eds. William F. Powers, Herbert E. Rauch, Byron D. Tapley, Carmelo E. Velez, Hard Cover $35 *(ISBN 0-87703-079-0)*; Microfiche Suppl. (Vol. 26 AAS Microfiche Series) $40 *(ISBN 0-87703-142-8)*

Vol. 34 Apollo Soyuz Mission Report, 1977, 336p., ed. Chester M. Lee, Hard Cover $35 *(ISBN 0-87703-089-8)*

Vol. 35 The Bicentennial Space Symposium - New Themes for Space: Mankind's Future Needs and Aspirations, 22nd AAS Meeting, Oct. 6-8, 1976, Washington, D.C., 1977, 242p., ed. William C. Schneider, Hard Cover $25 *(ISBN 0-87703-090-1)*

Vol. 36 The Industrialization of Space, 23rd Annual AAS Meeting, Oct. 18-20, 1977, San Francisco, CA, 1978, 1160p., eds. Richard A. Van Patten, Paul Siegler, Edward V.B. Stearns, Hard Cover, **Part 1**, 610p., $55 *(ISBN 0-87703-094-4)*; Hard Cover, **Part 2**, 550p., $45 *(ISBN 0-87703-095-2)*; Microfiche Suppl. (Vol. 28 AAS Microfiche Series) $15 *(ISBN 0-87703-121-5)*

Vol. 37 Space Shuttle and Spacelab Utilization, What are the Near-Term and Long-Term Benefits for Mankind?, 16th Goddard Memorial Symposium, 24th Annual AAS Meeting, March 8-10, 1978, Washington, D.C., 1978, 865p., eds. George W. Morgenthaler and Manfred Hollstein, Hard Cover, **Part 1**, 400p., $40 *(ISBN 0-87703-096-0)*; Hard Cover, **Part 2**, 465p., $45 *(ISBN 0-87703-097-9)*

Vol. 38 The Future U.S. Space Program, 25th Anniversary Conference, Oct. 20 - Nov. 2, 1978, Houston, TX, 1979, 880p., eds. Richard S. Johnston, Albert Naumann, Jr., Clay W. G. Fulcher, Hard Cover, **Part 1**, 444p., $45 *(ISBN 0-87703-098-7)*; Hard Cover, **Part 2**, 436p., $40 *(ISBN 0-87703-099-5)*; Microfiche Suppl. (Vol. 30 AAS Microfiche Series) $15 *(ISBN 0-87703-129-0)*

Vol. 39 Guidance and Control 1979, Feb. 24-28, 1979, Keystone, CO, 1979, 492p., ed. Robert D. Culp, Hard Cover $45 *(ISBN 0-87703-100-2)*; Microfiche Suppl. (Vol. 31 AAS Microfiche Series) $10 *(ISBN 0-87703-128-2)*

Vol. 40 AAS/AIAA Astrodynamics Conference, Jun. 25-27, 1979, Provincetown, MA, 1980, 996p., eds. Paul A. Penzo, Bernard Kaufman, Louis Friedman, Richard Battin, Hard Cover, **Part 1**, 494p., $45 *(ISBN 0-87703-107-X)*; Soft Cover $35 *(ISBN 0-87703-108-8)*; Hard Cover, **Part 2**, 502p., $45 *(ISBN 0-87703-109-6)*; Soft Cover $35 *(ISBN 0-87703-110-X)*; Microfiche Suppl. (Vol. 32 AAS Microfiche Series) $20 *(ISBN 0-87703-139-8)*

Vol. 41 Space Shuttle: Dawn of an Era, 26th Annual AAS Meeting, Oct. 29-Nov. 1, 1979, Los Angeles, CA, 1980, 980p., eds. William F. Rector, III and Paul A. Penzo, Hard Cover, **Part 1**, 452p., $45 *(ISBN 0-87703-111-8)*; Soft Cover $35 *(ISBN 0-87703-112-6)*; Hard Cover, **Part 2**, 528p., $55 *(ISBN 0-87703-113-4)*; Soft Cover $40 *(ISBN 0-87703-114-2)*; Microfiche Suppl. (Vol. 33 AAS Microfiche Series) $10 *(ISBN 0-87703-136-3)*

Vol. 42 Guidance and Control 1980, Feb. 17-21, 1980, Keystone, CO, 1980, 738p., ed. Louis A. Morine, Hard Cover $60 *(ISBN 0-87703-137-1)*; Soft Cover $45 *(ISBN 0-87703-138-X)*

Vol. 43 Shuttle/Spacelab - The New Transportation System and its Utilization, (3rd DGLR/AAS Symposium), Apr. 28-30, 1980, Hannover, Germany, 1981, 342p., eds. Dietrich E. Koelle and George V. Butler, Hard Cover $45 *(ISBN 0-87703-144-4)*; Soft Cover $35 *(ISBN 0-87703-146-0)*

Vol. 44 Space--Enhancing Technological Leadership, 27th Annual AAS Meeting, Oct. 20-23, 1980, Boston, MA, 1981, 580p., ed. Lawrence P. Greene, Hard Cover $65 *(ISBN 0-87703-147-9)*; Soft Cover $50 *(ISBN 0-87703-148-7)*; Microfiche Suppl. (Vol. 35 AAS Microfiche Series) $10 *(ISBN 0-87703-164-9)*

Vol. 45 Guidance and Control 1981, Jan. 31- Feb. 4, 1981, Keystone, CO, 1981, 506p., ed. Edward J. Bauman, Hard Cover $60 *(ISBN 0-87703-150-9)*; Soft Cover $50 *(ISBN 0-87703-151-7)*; Microfiche Suppl. (Vol. 36 AAS Microfiche Series) $15 *(ISBN 0-87703-156-8)*

Vol. 46 AAS/AIAA Astrodynamics Conference, Aug. 3-5, 1981, North Lake Tahoe, NV, 1982, 1124p., eds. Alan L. Friedlander, Paul J. Cefola, Bernard Kaufman, Walt Williamson, G.T. Tseng, Hard Cover, **Part 1**, 552p., $55 *(ISBN 0-87703-159-2)*; Soft Cover $45 *(ISBN 0-87703-160-6)*; Hard Cover, **Part 2**, 572p., $55 *(ISBN 0-87703-161-4)*; Soft Cover $45 *(ISBN 0-87703-162-2)*; Microfiche Suppl. (Vol. 37 AAS Microfiche Series) $40 *(ISBN 0-87703-163-0)*

Vol. 47 Leadership in Space - For Benefits on Earth, 28th Annual AAS Meeting, Oct. 26-29, 1981, San Diego, CA, 1982, 310p., ed. William F. Rector, III, Hard Cover $45 *(ISBN 0-87703-168 1)*; Soft Cover $35 *(ISBN 0-87703-169-X)*

Vol. 48 Guidance And Control 1982, Jan. 30 - Feb. 3, 1982, Keystone, CO, 1982, 558p., eds. Robert D. Culp, Edward J. Bauman, W. E. Dorroh, Jr., Hard Cover $65 *(ISBN 0-87703-170-3)*; Soft Cover $50 *(ISBN 0-87703-171-1)*; Microfiche Suppl. (Vol. 38 AAS Michrofiche Series) $10 *(ISBN 0-87703-180-0)*

Vol. 49 Spacelab, Space Platforms, and the Future, Fourth AAS/DGLR Symposium and 20th Goddard Memorial Symposium, Mar. 17-19, 1982, Greenbelt, MD, 1982, 502p., eds. Peter M. Bainum, Dietrich E. Koelle, Hard Cover $55 *(ISBN 0-87703-174-6)*; Soft Cover $45 *(ISBN 0-87703-175-4)*; Microfiche Suppl. (Vol. 42 AAS Microfiche Series) $15 *(ISBN 0-87703-181-9)*

Vol. 50 Proceedings on an International Symposium on Engineering Sciences and Mechanics, Dec. 29-31, Tainan, Taiwan, 1983, **two parts**, 1570p., eds. Han-Min Hsia, Richard W. Longman, You-Li Chou, Hard Cover $120 *(ISBN 0-87703-176-2)*; Microfiche Suppl. (Vol. 43 AAS Microfiche Series) $10 *(ISBN 0-87703-215-7)*

Vol. 51 Guidance and Control 1983, Feb. 5-9, 1983, Keystone, CO, 1983, 494p., eds. Edward J. Bauman, Zubin W. Emsley, Hard Cover $60 *(ISBN 0-87703-182-7)*; Soft Cover $50 *(ISBN 0-87703-183-5)*; Microfiche Suppl. (Vol. 44 AAS Microfiche Series) $10 *(ISBN 0-87703-214-9)*

Vol. 52 Developing the Space Frontier, 29th Annual AAS Meeting, Oct. 25-27, 1982, Houston, TX, 1983, 436p., eds. Albert Naumann, Grover Alexander, Microfiche Only $45 *(ISBN 0-87703-189-4)*

Vol. 53 Space Manufacturing 1983, May 9-12, 1983, Princeton, NJ, 1983, 496p., eds. James D. Burke, April S. Whitt, On Microfiche $50 *(ISBN 0-87703-188-6)*; Soft Cover $60 *(0-87703-189-4)*

Vol. 54 AAS/AIAA Astrodynamics Conference, Aug. 22-25, 1983, Lake Placid, NY, 1984, **two parts**, 1370p., eds. G.T. Tseng, Paul J. Cefola, Peter M. Bainum, David A. Levinson, Hard Cover $120 *(ISBN 0-87703-190-8)*; Soft Cover $90 *(ISBN 0-87703-191-6)*; Microfiche Suppl. (Vol. 45 AAS Microfiche Series) $40 *(ISBN 0-87703-192-4)*

Vol. 55 Guidance and Control 1984, Feb. 4-8, 1984, Keystone, CO, 1984, 500p., eds. Robert D. Culp, Parker S. Stafford, Hard Cover $60 *(ISBN 0-87703-199-1)*; Soft Cover $50 *(ISBN 0-87703-200-9)*; Microfiche Suppl. (Vol. 48 AAS Microfiche Series $15 *(ISBN 0-87703-201-7)*

Vol. 56 From Spacelab to Space Station, Fifth DGLR/AAS Symposium, Oct. 3-5, 1984, Hamburg, Germany, 1985, 270p., eds. H. Stoewer, Peter M. Bainum, Microfiche Only $30 *(ISBN 0-87703-209-2)*

Vol. 57 Guidance and Control 1985, Feb. 2-6, 1985, Keystone, CO, 1985, 618p., eds. Robert D. Culp, Edward J. Bauman, Charles A. Cullian, Hard Cover $65 *(ISBN 0-87703-211-4)*; Soft Cover $50 *(ISBN 0-87703-212-2)*; Microfiche Suppl. (Vol. 50 AAS Microfiche Series) $15 *(ISBN 0-87703-213-0)*

Vol. 58 AAS/AIAA Astrodynamics Conference, Aug. 12-15, 1985, Vail, CO, 1986, **two parts**, 1556p., eds. Bernard Kaufman, Joseph J.F. Liu, Robert A. Calico, Felix R. Hoots, Hard Cover $140 *(ISBN 0-87703-245-9)*; Soft Cover $110 *(ISBN 0-87703-246-7)*; Microfiche Suppl. (Vol. 51 AAS Microfiche Series); $60 *(ISBN 0-87703-247-5)*

Vol. 59 Space Station Beyond IOC, 32nd Annual AAS Meeting, Nov 6-7, 1985, Los Angeles, CA, 1986, 188p., ed. M. Jack Friedenthal, Hard Cover $40 *(ISBN 0-87703-252-1)*; Soft Cover $30 *(ISBN 0-87703-253-X)*

Vol. 60 Space Exploitation and Utilization, First AAS/JRS Symposium, Dec. 15-19, 1985, Honolulu, HI, 1986, 740p., eds. Gayle L. May, Peter M. Bainum, Kenji Ikeda, Tamiya Nomura, Tatsuo Yamanaka, Ryojiro Akiba, Hard Cover $70 *(ISBN 0-87703-254-8)*; Soft Cover $55 *(ISBN 0-87703-255-6)*; Microfiche Suppl. (Vol. 52 AAS Microfiche Series) $10 *(ISBN 0-87703-256-4)*

Vol. 61 Guidance and Control 1986, Feb. 1-5, 1986, Keystone, CO, 1986, 460p., eds. Robert D. Culp, John C. Durrett, Hard Cover $60 *(ISBN 0-87703-257-2)*; Soft Cover $50 *(ISBN 0-87703-258-0)*; Microfiche Suppl. (Vol. 53 AAS Microfiche Series) $10 *(ISBN 0-87703-259-9)*

Vol. 62 Tethers in Space, Proceedings of First International Conference on Tethers in Space (NASA & PSN Sponsors; AIAA, AAS, & AIDAA Co-Sponsors), Sept. 17-19, 1986, Arlington, VA, 1987, 784p., eds. Peter M. Bainum, Ivan Bekey, Luciano Guerriero, Paul A. Penzo, Hard Cover $80 *(ISBN 0-87703-264-5)*; Soft Cover $70 *(ISBN 0-87703-265-3)*

Vol. 63 Guidance and Control 1987, Jan. 31 - Feb. 4, 1987, Keystone, CO, 1987, 638p., eds. Robert D. Culp, Terry J. Kelly, Hard Cover $75 *(ISBN 0-87703-268-8)*; Soft Cover $60 *(ISBN 0-87703-269-6)*

Vol. 64 Aerospace Century XXI, 33rd AAS Annual Meeting, Oct. 26-29, 1986, Boulder, CO, 1987, **all three parts**, Hard Cover $225 *(ISBN 0-87703-276-9)*; Soft Cover $180 *(ISBN 0-87703-277-7)*; **Part I**, Space Missions and Policy, 686p., eds. George W. Morgenthaler, Gayle L. May, Hard Cover $75 *(ISBN 0-87703-279-3)*; Soft Cover $60 *(ISBN 0-87703-282-3)*; **Part II**, Space Flight Technologies, 608p., eds. George W. Morgenthaler, W. Kent Tobiska, Hard Cover $75 *(ISBN 0-87703-280-7)*; Soft Cover $60 *(ISBN 0-87703-283-1)*; **Part III**, Space Sciences, Applications, and Commercial Developments, 724p., eds. George W. Morgenthaler, Jean N. Koster, Hard Cover $75 *(ISBN 0-87703-281-5)*; Soft Cover $60 *(ISBN 0-87703-284-X)*; Microfiche Suppl. (Vol. 54 AAS Microfiche Series) $25 *(ISBN 0-87703-278-5)*

Vol. 65 AAS/AIAA Astrodynamics Conference, Aug. 10-13, 1987, Kalispell, MT, 1988, **two parts**, 1774p., eds. John K. Soldner, Arun K. Misra, Robert E. Lindberg, Walton Williamson, Hard Cover $180 *(ISBN 0-87703-285-8)*; Soft Cover $150 *(ISBN 0-87703-286-6)*; Microfiche Suppl. (Vol. 55 AAS Microfiche Series); $70 *(ISBN 0-87703-287-4)*

Vol. 66 Guidance and Control 1988, Jan. 30 - Feb. 3, 1988, Keystone, CO, 1988, 576p., eds. Robert D. Culp, Paul L. Shattuck, Hard Cover $75 *(ISBN 0-87703-288-2)*; Soft Cover $60 *(ISBN 0-87703-289-0)*; Microfiche Suppl. (Vol. 56 AAS Microfiche Series) $10 *(ISBN 0-87703-290-4)*

Vol. 67 Space - A New Community of Opportunity, 34th AAS Annual Meeting, Nov. 3-5, 1987, Houston, TX, 1989, 472p., eds. William G. Straight, Henry N. Bowes, Hard Cover $70 *(ISBN 0-87703-297-1)*; Soft Cover $55 *(ISBN 0-87703-298-X)*

Vol. 68 Guidance and Control 1989, Feb. 4-8, 1989, Keystone, CO, 1989, 708p., eds. Robert D. Culp, Robert A. Lewis, Hard Cover $85 *(ISBN 0-87703-299-8)*; Soft Cover $70 *(ISBN 0-87703-300-5)*

Vol. 69 Orbital Mechanics and Mission Design, Apr. 24-27, 1989, Greenbelt, MD, 1989, 862p., ed. Jerome Teles, Hard Cover $95 *(ISBN 0-87703-311-0)*; Soft Cover $80 *(ISBN 0-87703-312-9)*; Microfiche Suppl. (Vol. 57 AAS Microfiche Series) $10 *(ISBN 0-87703-313-7)*

Vol. 70 The 21st Century in Space, 35th AAS Annual Meeting, Oct. 24-26, 1988, St. Louis, MO, 1990, 446p., ed. George V. Butler, Hard Cover $90 *(ISBN 0-87703-314-5)*; Soft Cover $75 *(ISBN 0-87703-315-3)*; Microfiche Suppl. (Vol. 58 AAS Microfiche Series) $10 *(ISBN 0-87703-316-1)*

Vol. 71 AAS/AIAA Astrodynamics Conference, Aug. 7-10, 1989, Stowe, VT, 1990, **two parts**, 1472p., eds. Catherine L. Thornton, Ronald J. Proulx, John E. Prussing, Felix R. Hoots, Hard Cover $200 *(ISBN 0-87703-317-X)*; Soft Cover $170 *(ISBN 0-87703-318-8)*; Microfiche Suppl. (Vol. 59 AAS Microfiche Series) $50 *(ISBN 0-87703-319-6)*

Vol. 72 Guidance and Control 1990, Feb. 3-7, 1990, Keystone, CO, 1990, 676p., eds. Robert D. Culp, Arlo D. Gravseth, Hard Cover $95 *(ISBN 0-87703-320-X)*; Soft Cover $80 *(ISBN 0-87703-321-8)*

Vol. 73 Space Utilization and Applications in the Pacific, Third (PISSTA) AAS/JRS/CSA Symposium, Nov. 6-8, 1989, Los Angeles, CA, 1990, 764p., eds. Peter M. Bainum, Gayle L. May, Tatsuo Yamanaka, Yang Jiachi, Hard Cover $95 *(ISBN 0-87703-325-0)*; Soft Cover $80 *(ISBN 0-87703-326-9)*

Vol. 74 Guidance and Control 1991, Feb. 2-6, 1991, 730p., eds. Robert D. Culp, James P. McQuerry, Hard Cover $120 *(ISBN 0-87703-334-X)*; Soft Cover $90 *(ISBN 0-87703-335-8)*

Order from Univelt, Inc., P.O. Box 28130, San Diego, California 92198

SCIENCE AND TECHNOLOGY SERIES (1964-)

ISSN 0278-4017

A Supplement to *Advances in the Astronautical Sciences*. Proceedings and monographs, most of them based on AAS technical meetings.

Vol. 1 Manned Space Reliability Symposium, Jun. 9, 1964, Anaheim, CA, 1964, 112p., ed. Paul Horowitz, Hard Cover $20 *(ISBN 0-87703-029-4)*

Vol. 2 Towards Deeper Space Penetration (AAS/AAAS Symposium), Dec. 29, 1964, Montreal, Canada, 1964, 182p., ed. Edward R. Van Driest, Hard cover $20 *(ISBN 0-87703-030-8)*

Vol. 3 Orbital Hodograph Analysis, 1965, 150p., ed. Samuel P. Altman, Hard Cover $20 *(ISBN 0-87703-031-6)*

Vol. 4 Scientific Experiments for Manned Orbital Flight, 3rd Goddard Memorial Symposium, Mar. 18-19, 1965, Washington, D.C., 1965, 372p., ed. Peter C. Badgley, Hard Cover $30 *(ISBN 0-87703-032-4)*

Vol. 5 Physiological and Performance Determinants in Manned Space Systems (AAS/HFS Symposium, Apr. 14-15, 1965, Northridge, CA, 1965, 220p., ed. Paul Horowitz, Hard Cover $20 *(ISBN 0-87703-033-2)*

Vol. 6 Space Electronics Symposium (AAS/AES Meeting), May 25-27, 1965, Los Angeles, CA, 1965, 404p., ed. Chung-Ming Wong, Hard Cover $30 *(ISBN 0-87703-034-0)*

Vol. 7 Theodore von Karman Memorial Seminar, May 12, 1965, Los Angeles, CA, 1966, 140p., ed. Shirley Thomas, Hard Cover $30 *(ISBN 0-87703-035-9)*

Vol. 8 Impact of Space Exploration on Society, Aug. 18-20, 1965, San Francisco, CA, 1966, 382p., ed. William E. Frye, Hard Cover $30 *(ISBN 0-87703-O36-7)*

Vol. 9 Recent Developments in Space Flight Mechanics, (AAS/AAAS Symposium), Dec. 29, 1965, Berkeley, CA, 1966, 280p., ed. Paul B. Richards, Hard Cover $25 *(ISBN 0-87703-037-5)*

Vol. 10 Space in the Fiscal Year 2001, 4th Goddard Memorial Symposium, Mar. 15-16, 1966, Washington, D.C., 1967, 458p., eds. Eugene B. Konecci, Maxwell W. Hunter, II, Robert F. Trapp, Hard Cover $35 *(ISBN 0-87703-038-3)*

Vol. 11 Space Flight Specialist Conference, Jul. 6-8, 1966, Denver, CO, 1967, 618p., ed. Maurice L. Anthony, Hard Cover $45 *(ISBN 0-87703-039-1)*; Microfiche Suppl. (Vol. 2 AAS Microfiche Series) $15 *(ISBN 0-87703-221-1)*

Vol. 12 Management of Aerospace Programs Conference, Nov. 16-18, 1966, Columbia, MO, 1967, 392p., ed. Walter K. Johnson, Hard Cover $30 *(ISBN 0-87703-040-5)*

Vol. 13 Physics of the Moon (AAS/AAAS Symposium), Dec. 29, 1966, Washington, D.C., 1967, 260p., ed. S. Fred Singer, Hard Cover $25 *(ISBN 0-87703-041-3)*

Vol. 14 Interpretation of Lunar Probe Data, Sept. 17, 1966, Huntington Beach, CA, 1967, 270p., ed. Jack Green, Hard Cover $25 *(ISBN 0-87703-042-1)*

Vol. 15 Future Space Program and Impact on Range and Network Development Symposium, Mar. 22-24, 1967, Las Cruces, NM, 1967, 588p., ed. George W. Morgenthaler, Hard Cover $40 *(ISBN 0-87703-043-X)*

Vol. 16 Voyage to the Planets, 5th Goddard Memorial Symposium, Mar. 14-15, 1967, Washington, D.C., 1968, 184p., ed. S. Fred Singer, Hard Cover $20 *(0-87703-044-8)*

Vol. 17 Use of Space Systems for Planetary Geology and Geophysics Symposium, May 25-27, 1967, Boston, MA, 1968, 623p., ed. Robert D. Enzmann, Hard Cover $45 *(ISBN O-87703-045-6)*; Microfiche Suppl. (Vol. 5 AAS Microfiche Series) $15 *(ISBN 0-87703-135-5)*

Vol. 18 Technology and Social Progress, 6th Goddard Memorial Symposium, Mar. 12-13, 1968, Washington, D.C., 1969, 170p., ed. Philip K. Eckman, Hard Cover $20 *(ISBN 0-87703-046-4)*

Vol. 19 Exobiology - The Search for Extraterrestrial Life (AAS/AAAS Symposium) Dec. 30, 1967, New York, NY, 1969, 184p., eds. Martin M. Freundlich, Bernard W. Wagner, Hard Cover $20 *(ISBN 0-87703-047-2)*

Vol. 20 Bioengineering and Cabin Ecology (AAS/AAAS Symposium) Dec. 30, 1968, Dallas, TX, 1969, 162p., ed. William Cassidy, Hard Cover $20 *(ISBN 0-87703-048-0)*

Vol. 21 Reducing the Cost of Space Transportation, 7th Goddard Memorial Symposium, Mar. 4-5, 1969, Washington, D.C., 1969, 264p., ed. George K. Chacko, Microfiche only $25 *(ISBN 0-87703-049-9)*

Vol. 22 Planning Challenges of the 70's in the Public Domain, 15th Annual AAS Meeting, Jun. 17-20, 1969, Denver, CO, 1970, 504p., eds. William J. Burnsnall, George K. Chacko, George W. Morgenthaler, Hard Cover $40 *(ISBN 0-87703-050-2)*; Microfiche Suppl. (Vol. 13 AAS Microfiche Series) $20 *(ISBN 0-87703-131-2)*; See also Vols. 15-17, AAS Microfiche Series

Vol. 23 Space Technology and Earth Problems Symposium, Oct. 23-25, 1969, Las Cruces, NM, 1970, 418p., ed. C. Quentin Ford, Hard Cover $35 *(ISBN 0-87703-051-0)*; Microfiche Suppl. (Vol. 12 AAS Microfiche Series) $20 *(ISBN 0-87703-134-7)*

Vol. 24 Aerospace Research and Development, Jul. 14, 1966, Holloman AFB, NM, 1970, 500p., ed. Ernst A. Steinhoff, Hard Cover $40 *(ISBN 0-87703-052-9)*

Vol. 25 Geological Problems in Lunar and Planetary Research, Feb. 17-18, 1969, Huntington Beach, CA, 1971, 750p., ed. Jack Green, Hard Cover $45 *(ISBN 0-87703-056-1)*

Vol. 26 Technology Utilization Ideas for the 70s and Beyond, Oct. 30, 1970, Winrock, AR, 1971, 312p., eds. Fred W. Forbes, Paul Dergarabedian, Microfiche only $30 *(ISBN 0-87703-057-X)*

Vol. 27 International Cooperation in Space Operations and Exploration, 9th Goddard Memorial Symposium, Mar. 11, 1971, Washington, D.C. 1971, 194p., ed. Michael Cutler, Hard Cover $20 *(ISBN 0-88703-058-8)*

Vol. 28 Astronomy from a Space Platform (AAS/AAAS Symposium) Dec. 27-28, 1971, Philadelphia, PA, 1972, 416p., eds. George W. Morgenthaler, Howard D. Greyber, Hard Cover $35 *(ISBN 0-87703-061-8)*

Vol. 29 Space Technology Transfer to Community and Industry, 10th Goddard Memorial Symposium, 18th Annual AAS Meeting, Mar. 13-14, 1972, Washington, D.C., 1972, 196p., eds. Ralph H. Tripp, John K. Stotz, Jr., Hard Cover $20 *(ISBN 0-87703-062-6)*; on Microfiche $15

Vol. 30 Space Shuttle Payloads (AAS/AAAS Symposium) Dec. 27-28, 1972, Washington, D.C., 1973, 532p., eds. George W. Morgenthaler, William J. Bursnall, Hard Cover $40 *(ISBN 0-87703-063-4)*

Vol. 31 The Second Fifteen Years in Space, 11th Goddard Memorial Symposium, Mar. 8-9, 1973, Washington, D.C., 1973, 212p., ed. Saul Ferdman, Hard Cover $25 *(ISBN 0-87703-064-2)*

Vol. 32 Health Care Systems Conference, Nov. 21-22, 1972, Dallas, TX, 1974, 265p., ed. Eugene B. Konecci, Hard Cover $25 *(ISBN 0-87703-067-7)*

Vol. 33 Orbital International Laboratory, 3rd and 4th IAF/OIL Symposia, Oct. 5-6, 1970, Constance, Germany, Sept. 24-25, 1971, Brussels, Belgium, 1974, 322p., ed. Ernst A. Steinhoff, Hard Cover $30 *(ISBN 0-87703-068-5)*

Vol. 34 Management and Design of Long-Life Systems, Apr. 24-26, 1973, Denver, CO, 1974, 198p., ed. Harris M. Schurmeier, Hard Cover $20 *(ISBN 0-87703-069-3)*

Vol. 35 Energy Delta, Supply vs. Demand, (AAS/AAAS Symposium) Feb. 25-27, 1974, San Francisco, CA, 1975, 2nd Printing 1976, 604p., eds. George W. Morgenthaler, Aaron N. Silver, Hard Cover $35 *(ISBN 0-87703-070-7)*; Soft Cover $25 *(ISBN 0-87703-082-0)*; on Microfiche $20

Vol. 36 Skylab and Pioneer Report, 12th Goddard Memorial Symposium, Mar. 8, 1974, Washington, D.C., 1975, 160p., eds. Philip H. Bolger, Paul B. Richards, Hard Cover $20 *(ISBN 0-87703-071-5)*

Vol. 37 Space Rescue and Safety 1974, 7th International IAA Symposium, Sept. 30 - Oct. 5, 1974, Amsterdam, Netherlands, 1975, 294p., ed. Philip H. Bolger, Hard Cover $25 *(ISBN 0-87703-073-1)*

Vol. 38 Skylab Science Experiments, (AAS/AAAS Symposium) Feb. 28, 1974, San Francisco, CA, 1976, 274p., eds. George W. Morgenthaler, Gerald E. Simonson, Microfiche only $20 *(ISBN 0-87703-074-X)*

Vol. 39 Environmental Control and Agri-Technology, 1976, 346p., ed. Eugene B. Konecci, Microfiche only $20 *(ISBN 0-87703-075-8)*

Vol. 40 Future Space Activities, 13th Goddard Memorial Symposium, Apr. 11, 1975, Washington, D.C., 1976, 182p., ed. Carl H. Tross, Microfiche only $20 *(ISBN 0-87703-076-6)*

Vol. 41 Space Rescue and Safety 1975, 8th International IAA Symposium, Sept. 21-27, 1975, Lisbon, Portugal, 1976, 230p., ed. Philip H. Bolger, Hard Cover $25 *(ISBN 0-87703-077-4)*

Vol. 42 The End of an Era in Space Exploration, From International Rivalry to International Cooperation, 1976, 216p., by J.C.D. Blaine, Hard Cover $25 *(ISBN 0-87703-084-7)*; without volume number *(ISBN 0-87703-080-4)*

Vol. 43 The Eagle Has Returned, Part I, International Space Hall of Fame Dedication Conference, Oct. 5-9, 1976, Alamogordo, NM, 1976, 370p., ed. Ernst. A. Steinhoff, Hard Cover $30 *(ISBN 0-87703-086-3)*

Vol. 44 Satellite Communications in the Next Decade, 14th Goddard Memorial Symposium, Mar. 12, 1976, Washington, D.C., 1977, 188p., ed. Leonard Jaffe, Hard Cover $20 *(ISBN 0-87703-088-X)*

Vol. 45 The Eagle Has Returned, Part 2, International Space Hall of Fame Dedication Conference, Oct. 5-9, 1976, Alamogordo, NM, 1977, 454p., ed. Ernst A. Steinhoff, Hard Cover $35 *(ISBN 0-87703-092-8)*

Vol. 46 Export of Aerospace Technology, 15th Goddard Memorial Symposium, Mar. 31 - Apr. 1, 1977, Washington, D.C., 1978, 174p., ed. Carl H. Tross, Hard Cover $20 *(ISBN 0-87703-093-6)*

Vol. 47 Handbook of Soviet Lunar and Planetary Exploration, 1979, 276p., by Nicholas L. Johnson, Microfiche Only $25 *(ISBN 0-87703-105-3)*

Vol. 48 Handbook of Soviet Manned Space Flight, 2nd Edition, 1988, 474p., by Nicholas L. Johnson, Hard Cover $60 *(ISBN 0-87703-115-0)*; Soft Cover $45 *(ISBN 0-87703-116-9)*

Vol. 49 Space - New Opportunities for International Ventures, 17th Goddard Memorial Symposium, Mar. 28-30, 1979, Washington, D.C., 1980, 300p., ed. William C. Hayes, Jr., Hard Cover $35 *(ISBN 0-87703-124-X)*; Soft Cover $25 *(ISBN 0-87703-125-8)*; see also Vol. 2 AAS History Series

Vol. 50 Remember the Future - The Apollo Legacy, Jul. 20-21, 1979, San Francisco, CA, 1980, 218p., ed. Stan Kent, Hard Cover $25 *(ISBN 0-87703-126-6)*; Soft Cover $15 *(ISBN 0-87703-127-4)*

Vol. 51 Commercial Operations in Space 1980-2000, 18th Goddard Memorial Symposium, Mar. 27-28, 1980, Washington, D.C., 1981, 214p., eds. John L. McLucas, Charles Sheffield, Hard Cover $30 *(ISBN 0-87703-140-1)*; Soft Cover $20 *(ISBN 0-87703-141-X)*; Microfiche Suppl. (Vol. 34 AAS Microfiche Series) $10 *(ISBN 0-87703-165-7)*; see also Vols. 2 and 3, AAS History Series

Vol. 52 International Space Technical Applications, 19th Goddard Memorial Symposium, Mar. 26-27, 1981, Washington, D.C., 1981, 186p., eds. Andrew Adelman, Peter M. Bainum, Hard Cover $30 *(ISBN 0-87703-152-5)*; Soft Cover $20 *(ISBN 0-87703-153-3)*; see also Vol. 5, AAS History Series

Vol. 53 Space in the 1980's and Beyond, 17th European Space Symposium, Jun. 4-6, 1980, London, England, 1981, 302p., ed. Peter M. Bainum, Hard Cover $40 *(ISBN 0-87703-154-1)*; Soft Cover $30 *(ISBN 0-87703-155-X)*

Vol. 54 Space Safety and Rescue 1979-1981 (with abstracts 1976-1978), Proceedings of symposia of the International Academy of Astronautics held in conjunction with the 30th, 31st, and 32nd International Astronautical Federation Congresses, Munich, Germany, 1979, Tokyo, Japan, 1980, and Rome, Italy, 1981, 1983, 456p., ed. Jeri W. Brown, Hard Cover $45 *(ISBN 0-87703-177-0)*; Soft Cover $35 *(ISBN 0-87703-178-9)*; Microfiche Suppl. (Vols. 39-41 AAS Microfiche Series) $39 *(ISBN 0-87703-222-X)*; *(ISBN 0-87703-223-8)*; *(ISBN 0-87703-224-6)*

Vol. 55 Space Applications at the Crossroads, 21st Goddard Memorial Symposium, Mar. 24-25, 1983, Greenbelt, MD, 1983, 308p., eds. John H. McElroy, E. Larry Heacock, Hard Cover $45 *(ISBN 0-87703-186-X)*; Soft Cover $35 *(ISBN 0-87703-187-8)*

Vol. 56 Space: A Developing Role for Europe, 18th European Space Symposium, Jun. 6-9, 1983, London, England, 1984, 278p., eds. Len J. Carter, Peter M. Bainum, Hard Cover $45 *(ISBN 0-87703-193-2)*; Soft Cover $35 *(ISBN 0-87703-194-0)*; Microfiche Suppl. (Vol. 46 AAS Microfiche Series) $15 *(ISBN 0-87703-195-9)*

Vol. 57 The Case for Mars, Apr. 29 - May 2, 1981, Boulder, CO, 1984, Second Printing 1987, 348p., ed. Penelope J. Boston, Hard Cover $45 *(ISBN 0-87703-197-5)*; Soft Cover $25 *(ISBN 0-87703-198-3)*; on Microfiche $25

Vol. 58 Space Safety and Rescue 1982-1983, Proceedings of the International Academy of Astronautics held in conjunction with the 33rd and 34th International Astronautical Congresses, Paris, France, Sept. 27 - Oct. 2, 1982, and Budapest, Hungary, Oct. 10-15, 1983, 1984, 378p., ed. Gloria W. Heath, Hard Cover $50 *(ISBN 0-87703-202-5)*; Soft Cover $40 *(ISBN 0-87703-203-3)*

Vol. 59 Space and Society - Challenges and Choices, April 14-16, 1982, University of Texas at Austin, 1984, 442p., eds. Paul Anaejionu, Nathan C. Goldman, Philip J. Meeks, Hard Cover $55 *(ISBN 0-87703-204-1)*; Soft Cover $35 *(ISBN 0-87703-205-X)*

Vol. 60 Permanent Presence - Making It Work, 22nd Goddard Memorial Symposium, Mar. 15-16, 1984, Greenbelt, MD, 1985, 190p., ed. Ivan Bekey, Hard Cover $40 *(ISBN 0-87703-207-6)*; Soft Cover $30 *(ISBN 0-87703-208-4)*

Vol. 61 Europe/United States Space Activities - With a Space Propulsion Supplement, 23rd Goddard Memorial Symposium/19th European Space Symposium, Mar. 27-29, 1985, Greenbelt, MD, 31st Annual AAS Meeting, Oct. 22-24, 1984, Palo Alto, CA, 1985, 442p., eds. Peter M. Bainum, Friedrich von Bun, Hard Cover $55 *(ISBN 0-87703-217-3)*; Soft Cover $45 *(ISBN 0-87703-218-1)*

Vol. 62 The Case for Mars II, July 10-14, 1984, Boulder, CO, 1985, 730p., ed. Christopher P. McKay, Hard Cover $60 *(ISBN 0-87703-219-1)*; Soft Cover $40 *(ISBN 0-87703-220-3)*

Vol. 63 Proceedings of 4th International Conference on Applied Numerical Modeling, Dec. 27-29, 1984, Tainan, Taiwan, 1986, 800p., ed. Han-Min Hsia, You-Li Chou, Shu-Yi Wang, Sheng-Jii Hsieh, Hard Cover $70 (ISBN 0-87703-242-4)

Vol. 64 Space Safety and Rescue 1984-1985, Proceedings of the International Academy of Astronautics held in conjunction with the 35th and 36th International Astronautical Congresses, Lausanne, Switzerland, Oct. 7-13, 1984, and Stockholm, Sweden, Oct. 7-12, 1985, 1986, 400p., ed. Gloria W. Heath, Hard Cover $55 *(ISBN 0-87703-248-3)*; Soft Cover $45 *(ISBN 0-87703-249-1)*

Vol. 65 The Human Quest in Space, 24th Goddard Memorial Symposium, Mar. 20-21, 1986, Greenbelt, MD, 1987, 312p., ed. Gerald L. Burdett, Gerald A. Soffen, Hard Cover $55 *(ISBN 0-87703-262-9)*; Soft Cover $45 *(ISBN 0-87703-263-7)*

Vol. 66 Soviet Space Programs 1980-1985, 1987, 298p., by Nicholas L. Johnson, Microfiche Only $45 *(ISBN 0-87703-266-1)*

Vol. 67 Low-Gravity Sciences, Seminar Series 1986, University of Colorado at Boulder, 290p., ed. Jean N. Koster, Hard Cover $55 *(ISBN 0-87703-270-X)*; Soft Cover $45 *(ISBN 0-87703-271-8)*

Vol. 68 Proceedings of the Fourth Annual L5 Space Development Conference, Apr. 25-28, 1985, Washington, D.C., 1987, 268p., ed. Frank Hecker, Hard Cover $50 *(ISBN 0-87703-272-6)*; Soft Cover $35 *(ISBN 0-87703-273-4)*

Vol. 69 Visions of Tomorrow: A Focus on National Space Transportation Issues, 25th Goddard Memorial Symposium, Mar. 18-20, 1987, Greenbelt, MD, 1987, 338p., ed. Gerald A. Soffen, Hard Cover $55 *(ISBN 0-87703-274-2)*; Soft Cover $45 *(ISBN 0-87703-275-0)*

Vol. 70 Space Safety and Rescue 1986-1987, Proceedings of the International Academy of Astronautics held in conjunction with the 37th and 38th International Astronautical Congresses, Innsbruck, Austria, Oct. 4-11, 1986, and Brighton, England, Oct. 11-16, 1987, 1988, 360p., ed. Gloria W. Heath, Hard Cover $55 *(ISBN 0-87703-291-2)*; Soft Cover $45 *(ISBN 0-87703-292-0)*

Vol. 71 The NASA Mars Conference, Jul. 21-23, 1986, Washington, D.C., 1988, 570p., ed. Duke B. Reiber, Hard Cover $50 *(ISBN 0-87703-293-9)*; Soft Cover $30 *(ISBN 0-87703-294-7)*

Vol. 72 Working in Orbit and Beyond: The Challenges for Space Medicine, Jun. 20-21, 1987, Washington, D.C., 1989, 188p., ed. David Lorr, Victoria Garshnek, Hard Cover $45 *(ISBN 0-87703-295-5)*; Soft Cover $35 *(ISBN 0-87703-296-3)*

Vol. 73 Technology and the Civil Future in Space, 26th Goddard Memorial Symposium, Mar. 16-18, 1988, Greenbelt, MD, 1989, 246p., ed. Leonard A. Harris, Hard Cover $50 *(ISBN 0-87703-301-3)*; Soft Cover $35 *(ISBN 0-87703-302-1)*

Vol. 74 The Case for Mars III: Strategies for Exploration - General Interest and Overview, July 18-22, 1987, Boulder, CO, 1989, 744p., ed. Carol Stoker, Hard Cover $75 *(ISBN 0-87703-303-X)*; Soft Cover $55 *(0-87703-304-8)*

Vol. 75 The Case for Mars III: Strategies for Exploration - Technical, July 18-22, 1987, Boulder, CO, 1989, 646p., ed. Carol Stoker, Hard Cover $70 *(ISBN 0-87703-305-6)*; Soft Cover $50 *(ISBN 0-87703-306-4)*

Vol. 76 Global Environmental Change: The Role of Space in Understanding Earth, 27th Goddard Memorial Symposium, Mar. 8-10, 1989, Washington, D.C., 1990, 178p., ed. Richard G. Johnson, Hard Cover $50 *(ISBN 0-87703-322-6)*; Soft Cover $35 *(ISBN 0-87703-323-4)*; Microfiche Suppl. (Vol. 60 AAS Microfiche Series) $10 *(ISBN 0-87703-324-2)*

Vol. 77 Space Safety and Rescue 1988 - 1989, Proceedings of the International Academy of Astronautics held in conjunction with the 39th and 40th International Astronautical Congresses, Bangalore, India, Oct. 8-15, 1988, and Málaga, Spain, Oct. 7-12, 1989, 500p., ed. Gloria W. Heath, Hard Cover $70 *(ISBN 0-87703-327-7)*; Soft Cover $55 *(ISBN 0-87703-328-5)*

INDEX

INDEX TO ALL AMERICAN ASTRONAUTICAL SOCIETY PAPERS AND ARTICLES 1954 - 1985/86

This index is a numerical/chronological index (which also serves as a citation index) and an author index. (A subject index volume will be forthcoming.)

It covers all articles that appear in the following:

Advances in the Astronautical Sciences (1957 - August 1986)

Science and Technology Series (1964 - September 1986)

AAS History Series (1977 - 1986)

AAS Microfiche Series (1968 - August 1986)

Journal of the Astronautical Sciences (1954 - 1986)

Astronautical Sciences Review (1959 - 1962)

If you are in aerospace you will want this excellent reference tool which covers the first 30 years of the Space Age.

Numerical/Chronological/Author Index in two volumes, Library Binding (both volumes) $95.00; Soft Cover (both volumes) $80.00; Volume I (1954 - 1978) Library Binding $40.00; Soft Cover $30.00; Volume II (1979 - 1985/86) Library Binding $60.00; Soft Cover $45.00.

Order from Univelt, Inc., P.O. Box 28130, San Diego, California 92198.

NUMERICAL INDEX

VOLUME 74 **ADVANCES IN THE ASTRONAUTICAL SCIENCES,** *GUIDANCE AND CONTROL 1991*

(AAS Annual Rocky Mountain Guidance and Control Conference, February 2-6, 1991, Keystone, Colorado)

AAS 91-001 to -009 Not Assigned

AAS 91-010 Gravitational Effects on Low-Earth Orbiters, G. W. Rosborough

AAS 91-011 Atmospheric Drag, R. G. Roble

AAS 91-012 Orbital Debris, R. D. Culp

AAS 91-013 Radiation Effects on Spacecraft, J. C. Ritter

AAS 91-014 Not Available

AAS 91-015 Initial Results from Long Duration Exposure Facility (LDEF) Postretrieval Observations, W. H. Kinard, G. D. Martin, R. L. O'Neal

AAS 91-016 to -019 Not Assigned

AAS 91-020 Strategic Avionics Technology Planning, K. J. Cox, D. C. Brown

AAS 91-021 Not Available

AAS 91-022 Autonomous Guidance, Navigation and Control, A. J. Bordano, G. G. McSwain, S. T. Fernandes

AAS 91-023 A Scalable Architecture for an Operational Spaceborne Autonav System, L. J. Hansen

AAS 91-024 Mitigating Multipath Error in GPS Based Attitude Determination, C. E. Cohen, B. W. Parkinson

AAS 91-025 Autonomous, All-Stellar Attitude Determination Experiment: Ground Test Results, H. L. Fisher, T. E. Strikwerda, C. C. Kilgus, L. J. Frank, M. D. Shuster

AAS 91-026 The Minimum Control Authority of a System of Actuators with Applications to Gravity Probe-B, P. Wiktor, D. DeBra

AAS 91-027 to -029 Not Assigned

AAS 91-030 to -034 Not Available

AAS 91-035 Microgravity Flight Testing of a Laboratory Report, E. M. Hinman

AAS 91-036 to -037 Not Available

AAS 91-038	Reaction Wheel Control of the MSX Satellite, T. Marshall, T. Gunderman, F. Mobley
AAS 91-039 to -040	Not Available
AAS 91-041	Non-Collocated Control of an Inverted Flexible Structure, S. Clark, C. Fosha
AAS 91-042	A Magnetic Bearing Momentum Wheel for High Pointing Accuracy and Vibration Sensitive Space Applications, U. J. Bichler
AAS 91-043	Flight Telerobotic Servicer Prototype Simulator, R. Schein, L. Krauze, C. Hartley, A. Dickenson, T. LaVecchia, B. Working
AAS 91-044 to -045	Not Available
AAS 91-046	Space Telerobotic Research and Applications at Space Systems/Loral, S. W. Tilley, M. G. Hollars, C. M. Francis, A. H. Tadros
AAS 91-048	IntelliSTAR™ - An Emerging Spacecraft Control Architecture to Improve Autonomous Control and Operation, J. Barry, L. Raslavicius, T. P. Gathmann
AAS 91-049	Not Available
AAS 91-050	The Controls-Structures Interaction Guest Investigator Program - An Overview and Phase I Experimental Results, R. Smith-Taylor, S. E. Tanner
AAS 91-051	Demonstrations of LSS Active Vibration Control Technology on Representative Ground-Based Testbeds, D. C. Hyland, D. J. Phillips, E. G. Collins, Jr.
AAS 91-052	Experimental Verification of Nanometer Level Optical Pathlength Control on a Flexible Structure, M. O'Neal, D. Eldred, D. Liu, D. Redding
AAS 91-053	Acquisition, Tracking and Pointing Concept Solutions for an Independent Complementary Space Experiment (ICSE), S. M. Seltzer, R. W. Schock, N. Fitz-Coy, R. C. Francis, K. E. Weir, L. A. Morine
AAS 91-054	A Wavefront Compensation Approach to Segmented Mirror Figure Control, D. Redding, B. Breckenridge, G. Sevaston, K. Lau
AAS 91-055	A Class of Neural Networks and Their Application, N. Kermiche, R. Su
AAS 91-056	Thermal Elastic Shock and its Effect on TOPEX Spacecraft Attitude Control, D. F. Zimbelman
AAS 91-057 to -059	Not Assigned
AAS 91-060	Evaluation of Flight Results of the ROSAT Attitude Measurement and Control System, W. Schrempp, A. Böinghoff, E. Brüderle, N. Gradmann, H. Herbst, L. Kaffer, W. Oesterlin
AAS 91-061	Image Navigation and Registration of Second Generation METEOSAT, B. Blancke, J. L. Carr, P. Mauté
AAS 91-062	Control System Challenges Associated with the Mobile Servicing System for Space Station Freedom, M. E. Stieber, P. T. K. Fung

AAS 91-063	A Guidance, Navigation and Control System for Rendezvous and Proximity Operations Between Hermes, the Columbus Free Flyer and the Space Station Freedom, W. Fehse, A. Tobias, C. Champetier, J. M. Pairot, C. Pauvert
AAS 91-064	High Precision Control Systems for Telescopes and Interferometers, H. Jörck
AAS 91-065	Advanced GNC Concepts and Techniques for an Interplanetary Mission, C. Champetier, P. Régnier, J. de Lafontaine
AAS 91-066	Telecommunications, Navigation and Information Management Concept Overview for the Space Exploration Initiative Program, J. A. Bell, E. Stephens, G. Barton
AAS 91-067 to -069	Not Assigned
AAS 91-070	Attitude Control Experience with the Combined Release and Radiation Effects Satellite, W. Frazier, K. Saylors, F. Patton, K. Stakkestad
AAS 91-071	Initial Performance Improvements Due to Design Modifications for the Pointing Control System on the Hubble Space Telescope, G. S. Nurr, J. P. Sharkey, H. B. Waites
AAS 91-072	Magellan Star Scanner Experiences: What a Long, Strange Trip it's Been, E. H. Seale
AAS 91-073	GN&C for the Pegasus® Air-Launched Space Booster: Design and First Flight Results, D. Rovner
AAS 91-074	Execution of a Successful Rocket Plume Observation from the LACE Satellite, R. Campion, T. Tran, D. M. Horan, H. W. Smathers, R. E. Palma
AAS 91-075	Magnetic Attitude Determination Subsystem Performance on the LACE Spacecraft, S. M. Fox, P. K. Pal, M. L. Psiak, M. E. Mook
AAS 91-076	Discovery of BE-7 Accretion in Low Earth Orbit, G. W. Phillips, S. E. King, R. A. August, J. C. Ritter, J. H. Cutchin, P. S. Haskins, J. E. McKisson, D. W. Ely, A. G. Weisenberger, R. B. Piercey, T. Dybler
AAS 91-077 to -079	Not Assigned
AAS 91-080 to -082	Not Available
AAS 91-083 to -099	Not Assigned

AUTHOR INDEX*

* For each author the paper number is given. The page numbers refer to Volume 74, **Advances in the Astronautical Sciences**.